Silicon in Polymer Synthesis

Springer-Verlag Berlin Heidelberg GmbH

H.R. Kricheldorf (Ed.)

Silicon in Polymer Synthesis

With Contributions by
C. Burger   W.R. Hertler   P. Kochs
F.-H. Kreuzer   H.R. Kricheldorf   R. Mülhaupt

With 30 Figures and 73 Tables

 Springer

Editor

Professor H.R. Kricheldorf
Universität Hamburg
Institut für Technische und Makromolekulare Chemie
Bundesstr. 45
20146 Hamburg, Germany

Authors

Dr. W.R. Hertler
1375 Parkersville Road
Kennett Square, PA 19348-2114, USA

Dr. C. Burger
Dr. P. Kochs

Wacker Chemie
GBS-S
Hans-Seidel-Platz 4
81737 München, Germany

Dr. F.-H. Kreuzer
Consortium für Elektrochemische
Industrie
Zielstattstr. 20
81379 München, Germany

Professor Dr. H.R. Kricheldorf
see address above

Professor Dr. R. Mülhaupt
Universität Freiburg
Institut für Makromolekulare Chemie
Hermann-Staudinger-Haus
Stefan-Meier-Str. 31
79104 Freiburg, Germany

ISBN 978-3-642-79177-2   ISBN 978-3-642-79175-8 (eBook)
DOI 10.1007/978-3-642-79175-8

CIP data applied for
Die Deutsche Bibliothek – CIP-Einheitsaufnahme
Silicon in polymer synthesis: with 73 tables/H.R. Kricheldorf
(ed.). With contributions by W.R. Hertler... – Berlin; Heldelberg; New York; Barcelona; Budapest; Hong Kong;
London; Milan; Paris; Santa Clara; Singapore; Tokyo;
Springer, 1996

NE: Kricheldorf, Hans Rytger [Hrsg.]; Hertler, Walter R.

Cover design: Lewis + Leins, Berlin

© Springer-Verlag Berlin Heidelberg 1996
Originally published by Springer-Verlag Berlin Heidelberg New York in 1996
Softcover reprint of the hardcover 1st edition 1996

Typesetting: Thomson Press (India) Ltd., Madras

SPIN: 10109042   2/3020/SPS – 5 4 3 2 1 0 – Printed on acid-free paper

Preface

Over the past four decades, preparative silicon chemistry has taken two courses. Either silicon-containing compounds were the final goal of the syntheses and physico-chemical studies, or silicon compounds served as auxiliary agents in the synthesis of silicon-free organic chemicals. With a delay of approximately two decades the same trends are now observable in polymer science. On the one hand, silicon-containing polymers such as polysiloxanes, polysilazanes, poly-carbosilanes or polysilanes are attracting more and more interest; on the other hand, a rapidly increasing number of papers and patents report on various applications of silicon reagents for syntheses and modifications of silicon-free monomers and polymers. Reviews of smaller segments of this broad field, for instance, reviews on polysiloxanes, have been published by several authors. In this connection, it was the purpose of the present book to give the broadest possible overview on applications of silicon-containing monomers and reagents in polymer synthesis. Since this work was designed to have the character of a handbook and not that of a textbook, detailed discussions of reaction mechanisms were not included. A short description of some pecularities of silicon chemistry is given in Appendix B in connection with the reactivity of silylating agents. It is the sincere hope of the editor that the numerous examples of successful application of silicon compounds in polymer synthesis presented in this handbook will stimulate further research in this growing and promising field.

Finally the editor wishes to express his gratitude to the contributors who made this work possible and he wishes to thank Mrs. K. Griem and Mrs. I. Wasum for the preparation of the manuscript.

Contents

Chapter 3
Polysiloxanes and Polymers Containing Siloxane Groups 113
C. Burger and F.-H. Kreuzer

Chapter 6
Miscellaneous Applications of Silicon Reagents   355
H.R. Kricheldorf

Chapter 7
Chemical Modification of Polymers and Surfaces   404
H.R. Kricheldorf

Appendix A
Silicon-Based Thermoset Resins

R.Mülhaupt

Appendix B
Silylation and Silylating Agents

H.R. Kricheldorf

CHAPTER 1

Polymerization of Si-Containing Vinyl Monomers and Acetylenes

H.R. Kricheldorf

1.1
Si-Containing Styrenes

Almost all silicon-containing styrenes and the polymers derived from them possess one functional group in the *para* position relative to the vinyl group. The silicon in this functional group may exclusively form Si—C bonds but it may also involve Si—O, Si—S and Si—N bonds. All these Si-bonds are more or less sensitive to reactive cations, and thus, reports on cationic polymerizations of Si-containing styrenes are scarce. Albeit, Si—C and Si—O bonds do not poison Ziegler-Natta (ZN) catalysts, even reports on ZN-catalysed polymerizations of Si-containing styrenes are rare to the best of our knowledge. Therefore, this review will mainly deal with the synthesis of monomers and their polymerization by free radical mechanisms (discussed first)[1 –27] or by anionic mechanisms (discussed second).

The interest in Si-containing vinyl monomers and their polymers has rapidly increased since 1980 for several reason. Some of these polymers may serve as compatibilizers or "coupling agents" in polymer blends and composites. Polymers with a high content of silyl and siloxane groups may be useful as membranes with high permeability to oxygen. Blockcopolymers with well defined morphology allow a controlled preparation of micro porous membranes. Furthermore, silyl groups (particularly trimethylsilyl and *tert*-butyldimethylsilyl) were used as protecting groups for OH, SH and NH functions during the polymerization process and removed afterwards (see Appendix B). The removal of these silyl groups usually occurs under mild hydrolytic conditions (these and other chemical modifications will be discussed in chapter 7). Another important application of Si-containing styrenes and other vinyl monomers concerns the fabrication of resist materials for any kind of microlithography and semiconductor devices. The problems of a high-resolution, high-aspect ratio imaging technology are apparently best solved by bilayer or multilayer resists [4, 28]. In the simplest case, the bilayer system consists of a thin radiation-sensitive "imaging layer" on top of a oxidation sensitive thick "planarizing layer". The top layer receives the image in the standard fashion by UV- or X-ray radiation through a mask. After fixation the image of the thin top layer is then transferred to the lower "planarizing layer" by rapid oxidation with an O_2-plasma in a vacuum. Therefore, the lower layer must be an oxidation sensitive polymer with a relatively low carbon content [29, 30]. In contrast, the imaging layer needs to be stable against complete oxidation. This property is imparted by silicon (or tin and other metal atoms), because the initial oxidation with the O_2-plasma produces a non-volatile layer of SiO_2 which protects the lower layer against oxidation. Therefore, the material of a typical imaging layer consists of a copolymer or a blend of polymers containing silyl groups, on the one hand, and a radiation-sensitive group, on the other hand.

1.1.1
Radical Polymerizations

The syntheses of most Si-containing styrenes are based on the reaction pathways outlined in Eqs. (1–7). The Si—C bonds are typically formed via a Grignard

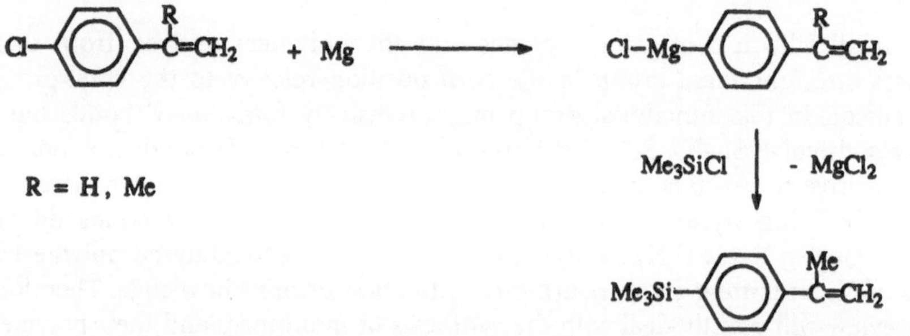

$$Cl-\!\!\!\bigcirc\!\!\!-\!\!\overset{R}{\underset{}{C}}\!\!=\!\!CH_2 \ + \ Mg \ \longrightarrow \ Cl-Mg-\!\!\!\bigcirc\!\!\!-\!\!\overset{R}{\underset{}{C}}\!\!=\!\!CH_2 \tag{1}$$

$$R = H, Me$$

$$\big\downarrow \ Me_3SiCl \quad -MgCl_2$$

$$Me_3Si-\!\!\!\bigcirc\!\!\!-\!\!\overset{Me}{\underset{}{C}}\!\!=\!\!CH_2 \tag{2}$$

$$X-(CH_2)_n-\!\!\!\bigcirc\!\!\!-\!\!\overset{R}{\underset{}{C}}\!\!=\!\!CH_2 \ + \ Mg \ \longrightarrow \ X-Mg-(CH_2)_n-\!\!\!\bigcirc\!\!\!-\!\!\overset{R}{\underset{}{C}}\!\!=\!\!CH_2 \tag{3}$$

$$R = H, Me$$
$$X = Cl, Br$$

$$\big\downarrow \ Me_3SiCl \quad -MgClX$$

$$Me_3Si-(CH_2)_n-\!\!\!\bigcirc\!\!\!-\!\!\overset{R}{\underset{}{C}}\!\!=\!\!CH_2 \tag{4}$$

$$HO-(CH_2)_n-\!\!\!\bigcirc\!\!\!-\!\!\overset{R}{\underset{}{C}}\!\!=\!\!CH_2 \ \xrightarrow[\ -HX\ ]{\ +XSiMe_3\ } \ Me_3SiO-(CH_2)_n-\!\!\!\bigcirc\!\!\!-\!\!\overset{R}{\underset{}{C}}\!\!=\!\!CH_2 \tag{5}$$

$$n = 0,1,2 \qquad R = H, Me \qquad X = Cl, NHSiMe_3$$

$$Cl-\!\!\overset{Me}{\underset{Me}{Si}}\!\!-(CH_2)_n-\!\!\!\bigcirc\!\!\!-\!\!\overset{R^1}{\underset{}{C}}\!\!=\!\!CH_2 \ \xrightarrow[\ -HCl\ ]{\ +R^2\!-\!OH\ } \ R^2\!-\!O-\!\!\overset{Me}{\underset{Me}{Si}}\!\!-(CH_2)_n-\!\!\!\bigcirc\!\!\!-CH\!\!=\!\!CH_2 \tag{6}$$

$$n = 0,1,2 \qquad R^1 = H, Me \qquad R^2 = Alkyl, Aryl$$

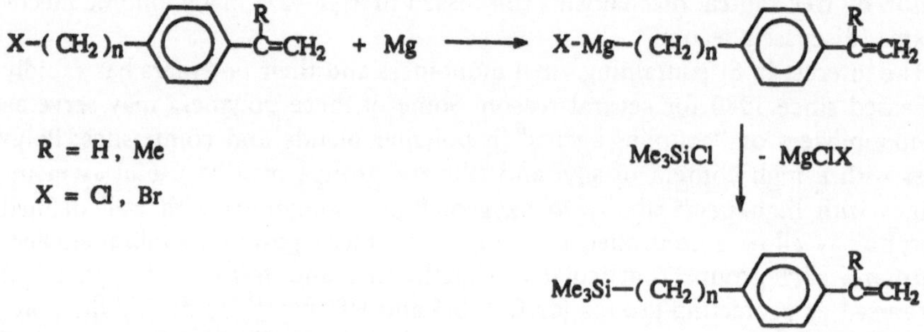

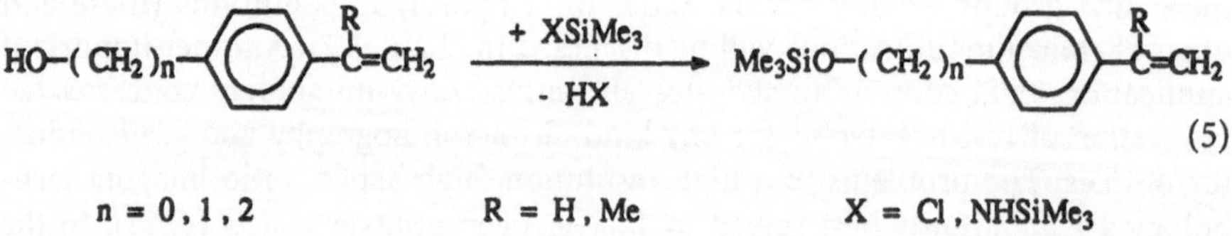

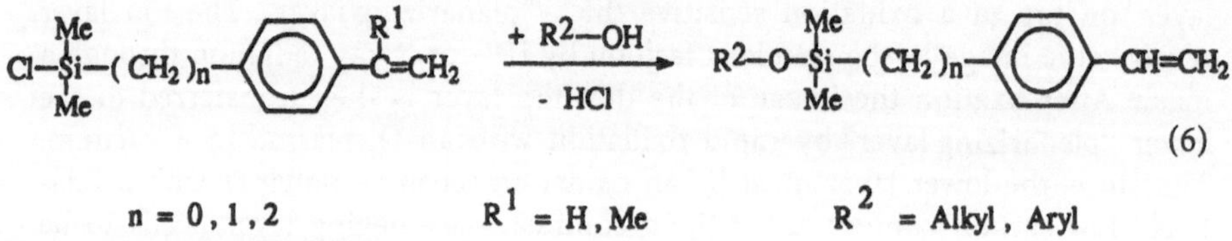

$$C_6H_5\text{-CO-CH}_3 \quad \xrightarrow[\text{- HCl}]{\text{+ ClSiMe}_3} \quad C_6H_5\text{-C}\begin{smallmatrix}\text{OSiMe}_3\\\text{CH}_2\end{smallmatrix} \qquad (7)$$

reagent of the styrene, Eqs. (1, 2) [1, 2, 31–33]. The silylation of OH groups is usually performed with chlorosilanes (in combination with a *tert*-amine) or with hexamethyldisilazane, Eq. (5). A *tert*-amine is again used as HCl-acceptor when styrenes with chlorosilane groups are reacted with alcohols or phenols, Eq. (6). The synthesis of α-trimethylsiloxystyrenes from acetophenone and chlorotrimethylsilane requires a strong base such as triethylamine or sodium hydride as HCl-acceptor, Eq. (7)[34].

The most frequently studied Si-containing styrene is the simplest one: 4-(trimethylsilyl)styrene[1–9]. In addition to the synthesis outlined in Eqs. (1, 2) [2, 9, 31–33] a more cumbersome synthetic route was described in one of the first papers dealing with this monomer, Eqs. (8–11) [1]. However, the 4-trichlorosilyl styrene prepared as intermediate Eq. (10) is useful for the preparation of numerous Si-containing styrenes such as the 4-(trimethoxysilyl)-styrene Eq. (12) [1]. The homopolymerization of 4-TMS-styrene and the properties of this resulting homopolymer have recently attracted increasing interest in connection with bilayer resists [2–6] MW's around $10\text{-}20 \times 10^3$ were reported, but no

$$\text{Br-}C_6H_4\text{-CH}_2\text{-CH}_3 + \text{SiCl}_4 \quad \xrightarrow[\text{- (MgClBr)}]{\text{+ (Mg)}} \quad \text{Cl}_3\text{Si-}C_6H_4\text{-CH}_2\text{-CH}_3 \qquad (8)$$

$$\text{Cl}_2\,(\text{h}\nu) \quad\quad\quad \text{- HCl} \qquad\qquad\qquad\qquad\qquad (9)$$

$$\text{Cl}_3\text{Si-}C_6H_4\text{-}\begin{smallmatrix}\text{Cl}\\\text{CH}\end{smallmatrix}\text{-CH}_3 \quad \xrightarrow[\text{- HCl}]{\Delta T} \quad \text{Cl}_3\text{Si-}C_6H_4\text{-CH=CH}_2 \qquad (10)$$

$$3\ \text{ClMgMe} \Big\downarrow\ \text{- 3 MgCl}_2 \qquad\qquad\qquad\qquad (11)$$

$$\text{Me}_3\text{Si-}C_6H_4\text{-CH=CH}_2$$

$$\text{Cl}_3\text{Si-}C_6H_4\text{-CH=CH}_2 \quad \xrightarrow[\text{- 3 HCl}]{\text{+ 3 MeOH}} \quad (\text{MeO})_3\text{Si-}C_6H_4\text{-CH=CH}_2 \qquad (12)$$

tacticity or T_g is available. The copolymerizations of styrene with both 4-TMS-styrene 11 and 4-(trimethoxysilyl)styrene were studied in detail [1]. The reactivity ratios determined for benzoylperoxide-initiated bulk polymerizations at 70 °C are listed in Table 1.1. Interestingly quite different reactivity ratios were obtained by another group [9] when the copolymerizations were conducted in toluene (Table 1.1). Copolymers with 4-chlorostyrene are of particular interest as imaging

Table 1. Reactivity ratios of radical copolymerizations of various styrene derivatives

Comonomers	Formula	r_1	r_2	Ref.
Styrene/Me$_3$SiC$_6$H$_4$CH = CH$_2$	(2)[a]	1.0	1.0	1
Styrene/(MeO)$_3$SiC$_6$H$_4$C = CH$_2$	(2)	0.48	0.28	9
Styrene/MeSiH$_2$C$_6$H$_4$CH = CH$_2$	(13)[a]	0.91	1.0	10
Styrene/HSiMe$_2$C$_6$H$_4$CH = CH$_2$	(13)[b]	0.56	1.0	14
Styrene/Me$_3$Si-SiMe$_2$C$_6$H$_4$CH = CH$_2$	(14)[a]	0.7	1.13	12
Styrene/Me$_3$Si-SiMe$_2$C$_6$H$_4$CH = CH$_2$	(14)[b]	0.88	0.04	12
Styrene/ClCH$_2$SiMe$_2$C$_6$H$_4$CH = CH$_2$	(15)[a]	0.86	0.69	13
Styrene/ClCH$_2$SiMe$_2$OSiMe$_2$C$_6$H$_4$CH = CH$_2$	(15)[b]	1.08	1.01	13
Styrene/C$_6$H$_5$SiMe$_2$C$_6$H$_4$CH = CH$_2$	(19)[a]	0.80	2.3	14
Styrene/CH$_2$ = CH-SiMe$_2$C$_6$H$_4$CH = CH$_2$	(19)[b]	0.90	1.8	14
Styrene/CH$_2$ = CH-CH$_2$-SiMe$_2$C$_6$H$_4$CH = CH$_2$	(19)[a]	0.69	0.96	14
4-Cl-Styrene/C$_6$H$_5$SiMe$_2$C$_6$H$_4$CH = CH$_2$	(19)[a]	0.51	0.51	14
4-Cl-Styrene/CH$_2$ = CHSiMe$_2$C$_6$H$_4$CH = CH$_2$	(19)[b]	0.76	0.84	14
MMA/C$_6$H5SiMe$_2$C$_6$H$_4$CH = CH$_2$	(19)[a]	0.50	0.05	14
MMA/CH$_2$ = CH-SiMe$_2$C$_6$H$_4$CH = CH$_2$	(19)[b]	0.76	0.18	14
Styrene/Me$_3$SiCH$_2$C$_6$H$_4$CH = CH$_2$	(20)	1.15	0.36	17
Styrene/(Me$_3$Si)$_2$CHC$_6$H$_4$CH = CH$_2$	(22)	1.15	0.31	17
MMA/Me$_3$SiCH$_2$C$_6$H$_4$CH = CH$_2$	(20)	0.17	0.06	17
MMA/(Me$_3$Si)$_2$CHC$_6$H$_4$CH = CH$_2$	(22)	0.20	0.10	17
Styrene/Me$_3$CSiMe$_2$OC$_6$H$_4$CH = CH$_2$	(31)[b]	0.90–0.95	0.90[c]	23
		1.02	1.00[d]	23
		1.12	0.87[e]	23
Styrene/(Me$_3$SiO)$_2$BC$_6$H$_4$-CH = CH$_2$		0.87	0.28	25
Styrene/C$_6$H$_5$—C(OSiMe$_3$)=CH$_2$		1.48	0	26
Acrylonitrile/C$_6$H$_5$—C(OSiMe$_2$)=CH$_2$		0.05	0	27

[a] in bulk; [b] in toluene; [c] 70°C low conversion, Fineman-Ross or Kelen-Tüdös method, [d] 70°C Mayo-Lewis method, [e] 80°C high conversion Mayo-Lewis method

layers of bilayer resists because the chlorostyrene units enable radiation-induced crosslinking [4–6]. For the copolymer of 4-TMS styrene and N-(4-hydroxyphenyl)-maleimide an alternating sequence was postulated and a M_w of 53×10^3 with M_w/M_n ratio near 3 was reported [7]. Finally, a terpolymer prepared by AiBN initiated polymerization of 4-TMS-styrene, 2-hydroxyethyl acrylate and N, N-diethylaminoethyl acrylate was described [8]. The resulting terpolymer seems to be useful as membrane for the encapsulation of insulin. Yet Ref [8] again provides little information about the basic properties of the terpolymer.

Synthesis and polymerizations of two styrenes with hydridosilyl groups 13a, b, were reported by three different research groups [10, 11, 14]. These monomers were prepared according to Eqs. (1, 2). The AiBN initiated copolymerization of 13a with styrene was conducted at 60 °C in bulk and gave the reactivity ratios listed in Table 1.1 [10]. The copolymerization of 13b with styrene was again

$$\text{Me-}\underset{\underset{\text{H}}{|}}{\overset{\overset{\text{H}}{|}}{\text{Si}}}\text{-}\langle\bigcirc\rangle\text{-CH=CH}_2 \qquad\qquad \text{H-}\underset{\underset{\text{Me}}{|}}{\overset{\overset{\text{Me}}{|}}{\text{Si}}}\text{-}\langle\bigcirc\rangle\text{-CH=CH}_2 \qquad (13)$$

$$\qquad\qquad\qquad\text{a}\qquad\qquad\qquad\qquad\qquad\qquad\text{b}$$

initiated with AiBN, but conducted in toluene at 80 °C [14]. The reactivity ratios (Table 1.1) differ somewhat from those of 13a.

Two monomers with a pentamethyldisilane group in the *para* position 14a, b, were synthesized according to Eqs. (1, 2) [12]. The homo- and copolymerizations were conducted either with AiBN at 60 °C or with dibenzoylperoxide at 80 °C in benzene. The homopolymers possess relatively high T_g's. The copolymerizations with styrene were analyzed with regard to the reactivity ratios (Table 1.1) [12].

$$\text{Me}_3\text{Si-}\underset{\underset{\text{Me}}{|}}{\overset{\overset{\text{Me}}{|}}{\text{Si}}}\text{-}\langle\bigcirc\rangle\text{-}\overset{\overset{\text{R}}{|}}{\text{C}}\text{=CH}_2 \qquad\qquad \begin{array}{l}\text{a : R = H} \\ \text{b : R = Me}\end{array} \qquad (14)$$

$$\text{ClCH}_2\text{-}\underset{\underset{\text{Me}}{|}}{\overset{\overset{\text{Me}}{|}}{\text{Si}}}\text{-}\left(\text{O-}\underset{\underset{\text{Me}}{|}}{\overset{\overset{\text{Me}}{|}}{\text{Si}}}\right)_n\text{-}\langle\bigcirc\rangle\text{-CH=CH}_2 \qquad\qquad \begin{array}{l}\text{a : n = 0} \\ \text{b : n = 1} \\ \text{c : n = 2} \\ \text{d : n = 3}\end{array} \qquad (15)$$

$$\text{ClCH}_2\text{-}\underset{\underset{\text{Me}}{|}}{\overset{\overset{\text{Me}}{|}}{\text{Si}}}\left(\text{O-}\underset{\underset{\text{Me}}{|}}{\overset{\overset{\text{Me}}{|}}{\text{Si}}}\right)_m\text{-Cl} \quad + \quad \text{HO-}\underset{\underset{\text{Me}}{|}}{\overset{\overset{\text{Me}}{|}}{\text{Si}}}\text{-}\langle\bigcirc\rangle\text{-CH=CH}_2$$

$$m = 0, 1, 2 \qquad\qquad\qquad\qquad\qquad\qquad\qquad\qquad (16)$$

$$+ \text{Py} \quad\Big\downarrow\quad - \text{Py} \cdot \text{HCl}$$

$$(\underline{15})\ \underline{b}, \underline{c}, \underline{d}$$

$$\text{H-}\underset{\underset{\text{Me}}{|}}{\overset{\overset{\text{Me}}{|}}{\text{Si}}}\left(\text{O-}\underset{\underset{\text{Me}}{|}}{\overset{\overset{\text{Me}}{|}}{\text{Si}}}\right)_n\text{-}\langle\bigcirc\rangle\text{-CH=CH}_2 \qquad\qquad \begin{array}{l}\text{a : n = 1} \\ \text{b : n = 2} \\ \text{c : n = 3} \\ \text{d : n = 4}\end{array} \qquad (17)$$

$$\text{Cl-}\underset{\underset{\text{Me}}{|}}{\overset{\overset{\text{Me}}{|}}{\text{Si}}}\left(\text{O-}\underset{\underset{\text{Me}}{|}}{\overset{\overset{\text{Me}}{|}}{\text{Si}}}\right)_n\text{-}\langle\bigcirc\rangle\text{-CH=CH}_2 \qquad\qquad \begin{array}{l}\text{a : n = 0} \\ \text{b : n = 1} \\ \text{c : n = 2} \\ \text{d : n = 3}\end{array} \qquad (18)$$

$$\text{R-}\underset{\underset{\text{Me}}{|}}{\overset{\overset{\text{Me}}{|}}{\text{Si}}}\text{-}\langle\bigcirc\rangle\text{-CH=CH}_2 \qquad\qquad \begin{array}{l}\text{a : R = C}_6\text{H}_5 \\ \text{b : R = CH=CH}_2 \\ \text{c : R = CH}_2\text{-CH=CH}_2 \\ \text{d : R = O-CH}_2\text{-CH=CH}_2\end{array} \qquad (19)$$

Three classes of styrenes with oligosiloxanes in p-position (**15b–d, 17a–d** and **18b–d**) were synthesized by Greber and coworkers [13, 14, 34, 35]. The syntheses of **15a, 18a** and **19a–d** is feasible according to Eqs. (1, 2). In the case of **15b–d** the route of Eq. (16) was more successful [13]. The homopolymers of **15a–d** were prepared by means of AiBN in toluene but with exception of elemental analyses no properties were described. The copolymerizations with styrene were nearly ideal (Table 1.1). The AiBN initiated copolymerizations of all other "oligosiloxanes" **17a–d** and **18b–d** yielded again reactivity ratios around 1.0 and are not listed in Table 1.1. In the case of monomers **19b–d** homopolymerizations were described, and the isolated homopolymers characterized by elem. analyses and IR-spectra. The reactivity ratios of copolymerizations of **19a–c** are compiled in Table 1.1. Taken together, it may be said the groups directly attached to the aromatic ring have little influence on the reactivity of the vinyl group, so that copolymerizations with styrene tend to take an ideal course.

Numerous styrenes with silyl groups attached to the α-carbon of 4-methylstyrene were prepared by Naga and Tsuruta [15–20] via lithiation reactions. 4-Methylstyrene is acidic enough to allow for lithiation with lithium diisopropylamide, Eq. (20). Addition of chlorotrimethyl silane gave the α-trimethylsilyl-4-methylstyrene (**21**) in high yields [15,17]. This monomer is less acidic than 4-methylstyrene, but allows again lithiation, Eq. (22). With addition of chlorotrimethyl silane the α,α-bis(trimethylsilyl)-4-methylstyrene (**23**) was obtained. Starting from lithiated 4-methylstyrene the monomers **24a–c** were prepared analogously to **21**. By means of **22** monomers **25a,b** and **26a,b** were synthesized in analogy to Eqs. (22, 23). Treatment of the lithiated α-trimethylsilyl-4-methylstyrene with oxiranes yielded the monomers **27a–c** [16]. However radical polymerizations of these monomers have not been reported yet. Another

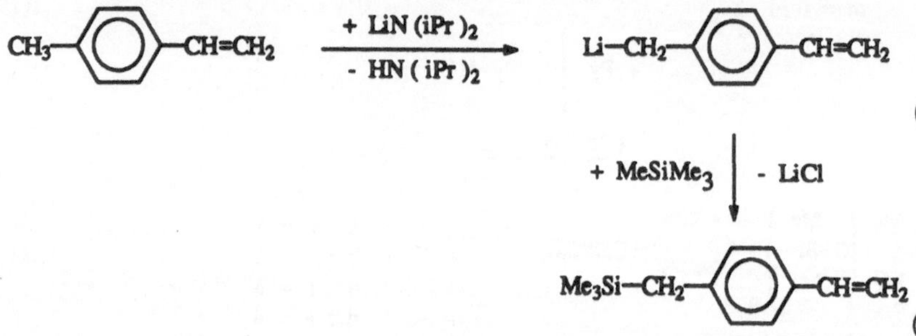

(20)

(21)

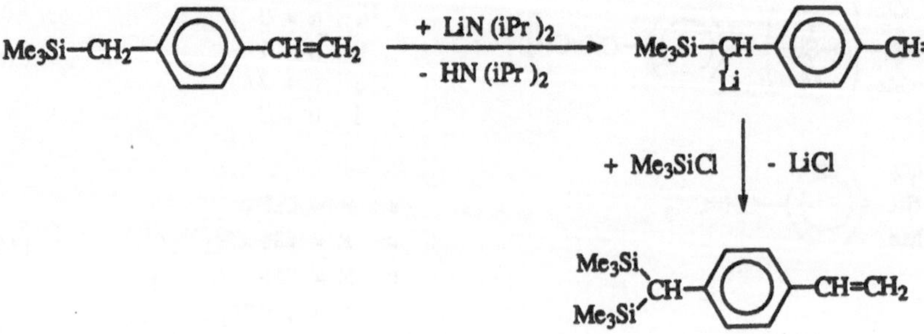

(22)

(23)

interesting property of the lithiated monomer (20) is its capability of initiating the anionic oligomerization of monomer 21. Depending on the reaction conditions either oligomers of structure 28 were formed by an polyaddition process, or copolymers consisting of the structural elements 28 and 29[19]. In other words the polyaddition and an anionic vinyl polymerization occurred simultaneously. Owing to the styrene-type endgroups these oligomers are macromers, however, details of their radical polymerization have not been reported so far [19].

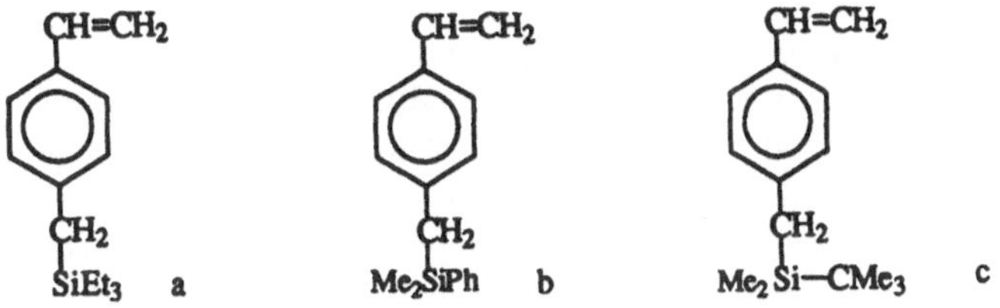

(24)

(25)

(26)

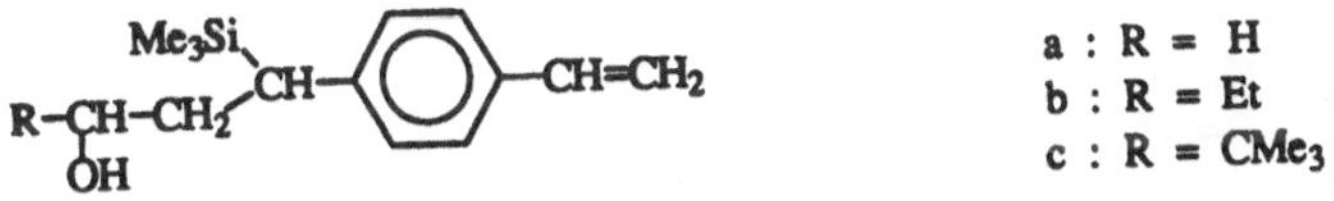

a : R = H
b : R = Et
c : R = CMe$_3$

(27)

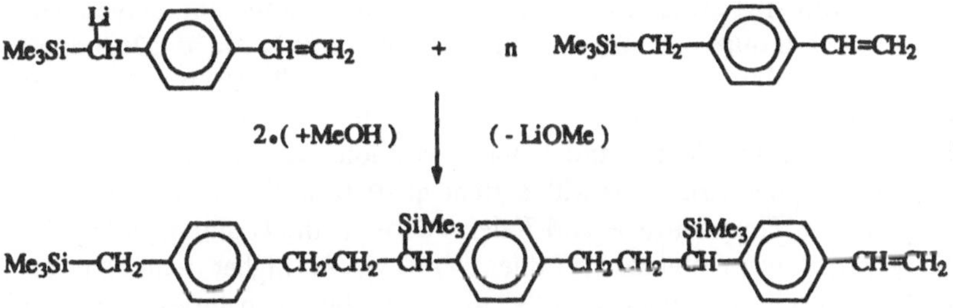

(28)

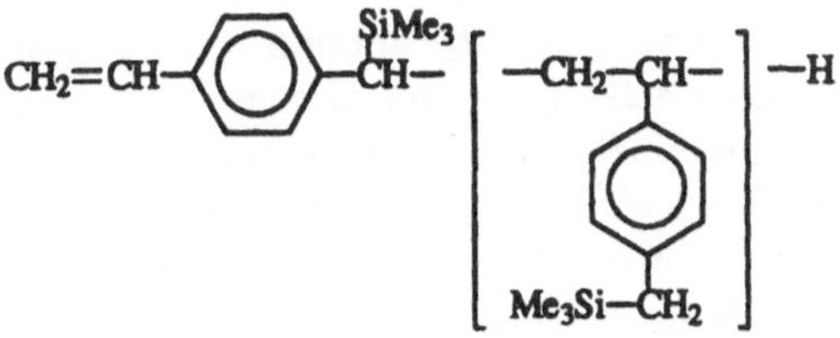

(29)

The reactivities of monomers 20 and 22 were studied in great detail and the Hammett σ-values were determined [17, 19]. AiBN-initiated polymerizations of both monomers were conducted at 60 °C in bulk. For the homopolymer of (22) molecular weights in the range of $400-500 \times 10^3$, a narrow MWD and a T_g of 148 °C was reported [18]. Furthermore, monomers 20 and 22 were copolymerized with styrene and MMA [17]. The reactivity ratios are given in Table 1.1. The properties of membranes with regard to gas permeation were also evaluated [18].

The monomer 30a was prepared according to Eqs. (3) and (4) from chloro-dimethylphenylsilane [21]. It was copolymerized with divinylbenzene and the resulting network was subjected to various chemical modifications based on the acid catalyzed cleavage of the Si-Ph bond [21]. The monomer 30b was prepared from 4-mercaptomethylstyrene, by silylation with *tert*-butyldimethylchlorosilane [22]. Its homopolymerization was initiated with AiBN at 80 °C in bulk. Afterwards the silyl protecting group was hydrolyzed and a polystyrene with free mercapto groups was isolated [22].

$$\text{Ph-}\underset{\underset{\text{Me}}{|}}{\overset{\overset{\text{Me}}{|}}{\text{Si}}}\text{-(CH}_2\text{)}_3\text{-}\langle\bigcirc\rangle\text{-CH=CH}_2 \qquad \text{Me}_3\text{C-}\underset{\underset{\text{Me}}{|}}{\overset{\overset{\text{Me}}{|}}{\text{Si}}}\text{-S-CH}_2\text{-}\langle\bigcirc\rangle\text{-CH=CH}_2 \qquad (30)$$

$$\qquad\qquad a \qquad\qquad\qquad\qquad\qquad\qquad\qquad b$$

$$\text{R}^2\text{-}\underset{\underset{\text{R}^1}{|}}{\overset{\overset{\text{R}^1}{|}}{\text{Si}}}\text{-O-(CH}_2\text{)}_n\text{-}\langle\bigcirc\rangle\text{-CH=CH}_2$$

$$
\begin{array}{lll}
a: & n = 0 & R^1, R^2 = Me \\
b: & n = 0 & R^1 = Me, R^2 = CMe_3 \\
c: & n = 2 & R^1, R^2 = Me \\
d: & n = 2 & R^1 = Me, R^2 = CMe_3
\end{array}
\qquad (31)
$$

$$\text{R}^2\text{-}\underset{\underset{\text{R}^1}{|}}{\overset{\overset{\text{R}^1}{|}}{\text{Si}}}\text{-O-}\langle\bigcirc\rangle\text{-}\overset{\overset{\text{Me}}{|}}{\text{C}}\text{=CH}_2$$

$$
\begin{array}{ll}
a: & R^1, R^2 = Me \\
b: & R^1 = Me, R^2 = CMe_3
\end{array}
\qquad (32)
$$

Styrenes with silyl-protected OH-groups were synthesized and polymerized by several research groups [23-27]. Among the monomers 7, 31a-d, 32a,b and 36 it was the styrene 31b which was studied first in more detail [23]. It was synthesized by silylation of 4-hydroxystyrene with *tert*-butyldimethylchloro-silane according to Eq.(5) [36, 37]. Both homo- and copolymerizations were initiated with AiBN in solution. The copolymerizations with styrene at 70 °C in dioxane was evaluated by the Mayo-Lewis, Fineman-Ross and Kelen-Tüdös methods and slightly differing reactivity ratios were obtained (Table 1.1) [23]. At higher temperatures the reactivity ratios were again slightly different, but nonetheless nearly ideal copolymerizations were observed under all circumstances.

The monomers 31a-c and 32a were synthesized by another research group [24] via silylation of 4-hydroxystyrene or 4-(2'-hydroxyethyl)styrene with hexa-methyldisilazane, Eq. (5). The monomers 31b and 32b were prepared by the novel approach outlined in Eqs. (33, 34). This approach is remarkable, because the vinyl group was formed after the silylation of the OH-group. Monomer 31d was obtained by silylation of 4-(2'-hydroxyethyl)styrene with t-BuSi(Me)$_2$Cl and imidazole.

$$HO-\text{C}_6H_4-\underset{R}{\overset{}{\text{C}}}=O \quad \xrightarrow[(\text{Imidazole})]{+ \text{ClSi (Me)}_2\text{CMe}_3} \quad Me_3C-\underset{Me}{\overset{Me}{\text{Si}}}-O-C_6H_4-\underset{R}{\overset{}{\text{C}}}=O \quad (33)$$

$$\downarrow \begin{array}{c} Me_3C\,O\,K \\ Ph_3\,P\,CH_2\,Br \end{array} \quad - \text{KBr}, - \text{HOCMe}_3$$

$$(34)$$

$$(\underline{31})\ \underline{b},\,(\underline{32})\ \underline{b}$$

R = H, Me

$$\left[\begin{array}{c} -CH_2-CH- \\ C_6H_4 \\ | \\ O \\ | \\ CO-O-CMe_3 \end{array} \middle| \begin{array}{c} -CH_2-CH- - \\ C_6H_4 \\ | \\ (CH_2)_2 \\ | \\ O-SiMe_3 \end{array} \right] \quad (35)$$

$$\underset{R-O}{\overset{R-O}{>}}B-C_6H_4-CH=CH_2 \qquad \begin{array}{l} a:R=H \\ b:R=SiMe_3 \end{array} \quad (36)$$

Unfortunately, the monomers and reaction intermediates were only characterized by IR- and ^{1}H NMR spectroscopic data. Radical homopolymerizations were reported for 31a, b and d. The resulting homopolymers were characterized by M_n, M_w and spectroscopic data. Several attempts were made to prepare the copolymer 35 from 31c. Whereas initiation with radicals only yielded oligomers, anionic copolymerizations with naphthalin/lithium in THF at -78 °C proved to be more successful.

Of particular interest is the successful cationic polymerization of monomers 32a, c, because cationic polymerizations of Si-containing styrenes have never been studied in detail. Using $BF_3.OEt_2$ as catalyst and H_2O as cocatalyst, M_n values around 400×10^3 were obtained for 32a or 500×10^3 for 32b [24]. IR- and ^{1}H-NMR spectra of the resulting polymers were reported.

Polymers containing benzene boronic acid groups are of analytical or preparative interest, because the boronic acid groups reversibly form complexes with saccharides which can be utilized for chromatographic purposes [38]. In this connection the paper was published describing the radical co-polymerization of monomers 36a and b with styrene [25]. Nearly identical reactivity ratios were found for the combinations styrene/36a and styrene/36b (Table 1.1). Hydrolysis of the silylester groups with various metal hydroxide solutions yielded copolymers with pending metal boronate groups. Furthermore, the influence of the boronic acid content on the T_g of the copolymers was investigated.

Last but not least, synthesis and polymerization of a trimethylsiloxystyrene need discussion. The synthesis of this monomer from acetophenone and chlorotrimethyl silane [26, 27, 39] is illustrated by Eq. (7). Apparently for thermodynamical reasons neither a radical nor an anionic homopolymerization of this monomer was successfull [26]. AiBN initiated copolymerizations with styrene confirmed the absence of homopropagation steps of α-trimethylsiloxystyrene [26]. The reactivity ratios are given in Table 1. Copolymerizations with acetonitrile [26], maleic anhydride or fumarodinitrile [27] yielded almost perfectly alternating copolymers (Form 37b, 38a,b). A terpolymer with styrene and maleic anhydride was also prepared. Furthermore the hydrolysis of the TMS-O groups was studied [26, 27] (chapter 8).

$$ \left[-CH_2-CH-\backslash-CH_2-\overset{\overset{\displaystyle O-TMS}{|}}{C}- \right]_a \qquad \left[-CH_2-\overset{\overset{\displaystyle }{|}}{\underset{\underset{\displaystyle CN}{|}}{CH}}-CH_2-\overset{\overset{\displaystyle O-TMS}{|}}{C}- \right]_b \qquad (37) $$

a   b

$$ \left[-\overset{\overset{\displaystyle }{|}}{\underset{\underset{\displaystyle OC_{}O_{}CO}{}}{CH}}-CH-CH_2-\overset{\overset{\displaystyle O-TMS}{|}}{C}- \right]_a \qquad \left[-\overset{\overset{\displaystyle CN}{|}}{CH}-\overset{\overset{\displaystyle CN}{|}}{CH}-CH_2-\overset{\overset{\displaystyle O-TMS}{|}}{C}- \right]_b \qquad (38) $$

a   b

1.1.2
Anionic Polymerizations

Quite recently Hirao and Nakahama published two reviews dealing with the anionic living polymerization of vinyl-monomers with functional silyl groups and some other protected functional groups [40, 41]. Owing to the availability of these reviews the discussion of this topic in the present section is somewhat shortened, in as much as most contributions to this field have been published by Hirao, Nakama and coworkers [42 – 54]. These authors conducted the anionic polymerizations of Si-containing styrenes under the following standard conditions. Lithium, sodium or potassium naphthalenide served as electron donors for a small amount of α-methylstyrene which played the role of a coinitiator, Eq.(39). The radical anion of α-methylstyrene dimerizes rapidly with subsequent polymerization yielding an oligomeric dianion, Eq.(40), which served as initiator in almost all polymerizations (e.g. Eq. 41). These mostly living polymerizations were stopped by addition of methanol or methyl iodide, so that polymers with stable endgroups were formed. These polymers were called homopolymers, albeit they contain a short central block of α-methylstyrene.

All polymerizations were conducted at −78 °C in THF or THF/pentane mixtures to minimize side reaction. Side reactions, if they occured at all, will be

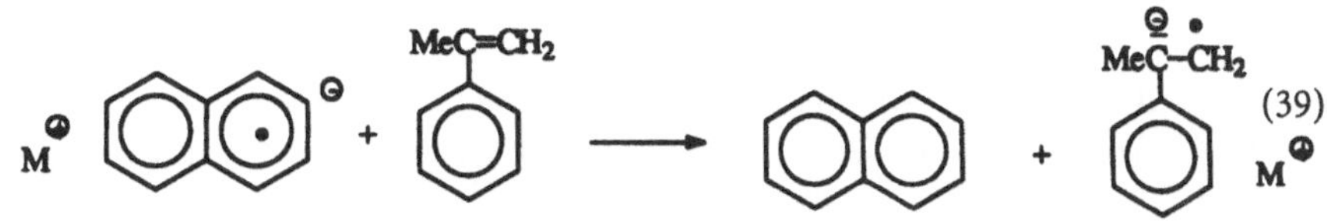

$$M^{\oplus} \text{(naphthalene)}^{\ominus} \cdot + \underset{MeC=CH_2}{\bigcirc} \longrightarrow \text{(naphthalene)} + \underset{MeC-CH_2}{\overset{\ominus \ \bullet}{\bigcirc}} M^{\oplus} \qquad (39)$$

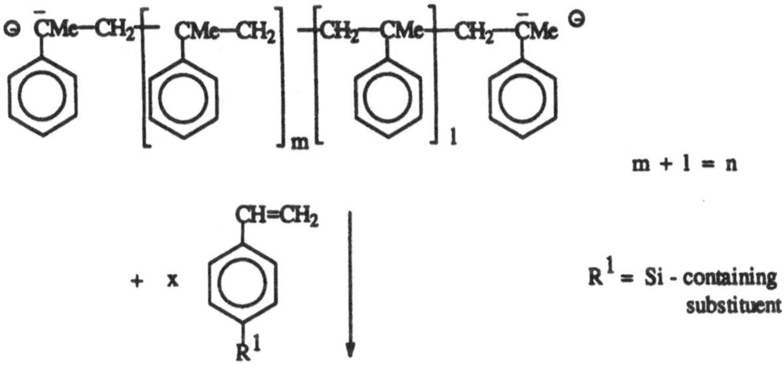

$$m + l = n \qquad (40)$$

$$+ \ x \ \underset{R^1}{\overset{CH=CH_2}{\bigcirc}} \qquad R^1 = Si - \text{containing substituent}$$

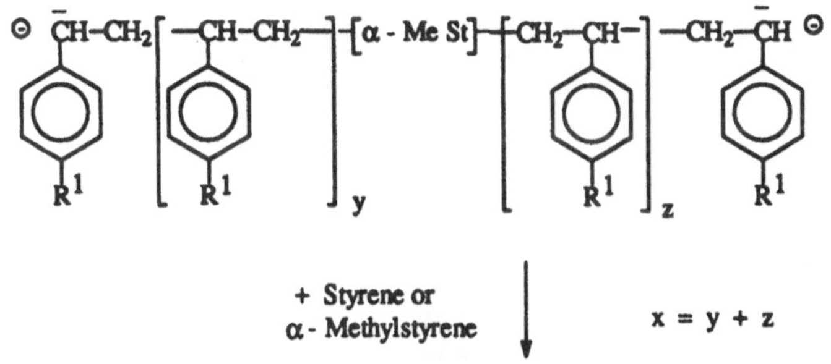

$$(41)$$

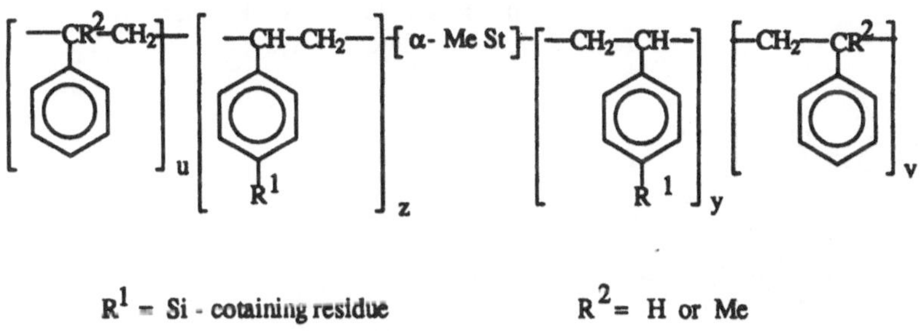

+ Styrene or
α - Methylstyrene

$$x = y + z$$

$$(42)$$

R^1 = Si - cotaining residue R^2 = H or Me

discussed below in connection with individual monomer. However, in most cases side reactions were nearly absent and the polymerizations showed a true living character. Accordingly, the DP's paralleled the M/I ratios and narrow MWD's with M_w/M_n ratios around 1.1 ± 0.05 were found. An important preparative aspect of these living polymerizations is the synthesis of tailor-made A-B-A triblock copolymers. Two versions of triblock copolymers were synthesized. Either the Si-containing styrene formed the B-block (in combination with the initiator fragment) and styrene or α-methylstryene the A-blocks, Eq.(42), or the inverse pattern was prepared with isoprene, styrene or α-methylstyrene as B-monomers **43**.

A broad variety of 4-(alkoxysilyl)styrenes, **44a–h**, **45a,b** and **46**, were synthesized according to Eqs. (1, 2) or from methoxysilanes according to Eq. (47) [42, 43]. Their anionic polymerizations were conducted under standard conditions and the stability of the coloured carbanions was studied in detail. Lithium as counterion showed a marked tendency to cause side reactions with the methoxy Si groups. Nonetheless most polymerizations followed the "living pattern" and molecular weights in the range of $13 \times 10^3 - 36 \times 10^3$ were obtained. The isopropyloxysilane **44d** was widely used to synthesize triblockcopolymers with isoprene **43a** [44, 45]. Films of these blockcopolymers were prepared under conditions leading to phase separation of the A and B-blocks. In order to obtain microporous membranes with well defined morphology and pore size, the isoprene blocks of the biphasic films were selectively oxidized with O_3 [44, 45].

Side reactions of the anionic chain end were also observed when the dimethylsilyl styrene **42a** was polymerized in neat THF [46]. Living polymerizations

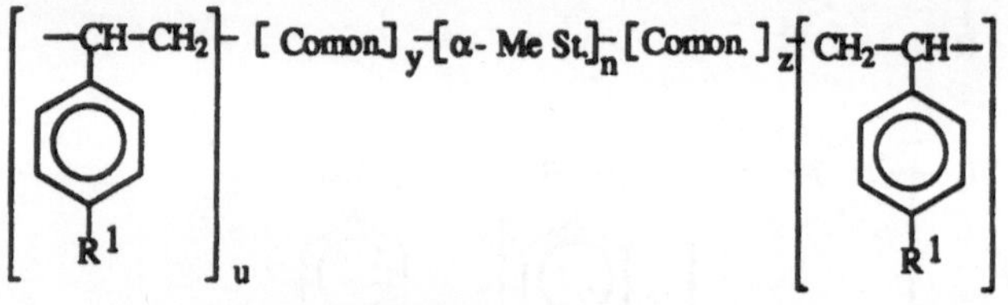

$$\left[\begin{array}{c} -CH-CH_2- \\ \bigcirc \\ R^1 \end{array}\right]_u [\, Comon\,]_y [\alpha\text{-}Me\,St\,]_n [\,Comon.\,]_z \left[\begin{array}{c} CH_2-CH- \\ \bigcirc \\ R^1 \end{array}\right]_v \qquad (43)$$

a : Comon. = isoprene
b : Comon. = styrene
c : Comon. = α- methylstyrene

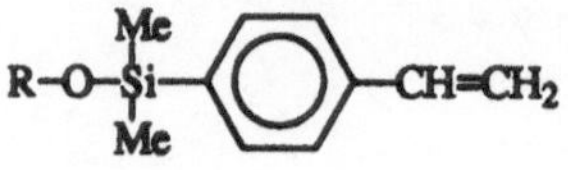

$$R-O-\underset{\underset{Me}{|}}{\overset{\overset{Me}{|}}{Si}}-\bigcirc-CH=CH_2 \qquad (44)$$

a : R = Me   e : R = n - Bu
b : R = Et   f : R = i - Bu
c : R = n - Pr g : R = sec. - Bu
d : R = i - Pr h : R = tert. - Bu

$$Me-\underset{\underset{RO}{|}}{\overset{\overset{RO}{|}}{Si}}-\bigcirc-CH=CH_2 \qquad (45)$$

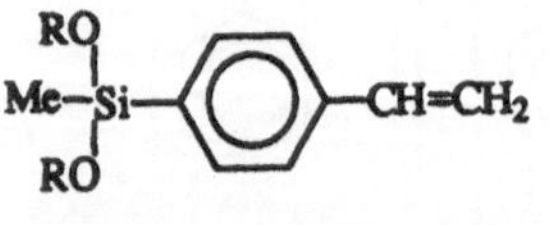

a : R = Et
b : R = i - Pr

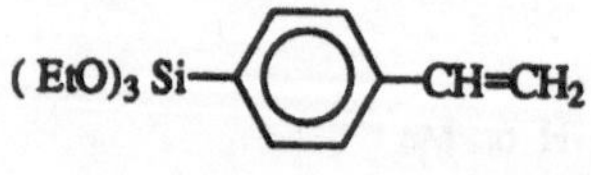

$$(EtO)_3 Si-\bigcirc-CH=CH_2 \qquad (46)$$

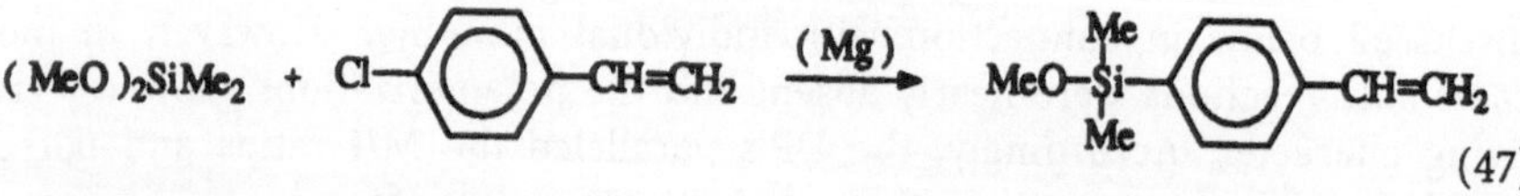

$$(MeO)_2SiMe_2 + Cl-\bigcirc-CH=CH_2 \xrightarrow{(Mg)} MeO-\underset{\underset{Me}{|}}{\overset{\overset{Me}{|}}{Si}}-\bigcirc-CH=CH_2 \qquad (47)$$

$$H-\underset{\underset{Me}{|}}{\overset{\overset{Me}{|}}{Si}}-\bigcirc-CH=CH_2 \qquad Et_2N-\underset{\underset{Me}{|}}{\overset{\overset{Me}{|}}{Si}}-\bigcirc-CH=CH_2 \qquad (48)$$

a   b

were obtained in mixtures of THF and pentane. A triblock copolymer of type **42** was prepared with styrene ($R^2 = H$) and two triblock copolymers of type **43** with isoprene and styrene. Monomer **48a** like **48b** was prepared from the corresponding chlorosilanes according to Eq. (1). The living polymerizations of **48b** proceeded with both Li and K counterions without side reactions [47].

In addition to styrenes with the silicon directly attached to the aromatic ring numerous monomers were synthesized with more than one σ-bond between Si and styrene unit. To the latter group of monomers belong the 4-(2'-siloxyethyl)styrenes **49a–e**. These monomers were prepared according to Eq. (5) by silylation of 4-(2'-hydroxyethyl)styrene with chlorosilanes and imidazole [48, 49]. The anionic polymerization of these monomers under standard conditions followed the "living pattern". In addition to homopolymers, numerous di- and triblock copolymers **42** and **43b,c** were prepared and characterized [49]. As already discussed in connection with radical polymerization, α-trimethylsiloxy styrene cannot homopolymerize for thermodynamical reasons. This severe steric hindrance along with the $+M$ effect of the siloxy group strongly reduces the reactivity of the siloxy vinyl groups against any nucleophilic attack and allows a selective anionic polymerization of the vinyl or methyl vinyl groups of monomers **50b,c**, **51b** and **52a,b**. However, attempted anionic polymerizations of **50a** and **51a** failed. In the case of **50b** and c triblock copolymers with styrene were synthesized. The monomers were synthesized from the corresponding aceto or propiophenones and *tert*-butyldimethylchlorosilane with triethylamine [55] or NaH [56] as proton acceptors.

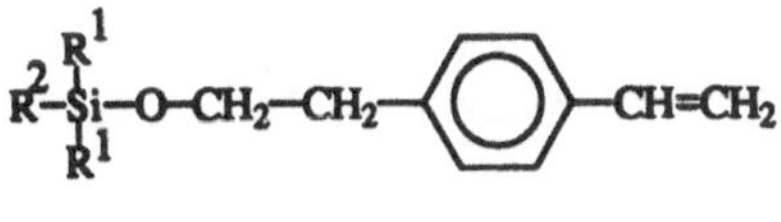

$$
\begin{array}{l}
a : R^1 = R^2 = Me \\
b : R^1 = Me, R^2 = Et \\
c : R^1 = Me, R^2 = i\text{-}Pr \\
d : R^1 = Me, R^2 = CMe_3 \\
e : R^1 = Me, R^2 = OMe
\end{array}
\qquad (49)
$$

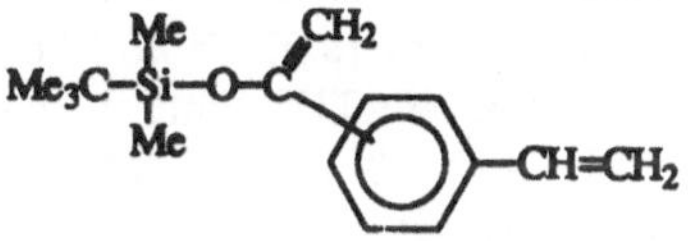

$$
\begin{array}{l}
a : \text{otho} \\
b : \text{meta} \\
c : \text{para}
\end{array}
\qquad (50)
$$

$$(51)$$

a   b

$$
\begin{array}{l}
a : R = H \\
b : R = Me
\end{array}
\qquad (52)
$$

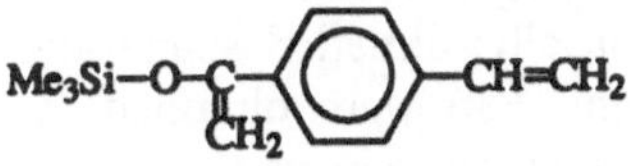

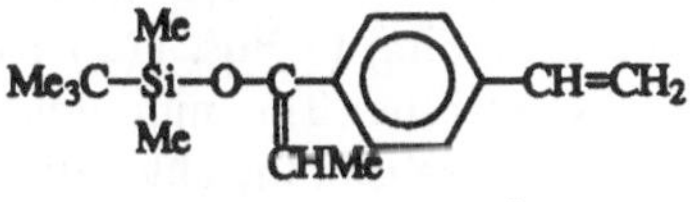

$$
\begin{array}{l}
a : n = 0 \\
b : n = 1 \\
c : n = 2
\end{array}
\qquad (53)
$$

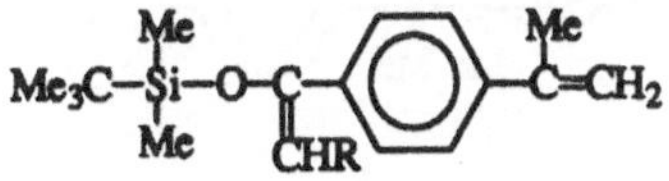

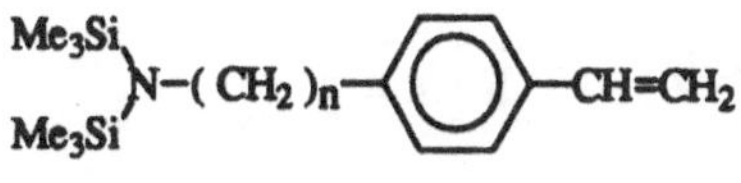

Protection with two TMS groups also allows the anionic polymerization of styrenes with primary amino groups in *para*-position, **53a–c** [51–53].
Monomer **53a** was prepared from 4-aminostyrene by stepwise silylation with
hexamethyldisilazane and chlorotrimethylsilane + EtMgBr [51, 52]. The anionic
polymerization showed a living character with all three counterions. With
styrene and isoprene triblock copolymers of structure **42** and **43** were prepared.
The chemical modification of the resulting homo- and copolymers is described
in chapter 8. Monomers **53b** and **c** were synthesized by the method outlined in
Eqs.(54, 55) [54]. The anionic polymerizations of both monomers under

$$\begin{array}{c} Me_3Si \\ N-CH_2-O-CH_3 \\ Me_3Si \end{array} \Bigg\langle \begin{array}{l} \xrightarrow[]{ClC_6H_4CH=CH_2\quad (Mg)} \quad (53)b \qquad (54) \\[2ex] \xrightarrow[]{ClCH_2C_6H_4CH=CH_2\quad (Mg)} \quad (53)c \qquad (55) \end{array}$$

standard conditions and with BuLi had a living character, and triblock
copolymers of structure **42** were prepared with styrene. Living polymerizations
with M_w/M_n ratios of 1.04 and M_n's around 20×10^3 were also found, when
4-(2'-trimethyl-silylethynyl) styrene was initiated with BuLi, or K-naphth. at
$-78\,°C$ [44]. The monomer itself was obtained from 4-bromostyrene according
to Eq. (56) [57].

$$Me_3SiC\equiv CH \;+\; Br-\!\!\raisebox{-1ex}{\bigcirc}\!\!-CH=CH_2 \xrightarrow[(-HBr)]{Pd} Me_3SiC\equiv C-\!\!\raisebox{-1ex}{\bigcirc}\!\!-CH=CH_2 \qquad (56)$$

Two papers of Saigo et al. [58, 59] describe four new polystyrenes suitable as
top(imaging) layers of bilayer resists for O_2-plasma etching. These poly-styrenes
were prepared by BuLi-initiated polymerization of monomers **57a,b** and **58a,b**.
These anionic polymerizations conducted at $-78\,°C$ in THF had a living
character and yielded M_w's in the range of $15 \times 10^3 - 45 \times 10^3$ with M_w/M_n ratios
< 1,1. The allyl groups did not participate in the polymerizations process, but
are responsible for the radiation sensitivity (imaging process). Monomer **57a**
was prepared according to Eqs. (1,2) from chloroallyl-dimethyl silane and 4-
chloro-α-methyl styrene. The monomers **57b** and **58b** were obtained in an
analogous way by means of the chlorosilanes **59a,b** which were synthesized from

$$CH_2=CH-CH_2-\underset{Me}{\overset{Me}{Si}}-\!\!\raisebox{-1ex}{\bigcirc}\!\!-\underset{}{\overset{Me}{C}}=CH_2 \qquad\qquad CH_2=CH-CH_2-\underset{Me\ Me}{\overset{Me\ Me}{Si-Si}}-\!\!\raisebox{-1ex}{\bigcirc}\!\!-\overset{Me}{C}=CH_2 \qquad (57)$$

$$\text{a} \qquad\qquad\qquad\qquad\qquad\qquad\qquad \text{b}$$

$$CH_2=CH-CH_2-\underset{Me}{\overset{Me}{Si}}-(CH_2)_n-\underset{Me}{\overset{Me}{Si}}-\!\!\raisebox{-1ex}{\bigcirc}\!\!-\overset{Me}{C}=CH_2 \qquad\qquad \begin{array}{l} a : n = 2 \\ b : n = 3 \end{array} \qquad (58)$$

1,2-dichlorotetramethyl silane. Monomer **58a** was obtained in a somewhat different way as outlined in Eqs.(60, 61).

$$CH_2{=}CH{-}CH_2{-}\underset{\underset{Me}{|}}{\overset{\overset{Me}{|}}{Si}}{-}\underset{\underset{Me}{|}}{\overset{\overset{Me}{|}}{Si}}{-}Cl \qquad CH_2{=}CH{-}CH_2{-}\underset{\underset{Me}{|}}{\overset{\overset{Me}{|}}{Si}}{-}(CH_2)_2{-}\underset{\underset{Me}{|}}{\overset{\overset{Me}{|}}{Si}}{-}Cl \qquad (59)$$

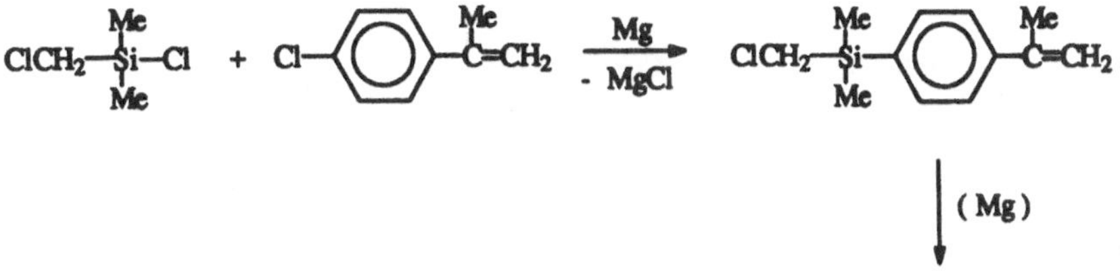

a   b

$$ClCH_2{-}\underset{\underset{Me}{|}}{\overset{\overset{Me}{|}}{Si}}{-}Cl \;+\; Cl{-}\!\!\bigcirc\!\!{-}\overset{\overset{Me}{|}}{C}{=}CH_2 \xrightarrow[-\,MgCl]{Mg} ClCH_2{-}\underset{\underset{Me}{|}}{\overset{\overset{Me}{|}}{Si}}{-}\!\!\bigcirc\!\!{-}\overset{\overset{Me}{|}}{C}{=}CH_2 \quad (60)$$

$$\Big\downarrow (Mg) \qquad (61)$$

$$(\,58\,)\,a$$

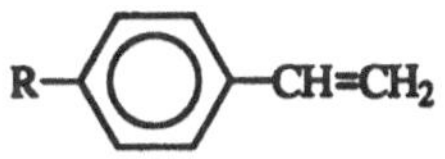

$$R{-}\!\!\bigcirc\!\!{-}CH{=}CH_2$$

a : R = SiMe₃   d : R = CH₂SiMe₃
b : R = H   e : R = CH(SiMe₃)₂   (62)
c : R = CH₃

Finally three studies of the anionic polymerization of several Si-containing styrenes conducted by Nagasaki et al. [60] should be mentioned. With lithium diisopropylamide the following order of decreasing reactivity was found:
(63) > (62)b > (62)c > (62)d > (62)e.

Furthermore, the synthesis of macromers by polyaddition of lithiated monomers, Eqs.(20, 28), was found, and the oligomer **63** was isolated and characterized [60].

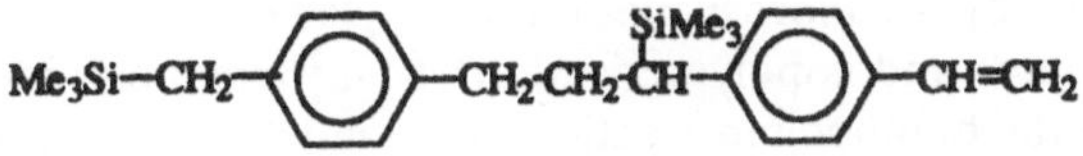

$$Me_3Si{-}CH_2{-}\!\!\bigcirc\!\!{-}CH_2CH_2\overset{\overset{SiMe_3}{|}}{C}H{-}\!\!\bigcirc\!\!{-}CH{=}CH_2 \qquad (63)$$

1.2
Various Vinyl Monomers

This subchapter deals with a variety of Si-containing vinyl monomers, which will be discussed in the following order: vinylsilanes, allylsilanes, methacrylates and fumarates, vinyl ethers and 1,3-dienes.

1.2.1
Vinyl Silanes

The vinyl silanes used as monomers may be subdivided into two groups:

A) Vinyl silanes with one, two or three Si—O—C bonds (alkoxy- or acyloxysilanes), **64a–c**, **65a–f**, **66a,b**, and

$$CH_2=CH-\underset{R}{\overset{R}{Si}}(OEt)_2 \qquad \begin{array}{l} a : R = Me \\ b : R = Et \\ c : R = Ph \end{array} \qquad\qquad (64)$$

$$CH_2=CH-Si(OR)_3 \qquad \begin{array}{ll} a : R = Me & e : R = SiMe_3 \\ b : R = Et & f : R = CO-CH_3 \\ c : R = iPr \end{array} \qquad (65)$$

$$\underset{a}{CH_2=CH-\underset{Me}{\overset{Me}{Si}}-O-SiMe_3} \qquad \underset{b}{CH_2=CH-\overset{Me}{Si}(OSiMe_3)_2} \qquad (66)$$

$$\underset{a}{CH_2=CH-SiMe_3} \qquad \underset{b}{CH_2=CH-SiH_2-Ph} \qquad\qquad (67)$$

$$CH_2=CH-Si-Cl_3 \ + ROH \xrightarrow[-\ HCl]{} CH_2=CH-Si(OR)_3 \qquad (68)$$

$$CH_2=CH-Si-Cl_3 \ + 3\ Ac_2O \xrightarrow[-\ 3\ AcCl]{} CH_2=CH-Si(OAc)_3 \qquad (69)$$

B) vinyl silanes exclusively containing Si—C and Si—H bonds, 67a,b.

Most studies and the oldest publications in this field deal with group A monomers [61–68], and thus, will be discussed first, and group B monomers second [16, 64, 69–71]. The di- and tri-alkoxysilanes are usually prepared by alcoholysis of the corresponding vinyl chlorosilanes in the presence of a tertiary amine, Eq.(68). The acyloxysilane 65f can be prepared by means of sodium acetate or by means of acetic anhydride, Eq.(69). The vinyl chlorosilanes may be synthesized by addition of the hydrochlorosilanes onto acetylene.

In a first study of the homo- and copolymerization of vinyltriethoxysilanes Wagner [61] concluded that the homopolymerization of the vinylsilane requires temperatures $> 130\,°C$ if it takes place at all. All further studies on radical copolymerization of vinyl silanes confirmed the homopolymerization steps are extremely rare, so that the reactivity ratios (r_2) of most vinylsilanes (M_2) are close to 0 [10, 62–68]. Therefore Table 1.2 exclusively summarizes the reactivity ratios (r_1) of the comonomers (M_1).

A comparison of these reactivity ratios needs, of course, to take into account that the reaction conditions selected by different research group vary to some extent. For instance acetyl peroxide was used as initiator in dry acetone at $50\,°C$ in Refs. 62 and 63. In Ref. 64, AiBN initiated bulk copolymerizations were conducted at $60\,°C$. Again bulk copolymerization were evaluated, in Ref. 65, but benzoylperoxide was used as initiator at $80\,°C$. Benzoyl peroxide was used in bulk copolymerizations at $70\,°C$ in Ref. 66, at $50\,°C$ in Ref. 67 and at $60\,°C$ in Ref. 68. In Ref. 65, the temperature dependence of the reactivity ratios in the range of $60–100\,°C$ was investigated. Furthermore, polymer properties such as solubilities and thermostabilities were determined in the work of Babu et al. [65–68].

Table 1.2. Reactivity ratios of vinyl-monomers (M_n) in radical copolymerizations with vinyl silanes (M_2)[a]

Comonomers	r_1	Ref.	Comonomers	r_1	Ref.
Acrylonitrile/$CH_2 = CH - SiMe(OEt)_2$	6.0	62	Styrene/$CH_2 = CH - SiMe_2OSiMe_3$	60.0[b]	63
Acrylonitrile/$CH_2 = CH - SiEt(OEt)_2$	9.0	62	Styrene/$CH_2 = CH - SiMe(OSiMe_3)_2$	60.0[b]	63
Acrylonitrile/$CH_2 = CH - SiPh(OEt)_2$	8.3	62	Styrene/$CH_2 = CH - Si(OSiMe_3)_3$	60.0[b]	63
Acrylonitrile/$CH_2 = CH - Si(OMe)_3$	6.0	62	Acrylonitrile/$CH_2 = CH - SiMe_2(OSiMe_3)$	8.0[b]	63
Acrylonitrile/$CH_2 = CH - Si(OEt)_3$	4.5	62	Acrylonitrile/$CH_2 = CH - SiMe(OSiMe_3)_2$	8.0[b]	63
Acrylonitrile/$CH_2 = CH - Si(OEt)_3$	5.0	64	Acrylonitrile/$CH_2 = CH - Si(OSiMe_3)_3$	8.0[b]	63
Acrylonitrile/$CH_2 = CH - Si(OEt)_3$	4.7	67	Vinylpyrrol/$CH_2 = CH - Si(OSiMe_3)_3$	4.0[b]	63
Acrylonitrile/$CH_2 = CH - Si(OiPr)_3$	6.5	62	Vinylacetate/$CH_2 = CH - SiMe(OSiMe_3)_2$	1.0[a]	63
Acrylonitrile/$CH_2 = CH - Si(OCH_2CH_2OMe)_3$	2.5	67	Vinylacetate/$CH_2 = CH - Si(OSiMe_3)_3$	1.0[a]	63
Vinylchloride/$CH_2 = CH - SiMe(OEt)_2$	1.2	62	Vinylchloride/$CH_2 = CH - SiMe(OSiMe_3)_2$	0.9[d]	63
Vinylchloride/$CH_2 = CH - SiEt(OEt)_2$	1.0	62	Vinylchloride/$CH_2 = CH - Si(OSiMe_3)_3$	0.9[d]	63
Vinylchloride/$CH_2 = CH - SiPh(OEt)_2$	0.7	62	$CF_2 = CFCl/CH_2 = CHSiMe(OSiMe_3)_2$	0.05[c]	63
Vinylchloride/$CH_2 = CH - Si(OMe)_3$	0.8	62	$CF_2 = CFCl/CH_2 = CH - Si(OSiMe_3)_3$	0.05[c]	63
Vinylchloride/$CH_2 = CH - Si(OEt)_3$	0.9	62	Styrene/$CH_2 = CH - Si(OEt)_3$	22.0	64
Vinylchloride/$CH_2 = CH - Si(OiPr)_3$	0.8	62	Styrene/$CH_2 = CH - Si(OEt)_3$	13.0	65
MMA/$CH_2 = CH - Si(OCH_2CH_2OMe)_3$	11.2[c]	68	Styrene/$CH_2 = CH - Si(OCH_2CH_2OMe)_3$	11.2[d]	68
MMA/$CH_2 = CH - Si(OAc)_3$	7.75	66	Styrene/$CH_2 = CH - Si(OAc)_3$	3.5	65
			BuMA/$CH_2 = CH - Si(OAc)_3$	4.62	66

[a] $r_2 = 0.01$ if not otherwise mentioned,　[b] $r_2 \approx 0.1$,　[c] $r_2 = 0.2$,　[d] $r_2 \approx 0.5$

A broader variety of initiators and polymerization mechanisms was reported for the vinyl silanes **67a** and **b**. AiBN initiated copolymerizations of **67a** with styrene were conducted in bulk, and a reactivity ratio (r_1) of 5.7 was found for styrene, but $r_2 = 0$ for the vinyl silane. This result fits well in with the data of Table 1.2. Albeit vinyl silanes are reluctant to undergo radical homopolymerization, monomer **67a** was successfully homopolymerized by *n*-BuLi or *sec*-. BuLi [69–71]. Obviously anionic polymerizations take place with the negative charge stabilized by the free *d*-orbitals of the Si-atom, Eq.(70). Nonetheless, these anionic polymerizations are plagued by termination steps and did neither show a "living pattern" nor did they yield high molecular weights. Addition of MeOCH$_2$CH$_2$OMe to the reaction mixture in cyclohexane accelerates all reactions but mostly the termination steps. Detailed kinetic studies and molecular weight measurements were reported [69–71]. Diblock copolymers with styrene, **71**, were prepared in such a way that styrene served as the A-monomer to overcome the problem of termination steps [70].

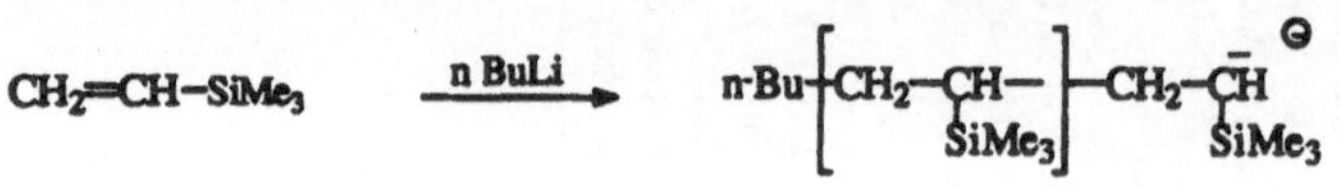

$$CH_2{=}CH{-}SiMe_3 \xrightarrow{\ n\,BuLi\ } n\text{-}Bu\!\left[CH_2{-}\underset{SiMe_3}{CH}{-}\right]\!\!\left[CH_2{-}\underset{SiMe_3}{\overset{\ominus}{CH}}\right] \qquad (70)$$

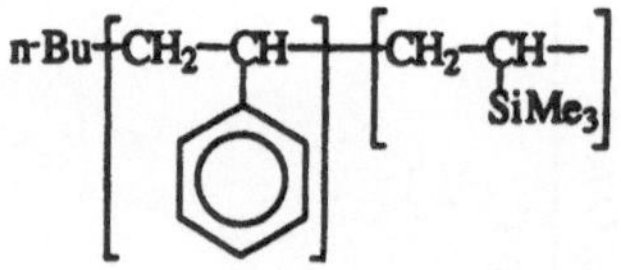

$$n\text{-}Bu\!\left[CH_2{-}\underset{\bigcirc}{CH}{-}\right]\!\!\left[CH_2{-}\underset{SiMe_3}{CH}{-}\right] \qquad (71)$$

1.2.2
Allyl Silanes

Allyl silanes can be subdivided into two groups:

A) monofunctional monomers bearing one allyl group attached to silicon **57a,b**, **58a,b**, **72a,b**, and
B) di- or trifunctional monomers with two or three allyl groups **73a–h**, **74a–c**, **75a,b**.

$$CH_2{=}CH{-}CH_2{-}SiH_2R \qquad\qquad\qquad\qquad (72)$$

a : R = Me
b : R = Ph

$$(CH_2{=}CH{-}CH_2)_2\!{-}SiR_2 \qquad\qquad\qquad\qquad (73)$$

a : R = H   f : R = OMe
b : R = Me   g : R = OEt
c : R = Et   h : R = NMe$_2$
d : R = Ph   i : R = NEt$_2$
e : R = p-MePh

$$(CH_2{=}CH{-}CH_2)_2\!{-}Si\!\begin{smallmatrix}R^1\\R^2\end{smallmatrix} \qquad\qquad\qquad\qquad (74)$$

a : R^1 = Me , R^2 = Ph
b : R^1 = Me , R^2 = p-MePh
c : R^1 = Ph , R^2 = p-MePh

$$\left(CH_2=CH-CH_2\right)_3 Si-R \qquad \begin{array}{l} a : R = Me \\ b : R = Ph \end{array} \qquad (75)$$

Most allyl silanes were synthesized by the reaction of an allyl Grignard agent with a suitable chlorosilane e.g. Eq.(76). The monomers **73f – h** were prepared by alcoholysis or aminolysis of bisallyldichlorosilane, Eq.(77) [74].

$$2\ CH_2=CH-CH_2-MgBr \xrightarrow[\ -\ 2\ MgClBr\ ]{+\ Cl_2SiR_2} \left(CH_2=CH-CH_2\right)_2 SiR_2 \qquad (76)$$

$$\left(CH_2=CH-CH_2\right)_2 SiCl_2 \xrightarrow[\ -\ 2\ HCl\ ]{+\ 2\ HOR} \left(CH_2=CH-CH_2\right)_2 Si\,(\,OR\,)_2 \qquad (77)$$

Relatively little is known about the homo- or copolymerization of monoallyl silanes. This is not surprising because the allyl group stabilizes neither radicals nor ions in a normal vinyl type chain growth reaction. On the other hand, the allyl group favors side reactions and termination steps by formation of resonance stabilized allyl radicals or allyl ions. The low reactivity of the $C = C$ double bond in polymerization processes is illustrated by the regioselective polymerization of monomers **57a,b** and **58a,b** [58]. It is also confirmed by copolymerizations of monomers **72a,b** with styrene. These copolymerization were initiated with AiBN at 60 °C and proved that both allyl silanes do not undergo homopolymerization steps ($r_2 = 0$) [16]. The reactivity ratios of styrene (r_1) were 36 and 29 respectively.

In contrast to monoallyl silanes, the di- and triallyl silanes are prone to homopolymerization, because cyclization with formation of six-membered rings can take place (e.g. **78a,b**). Depending on initiator and reaction conditions these cyclopolymerizations are not clean reactions, and in addition to termination steps crosslinking may occur. In the first two papers dealing with cyclopolymerizations of **73b** and **d** [72,73] Zielger-Natta-catalysts prepared from $TiCl_4$ and $AlEt_3$ in heptane were used as initiators. Insoluble fractions $< 10\%$ and low viscosity values were reported. No information on molecular weights or MWD was given and the structural characterization was exclusively based on IR-spectroscopy.

$$(78)$$

a   b

In a later, more comprehensive study of another research group [74] a variety of initiators was used and compared. In addition to the Ziegler-Natta catalyst mentioned above, AiBN was used at 60 °C as radical initiator and $AlBr_3$ in toluene at 25 °C as cationic initiator. Furthermore, numerous attempts were made to initiate anionic polymerizations by means of n-BuLi, n-BuMgBr or sodium metal, either in hexane, diethyl ether or THF at temperatures between −20 and +20 °C. However, all attempts with anionic initiators failed completely. Furthermore, a broad selection of monomers 73a–i and 74a–c was studied. Regardless of the monomer structure the highest yields were obtained with AlBr₃, but even this initiator failed in the case of 73a and 73f–i. Furthermore, only low viscosity values were obtained despite high yields [74].

Based on these results another group [76, 77] studied AlBr₃ initiated polymerizations of monomers 73d, e and 74a,b. Again high yields were obtained in several cases but all M_n's were below 5000 g/mol (mostly around 2000). The formation of six-membered rings and their stereo sequences were analysed by means of ^{1}H and ^{29}Si NMR spectroscopy [77].

Radical homopolymerizations of monomers 73b, 74a and 75a, b were studied by Saigo et al. [75]. Benzoylperoxide was used as initiator in benzene at 80 °C for 73b and 74a. Bis-tert-butylperoxide was prefered in ortho-dichlorobenzene at 130 °C for the polymerization of 75a,b. The formation of six membered rings in all cases was confirmed by IR- and ^{1}H-NMR spectroscopy studies including low-molar mass model compounds. Remarkable is the almost perfect formation of a polymer chain with pending allyl groups 78b when monomers 75a or b were polymerized [75]. These polymers were considered to be useful as imaging layers in bilayer resists for O_2-plasma etching.

1.2.3
Methacrylates and Fumarates

The Si-containing methacrylates studied as monomers by several research groups may be subdivided into the following two classes:

A) The silyl group is bound to the methacrylate via an oxygen, e.g. 79 and 80a–c [78–80].;

B) The silyl group is bound to the methacrylate via a σ-bond, e.g. 81a–d [81–85].

$$CH_2=\overset{\overset{\displaystyle Me}{|}}{C}-CO_2\text{-}CH_2\text{-}CH_2\text{-}O\text{-}Si\text{-}Me_3 \tag{79}$$

$$CH_2=\overset{\overset{\displaystyle Me}{|}}{C}-CO_2CH_2\text{-}\overset{\overset{\displaystyle Me}{|}}{C}H\text{-}O\text{-}\overset{\overset{\displaystyle R^1}{|}}{\underset{\underset{\displaystyle R^1}{|}}{Si}}\text{-}R^2 \qquad \begin{array}{l} a\ :\ R^1, R^2 = Me \\ b\ :\ R^1 = Me, R^2 = Ph \\ c\ :\ R^1 = Ph, R^2 = Me \end{array} \tag{80}$$

$$CH_2=\overset{\overset{\displaystyle Me}{|}}{C}-CO_2\text{-}(CH_2)_3\text{-}Si(OR)_3 \qquad \begin{array}{l} a\ :\ R = Me \\ b\ :\ R = Et \\ c\ :\ R = iPr \\ d\ :\ R = SiMe_3 \end{array} \tag{81}$$

Monomers of class A are typically synthesized by silylation of the 2-hydroxyalkylmethacrylate, Eq. (82). The class-B monomers are usually synthesized by acylation of the commercial γ-hydroxypropyltrialkoxysilanes, Eq. (83). An alternative approach is outlined in Eqs. (84–85) [81].

$$\underset{\underset{\displaystyle CH_2=C-CO_2-CH_2-CH_2-OH}{|}}{Me} \quad + \; ClSiMe_3 \quad \xrightarrow[-\;HCl]{} \quad (\,79\,) \tag{82}$$

$$HO-(CH_2)_3-Si(OR)_3$$
$$\xrightarrow[-\;HCl]{} \quad (\,81\,) \tag{83}$$
$$+ \quad CH_2=CMe-CO-Cl$$

$$\underset{\underset{\displaystyle CH_2=C-CO_2CH_2-CH=CH_2}{|}}{Me} \quad \xrightarrow[(Pt)]{+\;HSiCl_3} \quad \underset{\underset{\displaystyle CH_2=C-CO_2-(CH_2)_3-SiCl_3}{|}}{Me} \tag{84}$$

$$+\;3\,MeOH \;\left\downarrow\; -\;3\,HCl\right. \tag{85}$$
$$(\,81\,)\,a$$

In a relatively early study in this field [78] benzoylperoxide initiated copolymerizations of styrene with 79 or 80a were carried out in bulk. In the case of 79 the reactivity ratios were evaluated at different temperatures. The data listed in Table 1.3 indicate that increasing temperature favors an ideal character of such copolymerizations. Furthermore, the homopolymer of 79 and its copolymers were characterized by viscosity measurements different thermoanalysis and thermogravimetric analyses. For the homopolymer of 79 a T_g of 112 °C was found [78]. AiBN initiated copolymerizations of styrene with the monomers 80a–c yielded the reactivity ratios listed in Table 1.3 [79]. Again viscosities and thermogravimetric analyses were reported. Also copolymers of acrylonitrile and monomers 79 or 80a, prepared with AiBN at 60 °C in DMF, were characterized in much detail [80]. The crystallinity typical for neat polyacrylonitrile decreases with increasing molar fraction of the Si-containing comonomer. Also the thermostability decreases with higher contents of 79 or 80a. Furthermore, density and viscosity measurements were described.

Copolymers of monomer 81a with MMA were synthesized and studied because hydrolysis of the methoxy silyl group allows the preparation of networks with a controlled number of branching points [81]. Although numerous copolymers were prepared, reactivity ratios were not determined. The copolymers were characterized by viscosities, IR- and ^{1}H NMR spectra. The thermogravimetrical analyses were combined with mass-spectroscopy of the volatile degradation products. Benzoylperoxide initiated copolymerizations of the same monomer 81a with styrene were conducted in toluene or with acrylonitrile and MMA in bulk at 65 °C [82]. Reactivity ratios were evaluated (see Table 1.3), viscosities and T_g's were measured and thermogravimetric analyses were conducted.

Table 1.3. Reactivity ratios determined for radical copolymerizations of various Si-containing vinyl monomers

Comonomers	Temp. (°C)	r_1	r_2	Ref.
Styrene/CH_2 = CH-SiH_2Ph	60	5.7	0	10
Styrene/CH_2 = CH-CH_2SiH_2Ph	60	29.0	0	10
	60	1.18	0.18	78
	80	1.04	0.30	78
Styrene/CH_2 = CMe-CO_2-CH_2CHMe-$OSiMe_3$	100	1.02	0.35	78
Styrene/CH_2 = CMe-CO_2-CH_2CHMe-$OSiMe_3$	60	0.95	0.48	79
	80	0.85	0.58	79
Styrene/CH_2 = CMe-CO_2-CH_2CHMe-$OSiMe_2$Ph	60	0.97	0.39	79
Styrene/CH_2 = CMe-CO_2-CH_2CHMe-$OSiPh_2$Me	60	1.05	0.35	79
Styrene/CH_2 = CMe-$CO_2(CH_2)_3Si(OMe)_3$	65	0.45	0.90	82
Acrylonitrile/CH_2 = CMe-$CO_2(CH_2)_3Si(OMe)_3$	65	0.11	3.7	82
MMA/CH_2 = CMe-$CO_2(CH_2)_3Si(OMe)_3$	65	0.74	1.33	82
Bis-iPro-fumarate/(86)b	60	0.87	0.07	87
Bis-t-Bu-fumarate/(86)b	60	0.67	0.21	87
Styrene/CH_2 = C—CH = CH_3, with SiMe₂ branch	15	0.49	2.20	100
Styrene/CH_2 = C—CH = CH_2, with $SiMe_2$OiPr branch	15	0.54	3.35	100
Styrene/CH_2 = C—CH = CH_2, with $SiMe(OiPr)_2$ branch	15	0.41	3.70	100
Styrene/CH_2 = C—CH = CH_2, with $Si(OiPr)_3$ branch	15	0.39	4.10	100
Styrene/CH_2 = C—CH = CH_2, with $OSiMe_3$ branch	60	0.64	1.20	103
Acrylonitrile/CH_2 = C—CH = CH_2, with $OSiMe_3$ branch	60	0.04	0.06	103

Copolymers of monomer **81d** and *p*-vinyl *tert*-butylbenzoate were prepared by AiBN-initiated copolymerizations at 60 °C in bulk [83]. High yields and high molecular weights (Mw > 10^5) were found. The properties of these copolymers were studied with regard to gas separation and permeability of oxygen [83]. Concerning radical copolymerizations of monomer **81a** a patent describing emulsion copolymerizations with acrylonitrile or *n*-butylmethacrylate should be mentioned [84]. These copolymerizations were initiated with $K_2S_2O_8$ at 70 °C.

Finally a series of anionic homopolymerizations of monomers **81a–c** needs discussion [85]. All polymerizations were conducted at −78 °C in THF. Numerous initiators, such as *n*-BuLi, cumyl potassium, K-naphthalenide, benzylmagnesium chloride, K-*tert*-butoxide and LiAlEt$_4$ were used and compared. Particularly successful were polymerizations initiated with BuLi in combination with 1,1-diphenylethane. High yields, M_n's up to 13×10^3 and narrow MWD's ($M_w/M_n \leq 1.05$) were reported. These living polymerizations were also utilized to synthesize diblock copolymers with styrene or MMA as comonomers. Furthermore, hydrolysis and crosslinking of the resulting blockcopolymers were studied [85].

In addition to methacrylates two Si-containing fumarates should be mentioned; **86a** and **b**. Various attempts to polymerize **86a** anionically or cationically failed, whereas the AiBN initiated radical polymerization in THF at 60 °C was successful. The resulting polymer was characterized by ^{1}H-NMR spectroscopy and hydrolyzed to yield polyfumaric acid [86].

Monomer **86b** was subjected to radical homopolymerization and various conditions, but only low yields and viscosities were obtained [87]. More successful proved copolymerizations with either diisopropyl fumarate or bis-*tert*-butyl fumarate (see reactivity ratios in Table 1.3). Films cast from solution showed a

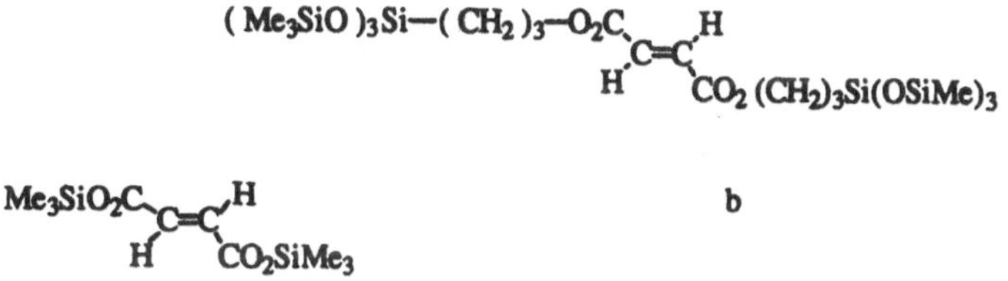

$$(Me_3SiO)_3Si-(CH_2)_3-O_2C \underset{H}{\overset{H}{\underset{}{C=C}}} CO_2 (CH_2)_3Si(OSiMe)_3$$

$$Me_3SiO_2C \underset{H}{\overset{H}{\underset{}{C=C}}} CO_2SiMe_3$$

b

a

(86)

good permeability to oxygen [87].

1.2.4
Vinyl Ethers

Relatively little work has been published about the polymerization of Si-containing vinyl ethers [88–93]. Again two classes of monomers may be distinguished

A) monomers with a C—O—Si ether group **87** [88–92]

B) monomers with C—O—C ether group and a silyl group part of the ether **88a,b** [93,94].

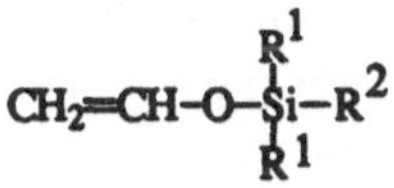

$$CH_2=CH-O-\underset{R^1}{\overset{R^1}{\underset{}{Si}}}-R^2$$

a : R^1, R^2 = Me
b : R^1, R^2 = n-Pr
c : R^1, R^2 = i-Pr
d : R^1, R^2 = n-Bu
e : R^1, R^2 = i-Bu
f : R^1 = Me, R^2 = CMe$_3$

(87)

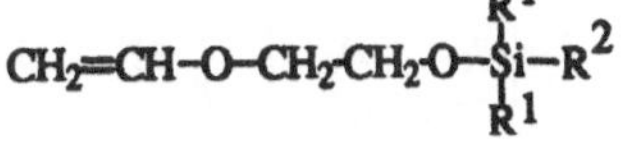

$$CH_2=CH-O-CH_2CH_2-O-\underset{R^1}{\overset{R^1}{\underset{}{Si}}}-R^2$$

a : R^1, R^2 = Me
b : R^1 = Me, R^2 = CMe$_3$

(88)

Monomers of structure **87a–f** were synthesized from the corresponding halosilanes and acetaldehyde using either triethylamine as HCl acceptor, Eq. (89) [90] or the mercury salt of acetaldehyde, Eq. (90) [89]. Various methods for the synthesis of enolethers are summarized in a monograph by Weber [95].

$$CH_3-CHO \; + \; ClSiMe_3 \quad \xrightarrow[\text{- NEt}_3\cdot\text{HCl}]{\text{+ NEt}_3} \quad CH_2=CHO-SiMe_3 \tag{89}$$

$$Hg\,(\,CH_2-CHO\,)_2 \; + \; 2\; BrSiR_3 \quad \xrightarrow[\text{- HgBr}_2]{} \quad CH_2=CH-O-SiR_3 \tag{90}$$

The first study dealing with the polymerization of a vinyl ether, **87a** [88], is based on acidic catalysts such as $SnCl_4$, Et_2CHCl or $EtAlCl_2$. All three catalysts were compared in three different solvents: hexane, toluene and CH_2Cl_2. The highest yields ($\sim 100\%$) and highest molecular weights (DP ~ 950) were obtained with Et_2AlCl in CH_2Cl_2. The silyl groups of the resulting polymers were hydrolyzed, Eq.(91), and the tacticity of the isolated polyvinyl alcohols was determined by IR spectroscopy. Mainly isotactic polymers were obtained in nonpolar solvents, predominantly syndiotactic poly(vinyl alcohol) in CH_2Cl_2. The influence of various silyl groups **87a–e** and different reaction media on the tacticity of polyvinyl alcohol was further studied by Noyakuva et al. [89]. Lewis acids, such as BF_3OEt_2, $SnCl_4$ or $EtAlCl_2$ were used at low temperatures as cationic initiators. It was found the polar solvents favor the formation of syndiotactic sequences. Increasing bulkiness of the trialkylsilyl group favors heterotactic stereosequences at the expense of isotactic ones.

$$\left(\begin{array}{c} -CH_2-CH- \\ \quad\quad\;\; OSiMe_3 \end{array}\right) \quad \xrightarrow{\hspace{2cm}} \quad \left(\begin{array}{c} -CH_2-CH- \\ \quad\quad\; OH \end{array}\right) \tag{91}$$

Another research group [90] used the same cationic initiators for copolymerizations of **87a** with several optically active vinyl ethers (free of silicon!). The resulting copolymers were desilylated and characterized.

A new kind of "Group-transfer-Polymerization" or in other words "Silyl-Aldol Polycondensation" of monomer **87f** was reported by Soga and Webster [91]. Aromatic aldehydes in combination with Lewis acids serve as initiators, Eq. (92). The best results were obtained with benzaldehyde/$ZnBr_2$ in dichloro-methane. 100% conversion can be obtained and the course of the polymerization follows the "living pattern" with relatively narrow MWD's. With α,α-dibromo-p-xylene as initiator chain growth in two directions was observed, Eq.(93). This "Silyl-Aldol Polycondensation" offers also a great potential for the synthesis of di- and triblock copolymers (see. Chapter 2).

$$Ph-\underset{O}{\overset{}{C}}-H \; + \; CH_2=CH-O-SiMe_2t.Bu \quad \xrightarrow{(\,ZnBr_2\,)} \quad Ph\left[CH-CH_2-\atop OSiMe_2t.Bu\right]_n\!\!-\underset{O}{\overset{}{C}}-H \tag{92}$$

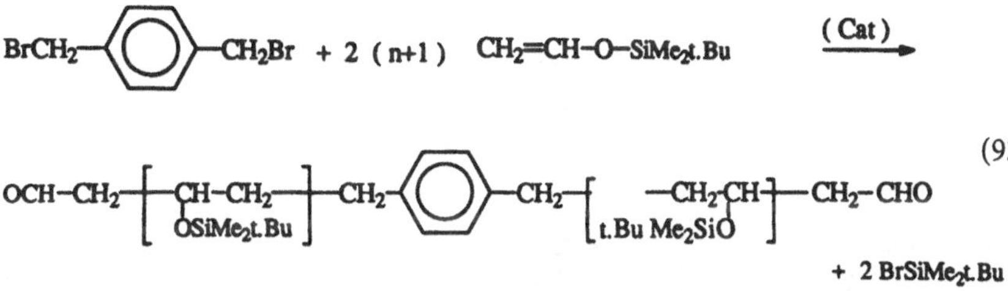

$$BrCH_2-\underset{}{\bigcirc}-CH_2Br \ + \ 2\ (n{+}1)\ \ CH_2{=}CH{-}O{-}SiMe_2t.Bu \quad \xrightarrow{(Cat)}$$

$$(93)$$

$$OCH{-}CH_2\left[\underset{\dot{O}SiMe_2t.Bu}{CH{-}CH_2}\right]CH_2{-}\underset{}{\bigcirc}{-}CH_2\left[\underset{t.Bu\ Me_2Si\dot{O}}{-CH_2{-}CH{-}}\right]CH_2{-}CHO$$

$$+\ 2\ BrSiMe_2t.Bu$$

A new route to the synthesis of a diblock copolymer involving the afore-mentioned polymerization method, was recently published by Risse and Grubbs [92]. The A-block was prepared by metathesis ROP of norbornene with a titanacyclobutane derivative, Eq. (94). The active endgroup was treated with a large excess of terephthaldehyde, so that a poly(norbornene) with aldehyde endgroup was formed, Eq. (95). This polymer served as coinitiator for the polymerization of monomer 87f, Eq. (96). Desilylation yielded an amphiphilic blockcopolymer.

With the vinyl ethers 88a and b Higashimura et al. [93] demonstrated that living cationic polymerizations are feasible when HJ/J$_2$ is used as initiator in

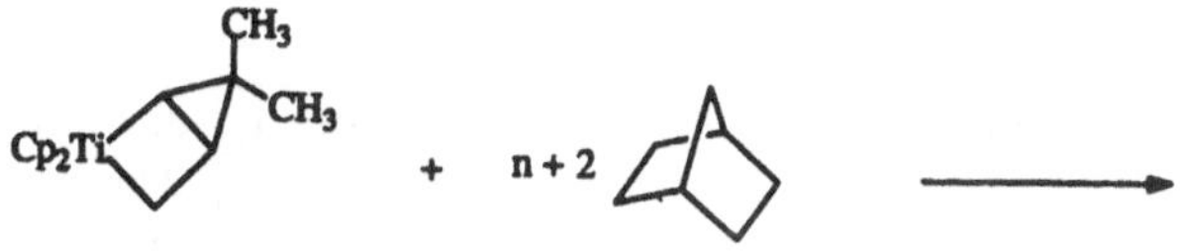

$$Cp_2Ti\overset{CH_3}{\underset{CH_3}{\diamond}} \ + \ n{+}2\ \diamond \ \longrightarrow$$

$$(94)$$

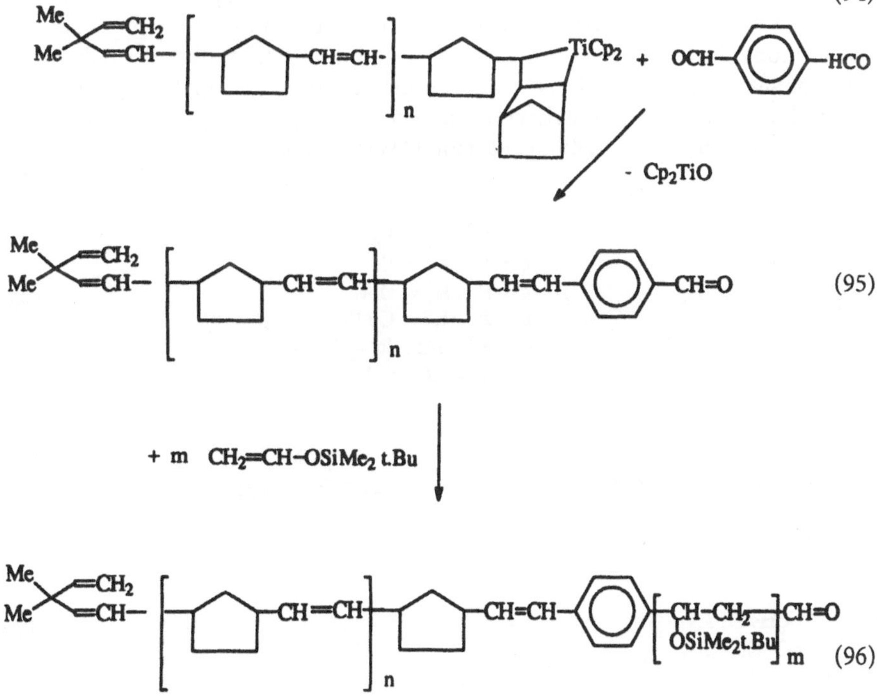

$$\underset{Me}{\overset{Me}{>}}{\overset{=CH_2}{\underset{=CH-}{}}}\left[\ {-}CH{=}CH{-}\ \right]_n \ {-}TiCp_2 \ + \ OCH{-}\underset{}{\bigcirc}{-}HCO$$

$$\bigg\downarrow \ {-}\ Cp_2TiO$$

$$\underset{Me}{\overset{Me}{>}}{\overset{=CH_2}{\underset{=CH-}{}}}\left[\ {-}CH{=}CH{-}\ \right]_n {-}CH{=}CH{-}\underset{}{\bigcirc}{-}CH{=}O \quad (95)$$

$$\bigg\downarrow \ +\ m\ \ CH_2{=}CH{-}OSiMe_2\ t.Bu$$

$$\underset{Me}{\overset{Me}{>}}{\overset{=CH_2}{\underset{=CH-}{}}}\left[\ {-}CH{=}CH{-}\ \right]_n {-}CH{=}CH{-}\underset{}{\bigcirc}\left[\underset{OSiMe_2t.Bu}{CH{-}CH_2}\right]_m CH{=}O \quad (96)$$

toluene at low temperature (e.g. $-40\,°C$). M_w/M_n ratios around 1.1 were found and diblockcopolymers were prepared by batchwise copolymerization with vinyl isobutyl ether, Eq. (97). Desilylation again yielded amphiphilic blockcopolymers.

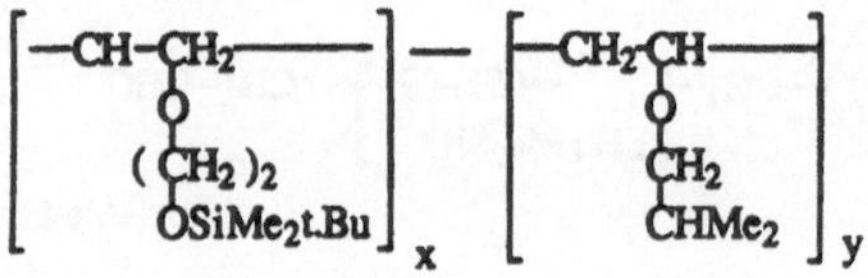

$$
\left[\begin{array}{c} -CH-CH_2- \\ O \\ (CH_2)_2 \\ OSiMe_2t.Bu \end{array}\right]_x - \left[\begin{array}{c} -CH_2-CH- \\ O \\ CH_2 \\ CHMe_2 \end{array}\right]_y \tag{97}
$$

1.2.5
1,3-Dienes

By analogy with other monomers (see above),
the Si-containing 1.3-dienes used as monomers may be subdivided into the following two classes

A) dienes with the silyl group directly attached to the diene **98a–f** [96–101],
B) dienes with the silyl group separated from diene structure by a C or O atom
   99a, b [102, 103] or **100** [101].

Most monomers of structure **98** were prepared from 1,4-dichlorobutyne-2 by hydrosilylation as illustrated in Eqs. (101–103) [96–98, 104]. The monomers **99 a** or **b** were obtained by silylation of croton aldehyde [102] or methyl vinyl ketone, Eq. (104) [103, 105] with chlorotrimethylsilane in the presence of a strong base (s. Ref. 95 and Appendix). Monomer **99** was prepared from 2-chlorobutadiene and the Grignard reagent of chlorotetramethylsilane [101]. The 1,3-dienes of structure **98a–f** were polymerized by a variety of methods, but most authors used anionic initiators [96–99]. Takenaka et al. [96–98] studied the polymerization of **98a**, and **c–f** in THF at $-78\,°C$ initiated with Li, Na or K.-naphthalenide in combination with α-methylstyrene as coinitiators in analogy to Eqs. (39–41). Living polymerizations yielding narrow MWD's were observed in most cases.

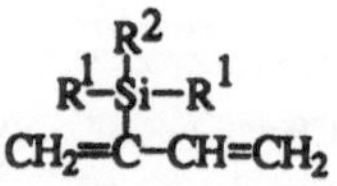

$$
\begin{array}{l}
R^2 \\
R^1-Si-R^1 \\
CH_2=C-CH=CH_2
\end{array}
\qquad
\begin{array}{l}
a : R^1, R^2 = Me \\
b : R^1, R^2 = Et \\
c : R^1, R^2 = OMe \\
d : R^1, R^2 = OiPr \\
e : R^1 = Me, R^2 = OiPr \\
f : R^1 = OiPr, R^2 = Me
\end{array}
\tag{98}
$$

$$
Me_3SiO-CH=CH-CH=CH_2 \qquad\qquad \begin{array}{l} OSiMe_3 \\ CH_2=C-CH=CH_2 \end{array} \tag{99}
$$

$$
\text{a} \qquad\qquad\qquad\qquad\qquad\qquad \text{b}
$$

$$
\begin{array}{l}
CH_2-SiMe_3 \\
CH_2=C-CH=CH_2
\end{array}
\tag{100}
$$

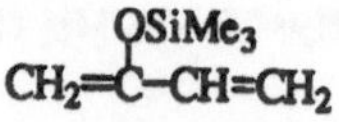

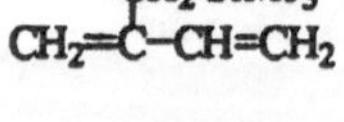

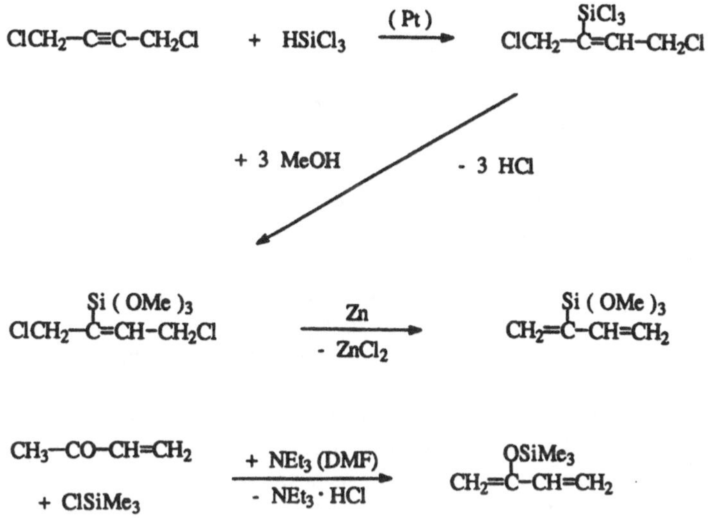

$$ClCH_2-C\equiv C-CH_2Cl \; + \; HSiCl_3 \; \xrightarrow{(Pt)} \; ClCH_2-\underset{SiCl_3}{C}=CH-CH_2Cl \qquad (101)$$

$$+\ 3\ MeOH \qquad\diagdown\qquad -\ 3\ HCl \qquad\qquad (102)$$

$$\underset{ClCH_2-C=CH-CH_2Cl}{\overset{Si(OMe)_3}{}} \; \xrightarrow[-\ ZnCl_2]{Zn} \; \underset{CH_2=C-CH=CH_2}{\overset{Si(OMe)_3}{}} \qquad (103)$$

$$\begin{aligned}&CH_3-CO-CH=CH_2\\ &+\ ClSiMe_3\end{aligned} \; \xrightarrow[-\ NEt_3\cdot HCl]{+\ NEt_3\ (DMF)} \; \underset{CH_2=C-CH=CH_2}{\overset{OSiMe_3}{}} \qquad (104)$$

The tendency of side reactions causing termination steps increased in the order:

K < Na < Li and (98)a < (98)d < (98)c.

The polymerization of 98a was faster than that of 98b–f. The chemical structures of all homopolymers were analyzed by means of ^1H and ^{29}Si-NMR spectroscopy. The data compiled in Table 1.4 indicate that 1,4-units with

Table 1.4. Microstructure of Poly 98a to Poly 98d[a]

polymer	SiR¹R²R³			condition	microstructure, %			
	R¹	R²	R³		1,4-E	1,4-Z	1,2	3,4
poly 98d	OPr^i	OPr^i	OPr^i	Li/THF	100	0	0	0
				Na/THF	100	0	0	0
				K/THF	100	0	0	0
poly 98c	OMe	OMe	OMe	Li/THF	90	10	0	0
				Na/THF	75	25	0	0
				K/THF	73	27	0	0
poly 98f	OPr^i	OPr^i	Me	Li/THF	91	3	6	0
				Na/THF	67	17	16	0
				K/THF	62	19	19	0
poly 98e	OPr^i	Me	Me	Li/THF	65	15	20	0
				Na/THF	45	30	25	0
				K/THF	47	30	23	0
poly 98a	Me	Me	Me	Li/THF	43	23	34	0
				Na/THF	44	27	29	0
				K/THF	44	28	28	0
				Li/hexane	84	4	12	0
				Li/hexane/ TMEDA	67	11	13	9

[a] data adapted from Ref. 98.

E-isomerism are the predominant repeating units under all circumstances [98]. Furthermore, the reactivities of the anionic chain ends in blockcopolymerizations with isoprene, styrene and 2-vinylpyridine were investigated [97, 98].

Another research group [99] studied the anionic polymerizations of **98b** initiated with *n*-BuLi in neat hexane at 25 °C or in the presence of HMPA at −25 and −78 °C. Under these conditions no "living character" was found, but *cis*-1,4-units were again prevalent. When AiBN initiated radical polymerizations of **98a** and **d**–**f** were conducted in bulk at 80 °C, Diels-Alder dimers were formed along with polymers [100]. Initiation with redox-initiators at 15 °C prevented the formation of Diels-Alder products, but the polymerizations were slow and gave low yields. Copolymerization conducted with styrene in bulk were evaluated with regard to reactivity ratios (Table 1.3). Relatively high reactivities were found for all dienes. Copolymers of **97c**–**f** are of general interest, because hydrolysis of the alkoxysilyl groups allows controlled crosslinking.

Monomere **100** was either polymerized by BuLi in hexane or by a Ziegler-Natta-catalyst prepared from $TiCl_4$ and $AlEt_3$ in hexane. The microstructures of the isolated polymers were different. The Ziegler-Natta-product consisted almost exclusively of *cis*-1,4-units, whereas the anionic polymer contained considerable amounts of *trans*-1,4 and 3,4-units. The chemical modification of these polymers is mentioned in chapter 7.

The cationic polymerization of 1-trimethylsiloxybutadiene **99a** with $SnCl_4$ or ACl_3 resulted in gelation due to crosslinking even at −78 °C. Soluble polymers predominantly consisting of *trans*-1,4-units were obtained when an aldol-type GTP was conducted in analogy to Eqs. (93, 94). A M_w/M_n ratio of 1.5 was determined by GPC. A poly(trimethylsiloxybutadiene) with a more complex and not fully elucidated microstructure was obtained from the AiBN initiated bulk polymerization of **99b**. IR-spectra, ^{1}H-NMR data and a molecular weight around 10×10^3 were reported for the characterization of this homopolymer. Copolymerizations of **99b** with styrene and acrylonitrile at 60 °C were analyzed with regard to the reactivity ratios (Table 1.3). Copolymers with quite different sequences were obtained [103].

Last but not least, radical homo- and copolymerization of monomers **105a** and **b** should be mentioned [106]. Due to steric hindrance the polymerizability of both monomers is extremely slow. However, in the case of **105b** copolymers with styrene or acrylonitrile were prepared. Regardless of comonomers and reaction conditions, no homopolymerization steps of **105b** were observed ($r_2 = 0$). In this regard **105b** resembles most allyl or vinyl silanes.

$$Me_2C=CH-CH_2-CH_2-\overset{\overset{\textstyle Me}{|}}{C}H-CH_2O-SiMe_3 \qquad\qquad Me_2C=CH-CH_2-CH_2-\overset{\overset{\textstyle Me}{|}}{\underset{\underset{\textstyle OSiMe_3}{|}}{C}}H-CH=CH_2$$

a   b   (105)

1.3
Polymerizations of Si-Containing Acetylenes

The polymerizations of Si-containing acetylenes described in this section are defined as Ziegler-Natta or Metathesis-type chain growth reactions involving the conversion of π-bonds into σ-bonds. Oligomerization and polymerization processes based on addition or condensation steps are reported in the subsequent section. The polymerization of silylated acetylene was stimulated by the finding that the resulting Si-containing polyacetylenes are useful as membranes because of their high permeabilities to oxygen. This useful property was in fact detected for the first time for polyacetylenes with alkyl substituents [107]. However, polyacetylenes containing silyl groups were found to be superior. Papers exclusively dealing with characterization and properties of films or membranes [107–116] will not be discussed in detail.

Si-containing acetylenes used in chain growth processes may be subdivided into two main groups with two subgroups.

A) acetylenes with a directly attached silyl group, and a hydrogen or another substituent bound to the alkyne group, for example:

$$HC \equiv C + SiR^1R^2R^3 \quad \text{and} \quad X - C \equiv C - SiR^1R^2R^3$$

B) acetylenes with the silyl group located somewhere in one of the substituents: for instance

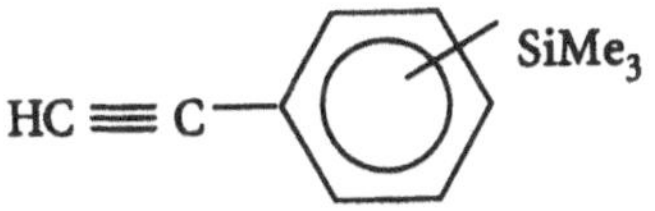

and

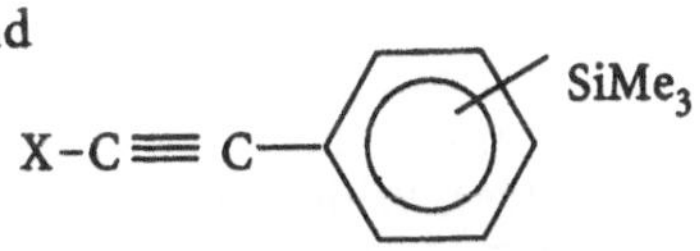

The simplest monomer, TMS-acetylene, **106a**, was first polymerized by Voronkov et al. [117] with $MoCl_5$ as catalyst and a yellow powder insoluble in all common solvents was obtained. Under the same conditions, even the alkoxy monomers **107a** and **b** yielded insoluble polyacetylenes. Masuda, Higashimura and coworker [118, 119], who studied numerous Si-containing acetylenes using $MoCl_5$ obtained soluble oligomers of TMS-acetylene and high molecular weight polymers using WCl_6 based catalysts, but these polymers were mainly insoluble. Other research groups [120–123] experienced the same results using $W(CO)_6/Cl_4$, WCl_6/nBuLi or $R^1R^2C = W(CO)_4$ as catalysts. Small soluble fractions with molecular weights $\leqslant 10^4$ were obtained in all cases but never a fully soluble material capable of forming tough films or membranes.

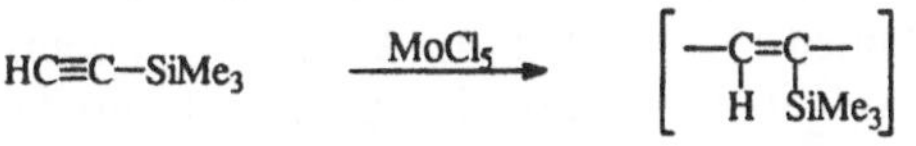

$$(106)$$

This failure prompted several research groups to vary the structure of **106a** and **b** over a broader range. For instance the monomers **108a–d** were prepared from sodium acetylide and the corresponding chlorosilanes, Eq.(108) [123]. Polymerizations conducted with $MoCl_5$ gave extremely low yields, whereas catalysts based on WCl_6 (e.g. in combination with Ph_4Sn, Et_3SiH, Ph_3Bi or (Cl_4) proved to be more successful. Partially insoluble polyacetylenes resulted from the monomers **108a** and **c**, whereas completely soluble poly(acetylene)s, were obtained from all other monomers. The influence of solvent and temperature on the polymerizations catalyzed with WCl_6/Ph_4Sn was studied and Mn's up to 18 000 were determined. The resulting polyacetylenes were characterized by IR, ^1H- and ^{13}C NMR spectroscopy [123].

$$HC≡C–SiMe_2OEt \qquad\qquad HC≡C–SiMe_2OPr \tag{107}$$

$$\text{a} \qquad\qquad\qquad\qquad \text{b}$$

$$HC≡C–Na \;+\; Cl–SiMe_2R \xrightarrow[\text{– NaCl}]{} HC≡C–SiMe_2R$$

$$
\begin{aligned}
&\text{a : R = t-Bu}\\
&\text{b : R = n-}C_6H_{13}\\
&\text{c : R = Ph}\\
&\text{d : R = }CH_2Ph\\
&\text{e : R = }CH_2CH_2Ph
\end{aligned}
\tag{108}
$$

Further studies concentrated on the polymerization of 1-trimethylsilyl propyne, Eq.(109) [124–130]. The polymerization of this monomer was studied with all Nb and Ta halogenides in solution [124–126]. With NbF_5 high yields of an insoluble material were obtained. With NbJ_5, TaF_5 or TaJ_5 no polymerization took place [125]. Yet, soluble polyacetylenes were isolated in high yields when $NbCl_5$, NBr_5, $TaCl_5$ or $TaBr_5$ were used as catalysts in toluene at 80 °C. The highest molecular weights (> 16^6) were finally obtained with combinations of $TaCl_5$ and a reducing cocatalyst such as Ph_3Bi Ph_3Sb, Ph_4Sn, n-Bu_4Sn, Ph_3SiH or Et_3SiH [126–128]. The influences of temperature, solvent, monomer and catalyst concentration on yield and molecular weight were studied in detail. Polymerization of TMS-propyne in cyclohexane with $NbCl_5$ was found to yield narrow MWD's ($M_w/M_n = 1,2$) and the M_n increased with conversion and monomer/initiator ratio as expected for a living polymerization [127]. The high molecular weight poly(acetylene) prepared according to Eq.(109) allowed the production of tough films or membranes with interesting properties [127]. Furthermore, numerous copolymers of TMS-propyne were synthesized **110a–d**, **111a,b** and their properties as membranes were evaluated [128, 129, 130]. All these copolymers were prepared with high molecular weights by means of $TaCl_5$-based initiators.

$$CH_3–C≡C–SiMe_3 \longrightarrow \left[-\underset{Me}{C}=\underset{SiMe_3}{C}- \right] \tag{109}$$

Masuda et al. [131] also synthesized the analogs of TMS-propyne **112a–c**, **113a,b** and **114a,b** from lithium salts of 1-alkynes and chlorosilanes. The poly-

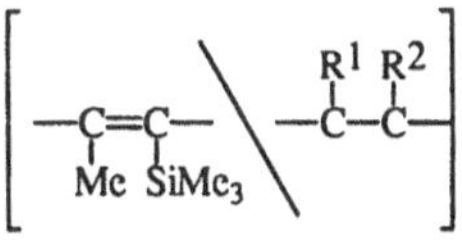

$$\left[-\underset{Me}{\overset{}{C}}=\underset{SiMe_3}{\overset{}{C}}- \diagdown -\underset{R^1}{\overset{}{C}}-\underset{R^2}{\overset{}{C}}- \right]$$

a : R^1 = Me, R^2 = C_5H_{11}

b : R^1 = Me, R^2 = Ph

c : R^1 = n-Pr, R^2 = n-Pr

d : R^1 = n-Bu, R^2 = Ph

(110)

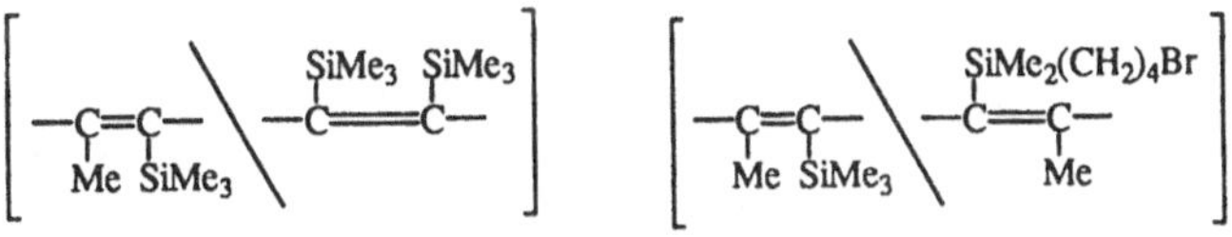

$$\left[-\underset{Me}{\overset{}{C}}=\underset{SiMe_3}{\overset{}{C}}- \diagdown -\underset{}{\overset{SiMe_3}{C}}=\underset{}{\overset{SiMe_3}{C}}- \right]$$
a

$$\left[-\underset{Me}{\overset{}{C}}=\underset{SiMe_3}{\overset{}{C}}- \diagdown -\underset{Me}{\overset{SiMe_2(CH_2)_4Br}{C}}=\underset{}{\overset{}{C}}- \right]$$
b

(111)

merizations were studied in toluene at 110 °C with $TaCl_5/Ph_4Sn$ or $TaCl_5/Ph_3Bi$ catalysts. Satisfactory yields and molecular weights (MW above 10^5) were found in the case of monomers 112a–d [131] and 113a [132]. Yet, apparently due to steric hindrance the Si-alkynes 113b,c and 114a,b did not polymerize [131]. The reluctance of these alkynes to polymerize was ascribed to steric hindrance [131]. If this is true, it is surprising that the aromatic monomers 115a,b and 116a,b were polymerizable [133–134]. A broad variety of catalysts were examined, for instance neat $NbCl_5$, $TaCl_5$, $MoCl_5$ and WCl_6 or combinations of either $MoCl_5$ or WCl_5 with cocatalysts such as Me_4Sn, Bu_4Sn Ph_4Sn, Et_3Al, Bu_3Al or Et_2AlCl. In all cases the yields were low (< 32%) most molecular weights (M_n) were below 10 000, but for a few polyacetylenes M_n's up to 27 000 were found [113, 134].

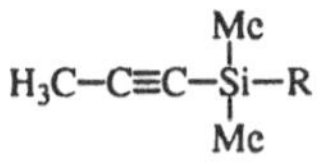

$$H_3C-C\equiv C-\underset{Me}{\overset{Me}{Si}}-R$$

a : R = n-C_6H_{13} b : R = Ph

c : R = CH_2SiMe_3 d : R = $CH_2CH_2SiMe_3$

(112)

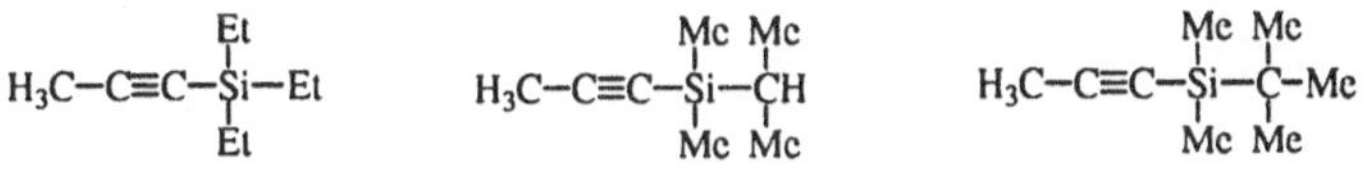

$$H_3C-C\equiv C-\underset{Et}{\overset{Et}{Si}}-Et$$
a

$$H_3C-C\equiv C-\underset{Me}{\overset{Me\ Me}{Si}}-\underset{Me}{\overset{}{CH}}$$
b

$$H_3C-C\equiv C-\underset{Me}{\overset{Me\ Me}{Si}}-\underset{Me}{\overset{}{C}}-Me$$
c

(113)

Et$-C\equiv C-$SiMe$_3$
a

n-Bu$-C\equiv C-$SiMe$_3$
b

(114)

a

b

(115)

Numerous monosubstituted acetylenes with the silyl group in α-position, 117a–d, 118a,b and 119 were synthesized and polymerized by several research groups [135–141]. The polymerizations of monomers 117a–d were conducted in toluene at 30 °C [135]. In addition to neat $NbCl_5$, $TaCl_5$, $MoCl_5$ and WCl_6

combinations of this Lewis-acids with Et_3SiH, Ph_4Sn or Ph_3Bi were used as catalysts. In most cases yields above 70% and molecular weights (MW) above 10^5 were obtained. All polymers were isolated in the form of yellow powders, but only the polymers of **117c** and **d** were completely soluble. The polyacetylene structure was characterized by IR-, ^{1}H-NMR and ^{13}C NMR spectroscopy. Analogous polymerizations of monomers **118a,b** were conducted in toluene at $0\,^\circ$C [136]. Again molecular weights (M_n and M_w) around 10^5 were achieved, and all polymers were soluble in several common solvents. The casting of films and their mechanical and thermal properties were discussed. The synthesis of the chiral, optically active monomer **119** is of particular interest because two silyl groups of different reactivities are involved, Eqs. (120–122) [137]. The

$$\text{(116)}$$

a   b

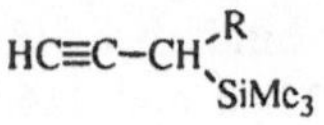

$$HC\equiv C-CH\begin{smallmatrix}R\\SiMe_3\end{smallmatrix} \qquad \begin{array}{ll} a : R = CH_3 & b : R = n\text{-}Pr \\ c : R = n\text{-}C_5H_{11} & d : R = n\text{-}C_7H_{15} \end{array} \qquad \text{(117)}$$

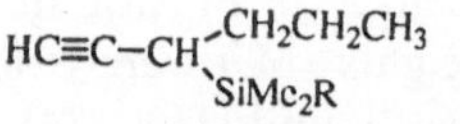

$$HC\equiv C-CH\begin{smallmatrix}CH_2CH_2CH_3\\SiMe_2R\end{smallmatrix} \qquad \begin{array}{l} a : R = n\text{-}C_6H_{13} \\[6pt] b : R = Ph \end{array} \qquad \text{(118)}$$

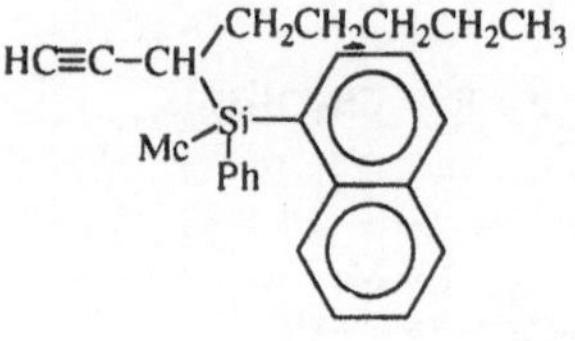

$$HC\equiv C-CH\begin{smallmatrix}CH_2CH_2CH_2CH_2CH_3\\ \\ Me-Si-\\ Ph\end{smallmatrix} \qquad \text{(119)}$$

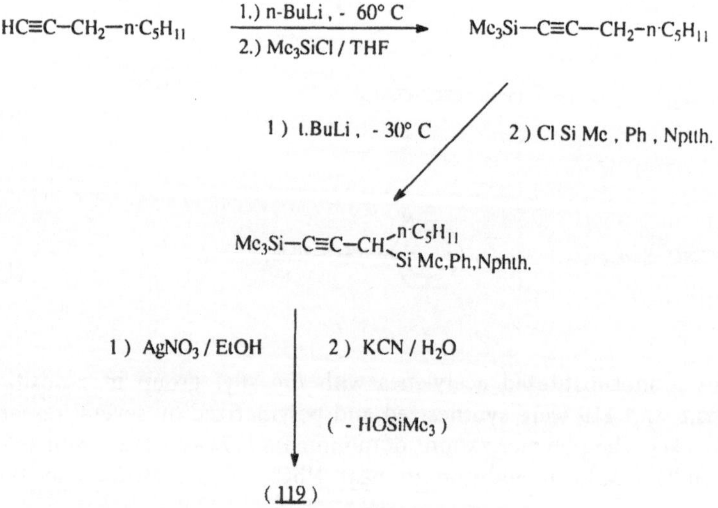

$$HC\equiv C-CH_2-n\text{-}C_5H_{11} \xrightarrow[\text{2.) Me}_3\text{SiCl / THF}]{\text{1.) n-BuLi, - 60}^\circ\text{ C}} Me_3Si-C\equiv C-CH_2-n\text{-}C_5H_{11} \qquad \text{(120)}$$

$$\begin{array}{c} \text{1) t.BuLi , - 30}^\circ\text{ C} \diagdown \quad \text{2) Cl Si Me , Ph , Npth.} \\ \text{(121)} \end{array}$$

$$Me_3Si-C\equiv C-CH\begin{smallmatrix}n\text{-}C_5H_{11}\\Si\ Me,Ph,Nphth.\end{smallmatrix}$$

$$\begin{array}{c} \text{1) AgNO}_3\text{ / EtOH} \quad | \quad \text{2) KCN / H}_2\text{O} \\ | \\ (- HOSiMe_3) \\ \\ (\underline{119}) \end{array} \qquad \text{(122)}$$

trimethylsilyl group plays the role of a protecting group for the introduction of the chiral silyl substituent. Polymerizations of monomer **119** with various Mo or W catalysts gave mixed results. The highest molecular weight (MW $\sim 10^4$) was obtained with $Mo(CO)_3 \cdot (CH_3CN)_3$ in CCl_4. The optical active polyacetylenes were characterized by optical rotation, ^{13}C NMR spectroscopy and thermogravimetric measurements.

Another highly interesting group of monomers are the diacetylenes **123a–e** [138] and **125a–c** [139,140]. The monomers **123a–e** were polymerized in toluene at 80 °C with WCl_6 and in a few cases with $MoCl_5$, $NbCl_5$ or $MoCl_5/Ph_4Sn$ and WCl_6/Ph_4Sn combinations. In all cases, soluble polymers were isolated but the molecular weights were relatively low ($M_n \sim 4000–15\ 000$). Films were cast, doped with J_2, and conductivities were measured ($\sigma \sim 10^{-2}$ to 10^{-4} Siemens/cm). Albeit, detailed structural analyses were not reported, both, solubility and electrical conductivity, together suggest that cyclopolymerizations yielding polymers of structure **124a–e**, took place. When the monomers **125a–c** were polymerized with $MoCl_5$ or WCl_6-based catalysts in chlorobenzene at 60 °C soluble polymers were isolated in the case of **125b** and **c** [139, 140]. Monomer **125a** yielded partially crosslinked polymers under all circumstances [139].

$$\left(HC\equiv C-SiMe_2\right)_2 X$$

$$\downarrow$$

a : X = σ-bond
b : X = CH_2 (123)
c : X = CH_2CH_2
d : X = O
e : X = $SiMe_2$

$$\left[\begin{array}{c} -\overset{C}{\underset{Me_2Si}{C}}\diagdown_{X}\diagup\overset{C=C-}{SiMe_2} \end{array} \right]$$

(124)

$$HC\equiv C-CH_2-\overset{\overset{R}{|}}{\underset{\underset{R'}{|}}{Si}}-CH_2-C\equiv CH$$

a : R , R' = Me
b : R , R' = Ph   (125)
c : R = Me , R' = Ph

The structure of the polymers based on cyclopolymerization was characterized by IR-, ^1H- and 13C NMR spectroscopy. Furthermore, thermogravimetric analyses were conducted, and films doped with J_2 were prepared. Again conductivities in the range of $10^{-5} - 10^{-2}$ Siemens/cm were found. A semiconductive polymer was also obtained by regioselective 1,2-polymerization of monomer **126a** [141]. Ti(OBu)4/AlEt$_3$ proved to be the most successful catalyst yielding a soluble polymer **126b**, with $M_n = 4600$ g/mol.

$$HC\equiv C-C\equiv C-SiMe_3 \longrightarrow \left[\begin{array}{c} -\overset{C=C}{\underset{H\ \ C\equiv C-SiMe_3}{|\ \ \ \ |}}- \end{array} \right]$$

(126)

Quite recently 1-(o-trimethylsilylphenyl)acetylene **129b** has found increasing interest [142–145]. This monomer was synthesized from phenylacetylene according to Eqs. 127–129. High molecular weight polyacetylenes ($M_w > 10^6$) were obtained in high yields when the polymerization was conducted in toluene

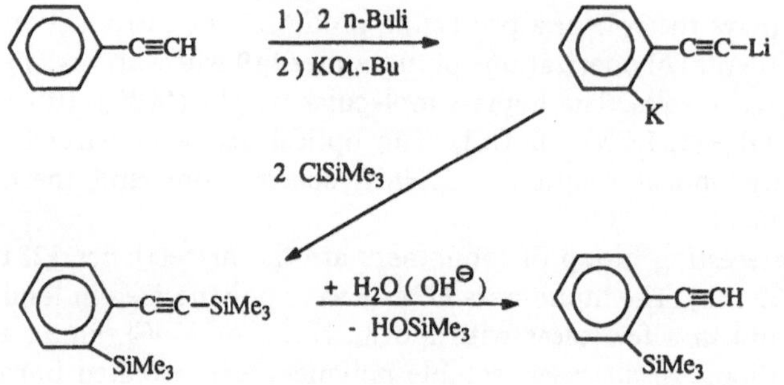

$$\text{(127)}$$

$$\text{(128)}$$

$$\text{(129)}$$

a   b

at 30 °C with catalysts such as: W(CO)6-hv, MoCl$_5$-Ph$_4$Sn, WCl$_6$-Ph$_4$Sn, WCl$_6$-PhSb WCl$_6$-Ph$_3$Bi, WCl$_6$-Bu Sn or WCl$_6$ Et$_3$SiH [143]. The influence of solvents and cocatalyst concentration on yield and molecular weight was studied extensively. Detailed studies of the reaction mechanism based on WCl$_6$-Me$_4$Sn revealed a W = CH$_2$ carbon complex as initiator [145]. The polymer was characterized by a variety of methods. In contrast to unsubstituted poly(phenylacetylene) the ortho-TMS-derivative forms tough, flexible films which are of interest due to their photoconductivity [142]. Yet, a high molecular weight polymer was obtained from the *meta* compound **130b** which was prepared from 1-(*m*-bromophenyl) propyne **130a** [146]. Whereas NbCl$_5$-based catalysts yielded partially crosslinked polymers TaCl$_5$ in combination with suitable co-catalysts (Ph$_4$Sn, Ph$_3$Bi, Et$_3$SiH etc) proved to be an excellent catalyst. Films of the resulting polyacetylene are of interest, because of their gas-permeability [146].

Finally, it should be mentioned, that attempts to obtain a vinylpolymer with a pendant TMS-acetylene group by radical polymerization failed, Eq. (131) [147].

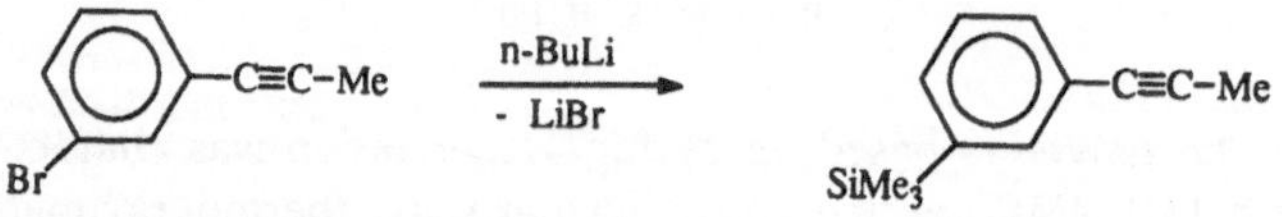

$$\text{(130)}$$

a   b

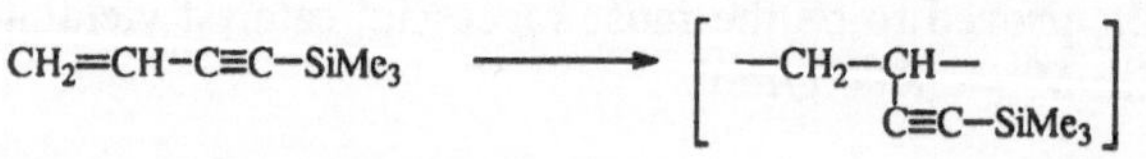

$$\text{(131)}$$

The acetylene group was somehow involved in the polymerization process and caused crosslinking. Taken together, the results of this section indicate that the polymerization of Si-containing acetylenes is a rapidly growing, highly interesting new field.

1.4
Silylated Acetylenes in Condensation Reactions

This subchapter deals with the following three aspects of acetylene chemistry:

I) Stepwise synthesis of oligoacetylenes from silylated acetylenes.
II) Synthesis of acetylene end capped oligomers designed as thermosetting resins.
III) Polyadditions and polycondensations of Si-containing acetylenes.

1.4.1
Stepwise Syntheses of Oligomers

The stepwise synthesis of well defined oligoacetylenes requires acetylenes with suitable protecting groups as starting materials. These protecting groups should be stable under the reaction conditions of the coupling step, but easy to remove without damage of the highly reactive oligoacetylenes. Trialkylsilyl groups, particularly the triethylsilyl group, proved to be most useful for this purpose. Walton and coworkers [148-150] were the first who developed a synthetic strategy based on triethylsilylacetylene, Scheme (132). Trialkylsilylalkines can be

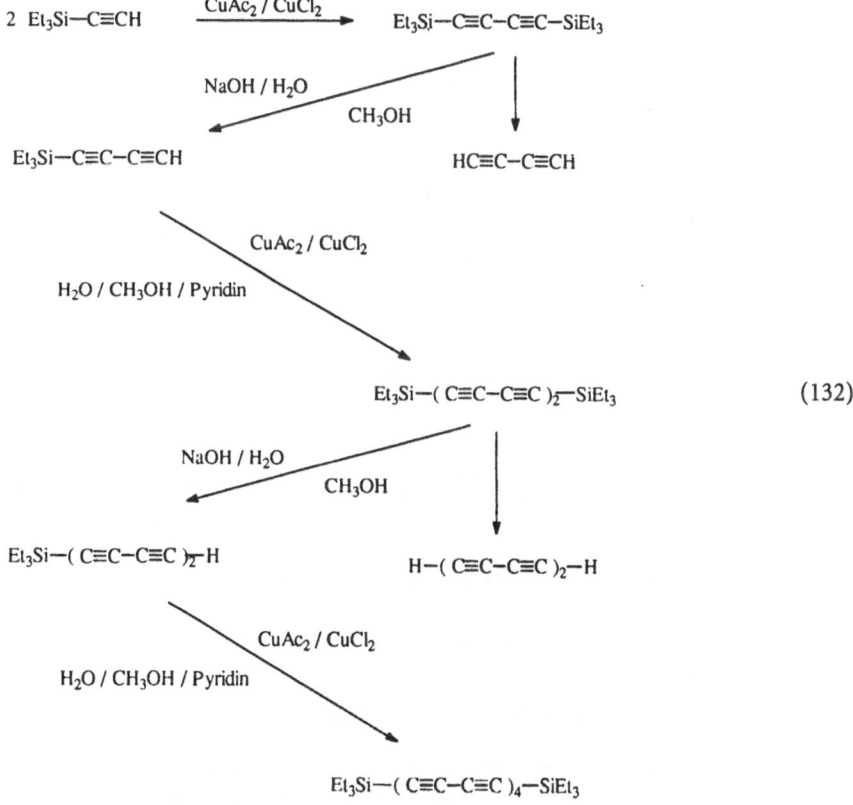

(132)

prepared by the reaction of chlorotrialkylsilanes with lithium sodium or Grignard derivatives of acetylene [150–153]. However, treatment of 1,3-butadiyne with EtMgBr followed by Me$_3$SiCl yielded a mixture of the mono- and disilylated products **133a, b** [151]. After coupling the triethylsilyl group is easily and selectively removed by alkaline methanol solutions [148]. The most satisfactory coupling method proved to be the "Hay version" [154] of the "Glaser coupling" as illustrated in Scheme (132). Typical for this procedure is a combination of *N, N'*-tetramethyl ethylene diamine and pyridine for the complexation of copper ions. By this strategy oligomers up to the hexadecayne (n = 16) were synthesized [155]. An alternative condensation method the "Cadiot-Chodkiewicz coupling" proved to be less favorable, because it requires selectively brominated triethylsilyl alkynes Eq. (134).

$$Me_3Si-C\equiv C-C\equiv CH \qquad\qquad Me_3Si-C\equiv C-C\equiv C-SiMe_3$$

$$a \qquad\qquad\qquad\qquad\qquad\qquad b \tag{133}$$

$$Ph-C\equiv C-Br \;+\; HC\equiv C-SiEt_3 \xrightarrow[-\,HBr]{Pd\,(\,II\,)} Ph-C\equiv C-C\equiv C-SiEt_3 \tag{134}$$

Several oligoalkynes and their α, ω-bistrimethylsilyl derivatives were prepared by Kloster Jensen [156] according to Scheme (135). However, yields were not reported and the oligomers were only characterized by ^{1}H-NMR and UV-spectra. A third approach is illustrated in Eq. (136). It is based on cyclobutenediones as thermally unstable precursors of an acetylene group. A special technique "Solution-Spray-Flash-Vacuum-Pyrolysis" (SS-FVP) is required to obtain acceptable yields [157]. Furthermore, the synthesis of the silyl protected, cross-conjugated oligoacetylenes **137a–c** and **138** should be mentioned [158]

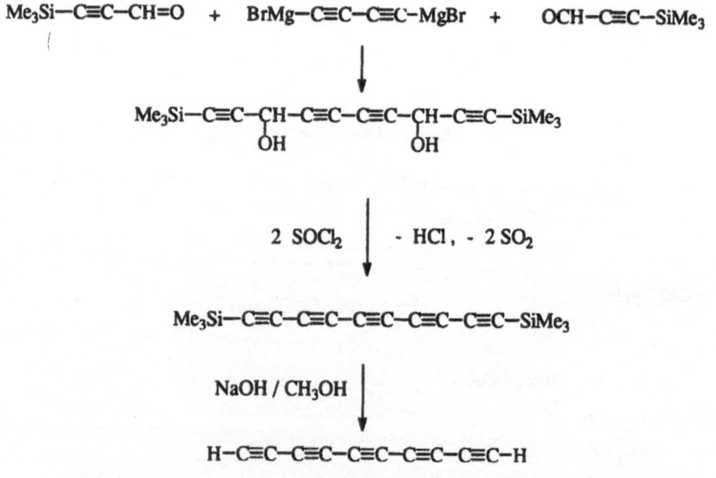

$$Me_3Si-C\equiv C-CH=O \;+\; BrMg-C\equiv C-C\equiv C-MgBr \;+\; OCH-C\equiv C-SiMe_3$$

$$\downarrow$$

$$Me_3Si-C\equiv C-\underset{\underset{OH}{|}}{C}H-C\equiv C-C\equiv C-\underset{\underset{OH}{|}}{C}H-C\equiv C-SiMe_3$$

$$2\,SOCl_2 \;\Big\downarrow\; -\,HCl,\,-\,2\,SO_2 \tag{135}$$

$$Me_3Si-C\equiv C-C\equiv C-C\equiv C-C\equiv C-C\equiv C-SiMe_3$$

$$NaOH\,/\,CH_3OH \;\Big\downarrow$$

$$H-C\equiv C-C\equiv C-C\equiv C-C\equiv C-C\equiv C-H$$

In addition to these quasi-aliphatic oligoacetylenes a synthesis of dendritic aromatic oligomers was recently described [159]. The reaction sequence begins with the Pd catalyzed condensation of phenylacetylene with silylated 3,5-dibromophenylacetylene, Eq. (139). Removal of the silylgroup generates a free acetylene group **140** which can be used for the same type of condensation yielding the monomer **141a**. Condensation of this monomer with **141b** finally yields the dendritic oligomer **142**.

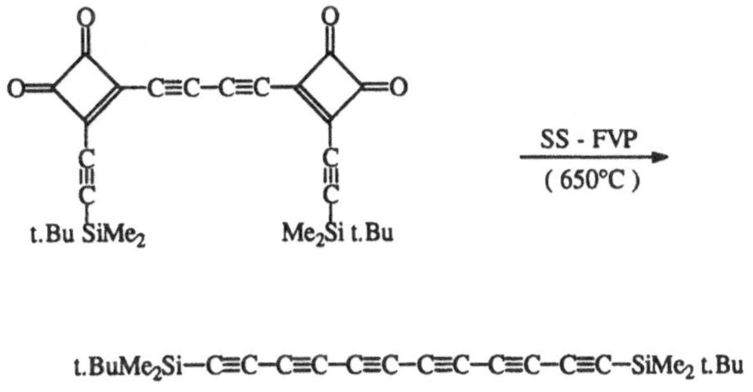

$$\xrightarrow[(650°C)]{SS - FVP}$$

(136)

$$t.BuMe_2Si-C\!\equiv\!C-C\!\equiv\!C-C\!\equiv\!C-C\!\equiv\!C-C\!\equiv\!C-C\!\equiv\!C-SiMe_2\, t.Bu$$

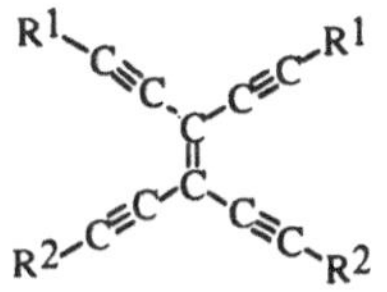

$\underline{a}$: R^1, R^2 = Me_3Si
$\underline{b}$: R^1 = Me_3Si, R^2= Ph
$\underline{c}$: R^1 = Me_3Si, R^2= Pr_3Si

(137)

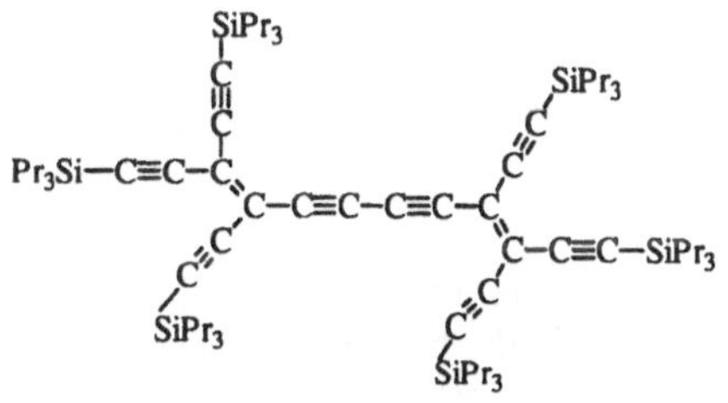

(138)

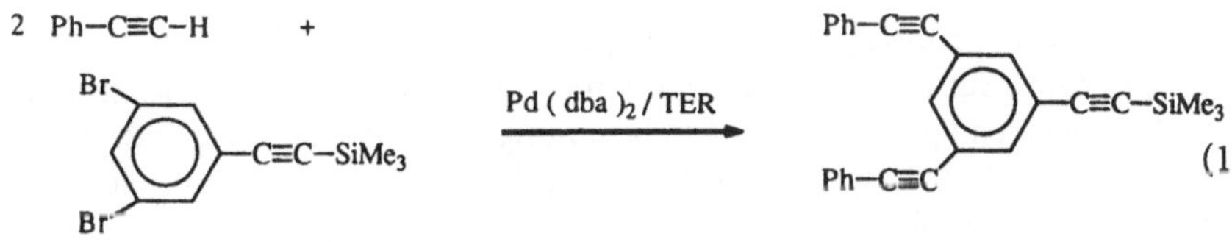

$$\xrightarrow{Pd\,(\,dba\,)_2\,/\,TER}$$

(139)

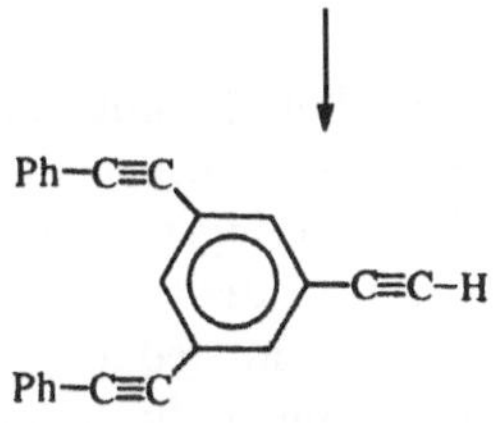

(140)

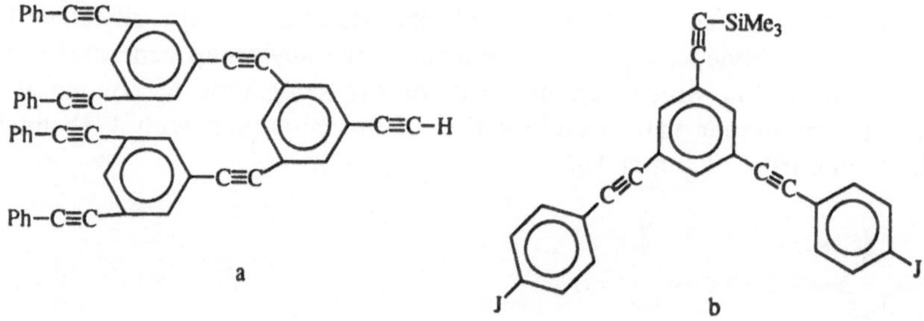

(141)

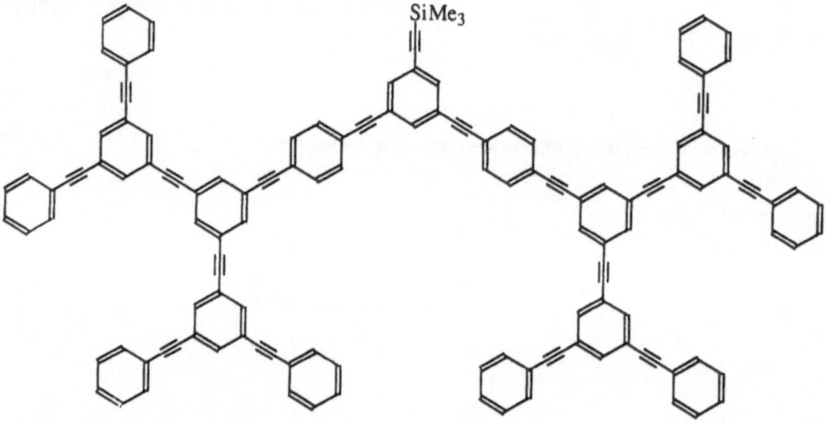

a

b

(142)

1.4.2
Acetylene-Terminated Oligomers

Acetylene-terminated telechelic oligomers have recently attracted increasing interest as thermally curable materials. Thermostable, high molecular weight aromatic polymers, such as poly(ether sulfone)s, poly(etherketone)s, polyimides, poly(benzimidazole)s etc., have the basc disadvantage that their processability is affected by a high melt viscosity. Oligomers of similar structure are easier to process, due to their lower melt viscosity, and if these oligomers possess functional end groups suited for thermal cure, they represent the ideal thermosetting resins. Acetylene end groups proved to be among the best functional end groups for this purpose [160–167], the main problem being the synthesis of telechelic, acetylene-terminated oligomers.

The most widely used method for synthesis of acetylene endgroups consist of a condensation of TMS-acetylene with brominated aromatic compounds catalyzed by Pd(II) complexes, Eq. (143). Functional groups, such as aldehyde, ester, CF_3, fluoro, nitro- and amino groups do not effect this condensation step [164]. The resulting functional phenylacetylenes are in turn used to build up acetylene-terminated oligomers. The alternative approach consists of the preparation of brominated oligomers and their reaction with TMS acetylene. A typical problem of this alternative approach may be a low solubility of the oligomers. Furthermore, it should be mentioned that instead of bromo-substituted aromatic compounds, triflate groups allow the incorporation of TMS-acetylene groups, Eq. (144) [166]. Typical acetylene-terminated oligomers prepared by one of these methods are the oligo(ether sulfone)s **145**, the bisimidazoles **146** [165] and the oligo(pyridinesulfide) **147** [167]. The pertinent literature also reports on thermal cure and thermogravimetric analyses.

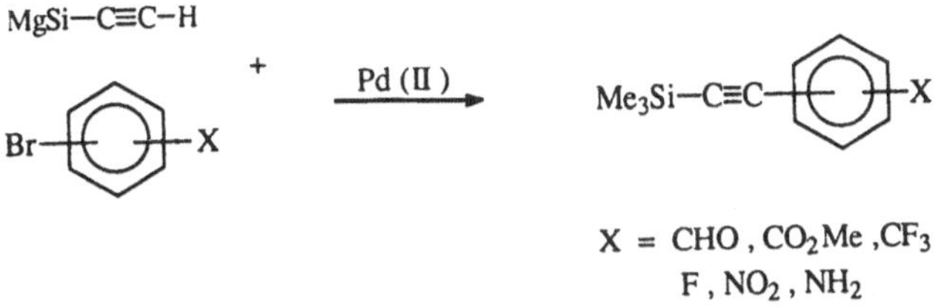

$$
\text{MgSi}-\text{C}\equiv\text{C}-\text{H}
$$

(143)

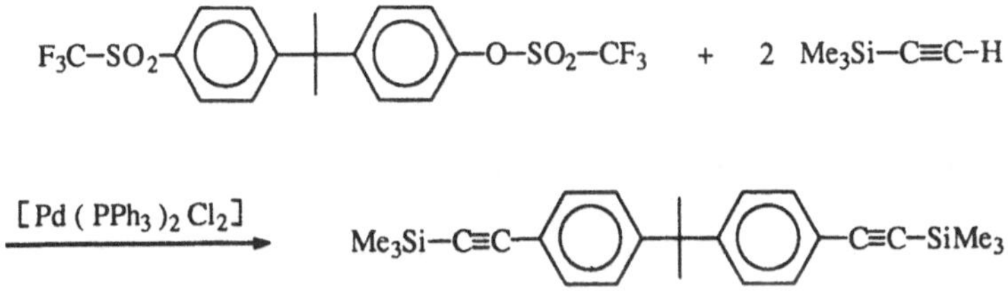

$$
X = \text{CHO}, \text{CO}_2\text{Me}, \text{CF}_3
$$
$$
\text{F}, \text{NO}_2, \text{NH}_2
$$

(144)

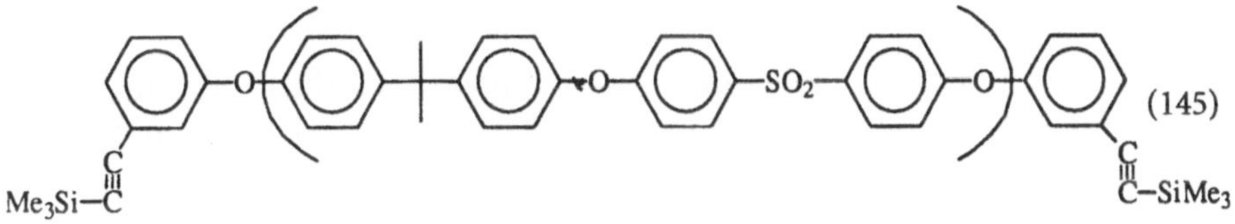

(145)

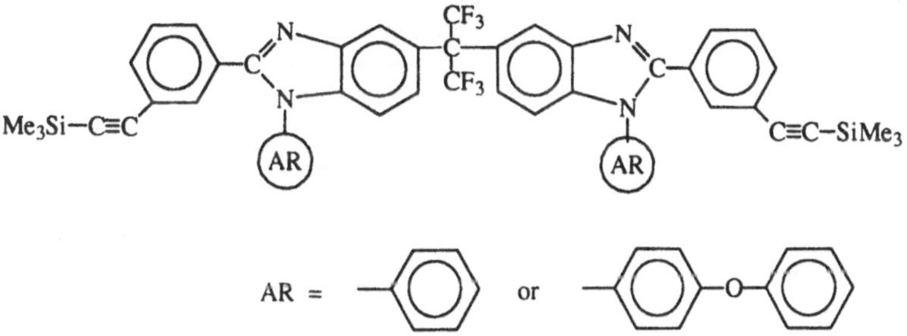

(146)

$$
\text{AR} =
$$

or

(147)

1.4.3
Polycondensations

The oxydative polycondensation of diacetylenes, Eq. (148), was first described by
Hay in 1969 [168, 169]. In order to control the molecular weight phenylacetylene
was added as a monofunctional chain stopper. More recently another research
group [170] described a similar approach for the synthesis of oligo(phenylene
diacetylene)s using trimethylsilyl-1, 3-diethynyl benzene as chain stopper 149.

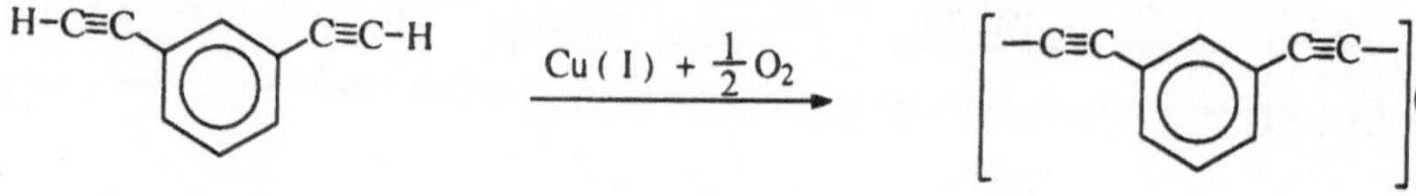

(148)

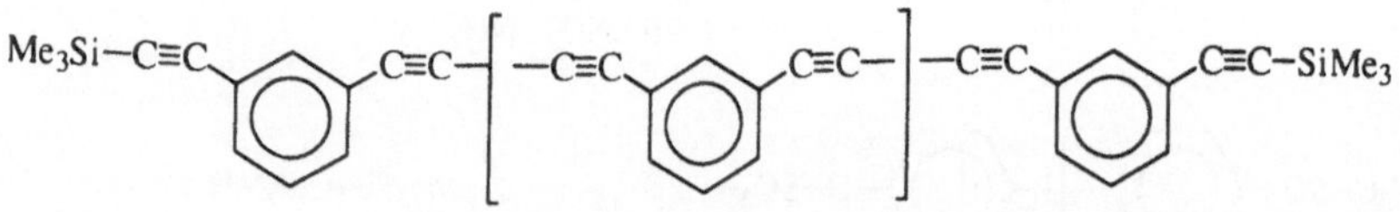

(149)

Detailed ^{13}C NMR spectroscopic studies were reported [170]. The oxydative
polycondensation of bisethynyl thiophenes was described more recently [171].
In that work, TMS-acetylene played the key role for the synthesis of the
monomers Eqs. (150, 151). Another class of polymers and polycondensation
processes studied by several research groups [172–175] concerns polymers containing both acetylene and silylene groups in their backbone. The first polymers
of this type reported in literature were prepared by polycondensation of the
disodium or dilithium salts of *meta* or *para* diethylnyl benzene with
dichlorosilanes, Eq. (152). The resulting polymers were yellow or brownish,
amorphous, soluble materials of relatively low molecular weight, but they were

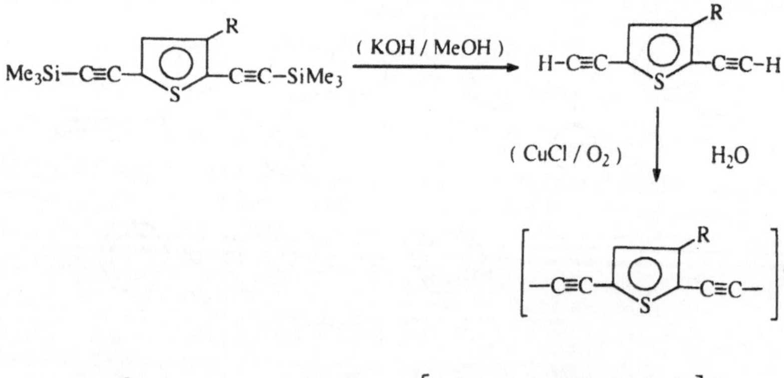

(150)

(151)

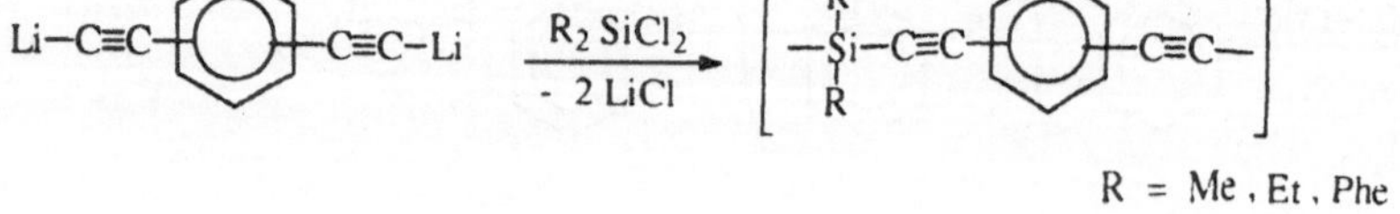

(152)

R = Me , Et , Phe

poorly characterized. An alternative synthetic approach which allows a broader variation of the aromatic building block was more recently reported by French authors [175]. Their polycondensation method is based on the Pd(II) catalyzed coupling of a diethynylsilane with a dibromoaromat, Eq. (153). Again amorphous, soluble polymers with low or moderate molecules weights ($M_n <$ 10 000) were obtained, which were well characterized.

Polymers with an alternating sequence of acetylene groups and disilane units were seemingly prepared by polycondensations of dilithium salts with dichlorodisilane, Eq. (154) [176]. Semicrystalline materials with $M_n \leqslant 7000$ g/mol were isolated and the ^{1}H- and ^{13}C-NMR spectra suggest the formation of alternating sequences, when R and R1 are different. An analogous polycondensation was studied by another research group again in THF at 20–25 °C [177]. Instead of disilane units simple silylgroups were used, Eq. (155). Interestingly, the ^{29}Si-NMR spectra of the isolated polymers clearly indicate that random sequences of the silylgroups were formed for R = Me and R^1 = Ph. The existence

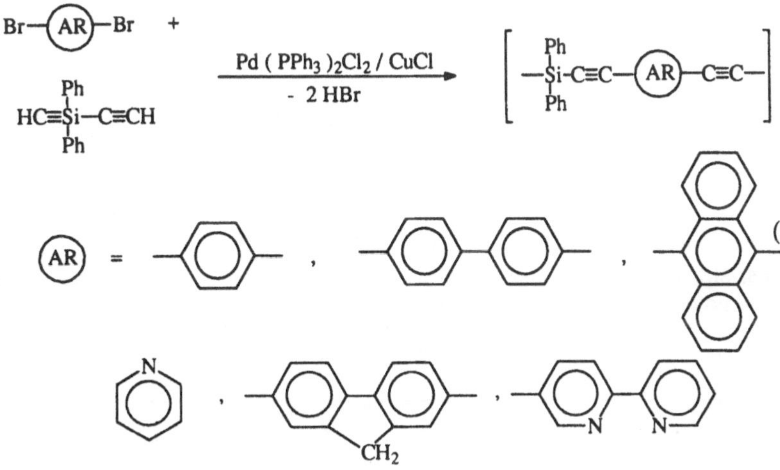

(153)

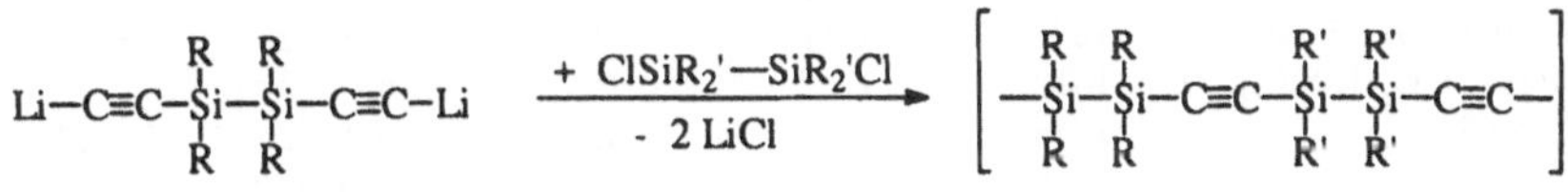

R , R' = Me or Bu

(154)

(155)

R = Me , R' = Ph

of equilibration reactions was also supported by the observation that cyclic oligomers were formed.

Several papers deal with synthesis and properties of polymers built up by 1, 3-butadiyne units connected by silyl groups (i.e. poly(silanylene-diethynylene)s) [178-180]. Their synthesis is based on the polycondensation of the diethynylene dilithium salt with various dichlorosilanes, Eq. (157). The dilithium salt is conveniently prepared from commercial 1,4-bis-TMS-1,3-butadiyne, Eq. (156), which is stable on storage in contrast to the free 1,3-butadiyne. The molecular weights of all these polymers **157** were low (M_n < 5000 g/l). They were characterized by ^1H, ^{13}C and ^{29}Si NMR spectroscopy, and their pyrolysis yielding β-Si-C ceramics was studied in detail. The data in Table 1.5 illustrate the variation of

$$Me_3Si-C\equiv C-C\equiv C-SiMe_3 \xrightarrow[- \ 2\,BuSiMe_3]{+\ 2\,BuLi} Li-C\equiv C-C\equiv C-Li$$

$$\Bigg\downarrow R_2SiCl_2$$

(156)

$$\left[\begin{array}{c} R \\ | \\ -Si-C\equiv C-C\equiv C- \\ | \\ R \end{array} \right]$$

(157)

Table 1.5. Structure of the dichlorosilanes used for the synthesis of poly(silanylene-diethynylene)s and properties of the resulting polymers[*]

dichlorosilane	MWc	MW/M_n	n_w	mp, °C	conductivity S/cm^{-1}
Me_2SiCl_2	4063	1.24	38	150	8×10^{-5}
$MePhSiCl_2$	3068	1.64	18	80-110	10^{-4}
$Me(CH_2=CH)SiCl_2$	2154	1.51	18	oil	
$Me(p\text{-}MeC_6H_4)SiCl_2$	2712	1.30	15	90-130	2×10^{-4}
$Me(p\text{-}FC_6H_4)SiCl_2$	1790	1.31	10	oil	
$Me(p\text{-}MeOC_6H_4)SiCl_2$	1822	1.24	9	90-120	3×10^{-4}
$Me(p\text{-}CF_3C_6H_4)SiCl_2$	1612	1.25	7	oil	
Ph_2SiCl_2	1826	1.31	8	80-120	3×10^{-4}
$(Me_2ClSi)_2$	3536	1.41	21	100-140	5×10^{-5}

[*] data adapted from Ref. 179.

the silyl group, the molecular weights and the electric conductivity found after doping with $FeCl_3$.

Alternating sequences of acetylene and disilane units (i.e. poly(disilanylene-ethynylene)s) were seemingly best prepared by anionic ring-opening polymerization, Eqs. (158, 159). Using BuLi as initiator MW's up to 10^5 were obtained which allowed the casting of tough films. High electric conductivities were found after doping with SbF_3. However the isolation of the cyclic monomers requires

$$
\begin{array}{c}
\text{Me}\ \text{Me}\\
\text{R}-\underset{\substack{\text{C}\\\|\\\text{C}}}{\text{Si}}-\underset{\substack{\text{C}\\\|\\\text{C}}}{\text{Si}}-\text{R}\\
\text{BrMg}-\text{C} \ +\ \text{C}-\text{MgBr}\\
\text{Me}\ \text{Me}\\
\text{Cl}-\underset{\text{R}}{\text{Si}}-\underset{\text{R}}{\text{Si}}-\text{Cl}
\end{array}
\quad\xrightarrow[-\ 2\ \text{MgClBr}]{\text{THF}}\quad
\begin{array}{c}
\text{RMe}-\underset{\big|}{\text{Si}}-\text{C}\equiv\text{C}-\underset{\big|}{\text{Si}}-\text{MeR}\\[4pt]
\text{RMe}-\text{Si}-\text{C}\equiv\text{C}-\text{Si}-\text{MeR}
\end{array}
\qquad(158)
$$

$$\big\downarrow\ \text{BuLi}$$

$$
R = \text{Me},\ \text{Et},\ \text{Ph}
\qquad\qquad
\left[\begin{array}{c}
\text{Me}\ \text{Me}\qquad\text{Me}\ \text{Me}\\
-\underset{\text{R}}{\text{Si}}-\underset{\text{R}}{\text{Si}}-\text{C}\equiv\text{C}-\underset{\text{R}}{\text{Si}}-\underset{\text{R}}{\text{Si}}-\text{C}\equiv\text{C}-
\end{array}\right]
\qquad(159)
$$

column chromatography. Polymers of structure **161** were also prepared by three direct polycondensation methods, Eqs. (160, 161, 162) [183]. The best result, namely a polymer with MW ~ 70 000, was obtained, when the dilithium acetylene was used as monomer, Eq. (161). For, this synthesis of alternating sequences of disilane and diacetylene units (i.e. poly(disilanylene-diethynylene)s) cyclic monomers were not available, and the normal polycondensation method based on Li_2C_4 was exclusively used, Eq. (163) [184]. Soluble polymers with MW's up to 44 000 were isolated and characterized by various spectroscopic methods. Furthermore, thermal properties including the formation of SiC ceramic were studied. In another paper of the same research group [185] the thermal polymerization of diethynyl diphenylsilane **164a** and its polymerization catalyzed by $MoCl_5$ were investigated. Both polymerization processes yielded polymers of complex

$$\text{ClMg}-\text{C}\equiv\text{C}-\text{MgCl}\ +\ \diagdown\text{ClSiMe}_2-\text{SiMe}_2\text{Cl} \qquad\qquad\qquad (160)$$

$$\text{Li}-\text{C}\equiv\text{C}-\text{Li}\ \xrightarrow{\ \text{ClSiMe}_2-\text{SiMe}_2\text{Cl}\ }
\left[\begin{array}{c}\text{Me}\ \text{Me}\\ -\underset{\text{Me}}{\overset{}{\text{Si}}}-\underset{\text{Me}}{\overset{}{\text{Si}}}-\text{C}\equiv\text{C}-\\ \text{Me}\ \text{Me}\end{array}\right] \qquad (161)$$

$$
\begin{array}{c}
\text{Me}\qquad\text{Me}\\
\text{Cl}-\underset{\text{Me}}{\text{Si}}-\text{C}\equiv\text{C}-\underset{\text{Me}}{\text{Si}}-\text{Cl}
\end{array}
\xrightarrow[-\ 2\ \text{NaCl}]{2\ \text{Na}}
\qquad\qquad\qquad (162)
$$

$$
\begin{array}{c}
\text{Li}-\text{C}\equiv\text{C}-\text{C}\equiv\text{C}-\text{Li}\ +\\[8pt]
\text{Me}\ \text{Me}\\
\text{Cl}-\underset{\text{R}}{\text{Si}}-\underset{\text{R}}{\text{Si}}-\text{Cl}
\end{array}
\quad\longrightarrow\quad
\left[\begin{array}{c}\text{Me}\ \text{Me}\\ -\underset{\text{R}}{\text{Si}}-\underset{\text{R}}{\text{Si}}-\text{C}\equiv\text{C}-\text{C}\equiv\text{C}-\end{array}\right]
\qquad(163)
$$

$$
\begin{array}{c}
\text{Ph}\\
\text{HC}\equiv\text{C}-\underset{\text{Ph}}{\text{Si}}-\text{C}\equiv\text{CH}
\end{array}
\qquad\qquad
\left[\begin{array}{c}\text{Ph}\\ -\underset{\text{Ph}}{\text{Si}}-\text{C}\equiv\text{C}-\text{C}\equiv\text{C}-\end{array}\right]
\qquad(164)
$$

$$\qquad\qquad\text{a}\qquad\qquad\qquad\qquad\qquad\qquad\text{b}$$

structure and not the polymer **164b**, which was synthesized by another route, Eqs. (156, 157).

Polymers containing siloxane groups alternating with acetylene were prepared in two closely related ways, Eqs. (165, 166) [186]. The soluble polymers having M_w values in the range of 6000–23 000 were characterized by ^1H and ^{29}Si NMR spectroscopy and their thermal properties were determined. On the basis of the

$$HOPh_2-Si-C\equiv C-Si-Ph_2OH$$
$$+ Ph_2Si(NMe-COMe)_2 \xrightarrow{-2\ HNMe-COMe} \left[\begin{array}{c} Ph \\ -Si-C\equiv C-Si-O-Si-O- \\ Ph \end{array} \begin{array}{c} Ph \\ \\ Ph \end{array} \begin{array}{c} Ph \\ \\ Ph \end{array}\right] (165)$$

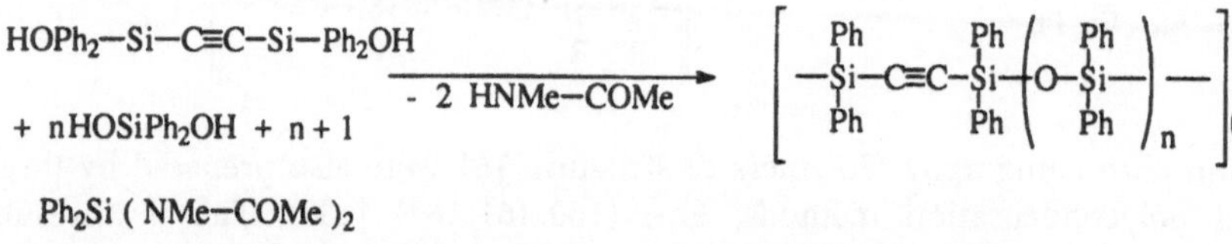

$$HOPh_2-Si-C\equiv C-Si-Ph_2OH$$
$$+ nHOSiPh_2OH + n+1 \xrightarrow{-2\ HNMe-COMe} \left[\begin{array}{c} Ph \\ -Si-C\equiv C-Si \\ Ph \end{array} \begin{array}{c} Ph \\ \\ Ph \end{array} \left(\begin{array}{c} Ph \\ O-Si- \\ Ph \end{array}\right)_n \right](166)$$
$$Ph_2Si(NMe-COMe)_2$$

results compiled in this section, it is easy to predict that polymers made up of silyl groups and acetylene units will attract increasing interest in the future.

1.5
Silicon-Containing Initiators and Catalysts

This section will deal exclusively with Si-containing initiators and catalysts used for the polymerization of vinyl monomers and olefins. Ring-opening polymerizations initiated by Si-containing initiators will be discussed in Chapter 6 (Sect. 3). Furthermore, this section does not include purely inorganic catalysts such as zeolites. First studies on Si-containing initiators useful for the polymerization of vinyl monomers go back to the late 1950s when the potential of peroxysilanes as radical initiators were explored [188, 189]. Since then, the field of Si-containing initiators and catalysts has found increasing interest, mainly over the past fifteen years. The initiator and catalyst systems discussed in this section will be subdivided into the following four groups:

1. Initiators of radical polymerizations.
2. Initiators of cationic polymerizations.
3. Silicon-modified Ziegler-Natta catalysts.
4. Polymerization initiated on silica or silicate surfaces.

1.5.1
Initiators of Radical Polymerizations

Si-containing initiators useful for the radical polymerization of olefins and vinyl compounds may be subdivided into the following three classes:

1) Peroxides with a C-O-O-Si group.
2) Hexasubstituted ethenes with a strained C-C bond.
3) Polysilanes with photosensitive Si-Si bonds.

The given order also represents the historic order in which these three classes of initiators appeared in the literature.

Peroxysilanes with C-O-O-Si structure were studied as initiators of radical polymerizations of styrene and alkyl methacrylate by several research groups [189–193]. The interest in these initiators has two main reasons. Firstly, their thermostability is unusually high when compared to other organic peroxides. Secondly, they may form complexes with the monomers and with the radical chain ends, and thus, influence the kinetic course of the polymerizations. All peroxysilanes reported so far were prepared by silylation of the corresponding hydroperoxides with chlorotrimethylsilane in the presence of a *tert* amine, Eq. (167). The structures and properties of the peroxysilanes described in Refs. 188 and 189 are compiled in Table 1.6. Applications of the peroxysilanes 168a and

$$Me_3COOH + Me_3SiCl \xrightarrow[- Py \cdot HCl]{+ Py} Me_3C-O-O-SiMe_3 \qquad (167)$$

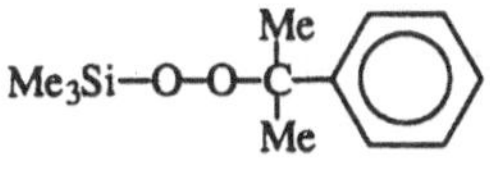

Et$_3$Si$-$O$-$O$-$CMe$_3$

(168)

a   b

Table 1.6. Structure and properties of selected peroxysilanes recommended as radical initiators in Ref. 188[*]

Chemical Structure	n_d^{25}	d_{25}	bip(°C/mm)
Me₃C—CO—OSiMe₃	1.3925	0.8219	79 °C/215 mm
EtMe₂C—O—O—SiMe₃	1.4032	0.8419	78 °C/95 mm
Ph-Me₂C—O—O—SiMe₃	1.4782	0.9501	43 °C/0.05 mm
Me₃C—O—O—SiEt₂—O—O—CMe₃	1.4149	0.9415	40 °C/1 mm
(Me₃C—O—O)₃SiMe	1.4097	–	50 °C/0,1 mm
Me₃C—O—O—SiPh₂—O—O—CMe₃	1.5103	1.033	110 °C/0,001 mm

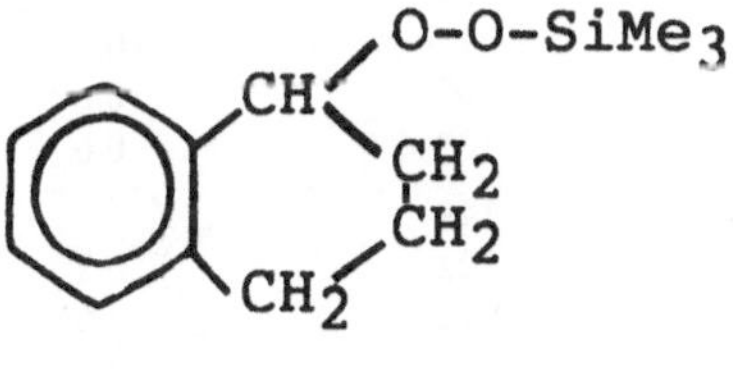	1.5102	–	53 °C/0,01 mm
Me₃C—O—O—SiPh₃	–	–	m.p. ~ 50 °C
Me₃C—O—C—Si(OMe)₃	–	–	49 °C/6 mm

[*]data adapted from Refs. 188 and 189.

168b were also reported by another groups [190–192]. All these peroxysilanes are thermostable enough to allow their purification by distillation in vacuum.

The high thermostability of the peroxides 168a and 168b also allows their application as radical crosslinking agents for polyolefins. Their half-lives has been reported [194] to be 6500–7500 h at 130 °C and 1.5–2.5 h at 190 °C. Therefore, they may be incorporated into polyolefins and their blends in the course of an extrusion or moulding process $\leqslant$ 150 °C. The crosslinking is then achieved by annealing of the doped products at temperatures $\geqslant$ 180 °C.

Concerning peroxysilane-initiated polymerizations of vinyl compounds, the most interesting mechanistic and preparative aspects were found, when alkyl methacrylates were used as monomers [193]. As evidenced by ^{1}H NMR spectroscopy the monomers form donor-acceptor complexes with the initiators in such a way that the silicon with its free d-orbital acts as acceptor (Form. 169).

$$CH_2 = \underset{\underset{Me}{|}}{\overset{}{C}} - \underset{\underset{OR}{|}}{\overset{}{C}} = \underline{O} | \cdots \underset{\underset{Me_3}{|}}{\overset{}{Si}} - O - O - CMe_3$$

(169)

The thermodynamical aspects of these complexation equilibria were studied in much detail [195–197]. The complexation reduces the stability of the peroxide group, and thus, lowers the initiation and polymerization temperature. For the decomposition of neat $tert$-butyl peroxytrimethylsilane rate constants of 2×10^{-9} s^{-1} and 5×10^{-9} s^{-1} were found in nonane or anisole, but 2.4×10^{-6} s^{-1} in butylmethacrylate at 373 K [195]. The activation energies determined for decomposition and polymerization processes of three peroxysilanes are listed in Table 1.7 [193]. Another interesting aspect of this chemistry is the variation of the reactivity ratios, when two vinylmonomers are copolymerized with peroxysilanes as initiators [193]. Complexation of the active chain ends by the silanes was discussed as speculative explanation.

Table 1.7. Overall activation energy of polymerization, E, activation energy of initiation stage, E_i, reaction order with respect to initiator, n. Data adapted from [193]

Peroxide	Monomer	E (kJ mol^{-1})	E_i (kJ mol^{-1})	n
Me$_3$C—O$_2$—SiMe$_3$	Anisole	—	159	
	MMA	77	112	0.5
	BMA	75	107	0.6
	St	82	109	0.6
(Me$_3$C—O$_2$) SiMe$_2$	Anisole	—	149	
	MMA	72	102	0.5
	BMA	70	99	0.6
	St	81	109	0.5
(MeC—O$_2$)$_3$ SiMe	Anisole	—	133	
	MMA	64	86	0.5
	BMA	70	98	0.3
	St	69	96	0.4

An interesting property of the peroxysilanes is their complexation with Lewis acids. In addition to $ZnCl_2$ various tin halogenides, such as $SnCl_4$, Me_2SnCl_2, Et_2SnCl_2, n Bu_2SnCl_2 and Bu_3SnCl were investigated. Stoichiometry and thermodynamical aspects of this complexation were examined by cryoscopy, calorimetry and ^1H NMR spectroscopy [198, 199]. In these complexes, the peroxysilanes play the role of the donor. Obviously the oxygen atom is coordinated with the metal ion. Whereas $SnCl_4$ is capable of complexing two peroxysilane molecules (e.g. Form. 170), butyltinchlorides form 1:1 complexes.

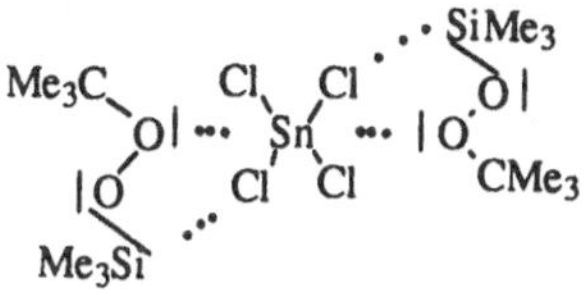

$$(170)$$

$$(171)$$

Addition of small amounts of Lewis acids to the mixture of peroxysilanes and alkylmethacrylates considerably lowers the activation energy of initiation and polymerization [193]. These results were speculatively explained by the intermediate formation of a complex consisting of Lewis acid, peroxysilane and monomer. Hypothetical formulations of such complexes are outlined in Forms 171a and 171b.

Bis(*tert*-Butylperoxy)dimethylsilane is particularly noteworthy as initiator of styrene, because under mild conditions, polystyrene chains bearing one peroxide group may be formed (Form. 172). When heated in the presence of another

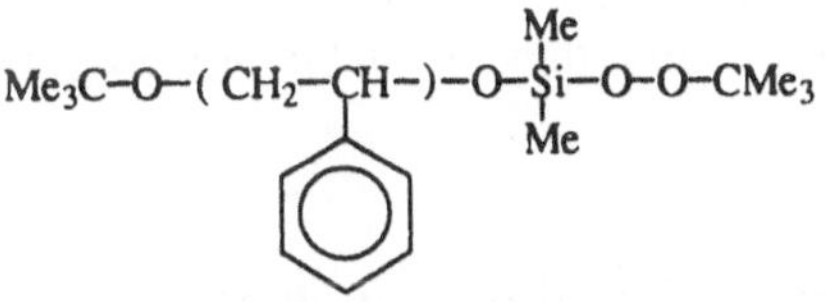

$$(172)$$

monomer, such as an alkyl methacrylate the peroxide endgroup can act as initiator resulting in the formation of two-block copolymers [193]. However, this method is, of course, not a clean process.

It has been well known for more than four decades that sterically hindered carbon-carbon bonds such as that of hexaphenyl ethene easily dissociate into radicals [200]. However, the steric effect caused by six bulky substituents is only one reason for the low energy of dissociation, the delocalization, and thus, stabilization of the resulting radicals is even more important. The excellent stabilization of the trityl radicals has, on the other hand, the consequence that they are not reactive enough to initiate the polymerization of most vinyl monomers. When two phenyl groups are replaced by hydroxy, alkoxy, aryloxy or siloxy groups the stabilization of the radicals is significantly reduced, the dissociation requires higher temperatures (e.g. $> 100\,°C$), but the radicals are now reactive enough to initiate polymerizations of styrene and other vinyl compounds.

This phenomenon was first observed for dihydroxytetraphenylethane (benzpinacol) [201, 202]. Braun and coworkers [202, 202] also found that in this particular case the initiation step consists of a transfer of H-radicals to the monomers, Eqs. (173, 174). This H-transfer and other side reactions are absent,

$$
\underset{\substack{\displaystyle OH \quad\ \ OH}}{C_6H_5-\overset{\displaystyle \overset{C_6H_5}{|}}{C}-\overset{\displaystyle \overset{C_6H_5}{|}}{C}-C_6H_5} \ \rightleftharpoons \ 2 \ \underset{\displaystyle OH}{C_6H_5-\overset{\displaystyle \overset{C_6H_5}{|}}{C}} \tag{173}
$$

$$+2 \ CH_2=CHR \ \Big\downarrow$$

$$2 \ H_3C-CHR \ + \ 2 \ (C_6H_5)_2C=O \tag{174}$$

when arylethers or silylethers of benzpinacol are used as initiators [203–205]. Therefore, a broad variety of silylbenzpinacolates were synthesized by numerous authors [205–212] and studied with regard to their usefulness as radical initiators. Two methods were reported for the synthesis of silylbenzpinacolates and related compounds. When benzpinacol is available, the silylation with chlorosilanes and a tertiary amine as HCl acceptor is the most convenient method, Eq. (175). However, the simultaneous condensation and silylation of ketones by means of elementary magnesium is the more flexible and most widely used approach, Eq. (176) [206–213]. This approach allows a broad variation of substituents including silylgroups directly attached to the central carbons 176 [209]. Even these diphenylethane derivatives are useful as radical initiators [209]. Three classes of siloxy initiators with cyclic structures are illustrated by Forms. 177a [211], 177c [211] and 178 [210]. Tables 1.8 and 1.9 summarize the details of their structures along with some properties.

The silylethers of benzpinacol and related compounds are of great interest as radical initiators for preparative purposes. Firstly, they are stable in storage and

$$
\underset{\substack{\displaystyle OH \ \ OH}}{(C_6H_5)_2-\overset{}{C}-\overset{}{C}-(C_6H_5)_2} \ \xrightarrow[{- \ 2 \ HCl}]{+ \ 2 \ ClSiMe_3} \ \underset{\substack{\displaystyle Me_3SiO \ \ OSiMe_3}}{(C_6H_5)_2-\overset{}{C}-\overset{}{C}-(C_6H_5)_2} \tag{175}
$$

$$
2 \ \underset{\displaystyle R}{C_6H_5-\overset{}{C}=O} \ + \ 2 \ ClSiMe_3 \ \xrightarrow{(Mg)} \ \underset{\substack{\displaystyle Me_3SiO \ \ OSiMe_3}}{C_6H_5-\overset{\displaystyle \overset{R}{|}}{C}-\overset{\displaystyle \overset{R}{|}}{C}-C_6H_5}
$$

$$\tag{176}$$

a : R = Alkyl   b : R = Aryl   c : R = Silyl

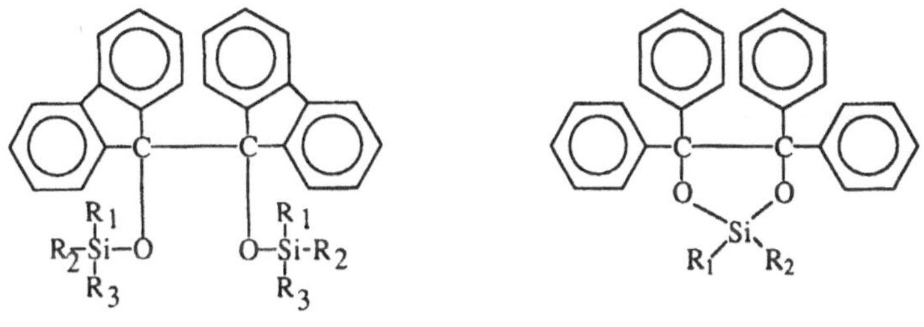

(177)

a   b

$(C_6H_5)_2-C-O\quad O-C-(C_6H_5)_2$
$\qquad\qquad Si$
$(C_6H_5)_2-C-O\quad O-C-(C_6H_5)_2$

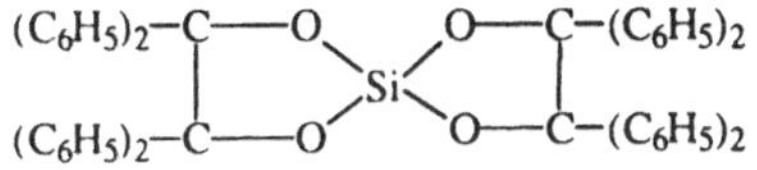

(178)

Table 1.8. Characteristics of cyclic silyl pinacole ether initiators. Data adapted from [210]

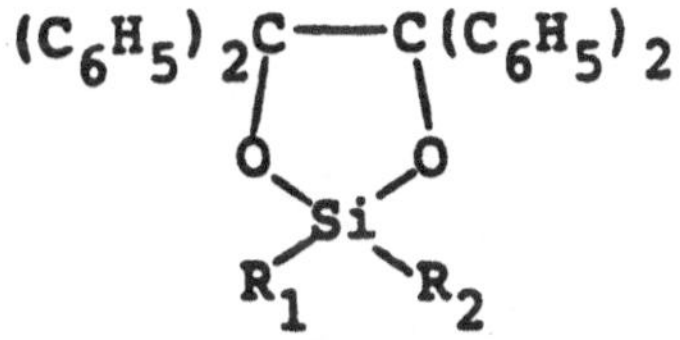

$(C_6H_5)_2C-C(C_6H_5)_2$
$\qquad O\quad O$
$\qquad\quad Si$
$\qquad R_1\quad R_2$

R_1	R_2	m.p. (°C)[a]	δ ^{29}Si (ppm)[b]	Elemental Analysis	% C	% H	% Si	Method (yield %)
CH$_3$	CH$_3$	134 − 136	20.82	Calc:	79.58	6.20	6.64	A(36),
				Fnd:	79.31	6.5	6.7	B(33),
								C(59)
C$_2$H$_5$	C$_2$H$_5$	123 − 125	19.16	Calc:	79.95	6.71	6.23	A(54)
				Fnd:	79.77	6.67	6.2	
C$_6$H$_5$	C$_6$H$_5$	> 200	−9.51	Calc:	83.48	5.53	5.14	A(35)
				Fnd:	83.38	5.47	4.97	
CH$_3$	CH$_2$=CH	135 − 137	4.68	Calc:	80.14	6.03	6.46	A(36)
				Fnd:	79.97	6.09	6.34	
CH$_3$	C$_6$H$_5$	150 − 151	5.22	Calc:	81.82	5.79	5.74	C(45)
				Fnd:	81.16	5.76	5.80	
(CH$_3$)$_3$C	(CH$_3$)$_3$C	185 − 199	12.59	Calc:	80.58	7.56	5.53	A(13)
				Fnd:	80.16	7.41	5.03	

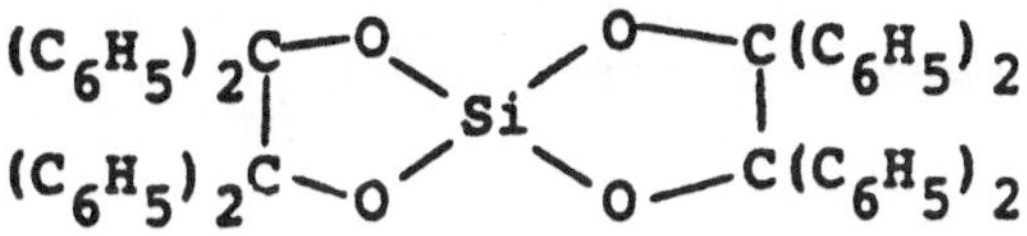

$(C_6H_5)_2C-O\quad O-C(C_6H_5)_2$
$\qquad\qquad\quad Si$
$(C_6H_5)_2C-O\quad O-C(C_6H_5)_2$

		240	−45.5	Calc:	82.45	5.29	3.70	A(24)
				Fnd:	79.2	5.30	3.65	

[a] Melting is accompanied by decomposition.   [b] ^{29}Si NMR run in CDCl$_3$.

Table 1.9. Characteristics of bissilyl pinacolate initiators. Data adapted from [211]

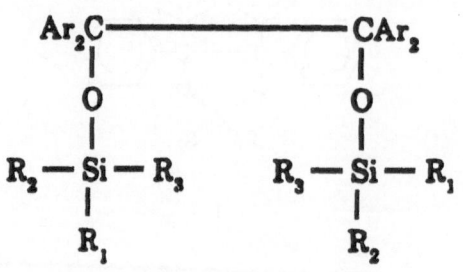

Ar	R_1	R_2	R_3	mp^a ($^\circ$C)	$\delta\ ^{29}Si^b$ (ppm)
C_6H_5	CH_3	CH_3	CH_3	132–134	9.22
C_6H_5	CH_3	CH_3	$CH=CH_2$	132–134	−1.75
C_6H_5	CH_3	CH_3	H	116–119	−6.24
C_6H_5	CH_3	C_6H_5	$CH=CH_2$	123–125	−11.06
C_6H_5	C_6H_5	C_6H_5	$CH=CH_2$	137–140	−18.64
C_6H_5	CH_3	CH_3	$n\text{-}C_3H_7$	64–66	9.794
C_6H_4					
C_6H_4	CH_3	CH_3	CH_3	247–253	14.82
C_6H_4					
C_6H_4	CH_3	CH_3	$CH=CH_2$	210–212	3.08

^{a} Melting is accompanied by decomposition. ^{29}Si NMR run in $CDCl_3$.

safe to handle in contrast to peroxides and azoisobutyronitrile. Secondly, the kinetics of the polymerization process initiated by these initiators is different from normal radical polymerizations. Since recombination of the radicals is fully reversible and other side reactions are almost absent these radicals act as initiators and terminators as illustrated by Eqs. (179–181) [203–205]. This reaction sequence has several interesting consequences for preparative applications. For instance, the resulting polymers possess two well-defined endgroups which may be varied at will by variation of the initiator Eq.(181). Furthermore, one of these endgroups is of low thermostability, and can dissociate upon heating into two radicals. Thus, the reversibility of the termination step may be used to reinitiate at higher temperatures either a polymerization of the same

$$(C_6H_5)_2-\underset{\underset{OR}{|}}{C}-\underset{\underset{OR}{|}}{C}-(C_6H_5)_2 \xrightarrow{\Delta T} 2\ (C_6H_5)_2-\underset{\underset{OR}{|}}{C}\cdot$$

$$+\ n\ \ CH_2=CHR' \qquad\qquad R = Ph\ ,\ SiMe_3$$

(179)

$$(C_6H_5)_2-\underset{\underset{OR}{|}}{C}-\underset{\underset{R'}{|}}{\left(CH_2\text{-}CH\right)}-CH_2\text{-}\underset{\underset{R'}{|}}{CH}\cdot \quad + \quad (C_6H_5)_2-\underset{\underset{OR}{|}}{C} \tag{180}$$

$$\Big\updownarrow$$

$$(C_6H_5)_2-\underset{\underset{OR}{|}}{C}-\underset{\underset{R'}{|}}{\left(CH_2\text{-}CH\right)}-CH_2\text{-}\underset{\underset{R'}{|}}{CH}-\underset{\underset{OR\,\cdot}{|}}{C}-(C_6H_5)_2 \tag{181}$$

monomer (with gradually increasing molecular weights) or of another monomer. In the latter case, two-block copolymers will be formed along with
homopolymers of the comonomer.

Of particular interest are cyclic silyl benzpinacolates such as **177b** (Table 1.8).
Dissociation of the labile C—C bond generates a diradical which in the presence
of a suitable vinyl monomer will grow in two directions. This chain growth will
be followed by recombination of the diradicals, and thus, segmented polymers
with oligosiloxane units and high molecular weight may be formed Eq. (182).
Segmented polymers with longer siloxane blocks can be prepared by an alternative approach outlined in Eqs. (183, 184). The first step of this approach consists
of the synthesis of silyl-benzpinacolates with two pendant vinylgroups
(Table 1.9).

Platinum-catalyzed polyaddition of a dihydrosiloxane yields oligo- or
polysiloxanes containing numerous thermolabile C—C bonds [211, 212]. When
solutions of such polysiloxanes in vinyl monomers are heated, the polysiloxane

$$Ph-\underset{\underset{\underset{Me}{\diagup Si\diagdown}Me}{O\diagup\;\diagdown O}}{\overset{\overset{Ph\quad Ph}{|\quad\;|}}{C-\!-\!-C}}-Ph \;+\; n\; CH_2\!=\!\underset{\underset{R}{|}}{CH} \longrightarrow \left(\underset{\underset{R}{|}}{CH}\text{-}CH_2\right)_n\!\underset{\underset{Ph}{|}}{\overset{\overset{Ph}{|}}{C}}-O-\underset{\underset{Me}{|}}{\overset{\overset{Me}{|}}{Si}}-O-\underset{\underset{Me}{|}}{\overset{\overset{Me}{|}}{Si}}-O-\underset{\underset{Ph}{|}}{\overset{\overset{Ph}{|}}{C}}\left(CH_2\text{-}\underset{\underset{R}{|}}{CH}\right)_m \tag{182}$$

$$\left(CH_2\!=\!CH-\underset{\underset{Me}{|}}{\overset{\overset{Me}{|}}{Si}}-O-\underset{\underset{Ph}{|}}{\overset{\overset{Ph}{|}}{C}}-\right)_2 \;+\; H-\left(\underset{\underset{Me}{|}}{\overset{\overset{Me}{|}}{Si}}-O\right)_n-\underset{\underset{Me}{|}}{\overset{\overset{Me}{|}}{Si}}-H \quad\xrightarrow{\;H_2PtCl_4\;} \tag{183}$$

$$\left[\left(\underset{\underset{Me}{|}}{\overset{\overset{Me}{|}}{Si}}-O\right)_n-\underset{\underset{Me}{|}}{\overset{\overset{Me}{|}}{Si}}-CH_2-CH_2-\underset{\underset{Me}{|}}{\overset{\overset{Me}{|}}{Si}}-O-\underset{\underset{Ph}{|}}{\overset{\overset{Ph}{|}}{C}}-\underset{\underset{Ph}{|}}{\overset{\overset{Ph}{|}}{C}}-O-\underset{\underset{Me}{|}}{\overset{\overset{Me}{|}}{Si}}-CH_2\text{-}CH_2-\right]_x$$

$$\left[\left(\underset{\underset{Me}{|}}{\overset{\overset{Me}{|}}{Si}}-O\right)_n-\underset{\underset{Me}{|}}{\overset{\overset{Me}{|}}{Si}}-CH_2-CH_2-\underset{\underset{Me}{|}}{\overset{\overset{Me}{|}}{Si}}-O-\underset{\underset{Ph}{|}}{\overset{\overset{Ph}{|}}{C}}-\left(CH_2\text{-}CH\right)_z-\underset{\underset{Ph}{|}}{\overset{\overset{Ph}{|}}{C}}-O-\underset{\underset{Me}{|}}{\overset{\overset{Me}{|}}{Si}}-CH_2\text{-}CH_2\right]_x \tag{184}$$

gradually dissociate into radicals, which initiate the polymerization of the vinyl monomer and recombine. In this way a controlled synthesis of multiblock copolymers is feasible [212].

More recently, another group of oligomers or polymers was found to be useful as radical initiators, namely polysilanes with a Si-Si backbone [213,214]. Syntheses and various structure-property relationships of polysilanes are discussed in Chapter 4. The present section concentrates on their potential application as radical initiators. These polysilanes possess the relatively rare property that the backbone itself absorbs UV-light. Table 1.10 summarizes some typical structures and their absorption maxima. The absorption of UV-light has the consequence that the Si—Si-bond breaks down and Si-radicals are formed along with silenes [215], Eqs. (185,186). The resulting Si-radicals are reactive enough to initiate radical polymerizations of various vinyl monomers, such as

Table 1.10. Polysilanes used to photoinitiate methylmethacrylate (MMA) in sunlight. ˙Data adapted from [213]

Polymer	λ_{max}(M)	$\in \times 10^{-3}$/Si	$\bar{M}_n$
$(PhMeSi)_n$	337	7.0	1×10^4
$(PhMeSi)_{0.56}$ $PhC_2H_4SiMe)_{1.0}$	327	5.8	4×10^5 8.4×10^4 (bimodal)
$(cyclo\text{-}HexSiMe)$ $(PhMeSi)_{1.0}$	334	7.8	8×10^4
$(PhC_2H_4SiMe)_n$	306	6.8	1.8×10^5
$(p\text{-}CH_3C_6H_6SiMe)_{0.91}$ $(Me_2Si)_{1.0}$	332	5.0	1.7×10^5
$(PhC_2H_4SiMe)_{0.67}$ $(cyclo\text{-}HexSiMe)_{1.0}$	310	6.7	1.5×10^5
$(PhC_2H_4SiMe)_{0.79}$ $(Me_2Si)_{1.0}$	305	5.0	3.3×10^5 1.1×10^4 (bimodal)
$(cyclo\text{-}HexSiMe)_{0.66}$ $(Me_2Si)_{1.0}$	309	4.3	9×10^5 7×10^4 (biomodal)
$(cyclo\text{-}HexSiMe)_n$	326	7.6	1.7×10^5
$(PhMeSi)_n$	254, (301t)a	1.5	3.5×10^3

ᵃ A UV absorption tail is observed to extend past 310 nm.

$$\text{Pol} \sim \underset{\underset{R}{|}}{\overset{\overset{R}{|}}{Si}} - \underset{\underset{R}{|}}{\overset{\overset{R}{|}}{Si}} \sim \text{Pol} \quad \overset{h\nu}{\rightleftharpoons} \quad \text{Pol} \sim \underset{\underset{R}{|}}{\overset{\overset{R}{|}}{Si}} \cdot \quad + \quad \cdot \underset{\underset{R}{|}}{\overset{\overset{R}{|}}{Si}} \sim \text{Pol} \qquad (185)$$

$$\text{Pol} \sim \underset{\underset{R}{|}}{\overset{\overset{R}{|}}{Si}} - \underset{\underset{R}{|}}{\overset{\overset{R}{|}}{Si}} - \underset{\underset{R}{|}}{\overset{\overset{R}{|}}{Si}} \sim \text{Pol} \quad \overset{h\nu}{\rightleftharpoons} \quad 2 \text{ Pol} \sim \underset{\underset{R}{|}}{\overset{\overset{R}{|}}{Si}} \cdot \quad + \quad \underset{\underset{R}{|}}{\overset{\overset{R}{|}}{Si}} : \qquad (186)$$

styrene, acrylic acid, acrylamide, methacrylic acid, methylmethacrylate, 2'-hy-droxyethylmethacrylate and vinylpyrrolidone [213, 214].

Polysilanes and particularly poly(methyl-phenylsilane) possess high extinction coefficients (Table 1.10) and the quantum efficiency of the photoscission is quite high (0.30–0.97). Surprisingly, the efficiency of the radical polymerization is by a factor of 10^3 lower [213]. The side reactions responsible for the loss of active radicals have not been elucidated in detail. A more positive aspect of the polysilanes as radical initiators is their low sensitivity to the presence of oxygen. The intermediately formed silenes and their rapid reaction with O_2 seem to be responsible for this effect. From the preparative point of view it is also worth noting that polymerizations with short irradiation times yield two-block and A-B-A three-block copolymers (Form. 187). Longer irradiation results in gradual degradation of the polysilane blocks. A partial problem of silane-initiated polymerizations may be the low solubility of these polymers in various vinyl monomers [213]. Due to their hydrophobic nature they are abso-lutely ineffective in combination with aqueous solutions of vinyl monomers. To overcome this difficulty, polysilanes with pendant quarternary ammonium groups were synthesized [216] (Form. 188) and successfully applied as initiators in aqueous systems [214].

$$R^2 \!-\!\!\left[\!\begin{array}{c} R^1 \\ | \\ Si \\ | \\ R^1 \end{array}\!\right]_n\!\!\left(\!CH_2\!-\!\underset{R^3}{CH}\!\right)_m\!\!\!-\!\!\left[\!\begin{array}{c} R^1 \\ | \\ Si \\ | \\ R^1 \end{array}\!\right]_n\!\!-\!R^2 \tag{187}$$

$$-\!\!\left[\!\begin{array}{c} Me \\ | \\ -Si- \\ | \\ \bigcirc \end{array}\!\right]_l\!\!-\!\!\left[\!\begin{array}{c} Me \\ | \\ -Si- \\ | \\ \bigcirc\!-\!CH_2\!-\!\overset{\oplus}{N}\!\!\bigcirc \\ \\ Cl^\ominus \end{array}\!\right]_m \tag{188}$$

1.5.2
Initiators of Cationic Polymerizations

Trialkylsilyl trifluoromethanesulfonates (triflates) are known in organic chemis-try to be highyl electrophilic compounds which react with a variety of electron-rich olefins or carbonyl compounds. Consequently, trimethylsilyl triflate initiated polymerizations of various vinyl monomers were reported [217]. In contrast, trimethylsilyl methane sulfonate and diphenyl trimethylsilyl phosphate were not reactive enough to initiate the polymerization of styrene or substituted styrenes [218]. However, the cationic polymerizations initiated with trimethyl-silyl or triisopropylsilyl triflate were not conducted in the presence of a proton scavenger such as 2,6-bis-*tert*-butylpyridine and the Me$_3$Si-endgroup which should be formed according to the initiation mechanism of Eq. (189) was not identified. It was demonstrated in a later reinvestigation of these polymerizations [219] that traces of triflic acid resulting from the hydrolysis of the silyl triflates is the true initiator and not the silyl triflates themselves.

$$Me_3Si-O-SO_2\text{-}CF_3 + CH_2{=}CH\text{-}Ph \longrightarrow Me_3Si-CH_2-\overset{\oplus}{\underset{Ph}{CH}} + CF_3SO_3^{\ominus} \quad (189)$$

Based on the finding that the $HJ \cdot ZnJ_2$ complexes allow a cationic polymerization of vinyl ethers and p-methoxystyrene with a living character [220], the combination of $Me_3Si\cdot J$ and ZnJ_2 was studied as potential initiator [221, 222]. In analogy to the chemistry of Me_3Si-triflate a direct attack of the silyl group on the olefin, Eq. (190), was never detected. The pure $Me_3SiJ\cdot ZnJ_2$ complex was found to be a poor initiator, but its reactivity markedly increased when small amounts of water or more than equimolar amounts of acetone were added as coinitiators. In the case of water, the hydrogen iodide resulting from the hydrolysis of Me_3Si-J was found to be the true initiator. When acetone is added, Me_3-J adds to the carbonyl group and generates a relatively well-stabilized tertiary oxocarbenium ion, Eq. (191). This cation is still reactive enough to initiate cationic polymerizations of vinyl ethers, Eq. (192). After precipitation into methanol and hydrolysis of the silyl group polymers with well-defined and functional end groups were isolated, Eqs. (193, 194). When these polymerizations are conducted at low temperatures the DP's parallel the M/I ratios and narrow molecular

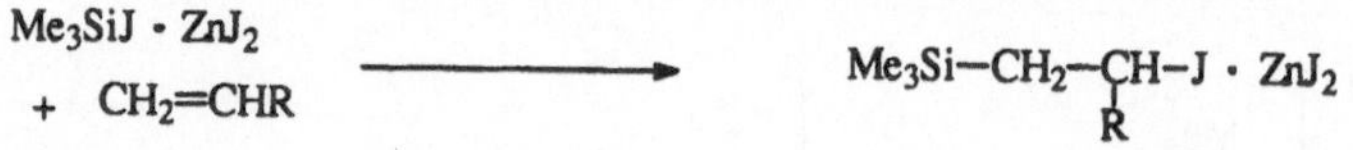

$$Me_3SiJ \cdot ZnJ_2 + CH_2{=}CHR \longrightarrow Me_3Si-CH_2-\underset{R}{CH}-J \cdot ZnJ_2 \quad (190)$$

$$Me_3SiJ \cdot ZnJ_2 + \overset{CH_3}{\underset{CH_3}{CO}} \rightleftharpoons Me_3Si-O-\overset{CH_3}{\underset{CH_3}{\overset{\oplus}{C}}} \quad ZnJ_3^{\ominus} \quad (191)$$

$$+ \; n \; CH_2{=}CHR \quad (192)$$

$$Me_3Si-O-\overset{Me}{\underset{Me}{C}}-\left(CH_2-\underset{R}{CH}\right)_n \cdot J \cdot ZnJ_2 \xrightarrow{CH_3OH} Me_3Si-O-\overset{Me}{\underset{Me}{C}}-\left(CH_2-\underset{R}{CH}\right)-OCH_3 \quad (193)$$

$$+ H_2O \quad / \quad - HOSiMe_3$$

$$HO-\overset{Me}{\underset{Me}{C}}-\left(CH_2-\underset{R}{CH}\right)-OCH_3 \quad (194)$$

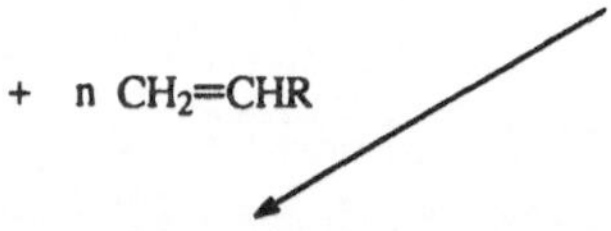

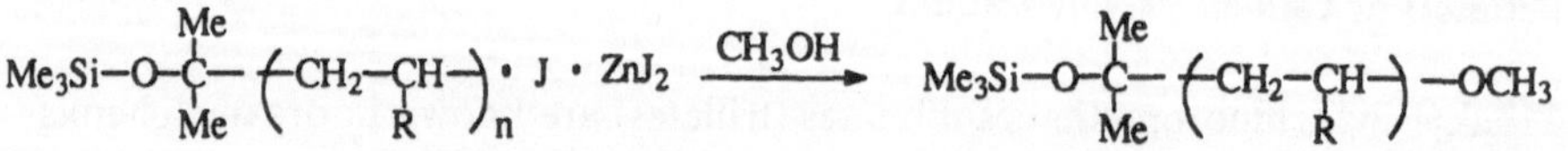

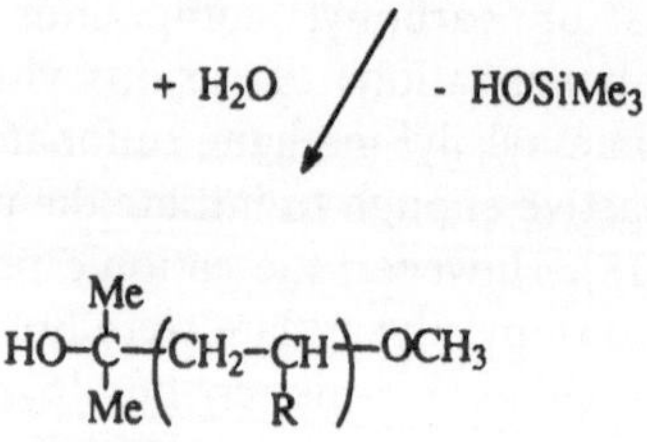

weight distributions were obtained [221, 222]. The kinetics of these polymerizations were studied in much detail.

Another approach based on $Me_3Si\text{-}J$ utilizes dioxolane as coinitiator. Again a highly reactive C—J bond is formed, Eq. (195), which in the presence of

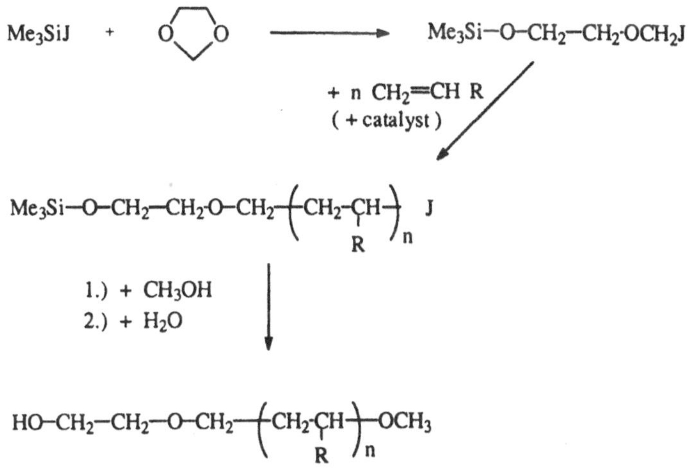

$$Me_3SiJ \quad + \quad \underset{O \diagdown \diagup O}{\bigcirc} \quad \longrightarrow \quad Me_3Si\text{-}O\text{-}CH_2\text{-}CH_2\text{-}OCH_2J$$

$$+ \; n \; CH_2{=}CH \; R \tag{195}$$
$$(+ catalyst)$$

$$Me_3Si\text{-}O\text{-}CH_2\text{-}CH_2\text{-}O\text{-}CH_2\text{-}\!\left(\!CH_2\text{-}\underset{R}{CH}\!\right)_{\!n}\!\! J \tag{196}$$

1.) + CH_3OH
2.) + H_2O

$$HO\text{-}CH_2\text{-}CH_2\text{-}O\text{-}CH_2\text{-}\!\left(\!CH_2\text{-}\underset{R}{CH}\!\right)_{\!n}\!\!\text{-}OCH_3 \tag{197}$$

tetrabutylammonium triflate enables a cationic polymerization of vinyl ethers [223]. These polymerizations show again a "living character" with relatively narrow molecular weight distributions. Hydroxy terminated polymers result from the hydrolysis of the silyl end group, Eqs. (196, 197).

1.5.3
Ziegler-Natta Catalysts and Related Systems

Highly active Ziegler-Natta (ZN) catalysts used in modern technical processes usually consist of $TiCl_4/AlEt_3$ combination on a $MgCl_2$ support: The reaction conditions used for the preparation of these $MgCl_2$-ZN catalysts have a considerable influence on their activity and stereoselectivity. Systematic studies by various research groups have revealed that addition of a Lewis base, such as ethers, esters or particularly alkoxysilanes, may enhance both activity and stereoselectivity of the catalyst systems [224–234]. When the Lewis base is added together with the triethylaluminium to the $MgCl_2/TiCl_4$ system it is called "external base" in contrast to the "internal base" which may be a constituent of the $MgCl_2/TiCl_4$ complex. Alkoxysilanes like other Lewis bases form 1:1 complexes with $AlEt_3$ [235], Eq. (198). These complexes are stable for several hours even at 60 °C. However, when an excess of $AlEt_3$ is present redox reactions take place more rapidly resulting in the formation of Al-alkoxide and Et—Si bonds, Eq. (199).

Of particular interest is the finding that alkoxysilanes are more effective in enhancing the stereoselectivity of the propylene polymerization. Polypropylene

$$\text{Et}_3\text{Al} \quad + \quad \overset{\text{SiMe}_3}{\underset{R}{\diagdown O \diagdown}} \quad \rightleftharpoons \quad \text{Et}_3\text{Al} \cdots \overset{\text{SiMe}_3}{| O \underset{\text{OR}}{\diagdown}}$$

(198)

$$\downarrow$$

$$\text{Et}_2\text{Al-OR} \quad + \quad \text{EtSiMe}_3 \tag{199}$$

with $\geqslant 95\%$ isotactic diads were isolated [229]. This stereoregulation is ascribed to two effects:

1) a selective poisoning of the active centers producing atactic polypropylene,
2) an activation of those active centers producing isotactic polypropylene.

Initially, PhSi(OMe)$_3$ was preferentially used as external base [224–227]. More recently a broad variety of alkoxysilanes (Forms. 200–207) were systematically studied [229, 231, 233, 234]. Only free alkoxysilanes or alkoxysilane-AlEt$_3$ complexes with an additional free Si—O—C group capable of deactivating active centers. The bulkiness of the organic residues directly attached to silicon favors the selectivity in such a way that the sterically more hindered "isotactic active centers" are less efficiently deactivated than the "atactic active centers [234, 235].

The role of silanes as cocatalysts (or catalyst modifying agents) was also studied in the field of metathetic polymerizations of 1, 2-disubstituted acetylenes [236–238]. The monomer which was firstly studied in this regard was 1-chloro-2-phenyl acetylene [236]. VCl$_4$, TaCl$_5$, ReCl$_5$ and MoCl$_5$ were found to catalyze the polymerization of this monomer. However, both yields and molecular weights were quite low. When MoCl$_5$ was combined with several silanes, Bu$_4$Sn, Ph$_3$Sb or Ph$_3$Bi significantly higher yields (up to 91%) and molecular weights (MW up to 122×10^4) were obtained [236]. In the case of silanes Et$_3$SiH and PhMe$_2$SiH were particularly useful. Somewhat lower yields and Mw's were found with (Me$_3$Si)$_2$O or Me$_3$SiOn-Bu. Combinations with PhMe$_3$Si, PhEt$_3$Si, Me$_3$SiCl or Ph$_2$SiCl$_2$ gave poorer results than neat MoCl$_5$.

In addition to 1-chloro-2-phenyl ethene two series of 1-(n-alkylthio)propynes 208a–f were successfully polymerized with the MoCl$_5$/Ph$_3$SiH catalyst [237]. In the case of 1-(-butyl)propyne, the catalyst was also varied. Various transition metal chlorides were studied with and without addition of Ph$_3$SiH. In all cases addition of the silane improved the yield. Analogous results were obtained, when monomer 209c was polymerized with various catalysts [238]. Furthermore, all members of series 209a–d were successfully polymerized with the MoCl$_5$/Ph$_3$SiH catalyst. No details of the polymerization mechanism were published but the indermediate formation of a Mo-carbene as reactive species was postulated.

Chromium oxides on a solid support such as silicon are well known to be highly effective catalysts for the polymerization of ethylene ("Phillips catalyst") [239, 240]. In this connection a research group of Union Carbide found [241, 242] that even neat bis(triphenylsilyl)chromate is a catalyst for the polymerization of ethylene. The catalytic activity of this soluble catalyst is considerably increased when the compound is deposited on silica-alumina. An additional activation results from treatment of the deposited catalyst with aluminum alkyls and particularly with diethylaluminum alkoxides [242]. The usefullness of this catalyst system was confirmed by another group [243]. Bis(triphenylsilyl) chromate itself is easy to synthesize from triphenylsilanol and CrO_3 [244].

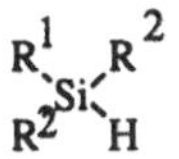

$$\begin{array}{ll} R^1 \quad R^2 \\ Si \\ R^2 \quad H \end{array} \qquad \begin{array}{l} a : R^1 = R^2 = Et \\ c : R^1 = Ph, R^2 = H \end{array} \qquad b : R^1 = R^2 = OSiMe_3 \qquad (200)$$

$$\begin{array}{ll} Me \quad Me \\ Si \\ Me \quad OR \end{array} \qquad \begin{array}{l} a : R = Me \\ b : R = Et \end{array} \qquad (201)$$

$$\begin{array}{l} R^1 \quad OMe \\ Si \\ R^2 \quad OMe \end{array} \qquad (202)$$

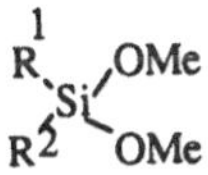

a : $R^1 = R^2 = Ph$ d : $R^1 = Me, R^2 = $ (structure with OMe groups)

b : $R^1 = H, R^2 = CH_3$

c : $R^1 = CH_3, R^2 = (CH_2)_7 CH_3$ e : $R^1 = Me, R^2 = $ (diphenyl ether structure)

$$\begin{array}{l} R^1 \quad OEt \\ Si \\ R^2 \quad OEt \end{array} \qquad (203)$$

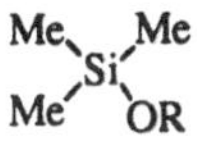

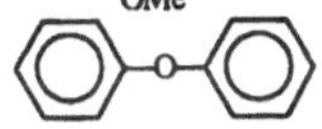

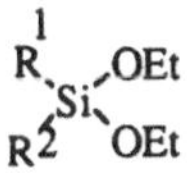

a : $R^1 = R^2 = Me$,   b : $R^1 = Me, R^2 = Ph$,   c : $R^1 = Me, R^2 = (CH_2)_n CH_3$

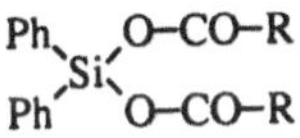

$$\begin{array}{ll} Ph \quad O-CO-R \\ Si \\ Ph \quad O-CO-R \end{array} \qquad \begin{array}{l} a : R = Me \\ b : R = -CMe=CH_2 \end{array} \qquad (204)$$

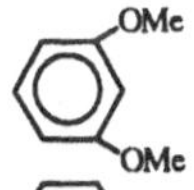

$$\begin{array}{ll} Me \quad O-CO-CMe=CH_2 \\ Si \\ Me \quad O-CO-CMe=CH_2 \end{array} \qquad \begin{array}{l} (CH_3)_2CH-CH_2 \quad OH \\ Si \\ (CH_3)_2CH-CH_2 \quad OH \end{array} \qquad (205)$$

a   b

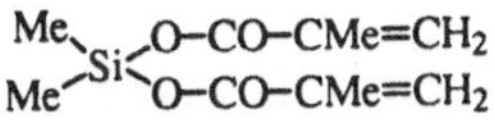

$R-Si(OMe)_3$

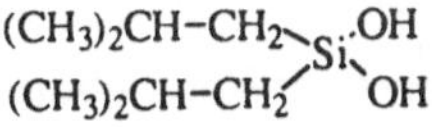

a : R = Me   b : R = Et   c : R = n-Pr   d : R = i-Bu

e : R = Ph   f : R = (structure)-CMe_3   g : R = (biphenyl structure)   (206)

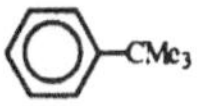

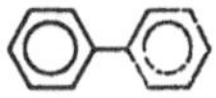

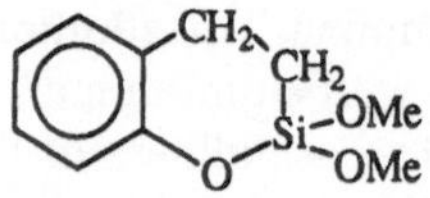

a 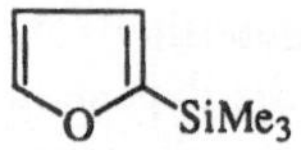   b

(207)

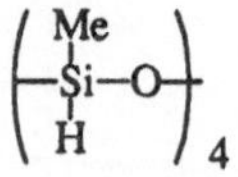

c   d

$$CH_3-C{\equiv}C-S-R$$

a : R = Me   d : R = n-C$_6$H$_{13}$
b : R = Et   e : R = n-C$_8$H$_{17}$
c : R = n-Bu   f : R = n-C$_{10}$H$_{21}$

(208)

$$R-C{\equiv}C-S-Me$$

a : R = Et   c : R = n-C$_6$H$_{13}$
b : R = n-Bu   d : R = n-C$_8$H$_{17}$

(209)

Finally, the polymerization of ethylene by a nickel catalyst with silylated ligand (Form. **210**) should be mentioned [245].

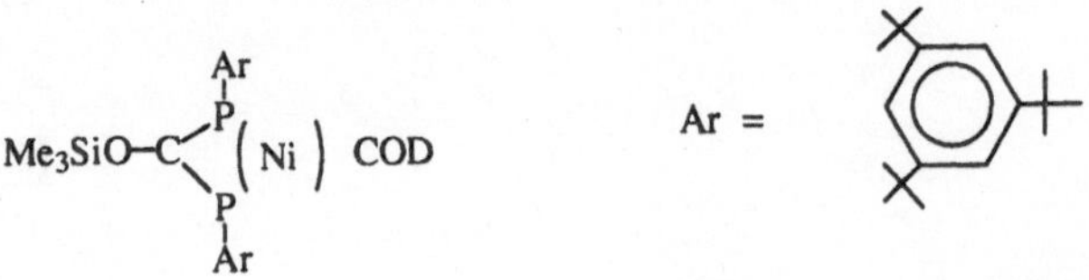

COD = 1,5-Cyclooctadiene

(210)

1.5.4
Polymerizations on Silica or Silicate Surfaces

Oligomers and polymers covalently bound to SiO$_2$ or silicate surfaces (e.g. glass-beads or glass fibers) are of interest for various reasons including, mechanical properties, compatibilization in composites catalysis or chromatographic applications. Three preparative strategies have been reported for a successful grafting of polymers onto silica or silicate surfaces:

1) Functional groups capable of initiating the polymerizations of vinyl monomers are covalently bound to the inorganic surface and finally used as initiator.
2) Functional groups (usually vinyl groups) are covalently bound to the inorganic surface and copolymerized with surrounding monomers.

3) Functional groups covalently attached to the inorganic surface serve as terminators of polymerizations in such a way that the termination step yields a stable, covalent bond.

Examples of these three strategies will be discussed in the order given above.

An example of strategy A already discussed above is possibly the polymerization of ethylene by bis(triphenylsilyl)chromate or a silica support. However, it is not clear, if in this case, a stable covalent bond between polyethylene and SiO_2 is really formed. Most methods based on strategy A involve the synthesis of a radical initiator, such as a thermolabile azocompound, on the SiO_2 surface [246–249]. In a first study dealing with this strategy two routes were described yielding phenyl diazo groups [246, 247]. Both routes have in common that p-nitrophenyl groups were generated on the surface of a nonporous silica powder (Aerosil of DEGUSSA AG). The first routes starts with the chlorination of OH-groups followed by condensation with phenyllithium and nitration of the phenyl group, Eqs. (211–213). The second route is more convenient, when 4-nitrophenyltrichlorosilan is available, Eq. (214). Regardless of how the nitrophenyl groups were attached to the silica, they were reduced with hydrazin, a step which either entails a low yield under mild conditions or a loss of phenyl groups under harsher conditions, Eq. (215). The resulting amino groups were treated with nitrous acid and the diazonium salts coupled with various C-H acidic compounds, Eqs. (216–218) such as, thiophenol, aromatic sulfinic acids, sulfonic acids, malonic acid esters, substituted malo dinitriles or alkyl cyanoacetates [246]. Most radical polymerizations were initiated with the diazophenylsulfide (Z = S-Ph in Eq. (218)) or diazonaphthylsulfide (Z = S-naphthyl). The following monomers were polymerized: acrylic acid, acrylamide, acrylonitrile, methylmethacrylate, vinylpyridin and particularly styrene. High grafting efficiencies and molecular weights were found [246, 247].

Later studies [248, 249] started with γ-aminopropyltrimethoxysilane as coupling agent (see chapter 8), Eq. (219). The γ-amino groups were acylated with the acid chloride of 4,4'-azobis(4-cyanopentanoic acid). Two structures were discussed for the reaction product, Eqs. (220, 221). In one paper [238] these initiators were used to polymerize butyl acrylate. The ester groups of the grafted poly(butylacrylate) were transformed in hydrazide and finally azide groups which in turn were used to immobilize lipase. The second paper

$$SiO_2\!\!-\!\!Si\!-\!OH + SOCl_2 \longrightarrow SiO_2\!\!-\!\!Si\!-\!Cl + SO_2 + HCl \quad (211)$$

$$Li\!-\!\!\bigcirc \qquad \qquad -\ LiCl \qquad \qquad (212)$$

$$SiO_2\!\!-\!\!Si\!-\!\!\bigcirc \xrightarrow{\text{Nitration}} SiO_2\!\!-\!\!\bigcirc\!-\!NO_2 \qquad (213)$$

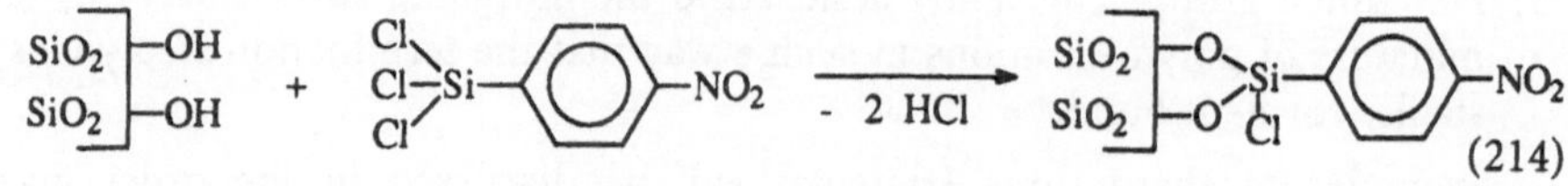

$$SiO_2\text{—OH} + Cl_3\text{—Si—}\langle\bigcirc\rangle\text{—NO}_2 \xrightarrow{-2\,HCl} SiO_2\text{—O—Si(Cl)—}\langle\bigcirc\rangle\text{—NO}_2 \quad (214)$$

$$\xrightarrow{NH_2NH_2 \cdot H_2O} \quad (215)$$

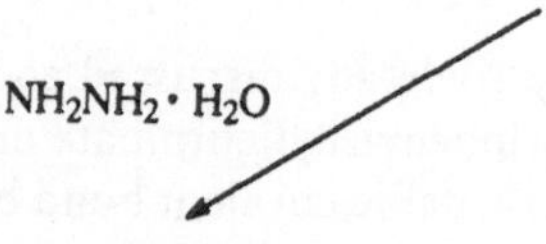

$$SiO_2\text{—O—Si(OH)—}\langle\bigcirc\rangle\text{—NH}_2 \xrightarrow{HNO_2} SiO_2\text{—O—Si(OH)—}\langle\bigcirc\rangle\text{—N}\equiv\overset{\oplus}{N} \quad X^{\ominus} \quad (216)$$

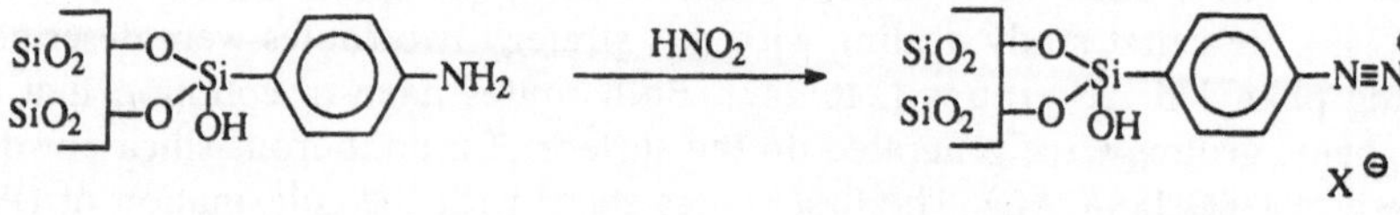

$$+ HZ \qquad - HX \qquad (217)$$

$$SiO_2\text{—O—Si(OH)—}\langle\bigcirc\rangle\text{—N=N—Z} \xrightarrow[-N_2]{\Delta T} SiO_2\text{—O—Si(OH)—}\langle\bigcirc\rangle + Z \quad (218)$$

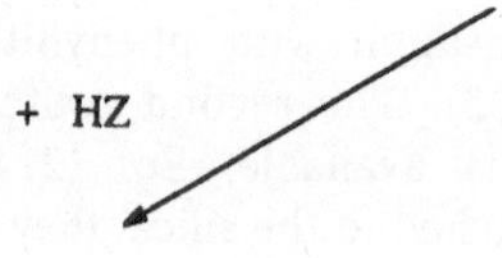

$$SiO_2\text{—OH} \xrightarrow[-2\,MeOH]{(MeO)_3Si(CH_2)_3NH_2} SiO_2\text{—O—Si(OMe)—(CH}_2)_3\text{—NH}_2 \quad (219)$$

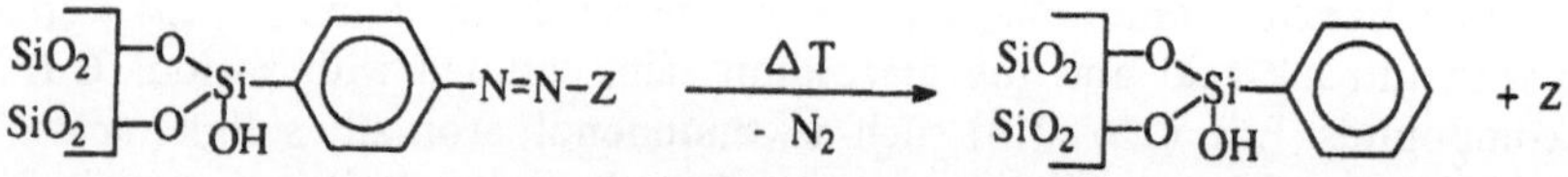

$$+ \left(\begin{array}{c}Me\\ =N\text{—C—CH}_2\text{—CH}_2\text{—COCl}\\ CN\end{array}\right)_2$$

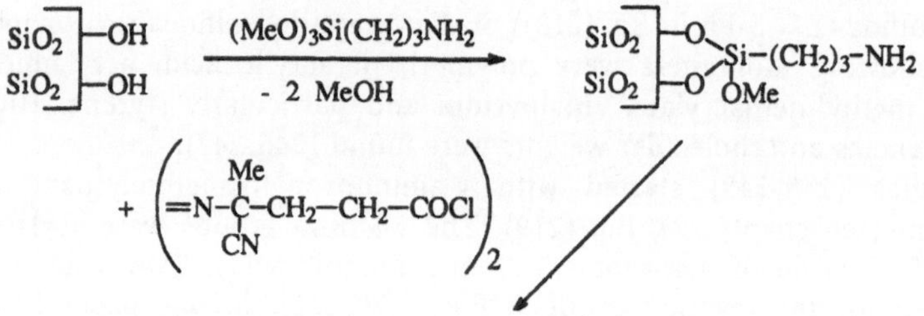

$$SiO_2\text{—O—Si(OMe)—(CH}_2)_3\text{—NH—CO—(CH}_2)_2\text{—C(Me)(CN)—N=N—C(Me)(CN)—(CH}_2)_2\text{—COCl} \quad (220)$$

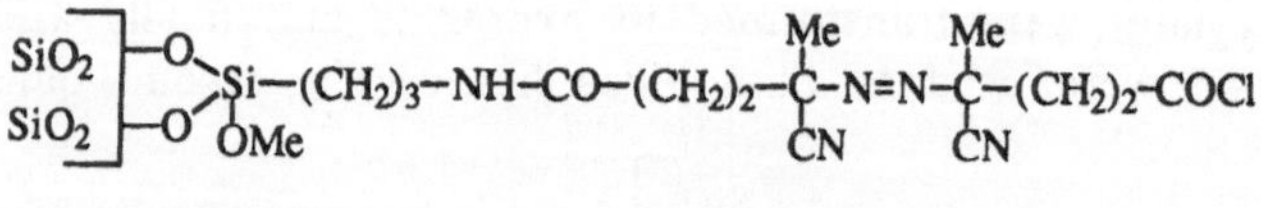

$$SiO_2\text{—O—Si(OMe)—(CH}_2)_3\text{—NH—CO—(CH}_2)_3\text{—C(Me)(CN)—N=}\\ \quad \text{O—Si(OMe)—(CH}_2)_3\text{—NH—CO—(CH}_2)_3\text{—C(Me)(CN)—N} \quad (221)$$

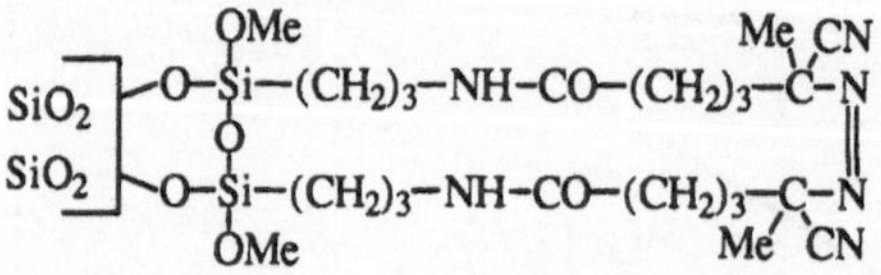

$$SiO_2\text{Si—C}\equiv N + HO_2C\text{—CCl}_3 \longrightarrow SiO_2\text{Si—CO—NH—CO—CCl}_3 \quad (222)$$

describes the grafting of methyl methacrylate in much detail. Another approach yielding radicals on silica surfaces is based on the reduction of trichloroacetyl groups. Both methods described in Ref. 250 started with the chlorination of silica. The Si-Cl groups were then treated with silver cyanide and trichloroacetic acid, Eq. (222) or with phenyl lithium and trichloroacetylchloride, Eq. (223). Radicals useful for the polymerization of styrene, Eq. (224), result from the treatment of the trichloroacetyl groups with Mo(CO)$_6$.

$$\begin{array}{c}SiO_2 \\ SiO_2\end{array}\!\Big]\!Si\!-\!\langle\bigcirc\rangle\!-\!Li \ + \ Cl\!-\!CO\!-\!CCl_3 \ \xrightarrow{\ AlCl_3\ } \ \begin{array}{c}SiO_2 \\ SiO_2\end{array}\!\Big]\!Si\!-\!\langle\bigcirc\rangle\!-\!COCCl_3 \qquad (223)$$

$$\begin{array}{c}SiO_2 \\ SiO_2\end{array}\!\Big]\!Si\!-\!\langle\bigcirc\rangle\!-\!{}_{COCl_2\cdot} \ \xrightarrow{\ +\ n\ styrene\ } \ \begin{array}{c}SiO_2 \\ SiO_2\end{array}\!\Big]\!Si\!-\!\langle\bigcirc\rangle\!-\!{}_{COCl_2}\!\Big[\!CH_2\!-\!\underset{Ph}{CH}\!\Big]_{n}\!H \qquad (224)$$

Several research groups [6251–254] used vinyl groups attached to the surface of glass-fiber or silica as anchors for grafting polystyrene or poly(methyl methacrylate). One group [251, 252] introduced the functional group via allyl glycidyl ether (Form. 225) and copolymerized the allyl group in a peroxide-initiated radical polymerization. Another group [253] started with a coupling agent (see Chapter 8) containing a methacrylate residue. The methacrylate group was activated by means of phenyl magnesium bromide, Eq. (226), and used as initiator for MMA. An insertion mechanism was postulated [253]. An anionic grafting process of styrene and isoprene was described by a third group. A special

$$\begin{array}{c}SiO_2 \\ SiO_2\end{array}\!\Big]\!Si\!-\!O\!-\!CH_2\!-\!\underset{\overset{|}{OH}}{CH}\!-\!CH_2\!-\!O\!-\!CH_2\!-\!CH\!=\!CH_2 \qquad (225)$$

$$\begin{array}{c}SiO_2 \\ SiO_2\end{array}\!\Big]\!\!\begin{array}{c}-O \\ -O\end{array}\!\!\underset{OH}{\overset{}{Si}}\!-\!(CH_2)_3\!-\!O\!-\!CO\!-\!CMe\!=\!CH_2 \ \xrightarrow{\ +\ PhMgBr\ }$$

$$\begin{array}{c}SiO_2 \\ SiO_2\end{array}\!\Big]\!\!\begin{array}{c}-O \\ -O\end{array}\!\!\underset{OH}{\overset{}{Si}}\!-\!(CH_2)_3\!-\!O\!-\!CO\!-\!\underset{CH_2\cdot Ph}{\overset{CH_3}{C}}\!-\!MgBr \qquad (226)$$

$$ClCH_2\!-\!\langle\bigcirc\rangle\!-\!CH\!=\!CH_2 \ \xrightarrow[-\ MgCl_2]{+\ SiCl_4\ +\ Mg} \ Cl_3Si\!-\!CH_2\!-\!\langle\bigcirc\rangle\!-\!CH\!=\!CH_2 \quad (227)$$

styrene derivative served as anchor of the functional group which was activated by *tert*-butyl lithium, Eqs. (227, 228). Finally, it should be noted that chlorinated silica surfaces may serve as terminators of anionic polymerizations as illustrated in Eq. (229) [255].

$$
\text{SiO}_2 \begin{bmatrix} -\text{O} \\ -\text{O} \end{bmatrix} \overset{\text{Si}-\text{CH}_2}{\underset{\text{OR}}{\big\rangle}} \text{—}\bigcirc\text{—CH=CH}_2 \quad \xrightarrow{\text{t.BuLi}}
$$

$$(228)$$

$$
\text{SiO}_2 \begin{bmatrix} -\text{O} \\ -\text{O} \end{bmatrix} \overset{\text{Si}-\text{CH}_2}{\underset{\text{OR}}{\big\rangle}} \text{—}\bigcirc\text{—} \overset{\overset{\ominus}{\text{CH}|}}{\underset{\text{CH}_2}{}} \quad \text{Li}^{\oplus}
$$

$$
\text{SiO}_2 \begin{bmatrix} \\ \end{bmatrix} \text{Si—Cl} \quad + \quad \text{Bu} \begin{bmatrix} \text{—CH}_2\text{—CH—} \\ \quad\quad\text{Ph} \end{bmatrix} \text{CH}_2\text{—CH}| \overset{\ominus}{} \quad \text{Li}^{\oplus} \quad \underset{\text{Ph}}{\longrightarrow}
$$

$$(229)$$

$$
\text{SiO}_2 \begin{bmatrix} \\ \end{bmatrix} \text{Si—CH—CH}_2 \begin{bmatrix} \text{—CH—CH}_2\text{—} \\ \quad\text{Ph} \end{bmatrix} \text{Bu}
$$
with Ph under the first CH.

1.6
References

1. Lewis CL, Lewis DW (1959) J Polym Sci 36: 325
2. Pike RM (1959) J Polym Sci 40: 577
3. Taylor GN, Wolf TM, Moran JM (1981) J Vac Sci Technol 19: 872
4. Mc Donald SA, Ito H, Wilson CG (1983) Macroelectronic Engin 1: 269
5. Mc Donald SA, Steinmann F, Ito H, Lu WY, Wilson CG (1983) Proc ACS Division of Polym Mat Sci Engin 49: 104
6. Suzuki M, Saigo K, Gokan H, Onishi Y (1983) I Electrochem Soc 130: 1962
7. Chiang WY, Lu JY (1991) J Polym Sci Part A, Polym Chem 29: 399
8. Ishihara K, Matsui K (1986) J Polym Sci Polym Letters Ed. 29: 413
9. Nagasaki Y, Bittau S, Kato M (1992) Makromol Chem 193: 1633
10. Iwakura Y, Toda F, Hattori K (1968) J Polym Sci Part A-1 6: 1633
11. Jenkins AD, Petrak K, Roberts GAF, Walton DRM (1975) Eur Polym J 11: 635
12. Saigo K (1989) J Polym Sci Part A Polym Chem 27: 2203
13. Greber G, Tölle J (1964) Makromol Chem 77: 98
14. Greber G, Reese E (1964) Makromol Chem 77: 7 and 13
15. Nagasaki Y, Tsuruta T (1986) Makromol Chem, Rapid Commun 7: 437
16. Nagasaki Y, Takahashi S, Tsuruta T (1990) Makromol Chem 191: 2297
17. Nagasaki Y, Tsuruta T (1989) Makromol Chem 190: 1855
18. Nagasaki N, Suda M, Tsuruta T (1989) Makromol Chem, Rapid Commun 10: 255
19. Nagasaki Y, Tsuruta T (1990) New Polym Mater 2: 357
20. Nagasaki Y, Han SB, Kato M (1992) Makromol Chem 193: 1633
21. Stover HDH, Lü P-Z, Fréchet JMJ (1991) Polym Bull 25: 575

22. Yamagouchi K, Kato T, Hirao A, Nakahama S (1987) Makromol Chem, Rapid Commun 8: 203
23. Packinisani E, Hirao A, Nakahama S (1989) J Polym Chem Part A, Polym Chem 27: 2811
24. Madit N, Bonfils F, Giral L, Montginoul L, Sagnes R, Schué F (1991) Makromol Chem 192: 1467
25. Hartmann M, Carlsohn H, Pauls J (1976) Makromol Chem 177: 131
26. Nagai K, Asadas K, Chiba K, Kuramoto N (1989) J Polym Sci Part A, Polym Chem 27: 3779
27. Nagai K, Chiba K, Asada K, Masui K, Kuramoto N (1990) J Polym Sci Part A, Polym Chem 28: 2195
28. Hu EL (1981) American Scientist 69: 517
29. Taylor GN, Wolf TM (1980) Polym Enge Sci 20: 1087
30. Tsuda M, Oikawa S, Ohnogi S, Suzuki A (1980) Proceedings of Microcircuit Engin Amsterdam NL, 533
31. Lewis DW (1958) J Org Chem 23: 1893
32. Leebrick JR, Ramsden HE (1960) J Org Chem 23: 935
33. Sennear AE, Wirth J, Neville RG (1960) J Am Chem Soc 25: 807
34. Greber G, Reese E (1962) Makromol Chem 55: 96
35. Greber G, Reese E (1964) Makromol Chem 77: 7
36. Hirao A, Yamagouchi K, Takenaka K, Suzuki K, Nakahama S (1982) Makromol Chem, Rapid Commun 3: 941
37. Hirao A, Takenaka K, Packirisamy S, Yamaguchi K, Nakahama S (1985) Makromol Chem 186: 1157
38. Barker SA, Hatt BW, Somers PJ, Woodburg RB (1973) Carbohydr Res 26: 55
39. Walsh NDA, Goodwing GBT, Smith GC, Woodward FE (1986) Org Synth 651
40. Nakahama S, Hirao A (1990) Progr Polym Sci 15: 299
41. Hirao A, Nakahama S (1992) Prog Polym Sci 17: 283
42. Hirao A, Nagawa T, Hatagama T, Yamaguchi K, Nakahama S (1985) Macromolecules 18: 2101
43. Hirao A, Hatagama T, Nagava T, Yamagouchi N, Yamagouchi K, Nakahama S (1987) Macromolecules 20: 242
44. Lee J-S, Hirao A, Nakahama S (1988) Macromolecules 21: 276
45. Lee J-S, Hirao A, Nakahama S (1989) Macromolecules 22: 2602
46. Hirao A, Hatayama T, Nakahama S (1987) Macromolecules 20: 1505
47. Taki T, Hirao A, Nakahama S (1991) Macromolecules 24: 1455
48. Hirao A, Takenaka K, Yamagouchi K, Nakahama N, Yamazaki N (1983) Polym Commun 24: 339
49. Hirao A, Yamamoto A, Takenaka K, Yamagouchi K, Nakahama S (1987) Polymer 28: 303
50. Hirao A, Kato K, Nakahama S (1992) Macromolecules 25: 535
51. Yamagouchi K, Hirao A, Suzuki K, Takenaka K, Nakahama S, Yamazaki N (1983) J Polym Sci, Polym Lett Ed 21: 395
52. Suzuki K, Yamagouchi K, Hirao A, Nakahama S (1989) Macromolecules 22: 2607
53. Suzuki K, Hirao A, Nakahama S (1989) Makromol Chem 190: 2893
54. Ishizone T, Hirao A, Nakahama S, Kakuchi T, Yokota K, Tsuda K (1991) Macromolecules 24: 5230
55. House HO, Czuba LJ, Gall M, Olnestead HP (1969) J Org Chem 34: 2324
56. Orban J, Turner JV, Twitehin B (1984) Tetrahedron Lett 25: 5099
57. Takahashi S, Kuroyama Y, Sonogashira K, Hagehara N (1990) Synthesis 627
58. Saigo K, Watanabe F, Onishi Y (1986) J Vac Sci Technol B 4: 692
59. Saigo K, Watanabe F (1989) J Polym Sci Part A, Polym Chem 27: 2611
60. Nagasaki Y, Tsuruta T (1989) Makromol Chem, Rapid Commun 10: 403
61. Wagner GH et al (1953) Ind Eng Chem 45: 367
62. Thompson BR (1956) J Polym Sci 19: 373
63. Pike RM, Bailey DL (1956) J Polym Sci 23: 55
64. Scott CE, Price CC (1959) J Am Chem Soc 81: 2670

65. Bajaj P, Khamma YP, Babu GN (1976) J Polym Sci, Polym Chem Ed 14: 465
66. Babu GN, Atodaria DR, Desphande A (1981) Eur Polym J 17: 427
67. Rao VL, Babu GN (1983) J Macromol Sci Chem A 20: 527
68. Rao VL, Eshwar MC, Babu GN (1986) J Macromol Sci Chem A 23: 1079
69. Richle GK (1986) J Macromol Sci Chem A 23: 1287
70. Richle GK (1987) J Macromol Sci Chem A 24: 93
71. Oku J, Hasegawa T, Takeuchi T, Takaki M (1991) Polym J 23: 1377
72. Marvel CS, Woolford RG (1960) J Org Chem 25: 1641
73. Butler GB, Stackman RW (1960) J Org Chem 25: 1643
74. Billingham NC, Jenkins AD, Kronfil EB, Walten DRM (1977) J Polym Sci, Polym Chem Ed 15: 675
75. Saigo K, Tateishi K, Adadi H (1988) J Polym Sci Part A, Polym Chem 26: 2085
76. Cragg RH, Jones RG, Swain AC (1991) Eur Polym J 27: 785
77. Jones RG, Cragg RH, Swain AC (1992) Eur Polym J 28: 651
78. Bajaj P, Nanabu G (1976) Eur Polym J 12: 601
79. Bajaj P, Gupta DC (1983) J Polym Sci, Polym Chem Ed 21: 1347
80. Bajaj P, Gupta DC, Gupta AK (1980) J Appl Polym Sci 25: 1673
81. Varma IK, Tomar AK, Anand RC (1987) J Appl Polym Sci 33: 1377
82. Rao VL, Babu GN (1989) Eur Polym 25: 605
83. Tsutsumi N, Nishikawa Y, Kigotsokuri T, Nayata M (1992) Polymer 33: 209
84. Isobe Y (1992) Jap Pat 03.227.313 (1991) to Toa Gosei Chem Ind Co C. A. 116: 1.52627
85. Osaki H, Hirao A, Nakahama S (1992) Macromolecules 25: 1391
86. Kitano T, Ishigaki A, Uematsu C-T, Kawaguchi S, Ito K (1987) J Polym Sci Part A, Polym Chem 25: 979
87. Otsu T, Yoshioka M (1992) Makromol Chem 193: 2283
88. Murahashi S, Nazakura S, Sunae M (1965) J Polym Sci, Polym Lett 3: 245
89. Nozakura S, Ishihara S, Inaba Y, Matsumura K (1973) J Polym Sci 11: 1053
90. Solaro R, Chiellini E (1973) Gazz Chim Ital 106: 1053
91. Sogah DY, Webster OW (1986) Macromolecules 19: 1775
92. Risse W, Grubbs RH (1989) Macromolecules 22: 1558
93. Higashimura T, Ebara K, Aoshima S (1989) J Polym Sci Part A, Polym Chem 27: 2937
94. Nesmejanov AN, Lutsenko IF, Brattsev VA (1959) Dokl Akad Nauk SSSR 128: 551
95. Weber PW (1983), "Silicon Reagents for Organic Synthesis", Springer Verlag, Berlin Heidelberg, Chapter 16
96. Takenaka K, Hirao A, Hattori T, Nakahama S (1987) Macromolecules 20: 2035
97. Takenaka K, Hirao A, Hattori T, Nakahama S (1989) Macromolecules 22: 1563
98. Takenaka K, Hattori T, Hirao A, Nakahama S (1992) Macromolecules 25: 96
99. Ding Y-X, Weber P (1988) Macromolecules 21: 530
100. Takenaka K, Hirao A, Nakahama S (1992) Makromol Chem 193: 1943
101. Ding Y-X, Weber P (1988) Macromolecule 21: 2672
102. Hirabayashi T, Itoh T, Yokota K (1988) Polym J 20: 1041
103. Nagai K, Asada K, Kuramoto N (1990) J Polym Sci, Part A Polym Chem 28: 2845
104. Sato F, Uchiyama H, Samaddar AK (1984) Chem Id 743
105. Jung HE, McLombs CA, Takeda Y, Pan Y (1981) J Am Soc 103: 6677
106. Babu GN, Bajaj P (1977) Angew Makromol Chem 64: 211
107. Higashimura T, Masuda T, Okada M (1983) Polym Bull 10: 114
108. Masuda T, Iguchi Y, Tang BZ, Higashimura T (1988) Polymer 29: 2041
109. Tang BZ, Masuda T, Higashimura T (1989) J Polym Sci Part B, Polym Phys 27: 1261
110. Kita H, Sakamoto T, Tanaka K, Okamoto J (1988) Polym Bull 20: 349
111. Kang ET, Neoh KG, Tan KL (1991) J Polym Sci, Part B, Polym Phys 29: 669
112. Nagase Y, Mori S, Matsui K (1989) J Appl Polym Sci 37: 1259
113. Tasaka S, Inagaki N, Igawa M (1991) J Polym Sci Part B, Polym Phys 29: 691
114. Nishide H, Kawakami H, Sasame Y, Ishiwata K, Tsuchida E (1992) J Polym Sci Part A, Polym Chem 30: 77
115. Aoki T, Shinohara K, Oikawa E (1992) Makromol Chem, Rapid Commun 13: 565

116. Tanaka A, Nitta K, Maekawa R, Masuda T, Higashimura T (1992) Polym J 24: 1173
117. Voronkov MG, Pukhnarevich VB, Sushchinskaya SP, Annekova VZ, Annenkova VM, Andreeva NJ (1980) J Polym Sci, Polym Chem Ed 18: 53
118. Higashimura T (1984) Jpn Kokai Tokkyo Koho 58.206.611 (1984) CA 110: 24114v
119. Okano Y, Masuda T, Higashimura T (1984) J Polym Sci, Polym Chem Ed 22: 1603
120. Zeigler JM (1984) ACS Polymer Prepr 25: 223
121. Liaw DJ, Soum A, Fondanille M, Parlier A, Rudler H (1985) Makromol Chem, Rapid Commun 6: 309
122. Otsubo M, Hirokawa Y (1992) Jpn Kokai Tokkyo Koho 0404207 (1992) to Nippon Zeon Co Ltd CA 116: 195107i
123. Tajima H, Masuda T, Higashimura T (1987) J Polym Sci Part A, Polym Chem 25: 2033
124. Masuda T, Isobe E, Higashimura T (1983) J Am Chem Soc 105: 7473
125. Masuda T, Isobe E, Higashimura T (1985) Macromolecules 18: 841
126. Masuda T, Isobe E, Hamano T, Higashimura T (1986) Macromolecules 19: 2448
127. Fujimori J, Masuda T, Higashimura T (1988) Polym Bull 20: 1
128. Kunzler J, Percec V (1990) New Polymeric Mater. 4: 271
129. Hamano T, Masuda T, Higashimura T (1988) J Polym Sci Part A, Polym Chem 26: 2603
130. Zheng G, Xia J, Nakagawa T, Higuchi A, Nagai K (1991) Yingyong Huaxue 8 43 CA 116: 1073.87x
131. Masuda T, Isobe E, Hamano T, Higashimura T (1987) J Polym Sci Part A, Polym Chem 25: 1353
132. Isobe E, Masuda T, Higashimura T, Yamamoto A (1986) J Polym Sci Part A, Polym Chem 24: 1839
133. Gal Y-S, Choi S-K (1987) J Polym Sci Part A, Polym Chem 25: 2323
134. Gal Y-S, Choi S-K (1989) J Polym Sci Part A, Polym Chem 27: 31
135. Masuda T, Tajima H, Yoshimura T, Higashimura T (1987) Macromolecules 20: 1467
136. Masuda T, Tsuchihara K, Ohmameuda K, Higashimura T (1989) Macromolecules 22: 1036
137. Tang B-Z, Kotera N (1989) Macromolecules 22: 4388
138. Kusumoto T, Hiyama T (1988) Chemistry Lett 1149
139. Kim Y-H, Gal Y-S, Kim U-Y, Choi S-K (1988) Macromolecules 21: 1991
140. Kim Y-H, Kwan S-K, Gal Y-S, Choi S-K Pure JMS (1992) Appl Chem A29: 589
141. Kobayashi N, Nakada M, Ohrio H, Tsuchida E, Matsuda H, Nakanishi H, Kato M (1987) New Polymeric Mater 1: 3
142. Kang ET, Neho KG, Masuda T, Higashimura T, Yamamoto M (1989) Polymer 30: 1328
143. Tang B-Z, Masuda T, Higashimura T (1989) J Polym Sci Part A, Polym Chem 27: 1197
144. Masuda T, Hamono T, Tsuchira K, Higashimura T (1990) Macromolecules 23: 1374
145. Percec V, Kü:nzler J (1991) Polym Bull 25: 483 (1991)
146. Tsuchira K, Oshita T, Masuda T, Higashimura T (1991) Polymer J 23: 1273
147. Hallensleben ML, Marchig C (1992) Polym Bull 27: 367
148. Eaborn C, Walton DRM, Organometal Chem J (1965) 4: 217
149. Eastmond R, Walton DRM (1968) Chem Commun 204
150. Eastmond R, Johnson TR, Walton DRM (1972) Tetrahedron 25: 460
151. Shakowski BG, Stadnichuk MD, Petrov AA (1965) Zh Obshch Khim 35: 1714 (1965) CA 64 2119
152. Jones ERH, Skattebol L, Whiting MC (1956) J Chem Soc 4765
153. Krüerke U, J Chem Organometal (1970) 21: 83
154. Hay S, (1962) J Org Chem 27: 3320
155. Cadiot P, Chodkiewicz W (1969) In: "Chemistry of Acetylenes" (Viehe HG, Ed) Marcel Dekker New York pp 597 – 647
156. Kloster-Jensen E (1972) Angew Chem 84: 483
157. Diederich F, Rubin Y (1992) Angew Chem 104 1123 Int Ed 31: 1101 (1992)
158. Boldi AM, Anthony J, Knobler CB, Diederich F (1992) Angew Chem 104: 1270 Ind Ed 31 1240 (1992)
159. Moore JS, Xu Z (1991) Macromolecules 24: 5894

66   Chapter 1: Polymerization of Vinyl Monomers

160. Hergenrother PM (1985) Acetylene-terminated Prepolymers In: Mark HF, Bikales NM, Overberger CG, Menges G (eds) Encyclopedia of polymer Sci and Vol 1 Engin Wiley J, New York, p 61
161. Hergenrother PM (1985) Reactive Oligomers In: Harris FW, Spinelli HJ (eds) ACS Symp Ser 282: 1
162. Sillion B (1987) Recent Advances in Mechanistic and Synthetic Aspects of Polymerization In: Fontanille M, Guyot A (eds) Reidel Dordrecht p 237
163. Hergenrother PM (1982) J Polym Sci Polym Chem Ed 20: 3131
164. Austin WB, Bilow N, Kelleghan WJ, Lau KSY (1981) J Org Chem 46: 2280
165. Lau KSY, Kelleghan WJ, Boschan RH, Bilow N (1983) J Polym Sci Polym Chem Ed 21: 3009
166. Sauvage N, Mereier R, Sillion B (1990) Polym Bull 23: 7
167. Ben Romdhane H, Leuze A De, Boileau S, Bartholin M, Sillion B (1992) Polym Prepr 33: 193
168. Hay AS (1962) J Org Chem 27: 3320
169. Hay AS (1969) J Polym Sci part A 17: 1625
170. White DM, Levy GC (1972) Macromolecules 5: 526
171. Rutherford DR, Stille JK, Elliot CM, Reichert VR (1992) Macromolecules 25: 2294
172. Korshak VV, Sladkov AM, Luneva LK (1962) Izv Akad Nauk CSSR Otd Khim Nauk 728 (1962) CA 57: 14970i
173. Luneva LK, Sladov AM, Korshak VV (1967) Vysokomol Soedin Ser A 9 910 (1967) CA 67: 5488 3m
174. Shim IW, Rissen WM, Organometal J (1984) Chem 269: 171
175. Corriu RJP, Douglas WE, Yang Z-X (1990) J Polym Sci Part C, Polym Letters 28: 431
176. Iwahara T, Hayase S, Wert R (1990) Macromolecules 23: 1298
177. Bortolin R, Brown SSD, Parbhoo B (1990) Macromolecules 23: 2465
178. Corriu RJP, Guerin C, Henner B, Kuhlmann Th, Jean A (1990) Chem Matter 2: 351
179. Bre´fort JL, Corriu RJP, Gerbier Ph, Henner BJL, Jean A, Kuhlmann Th (1992) Organometallics 11: 2500
180. Corriu RJP, Gerbier Ph, Guerin C, Henner BJL, Jean A, Mutin PH (1992) Organometallics 11: 2507
181. Ishikawa M, Hasegawa Y, Hatano T, Kunai A (1989) Organometallics 8: 2741
182. Ishikawa M, Hasegawa Y, Kunai A, Yamanoka R (1990) J Organomet Chem 381: C 578
183. Ijadhi-Maghsoodi S, Pang Y Barton TJ (1990) J Polym Sci Part A, Polym Chem 28: 955
184. Ijadhi-Maghsoodi S, Barton TJ (1990) Macromolecules 23: 4485
185. Barton TJ, Ijadhi-Maghsoodi S, Pang Y (1991) Macromolecules 24: 1257
186. Suzuki T, Mita I (1992) Eur Polym J 28: 1373
187. Cho O-K, Kim Y-H, Choi KY, Choi SK (1990) Macromolecules 23: 12
188. Buncel E, Davies A (1958) J Chem Soc 1550
189. Davies AG, Buncel E (1960) Brit Pat 827 366 (1960) CA 54: 14097f
190. Mageli OC, Light RE, Varnagy EJ (1967) US Pat 3.388.864 (1967) to Wallace & Tiernan Inc CA 67: 91391a
191. Harries AF (1968) US Pat 3.385.911 to Monsanto Co CA 69: 20020j (1968)
192. Harries AF (1968) US Pat 3.385.912 to Monsanto Co CA 69: 20021k (1968)
193. Semchikov YD, Kopylova NA, Yablokova NV, Nistratov LN (1986) Eur Polym J 22: 569
194. Matsumura S, Miyauchi H, Yoshida M, Nakashio Y (1972) to Sumitomo El Ind Ger.Offen 2.064.114 (1971) CA 16: 46885q
195. Nistratova LN, Kopylova NA, Sanchikov YD, Yablokova NV, Kabanova EG, Kurshyi YA, Alexandrov YA (1985) Vysokomol Soedin A27: 825
196. Kabanova EG, Yablokova NV, Alexandrov YA, Nistratova LN, Kopylova NA, Semchikov YD (1982) Khim.Elementorg Soedin Gorky State University Press
197. Sluchivskaya NP, Yablokov VA, Yablokov NV, Alexandrov YA (1976) Zh Obshch Khim 46: 1540
198. Gorbatov VV, Kurshyi YA, Alexandrov YA, Yablokova NA (1979) Zh Obshch Khim 49: 365
199. Hirai H, Komijama M (1974) J Polym Sci, Polym Chem Ed 12: 2701

200. Gilman H, Dunn GE (1951) J Am Chem Soc 73: 5077
201. Braun D, Becker KH (1969) Angew Makromol Chem 6: 136
202. Braun D, Becker KH (1971) Makromol Chem 147: 91
203. Bledzki A, Braun D (1981) Makromol Chem 182: 1047
204. Bledzki A, Braun D (1981) Makromol Chem 182: 1395
205. Bledzki A, Braun D, Titzschkau K (1983) Makromol Chem 184: 745
206. Vio L, (1974) US Ger Offen 2.131.623 (1970) or Pat 3.792.126 (1974) to Soc Nat Petr d' (1974) Aquitaine CA 76 P 127805r
207. Rudolph H, Traenckner HJ (1976) US Pat 3.431.355 to Bayer AG
208. Wolfers H, Rudolph H, Rosenkranz HJ (1978) Ger Offen 2.632.294 to Bayer AG
209. Reuter K, Dhein R (1983) Ger Offen 3.151.444 to Bayer AG
210. Crivello JV, Lee JL, Conion DA (1986) Polymer Bull 16: 95
211. Crivello JV, Conlon DA, Lee JL (1986) J Polym Sci Part A, Polym Chem 24: 1197
212. Crivello JV, Lee JL, Conlon DA (1986) J Polym Sci part A, Polym Chem 24: 1251
213. West R, Wolff AR, Peterson DJ (1986) Rad J Curring 35:
214. Krainek I, Yagci Y, Schnabel W (1992) Polym Bull 29: 277
215. Trefonas P, Müller R, West R (1985) J Am Chem Soc 107: 2737
216. Kminek I, Brynda E, Schabel W (1991) Eur Polym J 27: 1073
217. Gong MS, Hall HK jr (1986) Macromolecules 19: 3011
218. Hall HK jr, Padias AB, Atsumi M Way TF (1990) Macromolecules 23: 678
219. Lin C-H, Matyjaszewski K (1990) J Polym Sci Part A, Polym Chem 28: 1771
220. Sawamoto M, Okamoto C, Higashimura T (1987) Macromolecules 20: 2693
221. Sawamoto M, Kamigaito M, Kojima K, Highashimura T (1988) Polym Bull 19: 359
222. Kamigaito M, Sawamoto M, Higashimura T (1990) Macromolecules 23: 4896
223. Meirvenne DV, Haucourt N, Goethals EJ (1990) Polym Bull 23: 185
224. Pio P, Guastalla G, Rotzinger B, Mülhaupt R (1983) In: Quirk RP (ed) Transition Metal Catalyzed Polymerization: Alkenes and Dienes Harwood, New York, p 435
225. Kashiwa N (1983) In: Quirk RP Transition Metal Catalyzed Polymerization: Alkenes and Dienes Harwood Press, New York p 379
226. Parodi S, Nocci R, Giannini U, Barbe PC, Scata U (1984) Eur Pat Appl 45.975 (1982) to Montedison SPA CA 101: 38989w
227. Parodi S, Nocci R, Giannini U, Barbe CP, Scate U (1982) Eur Pat Appl 45.976 and 45.977 (1982) to Montedison SPA CA 96: 18808v (1982) and CA 96: 200 358s
228. Mülhaupt R, Klabunde U, Ittel SD (1985) J Chem Soc, Chem Commun 1945 [82]
229. Spitz R, Bobichon C, Guyot A (1989) Makromol Chem 190: 707
230. Sacchi MC, Shan C, Locatelli P, Tritto I (1990) Macromolecules 23: 383
231. Sacchi MC, Forlini F, Tritto I, Mendichi R, Zameoni G (1992) Macromolecules 25: 5914
232. Vähäsarja I, Pakkanen TT, Pakkanen TA, Iiskola E, Sormunen P (1987) J Polym Sci Part A, Polym Chem 25: 3241
233. Seppälä JV (1991) Härkönen M, Makromal Chem 190: 2535
234. Härkönen M, Seppälä JV (1991) Makromol Chem 192: 2857
235. Iiskola E, Sormunen P, Garoff T, Vähäsarja E, Pakkanen TT, Pakkanen TA (1988) In: "Transition Metals and Organometallics as Catalysts for Olefin Polymerization" (Kaminsky W, Sinn HJ Eds) Springer Verlag, Berlin, p 113
236. Masuda T, Yamagata M, Higashimura T (1984) Macromolecules 17: 126
237. Masuda T, Matsumoto T, Yoshimura T, Higashimura T (1990) Macromolecules 23: 4902
238. Matsumoto T, Masuda T, Higashimura T (1991) J Polym Sci Part A, Polym Chem 29: 295
239. Hogan JP (1970) J Polym Sci Part A 18: 2637
240. McDaniel MP (1982) J Catal 76: 37
241. Turbett RJ, Pollart DF (1971) Ger Offen 2.052.573 (1971) to Union Carbide, CA 75: 37058z
242. Carrick WL, Turbett RJ, Karol FJ, Karapinka GL, Fox AS, Johnson RN (1972) J Polym Sci Part A-1 10: 2609
243. Scheirs J, Bigger SW, Billingham NC (1992) J Polym Sci Part A, Polym Chem 30: 1773
244. Baker LM, Carrick WL (1970) J Org Chem 35: 774
245. Keim W, Appel R, Gruppe S, Knoch F (1987) Angew Chem 99 1042 (1987) Angew Chem Int Ed.

246. Fery N, Laible R, Hamann K (1973) Angew Makromol Chem 34: 81
247. Laible R, Hamann K (1975) Angew Makromol Chem 48: 97
248. Nakatsuka T (1987) J Appl Polym Sci 34: 2125
249. Boven G, Osterling MLCM, Challa G, Schouten AJ (1990) Polymer 31: 2377
250. Eastmand GC, Nguyen-Huu C, Piret WH (1980) Polymer 21: 598
251. Hashimoto K, Fujisawa T, Kobayashi M, Yosomiya R (1982) J Appl Polym Sci 27: 4529
252. Hashimoto K, Fujisawa T, Kobayashi M, Yosomiya R (1982) J Macromol Sci Chem A 18: 173
253. Schomaker E, Zwarteveen A-J, Challa G, Capka M (1988) Polymer Comm 29: 158
254. Osterling MLCM, Sein A, Schoutenn AJ (1992) Polymer 33: 4394
255. Fery N, Hoene NA, Hamann K (1972) Angew chem 84: 359

CHAPTER 2

Group Transfer Polymerization

W. R. Hertler

Introduction

The importance of silicon as a structural element of polymers has long been recognized. But, during the past ten years a new role for silicon which has been developed is that of a mediator for polymer synthesis. When silicon has completed its tasks in polymer synthesis, it can be discarded.

The historical methods for synthesis of acrylic polymers are free radical polymerization and low temperature living anionic polymerization. When acrylic polymers with narrow MWD, strictly controlled M_n, or block polymer structure were required, living anionic polymerization was the only reasonable method to use. However, because of rapid termination at temperatures above about $-40\ °C$, anionic polymerization could be conducted successfully only at very low temperatures, which rendered such polymers very costly in comparison with polymers prepared by free radical polymerization. The application of the silicon-mediated Michael addition reaction of Mukiyama [1] to polymer chemistry in 1983 [2, 3] provided, for the first time, a practical process to partake of all the benefits of living anionic polymerization (the ability to control polymer architecture, M_n, and MWD) at the reflux temperatures of common polymerization solvents. This silicon-mediated polymerization of acrylic monomers is termed group transfer polymerization (GTP). GTP is called a living polymerization because rates of transfer and termination can be diminishingly small relative to the rate of propagation [4,5]. This chapter describes the GTP process for acrylic monomers as well as the related aldol-GTP [6–8] of silyl vinyl ethers. There are several general reviews of GTP [9–13], reviews with particular emphasis on mechanism of GTP [14, 15], and a review of polymer architecture with emphasis on block polymers [16].

2.1
The GTP Process

The sequential Michael addition reaction of a silyl ketene acetal **1a** (the initiator) with an α,β-unsaturated ester, typically a methacrylate, in the presence of a catalyst, typically an onium salt of a nucleophilic anion, to give an acrylic polymer with a "living"silyl ketene acetal end group **3c** is the process known as GTP, Eqs. (1, 2) [2]. When the polymerization process is complete, the silyl ketene acetal end group is conveniently quenched by addition of proton-source, such as methanol, which introduces a hydrogen at the polymer terminus, Eq. (3), with concomitant formation of a silyl ether **3d**. Since the silyl ether **3d** is usually volatile, the polymer can readily be obtained free of impurities such as the alkali metal salts which accompany polymers prepared by conventional anionic polymerization. The only contaminant in a polymer prepared by anion-catalyzed GTP is the catalyst, which generally is present at a level of only about 1 mol % of the silyl ketene acetal initiator.

$$\tag{1}$$

$$\tag{2}$$

$$\tag{3}$$

GTP, in common with other living polymerization processes, is capable of preparing poly(methacrylates) with very narrow MWD and with DP equal to the molar ratio of monomer to initiator **1a**. A principal use of GTP, for which livingness is critical, is the preparation of block copolymers. Müller et al. [17] provided a careful analysis of the capability of GTP in synthesis of block co-polymers. Although GTP is not limited to methacrylates, this class of monomer gives the best control of M_n and the lowest values of MWD. Onium salts of nucleophilic anions, such as sulfonium bifluorides, quaternary ammonium carboxylates, and quaternary ammonium cyanide are the catalysts most commonly used for GTP, but Lewis acids, Lewis bases, Zeolites, and mercuric iodide have also been used (vide infra). It is important to note that silyl ketene acetals, including the living polymer **3c**, are stable over long periods of time in the absence of catalyst. But, in the presence of catalyst they are highly reactive, and in the absence of monomer undergo a variety of decomposition processes [18–20]. Replacement of the silicon of silyl ketene acetals with germanium [21], tin [21], titanium [22], zirconium [22, 23], hafnium [22], and lanthanides, notably samarium [24], has been shown to support GTP-like processes for polymerization of acrylic monomers. These metals have not been studied in as great detail as silicon, and, in keeping with the theme of this book, this review is limited to GTP with silicon as the transfer group.

2.2
Mechanism

In their first publication on GTP of MMA [2] in 1983, Webster and co-workers proposed an intramolecular transfer mechanism, Eq (4), in which the silyl group is transferred directly from the complex of the silyl ketene acetal initiator (or the corresponding living polymer end) and the catalyst anion–a pentacoordinate siliconate 4a–to the carbonyl oxygen of the monomer via 4b [25, 26]. This proposal of a direct transfer of silicon from the pentacoordinate siliconate 4a to the monomer has led to a great deal of controversy concerning the mechanism of GTP, and has perhaps served as a spur to several groups to carry out detailed mechanism studies. Interestingly, Yasuda [24] has found that in the uncatalyzed samarium-GTP of MMA, the eight-membered cyclic analog of 4b, in which $RSm(C_5Me_5)_2$ replaces R_3SiNu^-, is actually a stable, isolable intermediate characterized by X-ray analysis.

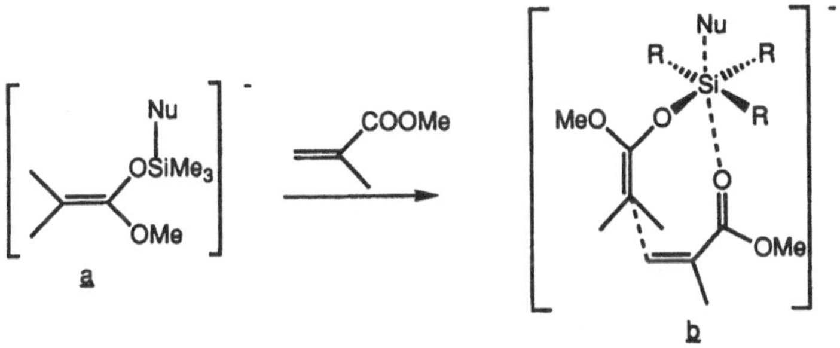

$$(4)$$

Sogah and Farnham [25, 26] showed with labelling experiments that in GTP of MMA catalyzed by trisdimethylaminosulfonium bifluoride 5a (TASHF2) at 25–55 °C or catalyzed by tris(dimethylamino)sulfonium difluorotrimethylsiliconate (TASF) at –95 °C, fluorosilane 5d is not produced in a reversible, dissociative step, Eq. (5), at a rate which is competitive with the rate of propagation.

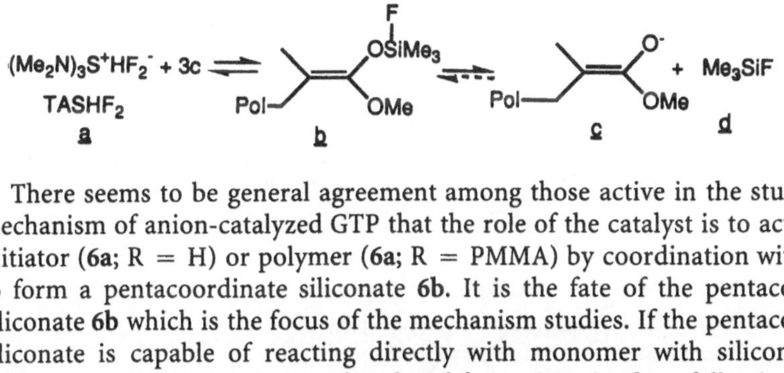

$$(5)$$

There seems to be general agreement among those active in the study of the mechanism of anion-catalyzed GTP that the role of the catalyst is to activate the initiator (6a; R = H) or polymer (6a; R = PMMA) by coordination with silicon to form a pentacoordinate siliconate 6b. It is the fate of the pentacoordinate siliconate 6b which is the focus of the mechanism studies. If the pentacoordinate siliconate is capable of reacting directly with monomer with silicon-transfer either accompanying carbon-carbon bond formation via 4b or following carbon-carbon bond formation, then it is possible that exchange of silicon groups among polymer chains could be slower than the rate of propagation. Sogah and Farnham [25, 26] carried out double-labelling experiments in GTP of MMA and n-BuMA catalyzed by $TASHF_2$ at ambient temperature or catalyzed by TASF at –95 °C, in which the polymer chains were labelled structurally to permit separation by solubility, and silicon was labelled with different substituents (trimethylsilyl vs triethylsilyl for the $TASHF_2$-catalyzed process, and dimethylphenylsilyl vs dimethyltolylsilyl for the TASF-catalyzed, low temperature process). These workers reported that, during the time frame of the addition of a few monomer units to polymer, there was no exchange of silicon between polymer chains. This lack of silicon exchange between growing polymer chains is consistent with transfer of silicon from the silyl ketene acetal group of the polymer to the carbonyl oxygen of the incoming monomer, Eq. (7). In order for this mechanism to be consistent with the observation that M_n is determined by the ratio of monomer to initiator rather than by the ratio of monomer to catalyst, it is necessary that the rate of catalyst exchange among polymer chains, Eq. (6), be fast relative to the rate of monomer addition, Eq. (7).

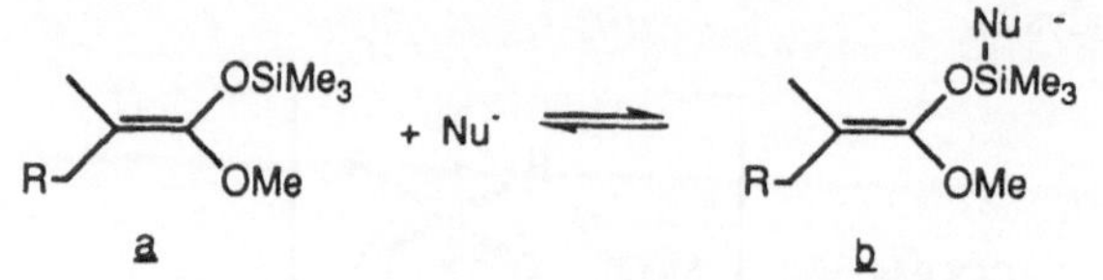

$$(6)$$

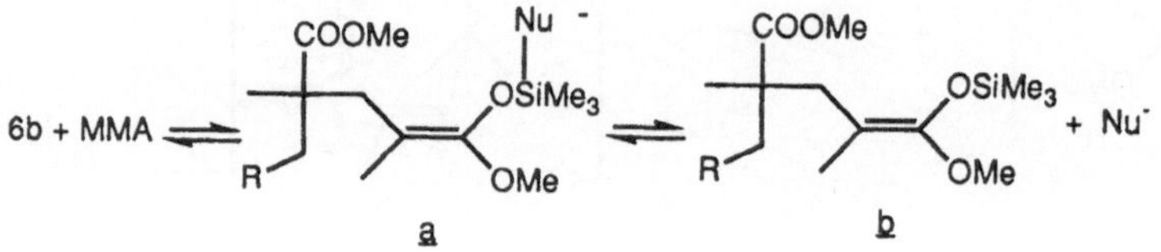

$$(7)$$

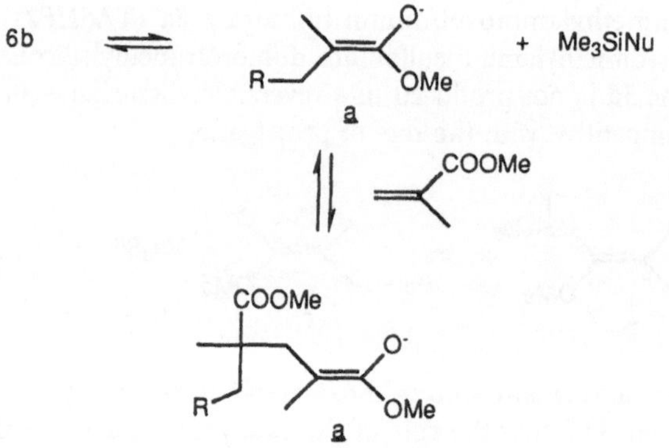

$$(8)$$

$$(9)$$

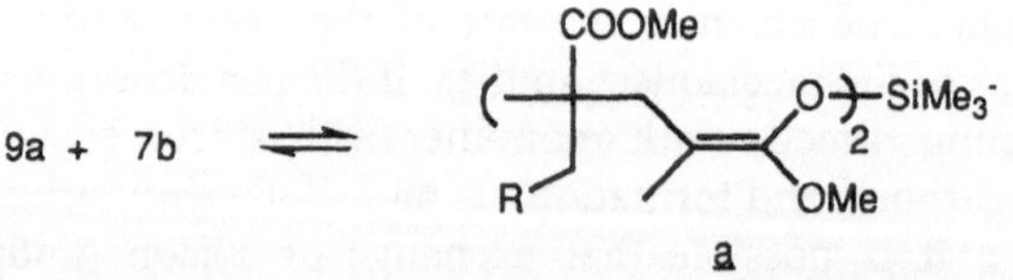

$$(10)$$

Quirk [27, 28] has challenged the proposal that the pentacoordinate siliconate **6b** reacts directly with monomer with silicon-transfer accompanying carbon-carbon bond-formation. He favors the mechanism, Eqs. (8–10), in which the pentacoordinate siliconate **6b** dissociates to ester enolate anion **8a**, followed by reaction with monomer to give **9a**. In order to account for control of M_n, by monomer:initiator ratio rather than monomer:catalyst ratio, the propagating ester enolate anions **9a** must undergo silicon-exchange reactions, Eq. (10), with "dormant" polymer chains **7b** (or initiator molecules) at a rate which is fast relative to monomer addition. Quirk and Bidinger [27], in fact, showed that tetrabutylammonium 9-methylfluorenide **11a**, which reacts with MMA to give the ester enolate **11b**, can be used with silyl ketene acetal **1a** (at a level of 0.043 mol % relative to **1a**) to catalyze polymerization of MMA. The M_n of the resulting PMMA (5000) is consistent with control of DP by the molar ratio of MMA to **1a**. This clearly implicates an intermediate such as **10a**. Burggraf and Davis [29] have postulated that a pentacoordinate siliconate with two electronegative axial substituents will have the greatest stability when the two electronegative substituents are identical. The silicon-exchange intermediate **10a** has just such a structure.

$$(11)$$

Quirk and Ren [28] carried out double labelling experiments in the GTP of MMA catalyzed by TASHF$_2$ in which the polymer chain label was molecular weight, permitting chromatographic separation, and the silicon was labelled with substituents (trimethylsilyl vs dimethylphenylsilyl). In these experiments varying amounts of silicon-exchange between polymer chains were found during the time period in which addition of about 60 monomer units took place. This result was interpreted as supporting the enolate ester anion mechanism, Eqs. (8–10), for TASHF$_2$-catalyzed GTP of MMA and disproving the direct silicon-transfer process, Eqs. (6, 7), proposed by Sogah and Farnum [25, 26]. However, the observation of (incomplete) silicon-exchange accompanying a large number of propagation steps does not address the relative rates of silicon-exchange and monomer-addition. A further caveat regarding the double labelling experiments of both research groups [25–28] is that, in the TASHF$_2$-catalyzed polymerizations, the substituents used as silicon "labels" (triethylsilyl/trimethylsilyl [25, 26] and dimethylphenylsilyl/trimethylsilyl [28]) would, in fact, be expected to introduce reactivity differences which could effect the rates and equilibrium constants of Eqs. (6–8) and Eq. (10). Martin and Bywater [20] have shown the substantial reactivity difference between a trimethylsilyl ketene acetal and a triethylsilyl ketene acetal in GTP. Moreover, Webster has reported [30] that in biacetate-catalyzed GTP of MMA using a mixture of two initiators, one having a dimethylphenylsilyl group and the other having a trimethylsilyl group, the MWD of the resulting polymer is broader than is obtained with either initiator alone. This suggests that there are two different intermediates which are not undergoing interchange on the time scale of monomer-addition. There is certainly opportunity for additional experimentation to clarify the relative rates of silicon-exchange and monomer addition.

Kinetics of GTP of MMA have been studied by Brittain [31] using stopped-flow FT-IR spectroscopy, by Mai and Müller [32–34] using gravimetric and gas chromatographic analysis, and by Bandermann and coworkers [18, 35, 36] using dilatometry. Brittain [31] determined that the pentacoordinate fluorosiliconate **12a** (which is a source of fluoride) catalyzes GTP of MMA with a first order dependence. With bifluoride as the catalyst, Bandermann [18] and Brittain [31] determined a reaction order of 2 for catalyst concentration, while Mai and Müller [32] working in a much lower concentration range found a reaction order of 1.17 for catalyst concentration and 0 order for initiator concentration. For GTP of MMA with bibenzoate catalyst, Brittain observed a reaction order of 0.3 for catalyst concentration, and Müller [34] reported a reaction order of 1 for initiator concentration. In the case of catalysis by benzoate, both Brittain [31] and Müller [34] found a reaction order about 1 for catalyst concentration, and, for the initiator, Müller [34] reported a reaction order of 1.2. Brittain's measurements indicated that, with benzoate- or bibenzoate-catalysis, the rates of initiation and the rates of early propagation steps were essentially identical. Müller [37] has calculated the limiting reaction orders for catalyst and initiator for the various mechanisms represented in Eqs. (6–10) (i.e. associative, reversible dissociative, and irreversible dissociative mechanisms). For catalysis by $Nu^- =$ benzoate, only the associative mechanism, Eqs. (6, 7), is consistent with the experimentally observed first order in catalyst and in initiator concentrations. For catalysis by bifluoride and bibenzoate, the data does not distinguish among the mechanisms.

$$(12)$$

Consideration of the mechanistic evidence available thus far for GTP suggests to this writer that both associative and dissociative processes occur depending upon the conditions. Enough factors are now understood that it is possible to control the variables of structure, catalyst, and concentration in order to favor either process. The model ab initio calculations of Dixon [38] for intermediates in the direct reaction of the pentavalent silyl enolate **6b** with monomer (associative process, Eq. (7)) help to make such a process understandable energetically. Dixon favors formation of the carbon-carbon bond before transfer of the silicon, and, if an eight-membered ring (Form **4b**) is involved in the transfer process, it is believed to be a transition state rather than an intermediate.

The tacticity of PMMA prepared by anion-catalyzed GTP has been studied by several groups, and found to be predominantly syndiotactic, with a pronounced inverse temperature-dependence, and with Bernoullian statistics prevailing [21, 39–43]. The tacticity resulting from anion-catalyzed GTP is similar to that from anionic polymerization of methacrylates with large counterions [27, 34]. Banerjee and Hogen-Esch [44] found that GTP of triphenylmethyl methacrylate results in 70–90% isotactic polymer, the isotactic content increasing with rising polymerization temperature. Diphenylmethyl methacrylate, in contrast, gives mostly syndiotactic polymer.

Several groups have measured reactivity ratios for random copolymerization of several methacrylates by GTP. The comonomer pairs include MMA (r = 4.59) and *tert*-butyl methacrylate (r = 0.16) [34]; MMA (r = 1.36) and ethyl methacrylate(r = 0.51) [17]; MMA (r = 1.66) and decyl methacrylate (r = 0.48) [17]; and MMA (r = 0.44) and *n*-butyl methacrylate (r = 0.26) [45]. Except for the last values, the large effect of the size of the alkyl groups is more characteristic of anionic polymerization than free radical polymerization.

2.3
Catalysis

2.3.1
Nucleophilic Anions

By far, the most widely used and studied class of catalysts for GTP is nucleophilic anions. One of the earliest, and most potent, catalysts to be used for GTP is TASHF$_2$ [2]. Similarly potent, but having much greater solubility in THF (the solvent nearly always used for GTP), are tetrabutylammonium fluoride [21] and tris(piperidino)sulfonium bifluoride (TPSHF$_2$, Form **12b**) [18, 31]. Chou and Niu [46] explored the use of potassium bifluoride and crown ether in catalysis of GTP of MMA. Catalysis of GTP by tetraalkylammonium cyanide has been studied in some detail by Bandermann [35, 41, 47–50]. A large number of oxyanion catalysts, such as tetrabutylammonium benzoates, acetate, phosphinates, phenolates, sulfinates, sulfonamidates, perfluoroalkoxides, nitrite, cyanate, and bicarboxylates (1:1 complexes of carboxylic acids with the corresponding carboxylate salt) were reported by Dicker and coworkers [51, 52]. The relative activity of these catalysts correlates well with the pK$_a$ (in DMSO) of the conjugate acids of the corresponding catalysts, and the pK$_a$ range of the conjugate acids of these catalysts is 4.5–23. The method used to rank the catalysts was to measure the time required to reach the maximum temperature resulting from the exothermic polymerization. The times to T$_{max}$ served as an approximation of the relative polymerization rates. The catalysts derived from the weakest acids (most basic oxyanions) had the greatest catalytic activity. All of the oxyanion catalysts were less active than bifluoride, and bicarboxylates were less active than the corresponding carboxylates. Among the oxyanion catalysts commonly used in GTP, acetate is more active than benzoate, which is more active than *m*-chlorobenzoate. Benzoate is more active than bibenzoate. Schmalbrock and Bandermann [19] have confirmed, with kinetics measurements, the activity order: bibenzoate > *m*-chlorobenzoate > bi-*m*-chlorobenzoate. In addition they have shown that cyanide is more active than bibenzoate. Müller [34] has found that the overall polymerization rate of MMA is two orders of magnitude greater with catalysis by bifluoride than with benzoate. Brittain [53] gives the order for relative rates of initiation for the TPS salts: bifluoride > benzoate > bibenzoate.

The studies by Dicker and coworkers [51, 52] also have shown that, with oxyanion catalysts, reducing the concentration of catalyst and using catalysts of lower relative activity reduces the rate of termination and narrows the MWD of the resulting polymers. These authors proposed that the relative activity of anion catalysts of GTP is determined by their effect on the equilibrium constant for Eq. (6). Thus, when Nu$^-$ is fluoride, the equilibrium is to the right, and with an oxyanion, the equilibrium is to the left. Müller [34, 37] has concluded from analysis of GTP initiator rate orders with various catalyst types that, indeed, with bifluoride the equilibrium constant is $\gg 1$, and for benzoate and bibenzoate it is $\ll 1$.

When bicarboxylates 13a are used to catalyze GTP, the carboxylic acid associated with the carboxylate salt probably reacts with silyl ketene acetal 13b to form the silyl carboxylate 13c, alkyl isobutyrate 13d, and mono-carboxylate salt 13e according to Eq. (13). The silyl carboxylate may then coordinate with the mono-carboxylate salt to form the pentacoordinate complex 14c according to the equilibrium of Eq. (14) [51, 52]. This will serve to reduce the concentration of active catalyst 14b, which generally increases livingness of the polymerization. In fact, Schneider and Dicker [54] have found that deliberate addition of a silyl ester, such as trimethylsilyl m-chlorobenzoate, to an oxyanion-catalyzed GTP reaction results in enhanced livingness characterized by lower polydispersity of the resulting polymer and reduced termination. Such "livingness enhancers" are believed to provide their beneficial effects by reducing the concentration of oxyanion catalyst, Eq. (14). This also leads to a reduction in polymerization rate.

Acetonitrile has also been recognized as a livingness enhancer for anion-catalyzed GTP [55]. This effect is seen at low concentrations of acetonitrile. There is a concommitant reduction in polymerization rate. When the catalyst is tetrabutylammonium cyanide, detailed studies have shown that the mechanism of livingness enhancement is complexation of the catalyst with acetonitrile, a Brønsted acid, by hydrogen-bonding. The complexed catalyst is inactive in GTP, but is in equilibrium with uncomplexed catalyst. The association constant for complex-formation between tetrabutylammonium cyanide and acetonitrile in THF solution was reported to be 4.2 M^{-1} [56]. The thermodynamic parameters for the equilibrium are $\Delta G = -0.86$ kcal/mol, $\Delta H = -6.8$ kcal/mol, $\Delta S = -20$ e.u. Low concentrations of acetonitrile in GTP reactions are often encountered when acetonitrile is used as a carrier for anion catalysts which have poor solubility in THF. When acetonitrile is used as the solvent for anion-catalyzed GTP, Bandermann and coworkers [36, 48–50] have shown that some termination occurs as a result of silylation of acetonitrile by the silyl ketene acetals.

Boettcher and coworkers studied GTP with anion catalysts which were supported on insoluble, lightly crosslinked polystyrene beads [56–58]. Two types of supported anion catalysts were prepared, 15a–c in which the quaternary ammonium ion is bound to the support ("cation-bound"), and 15d in which the

$$(RCOO)_2^{n\text{-Bu}_4N^+}H \; + \; \overset{OSiMe_3}{\underset{OMe}{=\!\!<}} \; \longrightarrow \; Me_3SiOOCR \; + \; Me_2CHCOOMe \; + \; \underset{RCOO^-}{\overset{n\text{-Bu}_4N^+}{}} \qquad (13)$$

$$\text{a} \qquad \qquad \text{b} \qquad \qquad \qquad \text{c} \qquad \qquad \text{d} \qquad \qquad \text{e}$$

$$Me_3SiOOCR \; + \; \begin{matrix} n\text{-}Bu_4N^+ \\ RCOO^- \end{matrix} \rightleftharpoons \underset{\underline{c}}{\overset{OOCR}{\underset{OOCR}{Me\diagdown \underset{|}{Si} - Me}}} \; n\text{-}Bu_4N^+ \tag{14}$$

<u>a</u>   <u>b</u>   <u>c</u>

nucleophilic anion is bound to the support ("anion-bound").Both the cation-bound- and the anion-bound-supported catalysts, catalyzed GTP of MMA in the presence of a silyl ketene acetal. In a typical procedure, the supported catalyst was allowed to equilibrate with a solution of monomer in THF. Then the initiator **1a** was added in one portion, and exothermic polymerization occured. With the cation-bound catalysts **15a–c** there was no induction period, and GPC of the resulting polymer showed a bimodal MWD. With the anion-bound catalyst **15d** there was an induction period followed by rapid, exothermic poly-merization, and GPC analysis of the polymer showed M_n close to theory for initiator-control, while $M_w/M_n = 1.6$. After removal of the supported catalyst **15d** by filtration, the filtrate from the reaction mixture still showed catalytic activity toward added monomer and initiator **1a**, giving quantitative conversion to polymer with about the theoretical M_n. This shows that the catalyst became solubilized, presumably during the induction period, Eq. (16), due to formation of silylated support **16a** and enolate **16b**, which resulted in GTP by the processes of Eqs. (9, 10) in homogeneous solution.

$$\left(\!\text{Sty}\!\right)\!\!-\!CH_2NBu_3^+ \; Nu^- \qquad\qquad \left(\!\text{Sty}\!\right)\!\!-\!CH_2COO^- \; Bu_4N^+$$

<u>a</u> Nu⁻ = acetate   <u>d</u>   (15)
<u>b</u> Nu⁻ = cyanide
<u>c</u> Nu⁻ = m-chlorobenzoate

$$\textbf{15d} \; + \; \overset{OSiMe_3}{\underset{OMe}{\diagup\!\!=\!\!\diagdown}} \; \longrightarrow \; \left(\!\text{Sty}\!\right)\!\!-\!CH_2COOSiMe_3 \; + \; \overset{O^- \; Bu_4N^+}{\underset{OMe}{\diagup\!\!=\!\!\diagdown}} \tag{16}$$

$$\underline{a} \qquad\qquad\qquad \underline{b}$$

2.3.2
Lewis Bases

Lewis bases were reported to catalyze GTP of methacrylates and acrylates [59]. Only a limited number of Lewis bases was effective, and those Lewis bases which catalyzed GTP were most effective in propylene carbonate solvent. Induction periods were often observed. The Lewis bases which were reported to be useful are quinuclidine **17a**, 1,4-diazabicyclo [2.2.2] octane **17b**, hexamethylphosphorous-triamide **17c**, bis(dimethylamino)methylphosphine **17d**, tributylphosphine, and triethylarsine. In some cases, addition of a second portion of monomer after a waiting period following the polymerization of the first batch of monomer produced the expected increase in M_n. The mechanism by which these Lewis bases catalyze GTP is not known.

$$(Me_2N)_3P \qquad (Me_2N)_2PMe \qquad\qquad (17)$$

a   b   c   d

2.3.3
Lewis Acids

Lewis acid catalysis of GTP was recognized early as a means for polymerizing acrylates, and, less effectively, methacrylates [2, 60]. Zinc halides, alkylaluminum oxides, and dialkylaluminum oxides were the most useful catalysts. However, the zinc halides, although effective at ambient temperature, were required at the impractically high levels of 10 mol % with respect to monomer to achieve complete conversion of monomer. The aluminum catalysts could be used at lower levels, but gave satisfactory results only at low temperatures, e.g. −78 °C. While alkyl acrylates could be quantitatively converted to polymer, MMA gave poor conversion. PMMA prepared by zinc halide-catalyzed GTP was more syndiotactic than PMMA prepared by anion-catalyzed GTP [60]. Liang et al. polymerized MMA with zinc chloride catalyst and a silyldienolate initiator and obtained polymer with MWD of 2.0 [61]. Both types of Lewis acid catalyst failed to catalyze GTP in basic solvents, so aromatic hydrocarbons or chlorinated alkanes, such as 1,2-dichloroethane, were most often used as solvent. The zinc halide catalysts are believed to activate the acrylate monomer for reaction with the silyl ketene acetal initiator by coordination with the carbonyl oxygen Eq. (18). Xu and Wang [62] studied the polymerization of ethyl acrylate in dichloroethane using zinc iodide catalyst in the presence of silyl ketene acetal **1a** to provide living polymers. Kinetics indicated that when zinc iodide was < 10 mol % of the ethyl acrylate, there was an induction period, and the polymerization rate was independent of monomer concentration. At > 10 mol % zinc iodide the polymerization rate was proportional to the monomer concentration. MWD of the resulting poly(ethyl acrylate) was about 1.2.

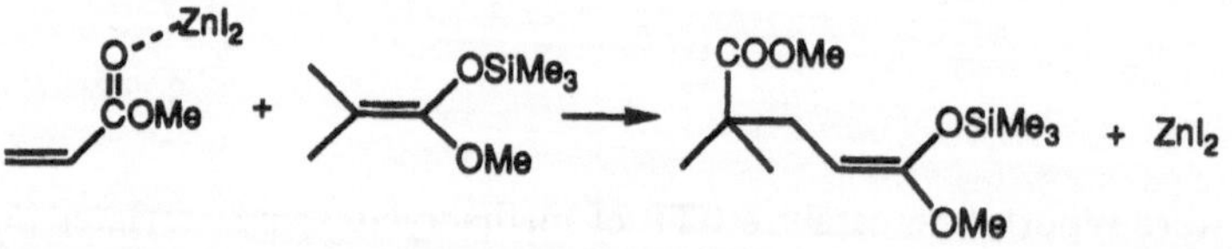

$$(18)$$

2.3.4
Zeolytes

Building on the clay-catalyzed Michael addition of silyl ketene acetals and silyl enol ethers reported by Kawai [63, 64], Corbin and Sormani [65] have found that Type Y zeolyte can be used as a Lewis acid catalyst for GTP of alkyl acrylates with silyl ketene acetal 1a, usually in toluene and at ambient temperature. Polymerization is usually rapid and exothermic. For best results, the initiator is usually added in one portion to a mixture of monomer, zeolyte, and solvent. In a typical polymerization the zeolyte is used at a level of about 25 wt % of initiator, a much lower level than is required with zinc halide catalysis of GTP. The resulting poly(alkyl acrylates) usually have a M_n somewhat higher than theory. Conversions are often quantitative, and values of M_w/M_n are as low as 1.02.

2.3.5
Mercuric Iodide

An important advance in GTP of acrylates was the finding that mercuric iodide at low levels could be used to catalyze their polymerization, preferably in less polar solvents [66–68]. Polymerization rates in THF are slow. *N,N*-dimethylacrylamide also undergoes mercuric iodide-catalyzed GTP [66]. The use of trimethylsilyl iodide in conjunction with the mercuric iodide gave greatly increased polymerization rates, permitted the polymerization of methacrylates, and greatly reduced the level of catalyst required [68, 69]. In a kinetics study, Zhuang and Müller [69] found that, in mercuric iodide-catalyzed GTP of *n*-butyl acrylate in toluene initiated with 1a, half-lives are in the range of hours, and the propagation reaction is first order with respect to initiator and catalyst, and 1.5 order with respect to monomer concentration after an induction period. In the presence of trimethylsilyl iodide, the induction period is eliminated, half-lives are in the range of minutes (in the range of seconds in dichloromethane solvent), the reaction is of first order in trimethylsilyl iodide, and Arrhenius plots indicate an apparently negative activation energy at higher temperatures. The authors explain the results by exothermic formation of a 1:1 complex of trimethylsilyl iodide with mercuric iodide, which acts as a nucleophilic catalyst.

2.3.6
High Pressure

In addition to anion-, Lewis base-, Lewis acid-, zeolyte-, and mercuric iodide-catalysis of GTP, uncatalyzed GTP was reported by Sogah and coworkers [70, 71]. When a solution of monomer and silyl ketene acetal 1a was subjected to high pressure of inert gas, polymerization occurred. Pressures of 1–3 kbar were used to polymerize ethyl acrylate and MMA. For ethyl acrylate, THF was the best solvent, while MMA polymerized best in dichloromethane. With both monomers, conversion was usually less than 100%, MWD was broader than in catalyzed GTP, and M_n was much higher than theory. Syndiotacticity of PMMA was higher than in anion-catalyzed GTP.

2.4
Initiators

2.4.1
Functional Initiators

Of the many initiators which have been used for GTP, the real workhorse initiator is the silyl ketene acetal, 1-methoxy-1-trimethylsiloxy-2-methyl-1-propene **1a**, which is commercially available. However, in order to achieve various objectives in polymer synthesis, such as terminal functional polymers, diblock polymers, ABA triblock polymers, graft polymers, and ladder polymers, synthetic chemists have designed and synthesized a wide variety of specialty initiators. Many of the silyl ketene acetals were synthesized by the classical method of Ainsworth [72], Eq. (19), or by catalytic hydrosilation, Eq. (20) [73, 74]. Dicker has reported an interesting method for homologation of silyl ketene acetals by mercuric iodide-catalyzed condensation of a silyl ketene acetal (in excess) with a methacrylate [75].

$$Me_2CHCOOMe + \textit{i-}Pr_2Li + Me_3SiCl \longrightarrow \begin{array}{c} Me \\ Me \end{array}\!\!\!=\!\!\!\begin{array}{c} OSiMe_3 \\ OMe \end{array} \tag{19}$$

$$\begin{array}{c} COOMe \\ Me \end{array}\!\!\!=\!\!\! + R_3SiH \xrightarrow{\text{cat.}} \begin{array}{c} Me \\ Me \end{array}\!\!\!=\!\!\!\begin{array}{c} OSiR_3 \\ OMe \end{array} \tag{20}$$

The use of functional initiators is the most efficient means to introduce exactly one functional group at the end of a polymer chain. Since many functional groups are not compatible with GTP, a protective group, often trimethylsilyl, may be required. Thus, a terminal carboxylic acid group on a poly(methacrylate) is readily obtained by initiating GTP with 1,1-bis(trimethyl-siloxy)-2-methyl-1-propene **21a** followed by deprotection with methanol, Eq. (21) [76]. And a terminal hydroxyl group is introduced by initiation with 1-(2-tri-methylsiloxyethoxy)-1-trimethyl-siloxy-2-methyl-1-propene **22a** followed by deprotection with methanol and acid catalyst, Eq. (22) [2, 21]. Similarly, a terminal phosphonic acid group is obtained from initiators **23a** and **23b** with a bis-(trimethylsilyl) phosphonate group Eq. (23) [21]. PMMA with a terminal dihydroxyaromatic group was prepared from an initiator with two protected phenolic groups **24a** for use in condensation polymerization to make graft polymers Eq. (24) [77].

$$\begin{array}{c} OSiMe_3 \\ OSiMe_3 \end{array}\!\!\!=\!\!\!\begin{array}{c} \\ \end{array} + \begin{array}{c} \\ \end{array}\!\!\!=\!\!\!\begin{array}{c} COOR \\ \end{array} \quad \begin{array}{c} 1) \text{ GTP} \\ \hline 2) \text{ MeOH} \end{array} \quad \begin{array}{c} COOH \\ \text{Polymer} \end{array} \tag{21}$$

a

Two different terminal functional groups are introduced to PMMA by the use of a vinyl silyl ketene acetal which contains a trimethylsilyl ester group (**25a**).

$$\text{a} \quad + \quad \text{COOR} \quad \xrightarrow[\text{2) MeOH}]{\text{1) GTP}} \quad \text{Polymer} \tag{22}$$

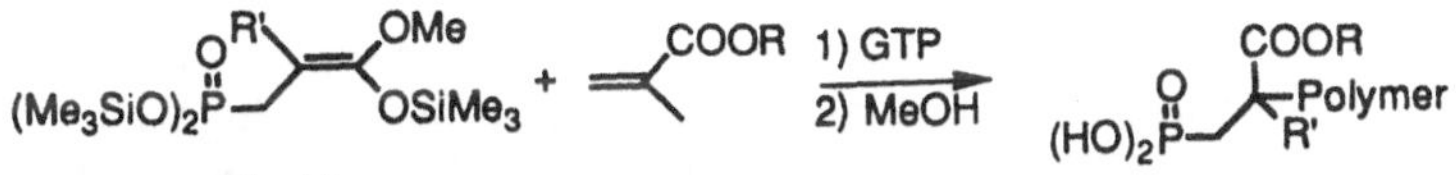

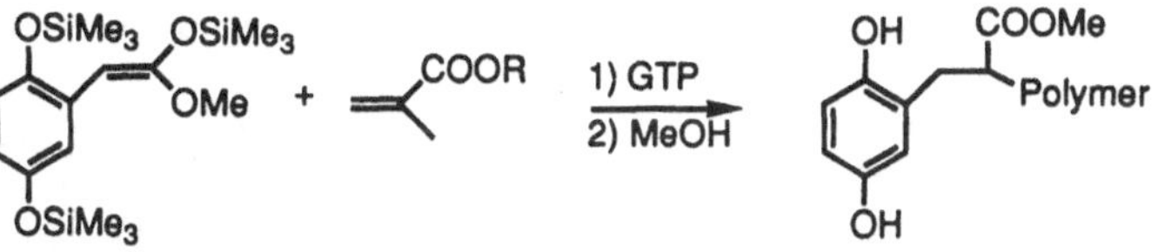

$$\text{a R' = Me} \quad \text{b R' = H} \quad + \quad \text{COOR} \quad \xrightarrow[\text{2) MeOH}]{\text{1) GTP}} \tag{23}$$

$$\text{a} \quad + \quad \text{COOR} \quad \xrightarrow[\text{2) MeOH}]{\text{1) GTP}} \tag{24}$$

The vinyl group does not require protection for GTP in contrast to radical polymerization, but the carboxylic acid group, of course, does need to be protected, and the butenoic acid **25c** is obtained Eq. (25) [78]. Two different terminal functions are also provided by tris(trimethysiloxyethylene **25b**, which gives PMMA with an α-hydroxycarboxylic acid group on one end [79].

$$+ \quad \text{COOMe} \quad \xrightarrow[\text{2) MeOH}]{\text{1) GTP}} \tag{25}$$

a R = CH$_2$=CH
b R = Me$_3$SiO

c R = CH$_2$=CH
d R = OH

Three initiators, which can be used for GTP of methacrylates to synthesize terminal macromonomers (polymers with a polymerizable functional group attached) are the styryl-, [80–83] the glycidyl-, [84] and trimethoxysilyl-substituted silyl ketene acetals **26a**, **26b**, and **26c**, respectively. The trimethoxysilyl end group of poly(methacrylates) prepared by GTP initiated by **26c** has been used to form the core of hybrid star-shaped polymers [85]. PMMA with a terminal phosphonic ester group was prepared from **26d**, [21] and block polymers with a terminal dimethylamino group were prepared from **26e** [86].

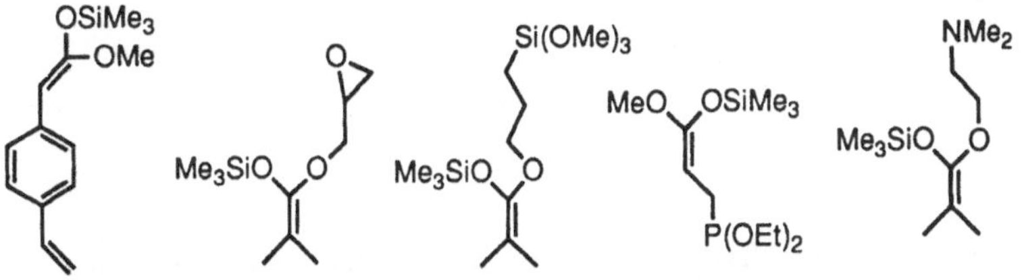

$$\tag{26}$$

a   b   c   d   e

The regiospecificity of initiation of GTP of MMA by silyl dienolates is dependent upon the substitution at C-2, Eqs. (27, 28). Thus, initiation by unsubstituted **27a** occurs only at C-2, while the 2-methyl-substituted homolog **28a** initiates at C-4 (28–29%) as well as C-2 (71–72%) [78]. The corresponding methyl ester **28b** also initiates GTP of MMA [80]. The silyltrienolate **29a** initiates GTP of MMA at C-2 to give a polymer with a terminal conjugated diene group, Eq. (29) [78]. All of these silylpolyenolates were reported to be highly reactive for initiation of GTP of MMA [78]. They provide examples of polymerization in which the rate of initiation is much faster than the rate of propagation. It has often been remarked that, in order to obtain narrow MWD in living polymerizations, the rate of initiation should be comparable to, or faster than, the rate of propagation.

Witkowski and Bandermann [80] used the unsaturated silyl ketene acetals **30a** and **30c** to initiate GTP of MMA to obtain polymers with terminal unsaturation, and Chou and Niu [46] obtained a similar result with **30b**. The PMMA macromonomer obtained from **30c** was found to undergo free radical homopolymerization [80].

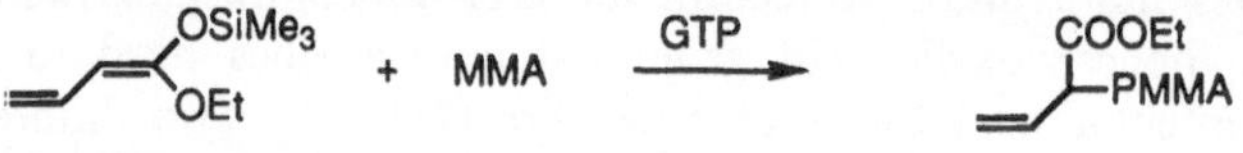

$$\text{(27)}$$

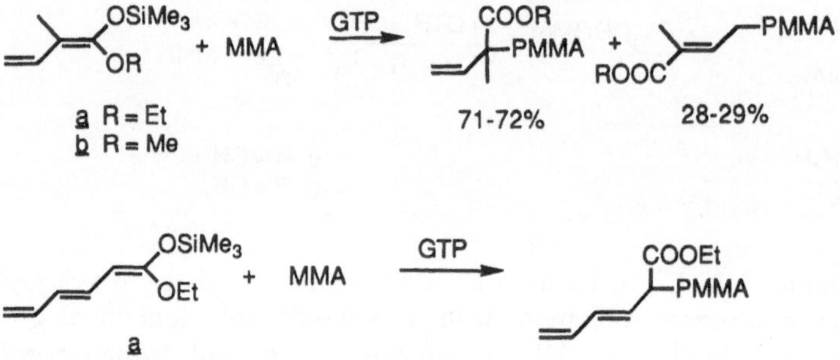

71-72%   28-29%

$$\text{(28)}$$

a R = Et
b R = Me

$$\text{(29)}$$

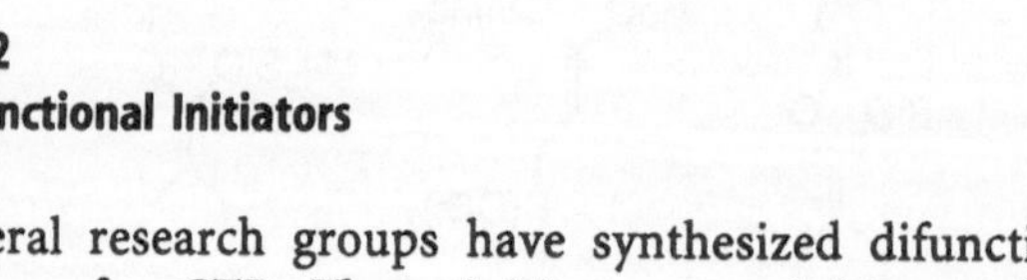

a R = Me
b R = Et

c

$$\text{(30)}$$

2.4.2
Difunctional Initiators

Several research groups have synthesized difunctional silyl ketene acetal initiators for GTP. These initiators can initiate polymer chain growth in two directions which facilitates synthesis of ABA triblock polymers in which the center B block is formed first. Choi et al. synthesized several interesting ABA triblock polymers with GTP using **31a** [87, 88] and **31c** [89] as initiators. The DuPont group prepared ABA block polymers with **31d** [21]. Steinbrecht and Bandermann reported the synthesis of **31a–e** and confirmed that they could be used to initiate GTP [90]. Initiator **31b** has the interesting ability to place two carboxylic acid groups in the center of the polymer chain. The conjugated difunctional initiator **32a** was reported to give poor conversions in GTP of MMA [78]. The fully substituted conjugated difunctional initiator **32b** was used by Manring et al. [91] to introduce a head-to-head linkage in the center of a PMMA chain for thermal degradation studies. The silyl bisenolate **32c** was used by Sogah [92–94] for the very special purpose of synthesizing ladder polymers from difunctional monomers (vide infra). The similar silylbisenolate **32d**, which is a silyl hydride, is reported to initiate GTP of MMA with good control of M_n and MWD [21].

$$ \text{Me}_3\text{SiO} \overset{\text{RO}}{\underset{}{\big|}} -(\text{CH}_2)_n - \overset{}{\underset{\text{OR}}{\big|}} -\text{OSiMe}_3 \qquad \overset{\text{OSiMe}_3}{\underset{\text{O(CH}_2\text{CH}_2)_n\text{O}}{}} \overset{}{\underset{\text{Me}_3\text{SiO}}{}} \qquad (31) $$

a n = 1; R = Me
b n = 1; R = Me₃Si
c n = 2; R = Me

d n = 1
e n = 2

$$ \underset{\text{MeO}}{\overset{\text{Me}_3\text{SiO}}{}} \underset{R}{\overset{R}{=}} \underset{\text{OMe}}{\overset{\text{OSiMe}_3}{}} \qquad \overset{\text{OMe}}{\underset{\text{OSiRMeO}}{=}} \underset{\text{MeO}}{\overset{}{}} \qquad (32) $$

a R = H
b R = Me

c R = Me
d R = H

2.4.3
Other -Silyl Ketene Acetal Initiators

Pan, Xia, and Zou used silyl ketene acetals **33a** and **33b** with conjugated ester groups [95–97] and **33c** with cyano groups [98] to initiate GTP. They found that **33a** selectively initiated GTP of methyl acrylate without polymerizing MMA, [97] although Dicker and Hertler reported that **33a** initiated GTP of MMA as well as ethyl acrylate [79]. In a comparison of the fully substituted silyl ketene acetal **1a** with the analog **33d**, in which one of the β-methyl groups is absent, it was found that GTP of MMA initiated with the latter produces polymer with M_n somewhat higher than theory and with broader MWD [21]. The lactone-derived silyl ketene acetal **33e** has also been used to initiate GTP [78]. Limited studies have been reported on the effects of substituents on silicon upon initiation of GTP. But, initiation of GTP of MMA with **33f**, which has a bulky *tert*-butyl group on silicon, resulted in a very slow rate of polymerization, much higher M_n than theory, and broad MWD [21]. In contrast, the ponderous, but less sterically demanding, n-octadecyl substituent on silicon in 33g gave good control of M_n, but broad MWD [21].

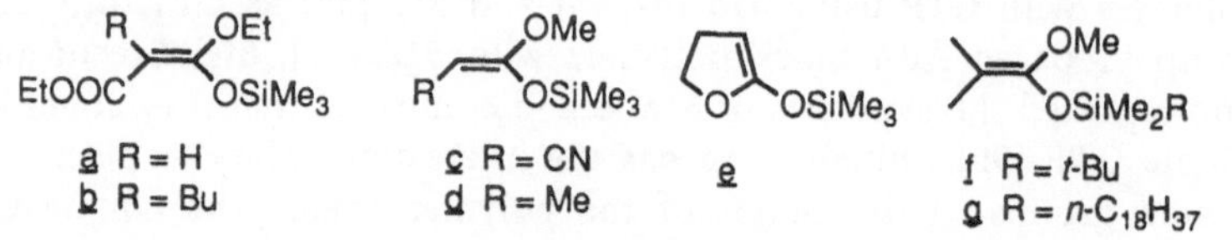

(33)

2.4.4
Macroinitiators

Polymers with silyl ketene acetal functionality are macroinitiators for GTP. Depending upon the number and placement of the silyl ketene acetal functionality, nonacrylic macroinitiators have been used to synthesize acrylic block polymers, graft polymers, star polymers, or gel-graft polymers. The DuPont group has attached multiple silyl ketene acetal groups to lightly crosslinked polystyrene beads to form a variety of insoluble, swellable macroinitiators (Forms 34a–c) for GTP [57, 58, 99]. A suspension of any of these macroinitiators in THF in the presence of an anion catalyst and MMA resulted in slow, exothermic polymerization of the monomer. Conversion was quantitative, and all of the resulting PMMA was grafted to the polystyrene beads – none was found in solution. Initiation of GTP of acrylonitrile in DMF solution with 34c in the presence of TASHF$_2$ catalyst at $-50\,°C$ gave 100% conversion to grafted poly(acrylonitrile) [99]. But at 25 °C conversion was incomplete, and both grafted and dissolved poly(acrylonitrile) was formed. This was interpreted as evidence that, at the higher temperature, chain transfer to monomer became important [99]. As expected, a stirred mixture of insoluble polystyrene-supported m-chloro-benzoate 15c, supported initiator 34b, and MMA in THF failed to form any PMMA [57, 58].

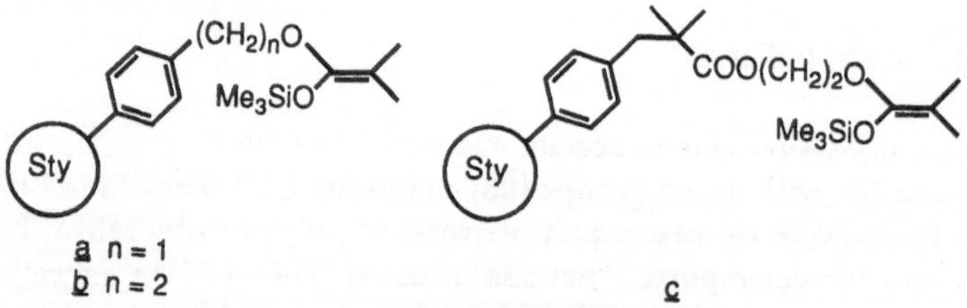

(34)

Wnek and Yang [100–103] synthesized a series of soluble poly-(dimethylsiloxanes) with pendant silyl ketene acetal functionality. This was accomplished by using poly(methylhydrosiloxane) 35a for platinum-catalyzed hydrosilation of the unsaturated silyl ketene acetals 35b and 36a. The hydrosilation reaction showed high specificity for the terminal olefin in preference to the highly substituted silyl ketene acetal group in both cases. Wnek and coworkers [103,104] also reported that when 35a was the cyclic tetramer, reaction with 36a gave a tetrafunctional silyl ketene acetal, which was then used to initiate GTP of MMA to give four-armed star-shaped PMMA with arms having from 20 to 150 MMA units. The MWD of these star polymers was ca. 1.2–1.3. Hellstern et al. [105] took another approach to synthesis of a poly(dimethylsiloxane) macroinitiator. They polymerized hexamethylcyclotrisiloxane by anionic ring-opening polymerization, and end-capped the living polymer with 3-chlorodimethysilylpropyl methacrylate to give a poly(dimethylsiloxane) with a single methacrylate end group. Hydrosilation of the methacrylate group with dimethylethylsilane gave poly(dimethylsiloxane) with a terminal silyl ketene acetal group. This was used to initiate GTP of MMA to give the AB diblock polymer. Matyjaszewski and coworkers [106] reported using polysilanes with silyl ketene acetal substituents to prepare PMMA-grafted polysilanes.

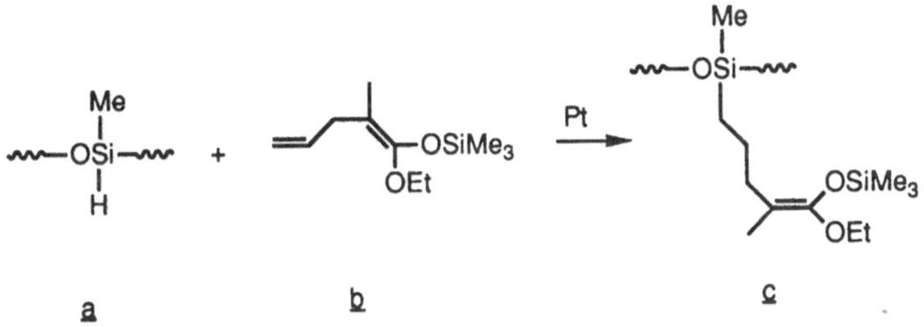

$$(35)$$

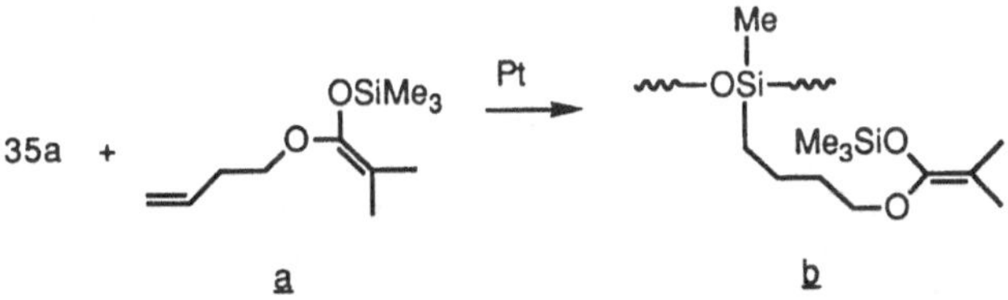

$$(36)$$

Jenkins et al. prepared soluble polystyrene copolymer macroinitiators for GTP [107–109]. The synthesis involved free radical copolymerization of styrene and 2-isobutyroxyethyl methacrylate **37b** to give the random copolymer **37b** with pendant isobutyrate ester groups, Eq. (37). The copolymer was then converted to macroinitiator **38a** with lithium diisopropylamide and chlorotrimethysilane Eq. (38). Finally, MMA was grafted to the macroinitiator by GTP using TASHF$_2$ in THF or zinc chloride in acetonitrile.

Kondo et al. [110] took a different approach for the synthesis of polystyrene with silyl ketene acetal functions as well as polystyrene with N-silylketeneimine

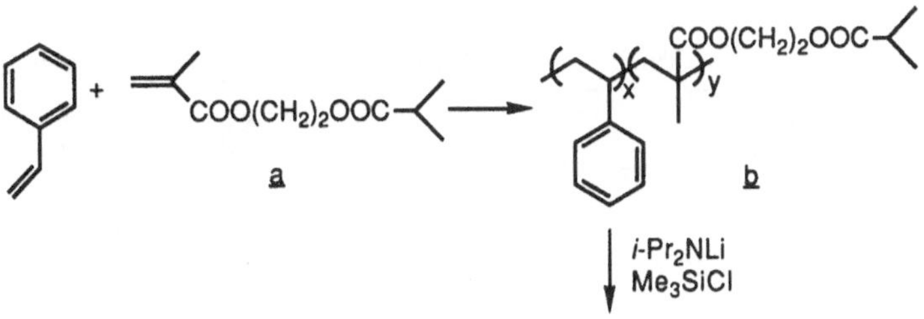

$$(37)$$

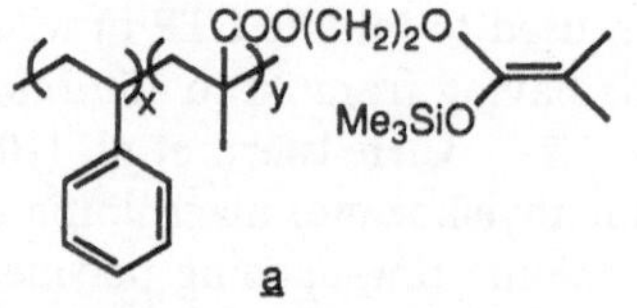

(38)

functions. The polymeric initiators were obtained by the copolymerization of styrene with *p*-vinylphenylketene methyl trimethylsilyl acetal **39a** or *N*-(trimethylsilyl)-*p*-vinyl-phenylketeneimine **39b**. Apparently at least some of the silyl ketene acetal functionality survived the radical polymerization. Initiation of GTP of MMA with the macroinitiator made from **39a** gave poly(styrene)-graft-PMMA with narrow MWD (1.4–1.8), and the initiator efficiency was nearly quantitative. In the case of the macroinitiator made from **39b**, however, the initiator efficiency was low and the MWD of the graft copolymer was broad. Witkowski and Bandermann [80] used the styryl initiator **39a** to initiate GTP of MMA to obtain a macromonomer which could be homopolymerized by radical polymerization to poly(styrene)-graft-PMMA, the conversion increasing with decreasing DP of the PMMA macromonomer. These authors found that radical polymerization of **39a** through the vinyl group resulted in loss of much of the silyl ketene acetal functionality.

Asami, Kondo et al. [81, 111] described the synthesis of silyl ketene acetal-ended polystyrene, which can be used to prepare AB diblock copolymers by GTP. Anionic living polystyrene was end-capped with 1,1-diphenylethylene, and the resulting carbanion **40a** was transformed into ester enolate anions **40e** by reaction with methyl cinnamate **40b**, methyl crotonate **40c**, or *tert*-butyl crotonate **40d**. Then each of the ester enolates (Form **40e**) was reacted with chlorotrimethylsilane to give the macroinitiators, Eq. (41). Block copolymers of polystyrene and PMMA were obtained by GTP of MMA using the macroinitiators. The macroinitiator derived from **40b** gave poly(styrene)-*block*-PMMA with MWD 1.2–1.4, while the macroinitiators derived from **40c** and **40d** led to block polymers with broad MWD.

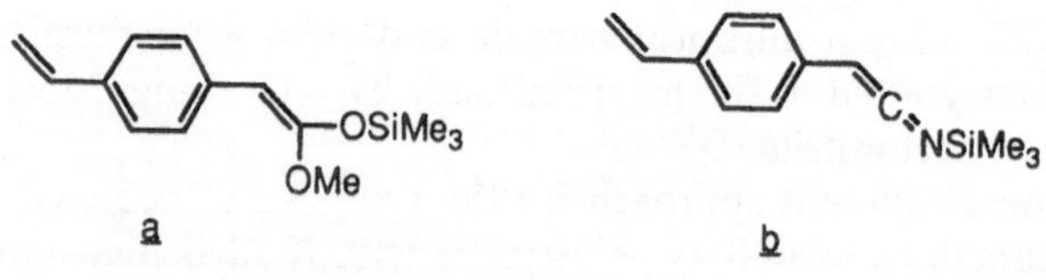

(39)

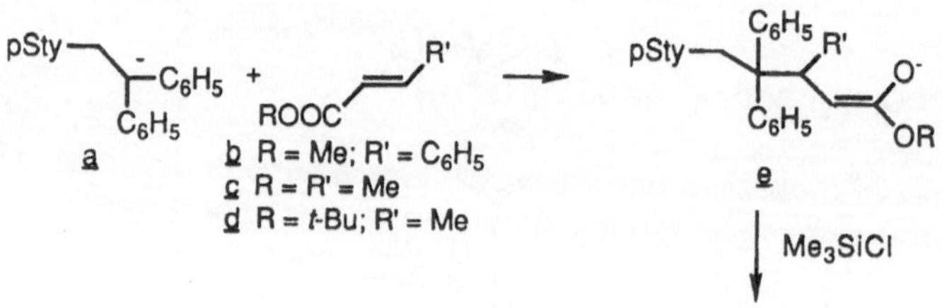

(40)

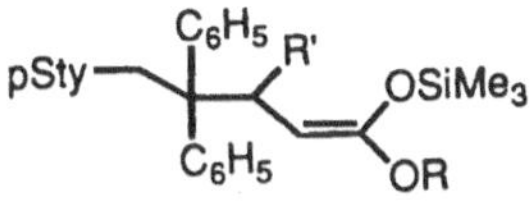

(41)

2.4.5
Other Silicon-Containing Initiators

Many organosilanes other than silyl ketene acetals have been found to initiate GTP of acrylic monomers. In many instances control of M_n is poor because the rate of initiation is slower than the rate of propagation. Examples of initiators for GTP of MMA which provide poor control of M_n and MWD are *n*-trimethylsilylisobutyronitrile 42a [21], methyl 2-methyl-4-trimethylsilyl-2-butenoate 42b [78], ethyl trimethylsilylacetate 42c [21], benzyltrimethylsilane 42d [112], and 1,4,4-trimethyl-2,5-bis(trimethylsiloxy)-1,4-dihydropyridine 42e [79]. Shen and Jin [113] found that zinc bromide, triphenylphosphine, and chlorotrimethyl-silane can be used directly to initiate GTP of acrylates. The resulting polyacrylates have terminal triphenylphosphonium chloride groups, which can be converted to the corresponding vinyl macromonomers by means of the Wittig reaction.

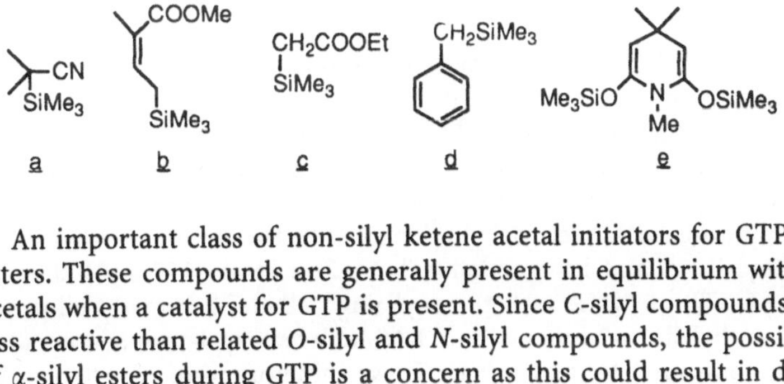

(42)

An important class of non-silyl ketene acetal initiators for GTP is the α-silyl esters. These compounds are generally present in equilibrium with silyl ketene acetals when a catalyst for GTP is present. Since *C*-silyl compounds are generally less reactive than related *O*-silyl and *N*-silyl compounds, the possible formation of α-silyl esters during GTP is a concern as this could result in degradation of control of M_n and MWD. For the anion-catalyzed equilibration of 1-methoxy-1-trimethylsiloxy-2-methyl-1-propene 43a and methyl α-trimethylsilylisobutyrate 43c, the equilibrium concentration of the ester 43c is reported to be 24–25% [18, 21, 41]. Perhaps more important than the position of the equilibrium is the rate of anion-catalyzed *O*-silyl to *C*-silyl isomerization, for, if this rate is comparable to the rate of polymerization, then the effects of the presence of the less reactive *C*-silyl isomer may be felt when it is generated during the polymerization. The rate of isomerization of 43a to 43c with catalysis by bifluoride or *m*-chlorobenzoate was determined by Sogah et al. [21] to be much slower than the apparent rates of polymerization of MMA with the respective catalysts. Brittain and Dicker [114] investigated the anion-catalyzed isomerization of a model compound for the living end of PMMA, 44a/44b. They found that the rate of bibenzoate-catalyzed isomerization of 44b (*E*-isomer) to 44a (*Z*-isomer) is 1260 times slower than the propagation rate in GTP of MMA, although the rates of reaction of MMA with **44a** and **44b** are identical. Moreover, no formation of *C*-silyl isomer **44c** was observed with several GTP catalysts [53, 115]. With bifluoride catalysis, the rate of isomerization of **44b** (E-isomer) to **44a** (Z-isomer) was comparable to the polymerization rate. Thus, intervention of *C*-silyl compounds during initiation by **43a** or during propagation of GTP of MMA is probably not important. Isomerization of **43b** to **43d** has been observed with anion catalysts [21, 49], and the equilibrium composition is reported to be 100% **43d** and 0% **43b** [114]. This O-C isomerization, added to an inherently lower reactivity of **43b** toward methacrylates, may contribute to the poorer control of M_n and MWD when GTP of MMA is initiated by **43b** rather than **43a** (vide supra).

$$
\begin{array}{ccc}
\text{(structure)} & \xrightleftharpoons{Nu^-} & \text{(structure)}
\end{array}
\tag{43}
$$

a R = Me   c R = Me
b R = H   d R = H

$$
\text{(structures)}
\tag{44}
$$

a   b   c

A large number of other silicon-containing compounds have been reported to initiate GTP of MMA, particularly when fluoride or bifluoride catalysts are used. In general, the M_n of the resulting PMMA is higher than theory indicative of a slow rate of initiation relative to the rate of propagation, or significant side reactions. Citron and Simms [116] reported initiation of GTP with 1-trimethylsilyl pyrazole, 1-trimethylsilylimidazole, diethylaminodimethylsilane, methoxytrimethylsilane, and hexamethyldisilane. Zou et al. [117, 118] claimed that initiation of GTP of MMA with the silyl vinyl ethers, 1-trimethylsiloxcyclohexene, 2-trimethylsiloxy-2-butene, and 1-trimethylsiloxy-1-propene resulted in PMMA with controlled M_n and MWD. These initiators were reported to be less active than silyl ketene acetals.

A series of silyl phosphites **45a–c** was reported by Sogah et al. [21] to initiate bifluoride-catalyzed GTP of MMA via catalyzed Michael addition to the monomer to form a silyl ketene acetal. Since the addition of the phosphites to monomer is slower than both the rate of initiation by silyl ketene acetal and the rate of propagation, the resulting polymers have higher M_n than theory and MWD ca. 2. Reetz and coworkers [119] studied the Lewis acid-catalyzed initiation of GTP of acrylates with the silyl sulfides **45d–g**, which also undergo Michael addition to the monomer to form a silyl ketene acetal initiator. With acrylate monomers there was good control of M_n and narrow MWD. With bifluoride catalyst, initiation of GTP of butyl acrylate with **45d** gave somewhat broader MWD. Cyanide-catalyzed initiation of GTP of MMA with **45d** gave M_n much higher than theory and MWD 1.6 [21]. The *N*-silylketeneimines, tris(trimethylsilyl)acetonitrile **45h**, and 1,3-bis(trimethylsilyl)-3-phenylketeneimine **45i**, were found by Dicker and Hertler [79] to initiate GTP of MMA with good control of M_n and MWD.

$R_2POSiMe_3$ 　　　　 $RSSiMe_3$

$$\begin{array}{c} Me_3Si \\[-4pt] R \end{array}\!\!\Big\rangle\!\!=C=NSiMe_3$$

　 a R = EtO 　　　　 **d** R = Me
　 b R = Me_3SiO 　　 **e** R = C_6H_5 　　　　　　　　　　　　　 (45)
　 c R = Me_2N 　　　 **f** R = Me_3SiO(CH$_2$)$_2$
　　　　　　　　　　 g R = Me_3SiOOCCH$_2$ 　　　 **h** R = Me_3Si
　　　　　　　　　　　　　　　　　　　　　　　 i R = C_6H_5

The most thoroughly studied of the silicon compounds which initiate GTP by catalyzed Michael addition to monomer is trimethylsilyl cyanide. Initiation of GTP with trimethylsilyl cyanide and a tetraalkylammonium cyanide catalyst [2, 21, 48, 56, 120] or TASF catalyst [121] is characterized by a substantial induction period. Nevertheless, it is possible to obtain PMMA with controlled M_n and narrow MWD. ^{1}H NMR was used to follow the concentration changes of MMA and trimethylsilyl cyanide during the induction period in tetraethylammonium cyanide-catalyzed GTP of MMA initiated with trimethylsilyl cyanide in acetonitrile solution [56, 122]. The rate of consumption of MMA was less than twice the rate of loss of trimethylsilyl cyanide until the last of the trimethylsilyl cyanide was consumed, at which point, the rate of MMA loss increased dramatically as rapid polymerization began. Thus, the induction period ended when all of the initiator had reacted. The probable mechanism for the reaction of trimethylsilyl cyanide with MMA, which is analogous to the mechanism proposed by Evans and Truesdale [123] for the cyanide-catalyzed addition of trimethylsilyl cyanide to carbonyl compounds, involves the Michael addition of cyanide ion to MMA, Eq. (46), to give the cyano ester enolate **46a**. The enolate **46a** then undergoes rapid silation by trimethylsilyl cyanide to give the cyano-substituted silyl ketene acetal **47b** with regeneration of cyanide ion, Eq. (47). The silyl ketene acetal, when activated by cyanide catalyst will initiate GTP of MMA. The explanatation proposed [48, 56, 122] for the failure of significant initiation of GTP of MMA to occur as long as trimethylsilyl cyanide is present is that trimethylsilyl cyanide complexes cyanide ion strongly by reversible formation, Eq. (48), of the pentacovalent siliconate, dicyanotrimethylsiliconate **48a**. If the equilibrium constant K_{eq} for **48a** is > 1, then the concentration of cyanide ion may be too low for GTP to occur until, with the consumption of all of the trimethylsilyl cyanide, the cyanide ion concentration rises and the propagation rate also increases. Dixon and coworkers [124] have provided support for the existence of **48a** by isolating and characterizing tetrabutylammonium dicyanotrimethylsiliconate. Using IR and ^{29}Si NMR techniques they found that in THF, $K_{eq} = 2.3$ M^{-1} at 24 °C Eq. (48). The thermodynamic parameters for the equilibrium were determined to be $\Delta H = -9$ kcal/mol, $\Delta S = -28$ e.u., and $\Delta G^{24 °C} = -0.5 \pm 0.2$ kcal/mol. Since trimethylsilyl cyanide competges successfully with the silyl ketene acetal **47a** for cyanide ion, it is reasonable to assume that the association constant for coordination of cyanide to a silyl ketene acetal is $\ll 2.3$ M^{-1}. This is consistent with Müller's estimate that the association constants of weaker (than bifluoride) GTP catalysts with silyl ketene acetals is $\ll 1$ [37, 115].

$$\text{CN}^- \ + \ \text{MMA} \ \longrightarrow \ \underset{\underset{\textbf{a}}{}}{\text{NC}-\overset{\displaystyle \text{O}^-}{\underset{\displaystyle \text{OMe}}{\diagup\diagdown}}} \tag{46}$$

$$\textbf{46a} \ + \ \text{Me}_3\text{SiCN} \ \longrightarrow \ \underset{\underset{\textbf{a}}{}}{\text{NC}-\overset{\displaystyle \text{OMe}}{\underset{\displaystyle \text{OSiMe}_3}{\diagup\diagdown}}} \ + \ \text{CN}^- \tag{47}$$

$$\text{Me}_3\text{SiCN} \ + \ \text{CN}^- \ \overset{K_{eq}}{\underset{}{\rightleftharpoons}} \ \left[\ \text{Me}-\overset{\displaystyle \text{CN}}{\underset{\displaystyle \text{CN}}{\overset{|}{\underset{|}{\text{Si}}}}}\overset{\text{Me}}{\diagdown}\,\overset{}{\underset{\text{Me}}{}} \ \right]^- \tag{48}$$

$$\textbf{a}$$

2.5
Monomers

2.5.1
α, β-Unsaturated Esters

The class of monomers to which GTP has been most successfully applied, and which has been most thoroughly studied, is methacrylates. Acrylates may also be polymerized by GTP with anion catalysts, but MWD is always broader than with the corresponding methacrylate. There appears to be a slow termination process which leads to broadening of the MWD. This will be discussed in the section on termination. When good control of M_n and narrow MWD are desired in GTP of acrylates, use of Lewis acid or mercuric iodide catalyst gives better results. Since the reactive end of a poly(acrylate) undergoing anion-catalyzed GTP resembles the initiator **33d** rather than the initiator **1a** (which resembles the reactive end of PMMA undergoing GTP), efficient addition of a methacrylate block onto a living poly(acrylate) is not possible. The rate of propagation of the methacrylate is greater than the rate of initiation of methacrylate by the living poly(acrylate). Therefore, to prepare block copolymers of acrylates and methacrylates, it is necessary to build the methacrylate block first and the acrylate block second. The rate of polymerization of acrylates is much faster than that of methacrylates as shown in a competition experiment [21]. The many methacrylates and acrylates which have been polymerized by GTP to give homopolymers, random copolymers or block copolymers are listed in Table 2.1.

Of particular interest are functional monomers, since these have been widely used to synthesize functional polymers. Functional monomers which do not require the use of a protective group for use in GTP include glycidyl methacrylate [2, 21, 125, 126], 2-dimethylaminoethyl methacrylate [17, 57, 58, 86, 127–132], and allyl methacrylate [2, 133–135]. The unusual N-metha-crylaxyalkyl-*p*-phenylenediamines **49a–c** and the related sorbates **49d–f** were synthesized by Laschewsky

Table 2.1. Monomers polymerized by group transfer polymerization[a]

Monomer	Literature References for Homopolymer, Random Copolymer (co), and Block Copolymer(b)
Acrylonitrile	[21, 59, 96–99]
Allyl methacrylate	[133] co: [2, 135] b: [2, 133, 134]
Benzyl methacrylate	co: [151–153] b: [132, 149, 153]
n-Butyl acrylate	[19, 21, 60, 66, 67, 113, 202] co: [137, 147] b: [67, 113, 142, 147]
tert-Butyl acrylate	co: [147] b: [147]
n-Butyl methacrylate	[133] co: [2, 21, 45, 127, 128, 134, 135, 147, 153] b: [2, 21, 86–89, 92, 93, 125, 127, 128, 131, 133, 147, 149, 154, 165, 172, 173]
tert-Butyl methacrylate	[42, 147, 148] co: [147] b: [87, 147, 149, 150]
n-Decyl methacrylate	co: [17, 203] b: [17, 203]
Diethylene glycol diMA	[93] b: [93]
Diethyl vinylphosphonate	[186]
Dimethylacrylamide	[2, 21, 57, 58, 67]
Dimethylaminoethyl MA	co: [127, 128] b: [17, 57, 58, 86, 127–132]
Diphenylmethyl MA	[44]
Ethyl acrylate	[6, 21, 59, 60, 66–68, 113, 204] co: [137] b: [2, 156, 205]
Ethylene dimethacrylate	co: [126] b: [126, 142, 146, 206]
2-Ethylhexyl MA	co: [153] b: [93, 126, 142, 154]
Ethyl methacrylate	co: [17, 146, 160] b: [146]
Ethyl muconate	[78]
Ethyl sorbate	[78] b: [78]
Furfuryl methacrylate	co: [156]
Glycidyl methacrylate	co: [2, 21] b: [125, 126, 172, 173]
Hexamethylene diMA	[92, 93] b: [92, 93, 126, 142]
Lauryl methacrylate	b: [21]
Methacrylonitrile	[21, 59, 159] co: [160]
2-Methoxyethyl MA	[147] co: [147] b: [147]
Methyl acrylate	[18, 35, 42, 49, 66, 67, 113, 207] b: [67, 161]
Methyl pentadienoate	[78]
Methyl sorbate	[78]
Morpholinoethyl MA	b: [17]
2-Phenylethyl acrylate	[157]
2-Phenylethyl MA	co: [132] b: [86, 131, 132]
N-Phenylmaleimide	[208] co: [208]
Phenyl vinylsulfone	[57, 58]
Propylene sulfide	b: [161]
Tetrahydrofuranyl A	[157]
Tetrahydrofuranyl MA	[157, 158] co: [153]
Tetrahydropyranyl A	[156]
Tetrahydropyranyl MA	[153, 156, 157] co: [151–153] b: [86, 131, 153, 156]
Tetramethylene diMA	[93]
Trimethylsilyl MA	[42] co: [132] b: [42, 86, 129–132, 154]
Triphenylmethyl MA	[44]
p-Vinyl benzyl MA	[138, 139] co: [138, 139]
49a	[136] co: [136] b: [136]
49b	[136]
49c	[136] co: [136] b: [136]
49d	[136]
49e	[136]
49f	[136]

Table 2.1. (*Continued*)

50a	[137]
50c	*co*: [140] *b*: [140]
51a	[150] *b*: [150]
51b	*b*: [209]
51c	*co*: [210]
52a	*co*: [142] *b*: [142]
52b	*b*: [142]
52c	*b*: [142]
52d	*b*: [142]
52e	*b*: [142]
53a	[141]
53b	*b*: [142]
53c	*b*: [142]
53d	*b*: [142]
54a	[21]
54b	[143]
54c	*co*: [132] *b*: [132]
54d	*b*: [132]
55b	[78] *b*: [78]
55c	[78]
56a	[78]
57a	*co*: [21, 146] *b*: [21, 88, 133, 172, 173]
57b	*co*: [137]
58c	[152, 153]
61b	*b*: [211]
62a	[164]
63a	[89] *b*: [89]
64a	[74]
65a	[165] *co*: [165] *b*: [165]
66a	[166]
68b	[93]
69a	[93]
69b	[93]

[a] Because of the large number of literature references, there is no entry for MMA homopolymer, but copolymers of MMA are included under the comonomer entries. Silyl vinyl ethers for aldol-GTP are not included in Table.

and Ward [136] and polymerized by GTP to give soluble, redox-active polymers. In these cases, free radical polymerization failed to give soluble polymers. GTP was used by Epple and Schneider to polymerize *N*-(2-acryloxyethyl)carbazole **50a** to obtain an electron-donor copolymer [137]. p-Vinylbenzyl methacrylate **50b** undergoes GTP of the methacrylate functionality, leaving the styryl group intact to provide macromonomers [138,139]. GTP of 3-trimethoxysilylpropyl methacrylate **50c** gives a polymer with useful crosslinking sites [140].

$$Me_2N\!-\!\langle\rangle\!-\!\overset{Me}{\underset{.}{N}}(CH_2)_n OOC\!-\!\!\diagdown \qquad Me_2N\!-\!\langle\rangle\!-\!\overset{Me}{\underset{.}{N}}(CH_2)_n OOC \diagup\!\diagdown\!\diagup\!\diagdown \tag{49}$$

    a n = 2   **d** n = 2
    b n = 3   **e** n = 3
    c n = 6   **f** n = 6

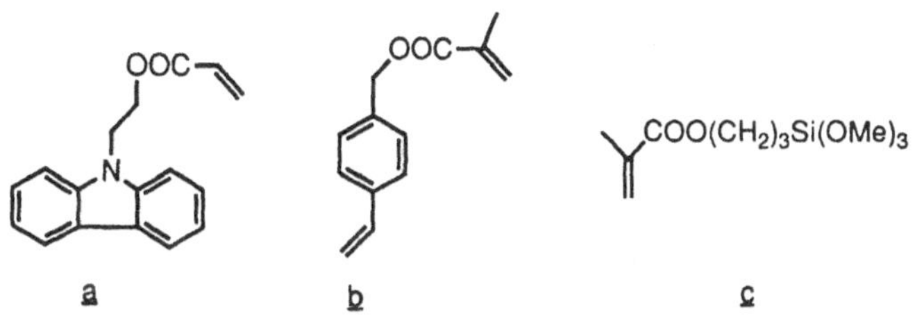

(50)

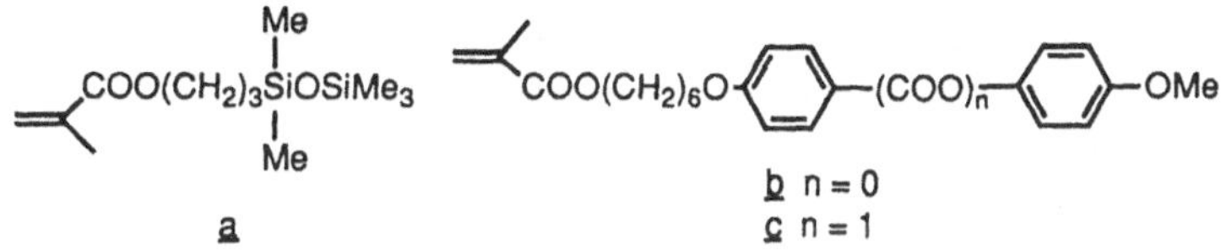

(51)

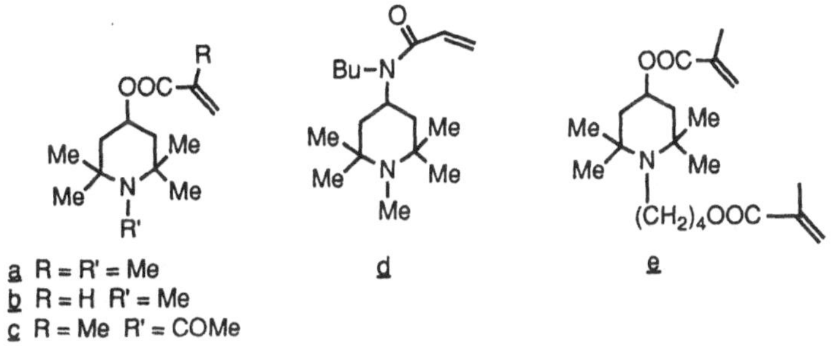

(52)

Gomez and Neidlinger [141] were able to use GTP to copolymerize the UV light stabilizer based on 2-(2-hydroxyphenyl)-2H-benzotriazole **53a** which contains an unmasked, intramolecularly hydrogen-bonded phenolic hydroxyl group. Slongo [142] used GTP to copolymerize the benzotriazoles **53b–d** with similarly unprotected, intramolecularly hydrogen-bonded phenolic hydroxyl groups, and long spacers. GTP of α-methylene-γ-butyrolactone **54a** was reported by Sogah et al. [21] to give polymer with much higher isotactic content than is characteristic of PMMA prepared by GTP. Stille and coworkers [143] used GTP to polymerize racemic α-methylene-γ-methyl-γ-butyrolactone **54b**. Comb-like polymers with polyethylene oxide oligomers attached to an acrylic backbone have been prepared by the use of methacrylates **54c** and **54d** [132].

GTP of a series of α,β-unsaturated esters with extended conjugated unsaturation was studied in some detail by Hertler et al. [78] In the absence of an

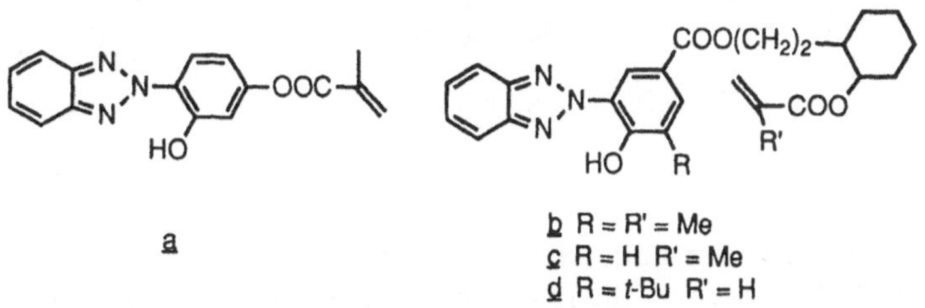

(53)

$$
\begin{array}{ll}
\underline{a} \ R = H & \underline{c} \ n = 3 \ R = Et \\
\underline{b} \ R = Me & \underline{d} \ n = 8 \ R = Me
\end{array}
\tag{54}
$$

ω-substituent, these polyenoates are very reactive in anion-catalyzed GTP, and propagation rates far exceed those of methacrylates, just as the related polyunsaturated initiators are much more reactive in GTP than are silyl ketene acetals (vide supra). All of the polyenoate monomers 55a–e undergo anion-catalyzed GTP with 1,4-regiospecificity to give polymers 55f–j Eq. (55). There is no evidence for any 1,2-addition during polymerization. In the case of the 2- methylpentadienoates 55b and 55c, this 1,4-regiospecificity contrasts with the lack of regiospecificity for the addition of the related 2-methylsilyldienolates 27a to MMA Eq. (27), and their anion-catalyzed reaction with other electrophiles [144]. While the regiospecificity is high, stereospecificity in anion-catalyzed GTP of the polyenoates is low with respect to geometry of the double bond, diastereoselectivity, and tacticity. Thus, both poly(ethyl sorbate) 55i and poly(diethyl muconate) 55j prepared by GTP have 75% trans and 25% cis backbone double bonds. Anionic polymerization of sorbates is known to form pure trans polymer [145]. Moreover, the diastereoselectivity in GTP of ethyl sorbate 55d and diethyl muconate 55e is the same, producing 2:1 erythro:threo 55i and 2:1 meso:racemic 55j. The tacticity appears to be irregular, in contrast to the diisotacticity in anionic poly(sorbates) [145]. The lower stereoregularity in the polydienoates prepared by GTP when compared with anionic polymerization is reflected in the low T_g of the polymers prepared by GTP. Anion-catalyzed GTP of the lactonetrienoate 56a gives exclusively 1,6-addition to form 56b with two all trans double bonds in the backbone. There is no diastereoselectivity in formation of 56b, however, and the erythro:threo ratio is nearly 1:1 [78].

$$
\begin{array}{ll}
\underline{a} \ R = R' = H \ R'' = Me & \underline{f} \ R = R' = H \ R'' = Me \ (T_g \ 2°) \\
\underline{b} \ R = H \ R' = R'' = Me & \underline{g} \ R = H \ R' = R'' = Me \ (T_g \ 3°) \\
\underline{c} \ R = H \ R' = Me \ R'' = Et & \underline{h} \ R = H \ R' = Me \ R'' = Et \ (T_g \ -27°) \\
\underline{d} \ R = Me \ R' = H \ R'' = Et & \underline{i} \ R = Me \ R' = H \ R'' = Et \ (T_g \ 7°) \\
\underline{e} \ R = EtOOC \ R' = H \ R'' = Et & \underline{j} \ R = EtOOC \ R' = H \ R'' = Et \ (T_g \ 15°)
\end{array}
\tag{55}
$$

$$
\underline{b} \ (T_g \ 150°)
\tag{56}
$$

Silyl ketene acetal 1a, the most broadly used initiator for anion-catalyzed GTP of methacrylates, served quite poorly in initiation of 55a–c and 56a with respect to control of M_n because the rate of propagation of the polyunsaturated monomers is much greater than the rate of initiation. However, when the silyldienolate **27a** was used for initiation of **55a-c**, there was much better control of M_n [78]. The inherent rate of polymerization of ethyl sorbate **55d** and diethyl muconate **55e** is slower than **55a-c** because of the ω-substitution, and initiation of GTP of **55d** with silyl ketene acetal **1a** resulted in good control of M_n.

2.5.2
Monomers with Protected Functional Groups

As with protected functional GTP initiators, the common functional groups which must be protected in monomers to be used in GTP are phenol, alcohol, and carboxylic acid. Other functional groups which are not compatible with GTP are unhindered primary and secondary amine, primary and secondary amide, aldehyde, ketone, and nitro group. The trimethylsilyl group is commonly used for protection of functionality in monomers just as it is with functional initiators. Thus, 2-trimethylsiloxyethyl methacrylate **57a** is the masked equivalent of 2-hydroxyethyl methacrylate [21, 88, 133, 146]. The resulting polymer is deprotected by heating with methanol and acid catalyst or tetrabutylammonium fluoride catalyst. 2-Trimethylsiloxyethyl acrylate **57b** has been used as a masked 2-hydroxyethyl acrylate [137]. Several research groups have used *tert*-butyl methacrylate **57c** [42, 87, 147–150] and *tert*-butyl acrylate **57d** [147] in GTP to obtain polymers with carboxylic acid functionality. The protecting group may be removed from the polymer with dilute HCl in isopropanol at 85 °C. Benzyl methacrylate **57e** undergoes GTP without the sluggishness of *tert*-butyl methacrylate, and the resulting polymer is deprotected by catalytic hydrogenolysis [132, 149, 151–153]. Trimethylsilyl methacrylate **57f** undergoes anion-catalyzed GTP, but because both the monomer and the polymer are, in fact, livingness enhancing agents (vide supra) the polymerization runs slowly and requires larger than normal amounts of catalyst [42, 86, 129–132, 154, 155]. Tetrahydropyranyl methacrylate **57g** [86, 131, 151–153, 155–157], tetrahydropyranyl acrylate **57h** [156], tetrahydrofuranyl methacrylate **58a** [153, 157, 158], and tetrahydrofuranyl acrylate **58b** [157] behave well in GTP, and the corresponding polymers can be deprotected with acid catalyst, or they may be heated neat for a few hours in a vacuum oven at 140 °C. A phenolic monomer, 4-tetrahydropyranyloxybenzyl methacrylate **58c**, has also been protected with a tetrahydropyranyl group for polymerization by GTP [152, 153].

$$
\begin{array}{ccc}
\text{COO(CH}_2)_2\text{OSiMe}_3 & \text{COOR'} & \text{COO} \\
\text{R} & \text{R} & \text{R}
\end{array}
\tag{57}
$$

a R = Me
b R = H

c R = Me R' = *t*-C$_4$H$_9$
d R = H   R' = *t*-C$_4$H$_9$
e R = Me R' = benzyl
f R = Me R' = SiMe$_3$

g R = Me
h R = H

(58)

a R = Me
b R = H   c

2.5.3
Other Monomers

Acrylonitrile undergoes rapid polymerization when silyl ketene acetal 1a is added to a solution of monomer and TASHF$_2$ in DMF at -50 °C. The MWD of the resulting poly(acrylonitrile) is broad (3.8) and M$_n$ is somewhat lower than theory [21]. This procedure avoids the gel-formation that occurs when monomer or catalyst is added to the other components due to the much greater rate of polymerization than of initiation. Hertler et al. found that initiation of GTP of acrylonitrile in DMF at -50 °C with the insoluble crosslinked polystyrene-supported silyl ketene acetal 34c resulted in 100% conversion to poly(acrylonitrile) all of which was grafted to the insoluble initiator. No poly(acrylonitrile) was found in solution. When the same experiment was performed at 25 °C, conversion to polymer was incomplete, and 42% of the poly(acrylonitrile) was isolated from solution while the remainder was grafted to the initiator [99]. This was interpreted as evidence that chain transfer to monomer, which is important at 25 °C, is suppressed at -50 °C.

Bandermann and Witkowski [159] studied GTP of methacrylonitrile in acetonitrile in some detail using α-trimethylsilyl-*iso*-butyronitrile 42a as initiator and cyanide, fluoride, or TASF as catalyst. The resulting polymers had M$_n$ much lower than theory, and only the last catalyst did not cause initiation of polymerization in the absence of initiator 42a. Conversion to polymer was often less than quantitative. Side reactions were credited for the results. Catalgil and Jenkins [160] determined the reactivity ratios for copolymerization of methacrylonitrile with MMA and with ethyl methacrylate (EMA) in THF with initiation by silyl ketene acetal 1a and catalysis by TASHF$_2$. The results showed methacrylonitrile (MAN) to be far more reactive in GTP than the methacrylates, with $r_{MMA} = 0.12$, $r_{MAN} = 2.69$ and $r_{EMA} = 0.16$, $r_{MAN} = 3.75$.

N,N-Dimethylacrylamide was polymerized by anion-catalyzed GTP in THF using silyl ketene acetal 1a, [21] or ethyl trimethylsilylacetate, as initiator [2]. Smooth polymerization was also obtained by initiation with 1a in acetonitrile solution with catalysis by mercuric iodide [67]. In all cases the M$_n$ as determined by GPC was lower than theory, but vapor phase osmometry gave values nearer to theory [21]. GTP of phenyl vinyl sulfone initiated by the polymer-supported silyl ketene acetal 34c was reported by Boettcher et al. [57, 58].

Quirk and Bidinger [161] used GTP to prepare an interesting hybrid polymer of propylene sulfide and methyl acrylate Eqs. (59, 60) [161]. The trimethylsilyl thioether 45f initiated ring-opening polymerization of propylene sulfide 59a in the presence of TASHF$_2$ catalyst to give poly(propylene sulfide **59b**. The living poly(propylene sulfide) **59b**, upon treatment with methyl acrylate and additional TASHF$_2$ catalyst, initiated GTP of the acrylate to give the hybrid polymer, poly(propylene sulfide)-*block*-poly(methyl acrylate) **60a**. Alternatively, the living poly(propylene sulfide) **59b** could be functionalized at the terminus by reaction with allyl bromide. Hovestadt et al. [162, 163] synthesized hybrid block copolymers from MMA and the cyclic carbonate, 2, 2-dimethyltrimethylene carbonate **61b**, by first using GTP to polymerize MMA with silyl ketene acetal **1a** as initiator and tetrabutylammonium cyanide as catalyst to give the living PMMA **61a**. Then, one equivalent of TASF and cyclic carbonate **61b** were added. The fluorosiliconate desilylated the silyl ketene acetal end group of the PMMA to form the ester enolate **5c** and fluorotrimethylsilane **5d**. Polymerization of **61b** was initiated by the ester enolate to give PMMA-*block*-poly(2,2-dimethyl-trimethylene carbonate) **61c**.

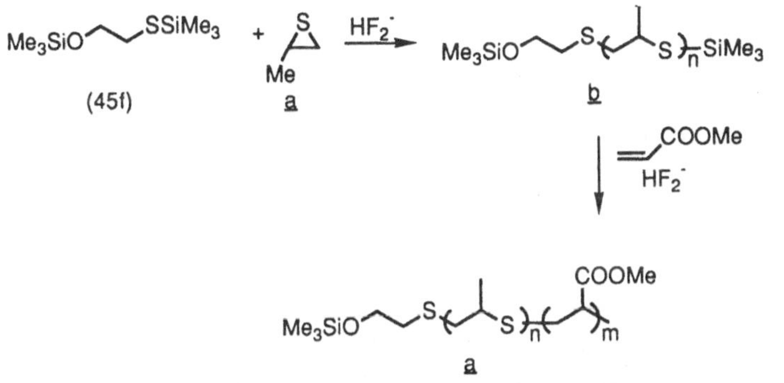

$$(59)$$

$$(60)$$

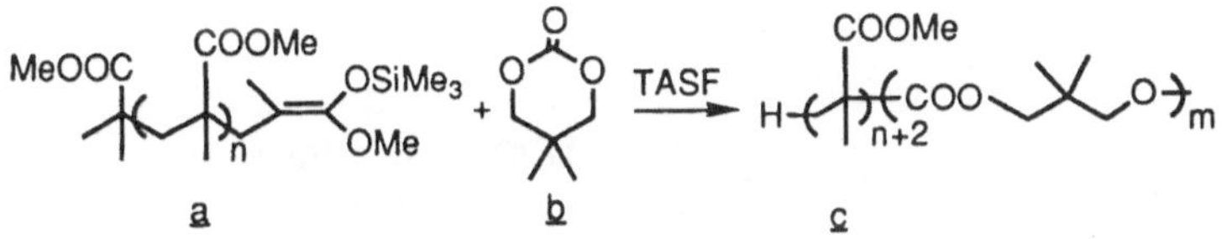

$$(61)$$

2.5.4
Difunctional Monomers

Difunctional monomers have provided an especially fertile research area of GTP. Beside crosslinked gels, three distinct types of polymer architectures have been reported based on difunctional monomers, cyclopolymers, star polymers, and ladder polymers. When the spatial arrangement of the two polymeriz-able functional groups is favorable, efficient cyclopolymerization can occur. Several research groups have demonstrated cyclopolymerization of a variety of monomer types by GTP. Kim et al. [164] found that GTP of

2,5-bis(methoxycarbonyl)-1,5-hexadiene **62a** initiated by silyl ketene acetal **1a** and catalyzed by bibenzoate or TASHF$_2$ gave exclusively the 5-membered ring cyclopolymer **62b** resulting from alternating intramolecular- and intermolecular polymerization Eq. (62). Significantly, these workers found that no polymer was obtained when anionic polymerization initiated by 9-fluorenyllithium at −78 °C was attempted. Kim et al. [89] also found that GTP of 2,6-bis(methoxycarbonyl)-1,6-heptadiene **63a** gave the 6-membered ring cyclopolymer **63b**. A very interesting ABA triblock polymer was prepared by using the difunctional initiator **31c** to prepare the center B block with 2,6-bis-(methoxycarbonyl)-1,6-heptadiene **63a**. Then n-butyl methacrylate was added to form the A blocks.

Carothers and Mathias [74] showed that n-butyl α-hydroxymethylacrylate ether **64a** underwent clean cyclopolymerization by anion-catalyzed GTP to give the polymeric tetrahydropyran **64b**. Choi et al. [165] found that bis(2-carbomethoxyallyl)methylamine **65a** could be converted to the polymeric piperidine **65b** by GTP. An ABA triblock polymer was synthesized by using the difunctional initiator **31c** to form the center B block with **65a**, followed by n-butyl methacrylate to form the two A blocks. Kozakiewicz et al. [166] used GTP to cyclopolymerize N-phenyldimethacrylamide **66a** to the polymeric 6-membered ring imide **66b**. In contrast, free radical polymerization gave the polymeric 5-membered ring imide.

$$\text{MeOOC} \overset{\|}{-}(CH_2)_2 \overset{\|}{-} \text{COOMe} \quad \xrightarrow{\text{GTP}} \quad \underset{\mathbf{b}}{\text{(cyclopentane ring structure)}} \qquad (62)$$

with **a** below the left structure.

$$\text{MeOOC} \overset{\|}{-}(CH_2)_3 \overset{\|}{-} \text{COOMe} \quad \xrightarrow{\text{GTP}} \quad \underset{\mathbf{b}}{\text{(cyclohexane ring structure)}} \qquad (63)$$

$$\text{BuOOC} \diagdown O \diagup \text{COOBu} \quad \xrightarrow{\text{GTP}} \quad \underset{\mathbf{b}}{\text{(tetrahydropyran ring structure)}} \qquad (64)$$

$$\text{MeOOC} \overset{\|}{-}CH_2\overset{Me}{\underset{N}{N}}CH_2\overset{\|}{-}\text{COOMe} \quad \xrightarrow{\text{GTP}} \quad \underset{\mathbf{b}}{\text{(piperidine ring structure, N-Me)}} \qquad (65)$$

$$\text{EtOOCCH}_2\text{SiMe}_3 \; + \; \text{(N-phenyl bis-methacrylamide)} \quad \xrightarrow{\text{GTP}} \quad \underset{\mathbf{b}}{\text{(6-membered ring imide, } C_6H_5)} \qquad (66)$$

A second, important class of polymers prepared from difunctional monomers is that of star-shaped polymers. These polymers can be formed by polymerizing a monofunctional methacrylate by GTP, and then adding a second, difunctional monomer which crosslinks the linear polymers ("arms") into a star-shaped polymer in which the "core" of the star is formed from the difunctional monomer [126, 146, 167, 168]. Ethylene dimethacrylate is the most common difunctional monomer used in star polymer synthesis by GTP, but tetraethylene glycol dimethacrylate has also been used [126]. Lang et al. [169, 170] have characterized these polymers and pointed out that they differ from common regular star-branched polymers in that the core is composed of a densely crosslinked microgel, which has a certain size distribution. Since the arms have a narrow MWD [146, 167], the nucleus distribution produces a corresponding distribution of the number of arms. Star polymer-formation has also been claimed when the difunctional monomer is polymerized first, followed by the monofunctional methacrylate. These core-first stars are much larger in size than those prepared arms-first [126].

Hutchins and Spinelli have prepared star polymers by synthesis of AB diblock polymers in which one of the blocks contains functional groups which can be crosslinked to form the core material of the star polymer. [171–173] Thus, the epoxy groups of PMMA-*block*-poly(glycidyl methacrylate) prepared by GTP were crosslinked with a diamine or with trifluoroacetic acid to form star polymers with crosslinked epoxy cores. Similarly, PMMA-*block*-poly(2-hydroxyethyl methacrylate) prepared by GTP was crosslinked with a diisocyanate to form star polymer.

Sogah [92–94] has reported that ladder polymers can be synthesized from difunctional methacrylates (with sufficiently long spacers between the functional groups) by GTP under conditions of high dilution, provided that the difunctional silyl ketene acetal 32b is used Eq. (67). These remarkable polymers are soluble in common solvents, have no discernible T_g, tenaciously retain solvent, and have high decomposition temperatures. Among other difunctional monomers which have been used are the binaphthyl

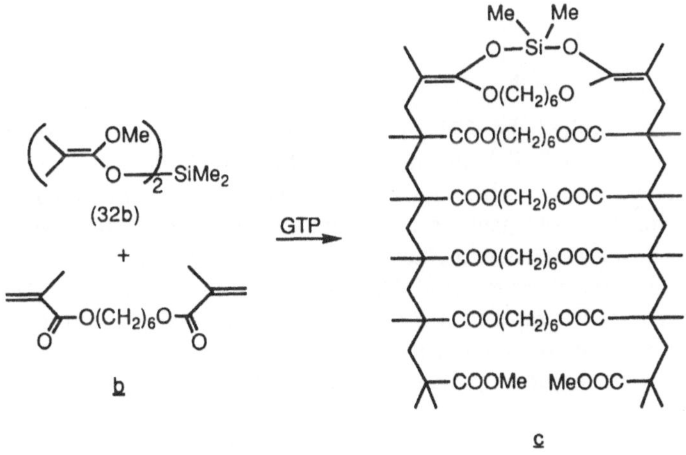

(67)

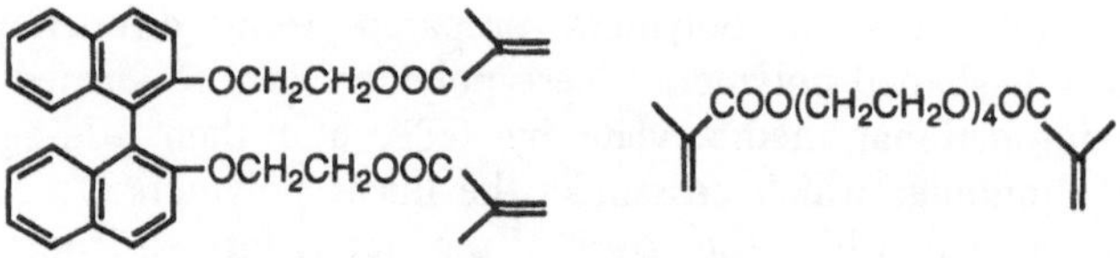

$$\text{(68)}$$

a A = O n = 1
b A = CH$_2$ n = 2

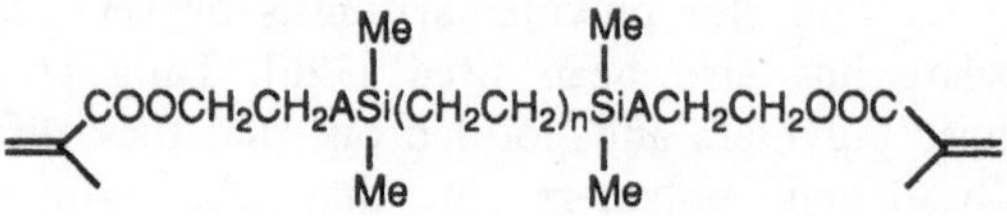

$$\text{(69)}$$

dimethacrylate **68a**, diethylene glycol dimethacrylate, tetraethylene glycol dimethacrylate **68b**, and the silyl dimethacrylates **69a** and **69b** [93].

2.5.5
Macromonomers

Macromonomers which can undergo GTP have been synthesized by several groups [174]. Cohen [175,176] has worked out procedures for efficiently converting the terminal trimethylsilyl ether group of a poly(methacrylate) **70a** prepared by GTP initiated by the silyl ketene acetal **22a** (which contains a hydroxyl group protected as a trimethylsilyl ether) to a methacrylate group, Eq. (70). The preferred reagent for this is methacrylyl fluoride **70b**. DeSimone et al. [177] prepared similar macromonomers by first deprotecting the trimethylsilyl ether-protected polymer **70a**, and then esterifying the hydroxyl group with methacrylyl chloride to obtain macromonomer **70c**. Radke and Müller [178] evaluated several procedures for preparing macromonomer **70c**. Sheridan and McGrath [179] prepared the corresponding acrylate macromonomer by reaction of **70a** with acryloyl chloride.

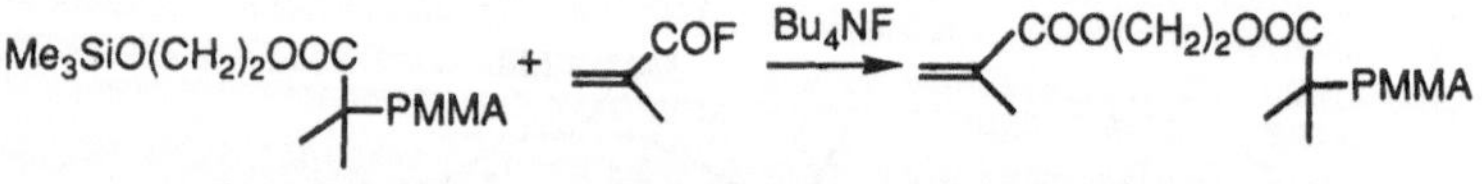

$$\text{(70)}$$

Asami et al. [180] studied the GTP of methacrylate-terminated polystyrene macromonomer **71a** using silyl ketene acetal **1a** and TASF as initiator and catalyst, respectively. Although poor results were obtained at 0 °C, oligomerization at −78 ° gave near quantitative oligomer of **71a** in which the DP was in good agreement with theory. The polymerizability of **71a** was markedly enhanced in the presence of MMA comonomer. DeSimone et al. [181] and Hellstern et al. [182] prepared a methacrylate-terminated poly(dimethylsiloxane) **71b** and studied copolymerization of the macromonomer with MMA by GTP, radical, and anionic polymerization to form PMMA-*graft*-poly-(dimethylsiloxane).

$$\text{COOCH}_2\text{CH}_2\text{-PStyr} \qquad \text{COO(CH}_2)_3\text{Me}_2\text{Si-PDMS} \qquad (71)$$

a   b

2.6
Termination

Termination of GTP may occur in many ways. GTP may be deliberately terminated, or quenched, by addition of a protonic compound, such as methanol, to remove the trimethylsilyl group from the living end. Or, termination may be an undesired side reaction which serves to limit the ultimate DP or MWD which can be attained by GTP. Termination may be a synthetic tool by which reagents are used to react with the living end of the polymer to cause coupling of chains, functionalization of chains, or transfer of livingness to begin a new chain (chain transfer).

Brittain and Dicker et al. [53, 115, 183] have carried out detailed studies on the major pathway for termination in GTP of MMA. They found that this process is a back-biting cyclization reaction which involves attack of the terminal ester enolate 72a on the penultimate methoxycarbonyl group with displacement of methoxide and formation of a ketone 72b. This cyclization only occurs when catalyst is present. The same termination process, leading to formation of a cyclohexanone, is characteristic of anionic polymerization of methacrylates. In kinetics studies of the catalyzed termination reaction of 72a (n = 3), Brittain and Dicker found that bifluoride catalyst was much more effective than bibenzoate in promoting termination, as well as in catalyzing polymerization. They found that the rate of cyclization of the oligomer 72a where n = 3 is greater than the rate of cyclization of all the higher oligomers investigated. Interestingly, with time they observed depolymerization of the higher oligomers 72a (n > 3) with a disproportionate amount of cyclic 72b (n = 3) formed. An important finding was the fact that, with both catalysts studied, the rate of termination is much lower than the rate of polymerization.

In the anion-catalyzed GTP of acrylates, a slow termination process leads to broadening of the MWD. Sogah et al. [21] have presented evidence that, due to the presence of hydrogen atoms on the backbone α- to the ester carbonyl group in poly(acrylates), silicon-migration occurs from the chain end to internal positions under the influence of the catalyst. The resulting internal silyl groups are too unreactive (possibly because they are C-silyl groups) to continue propagating. O-silyl to C-silyl isomerization of silyl ketene acetal 33d, a model of the propagating end group in GTP of acrylates, was shown to be catalyzed by anions [21, 49] and by mercuric iodide [114]. The C-silyl isomer is a much less reactive participant in GTP than the corresponding O-silyl isomer.

An important method for synthesis of acrylic polymers with a terminal functional group is to allow the silyl ketene acetal end group to react with a functionalized end-capping reagent. Thus, Quirk and Ren [184, 185] found that α-phenylacrylates do not homopolymerize under anion-catalyzed GTP

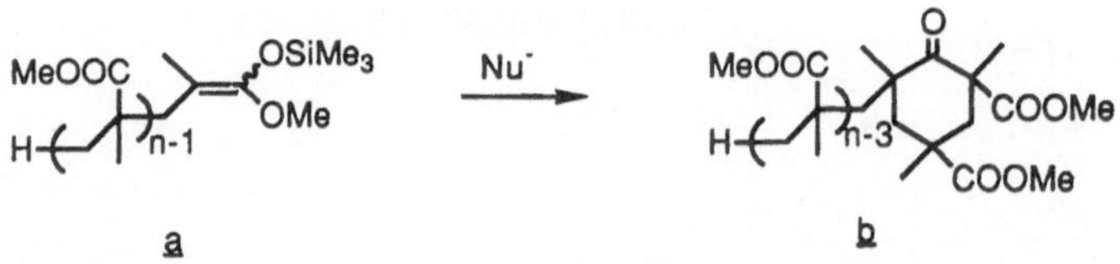

$$(72)$$

conditions, but that they are able to react with the silyl ketene acetal group at the end of a poly(methacrylate) prepared by GTP. They showed that the aromatic amine-containing α-phenylacrylate end-capper **73a** reacts efficiently with PMMA prepared by GTP in the presence of bifluoride catalyst to give very cleanly an aromatic amine-ended PMMA, Eq. (74). In similar fashion, they used the difunctional α-phenylacrylate **73b** to efficiently couple two chains of PMMA prepared by GTP. Diethyl vinylphosphonate **73c**, which was also found to resist homopolymerization under GTP conditions while remaining reactive toward silyl ketene acetals, was used to end-cap PMMA prepared by GTP to give a terminal diethyl phosphonate group [186]. The terminal phosphonic ester was converted to a phosphonic acid with bromotrimethylsilane followed by hydrolysis.

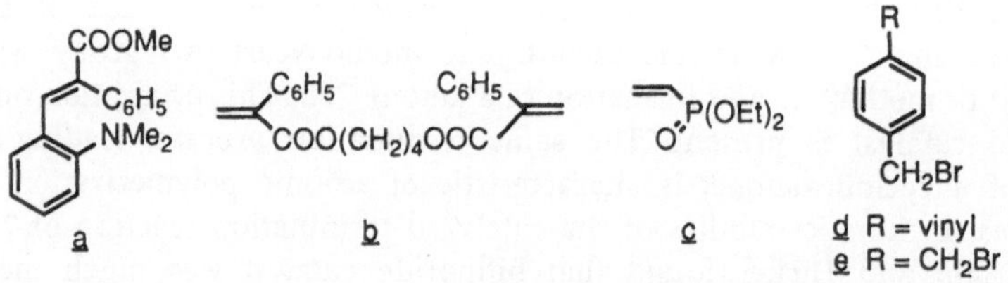

$$(73)$$

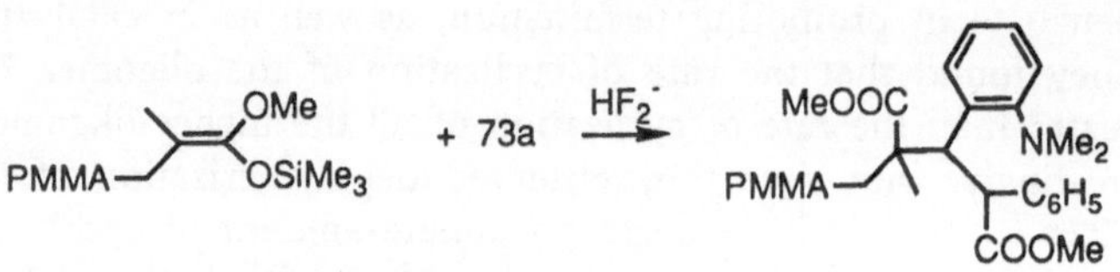

$$(74)$$

Asami et al. [81,187] used p-vinylbenzyl bromide **73d** to end-cap PMMA prepared by GTP, to give a macromonomer. For the end-capping reaction, stoichiometric TASF catalyst was required, and the best functionality achieved was 0.83. Sogah and Webster [76] used p-xylylenedibromide **73e** with stoichiometric amounts of the same catalyst to couple two chains of PMMA. This led to the telechelic polymers, α,ω-dihydroxy-PMMA and α,ω-dicarboxy-PMMA, when the protected initiators **22a** and **21a**, respectively, were used. High yields of functionality were claimed. These authors also reported very efficient coupling of the silyl ketene acetal end groups of PMMA with a combination of bromine and TiCl4 to produce a head-to-head linkage in the center of the PMMA chain where coupling occurred [76]. Banerjee and Hogen-Esch [43] used methyl iodide-^{13}C along with stoichiometric TASF to methylate PMMA prepared by GTP in order to determine chain end stereochemistry by NMR. Chou and Niu used an interesting method of chain-extending living PMMA. After MMA was polymerized by GTP using the difunctional initiator **32a**, p-xylylene-dibromide **73b** was added to couple the doubly-living-ended PMMA, resulting in a large increase in molecular weight [46].

Cohen reported a detailed study of reagents and conditions for end-capping and coupling living PMMA prepared by GTP [176]. Benzoyl fluoride **75a**, in the presence of 0.02–0.05 mol% of tetrabutylammonium biacetate catalyst, reacted with the silyl ketene acetal end group of PMMA to give a high yield of the C-acylated polymer **75b**. When acetyl fluoride, phenyl benzoate, or benzoic anhydride were used instead of **75a**, similarly high yields of acylated polymer were obtained. Efficient coupling of living PMMA was achieved with the analogous difunctional reagents, terephthaloyl fluoride and diphenyl tere-phthalate.

$$(75)$$

A special kind of termination, involving protonation-silicon exchange between carbon acids with pKa values 18–25 (measured in DMSO) and the silyl ketene acetal end group of a polymer undergoing GTP, is termed chain transfer when the silylated carbon acid is able to interact with monomer to initiate GTP, forming a new polymer chain. A variety of such carbon acids was found to exhibit a range of chain transfer constants in GTP in the presence of anion catalysts ranging in activity from bifluoride to m-chlorobenzoate and bibenzoate [188-190]. One of the most efficient chain transfer agents for GTP was reported to be 2-phenylpropionitrile **76a**, which in the presence of an anion catalyst for GTP, efficiently terminated a living PMMA chain by transferring a proton to the silyl ketene acetal, becoming silylated in the process to give the N-silylketeneimine **76c** and the terminated PMMA with an isobutyrate-like end group **76b**. Since the N-trimethylsilylketeneimine **76c** is an efficient initiator of GTP, a new polymer chain was initiated, probably at a rate of reaction which is greater than the rate of propagation of the newly formed silyl ketene acetal **77a** Eq. (77). Other chain transfer agents which were described are benzyl cyanide, methyl phenylacetate, methyl 2-phenylpropionate, indene, fluorene and γ-thiobutyrolactone. Concerning the mechanism of the proton-silicon exchange process involved in chain transfer, no experimental data has been reported. Hertler et al. have used as a chain transfer agent for GTP of MMA, a benzyl cyanide group bonded to an insoluble, crosslinked polystyrene support [99]. This resulted in part of the resulting PMMA being grafted to the support, and part being soluble in the polymerization medium.

$$(76)$$

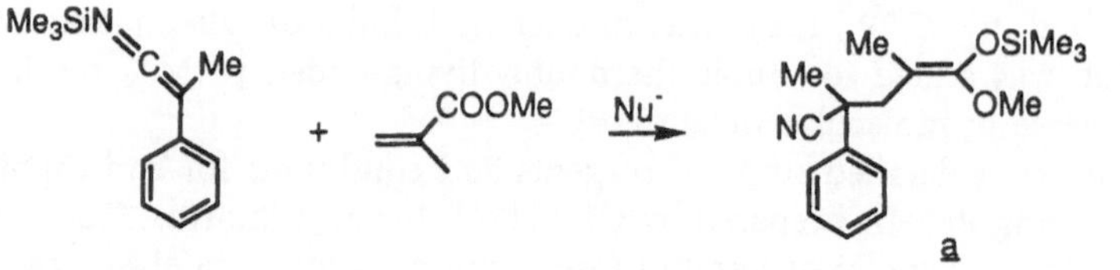

(77)

2.7
Aldol-GTP

Sogah and Webster [6–8, 191] applied the silyl aldol condensation reaction to the formation of polymers by the sequential addition of silyl vinyl ethers to aldehydes, Eq. (78). They termed this process aldol-GTP, since the silyl group of the silyl vinyl ether monomer 78b,c transfers to the carbonyl oxygen of the aldehyde initiator 78a or polymer end group 78f,g as the carbon-carbon bond forms. While the silyl ketene acetal end group of an acrylic polymer in GTP is nucleophilic in nature (reacting with electrophilic monomers), the aldehyde end group of a poly(silyl vinyl ether) in aldol-GTP is electrophilic in nature (reacting with a nucleophilic monomer). Although nucleophilic anions as well as Lewis acids catalyze aldol-GTP, Lewis acids are much preferred. The catalyst, typically a zinc halide, may be used at levels of 0.0001–0.01 mol% relative to monomer in chlorinated alkane or aromatic hydrocarbon solvents. The catalyst activates the aldehyde group so that it can react with the silyl vinyl ether [8]. Better control of molecular weight was obtained with the more hydrolytically stable *tert*-butyldimethylsilyl group 78b than with a trimethylsilyl group 78c in the monomer. Good control of M_n and narrow MWD were achieved. Catalyzed hydrolysis of the silyl groups in the polymer gave hydroxyl groups, i.e. poly-(vinyl alcohol). Besides aldehydes, benzyl halides and acetals were used as initiators for aldol-GTP. Initiation of aldol-GTP with *p*-xylylenedibromide resulted in the poly(silyl vinyl ether) growing in two directions [6]. Penelle and Verraver [192] reported an interesting instance in which methyl 2-trimethylsiloxyacrylate may undergo uncatalyzed aldol-GTP initiated by methyl pyruvate.

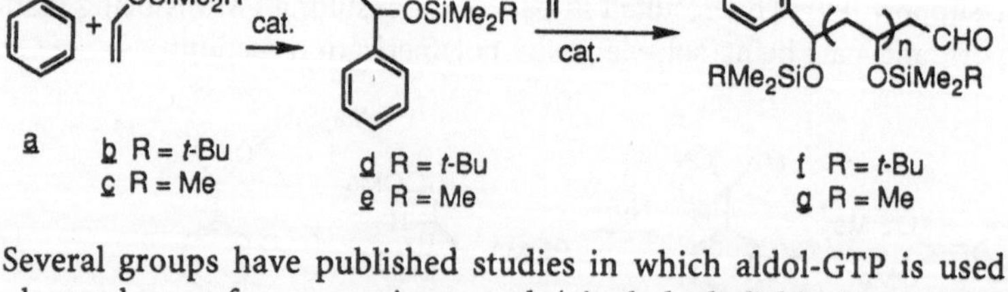

(78)

Several groups have published studies in which aldol-GTP is used to make block copolymers for conversion to poly(vinyl alcohol) block copolymers. Sogah and Webster [6] coupled poly(*tert*-butyldimethylsilyl vinyl ether) 79a prepared by aldol-GTP with PMMA **79b** prepared by GTP using TASHF$_2$ catalyst to obtain a diblock polymer **79c**, PMMA-*block*-poly(*tert*-butyldimethylsilyl vinyl ether), Eq. (79). Risse and Grubbs [193] used living ring-opening metathesis polymerization to prepare poly(norbornene) and poly(*exo*-dicyclopentadiene), which were then end-capped with excess terephthalaldehyde. The resulting aldehyde-ended hydrocarbon polymers 80a were then used to initiate zinc chloride-catalyzed aldol-GTP of *tert*-butyldimethylsilyl vinyl ether **78b** to obtain block copolymers 80b. The base-instability of **80b** was removed by reducing the terminal aldehyde group with sodium borohydride to give the terminal alcohol **81a**. Treatment with fluoride and methanol converted the silyl ethers to hydroxyl groups to obtain the desired hydrocarbon-poly(vinyl alcohol) block copolymers. Ruth et al. [194] used aldehyde-terminated poly(isobutylene) and aldehyde-terminated poly(styrene) to initiate aldol-GTP of *tert*-butyldimethylsilylvinyl ether with zinc bromide catalysis to prepare poly(isobutylene)-*block*-poly(*tert*-butyldimethylsilylvinyl ether) and poly(styrene)-*block*-poly(*tert*-butyldimethylsilylvinyl ether) with the objective of converting the copolymers to the corresponding poly(vinyl alcohol) copolymers. Macromonomers were prepared by initiation of aldol-GTP of *tert*-butyldimethylsilyl vinyl ether with *p*-vinylbenzaldehyde to obtain the poly(sily vinyl ether) with a terminal styryl group, which could then be used to prepare graft copolymers [195, 196].

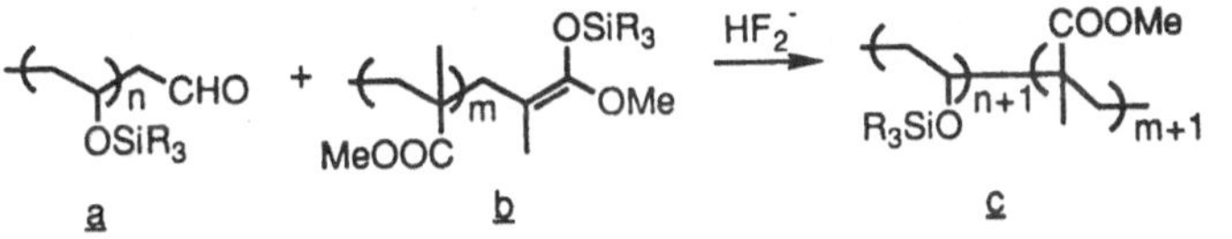

$$(79)$$

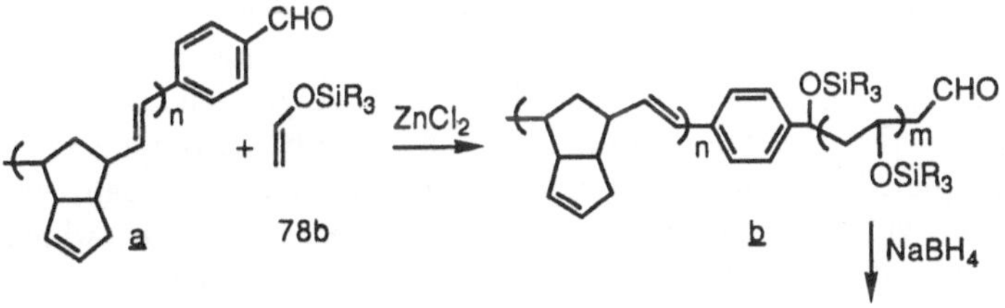

$$(80)$$

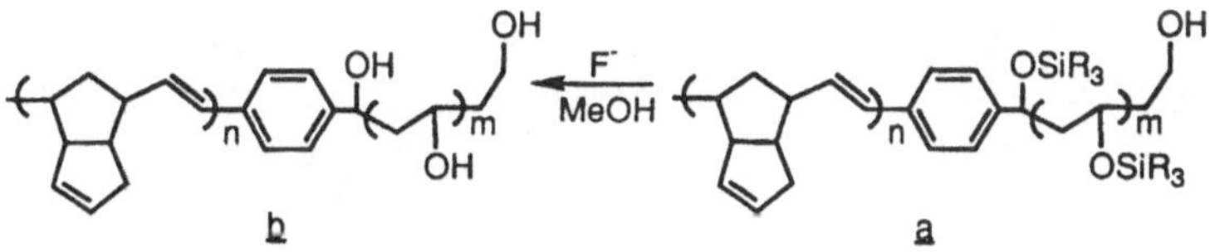

$$(81)$$

An interesting extension of aldol-GTP to silyldienolates was independently reported by Hirabayashi et al. [197,198] and Sogah and Harris [199–201]. Hirabayashi polymerized 1-trimethylsiloxybutadiene-1,3 **82a** with a variety of aromatic aldehyde initiators using one equivalent (on aldehyde) of zinc chloride catalyst and obtained the unsaturated polymer **82h** resulting exclusively from 1,4-addition, and with *trans* geometry. With benzaldehyde as initiator in benzene or toluene solvent at ambient temperature, the MWD ranged from 1.6–1.8 [197].

In a study of initiator effects, Hirabayashi found that electron-withdrawing substituents in the aromatic aldehyde increased the induction period and broadened the MWD. Initiation with *p*-anisaldehyde or cinnamaldehyde shortened the induction period and narrowed the MWD to about 1.3 [198]. Sogah and Harris [201] studied the effects of substituents on silicon in the silyldienolates 82b–g using zinc iodide catalyst and benzaldehyde initiator in dichloromethane, and obtained most satisfactory results with the sterically hindered *tert*-butyldimethylsiloxybutadiene **82d**. M_n was generally rather well controlled by the ratio of monomer to aldehyde, and MWD was in the range 1.5–2.8. The silyltrienolate **83a** underwent 1,6-aldol GTP to give the conjugated diene polymer **83b** [200, 201].

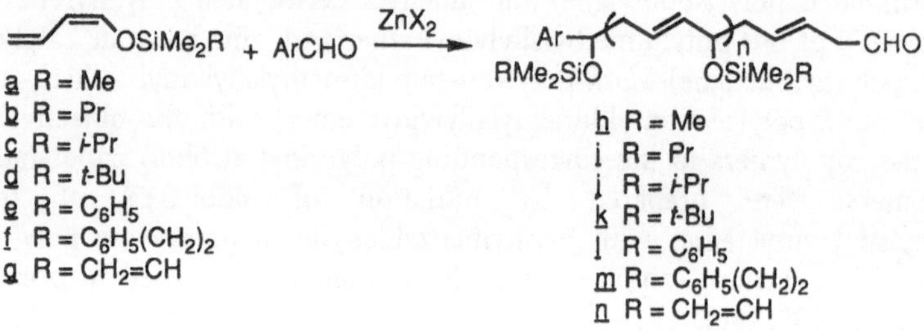

a R = Me	
b R = Pr	**h** R = Me
c R = *i*-Pr	**i** R = Pr
d R = *t*-Bu	**j** R = *i*-Pr
e R = C_6H_5	**k** R = *t*-Bu
f R = $C_6H_5(CH_2)_2$	**l** R = C_6H_5
g R = $CH_2=CH$	**m** R = $C_6H_5(CH_2)_2$
	n R = $CH_2=CH$

(82)

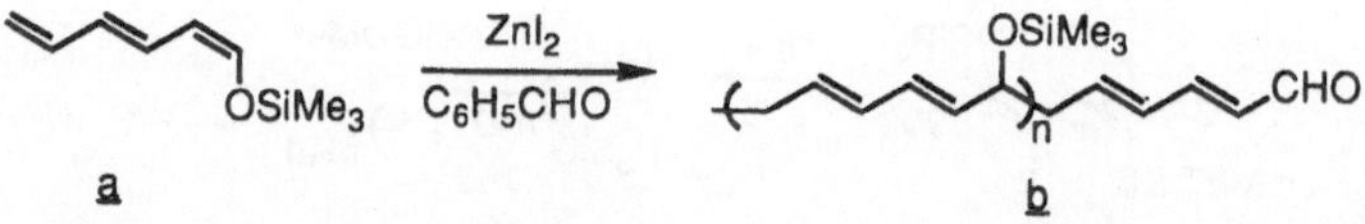

(83)

Sogah and Harris [201] used fluoride-catalyzed desilylation of polymers **82h–n** to convert them to the corresponding unsaturated polyols, which are formally alternating copolymers of acetylene and vinyl alcohol. Catalytic hydrogenation of the poly(siloxybutadienes) **84a** with tris(triphenylphosphine)rhodium chloride to remove the backbone unsaturation gave **84b**, which, upon desilylation with fluoride ion, gave the saturated polyol **84c**, which is formally a 1:1 alternating copolymer of vinyl alcohol and ethylene.

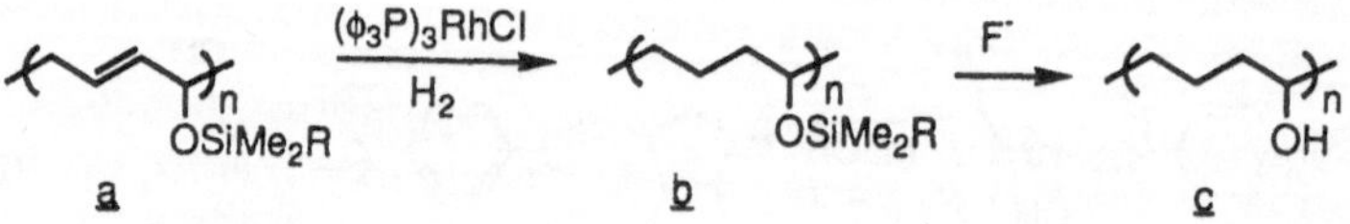

(84)

2.8
Conclusions

GTP has proven to be a powerful silicon-based methodology for the synthesis of living polymers of methacrylates, acrylates, silyl vinyl ethers, and several related monomers. It provides a practical way to prepare such polymers with controlled molecular weight, narrow MWD, and with as many variations in polymer architecture as the fertile imaginations of synthetic chemists are able to conjecture. It has also provided stimulus for mechanistic studies to better understand the complex series of equilibria involved in the polymer-forming process.

Abbreviations

TASF = tris(dimethylamino)sulfonium difluorotrimethylsiliconate
$TASHF_2$ = tris(dimethylamino)sulfonium bifluoerid
TPS = tris(piperidino)sulfonium
A = acrylate
MA = methacrylate
PMMA = poly(methyl methacrylate)
GTP = group-transfer polymerization

2.9
References

1. Narasaka K, Saigo K, Aikawa Y, Mukaiyama T (1976) Bull Chem Soc Jpn 49: 779
2. Webster OW, Hertler WR, Sogah DY, Farnham WB, RajanBabu TV (1983) J Am Chem Soc 105: 5706
3. Webster OW (1983) US Patent 4,417,034, to DuPont; (1984) Chem Abstr 100: 86327e
4. Matyjaszewski K (1993) Proc Am Chem Soc Div Polym Mater: Sci Eng 68: 58
5. Matyjaszewski K (1993) Macromolecules 26: 1787
6. Sogah DY, Webster OW (1986) Macromolecules 19: 1775
7. Sogah DY (1986) Polym Prepr, Am Chem Soc Div Polym Chem 27(1): 163
8. Sogah DY, Webster OW (1987) In: Fontanille M, Guyot A (eds) Recent advances in mechanistic and synthetic aspects of polymerization, Reidel, Dordrecht, Holland, p 61
9. Webster OW (1987) In: Kroschwitz JI (ed) Encyclopedia of Polymer Science and Engineering, vol 7, Wiley-Interscience, NY, p 580
10. Eastmond GC, Webster OW (1991) In: Ebdon JR (ed) New Methods of Polymer Synthesis, Blackie and Son, Ltd, Glasgow, p 22
11. Webster OW (1990) Makromol Chem, Macromol Symp 33: 133
12. Webster OW, Anderson BC (1992) In: Mijs WJ (ed) New Methods for Polymer Synthesis, Plenum, NY, p 1
13. Webster OW, Sogah DY (1987) In: Fontanille M, Guyot A (eds) Recent Advances in Mechanistic and Synthetic Aspects of Polymerization, D Reidel, Dordrecht, Holland, p 3
14. Brittain WJ (1992) Rubber Chem Technol 65: 580
15. Bywater S (1993) Makromol Chem, Macromol Symp 67: 339
16. Hertler WR In: Hatada K, Kitayama T, Vogl O (eds) Macromolecular Design of Polymeric Materials, M Dekker, NY (in press)
17. Müller MA, Augenstein M, Dumont E, Pennewiss H (1991) New Polym Mater 2: 315
18. Schubert W, Sitz H-D, Bandermann F (1989) Makromol Chem 190: 2193
19. Schmalbrock U, Bandermann F (1993) Makromol Chem 194: 2543
20. Martin DT, Bywater S (1992) Makromol Chem 193: 1011
21. Sogah DY, Hertler WR, Webster OW, Cohen GM (1987) Macromolecules 20: 1473
22. Farnham WB, Hertler WR (1988) US Patent 4,728,706, to DuPont; (1988) Chem Abstr 109: 55438y
23. Collins S, Ward DG (1992) J Am Chem Soc 114: 5460
24. Yasuda H, Yamamoto H, Yokota K, Miyake S, Nakamura A (1992) J Am Chem Soc 114: 4908
25. Sogah DY, Farnham WB (1985) In: Sakurai H (ed) Organosilicon and Bioorganosilicon Chemistry: Structure, Bonding, Reactivity and Synthetic Application, Ellis Horwood, Chichester, Chapter 20
26. Farnham WB, Sogah DY (1986) Polym Prepr Am Chem Soc Div Polym Chem 27(1): 167
27. Quirk RP, Bidinger GP (1989) Polym Bull 22: 63

28. Quirk RP, Ren J (1992) Macromolecules 25: 6612
29. Burggraf L, Davis LP (1987) In: Abstracts of Papers, 8th International Symposium on Organosilicon Chemistry, St Louis, MO, p 74
30. Webster OW (1994) J Macromol Sci Pure Appl Chem A31(8): 927
31. Brittain WJ (1988) J Am Chem Soc 110: 7440
32. Mai PM, Müller AHE (1987) Makromol Chem, Rapid Commun 8: 99 AH
33. Mai PM, Müller AHE (1987) Makromol Chem, Rapid Commun 8: 247
34. Müller AHE (1990) Makromol Chem, Macromol Symp 32: 87
35. Schubert W, Bandermann F (1989) Makromol Chem 190: 2721
36. Sitz HD, Speikamp HD, Bandermann F (1988) Makromol Chem 189: 429
37. Müller AHE (1994) Macromolecules 27: 1685
38. Dixon DA (1988) Polym Mater Sci Eng 58: 590
39. Müller MA, Stickler M (1986) Makromol Chem, Rapid Commun 7: 575
40. Konishi T, Tamai Y, Fujii M, Einaga Y, Yamakawa H (1989) Polym J 21: 329
41. Schmalbrock U, Sitz H-D, Bandermann F (1989) Makromol Chem 190: 2713
42. Wei Y, Wnek GE (1987) Polym Prepr, Am Chem Soc Div Polym Chem 28(1): 252
43. Banerjee KG, Hogen-Esch TE, (1987) Polym Prepr, Am Chem Soc Div Polym Chem 28(2): 320
44. Bannerjee KG, Hogen-Esch TE (1993) Macromolecules 26: 926
45. Jenkins AD, Tsartolia E, Walton DRM, Stejskal J, Kratochvil P (1988) Polym Bull 20: 97
46. Chou SSP, Niu CW (1987) MRL Bull Res Dev 1(2): 33
47. Sitz HD, Bandermann F (1987) NATO ASI Ser, Ser C, 215 (Recent Adv Mech Synth Aspects Polym): 41
48. Speikamp HD, Bandermann F (1988) Makromol Chem 189: 437
49. Schubert W, Bandermann F (1989) Makromol Chem 190: 2161 (1989)
50. Bandermann F, Sitz HD, Speikamp HD (1986) Polym Prepr, Am Chem Soc Div Polym Chem 27(1): 169
51. Dicker IB, Cohen GM, Farnham WB, Hertler WR, Laganis ED, Sogah DY (1990) Macromolecules 23: 4034
52. Dicker IB, Cohen GM, Farnham WB, Hertler WR, Laganis ED, Sogah DY (1987) Polym Prepr, Am Chem Soc Div Polym Chem 28(1): 106
53. Brittain WJ, Aquino EC, Dicker IB, Brunelle DJ (1993) Makromol Chem 194: 1249
54. Schneider LV, Dicker IB (1988) US Patent 4,736,003, to DuPont
55. Dicker IB, Hutchins CS, Spinelli HJ (1986) US Patent 4,622,372, to DuPont
56. Hertler WR (1994) Macromol Symp 88: 55
57. Boettcher FP, Dicker IB, Ebersole RC, Hertler WR (1991) US Patent 5,019,634, to DuPont; (1991) Chem Abstr 114: 24781d
58. Boettcher FP, Dicker IB, Ebersole RC, Hertler WR (1990) US Patent 4,940,760, to DuPont; (1991) Chem Abstr 114: 24781d
59. Hertler WR (1986) US Patent 4,605,716, to DuPont; (1986) Chem Abstr 105: 227555v
60. Hertler WR, Sogah DY, Webster OW, Trost BM (1984) Macromolecules 17: 1415
61. Liang L, Ren L, Ying S (1991) Hecheng Xiangjiao Gongye 14(5): 347
62. Xu L, Wang L (1989) Chin J Polym Sci 7(4): 299
63. Kawai F, Onaka M, Izumi T (1988) Bull Chem Soc Jpn 61: 2157
64. Kawai M, Onaka M, Izumi Y (1987) J Chem Soc, Chem Commun: 1203
65. Corbin DR, Sormani PME (1992) US Patent 5,162,467, to DuPont; (1993) Chem Abstr 118: 39608
66. Dicker IB (1988) Polym Prepr, Am Chem Soc Div Polym Chem 29(2): 114
67. Dicker IB (1988) US Patent 4,732,955, to DuPont; (1988) Chem Abstr 109: 38426a
68. Dicker IB (1989) US Patent 4,866,145, to DuPont
69. Zhuang R, Müller AHE (1993) In: Book of Abstracts, International Conference "Frontiers in Polymerization", Liege, Belgium, p 41
70. Sogah DY, Hertler WR, Dicker IB, DePra PA, Butera JR (1990) Makromol Chem, Macromol Symp 32: 75
71. Sogah DY (1990) U S Patent 4,957,973, to DuPont; (1991) Chem Abstr 114: 43757s
72. Ainsworth C, Chen F, Kuo YN (1972) J Organomet Chem 46: 59

73. Lukevics E, Belyakova ZV, Pomerantseva MG, Voronkov MG (1977) J Organomet Chem Library 5: 1
74. Carothers TW, Mathias LJ (1992) Polym Prepr, Am Chem Soc Div Polym Chem 33(2): 150
75. Dicker IB (1990) U S Patent 4,943,648, to DuPont; (1990) Chem Abstr 113: 212834b
76. Sogah DY, Webster OW (1983) J Polym Sci, Polym Lett Ed 21: 927
77. Heitz T, Webster OW (1991) Makromol Chem 192: 2463
78. Hertler WR, RajanBabu TV, Ovenall DW, Reddy GS, Sogah DY (1988) J Am Chem Soc 110: 5841
79. Dicker IB, Hertler WR (1991) US Patent 5,021,524, to DuPont; (1991) Chem Abstr 115: 136997
80. Witkowski R, Bandermann F (1989) Makromol Chem 190: 2173
81. Asami R, Kondo Y, Takaki M (1987) In: (Hogen-Esch TE, Smid J (eds)) Recent Adv Anionic Polym, Proc Int Symp, Elsevier, NY p 381
82. Asami R (1987) Jp Patent 62063595, to Toa Gosei Chemical Industry Co, Ltd
83. Asami R (1987) Jp Patent 62062801, to Toa Gosei Chemical Industry Co, Ltd
84. Spinelli HJ, Costello TD, Cohen GM (1988) Eur Patent Appl 248596, to DuPont; (1987) Chem Abstr 109: 7134z
85. Spinelli HJ (1991) US Patent 5,036,139, to DuPont
86. Ma SH, Hertler WR, Shor AC, Spinelli HJ (1993) Eur Patent Appl 556650, to DuPont; (1994) Chem Abstr 120: 33116s
87. Choi WJ, Kim YB, Kwon SK, Lim KT, Choi SK (1992) J Polym Sci, Polym Chem Ed 30: 2143
88. Choi WJ, Lim KT, Kwon SK, Choi SK (1993) Polym Bull 30: 401
89. Kim YB, Choi WJ, Choi BS, Choi SK (1991) Macromolecules 24: 5006
90. Steinbrecht K, Bandermann F (1989) Makromol Chem 190: 2183
91. Manring LE, Sogah DY, Cohen GM (1989) Macromolecules 22: 4652
92. Sogah DY (1988) Polym Prepr, Am Chem Soc Div Polym Chem 29(2): 3
93. Sogah DY (1990) US Patent 4,906,713, to DuPont; (1990) Chem Abstr 113: 60098z
94. Sogah DY (1991) Polym Prepr, Am Chem Soc Div Polym Chem 32(1): 307
95. Zou Y, Pan R (1988) Gaofenzi Xuebao (4): 301
96. Pan R, Xia H (1987) Xiamen Daxue Xuebao Ziran Kexueban 26(4): 474
97. Xia H, Zou Y, Pan R (1991) Gaofenzi Xuebao (2): 225
98. Xia H, Zou Y, Pan R (1991) Gaodeng Xuexiao Huaxue Xuebao 12(5): 699
99. Hertler WR, Sogah DY, Boettcher FP (1990) Macromolecules 23: 1264
100. Yang CY, Wnek GE (1989) Polym Prepr ACS Div Polym Chem 30(2): 177
101. Yang C-Y, Wnek GE (1990) Proc ACS Div Polym Mater Sci Eng 62: 601
102. Yang C-Y, Wnek GE (1992) Polymer 33: 4191
103. Wnek GE (1993) Polym Prepr, Am Chem Soc Div Polym Chem 34(2): 773
104. Zhu Z, Rider J, Yang CY, Gilmartin ME, Wnek GE (1992) Macromolecules 25: 7330
105. Hellstern AM, DeSimone JM, McGrath JE (1988) Polym Prepr, Am Chem Soc Div Polym Chem 29(1): 148 (1988)
106. Hrkach J, Ruehl K, Matyjaszewski K (1988) Report, Order No AD-A196645, 4 pp Avail NTIS From: Gov Rep Announce Index (U S) 1988, 88(23), Abstr No 857,741, "Grafting of living polymers from activated polysilanes"
107. Jenkins AD, Tsartolia E, Walton DRM, Horska-Jenkins J, Kratochvil P, Stejskal J (1990) Makromol Chem 191: 2511
108. Jenkins AD, Makromol Chem (1992) Macromol Symp 53: 267
109. Jenkins AD (1993) Makromol Chem, Macromol Symp 70/71: 67
110. Kondo Y, Takaki M, Oku J, Uemura K, Usui K, Asami R (1989) Kobunshi Ronbunshu 46(12): 789
111. Kondo Y, Kojima K, Shiraishi T, Takaki M, Asami R (1989) Kobunshi Ronbunshu 46(6): 367
112. Ballard DGH, Pickering A, Richards S, Runciman PJI (1988) Polym Prepr, Am Chem Soc Div Polym Chem 29(2): 35
113. Shen W, Jin H (1992) Makromol Chem 193: 743
114. Brittain WJ, Dicker IB (1993) Polym Intern 30: 101

115. Brittain WJ, Dicker IB (1993) Makromol Chem, Macromol Symp 67: 373
116. Citron JD, Simms JA (1988) U S Patent 4,771,117, to DuPont; (1989) Chem Abstr 110: 115514c
117. Zou Y, Xia H, Pan R (1990) Chin Chem Lett 1(3): 265
118. Zou Y, Pan R (1991) Gaofenzi Xuebao (4): 398
119. Reetz MT, Ostarek R, Piejko K-E, Arlt D, Bomer B (1986) Angew Chem Intl Ed Engl 25(12): 1108
120. Hertler WR, Webster OW (1984) In: Abstracts, 1984 International Chemical Congress of Pacific Basin Societies, Honolulu, Hawaii, p 09059
121. Bandermann F, Speikamp H-D (1985) Makromol Chem , Rapid Commun 6: 335
122. Webster OW, Hertler WR, Sogah DY, Farnham WB, RajanBabu TV (1984) J Macromol Sci, Chem A21(8 and 9): 943
123. Evans DA, Truesdale LK (1973) Tet Lett 1973: 4929
124. Dixon DA, Hertler WR, Chase DB, Farnham WB, Davidson F (1988) Inorg Chem 27: 4012
125. Hutchins CS, Shor AC (1987) U S Patent 4,656,226, to DuPont; (1987) Chem Abstr 107: 23910p
126. Spinelli HJ (1986) Int Patent Appl 8600626, to DuPont; (1986) Chem Abstr 105: 24811c
127. West NJW (1988) U S Patent 4,755,563, to DuPont; (1989) Chem Abstr 110: 39568f
128. Beadle PM, Rowan L, Mykytiuk J, Billingham NC, Armes SP (1993) Polymer 34: 1561
129. Patrickios CS, Jang CJ, Hertler WR, Hatton TA (1993) Polym Prepr, Am Chem Soc Div Polym Chem 34(1): 954, 2364
130. Patrickios CS, Gadam SD, Cramer SM, Hertler WR, Hatton TA (1993) Polym Prepr, Am Chem Soc Div Polym Chem 34(1): 1073
131. Dicker IB, Hertler WR, Ma S-H (1993) U S Patent 5,219,945, to DuPont; (1993) Chem Abstr 119: 250807b
132. Ma S-H, Dicker IB, Hertler WR (1993) Eur Patent Appl 556649, to DuPont; (1994) Chem Abstr 120: 79539k (1994)
133. Yu H-S, Choi W-J, Lim K-T, Choi S-K (1988) Macromolecules 21: 2893
134. Sogah DY, Hertler WR, Webster OW (1984) Polym Prepr, Am Chem Soc Div Polym Chem 25(2): 3
135. Catalgil-Giz H, Uyanik N, Erbil C (1992) Polymer 33: 655
136. Laschewsky A, Ward MD (1991) Polymer 32: 146
137. Epple U, Schneider HA (1990) Polymer 31: 961
138. Pugh C, Percec V (1985) Polym Bull 14(2): 109
139. Pugh C, Percec V (1985) Polym Prepr, Am Chem Soc Div Polym Chem 26(2): 303
140. Spinelli HJ (1991) Eur Patent Appl 420679, to DuPont; (1991) Chem Abstr 115: 115307r
141. Gomez PM, Neidlinger HH (1987) Polym Prepr , Am Chem Soc Div Polym Chem 28(1): 209
142. Slongo M, Rody J, Sitek F, Valet A (1988) Eur Pat Appl 293871, to Ciba-Geigy
143. Suenaga J, Sutherlin DM, Stille JK (1984) Macromolecules 17: 2913
144. Hertler WR, Reddy GS, Sogah DY (1988) J Org Chem 53: 3532
145. Farina M, Grassi M, DiSilvestro G, Zetta L (1985) Eur Polym J 21: 71
146. Simms JA (1991) Rubber Chem Technol 64: 139
147. Öner M, Calvert P (1993) Proc Am Chem Soc Div Polym Mater: Sci Eng 69: 166
148. Doherty MA, Müller AHE (1989) Makromol Chem 190: 527
149. Rannard SP, Billingham NC, Armes SP, Mykytiuk J (1993) Eur Polym J 29: 407
150. Gabor AH, Ober CK (1993) Polym Prepr, Am Chem Soc Div Polym Chem 34(2): 576
151. Taylor GN, Stillwagon LE, Houlihan FM, Wolf TM, Sogah DY, Hertler WR (1991) J Vac Sci Tech B 9: 3348
152. Taylor GN, Stillwagon LE, Houlihan FM, Wolf TM, Sogah DY, Hertler WR (1991) Chem Mater 3: 1031
153. Bauer RD, Chen GYY, Hertler WR, Wheland RC (1992) U S Patent 5,145,764, to DuPont
154. Ma S-H, Matrick H, Shor AC, Spinelli HJ (1992) U S Patent 5,085,698, to DuPont; (1992) Chem Abstr 116: 237559r
155. Patrickios CS, Hertler WR, Abbott NL, Hatton TA (1994) Macromolecules 27: 930
156. Raymond FA, Hertler WR (1992) J Imaging Sci Technol 36: 243

157. Simmons HE, III, Hertler WR, Sauer BB (1994) J Appl Polym Sci 52: 727
158. Hertler WR, Simmons HE, III (1993) U S Patent 5,229,244, to DuPont; (1994) Chem Abstr 120: 257549y
159. Bandermann F, Witkowski R (1986) Makromol Chem 187: 2691
160. Catalgil H, Jenkins AD (1991) Eur Polym J 27: 651
161. Quirk RP, Bidinger GP (1988) Polym Prepr, Am Chem Soc Div Polym Chem 29(2): 120
162. Hovestadt W, Keul H, Höcker H (1991) Makromol Chem 192: 1409
163. Höcker H, Keul H, Hovestadt W, Leitz E (1992) Eur Patent Appl 472079, to Bayer A-G; (1992) Chem Abstr 116: 256273j
164. Kim Y-B, Oh J-M, Choi W-J, Choi S-K (1993) Macromolecules 26: 2383
165. Choi W-J, Kim Y-B, Lim K-T, Choi S-K (1990) Macromolecules 23: 5365
166. Kozakiewicz J, Kurose NS, Draney DR, Huang SY, Falzone J (1987) Polym Prepr, Am Chem Soc Div Polym Chem 28(2): 347
167. Simms JA (1992) Polym Prepr, Am Chem Soc Div Polym Chem 33(1): 164
168. Simms JA, Spinelli HJ In: (Hatada K, Kitayama T, Vogl O (eds)) Macromolecular Design of Polymeric Materials, M Dekker, NY (in press)
169. Lang P, Burchard W, Wolfe MS, Spinelli HJ, Page L (1991) Macromolecules 24: 1306
170. Lang P, Burchard W (1987) Makromol Chem, Rapid Commun 8: 451
171. Hutchins CS, Spinelli HJ (1988) Eur Patent Appl 258065, to DuPont; (1988) Chem Abstr 109: 129894t
172. Hutchins CS, Spinelli HJ (1989) U S Patent 4,487,328, to DuPont
173. Hutchins CS, Spinelli HJ (1989) U S Patent 4,851,477, to DuPont
174. Spinelli HJ (1990) Adv Org Coat Sci Technol Ser 12: 34
175. Cohen GM (1989) U S Patent 4,845,156, to DuPont; (1990) Chem Abstr 112: 119613y
176. Cohen GM (1988) Polym Prepr, Am Chem Soc Div Polym Chem 29(2): 46
177. DeSimone JM, Hellstern AM, Siochi EJ, Ward TC, McGrath JE (1989) Polym Prepr, Am Chem Soc Div Polym Chem 30(1): 137
178. Radke W, Müller AHE (1992) Makromol Chem, Macromol Symp 54/55: 583
179. Sheridan MS, McGrath JE (1992) Makromol Chem, Macromol Symp 64: 85
180. Asami R, Takaki M, Moriyama Y (1986) Polym Bull 16: 125
181. DeSimone JM, Hellstern AM, Ward TC, McGrath JE, Smith SD, Gallagher PM, Krukonis VJ, Stejskal J, Strakova D, Kratochvil P (1988) Polym Prepr, Am Chem Soc Div Polym Chem 29(2): 116
182. Hellstern AM, Smith SD, McGrath JE (1987) Polym Prepr, Am Chem Soc Div Polym Chem 28(2): 328
183. Brittain WJ, Dicker IB (1989) Macromolecules 22: 1054
184. Quirk RP, Ren J (1992) Polym Prepr, Am Chem Soc Div Polym Chem 33(1): 978
185. Quirk RP, Ren J (1993) Polym Intern 32: 205
186. Hertler WR (1991) J Polym Sci, Polym Chem Ed 29: 869
187. Asami R, Kondo Y, Takaki M (1986) Polym Prepr, Am Chem Soc Div Polym Chem 27(1): 186
188. Hertler WR (1987) Macromolecules 20: 2976
189. Hertler WR (1987) Polym Prepr, Am Chem Soc Div Polym Chem 28(1): 108
190. Hertler WR, Sogah DY (1987) U S Patent 4,656,233, to DuPont
191. Sogah DY, Webster OW (1985) U S Patent 4,544,724, to DuPont; (1986) Chem Abstr 104: 89196e
192. Penelle J, Verraver S (1993) Makromol Chem, Rapid Commun 14: 563
193. Risse W, Grubbs RH (1989) Macromolecules 22: 1558
194. Ruth WG, Brittain WJ, Lubnin AV, Kuang J, Kennedy JP, Quirk RP (1993) Polym Prepr, Am Chem Soc Div Polym Chem 34(2): 584
195. Kawakamie Y, Aoki T, Yamashita Y (1987) Polym Bull 18: 473
196. Charleux B, Pichot C (1993) Polymer 34: 195
197. Hirabayashi T, Itoh T, Yokota K (1988) Polym J 20: 1041
198. Hirabayashi T, Kawasaki T, Yokota K (1990) Polym J 22: 287
199. Harris JF (1987) Eur Patent Appl 246053, to DuPont; (1988) Chem Abstr 109: 74129q
200. Harris JF (1988) U S Patent 4,739,021, to DuPont

201. Sogah DY, Harris JF (1991) Polym Prepr, Am Chem Soc Div Polym Chem 32(1): 435
202. Fleischmann G, Eck H (1991) U S Patent 5,003,015, to Wacker-Chemie
203. Augenstein M, Müller MA (1990) Makromol Chem 191: 2151
204. Shen W-P, Zhu W-D, Yang N-F, Wang L (1989) Makromol Chem 190: 3061
205. Kriz J, Masar B, Vlcek P (1993) Makromol Chem 194: 1435
206. Simms JA, Spinelli HJ (1987) J Coatings Technol 59: 125
207. Clever HA, Hall HK (1989) Polym Bull 21: 179
208. Saito A, Tirrell DA (1993) Polym Prepr, Am Chem Soc Div Polym Chem 34(1): 152
209. Hefft M, Springer J (1990) Makromol Chem, Rapid Commun 11: 397
210. Kreuder W, Webster OW (1986) Makromol Chem, Rapid Commun 7: 5
211. Hovestadt W, Müller AJ, Keul H, Höcker H (1990) Makromol Chem, Rapid Commun 11: 271

CHAPTER 3

Polysiloxanes and Polymers Containing Siloxane Groups

C. Burger and F.-H. Kreuzer

3.1
Polysiloxanes

3.1.1
Definition

Polysiloxanes or, to be more precise, poly (organosiloxane)s are polymers made up of the following structural units:

$R_2SiO_{1/2}$	R_2SiO	$RSiO_{3/2}$	$SiO_{4/2}$
M-	D-	T-	Q-unit
(mono-)	(di-)	(tri-)	(quadro-functional)

R is a monovalent organic alkyl and/or aryl group which can be substituted with functional groups X as necessary.

Linear (T and Q = 0) or branched (T and/or Q > 0) polymers are obtained [1], depending on how these units are combined.

Linear polyorganosiloxanes, i.e. those consisting only of D or DM units, are referred to in the literature as silicone fluids [2] whilst those consisting of T and/or Q-units, also combined with M- and/or D-units, are called silicone resins [3]. The M-function here serves as a polymer chain terminator.

3.1.2
Synthesis of Polydiorganosiloxanes

3.1.2.1
Synthesis of Polydiorganosiloxanes from Silanes

Unbranched polydiorganosiloxanes are prepared by the hydrolysis of the appropriate diorganodichlorosilanes, Eq. (1) [1, 4, 5, 6, 20, 241].

The ratio of linear α, ω-dihydroxypolydimethylsiloxanes 1a to cyclic compounds 1b is dependent upon reaction conditions. The formation of cyclic

$$n\,Me_2SiCl_2 \;+\; H_2O \xrightarrow[-\,2\,n\,HCl]{} HO\text{--}(Me_2SiO)_n\text{-}_m H \;+\; (Me_2SiO)_m \qquad (1)$$

$$\qquad\qquad\qquad\qquad\qquad\qquad\quad \text{a} \qquad\qquad\qquad \text{b}$$

compounds, for example, is favored by the use of ethers as solvent [7], by the addition of amines and ammonium [8] or phosphonium salts [9], alkali salts of organic sulphonic acids [10, 11, 12, 13] or by controlled low temperature hydrolysis [14, 15]. The reaction shown in Eq. (1) proceeds via dihydroxydimethyl silane as intermediate stage, which can only be isolated in neutral [21], but not in acidic or alkaline solution. Dihydroxydimethylsilane immediately reacts with further SiCl linkages or condenses in the presence of an acid catalyst to form α,ω-dihydroxypolydimethylsiloxanes 2a [16] (Eq. 2). If not enough water is present, α,ω-dichloro- 2c or α-chloro-ω-hydroxypolydimethylsiloxanes 2b are formed. Dichloro-polymers 2c can be prepared especially in a mixture of 1, 4-dioxane and water [5].

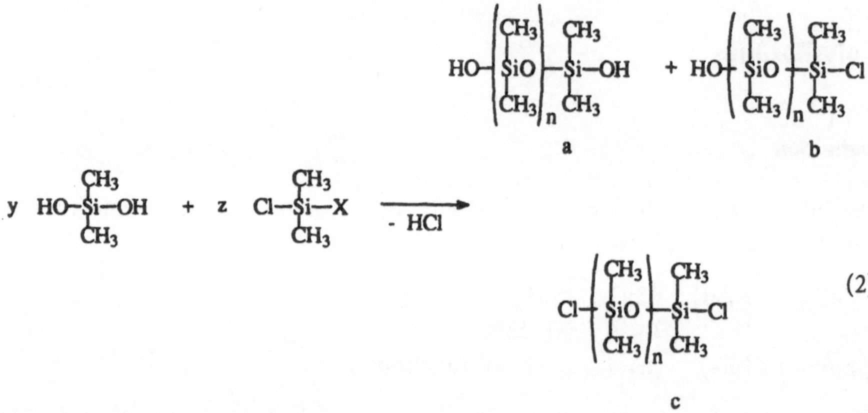

$$
HO\!\left[\!\begin{array}{c}CH_3\\SiO\\CH_3\end{array}\!\right]_n\!\!-\!\!\begin{array}{c}CH_3\\Si-OH\\CH_3\end{array} \quad + \quad HO\!\left[\!\begin{array}{c}CH_3\\SiO\\CH_3\end{array}\!\right]_n\!\!\begin{array}{c}CH_3\\Si-Cl\\CH_3\end{array}
$$

$$
\qquad\qquad\qquad\qquad a \qquad\qquad\qquad\qquad\qquad\qquad b
$$

$$
y \;\; HO\!-\!\begin{array}{c}CH_3\\Si\\CH_3\end{array}\!-\!OH \quad + \quad z \;\; Cl\!-\!\begin{array}{c}CH_3\\Si\\CH_3\end{array}\!-\!X \quad \xrightarrow[-\,HCl]{}
$$

$$
Cl\!\left[\!\begin{array}{c}CH_3\\SiO\\CH_3\end{array}\!\right]_n\!\!\begin{array}{c}CH_3\\Si-Cl\\CH_3\end{array} \tag{2}
$$

$$
\qquad\qquad\qquad\qquad\qquad c
$$

$$
X = Cl, \;\; -O\!-\!(\,Si\,Me_2O\,)_x H
$$
$$
\text{or} \quad -\!(\,OSiMe_2\,)_x Cl
$$

In its oligomeric form, α, ω-di-hydroxypolydimethylsiloxanes 2a can set free water intramolecularly, and thus form the above mentioned cyclic polydimethylsiloxanes, Eq. (3). Octamethylcyclotetrasiloxane (D4) is generally formed as the main component. Smaller amounts of decamethylcyclopentasiloxane (D5) and hexamethylcyclotrisiloxane (D3) are produced, and only negligible amounts of the higher cyclic compounds such as D6, D7 etc. are found in the end product, due to theromodynamic instability of the products with increasing ring size.

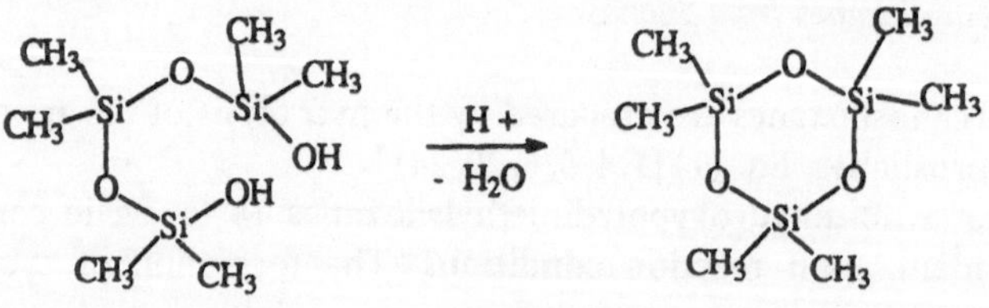

$$
\tag{3}
$$

Trimethylsilyl groups as chain terminators [18], SiH groups or other organic radicals (R) can be introduced by co-hydrolysis [19, 20], the reaction rate varying with the radical R, since the stability of intermediately formed diols varies. The polymerization rate decreases in the following order:

$$H(CH_3) = > (CH_3)_2Si = > (C_6H_5)(CH_3)Si = > (C_6H_5)_2Si =$$

Industrial scale hydrolysis is carried out in the liquid phase [22, 23], especially by the Loop process [24], as well as by gas phase reaction [24, 25, 26]. Since large amounts of hydrochloric acid are formed in these two processes, which must be converted into methyl chloride for environmental as well as financial reasons (use for the Rochow process), another process, referred to as methanolysis [17, 27], has been developed, in which dimethyldichlorosilane is reacted with methanol and hydrolyzed to form mainly linear polydimethylsiloxanes [4]. As is evident from Scheme (4), methyl chloride is formed directly in this reaction. This can be used again in the Rochow process so that, in contrast to hydrolysis, one stage is obsolete.

$$n\,(CH_3)_2SiCl_2 + 2\,n\,CH_3OH \longrightarrow (CH_3)_2{-}Si{-}(OCH_3)_2 + 2\,n\,HCl$$

$$\swarrow\; -\,2\,n\,CH_3Cl \tag{4}$$

$$n\,(CH_3)_2{-}Si{-}(OH)_2 \xrightarrow[-\,n\,H_2O]{} HO{-}[Si\,(CH_3)_2{-}O]_n{-}H$$

3.1.2.2
Synthesis of Polydiorganosiloxanes from Cyclic Compounds

The polymerization of cyclic polydimethylsiloxanes may be anionically or cationically catalyzed and also radiation initiated, as described below.

a) Cationic polymerization

Many acidic compounds such as sulfuric acid, trifluoromethanesulfonic acid as well as metal cations or suitable activated silicates such as clay or montmorillonite may be used as catalysts for the cationic polymerization of cyclic polydimethylsiloxanes, to mention only some of the possibilities described in the literature [28, 29, 30, 31, 32, 33, 34]. According to Voronkov [35], the first step in cationic polymerization is that the Lewis acid attacks the oxygen of the ring 5a, resulting in the formation of the linear cationic intermediate product 5b, as is shown by the example of hexamethyltrisiloxane Eq. (5). The silicenium ion 5b either reacts with further rings or with water with the formation of oligomeric α, ω-dihydroxydimethylsilanes which (as described in sect. 3.1.2 using dihydroxydimethyl silane as example) condense further under acidic conditions yielding the OH-terminated polymer, Scheme (5).

Other reactive intermediate stages have recently been described, e.g. nonionic structures which form a covalent linkage between the silicon atom and the alkali B [30, 32], resulting in a penta-coordinated silicon atom as transitional state. Sigwalt [37] has postulated the siloxonium ion (see Eq. 6). Here, however, a

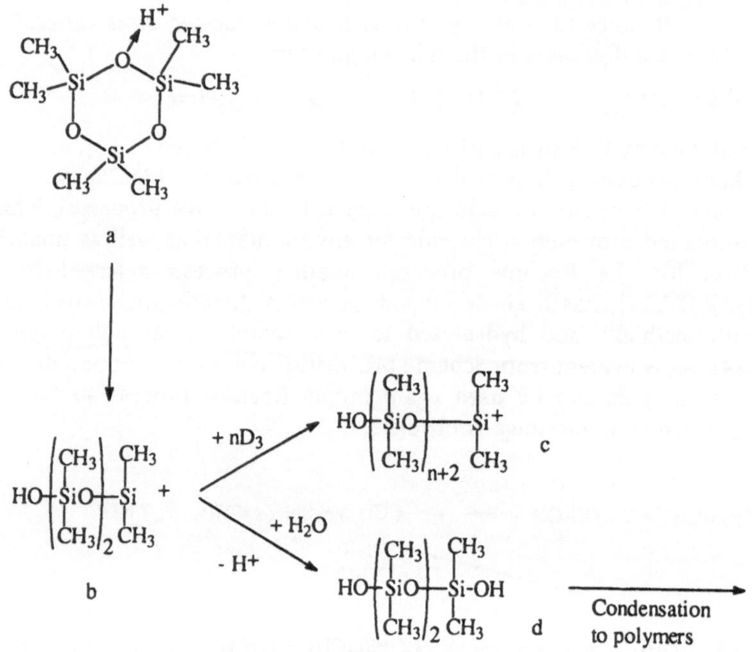

(5)

dual reaction mechanism which superimposes both processes cannot be excluded. Basically, all these reactions are equilibrium reactions, so that cyclic compounds are in a thermodynamic equilibrium with open-chain products, depending on the catalyst, temperature and, possibly, the substituent R.

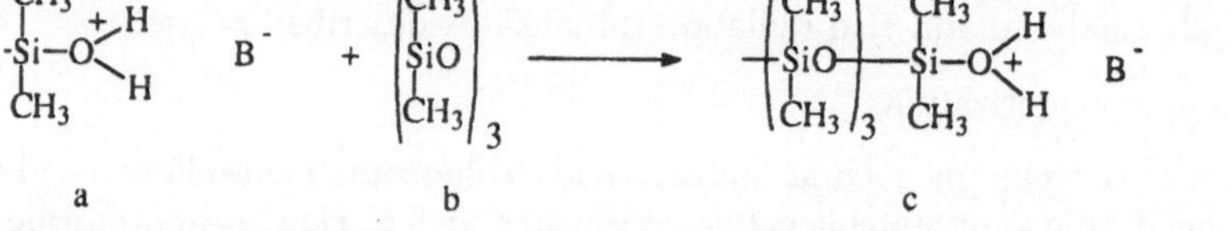

(6)

From this complex reaction behaviour it is evident that this cannot be a simple first order reaction. There is a dependence of reaction rate on the acid concentration, along with the expected dependence on the cyclic compounds concentration as expressed by Eq. (7)[36].

$$R_{polym} = (HA)^x (D_n)^y$$

(7)

$n = 3 - 6$; $x(CF_3SO_3H) = 2.2 - 2.7$; $y = -1$ to $+1$

The introduction of water as co-catalyst into the reaction medium additionally complicate matters. The polymerization of D3 is speeded up or slowed down, depending on the water/acid ratio, whereas the polymerization of D4 is retarded [37, 38].

Trimethylsilyl-terminated polydimethylsiloxanes can be prepared by reacting the cyclic compounds in the presence of hexamethyldisiloxane as end stopper

[17]. Oligomers $MD_{3n}M$ are formed with great regularity, particularly during the reaction of D3, like in Eq. (6). This indicates that under the given conditions, the redistribution of individual $=SiO$ groups only plays a subordinate role [39]. In the analogous reaction with D4, this regularity only occurs under very specific conditions [40, 30], which is very likely due to the lower reactivity towards acids compared with other D units. The reaction rate of the Si-O linkage in an acid medium decreases in the following order: $D_3 > MM > MDM > MD_2M > D_4$ [41]. By using carbofunctional disiloxanes $X\text{-}(CH_2)_nMe_2Si-O-SiMe_2(CH_2)_n\text{-}X$, where X is a CN-, OH-, (meth) acryloxy or other functional groups, the correspondingly terminated carbofunctional polydimethylsiloxanes are likewise made accessible [42].

b) Anionic Polymerization

The use of alkali metal hydroxides as initiators for the anionic polymerization of cyclic polydimethylsiloxanes was patented as early as 1949 by Hyde [43] and almost at the same time, but independently, by Warrick [44]. The most widely used base is potassium hydroxide, because it is freely available and because it is more reactive than sodium or lithium hydroxide, as the following order of reactivity demonstrates: $Cs > Rb > Na > Li$ [45].

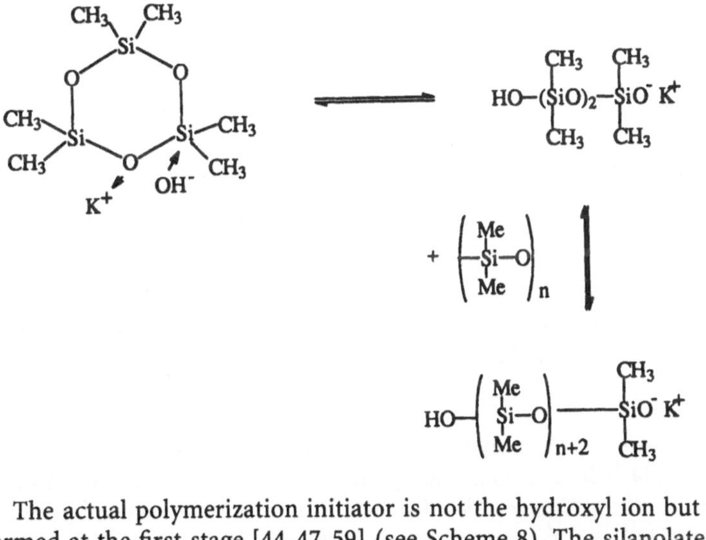

$$(8)$$

The actual polymerization initiator is not the hydroxyl ion but the silanolate formed at the first stage [44, 47, 59] (see Scheme 8). The silanolate is thought to possess cryptate or crown ether structures, which are known to accelerate polymerization [48, 49, 50]. Recently published information states that tetramethylammonium silanonates and siloxanolates are faster reacting than the alkali metal silanolates [51]. Other catalysts are tert-butyl lithium [50], alkali metal alcoholates and mercaptides [36]. In anionic polymerization, too, the structurally more strained D_3 ring reacts more rapidly than D_4 [53], whose reaction rate is, however, lower than that of the higher rings: $D_4 < D_6 < D_7 < D_8$. The cause is thought to be the already mentioned tendency to form crown ether-like structures, a tendency that is increasingly present in rings with more than five

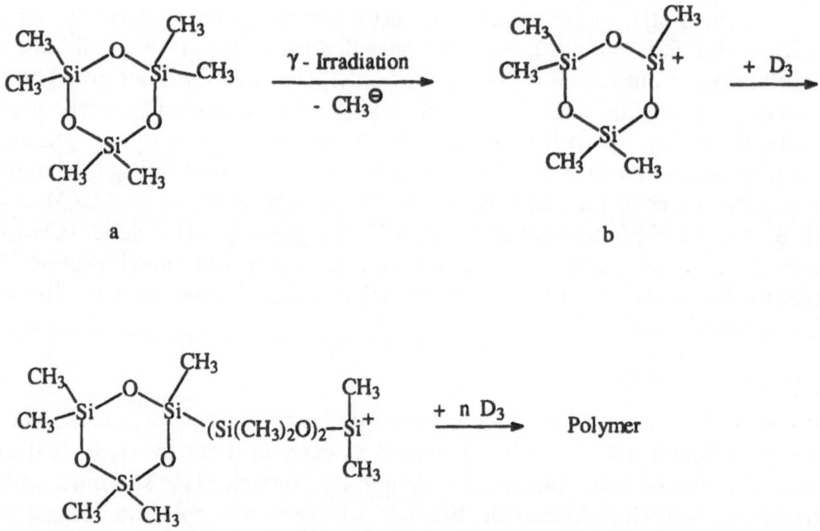

$$
\text{(a)} \quad \xrightarrow[\;-\,CH_3^{\ominus}\;]{\gamma\text{-Irradiation}} \quad \text{(b)} \quad +\ D_3 \ \longrightarrow
$$

(9)

$$
\xrightarrow{\;+\ n\ D_3\;} \quad \text{Polymer}
$$

SiO units. The reaction rate is almost of the first order, in relation to the ring, and is approximately given by the equation [36]: $R_{polym} = (\equiv SiO^-)^{1/n}\,(D_m)$ (the exponent $1/n$ varies with the chemical structure of the catalyst).

Since reaction sequence (8) likewise involves a chemical equilibrium, the catalyst must be destroyed if the cyclic compounds are to be separated from the open-chain, higher polymers by distillation. This is achieved either by adding traces of acid [44] or, more smoothly, by adding chlorosilanes [52], making it possible to prepare oligomeric compounds with organo-functional terminal groups. Chain termination by trimethylsilyl groups is generally achieved by adding hexamethyl disiloxane, as in the case of acid catalysed polymerization The method can also be used, however, to introduce carbofunctional terminal groups, especially aminopropyl groups [51, 54, 55].

c) Radiation-Induced Polymerization

The polymerization of cyclic polydimethylsiloxanes by radiation is possible in the liquid state only if they are extremely pure and after they have been dried thoroughly over sodium. The reaction is carried out in a high vacuum in sealed ampoules [33, 56, 57]. As far as the reaction mechanism is concerned, it is generally agreed that, in the first stage, a CH_3^- ion is split off by the γ radiation. The silicenium ion 9b formed as a result acts as a polymerization initiator (see Scheme 9) [33, 56, 58].

In contrast to cationically and anionically initiated polymerization, where D_3 reacts much faster than the higher cyclic compounds, there is not major difference in reaction rate between D_3, D_4 and D_5 in the case of radiation-induced polymerization [33, 56]. As far as product distribution is concerned, no higher (> 6 Si atoms) rings formed through intramolecular cyclization are evident, in contrast to ionically induced polymerization. Instead, one finds mainly (>50%) linear polymers, residues of the starting product and, depending on the educt, different cyclic compounds (D_3, D_4 D_5) in varying degrees.

3.1.3
Chemical Properties of Polydiorganosiloxanes

The most important chemical property of polydimethylsiloxanes is their chemical inertness, which is due to the high bond energy of the Si—O linkage [60].

Bond energy Si—O: 424–495 kJ/mol
C—O: approx. 357 kJ/mol

The bond angle Si—O—Si is 120° to 150°, depending on the substituent [60].

Opinions about the structure of the Si — O linkage and the resultant high bond strength vary. Earlier literature [60, 61, 62] assumes that there is a covalent bond in which the free electron pairs of the oxygen atom form a (p- >d)π bond with the empty d-orbitals of the silicon atom. This theory has recently been questioned [63, 64] because the basicity of oxygen and recent calculations speak against it. The presence of an inonic structure [63] and dipole character [64] are under discussion.

3.1.3.1
Hydrophobic Properties

Because of the free rotatability [65, 66] and polarizability of the Si — O bond, the siloxane chain is able to align itself accordingly, resulting in hydrophobic properties.

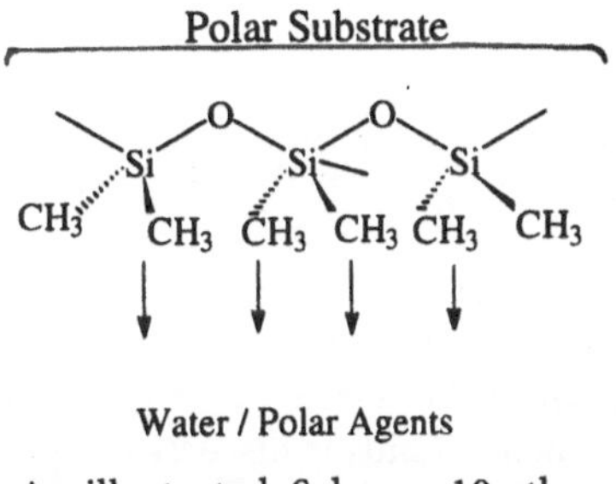

$$(10)$$

Water / Polar Agents

As illustrated Scheme 10, the oxygen atoms become aligned on the polar substrate, whilst the non-polar Si—CH$_3$ groups repel water, adhesives and polar substances generally.

3.1.3.2
Behavior towards Acids and Alkalis

One of the consequences of the dipole character of the Si—O bond is the ease with which these bonds are broken by acids and alkalis, as already mentioned in the preceding section. This splitting up and re-formation of Si—O linkages is referred to as equilibration and is used to introduce other substituents such as hydrogen, phenyl, vinyl and other organic groups which serve either as crosslinking points for the resultant polymer (see Sect. 3.1.5) or as a means of modifying the physical properties (see Sect. 3.1.4) of the pure polydimethyl-siloxane [67] (see Eq. 11):

$$(H_3C)_3Si-O-Si(CH_3)_3 \ + \ HO\left[\underset{CH_3}{\overset{CH_3}{Si}O}\right]_n H \ + \ HO\left[\underset{R'}{\overset{CH_3}{Si}O}\right]_m H \quad \xrightarrow{\ H+/OH^-\ }$$

$$\quad\quad a \quad\quad\quad\quad\quad\quad b \quad\quad\quad\quad\quad\quad\quad c$$

$$MM \quad + \quad HO-D_n-OH \quad + \quad HO-D'_m-OH \quad\quad\quad\quad (11)$$

$$(H_3C)_3SiO\left[\underset{CH_3}{\overset{CH_3}{Si}O}\right]_n\left[\underset{R'}{\overset{CH_3}{Si}O}\right]_m Si(CH_3)_3 \quad\quad d$$

$$MDDD'\ DDDD'\ DDDD'\ DDD. \ldots M$$

The general formula **d** is intended to symbolise that when $n \gg m$, there is a largely statistical distribution of R'SiO groups within the polymer. It should, however, be noted that the linear compounds are always in equilibrium with the appropriate cyclic ones, this equilibrium being characteristic for a specific reaction temperature. In the case of sterically hindered constituents such as the trifluoropropyl group, the equilibrium is shifted towards cyclic compounds, i.e. if these are to be equilibrated, the choice of reaction conditions, especially temperature, is of decisive importance.

3.1.3.3
Heat Resistance-Thermal Depolymerization

The excellent heat resistance of silicones is decisive for many applications. Silicones start to depolymerize very slowly (after about 20 h) only above about 200°C in the absence of oxygen and catalysts [68]. In the review by Kendrick and co-workers [69], the thermal decomposition of silicone fluids is discussed in great detail. So that we need only describe the essential features of the process here.

The thermal degradation of linear polydimethylsiloxanes is characterized by the cleavage of the Si—O bond, i.e. cyclic compounds are formed exclusively. Although, theoretically, methane or hydrogen could be formed, this has not been found to be the case in the absence of catalysts [69]. Comparison of bond energies Si—O (average about 460 kJ/mol), Si-C (315 kJ/mol) and C—H (411 kJ/mol), as well as consideration of the activation energy $E_a = 167$ kJ/mol of the depolymerization, make this unlikely. This apparent contradiction may be explained by the formation of cyclic, low-energy transitional states, which is made possible by the free **d**-orbitals on the silicon atom, Eq. (12).

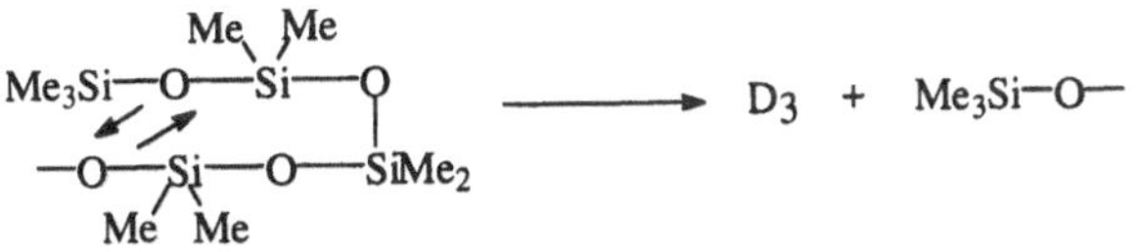

$$\underset{\underset{Me}{|}}{\overset{\overset{Me\;\;Me}{|\;\;\;|}}{Me_3Si-O-Si-O}}\quad\nwarrow\quad -O-\underset{\underset{Me}{|}}{\overset{\overset{}{|}}{Si}}-O-SiMe_2 \qquad\longrightarrow\qquad D_3\;+\;Me_3Si-O- \qquad (12)$$

The main depolymerization product is D_3. D_4 is produced in smaller quantities and D_5, D_6 etc. in only insignificant amounts, obviously depending on the reaction conditions. The prerequisite for complete depolymerization is that one is working in vacuo since otherwise an equilibrium will be formed. In the presence of traces of cationic or anionic polymerization catalysts however, different mechanisms for depolymerization are discussed and also the formation of methane has been observed [81]. If one compares the thermal stability of polydimethylsiloxanes which have been terminally blocked by a trimethylsilyl group with that of polymers containing terminal hydroxyl groups, one finds that the latter have poorer thermal stability, because the cyclic transition state is favored by the terminal hydroxyl group.

Thermal stability can be increased still further by adding phenyl groups to the silicon atom. With these polymers, benzene is produced at high temperatures via four-center transitional states. During the depolymerization of a mixture of polydimethylsiloxane and polymethylphenylsiloxane, cyclic compounds are formed which contain dimethyl as well as methylphenyl groups. This indicates that the cyclization explained in Eq. (12) can, in this case, proceed intra-, as well as intermolecularly.

In the presence of oxygen, thermal depolymerization is not the only process taking place. There is also oxidative cleavage of the Si—C bond, resulting in transverse crosslinking and, subsequently, the liberation of carbon dioxide, water and, in the case of incomplete oxidation, also formaldehyde and carbon monoxide.

3.1.3.4
Radiation Resistance

Silicone fluids are generally held to have good resistance to radiation although signs of degradation or crosslinking are apparent under certain conditions.

Tests have recently been carried out to find out more about the photodegradation of silicone fluids in the aqueous phase and the effect on the environment [70]). Here it was found that low molecular weight polydimethylsiloxanes $HO(Me_2SiO)_nH$ ($n = 3-9$) with a half life period of about 9 days are degraded into SiO_2 when subjected to UV light (xenon lamp $\lambda > 290\,nm$). Polydimethylsiloxanes with higher molecular weights are likewise degraded but at a very much slower rate. Degradation is accelerated by nitrate ions and ozone, which react with water, forming hydroxyl groups which are the true initiators of the degradation reaction (see Scheme 13) [71].

Carbon dioxide is formed as decomposition product, as well as formaldehyde. When polydimethylsiloxanes are subjected to energy-rich radiation, they crosslink, depending on the viscosity and radiation dose, see Table 3.1 [72]. The radiation source in this test was a ^{60}Co-γ-radiator.

$$NO_3^- + H_2O \xrightarrow{h\nu} NO_2^- + OH^- + OH\bullet$$

bzw..

$$O_3 + H_2O \xrightarrow{h\nu} O_2 + 2\ OH\bullet$$

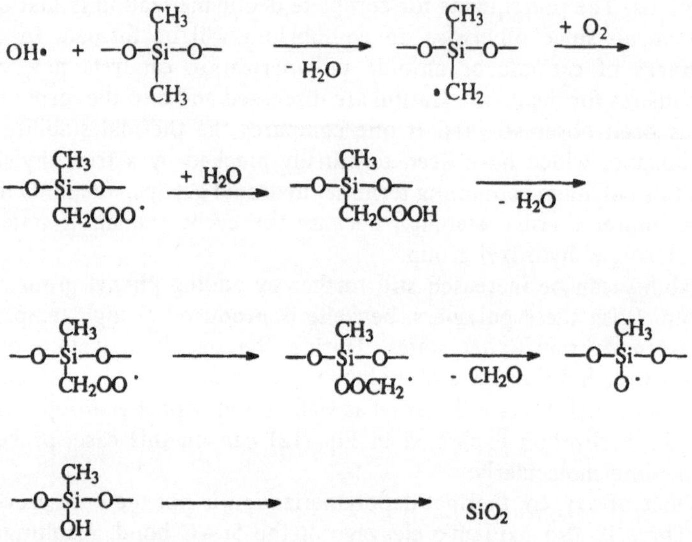

$$OH\bullet\ +\ -O-\underset{\underset{CH_3}{|}}{\overset{\overset{CH_3}{|}}{Si}}-O- \xrightarrow{-\ H_2O}\ -O-\underset{\underset{\bullet CH_2}{|}}{\overset{\overset{CH_3}{|}}{Si}}-O- \xrightarrow{+\ O_2}$$

$$-O-\underset{\underset{CH_2COO\bullet}{|}}{\overset{\overset{CH_3}{|}}{Si}}-O- \xrightarrow{+\ H_2O}\ -O-\underset{\underset{CH_2COOH}{|}}{\overset{\overset{CH_3}{|}}{Si}}-O- \xrightarrow{-\ H_2O} \tag{13}$$

$$-O-\underset{\underset{CH_2OO\bullet}{|}}{\overset{\overset{CH_3}{|}}{Si}}-O- \longrightarrow\ -O-\underset{\underset{OOCH_2\bullet}{|}}{\overset{\overset{CH_3}{|}}{Si}}-O- \xrightarrow{-\ CH_2O}\ -O-\underset{\underset{O\bullet}{|}}{\overset{\overset{CH_3}{|}}{Si}}-O-$$

$$-O-\underset{\underset{OH}{|}}{\overset{\overset{CH_3}{|}}{Si}}-O- \longrightarrow\ \longrightarrow\ SiO_2$$

Table 3.1. Influence of ^{60}Co-γ-irradiation on the viscosity of polydimethylsiloxanes

Initial viscosity in mm²/2	Dose of irradiation in Mrad				
	5	10	20	40	60
53	54	59	69	98	230
350	398	627	gel		
1000	1525	5973	gel		

3.1.4
Physical Properties of Polydiorganosiloxanes

3.1.4.1
Surface Tension

Certain physical properties of polydimethylsiloxanes are governed by their structure, described in Sect. 3.1.4. The flexibility of the siloxane chain in particular, and the low intermolecular forces between the methyl groups are the reason for the extremely low surface tension of polydimethylsiloxanes. Only that of fluoropolymers is lower [73] (Table 3.2).

Table 3.2. Surface tension of different polymers

Polymer	Surface tension (mN/m)
Polydimethylsiloxane	21–22
Poly(tetrafluorethylene)	14
Paraffinwax	25
Polyethylene	32
Poly (3, 3, 3-trifluoropropyl-methylsiloxane)	14–15

Whereas, in the case of unsubstituted polydimethylsiloxanes, liquid and solid polymers have the same surface tension from a chain length of 80 Si units upwards, there are major differences in the case of those containing trifluoro-propyl groups as substituents. In the solid state, the surface is 14–15 mN/m (see Table 3.2), but in the liquid state it is 24 mN/m, which is higher than the surface tension of polydimethylsiloxanes. This may be due to the fact that there is increased intramolecular and intermolecular interaction between the bigger trifluoroproyl group and neighbouring groups, which has no effect in the solid because of reduced mobility. Groups which are sterically more voluminous lead to increased surface tension because of the higher interactive forces [73]. The effect of intermolecular interaction is also evident in the dependence of the surface tension of polydimethylsiloxanes on their chain length in the region of up to about 50 Si units: the constant value of 21 mN/m is achieved only from a viscosity of about 100 mm^2/s (= about 80 Si units). Below this viscosity, the surface tension is lower and depends on chain length and viscosity (see Fig. 3. 1). This is due to the fact that the surface tension of a given substance reaches a minimum at the boiling point, due to the elimination of intermolecular forces, and because short chain polydimethylsiloxanes–unlike long chain ones–are still volatile.

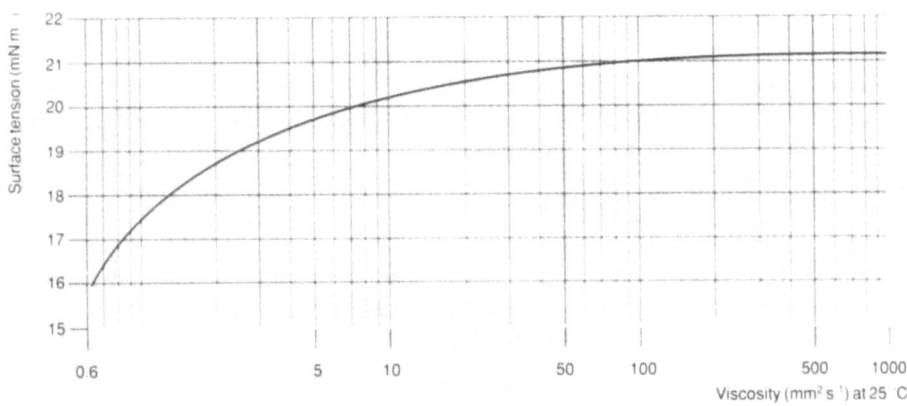

Fig. 3.1

The low surface tension is the reason why silicone fluids exhibit good spreading characteristics on nearly all organic and inorganic materials, except fluoropolymers. This is why they are used as antifoams [74] as well as wetting agents, e.g. in polishes [75] and waxes, to impart water repellent properties.

3.1.4.2
Thermal Properties

Another consequence of the high mobility of the siloxane chain is the extremely low melting point (about $-50\ °C$) and glass transition temperature (about $-120\ °C$) [73, 76]. The introduction of only small amounts of other substituents such as phenyl groups disturb the molecular order to such an extent that no crystallization is possible and the glass transition temperature of around $-100\ °C$ is alone responsible for low temperature properties.

The behaviour of polydimethyl-polymethylalkylsiloxane copolymers is interesting. Up to a side chain length of about 12 C units, the same behavior is evident as in the case of the incorporation of phenyl groups, i.e. crystallization of the siloxane chain is stopped and it is not possible to determine a crystallization temperature. If, on the other hand, the side chain consists of more than 12 C atoms, it will crystallize separately, which no longer has any appreciable effect on the crystallization temperature of the siloxane chain (approx. $-60\ °C$. The glass transition temperature, on the other hand, remains constant at $-120\ °C$ irrespective of the length of the side chain. The melting points of the side chain depend on the number of C units and, as expected, increase with increasing chain length [79]). The overall thermal properties enable these products to be used in applications where stability over a wide range of temperatures is an important requirement, e.g. hydraulic fluids, low-temperature lubricants and, especially in the case of polymethylphenylsiloxanes, as coolants.

3.1.4.3
Rheology

The special rheological behavior of polydimethylsiloxanes, which differs from that of organic mineral oils, has already been pointed out by Stark [76] and Flanigam [77]. We should, however, like to summarize the effect of pressure, temperature and shear force on viscosity.

a) Dependence of viscosity on pressure

The viscosity of polydimethylsiloxanes increases with increasing pressure, although to a smaller degree than in the case of mineral oils. For oils with a viscosity of >1000 mPas at 1.013 bar, increasing the pressure to 450 bar causes the viscosity to increase by a factor of 2 (see Fig. 3.2).

b) Dependence of viscosity on shear force

Polydimethylsiloxanes up to a viscosity of around 1000 mPas exhibit near-Newtonian flow, i.e. their viscosity remains constant with increasing shear force.

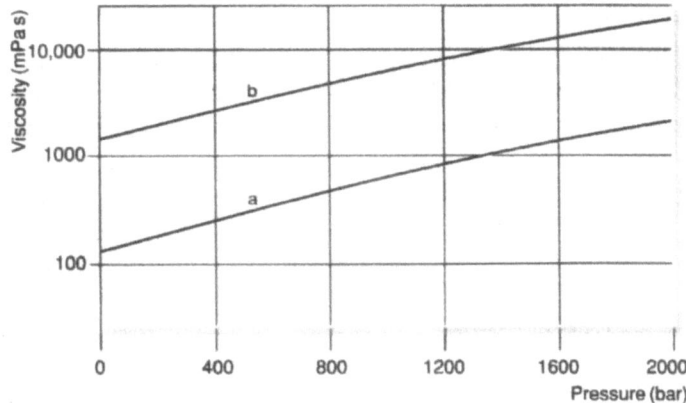

Fig. 3.2 Example: Effect of pressure on the viscosity of two silicone fluids at 25 °C, according to KUSS: a) 130 mPaS b) 1400 mPaS

High viscosity silicone fluids on the other hand show a high degree of non-Newtonian flow, or pseudoplasticity, which means that their viscosity decreases with increasing shear rate (see Fig. 3.3).

This behavior is thought to be due to the disentanglement and alignment of the individual polydimethylsiloxane chains in the direction of shear [78].

Polydimethyl-polymethylalkylsiloxane copolymers with side chains up to 14 C units long, and with an initial viscosity of max. 1000 mPas, exhibit Newtonian flow, just like pure polydimethylsiloxanes, Non-Newtonian flow is observed only if the side chains have 16 or 18 C atoms or more. The explanation is that longer side chains tend to crystallize and therefore tend to form network-like structures which are destroyed by shear stresses [79].

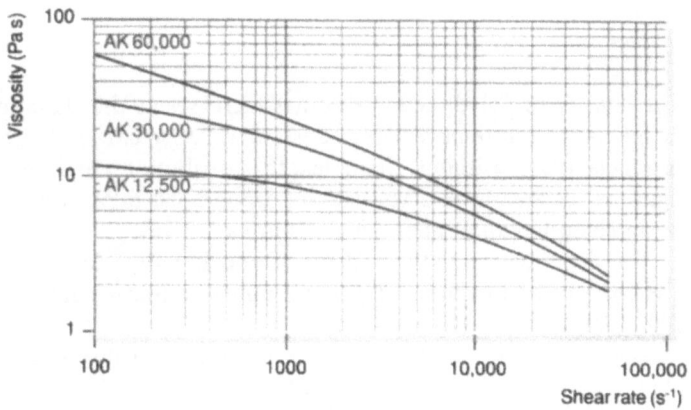

Fig. 3.3

c) Dependence of viscosity on temperature

Although the viscosity of polydimethylsiloxanes also shows a certain temperature dependence which is slightly increased by the incorporation of phenyl
groups, this is far less than in the case of mineral oils for example (see Fig. 3.4).

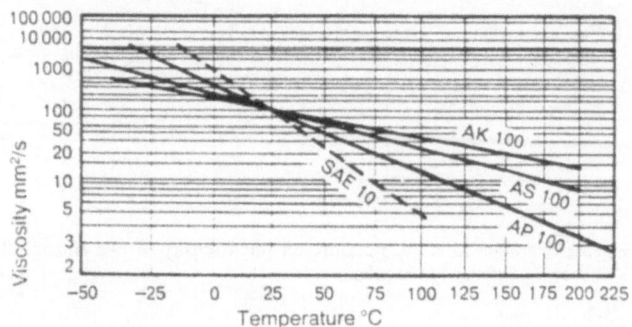

Fig. 3.4

The viscosity-temperature coefficient (VTC) represents a measure of the
dependence of viscosity on temperature, see Eq. (14). The VTC for polydimethylsiloxanes with a viscosity of $\geqslant 100$ mm²/s is 0.6, that of motor oil SAE 10 as a
typical representative of mineral oils is 0.9. From Eq. (3) one can see that the
temperature dependence of viscosity increases with increasing VTC values.

$$\text{VTC} = 1 - \frac{\text{viscosity } (\text{mm}^2/\text{s}^{-1}) \text{at } 99°\text{ C}}{\text{viscosity } (\text{mm}^2/\text{s}^{-1}) \text{at } 38°\text{ C}} \tag{14}$$

3.1.4.4
Density

The density of polydimethylsiloxanes increases as the chain length increases to
22 SiO units and then remains constant at 0.97 g/cm³, [77]. These are very pure
products (95%) without chain length distribution. In practice, however, this
figure is reached from a viscosity of around 350 mPas, which corresponds to a
mean chain length of about 200 SiO units [72, 76].

The dependence of the density of polydimethylsiloxanes on temperature is
greater than that of mineral oils and follows a linear course from a viscosity of
350 mPas upwards (see Fig. 3.5).

The dependence of the density of polydimethylsiloxanes on pressure is very
high compared with mineral oils (see Fig. 3.6).

These physical properties and the ones described in Sect. 3.2.4 favour the use
of polydimethylsiloxanes as damping media and hydraulic fluids.

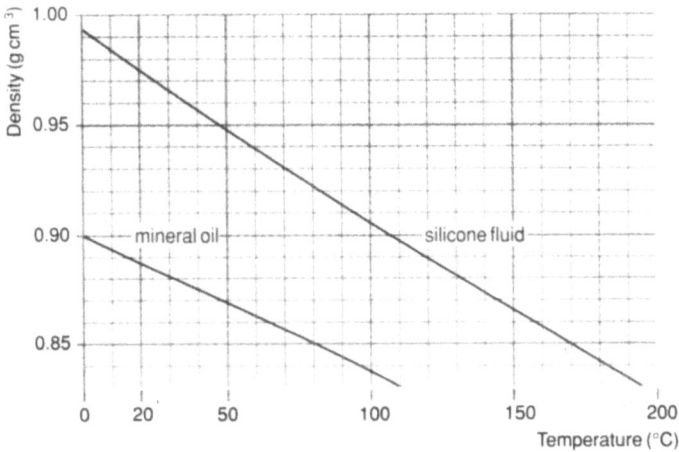

Fig. 3.5

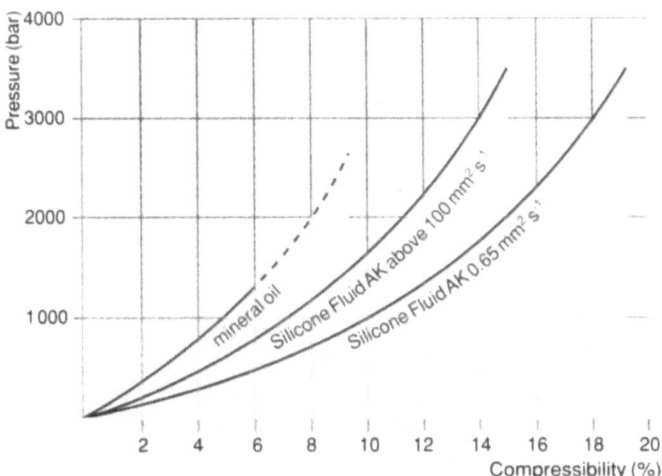

Fig. 3.6

3.1.4.5
Electrical Properties

Polydimethylsiloxanes have outstanding electrical isolating properties due to the fact that the SiO bond is protected by methyl groups and also due to their hydrophobic properties [76]. The dielectric constant ε for short chain lengths depends on the viscosity, see Fig. 3.7.

Table 3.3 lists further important electrical properties of polydimethyl and polydimethylmethylphenylsiloxanes.

Fig. 3.7

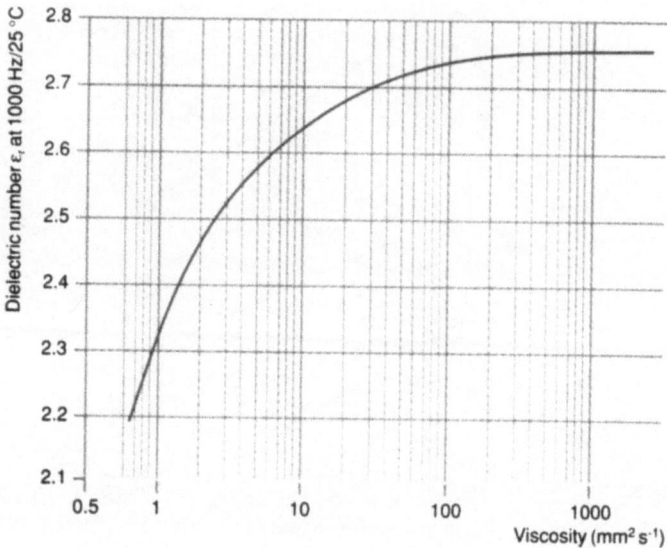

Table 3.3. Electrical properties of polydimethylsiloxane and polydimethylmethylphenylsiloxanes

R	Dielectric const. (ε) at 25 °C/10^2Hz	Dielectric strength (kV/mm)	Loss factor at 25 °C/10^3Hz
CH_3	2.76	15 − 16	$< 1 \times 10^{-5}$
C_6H_5	2.90	> 15	4×10^{-4}

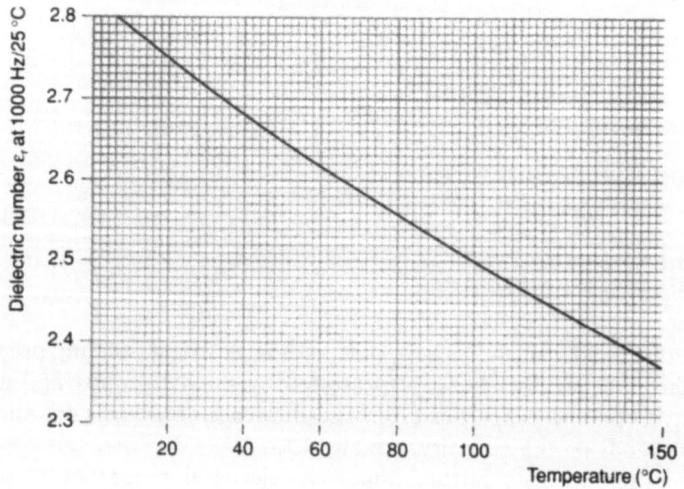

Fig. 3.8

As Table 3.3 shows, the electrical properties do not – surprisingly – change appreciably by the introduction of phenyl groups.

Trifluoropropyl groups, on the other hand, cause the dielectric constant to increase by a factor of about 2.5, resulting in a reduction of the dielectric strength by about 1 kV/mm [76]. The effects of substituents on electrical properties are not always accurately predictable and are evidently not yet fully understood. Since the electrical properties are affected also by the residual moisture content and, probably, catalyst residues from the manufacturing process, it is difficult to compare figures accurately for products made by different companies. The dielectric constant is but little affected by temperature when the products are used as liquid dielectrics. Other applications based on the electrical properties of these materials include insulating liquids in transformers and capacitors (see Fig. 3.8).

3.1.5
Polydimethylsiloxanes with Functional Groups

3.1.5.1
General

Polydimethylsiloxanes with functional groups may be divided into two main categories:

1. those which bear the functional group direct on the silicon atom (Si—X), i.e. which are silicon-functional, and
2. those in which the functionality is fixed via a SiC-bound alkylene or arylene group (Si—R—X), i.e. which are organofunctional. The most important functional groups X include hydrogen, vinyl, chloro, hydroxy, mercapto, alkoxy, cyano, (meth)acryloxy and amino groups.

A further differentiation is that between organo-modified and organo-functional polydimethylsiloxanes. The former are understood to be those polydimethylsiloxanes in which the methyl group has been replaced by another inert hydrocarbon group, e.g. a phenyl, alkyl, aralkyl or fluoro-alkyl group. Organo-functional polydimethylsiloxanes are substituted reactive organic group at the silicon atom. Four types of structure are possible for organo-functional polydimethylsiloxanes:

$Me_3SiO(Me_2SiO)_n (MeSiO)_nSiMe_3$ statistical or block-like distribution
$\quad\quad\quad\quad\quad\quad |$ of the functional group
$\quad\quad\quad\quad\quad\quad R$

$R\text{-}SiMe_2O(Me_2SiO)_n SiMe_2\text{-}R$ α, ω-termination of the functional group

$R\text{-}SiMe_2O(Me_2SiO)_n (MeSio)_m SiMe_2\text{-}R$ combination of 1 and 2
$\quad\quad\quad\quad\quad\quad\quad |$
$\quad\quad\quad\quad\quad\quad\quad R$

$Me_3SiO(Me_2SiO)_n SiMe_2\text{-}R$ monofunctional

3.1.5.2
Polydimethylsiloxanes with Si-Functional Groups

The Si-functional polydimethylsiloxanes are important for the production of silicone elastomers as well as being starting products for copolymers of siloxanes and organic polymers (see Sect. 3.2). The most important functional groups, which are attached directly to the silicon atom, are:

$$Si—OH, \ Si—H, \ Si—Vi, \ Si—Cl, \ Si—OR, \ Si—NR_2$$

Polydimethylsiloxanes with vinyl groups in the chain are cured with peroxides, whilst those with α, ω-terminal vinyl groups are cured with Si-H-functional polydimethyl or polymethylsiloxanes. α, ω-OH terminated polydimethyl-siloxanes are converted into elastomeric silicone networks using tri- or tetrafunctional alkoxy, acetoxy, alkylamino or oximosilanes. All these systems have been exhaustively described in the literature [1, 82, 83, 84, 85, 86, 87]. Polymers with vinyl or Si-H groups in the chain are synthesized either via vinyl-containing or Si-H cyclic compounds 15 b (see Scheme 15) and their polymerization, through co-hydrolysis of the silanes or via equilibration, through co-hydrolysis of the silanes or via equilibration of α, ω-hydroxypolydimethylsiloxanes 15 c with α, ω-hydroxy-polyhydrogenmethylsiloxanes 15 d [1, 20, 67, 88] or with α, ω-hydroxypolymethyl-vinylsiloxanes (see Scheme 15). Chain termination of the polymers is achieved using trimethylsilyl or dimethylvinylsilyl groups, or hydrogendimethylsilyl end groups.

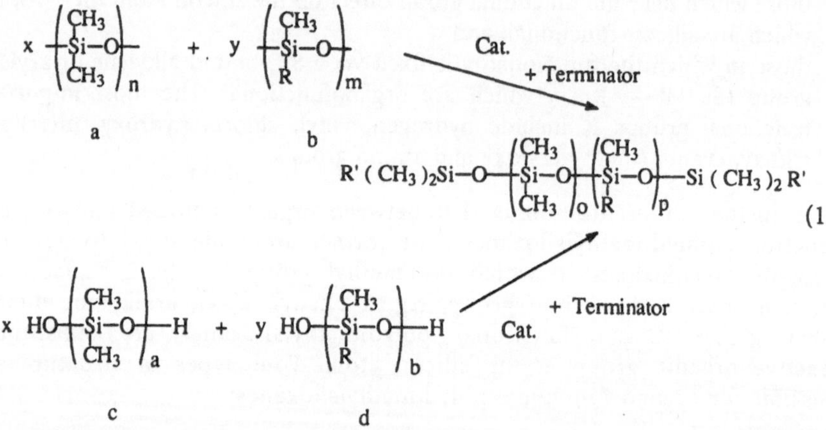

$$\tag{15}$$

The reaction mechanisms have been explained in great detail in Sect. 3.1.2. The disadvantage of silane co-hydrolysis compared with these two methods is that large amounts of cyclic products are produced and that it is difficult to achieve statistical distribution because of the different reactivities. Polydimethyl-siloxanes with Si—H- or Si—Vi- functionalities at the chain ends are prepared in the same way, only the dimethyl cyclic compounds 15a and dimethyl hydrolysates 15c being reacted with a terminator.

Polydimethylsiloxanes with terminal Si—Cl or Si—OR functionality are prepared through the controlled hydrolysis of the appropriate dichlorosilanes (see Sect.3.1.2) or dialkoxysilanes [89]. The formation of cyclic by-products is, however, unavoidable and it is for this reason that another synthetic approach is normally preferred, namely the polymerization of cyclic compounds in the presence of dichloro- and dialkoxysilanes [90, 91] or siloxanes (see scheme 16).

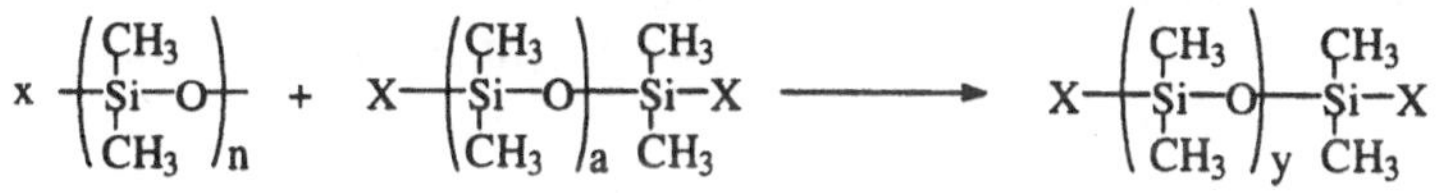

$$x \left(\begin{matrix} CH_3 \\ Si-O \\ CH_3 \end{matrix} \right)_n + X \left(\begin{matrix} CH_3 \\ Si-O \\ CH_3 \end{matrix} \right)_a \begin{matrix} CH_3 \\ Si-X \\ CH_3 \end{matrix} \longrightarrow X \left(\begin{matrix} CH_3 \\ Si-O \\ CH_3 \end{matrix} \right)_y \begin{matrix} CH_3 \\ Si-X \\ CH_3 \end{matrix}$$

(16)

X = Cl , OR or NR$_2$ a = 0 - 4 ;   n : preferably 4

By analogy with this, α, ω-bis (dialkylamino) polydimethylsiloxanes are also accessible [92]. Although it is basically possible to react α,ω-dichloro-polydimethylsiloxanes with dialkylamines to prepare polymers, the liberated hydrogen chloride needs to be neutralized and the salt formed needs to be removed from the polymer [93]. This is not so easy in view of the sensitivity of these polymers to hydrolysis.

Industrial applications of the three last-named polymers are mainly that of starting materials for the synthesis of polydimethylsiloxane copolymers described in Sect. 3.2.

3.1.5.3
Organo-Functional Polydimethylsiloxanes

The most important functional groups for organo-functional polydimethyl siloxanes are:

$$-R-OH, \quad -arylene-OH, \quad -R-SH, \quad -R-NR'_2, \quad -R-CH\overset{O}{\underset{}{\diagup\diagdown}}CH_2,$$

$$-R-O-\overset{O}{\overset{\|}{C}}-CR''{=}CH_2 \text{ and } R-CO_2H$$

R = divalent, substituted or unsubstituted alkyl group
R' = monovalent alkyl group
R'' = H or methyl

The most widely used methods of preparing organo-functional poly-dimethylsiloxanes are based either on hydrosilylation or on the already mentioned equilibration. In the first case, an unsaturated organic compound with the appropriate functional groups is added to a polydimethylsiloxane with Si-H bounds in the chain or at the chain ends. In the second case – with organo-functionality within the chain – a cyclic or linear prepolymer is prepared from a silane with the desired functional groups, which is then copolymerized in the usual manner with cyclic or linear polydimethylsiloxane. However, if the functional groups are at the end of the chain, the polydimethylsiloxane prepolymer with an organo-functional terminator is usually reacted with a di- or oligosiloxane. All these methods are illustrated below, with examples.

Polydimethylsiloxanes with hydroxyalkyl or hydroxyaryl groups. As in the acid or alkali-catalysed reaction of a chain terminator with alcoholic OH groups, e.g. 1, 3-bis (hydroxybutyl) tetramethyldisiloxane with polydimethylsiloxanes, the introduction of C—OH groups into the chain by equilibration cannot be used to prepare α, ω-bis(hydroxyalkyl)polydimethylsiloxanes, because of undesired cyclization reactions or because of the insertion of the C—O bond into the siloxane chain. The exception is sterically hindered secondary alcohols [95].

The platinum catalysed addition of unsaturated alcohols to Si-H bounds, e.g. allyl alcohol, is plagued by numerous side reactions such as the liberation of hydrogen and the associated formation of a silyl ether, or the β-addition to the C=C double bond [94]. Although the liberation of hydrogen can be limited by making use of buffered systems [99], the OH group of the unsaturated alcohol is preferentially provided with a protective group which is removed again after the addition reaction (Scheme 17) [96, 97, 98, 100, 134]:

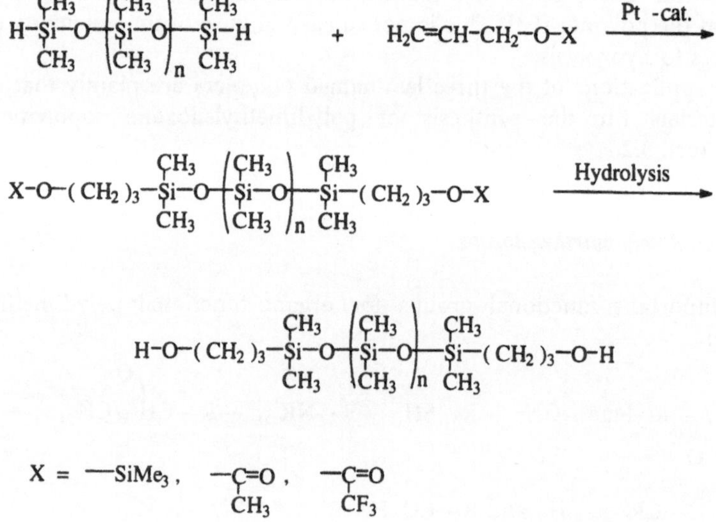

$$
\text{H-}\underset{\underset{CH_3}{|}}{\overset{\overset{CH_3}{|}}{Si}}\text{-O}\left(\underset{\underset{CH_3}{|}}{\overset{\overset{CH_3}{|}}{Si}}\text{-O}\right)_n\underset{\underset{CH_3}{|}}{\overset{\overset{CH_3}{|}}{Si}}\text{-H} \quad + \quad H_2C\text{=CH-}CH_2\text{-O-X} \xrightarrow{\text{Pt -cat.}}
$$

$$
\text{X-O-}(CH_2)_3\text{-}\underset{\underset{CH_3}{|}}{\overset{\overset{CH_3}{|}}{Si}}\text{-O}\left(\underset{\underset{CH_3}{|}}{\overset{\overset{CH_3}{|}}{Si}}\text{-O}\right)_n\underset{\underset{CH_3}{|}}{\overset{\overset{CH_3}{|}}{Si}}\text{-}(CH_2)_3\text{-O-X} \xrightarrow{\text{Hydrolysis}}
$$

(17)

$$
\text{H-O-}(CH_2)_3\text{-}\underset{\underset{CH_3}{|}}{\overset{\overset{CH_3}{|}}{Si}}\text{-O}\left(\underset{\underset{CH_3}{|}}{\overset{\overset{CH_3}{|}}{Si}}\text{-O}\right)_n\underset{\underset{CH_3}{|}}{\overset{\overset{CH_3}{|}}{Si}}\text{-}(CH_2)_3\text{-O-H}
$$

$$
X = \text{—SiMe}_3, \quad \underset{\underset{CH_3}{|}}{-C}\text{=O}, \quad \underset{\underset{CF_3}{|}}{-C}\text{=O}
$$

Another method starts out from cyclic compounds which are dissociated with diethoxydimethyl silane and then hydrolysed to form the polymer (Scheme 18) [101]:

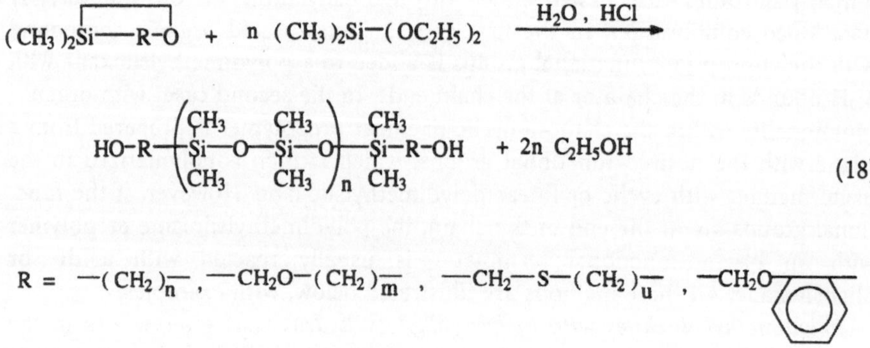

$$
\overline{(CH_3)_2Si\text{—R-O}} \quad + \quad n \ (CH_3)_2Si\text{-}(OC_2H_5)_2 \xrightarrow{\text{H}_2\text{O , HCl}}
$$

$$
\text{HO-R-}\left(\underset{\underset{CH_3}{|}}{\overset{\overset{CH_3}{|}}{Si}}\text{-O-}\underset{\underset{CH_3}{|}}{\overset{\overset{CH_3}{|}}{Si}}\text{-O}\right)_n\underset{\underset{CH_3}{|}}{\overset{\overset{CH_3}{|}}{Si}}\text{-R-OH} \quad + \text{ 2n } C_2H_5OH
$$

(18)

$$
R = \text{—}(CH_2)_{\overline{n}}, \text{—}CH_2O\text{-}(CH_2)_{\overline{m}}, \text{—}CH_2\text{-S-}(CH_2)_{\overline{u}}, \text{—}CH_2O\text{—}\bigcirc
$$

This method also makes α, ω-hydroxyaryl-terminated polydimethylsiloxanes accessible which are otherwise prepared like hydroxyalkylfunctional polydimethylsiloxanes by the addition of a stearyl ether to the Si — H bond, followed by the hydrolytic decomposition of the organic ether linkage [102].

Polydimethylsiloxanes with carboxyl groups. The synthesis of polydimethylsiloxanes with carboxyalkyl groups in the chain or at the chain end by equilibration with sulfuric acid [103] or trifluorosulfonic acid [104, 135] as catalysts takes place according to (Scheme 19):

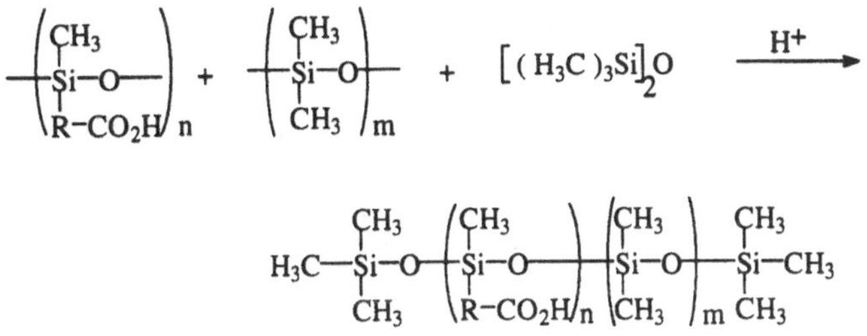

$$
\left(\begin{array}{c} CH_3 \\ | \\ -Si-O- \\ | \\ R-CO_2H \end{array}\right)_n + \left(\begin{array}{c} CH_3 \\ | \\ -Si-O- \\ | \\ CH_3 \end{array}\right)_m + \left[(H_3C)_3Si\right]_2 O \xrightarrow{\ H^+\ }
$$

$$
H_3C-\underset{\underset{CH_3}{|}}{\overset{\overset{CH_3}{|}}{Si}}-O\left(\underset{\underset{R-CO_2H}{|}}{\overset{\overset{CH_3}{|}}{Si}}-O\right)_n\left(\underset{\underset{CH_3}{|}}{\overset{\overset{CH_3}{|}}{Si}}-O\right)_m\underset{\underset{CH_3}{|}}{\overset{\overset{CH_3}{|}}{Si}}-CH_3 \tag{19}
$$

R = branched or nonbranched alkylene group

Another possibility consists of initially preparing cyclic poly(cyanoalkyl) methylsiloxanes and then to polymerize in one step and also to hydrolyze the cyano group (Scheme 20) [105, 106]:

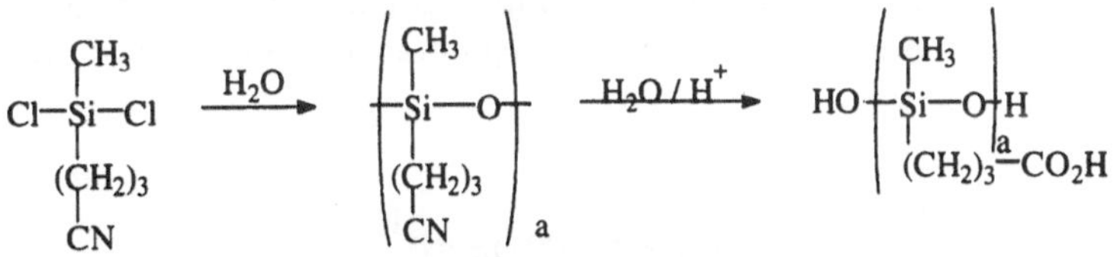

$$
\begin{array}{c} CH_3 \\ | \\ Cl-Si-Cl \\ | \\ (CH_2)_3 \\ | \\ CN \end{array} \xrightarrow{H_2O} \left(\begin{array}{c} CH_3 \\ | \\ -Si-O- \\ | \\ (CH_2)_3 \\ | \\ CN \end{array}\right)_a \xrightarrow{H_2O/H^+} HO\left(\begin{array}{c} CH_3 \\ | \\ -Si-O- \\ | \\ (CH_2)_3-CO_2H \end{array}\right)_a H \tag{20}
$$

a = 3 - 6

A third possibility–(which is only suitable for polydimethylsiloxanes with terminal carboxyaryl groups)–consists of reacting α, ω-dichloropolydimethylsiloxanes with bromophenyl lithium, followed by the introduction of the carboxyl group into the aromatic compound (Scheme 21) [107]:

Polydimethylsiloxanes with terminal carboxyalkyl or carboxyaryl groups are used as soft segments in polydimethyl siloxane-polycarbonate (see Sect. 3.2.3) or in polydimethyl siloxane-polyimide copolymers [107]. The textile industry uses metal carboxy-functional polydimethylsiloxanes to produce water repellent fabrics [104].

Polydimethylsiloxanes with aminoalkyl groups. Aminoalkyl-functional polydimethylsiloxanes have become very popular in the textile industry, where they are used as plasticizers and finishes [108], as well as in the production of

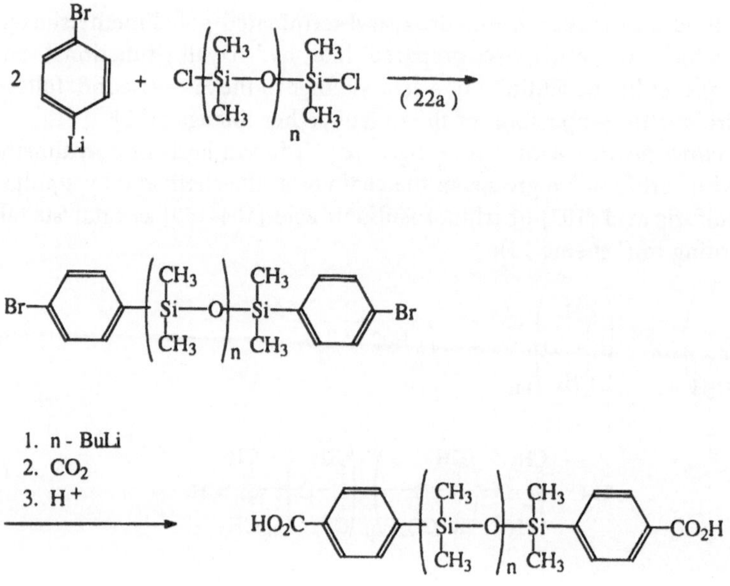

(21)

segmented copolymers (see Sect. 3.2). Ammonium-functional polydimethyl-siloxanes, which do not need to be discussed here, exhibit bactericidal properties [109].

Special preference is given to amino groups with the following structures (Scheme 22). Further variants are possible through branching of the alkyl group-$(CH_2)_n - CH_3$.

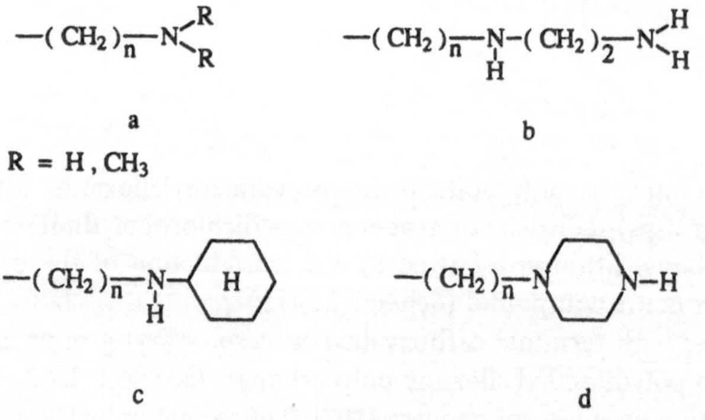

(22)

Two main methods are used for preparing aminoalkyl-functional poly-dimethylsiloxanes with a pendant aminoalkyl group in the main chain. One method consists of reacting cyclic (aminoalkyl)methylsiloxane with, preferably, four Si—O units, with cyclic dimethylsiloxane in the presence of a chain terminator such as hexamethyldisiloxane [110, 111], in analogy with carboxy-functional polydimethylsiloxanes (see Scheme 20). In the second method, an aminofunctional dialkoxy silane is reacted either with cyclic dimethylsiloxanes [112], or with trimethylsilyl-termined polydimethylsiloxanes [113] or with hydroxy-terminated polydimethylsiloxanes [114, 115]. The advantage of the silane route for preparing so-called "basic oils" is that mainly linear poly-dimethylsiloxanes with a well defined amine content are obtained. Because of the distinctly lower reactivity of the second Si—OMe bond, methoxy-terminated intermediates c are formed, which will react with further Si—OH groups only under more stringent conditions [115] (Scheme 23):

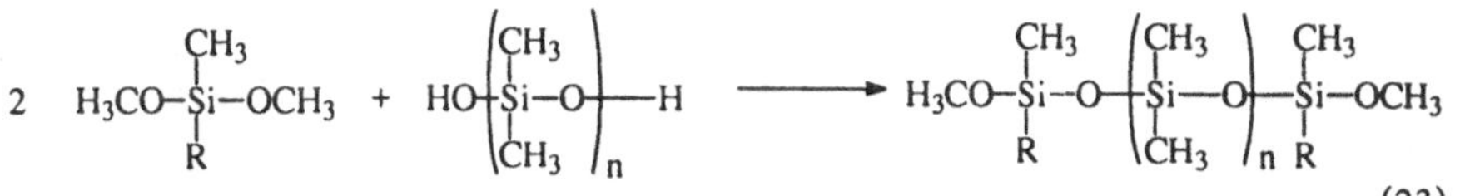

$$
2 \quad \text{H}_3\text{CO}-\underset{\underset{R}{|}}{\overset{\overset{CH_3}{|}}{Si}}-\text{OCH}_3 \quad + \quad \text{HO}\left[\underset{\underset{CH_3}{|}}{\overset{\overset{CH_3}{|}}{Si}}-O\right]_n\!\!-\text{H} \quad \longrightarrow \quad \text{H}_3\text{CO}-\underset{\underset{R}{|}}{\overset{\overset{CH_3}{|}}{Si}}-O\left[\underset{\underset{CH_3}{|}}{\overset{\overset{CH_3}{|}}{Si}}-O\right]_n\!\!\underset{\underset{R}{|}}{\overset{\overset{CH_3}{|}}{Si}}-\text{OCH}_3
$$

$$\hspace{3cm} a \hspace{3.5cm} b \hspace{6cm} c \hspace{5cm} (23)$$

$$R \;:\; -(\text{CH}_2)_3\text{NH}-(\text{CH}_2)_2\text{NH}_2$$

Further equilibration with pure polydimethylsiloxane produces polymers with excellent distribution of the amino group along the chain. When pure cyclic poly (aminoalkyl)methylsiloxanes are equilibrated with cyclic dimethyl compounds, the proportion of linear products in thermal equilibrium at 8.2 mol% CH_3RSiO units is 71% [111]. At higher amine contents, the cyclic compound content increases to >95% at 22.0 mol% CH_3RSiO units.

A third method of preparation which is not well established, involves a so-called polymer-analogous reaction. In this case, the chloride of a polydimethyl-siloxane with chloroalkyl groups in the chain is replaced by an appropriate amine [116].

Polydimethylsiloxanes with α, ω-terminal aminoalkyl groups are synthesized almost without exception via the alkali-catalysed polymerization of cyclic polydimethylsiloxanes, predominantly octamethylcyclotetra-siloxane (D_4) in the presence of 1,3-bis (aminoalkyl)tetramethyldisiloxanes as terminator, as has already been mentioned in Sect. 3.2.2 [51, 54, 55, 97, 117, 118] (Scheme 24):

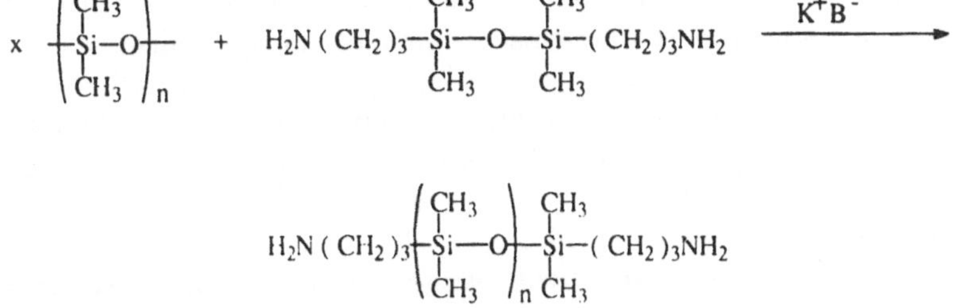

$$
x \left[\underset{\underset{CH_3}{|}}{\overset{\overset{CH_3}{|}}{Si}}-O\right]_n \;+\; \text{H}_2\text{N}(\text{CH}_2)_3-\underset{\underset{CH_3}{|}}{\overset{\overset{CH_3}{|}}{Si}}-O-\underset{\underset{CH_3}{|}}{\overset{\overset{CH_3}{|}}{Si}}-(\text{CH}_2)_3\text{NH}_2 \quad \xrightarrow{\;\;K^+B^-\;\;}
$$

$$(24)$$

$$
\text{H}_2\text{N}(\text{CH}_2)_3\left[\underset{\underset{CH_3}{|}}{\overset{\overset{CH_3}{|}}{Si}}-O\right]_n\!\!\underset{\underset{CH_3}{|}}{\overset{\overset{CH_3}{|}}{Si}}-(\text{CH}_2)_3\text{NH}_2
$$

Completion of the reaction with a view to the required chain length with optimum chain termination is decisively influenced by the choice of the base. Siloxanolates, which should be as pure as possible, are preferred to pure alkali hydroxides because of their better solubility in the mixture. The counter-ion K^+ is also of importance. Kinetic measurements carried out by McGrath et al. [51, 55, 117] have shown that the reaction with tetramethylammonium or tetrabutylphosphonium siloxanolate as catalyst is not only faster but also more efficient with regard to the incorporation of Bis-(aminopropyl)tetramethyldisiloxane as terminator.

Polydimethylsiloxanes with epoxy groups. Epoxy-functional polydimethylsiloxanes are used as cationically radiation-curing coating compounds [119, 120, 125–127, 129, 136–138] in the paper and electronics industries. Amine cured epoxysilicones are used in the textile industry as finishes for wool [121]. They are also incorporated as block segments in organic copolymers, as explained in Sect. 3.2. Epoxy-functional polydimethylsiloxanes may be synthesized by the addition of unsaturated organic epoxides to Si—H bonds **25 a**, or by the epoxidation of unsaturated substituents on the silicon atom [122, 123, 128] **25 b**. Another method consists of reacting α, ω-hydroxypolydimethylsiloxanes with epichlorohydrin [124] **25 c** or epoxyalkyldimethylchlorosilanes [98] **25 d**.

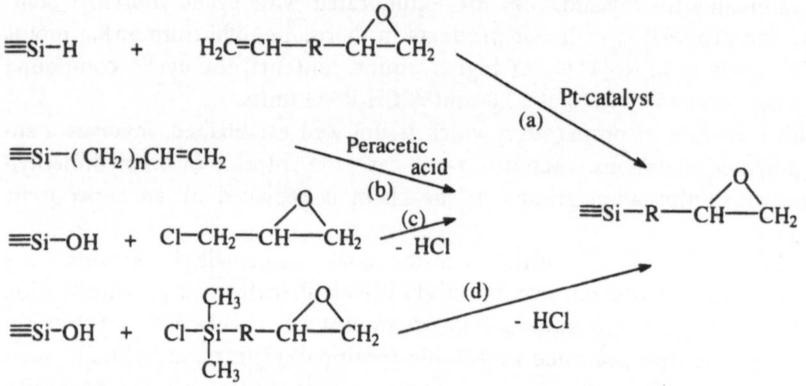

R', R = divalent, substituted or unsubstituted alkylene or alkylether group

The platinum catalyzed synthesis is probably the most elegant of methods, because this reaction produces the smallest amounts of by-products such as HCl, which needs to be eliminated from the end product. On the other hand, methods (25a) and (b) may be used to prepare epoxy-functional polydimethylsiloxanes with the epoxy group in the chain as well as at the end, what is not possible by methods (25c, d).

Polydimethylsiloxanes with (meth)acryloxyalkyl groups. (Meth)acryl functional polydimethylsiloxanes are used as soft segments in polyacrylate-polysiloxane copolymers (see Sect. 3.2) as well as radiation-curing coating compounds [139–141]. Mercapto groups need to be incorporated in order to achieve a non-smearing surface after UV curing in the presence of oxygen [130, 131].

Numerous ways of synthesizing acryloxyalkyl-functional polydimethylsiloxanes have been described in the literature. The article by P. J. Varaprath [132] provides a brief survey. Another possibility, which has not been mentioned in this article, is the ester interchange of (meth)acrylic acid esters with hydroxyalkyl-functional polydimethylsiloxanes [133] (Scheme 26):

$$\equiv Si-R-OH + R'-O-\overset{\overset{\displaystyle O}{\|}}{C}-CH=CH_2 \quad \xrightarrow[\text{- R'OH}]{} \quad \equiv Si-R-O-\overset{\overset{\displaystyle O}{\|}}{C}-CH=CH_2 \quad (26)$$

(Meth)acrylic-functional polydimethylsiloxanes with the (meth)acrylate groups in the chain or at the chain ends, can be prepared according to Scheme (26), but these highly reactive polymers need to be inhibited by means of radical interceptors (e.g. BHT) and protected against light, to prevent premature free-radical polymerization via the (meth)acrylate groups.

Polydimethylsiloxanes with mercaptoalkyl groups. Polydimethylsiloxanes with mercaptoalkyl groups are used industrially as release agents [142]. They are made either by equilibration of dimethyl and mercapto-functional, cyclic compounds or by the reaction of mercaptoalkyl methylalkoxysilanes with cyclic compounds (e.g. D_3, D_4) [142] in the presence of a chain terminator. Since these products are mentioned as starting materials in the next chapter about polysiloxane copolymers and since they are rarely used in the production of copolymers, they need not be discussed here.

3.2
Polysiloxane Copolymers

3.2.1
Fundamental Principles

Polysiloxane copolymers may be divided into block copolymers of type I, and into graft copolymers of type II. In type II polymers the substituent R attached to the silicon atom can either be a methyl group or another block B. The distribution of the grafted Si—O-groups in the copolymer is mostly statistical and not blocklike and is better represented by the comb-like structures of type II.

$$-(\, SiMe_2O)_n - B_x - (\, SiMe_2O\,)_m - \qquad\qquad I$$

$$-(\, SiMe_2O\,)_n - (\underset{\underset{\displaystyle B}{|}}{SiRO}\,)_m - (\, SiMe_2O\,)_n - (\underset{\underset{\displaystyle B}{|}}{SiRO}\,)_n \qquad\qquad II$$

Further branchings can be achieved by the introduction of $CH_3SiO_{3/2}$ groups (structure III), but specific comb-like structures can also be built up via an appropriate organic matrix structure (IV) or via cyclic compounds by addition of the organic component to silicon cycles containing Si-H groups (V).

Copolymers of polydiorganosiloxanes with substituents other than methyl groups can in principle be synthesized. However, such copolymers are of minor-technical importance for two reasons. Firstly, they are very expensive to prepare, and secondly, their properties depend mainly on the nature of the organic (on carbon based) blocks (B in structures I–V), so that a replacement of methyl groups by other substituents (e.g. Ph) will have little influence.

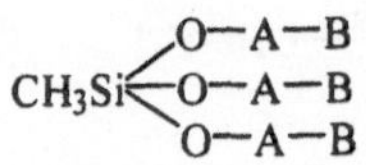

$$CH_3Si \Big\langle \begin{matrix} O-A-B \\ O-A-B \\ O-A-B \end{matrix}$$

III

$$\begin{matrix} | & | & | & | \\ B & B & B & B \\ | & | & | & | \\ A & A & A & A \end{matrix}$$

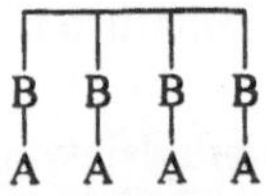

IV

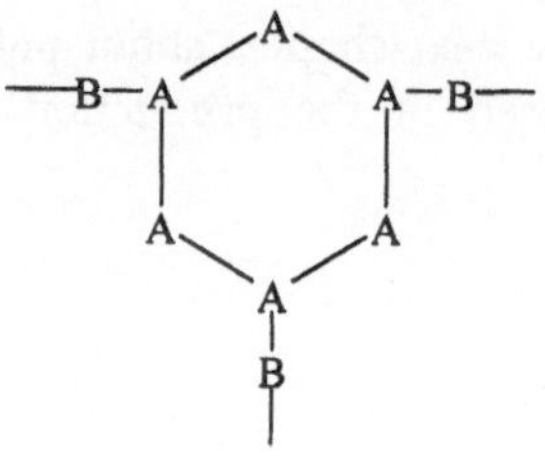

V

3.2.2
Siloxane-Alkylene Oxide Copolymers

Siloxane-alkylene oxide, also called siloxane polyether copolymers, form a very important group of compounds, being widely used as emulsifiers and foam stabilizers, e.g. in polyurethane foams [143, 155, 156].

As far as the linkage of the alkyleneoxide chain and siloxane is concerned, these copolymers may be divided into two groups: the hydrolytically more stable Si—C- and the hydrolytically less stable Si—O-linked copolymers. The structures of the various polymers can be subdivided into linear and branched types. Linear polymers are differentiated into (alkoxy)-(siloxane)-(alkoxy)-ABA, (siloxane)-(alkoxy)-(siloxane) = BAB- and ((siloxane)-(alkoxy))$_n$ = AB$_n$ types. The branched copolymers may consist of linear polydimethylsiloxanes to which the polyalkyleneoxide component (Form. 27) has been grafted

$$
\begin{array}{ll}
\underset{\displaystyle \overset{\displaystyle |}{\underset{\displaystyle |}{\text{R}'}}}{-\text{SiO}}\!\!-\!\!\overset{\displaystyle \overset{\text{CH}_3}{|}}{\underset{\displaystyle \underset{\text{CH}_3}{|}}{\text{Si}}}\!\!-\!\!\text{O}-
&
\end{array}
$$

CH$_3$ CH$_3$

—SiO—Si—O—

R'   CH$_3$

O (RO)$_n$ R'''

R', R = divalent alkylene group

R'' = H or monovalent alkyl or acyl group   (27)

They may also consist of products with T units which carry two or three poly (oxyalkylene) groups on the silicon atom. The different kinds of structures and their preparation have been very clearly and fully described in the review by Plumb and Asherton [144], so that we can confine ourselves to only brief descriptions of the basic synthetic principles.

Essentially, the methods of preparation depend upon the kind of linkage in the end product. Copolymers with SiO linkages are prepared by condensation or ester interchange, those with SiC linkages by hydrosilylation.

Preparation of Si—O—C-bound copolymers. Ester interchange is the preferred method of preparing polymers with Si—O—C linkages, since this takes place under milder conditions than condensation between SiOH or SiCl groups and poly(oxyalkylene) glycol. Hydrogen chloride is formed particularly when chlorine-functional siloxanes are used. The HCl can split the Si—O—C linkage and is very difficult to remove from the polymer. Although, ester interchange is carried out in the presence of acid or alkaline catalysts, these small amounts of acids and bases can easily be neutralized, and the liberated alcohol can be distilled off, Eq. (28) [146–148].

$$
x \;\; \text{C}_2\text{H}_5\text{-O}\!-\!\!\left(\!\!\overset{\text{CH}_3}{\underset{\text{CH}_3}{\text{SiO}}}\!\!\right)_{\!\!n}\!\!\overset{\text{CH}_3}{\underset{\text{CH}_3}{\text{Si-O-C}_2\text{H}_5}} \;+\; \text{HO-(C}_2\text{H}_4\text{O})_m\text{H} \;\xrightarrow[-\,\text{C}_2\text{H}_5\text{OH}]{\text{H}^+ \text{ o. B}^-\text{ Kat.}}
$$

$$
(28)
$$

$$
\text{C}_2\text{H}_5\text{-O}\!-\!\!\left[\!\!\left(\!\!\overset{\text{CH}_3}{\underset{\text{CH}_3}{\text{Si-O}}}\!\!\right)_{\!\!n}\!\!\overset{\text{CH}_3}{\underset{\text{CH}_3}{\text{Si-O}}}\text{---(C}_2\text{H}_4\text{O})_m\text{H}\!\right]_{\!x}
$$

The linear poly(oxyethylene)-polydimethylsiloxane copolymers formed according to Eq. (28) are relatively easily hydrolyzed, but their resistance to hydrolysis can be improved by introducing poly(1,2-oxypropylene) segments resulting in the formation of "silicone ethers" of secondary alcohols which are less sensitive to hydrolysis than the ethers of primary alcohols derived from poly(oxyethylene).

Branched structures are prepared by the same method as linear ones, branching taking place via the siloxane block by using oligomers with TD units (Form. 29a) or via the organic part, if etherified with a polyol (Form. 29b) [149].

$$
\begin{array}{cc}
\begin{array}{l}
\quad\quad O\,(\,SiMe_2O\,)_n\,R' \\
R{-}Si{\Large\langle}O\,(\,SiMe_2O\,)_n\,R' \\
\quad\quad O\,(\,SiMe_2O\,)_n\,R'
\end{array}
&
\begin{array}{ccc}
CH_2 & CH & CH_2 \\
| & | & | \\
O & O & O \\
| & | & | \\
R & R & R \\
| & | & | \\
OR' & OR' & OR'
\end{array}
\\[2em]
a & b
\end{array}
\qquad (29)
$$

R , R' = alkyl and alkylene groups respectively

Etherification of compound 29b takes place simultaneously with that of the siloxane and poly(oxyalkylene) component [149]. In contrast to Eq. (28) poly-(oxyalkylene) mono ethers are preferentially used. The advantage of branched copolymers is that they are more resistant to hydrolysis and chemical attack than linear ones, although the application decides which of the two types of product will to be used. Another very elegent method of preparing Si—O—C-linked poly(oxyalkylene)-polydimethylsiloxane copolymers is the alkali-[150] or tin-catalyzed [151] condensation of Si-H functions with the terminal OH groups of the poly(oxyalkylene). The hydrogen formed escapes from the system without the need to purify the end product, but the reaction needs careful control to prevent explosions. The evolution of H_2 has the additional advantage that no cleavage of Si—O—C bonds takes place.

Preparation of Si-C-linked copolymers. There are two different ways of preparing poly(oxyalkylene)-polydimethylsiloxane copolymers. The first way consists of condensing a carbo -functional polydimethylsiloxane with the hydroxyl group of the poly(oxyalkylene). The second way is based on the hydrosilylation of poly(oxyalkylene) with a terminal unsaturated group. Hydrosilylation, i.e. the addition of Si—H-functional polysiloxanes to the terminal C=C double bond of a poly(oxyalkylene) is a synthesis which produces no decomposition products. It may be used to prepare all the structural types tabulated in the review by Plumb and Atherton [144] and has been described in detail in the literature [144, 145, 157].

To explain the course of the reaction let us deal with an example, Eq. (30). Polymer 30b is a graft copolymer because the organic part has been added to individual chain segments of the polysiloxane by hydrosilylation. Linear analogues have been prepared by using α, ω-hydrogenpolysiloxanes as starting products.

"Inverse" polydimethylsiloxane-poly(oxyalkylene) copolymers, prepared by the hydrosilylation and, in part, subsequent condensation of hydrolyzable

$$(CH_3)_3\,SiO\left(\begin{matrix}CH_3\\|\\SiO\\|\\CH_3\end{matrix}\right)_n\left(\begin{matrix}CH_3\\|\\SiO\\|\\H\end{matrix}\right)_m Si(CH_3)_3 \quad + \;m\; H_2C{=}CH{-}CH_2{-}O\,(RO)_x R' \quad\longrightarrow$$

a

$$(CH_3)_3\,SiO\left(\begin{matrix}CH_3\\|\\SiO\\|\\CH_3\end{matrix}\right)_n\left(\begin{matrix}CH_3\\|\\SiO\\|\\(CH_2)_3{-}O\,(RO)\,R'_x\end{matrix}\right)_m Si(CH_3)_3 \tag{30}$$

$$R = \text{Alkylen}$$
$$R' = \text{Alkyl}$$

b

groups at the silicon atom, have been mentioned in recent patents [158]. In the case of the "inverse"copolymers of type ABA, A represents the siloxane fragment, just as in the graft copolymers of type -B-B-B-. In normal copolymers,

$$\begin{matrix}|&|\\A&A\end{matrix}$$

on the other hand, A represents the poly (oxyalkylene) segment.

Another interesting variant of the synthesis is the addition of the poly(oxy)alkylene component to epoxy-functional siloxanes, Eq. (31) [152]. The advantage of this method is that no by-products need to be removed, although, it is not frequently used because of its high cost.

$$\underset{CH_2}{\overset{O}{\triangle}}CH{-}CH_2{-}OR{-}\underset{CH_3}{\overset{CH_3}{Si}O{-}}(\underset{CH_3}{\overset{CH_3}{Si}O})_{\overline{n}}\underset{CH_3}{\overset{CH_3}{Si}O}{-}RO{-}CH_2{-}\underset{}{CH}{-}\underset{}{\overset{O}{\triangle}}CH_2 \quad + \;2\; HO\,(RO)_m R' \quad\longrightarrow$$

$$\tag{31}$$

$$R'\,(RO)_m O{-}CH_2{-}\underset{}{\overset{OH}{CH}}{-}CH_2{-}OR{-}\underset{CH_3}{\overset{CH_3}{Si}O}\left(\underset{CH_3}{\overset{CH_3}{Si}O}\right)_n\underset{CH_3}{\overset{CH_3}{Si}O}{-}RO{-}CH_2{-}\underset{}{\overset{OH}{CH}}{-}CH_2{-}O\,(RO)_m R'$$

$$R= \text{Alkylen} \quad R' = \text{Alkyl}$$

The physical properties of poly(oxyalkylene)-polysiloxane copolymers are due to their amphiphilic character, so that the interfacial properties of these substances were investigated in-depth. Polydimethylsiloxane-poly(oxyalkylene) block copolymers, unlike polydimethylsiloxanes, are water soluble and form micelles above a critical concentration. Above this so called critical micelle concentration, the surface tension of water is reduced from 72 mN/m to 20–30 mN/m, depending on the structure of the block copolymer [153, 154]. Linear copolymers apparently reduce the surface tension more than branched ones. Similarly, the surface tension of polyethylene glycol is also reduced. In this case, the surface tension of the solvent is 32 mN/m, so that a reduction to 20–30 mN/m is not as drastic as in the case of the aqueous solution [153, 154]. All figures quoted here refer to a 1% aqueous solution. Because of the hydrophilic properties of polysiloxane-poly(oxyalkylene) copolymers, these may be used for medical applications, (e.g. contact lenses), after crosslinking [229].

3.2.3
Polysiloxane-Polyester Copolymers

Polysiloxane-polyester copolymers may be subdivided into polysiloxane-polycarbonate, polysiloxane-polyphthalate and polysiloxane-poly(alkyl)esters as block copolymers, as well as polysiloxane-polymethacrylate copolymers.

3.2.3.1
Polysiloxane-Polycarbonate Copolymers

The polycarbonate blocks of these copolymers essentially consist of either bisphenol A or cyclobutylene carbonate segments [145]. The preparation of polysiloxane-bisphenol A-carbonate copolymers by phosgenation of bisphenol A and α, ω-dichloro-terminated polydimethyl siloxanes in the presence of pyridine was first described by Vaughn [159, 160]. The reaction mechanism has been described in detail in articles by Noshay and McGrath [145] as well as by Plumb and Atherton [144], so that it is sufficient to quote literature references. Instead of α, ω-dichloropolydimethylsiloxanes it is also possible to use α, ω-bis(dimethylamino)polydimethylsiloxanes as starting materials [161].

More recent methods consist of preparing the polydimethylsiloxanes with hydroxyaryloxy end groups (produced as intermediate stages in the above described reactions), followed by phosgenation after addition of other diphenols Eq. (32) [162–165].

$$HO-Ar-O-(\overset{\overset{\displaystyle R}{|}}{\underset{\underset{\displaystyle R^I}{|}}{Si}O})_{\overline{n}}-Ar-OH \quad + \quad HO-Ar^1-OH \quad \xrightarrow[\text{NaOH}]{COCl_2}$$

$$\text{excess}$$

$$H-\left[(O-Ar^1-O-\overset{\overset{\displaystyle O}{||}}{C}-)_m O-Ar-O-\left(\overset{\overset{\displaystyle R}{|}}{\underset{\underset{\displaystyle R^I}{|}}{Si}O}\right)_n Ar-OH\right]_x \tag{32}$$

The substituents R and R^1 can be identical or different and consist of monovalent substituted or unsubstituted alkyl or aryl groups. Ar and Ar^1 can likewise be identical or different, and are often of the bisphenol-A type, many different substitutions being possible.

The most elegant procedure is an interfacial polycondensation in which the two-phase system consists of an aqueous alkali hydroxide solution and a solvent which is immiscible with water. The advantage of this method is that the copolymers isolated from the organic phase are very pure and contain very well defined blocks.

The above-described copolymers have the disadvantage that they contain hydrolytically unstable Si—O—C-linkages. Therefore, reactions similar to Eq. (32), with α, ω-bis(hydroxyalkyl)polydimethylsiloxanes have also been described [166]. Another procedure is based on the polycondensation (esterification) of carboxyalkyl or carboxyaryl-functional polydimethylsiloxanes with the appropriate diols [168, 169]. This can be done either by the interfacial method [168] or in bulk in the molten state at temperatures between 300 and 400 °C [169]. Block copolymers are prepared from polydimethylsiloxanes which contain terminal carboxyl groups. Polysiloxanes in which the carboxyl groups are pending from the main chain, will produce graft polymers.

The physical properties of bisphenol A-polycarbonate-polydimethylsiloxane copolymers largely depend upon the bisphenol-A/polycarbonate/polysiloxane ratio. Neat polycarbonates show low elongation but high tensile strength. Incorporation of polydimethylsiloxane units increases their elongation, but reduces their tensile strength [170]. This behavior may be explained by the fact that block copolymers must generally be regarded as two-phase systems consisting of a hard, glassy phase and a soft elastic phase. In the above-described copolymers this soft phase is made up by the silicone blocks. In such two-phase systems the glassy or semicrystalline content acts as reinforcing material, rather like fillers, without which polydimethylsiloxanes, even in their crosslinked form, have hardly any mechanical strength [171]. The existence of a two-phase system is reflected by the fact that two glass transition temperatures can be detected by thermoanalysis one for the polydimethylsiloxane block, (around − 110 °C), and the second one between 60 and 140 °C, increasing with the number of carbonate units [145, 172]. The two-phase structure of the copolymer has also been confirmed by X-ray and electron microscopic examination [145].

As expected, the surface tension of the block copolymer is higher than that of neat polydimethylsiloxanes, with values in the range of 32–34 mN/m. These values depend little on the chain lengths of the siloxane blocks. On the other hand, the surface tension can be increased to > 40 mN/m by suitable end groups of the polysiloxane chain, i.e. by increasing the aromatic content of the organic substituents. This is desirable for certain applications, for instance, if a part made from poly(dimethylsiloxane-polycarbonate)s has to be painted [162].

The crosslinking of block copolymers by introducing reactive terminals groups such as acetoxy or allyl groups has been described. This measure leads to the formation of elastomers with high mechanical strength, which cannot be deformed even at high temperatures [174, 175].

Polydimethylsiloxane-polycarbonate block copolymers with tetramethyl-1, 3-cyclobutylene segments are less widely used in practice than the above described bisphenol-A products. They are prepared in a similar way by reacting a cyclobutylene carbonate prepolymer with an α, ω-bis (dimethylamino)poly-dimethylsiloxane. The reaction mechanism has been described in detail in review articles [145, 176], so that it is sufficient to illustrate the structure of the resultant block copolymers 33. As far as physical properties are concerned, there are no major differences compared to bisphenol-A analogs, except that they are easier to process from the molten state [177].

3.2.3.2
Polysiloxane-Polyphthalate Copolymers

Polysiloxane-polyphthalate copolymers can be prepared by condensation of the appropriate polyesters with α, ω-bis (dimethylamino)polydimethylsiloxanes. The

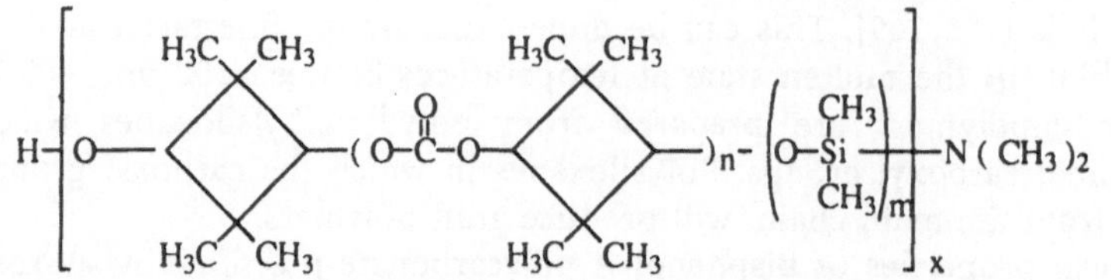

$$\text{(structure 33)} \tag{33}$$

polyester can also be formed in situ by reacting phthaloyl or isophthaloyl dichloride with alkyl or aryl diols and with the amino group-terminated poly-dimethylsiloxane. However, this method has the disadvantage that the blocks are not well defined [145]. Analogous reactions with polydimethylsiloxanes whose amino groups are linked with the polydimethylsiloxane via alkylene bridges have also been described [178]. Therefore, resultant copolymers no longer have a hydrolytically sensitive Si—O—C linkage, but the linkage between of polydimethylsiloxane and polyester block has an amide structure (34).

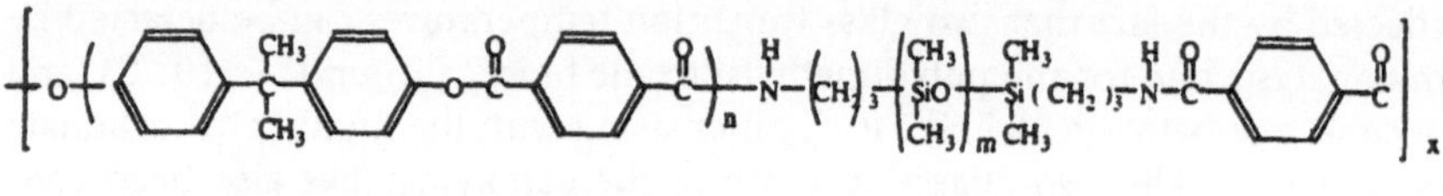

$$\text{(structure 34)} \tag{34}$$

Inoue [167] has likewise described polysiloxane-polyester copolymers with hydrolytically stable Si—C linkages obtained by reacting 1,3-bis (p-hydroxyphenyl) disiloxanes **35a** with iso- or terephthalic acid or their acid chlorides, Eq. (35). As it is evident from structure **35b** these copolymers possess a strictly alternating sequence and not a blocky structure with variable block lengths. They are remarkable due to their outstanding thermal stability. A loss of weight of 5% was only observed at temperatures of 436 °C and above [167].

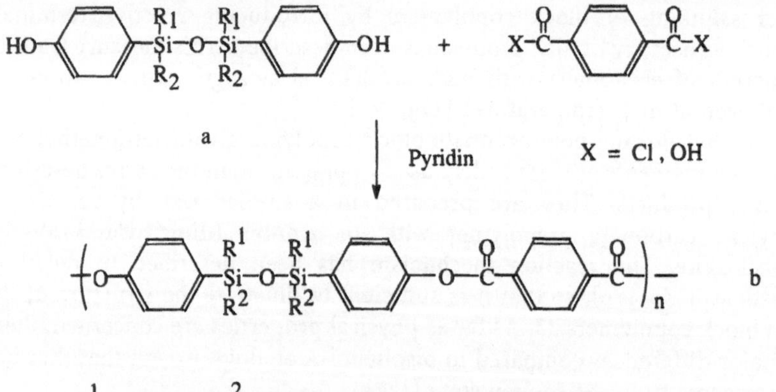

$$\text{(structure 35)} \tag{35}$$

R^1 = Methyl, R^2 = Methyl or Phenyl, X = Cl,OH

The morphology and the resultant physical properties of polysiloxane-phthalic acid polyester block copolymers have been exhaustively described in the literature [145, 173, 178]. They possess a two-phase structure with different glass transition temperatures (T_g's) like the previously described polysiloxane-bisphenol-A polycarbonates. The T_g of the polysiloxane segment is approximately

-110 to $-120\ °C$, that of the polyester segment around $140-210\ °C$ [173, 178]. The figures for the second glass transition temperature refer to the polyester fragments with bisphenol-A and tetramethyl-substituted bisphenol-A as the diol component. The tetramethyl substitution results in an increase of the T_g by about 30 °C compared to that of the unsubstituted bisphenol-A.

With alternating copolymers 35b, on the other hand, the T_g will only fall into the range of 62–79 °C depending on whether R_2 is a methyl or a phenyl group and depending on whether the organic fragment is an iso- or terephthalic acid ester. In totally similar polyesters, where bisphenol-A is the diol component instead of 1,3-bis (hydroxyphenyl)disiloxanes 35a, the T_g is twice as high. The lowering of the glass transition temperature in copolymers containing disiloxane fragments is thought to be due to the greater flexibility of the Si—O linkages [167].

Graft copolymers are a more recent group of phthalic acid ester copolymers containing siloxanes. In this case, a polyester with free allyl groups 36a is first prepared (e.g. from terephthalic acid dichloride and trimethylol propane mono-allyl ether) and the polysiloxane 36b is grafted on it by hydrosilylation (Scheme 36) [186].

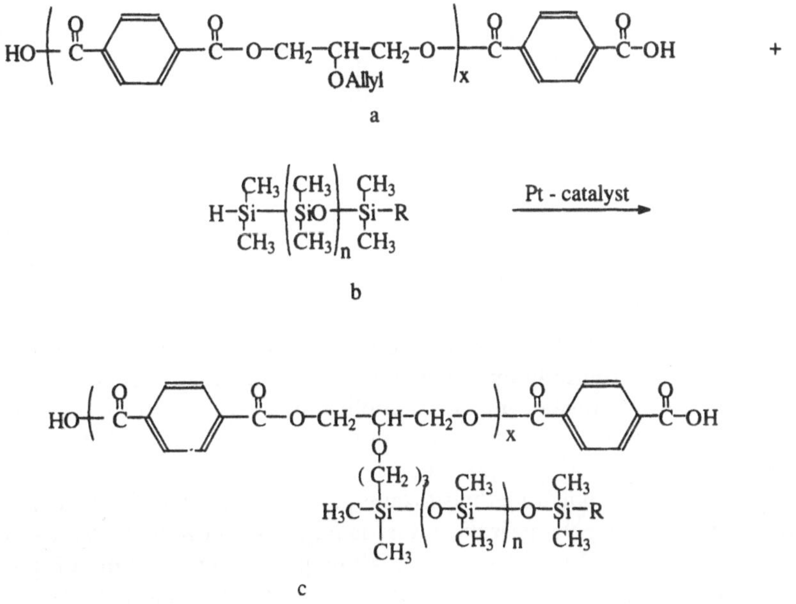

(36)

3.2.3.3
Polysiloxane-Poly(alkyl) Ester Copolymers

Besides the syntheses already mentioned for other groups of polymers the addition of carboxyl-terminated polyesters to α, ω-bis-(glycidoxy)polydimethyl siloxanes has been described for this type of copolymer [176, 179]. Products with hydrolytically stable Si—C linkages are obtained in this way.

Polysiloxane-poly(alkyl)ester copolymers have less economic importance than the products described in Sect. 3.2.3. However mention should be made of a special type of copolymer which is used particularly in medicine as implant material because of its hydrophilic properties and high flexibility, since it also differs structurally from the products mentioned earlier. In the copolymers described so far, the polyester fragment consists of a dicarboxylic acid which has been esterified with a diol. In this case, an α-hydroxycarboxylic acid (e.g. lactic or glycolic acid) is reacted with an hydroxy-terminated polydimethylsiloxane. The hydroxyl group can be linked direct to the silicon atom or can be linked via an alkylene group, Eq. (37) [180]. Since the α-hydroxycarboxylic acid is used in the form of lactide or glycolide, this procedure is a ring-opening polymerization of the organic component, and the hydroxy-terminated polysiloxane block serves as initiator.

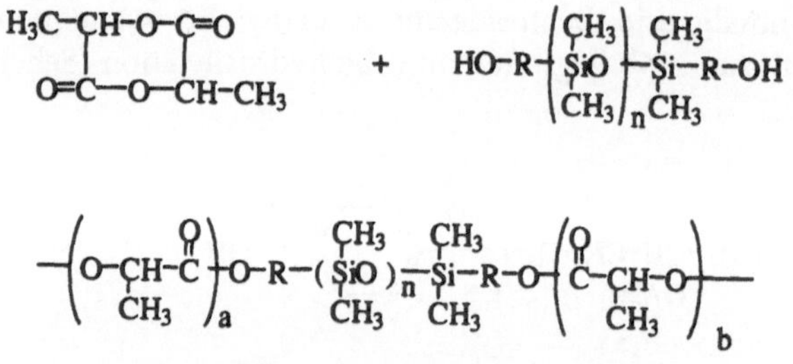

$$(37)$$

R = substituted or unsubstituted alkylene group

3.2.3.4
Polysiloxane-Poly(meth)acrylate Copolymers

Polysiloxane-poly(meth)acrylate copolymers are fairly new and their physical properties have not been fully investigated yet. However, they were found to be of interest for a wide range of industrial applications because silicone-modified acrylates are less brittle, making them suitable for applications requiring great flexibility such as the coating of glass fibers.

The production of polydimethylsiloxane-poly(meth)acrylate block copolymers by group transfer polymerization was, as far as we know, first described in 1989 in a patent by Hilty et al. [181], although the preparation of poly-(meth)acrylates by this technique was already described in the patents by Webster [182] and Farnham [183]. In the case of copolymers with polydimethyl-siloxanes, (meth)acrylate-terminated polydimethylsiloxanes are reacted with (meth)acrylic acid esters, preferably methyl methacrylate, Eq. (38) [181].

In this reaction, the product distribution depends on the ratio of the constituents and on the sequence in which they were added. Accordingly, the siloxane blocks can be distributed statistically as well as block-like between the polyacrylate segments. If polysiloxanes which have acrylate groups at both ends of the chain are used, polyacrylates crosslinked via siloxane blocks are produced.

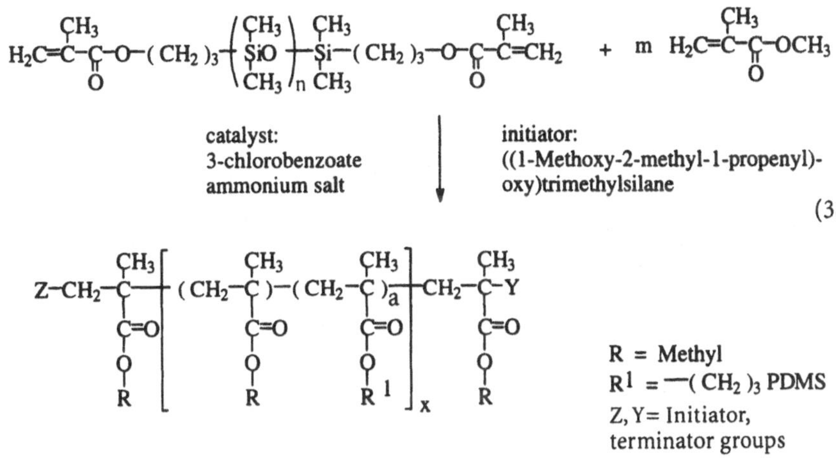

$$
\begin{array}{c}
\underset{\substack{\displaystyle \\ O}}{H_2C{=}C{-}C{-}O{-}(CH_2)_3}\!\!\!\overset{CH_3}{\underset{}{}}\!\!\!\left(\!\underset{CH_3}{\overset{CH_3}{Si}O}\!\right)_{\!\!n}\!\!\underset{CH_3}{\overset{CH_3}{Si}}\!{-}(CH_2)_3{-}O{-}\underset{\substack{\displaystyle \\ O}}{C{-}C{=}CH_2}\overset{CH_3}{} \quad + \quad m \ \underset{\substack{\displaystyle \\ O}}{H_2C{=}C{-}C{-}OCH_3}\overset{CH_3}{}
\end{array}
$$

catalyst:	initiator:
3-chlorobenzoate ammonium salt	((1-Methoxy-2-methyl-1-propenyl)-oxy)trimethylsilane

(38)

$$
Z{-}CH_2{-}\underset{\substack{C{=}O \\ \underset{R}{O}}}{\overset{CH_3}{C}}\!\!\!\left[{-}(CH_2{-}\underset{\substack{C{=}O \\ \underset{R}{O}}}{\overset{CH_3}{C}}){-}(CH_2{-}\underset{\substack{C{=}O \\ \underset{R^1}{O}}}{\overset{CH_3}{C}})_a\right]_{\!\!x}\!\!{-}CH_2{-}\underset{\substack{C{=}O \\ \underset{R}{O}}}{\overset{CH_3}{C}}{-}Y
$$

R = Methyl
R^1 = —(CH$_2$)$_3$ PDMS
Z, Y = Initiator,
terminator groups

Also the so-called "hybrid star" polymers are a siloxane-crosslinked with a siloxane nucleus from which polyacrylate "arms" branch off. The polysiloxane nucleus is formed by polycondensation of alkoxysilyl groups. The basic polymer to be crosslinked initially consists of a neat polyacrylate block on to which a mixture of acrylic monomer and silane-modified acrylate (e.g. methyl methacrylate and (trimethoxy)-silylpropyl methacrylate) is polymerized [184, 185]. The distribution of the silicon units in the second block is statistical. The crosslinking is then a consequence of the trifunctionality of the silicon quite analogous to that of silicon resins.

Polydimethylsiloxane-poly(meth)acrylate block copolymers can also be prepared by free-radical polymerization. This can be initiated by heat as well as by high energy radiation. The review by Yilgör and McGrath [178] gives an overview as well as a comprehensive bibliography of various methods of preparing graft, block and statistical copolymers. Inoue [207] reported on the thermal freeradical polymerization, in which the (meth)acrylic acid ester monomer is first polymerized using a radical initiator. In this case, a macro-initiator is used, an azo-bis(cyanopentamide)-substituted polysiloxane which forms the appropriate initiator radicals when heated. A patent [208] by Rhône Poulenc likewise reports on the high temperature polymerization with disulfur thiuram-coupled polysiloxanes, which is described in Sec 3.2.6. However, the method yielding the best defined block copolymers is said to be the radiations-induced radical polymerization using "polysiloxane-iniferter compounds" [206]. "Iniferter" compounds are those which have three functions: that of *initiator*, trans*fer* and *ter*minator. Such compounds include mono- and bis-xanthane or dithiocarbamate-terminated polydimethyl siloxanes 39c, which are obtained by reacting alkyl dithiocarbamates 39a (or xanthans) with α, ω-bis(chloroalkyl)polydimethylsiloxanes 39b [206]. When 39c is irradiated with UV light or other high energy radiation, it decomposes into the terminator radical and the polysiloxane macro-initiator which simultaneously with initiation of the methyl methacrylate polymerization transfers the block B (siloxane block) to the copolymer (Scheme 40).

In contrast to anionic polymerization, there is no equilibration in this type of reaction, so that the siloxane block will be incorporated into the block polymer

$$\underset{a}{Et_2N-\overset{\overset{S}{\|}}{C}-S^-\,Na^+} \;+\; \underset{b}{Cl-R-\overset{\overset{Me}{|}}{\underset{\underset{Me}{|}}{Si}}-\left(O-\overset{\overset{Me}{|}}{\underset{\underset{Me}{|}}{Si}}\right)_n O-\overset{\overset{Me}{|}}{\underset{\underset{Me}{|}}{Si}}-X} \quad\longrightarrow$$

R = alkylen   (39)
X = alkyl or a−R−alkoxy
Y = X or alkylene dithiocarbamate

$$\underset{c}{Et_2N-\overset{\overset{S}{\|}}{C}-S-R-\overset{\overset{Me}{|}}{\underset{\underset{Me}{|}}{Si}}-\left(O-\overset{\overset{Me}{|}}{\underset{\underset{Me}{|}}{Si}}\right)_n O-\overset{\overset{Me}{|}}{\underset{\underset{Me}{|}}{Si}}-Y}$$

$$39c \xrightarrow{\;h\nu\;} Et_2N-\overset{\overset{S}{\|}}{C}-S^\bullet \;+\; {}^\bullet R-\overset{\overset{Me}{|}}{\underset{\underset{Me}{|}}{Si}}-\left(O-\overset{\overset{Me}{|}}{\underset{\underset{Me}{|}}{Si}}\right)_n O-\overset{\overset{Me}{|}}{\underset{\underset{Me}{|}}{Si}}-Y$$

$$+\; aCH_2=\overset{\overset{Me}{|}}{C}-CO_2-Me \qquad\Big\downarrow$$

$$\underset{B}{Y-\overset{\overset{Me}{|}}{\underset{\underset{Me}{|}}{Si}}-O\left(\overset{\overset{Me}{|}}{\underset{\underset{Me}{|}}{Si}}-O\right)_n \overset{\overset{Me}{|}}{\underset{\underset{Me}{|}}{Si}}-R}-\underset{A}{\left(CH_2-\overset{\overset{Me}{|}}{\underset{\underset{\underset{OMe}{|}}{\overset{|}{C}=O}}{C}}\right)_b}CH_2-\overset{\overset{Me}{|}}{\underset{\underset{OMe}{\overset{|}{C}=O}}{C}}{}^\bullet$$

b = a - 1   (40)

$$+\; Et_2N-\overset{\overset{S}{\|}}{C}-S^\bullet \qquad\Big\downarrow$$

$$\underset{B}{Y-\overset{\overset{Me}{|}}{\underset{\underset{Me}{|}}{Si}}-O\left(\overset{\overset{Me}{|}}{\underset{\underset{Me}{|}}{Si}}-O\right)_n \overset{\overset{Me}{|}}{\underset{\underset{Me}{|}}{Si}}-R}-\underset{A}{\left(CH_2-\overset{\overset{Me}{|}}{\underset{\underset{\underset{OMe}{|}}{\overset{|}{C}=O}}{C}}\right)_b}CH_2-\overset{\overset{Me}{|}}{\underset{\underset{OMe}{\overset{|}{C}=O}}{C}}-S-\overset{\overset{S}{\|}}{C}-N-Et_2$$

b = a - 1

in the same form in which it is contained in the "iniferter"component. It is, thus, feasible to obtain copolymers with specific block of type AB, ABA as well as CAB and CABAC. According to the inventors [206], photo-"iniferter" polymerization is superior to all thermal processes since the technique produces neat block copolymers which do not contain any of the corresponding homopolymers as impurities. It is, thus, possible to synthesize tailor-made copolymers with low as well as high average molelcular weights. In the case of

Table 3.4. T_g of various poly dimethylsiloxane poly(methyl)methacrylate blockcopolymers

Type of Blockcopolymer	$T_G/°C$		Weight % Siloxane
	Siloxane	Organoblock	
MMA-PDMS-MMA	−124	+105	30%
A—B—A			
MMA-PDMS	−124	+106	46%
A—B			
HEA-MMA-PDMS-MMA			
-HEA	−120	+107	29%
C—A—B— A—C			

MMA: Methylmethacrylate.
HEA: Hydroxyethylacrylate.

ABA and CABAC block copolymers, difunctional siloxane "iniferter" 39, socalled "biniferter"compounds, are used (in this case $Y = Et_2NCS_2R$- in the structure 39c). Block polymers of this kind all exhibit microphase separation, as can be seen from the two different glass transition temperature (see Table 3.4) [206]. As is evident from Table 3.4, the glass transition temperature is unaffected by the sequence, sequence type and sequence length.

3.2.4
Polysiloxane-Polysulfone Block Copolymers

The simplest polymers of this group of products, chemically speaking, are built up from polydimethylsiloxane and 4,4′-dihydroxydiphenyl sulfone-polyether blocks 41.

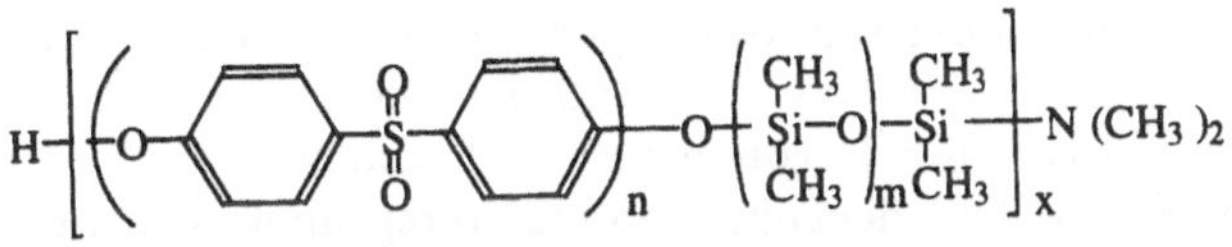

$$H-\left[\left(O-\!\!\left\langle\!\!\!\bigcirc\!\!\!\right\rangle\!\!-\!\!\overset{\overset{O}{\|}}{S}\!\!-\!\!\left\langle\!\!\!\bigcirc\!\!\!\right\rangle\!\!\right)_n\!\!-O\!-\!\!\left(\!\!\begin{array}{c}CH_3\\|\\Si-O\\|\\CH_3\end{array}\!\!\right)_{\!m}\!\!\begin{array}{c}CH_3\\|\\Si\\|\\CH_3\end{array}\!\!\right]_{\!x}\!\!\!-N\,(CH_3)_2 \qquad (41)$$

These block copolymers are prepared in a similar manner as the polysiloxane-polyester copolymers described in Sec 3.2.3, by condensing oligomeric 4,4′-dihydroxydiphenyl sulfone polyethers and α, ω-bis (dimethylamino) polydimethylsiloxane [187]. The polymers with formula 41 are far less hydrolytically stable than the copolymers 42 described earlier by Noshay [188].

The greater sensitivity of polymers 41 to hydrolysis compared to 42 is thought to be due to the fact that the electron-attracting effect of the sulfone group has a direct effect upon the Si—O—C linkage. In contrast, the opposite effect is operating in polymers of structure 42 due to the incorporation in bisphenol A. The isopropylidene group makes splitting the Si—O—C linkage by nucleophilic attack difficult, due to its +I effect [188].

$$\text{HO}\!\!\left[\text{R}\!\!\left(\underset{O}{\overset{O}{\underset{\|}{\underset{\|}{S}}}}\right)\!\!-\!\!O\!\!-\!\!\left(\underset{CH_3}{\overset{CH_3}{\underset{\|}{C}}}\right)_m\!\!-\!\!O\!\!\left(\underset{CH_3}{\overset{CH_3}{\underset{\|}{Si}}}\!\!-\!\!O\right)_n\!\!\underset{CH_3}{\overset{CH_3}{\underset{\|}{Si}}}\!\!-\!\!\right]_x\!\!N(CH_3)_2$$

(42)

$$R = \quad \left(\bigcirc\right)\!\!-\!\!\underset{CH_3}{\overset{CH_3}{\underset{\|}{C}}}\!\!-\!\!\left(\bigcirc\right)$$

Polysiloxane-polyaryl (alkyl) sulfone carbonate copolymers 43 [189] are a third variant. The polymers 43 are prepared in two stages—first, the poly-(arylsulfonecarbonate) copolymer is formed by phosgenation of a substituted or unsubstituted 4,4'- dihydroxyphenylsulfone (bisphenol-S). Copolymerisation can also take place in a mixture with bisphenol-A, followed by reaction with a polydimethylsiloxane containing bisphenol-A units at the end of the chain [189].

$$\text{H}\!\!\left[\!\!\left(\!\!-\!\!O\!\!-\!\!\left(\bigcirc\right)\!\!\underset{O}{\overset{O}{\underset{\|}{\underset{\|}{S}}}}\!\!\left(\bigcirc\right)\!\!-\!\!O\!\!-\!\!\underset{}{\overset{O}{\underset{\|}{C}}}\!\!\right)_m\!\!-\!\!O\!\!-\!\!R\!\!-\!\!O\!\!\left(\underset{CH_3}{\overset{CH_3}{\underset{\|}{Si}}}\!\!-\!\!O\right)_n\!\!\right]_x\!\!-\!\!R\!\!-\!\!OH$$

(43)

$$R = \quad \left(\bigcirc\right)\!\!-\!\!\underset{CH_3}{\overset{CH_3}{\underset{\|}{C}}}\!\!-\!\!\left(\bigcirc\right)$$

As expected, all these copolymers exhibit microphase separation, (i.e. two-phase morphology). In polymers with structures 42 and 43 the biphasic morphology is indicated by different glass transition temperatures from a critical siloxane content on [145, 189]. In contrast, only one T_g ($-120\,^\circ$C) is detectable in the case of copolymers 41 namely that of the polysiloxane segment. According to the authors [187], this finding is not surprising because only a broad diffuse glass-transition was previously observed for the polysulfone segment. Nonetheless, it was feasible to confirm the two-phase morphology of polymers with structure 41 by electron microscopy. In copolymers with structure 42, the T_g of the organic fragment is 160–170 °C, depending on the average molecular weight, i.e. the weight percentage of polydimethylsiloxane block, which should be at least 40% (corresponding to a M_n of 4900 g/mol). If the silicone content is <28% ($M_n = 1700$ g/mol), the glass transition temperature is observed at 140 °C [203]. The T_g values for the polysulfone fragment of copolymers with structure 43 are very high (around 200 °C).

It is interesting to note that, if the siloxane content is 60%, the glass transition temperature will be lower, namely 172 °C because this effect has not been described in the case of pure polydimethylsiloxane-polysulfone block copolymers (see Table 3.5) [187].

It is not clear how much siloxane has to be present for two glass transition temperatures to become detectable. However, it can be assumed that this would be along much the same lines as for copolymers with structure 42.

Table 3.5. T_g's of the poly(ether-sulfone) blocks in block-copolymers of structure (**43**)

Weight % PDMS[a]	T_G (°C)
5	208
40	192
60	172

[a]polydimethylsiloxane

3.2.5
Polysiloxane-Polyimide, -Polyamide, -Polyurea and -Polyurethane Copolymers

The synthesis of polydimethylsiloxane-polyamide copolymers from α, ω-bis(aminoalkyl)polydimethylsiloxanes and hydroxy-terminated polyamide prepolymers has been known for some time [190, 191]. However, Polysiloxane-polyimide copolymers have achieved greater industrial importance because of their excellent electrical properties, high impact strength and extremely high resistance to oxidation by atomic oxygen. These copolymers can also be synthesised from α, ω-bis(aminoalkyl)polydimethylsiloxanes which are condensed with organic diamines and dianhydrides to form imide copolymers. The numerous variants of these copolymers have been described in an excellent review by Yilgör and McGrath [178] which also contains a detailed bibliography. It is also possible to use polydimethylsiloxanes with terminal anhydride groups as the siloxane starting component [192–194]. A further possibility for the preparation of these copolymers is the platinum-catalysed addition of N,N'-diakylene diimides to the Si—H groups of polydimethylsiloxane [195]. Because poly(imide-siloxane)s are described elsewhere in this book due to their liquid crystalline properties, and because the above mentioned review [178] gives an excellent survey of their physical properties, there is no need to describe here these important products in detail.

$$\left[(-\overset{\underset{\displaystyle H}{|}}{N}-(CH_2)_3 - \left(\overset{\underset{\displaystyle CH_3}{|}}{\underset{\underset{\displaystyle CH_3}{|}}{Si}}-O \right)_n \overset{\underset{\displaystyle CH_3}{|}}{\underset{\underset{\displaystyle CH_3}{|}}{Si}}-(CH_2)_3 - \overset{H}{N}-\overset{O}{C}-\overset{H}{N}-R-\overset{H}{N}-\overset{O}{C}-) \right]_x \qquad (44)$$

R=substituted or unsubstituted alkylene or arylene group

Reaction of α, ω-bis (aminoalkyl)polydimethylsiloxanes with organic diisocyanates produces poly(urea-siloxane)s **44** [178, 196, 230]. With poly(urea-siloxane)s the strong hydrogen bridge linkages between the urea segments produce elastomer-like properties, i.e. high mechanical strength even at high temperatures [231].

The chemical structure of polysiloxane-polyurethane copolymers is very similar to that of poly(urea-siloxane)s. Here the siloxane consists of

α, ω-bis(hydroxyalkyl)polydimethylsiloxanes [232]. Analogous to poly(urea-siloxane)s only copolymers with short chains (average molecular weight about 1500–3000 g/mol) are usually obtained. This is thought to be due to the different solubility parameters of the starting product [178], which result in a 95% phase separation also in the copolymer [212]. In order to prepare high molecular weight copolymers with good mechanical properties, it is necessary to react polysiloxane-polyurethane prepolymers containing isocyanate endgroups, with chain-extending diols or polyols [178, 197]. Reacations with polyols result in crosslinked products which can be used for making contact lenses because of their high oxygen permeability and hydrophilic properties [197, 227]. Block copolymers containing ammonium groups in their skeleton structure possess "physical crosslinks" because of the strong ionic interactions. Therefore, they exhibit elastomeric behaviour (elongation up to 800%, tensile strength up to 880 psi), as well as good heat resistance (up to 250 °C) [198].

Polarizable, non-linear, optically active polysiloxane-polyurethane block copolymers are obtained by reaction of isocyanatearyloxyalkyl-terminated polydimethylsiloxanes with the corresponding polarizable organic diols. These hyperpolarizable copolymers, which can be excited by an electrical field, are used for highly specialized electro-optical applications [209].

3.2.6
Polysiloxane-Polyalkene Copolymers

The main alkene constituents of polysiloxane-polyalkylene copolymers are isoprene, styrene or α-methyl styrene. The synthesis of copolymers 45 with isoprene was first described by Morton in 1964, those with styrene as the polyalkylene component in 1970 by Dean [205]. In both cases the alkene is polymerized anionically to form a prepolymer which is then reacted with cyclic polydimethylsiloxanes (mainly octamethylcyclotetrasiloxane) via anionic ring opening. Coupling of the resulting block copolymer dianion with dichloro-dimethyl silane, yields a multiblock copolymer with $(ABA)_x$ structure 45 [233]. Details of the procedure were described in previous reviews [144, 145]. An other method is free-radical polymerization which has already been mentioned in

$$\left[-\left(O-\underset{\underset{CH_3}{|}}{\overset{\overset{CH_3}{|}}{Si}}\right)_n\left(CH_2-\underset{\underset{R}{|}}{CH}\right)_a\left(\underset{\underset{CH_3}{|}}{\overset{\overset{CH_3}{|}}{Si}}-O\right)_m\underset{\underset{CH_3}{|}}{\overset{\overset{CH_3}{|}}{Si}}-\right]_x \tag{45}$$

R = Phenyl oder
Iso-propenyl

Sect, 3.2.3., involving the polymerization of (meth)acrylalkyl- or styryl-terminated polydimethylsiloxanes with styrene in the presence of organic initiators [210, 211]. According to the very lates methods, "iniferter" polymers are used as initiators in preference to standard organic radical initiators. In the patent of Clouet [208], "mono-iniferter"compounds are used which enable specific block copolymers of the ABA type to be prepared. As explained in Sect. 3.2.3. such a synthesis normally requires "di-iniferter"compounds. However, in this special case the "mono-iniferter" is split symmetrically, so that initiator and terminator radicals are chemically identical, Eq. (46) [208].

$$H_3C\left(\begin{array}{c}CH_3\\ Si-O\\ CH_3\end{array}\right)_n \begin{array}{c}CH_3\\ Si\\ CH_3\end{array}-(CH_2)_3-\begin{array}{c}CH_3\\ N\end{array}-\begin{array}{c}C\\ \parallel\\ S\end{array}-S \quad - \quad S-\begin{array}{c}C\\ \parallel\\ S\end{array}-\begin{array}{c}CH_3\\ N\end{array}-(CH_2)_3-\begin{array}{c}CH_3\\ Si\\ CH_3\end{array}\left(\begin{array}{c}CH_3\\ O-Si\\ CH_3\end{array}\right)_n CH_3$$

$$\Delta T \quad \begin{array}{c} a \quad HC=CH_2\\ + \\ \bigcirc \end{array}$$

(46)

$$H_3C\left(\begin{array}{c}CH_3\\ Si-O\\ CH_3\end{array}\right)_n \begin{array}{c}CH_3\\ Si\\ CH_3\end{array}-(CH_2)_3-\begin{array}{c}CH_3\\ N\end{array}-\begin{array}{c}C\\ \parallel\\ S\end{array}-S-\left(\begin{array}{c}CH_2-CH\\ Ph\end{array}\right)_a -S-\begin{array}{c}C\\ \parallel\\ S\end{array}-\begin{array}{c}CH_3\\ N\end{array}-(CH_2)_3-\begin{array}{c}CH_3\\ Si\\ CH_3\end{array}\left(\begin{array}{c}CH_3\\ O-Si\\ CH_3\end{array}\right)_n CH_3$$

b

The physical properties of polydimethylsiloxane-polystyrene block copolymers largely depend on the total chain length as well as on the styrene/siloxane ratio. Only relatively short-chain variants or copolymers with structure **46b** (ABA type) were prepared, but, as far as we know, no physical constants such as glass transition temperature or mechanical properties have been published. According to investigations by Saam [233], x in structure **45** ((ABA)$_x$ type) must be higher than 2, and block A and B should have a high average molecular weight, if useful mechanical properties such as tensile strength or elongation are required. Copolymers with $M_n > 100\ 000$ and a silicone content of 80% exhibit elastomeric properties similar to those of cured silicones, i.e. relatively low tensile strength but high elongation. Those with a silicone content of 50% behave like polystyrene, i.e. they have high tensile strength and low elongation [234]. In this case it is the total degree of polymerization which is of decisive importance. Polydimethylsiloxane block copolymers with poly(α-methyl styrene) containing 73% siloxane and with $M_n \leqslant 20\ 000$ yield copolymers with extraordinarily good mechanical properties, such as a tensile strength of 1300 psi and elongation of 935% [145].

Polysiloxane-polystyrene block copolymers have two glass transition temperatures, that of the siloxane block around $-120\ °C$ and that of the polystyrene block around 50 °C [145]. The glass transition temperature of α-methyl styrene copolymers is between 90 and 140 °C, and even the slightest changes in block copolymer composition will cause it to fluctuate considerably [203].

New kinds of siloxane-polyene copolymers have in recent years found applications in microelectronics. The combination of physical and chemical properties, including a high glass transition temperature, low coefficient of thermal expansion and dielectric constant, as well as good moisture resistance, make these copolymers suitable for this application [218, 219], and they are, thus, a promising new development. They were synthesized by the platinum catalyzed hydrosilylation of the polyene double bonds. Dicyclopentodiene and its oligomers are preferred starting products for the polycyclic polyene component, vinyl norbornes are used for the polyene component and chain extender. Linear Si-H-terminated polydimethylsiloxanes or cyclic polyhydrogensiloxanes are used for the siloxane moiety. Hydrocarbon and siloxane segments alternate in the resultant copolymer molecules. The prepolymer will crosslink when heated. Glass transition temperatures of 132–138 °C and flexural strengths up to 28 Kpsi have been reported [218].

3.2.7
Polysiloxane-Polyalkine Copolymers

A great deal of information has been published in recent years about trialkylsilyl-substituted polyalkenes [221–224] and alkines [225] because their high oxygen permeability and selectivity towards nitrogen makes them the ideal materials for gas separation membranes. However, these polymers are not the subject under discussion, because the polyolefin skeleton does not contain a siloxane fragment but only a silyl substituent, as discussed in Chapter 1.

However, polysiloxane(1-trialkylsilyl-1-propine)graft copolymers are relevant to this section. They have far better mechanical properties and retain their gas permeability for a longer time than the homopolymers. The gas permeability of neat poly(1-trialkylsilyl propine), for example, decreases with time [226]. Polydimethylsiloxane-poly(1-trimethylsilyl-1-propine)graft copolymers can be prepared by various methods. Nagase et al. [226] consider the method outlined in Scheme (47) to be the best, because the starting polymers are resistant to hydrolysis and oxygen, and because the propine chain is not cleaved under the reaction conditions. Futhermore, D3 is not preferentially homopolymerised. It would be extremely difficult to separate these homopolymers from a mixture with the graft copolymer.

Graft copolymers with polydimethylsiloxane contents of 4 to 85 mol% were prepared and the oxygen permeability as well as the selectivity towards nitrogen

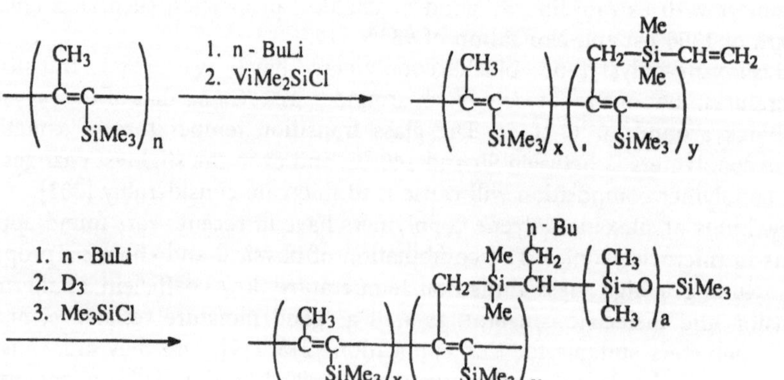

(47)

were determined. At a siloxane content of 55 mol % the oxygen permeability is at its minimum, whereas the selectivity towards nitrogen is at its maximum. The effect of time on the gas permeability of copolymers decreases with siloxane contents < 55 mol %, quite analogous to homopolymers of substituted propine. On the other hand, the oxygen permeability of copolymers with higher siloxane contents on the other hand does not decrease with time [226].

Another completely new group of copolymers are silethinylsiloxane copolymers in which the triple bond is retained in the end product (see structure 48c). These products are prepared by the anionically catalyzed ring-opening of hexamethyltrisiloxane and cyclic silylethinyl polymers, Eq. (48) [220]. This method may be used to prepare block-like as well as random copolymers. Unfortunately, no information about the physical properties of these products is available yet as far as we know.

$$\left(\begin{array}{c} CH_3 \\ Si-O \\ CH_3 \end{array}\right)_3 + \left(\begin{array}{c} CH_3 \\ Si-C\equiv C \\ CH_3 \end{array}\right)_n \xrightarrow{\text{LiB}^- / R'R_2SiCl} R'-\overset{R}{\underset{R}{Si}}\left(O-\overset{CH_3}{\underset{CH_3}{Si}}\right)_m\left(C\equiv C-\overset{CH_3}{\underset{CH_3}{Si}}\right)_n R' \quad (48)$$

a   b

R = Alkyl or Aryl, preferably Methyl
R' = May be same or different to R, Alkyl or Aryl

3.2.8
Polysiloxane-Silylarylene Copolymers

Poly (1,4 phenylene-bis (dimethylsiloxane))49, prepared by the self-condensation of 1,4-bis(dimethylhydroxysilyl)-benzene was first described by Sveda in 1946 [199]. As illustrated by structure 49, these copolymers consist of alternating tetramethyldisiloxy and phenylene units, they are largely crystalline and do not exhibit a two-phase morphology.

$$\left(\begin{array}{c} CH_3 \\ Si \\ CH_3 \end{array} - \underset{}{\bigcirc} - \begin{array}{c} CH_3 \\ Si-O \\ CH_3 \end{array}\right)_n \quad (49)$$

The synthesis of copolymers (49), with one additional Si—O unit between the aryl rings is feasible, by reacting 1,4-bis (dimethylhydroxysilyl) benzene with dichlorodimethyl silane [216]. This reaction produces only very short-chain copolymers apparently due to a desilylation reaction, i.e. splitting of the Si-Ar linkage by the liberated hydrogen chloride [216]. Copolymers with a much higher degree of polymerization were obtained were obtained from bis (N, N-tetramethylene-N'-phenylureido) dimethyl silane [217].

Details of copolymers with structure (50) were published in 1964 by Merker et al [200]. These copolymers contain longer polydimethylsiloxane segments

$$\left[\left(\begin{array}{c} CH_3 \\ Si \\ CH_3 \end{array} - \underset{}{\bigcirc} - \begin{array}{c} CH_3 \\ Si-O \\ CH_3 \end{array}\right)_n\left(\begin{array}{c} CH_3 \\ Si-O \\ CH_3 \end{array}\right)_{\dot{m}}\right]_x \quad (50)$$

and can be prepared by the co-condensation of α, ω-hydroxypolydi-methylsiloxanes with 1,4-bis (dimethylhydroxysilyl) benzene in the absence of equilibration. Details of this synthesis and of the physical properties have been given in several articles [144, 145, 178, 229]. The effect of water on the polymerization and the redistribution of siloxy units in the end product have been investigated by Williams using [29] Si-NMR spectroscopy [201]. As expected, water delays the polymerization reaction and causes the final polymers to degrade. A certain re-distribution within the siloxane chain also takes place. This effect is evident from a comparison of the chain lengths of the hydroxy-terminated polydimethylsiloxane oligomers with the corresponding sequences in the copolymer, as well as by the observation of small numbers of $(SiO)_3$ segments between the phenylene groups [201]. Other syntheses proceed via polymerization of cyclic dimethyl compounds (e.g. octamethylcyclotetra-siloxane), condensation of chloro-, amino-, acetoxy- or ureidosilanes or-siloxanes with bis (hydroxysilyl) arylenes. These compounds were described in detail by Dvornic and Lenz [229]. It is possible to prepare polymers with widely different molecular weights and widely different dimethylsiloxy/phenylene segment ratios and segment lengths, which influence morphology and physical properties. A two-phase morphology, which is reflected in two different glass transition tem-peratures, was only observed in copolymers containing $> 80\%$ silicone [178].

Otherwise, only one glass transition temperature is observed which, as ex-pected, is higher than that of neat polydimethylsiloxanes. This property depends on the type (e.g. monophenylene or diphenylene ether) and substitution of the phenylene nucleus as well as on the ratio of siloxy to phenylene units. Values of < -100 to about -25 °C have been reported [229]. Polysiloxane-polyarylene copolymers are known for their far better thermostability compared to that of neat polydimethylsiloxanes, provided that their sequence is alternating and not blocky. Therefore, the alternating copolymers are more important for industrial applications. Polydimethylsiloxanes start to degrade at 200 °C, although slowly, but above about 350 °C degradation acceleratates tremendously. Alternating polysiloxane-polyarylene copolmers on the other hand remain largely unaffected at these temperatures. Substantial weight loss as a result of thermal decomposi-tion occurs only around 450 °C or higher temperatures [229]. This property is thought to be due to the formation of cyclic, low-energy transitional states (see Sect. 3.1.3) which are of importance in the thermal decomposition of polydimethylsiloxanes. If the formation of these transitional states is inhibited by the presence of "stiffening"groups such as phenylene or p,p'-diphenylene ether groups, the material's heat resistance–which is characteristic for alternating copolymers but not for block copolymers–increases. Arylene-bridged polysilsesquioxanes were described which can be obtained by hydrolysis of 1,4-bis (triethoxysilyl) arylenes, Eq. (51) [202] quite analogous to linear copolymers.

(51)

The arylene groups serve as stiff spacers and lead to the formations of a network which is more expanded than that of silicic acid and more ordered than that of normal silsesquioxanes. Here, the distance between the bridged Si atoms can be accurately controlled by the type of spacer used (i.e. phenylene, di- or polyphenylenes). The heat resistance of arylene-bridged polysilsesquioxanes is extremely high compared to that of unbridged polysilsesquioxanes (500 °C against approx. 350 °C). These products can be used to prepare microporous xerogels, which can serve as starting materials for high purity glass and ceramics.

Polysiloxane-polyarylene copolymers which have a methylene group between the silicon atom and the arylene fragment, referred to as poly (m-silxylene siloxane)s and poly (siloxane-alkarylene) copolymers have been described. They are of little technical importance and difficult to synthesize. A summary of methods of synthesis and physical properties is given in the book of Dvornic and Lenz [229]. The more easily prepared products having the structure **52d** with methyleneoxy or alkylenoxy groups between the aromatic group and the silicon atom are, of greater interest [221].

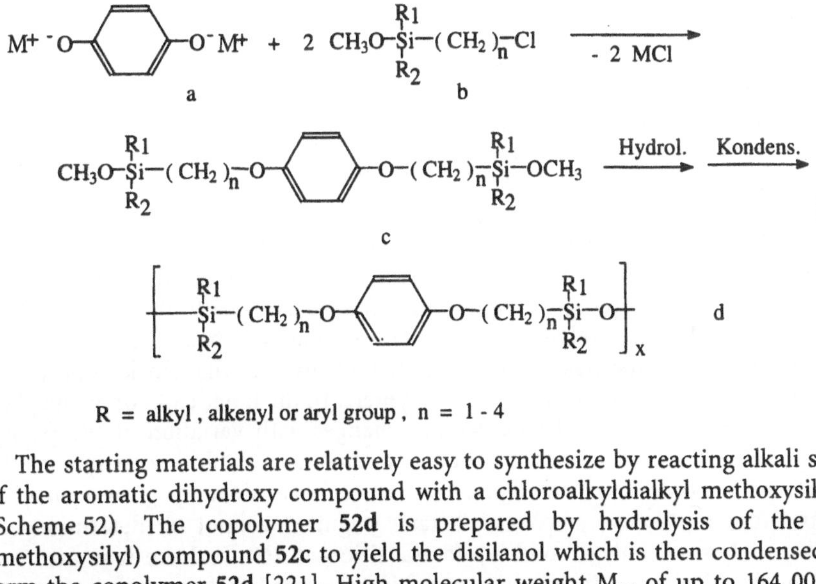

R = alkyl , alkenyl or aryl group , n = 1 - 4

The starting materials are relatively easy to synthesize by reacting alkali salts of the aromatic dihydroxy compound with a chloroalkyldialkyl methoxysilane (Scheme 52). The copolymer **52d** is prepared by hydrolysis of the bis (methoxysilyl) compound **52c** to yield the disilanol which is then condensed to form the copolymer **52d** [221]. High molecular weight M_w of up to 164 000 is obtained.

3.2.9
Polydimethylsiloxane-Polydiphenylsiloxane and Polymethylphenylsiloxane Copolymers

Polydimethylsiloxane copolymers in which the methyl groups have been partly or entirely replaced by phenyl groups, can exist in the form of random or block copolymers. The first class can be synthesized by co-condensation of the silanes or via anionically catalyzed ring-opening of the appropriate cyclic methyl or phenyl compounds (see Sect. 3.1.2). The physical and chemical properties have already been mentioned in Sects. 3.1.4 and 3.1.3, so that, the discussion here is limited to neat, block copolymers. Di-block (AB), tri-block (ABA) and star polymers $(AB)_n - C_m$ are synthesized sequentially [145, 213], cyclic dimethyl or diphenyl compounds are polymerized by anionic ring opening, followed by polymerization of the second component, as outlined in Scheme (53).

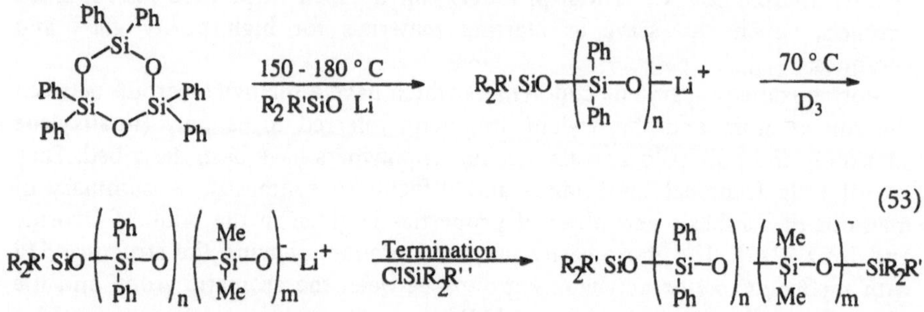

$$R = Me, \quad R' = Bu, \quad R'' = Ph \ o. \ Vi$$

If further cyclic methylphenyl or methylvinyl compounds are added, this method can also be used to prepare tri-block copolymers of type ABC. If the block C consists of methylvinylsiloxy units, the vinyl groups can be crosslinked to produce star polymers with high flexibility and strength [214]. Triblock copolymers of type ABA are prepared using a dianion initiator such as lithium diphenyldisilanolate which attacks the block B on both sides and adds the blocks A, at the chain ends [145, 213] analogous to the mechanism described in Scheme (53). The coupling of the anionic intermediates with diacetoxy disilane yielding multi-block copolymers of type $(ABA)_n$ had also been described [215].

The thermal properties of polydimethyl phenylsiloxane block copolymers clearly differ from those of random copolymers. In the latter case, only one glass transition temperature is detectable and changes with variation of the phenyl group content and distribution, whereas in the case of block copolymers the crystalline characters of polydiphenylsiloxanes is evident. A glass transition temperature for the polydiphenylsiloxane fragment cannot be determined in block copolymers, only a solid-liquid transition at 260 °C. Here it is assumed that the liquid phase is in a liquid crystalline state and that the transition to an isotropic liquid takes place only at higher temperatures. The glass transition temperature for the polydimethylsiloxane block was found at −125 °C, as expected [213].

3.3
Polysiloxanes Forming Mesophases

One of the reasons for the unique properties of silicones (poly-dimethyl-siloxanes) is the extreme flexibility of the siloxane chain [235]. This property is contradictory to requirements of the liquid crystalline state as rigid elongated or cyclic structures are a molecular prerequisite of liquid crystals [236]. Nevertheless different approaches have been found to obtain silicon containing liquid crystalline polymers:

The first one favours arrangements of the silicon substituents surrounding the siloxane chain resulting in a hexagonal symmetry in the shape of columns or stacks.. Beside the rigidity of the polymer structure, packing factors are also responsible for the existence of mesophases especially in the case of dialkylsiloxanes [237]. So siloxanes with alkyl substituents with more than two C-atoms, with phenyl groups or phthalocyaninato-complexes exhibit lc behaviour. Also the introduction of ring structures of hydrogen bonds [238] favours lc properties.

The second approach is characterized by the connection of relatively small siloxane blocks or a siloxane backbone with mesogenic groups dominating the structure. Thus lc siloxanes may be obtained either as main chain or side chain polymers.

3.3.1
Mesomorphic Siloxanes with Alkyl and Aryl Substituents

3.3.1.1
Linear Polyorganosiloxanes

Alkyl- and arylsilanes as well as -siloxanes were the subject of the classical work of Friedel, Crafts, Ladenburg, Dilthey [235] and Kipping [239]. However, decades passed before the character of the phase transitions of mesomorphic siloxanes was recognized. With the notable exception of polydimethylsiloxane, polydialkylsiloxanes with linear alkyl substituents exhibit mesophases [240, 241]. The mesophases of these compounds were classified as plastic or condis (*confor-mational diso*rdered) crystals [242–244]. The mesophases of these linear polymers show a hexagonal structure with the siloxane chain within a cyclic channel of alkyl groups stabilizing the structure. The synthesis of these polydialkylsiloxanes is started from the corresponding dialkylsilanes which are hydrolyzed to cyclic siloxanes e.g. hexaalkyltrisiloxane [245]. The cyclotrisiloxanes are then converted by anionic polymerization to linear polymers, employing KOH as catalyst [246].

The temperature range of the mesophase strongly depends on the molecular weight of the siloxane. As an example the increase of the temperature of isotropization of polydipropylsiloxane with increasing molecular weight is shown in Table 3.6:

Due to the strong dependence of the mesomorphic behavior on the molecular weight the polydispersity of these polymers has carefully be controlled. Recently synthetic routes using CsOH or *t*-BuLi were developed to obtain polymers with narrow molecular weight distributions and high DPs of up to 2050 so the tedious isolation of fractions from mixtures with broad distributions is no longer necessary [248].

· The phase transitions of several polydialkylsiloxanes are collected in Table 3.7.

Table 3.6. Influence of the molecular weight of polydipropylsiloxanes on the temperature of isotropization. Data adapted from [247]

$[\eta]$/dl/g (25°C, toluene)	$M_w^a \cdot 10^{-3}$	DP	G/°C	$C \rightarrow C^a$/°C	LC^b/°C	I/°C
0.04	8.0	61	−109	−76	–	19
0.06	27.0	207	−109	−76	46	69
0.09	36.0	277	−109	−73	52	122
0.11	43.0	330	−109	−73	43	145
0.12	51.0	390	−109	−70	53	172
0.16	68.0	523	−109	−62	55	177
0.26	82.0	630	−109	−57	58	206
0.28	87.0	670	−109	−55	60	207

[a] The transition $C \rightarrow C$ means a crystalline-crystalline transition.
[b] LC is the temperature of the transition crystalline $\rightarrow$ liquid crystalline, I: temperature of isotropization, G: temperature of glass transition.

Table 3.7. Phase transitions of polydialkylsiloxanes. Data adapted from [248]

R	$M_w^* \cdot 10^{-3}$	DP	T_{d1}/°C	T_{d2}/°C	I/°C
Ethyl	84	1,760	−67/−61	17/6	36
n-Propyl	170	930	−48	62	115
n-Butyl	62	400	−56	–	220
n-Pentyl	382	2,050	−44	−22	330
n-Hexyl	270	1,260	−25	25	310

Crosslinking either of linear polydialkylsiloxanes with peroxides or of hydroxyl terminated polydialkylsiloxanes with partially hydrolyzed ethylsilicate in the presence of a metallo-organic catalyst yields mesomorphic elastomers [249].

The temperature of isotropization of these compounds depends on the crosslink density (see Table 3.8) and on the extent of stretching. So the temperature of isotropization of a polymer sample ($M_w = 400\,000$) with 2 % crosslinker increases linearly from 17 °C without load to 150 °C at an extension ratio of 4.5 [249].

Likewise the optical properties depend on the extension rate: So the birefringence of the sample with 3% crosslinker increases spontaneously at RT when the extension ratio becomes higher than 1.5.

Table 3.8. Influence of the crosslink density on the phase transition temperature of a RT vulcanized polydiethylsiloxane with ethylsilicate as crosslinker. (The starting polymer has a Mw of 400 000). Data adapted from [249]

Crosslinker %	G/°C	C/°C	I/°C	Amount of LC phase
0	−139	17	52	100
1	−139	11	43	73
2	−139	10	15	10
3	−139	0	12	24
4	−139	0	–	–
5	−139	0	–	–

3.3.1.2
Cyclic and Cyclolinear Polyorganosiloxanes

Cyclic siloxanes with methyl substituents do not exhibit mesophases, but a high degree of phenyl substitution yields materials with mesomorphic properties. Although octaphenyltetrasiloxane (54) is not an elongated molecule a mesophase between 188 °C and 205 °C has been suggested [250, 251].

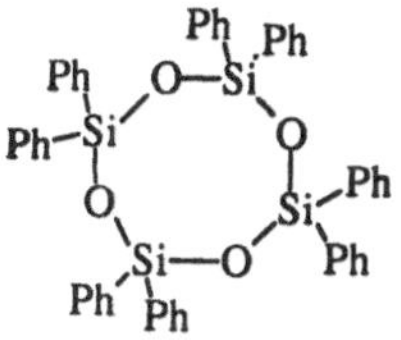

$$(54)$$

Octaphenyltetrasiloxane shows, in neutron diffraction experiments in this temperature range, a sharp reflection at 9.5 Å and a broad at about 4.8 Å. It is therefore assumed that the siloxane rings form a mesophase similar to a smectic A phase with the siloxane backbone in the chair conformation.

The synthesis of cyclic diphenylsiloxane starts from diphenyldichlorosilane [252, 253], which is hydrolyzed to diphenylsilandiol. Basic catalysis yields oligomeric cyclosiloxanes Eq. (55).

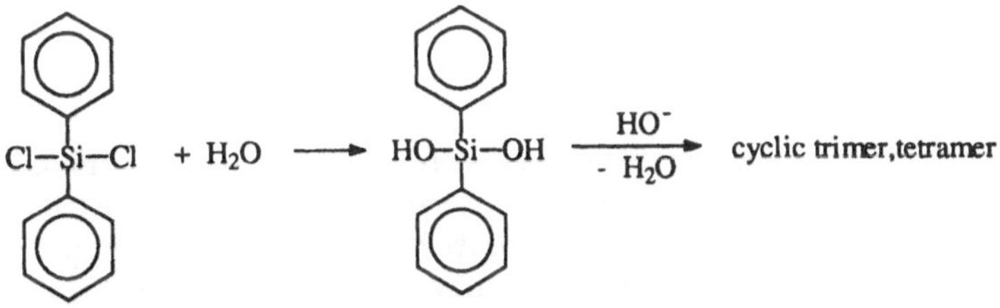

$$(55)$$

The conditions for the existence of mesophases are improved by linearly connected ring structures so they can form molecules with elongatead segments. The substitution patterns of such cyclolinear polysiloxanes exhibit a wide variety of ring systems with different degrees or rigidity (Form. 56). Thus even a

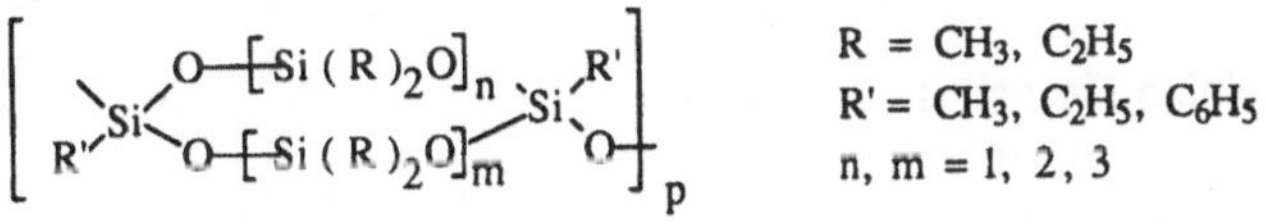

$$R = CH_3, C_2H_5$$
$$R' = CH_3, C_2H_5, C_6H_5 \qquad (56)$$
$$n, m = 1, 2, 3$$

cyclolinear system bearing only methyl groups exhibits mesophases [254]. If R' = phenyl and n, m = 1, 2, mesomorphic polymers with broad mesophases are obtained. The identification of the mesophases is sometimes difficult but the temperature range of the existence of the mesophase can be greater than 350 K as found e.g. for polydecamethylcyclohexasiloxane. In the case of cyclolinear polyhexamethylcyclotetrasiloxane this interval is only 10 K [255]. Furthermore the mesophase behavior is strongly influenced by the tacticity of the polymer chain [256].

In addition to cyclolinear siloxanes with rings linked by oxygen atoms several materials have been synthesized with siloxane spacers between the siloxane rings (Form. 57) also exhibiting mesomorphic properties [257].

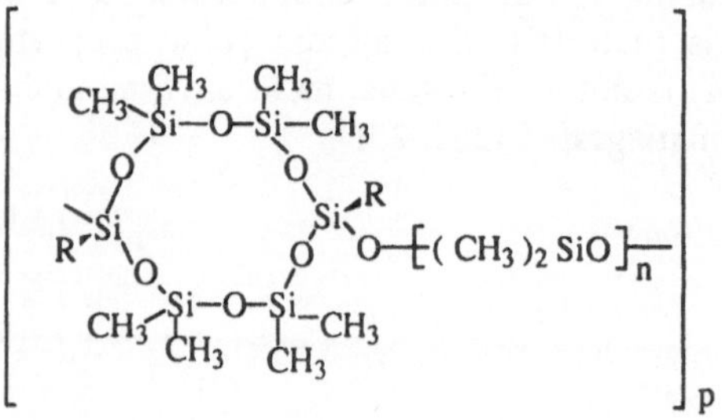

$$
\left[
\begin{array}{c}
\underset{CH_3}{\overset{CH_3\quad CH_3}{\underset{|}{Si}-O-\underset{|}{Si}-CH_3}} \\
\end{array}
\right]_p
\qquad (57)
$$

R = CH₃, C₆H₅

n = 1 - 4

(Structure 57: cyclic siloxane ring with R groups and $-O-[(CH_3)_2 SiO]_{n}-$ spacer; R = CH_3, C_6H_5; n = 1 - 4)

3.3.1.3
Ladder Polyorganosiloxanes

Ladder polymers consisting of siloxane rings (58) are very intersting candidates for mesomorphic materials:

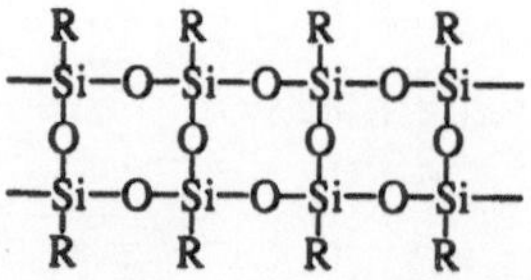

$$
\begin{array}{c}
\overset{R}{\underset{|}{Si}}-O-\overset{R}{\underset{|}{Si}}-O-\overset{R}{\underset{|}{Si}}-O-\overset{R}{\underset{|}{Si}}- \\
| \quad\quad | \quad\quad | \quad\quad | \\
O \quad\quad O \quad\quad O \quad\quad O \\
-\underset{R}{\overset{|}{Si}}-O-\underset{R}{\overset{|}{Si}}-O-\underset{R}{\overset{|}{Si}}-O-\underset{R}{\overset{|}{Si}}-
\end{array}
\qquad (58)
$$

In spite of the difficulties to isolate pure samples of the mesomorphic material from the reaction mixtures there is no doubt about the existence of mesophases of these materials. Several examples of double chain polyorganosilsesquioxanes with *cis*-syndiotactic conformation are reported [258, 259]. The investigation of these mesophases by DSC is difficult because partial decomposition is observed below or at the temperature of isotropization. X-ray experiments show a sharp reflection and an amorphous halo.

Depending on the degree of polymerization, mesophases are observed similar to nematic or smectic phases.

3.3.2
LC-Polysiloxanes Containing Mesogens – Syntheses

3.3.2.1
General Structure Aspects

The basic concept for a very rich development of siloxanes and other polymers with liquid crystalline properties was established by Ringsdorf, Finkelmann, Blumstein, Shibaev and others in the late 1970s.

Low molecular thermotropic liquid crystals consist of elongated or disc-like molecules with flexible end groups. Lc-polymers may be constructed by connecting these entities. Depending on the different patterns of connection, this leads to main chain and side chain LCPs or to lc networks. To obtain liquid crystalline siloxanes appropriately functionalized end groups have to be connected with mono-, di-, tri- or tetra functional siloxanes.

For systematic reasons the discussion should not be restricted to really high molar mass systems. Therefore, we will start with small molecules containing only one repeating-unit. The highest degrees of polymerization (DP) of linear main chain or side chain liquid crystalline polymers(LCPs) reported in the literature are lower than 100, most of them having DPs in the order of 30−50. Of special interest are glasses and networks with liquid crystalline structures because as they may become materials for promising technical applications.

Considering that that the number of mesogens already synthesized exceeds more than 20000, which further may be combined to copolymers and that additionally the geometry and composition of the siloxane component can also be varied, the resulting number of possible structures is enormous [260]. Furthermore combinations of different backbones or connecting groups are possible e.g. networks containing only siloxane units but also polyacrylate, polymethacrylate, polyvinyl or polyether structures. Fig 3.9 shows a selection of possible topological combinations of siloxane components and mesogens with functionalities between 1 and 3. In this context functionality is defined as the number of connections to the siloxane backbone resp. to mesogenic units. Systems with functionality 4 are possible as well. The quadratic symbol for the mesogenic group was chosen to cover both elongated, calamitic structures (calamitic from Greek $\kappa\alpha\lambda\alpha\mu\infty$ = stalk) and discotic mesogens.

The resulting LCP structures depend on the functionalities of the mesogens, of the siloxanes and on the combination of different components and on the overall geometrical arrangement of all basic units.

The aliphatic group at the end of a mesogen is often of advantage for the stability of the mesophase but is not necessary in all cases. Instead of this group often other substituents may be attached to the mesogenic core providing special physical properties, e.g. optical activity, high dipole moments, photochemical activity or NLO-properties. The mesogens can bear more than two side groups, e.g. disc like molecules with 3, 4, 5, 6 . . . .-fold symmetry. Lc-siloxanes of this structure need at least one of the end groups to provide the required attachment to the siloxane, scheme (59).

Some representative examples of siloxane structures, which have been preferentially used as starting materials for LCPs are listed below (59).

3.3.2.2
Synthesis of LC-Siloxanes

The general strategy to synthesize lc siloxanes requires three steps:

A) Synthesis of mesogens
B) Synthesis of siloxanes
C) Linking mesogens and siloxanes

Mesogens useful for LCPs are synthesized by the well known procedures applied to low molar mass lc chemistry. As an example for a synthesis of mesogenic

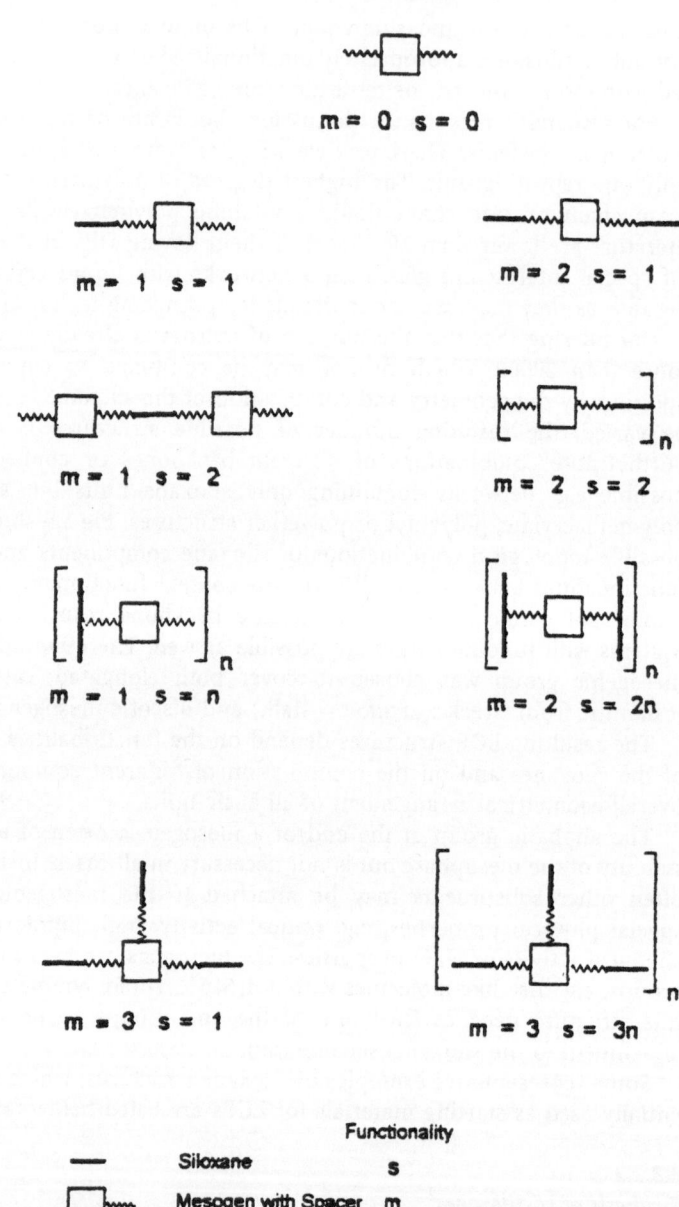

Fig. 3.9. Examples of basic topological combinations of mesogenic units, spacers and siloxanes with different functionalities. S is the functionality of the siloxane component against the mesogenic group, m the functionality of the mesogenic group against the siloxane component

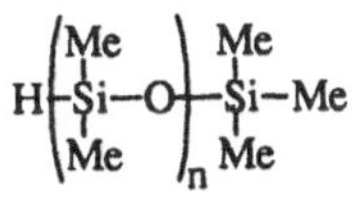

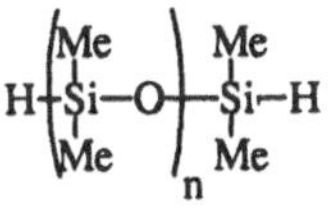

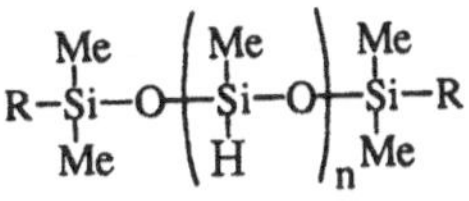

$$H\!-\!\begin{pmatrix}Me\\|\\ \underset{|}{Si}\!-\!O\\|\\Me\end{pmatrix}_{\!n}\!\begin{matrix}Me\\|\\ \underset{|}{Si}\!-\!Me\\|\\Me\end{matrix} \qquad H\!-\!\begin{pmatrix}Me\\|\\ \underset{|}{Si}\!-\!O\\|\\Me\end{pmatrix}_{\!n}\!\begin{matrix}Me\\|\\ \underset{|}{Si}\!-\!H\\|\\Me\end{matrix}$$

$$R\!-\!\begin{matrix}Me\\|\\ \underset{|}{Si}\!-\!O\\|\\Me\end{matrix}\!\begin{pmatrix}Me\\|\\ \underset{|}{Si}\!-\!O\\|\\H\end{pmatrix}_{\!n}\!\begin{matrix}Me\\|\\ \underset{|}{Si}\!-\!R\\|\\Me\end{matrix}$$

R = H, Me

n = 1, 2, ...

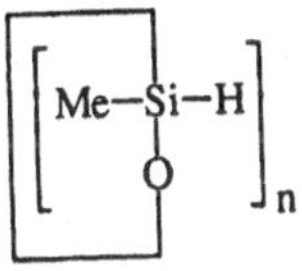

$$\left[\begin{matrix}Me\!-\!\underset{|}{\overset{|}{Si}}\!-\!H\\|\\O\end{matrix}\right]_{n}$$

n = 3, 4, 5, 6, 7, ...   (59)

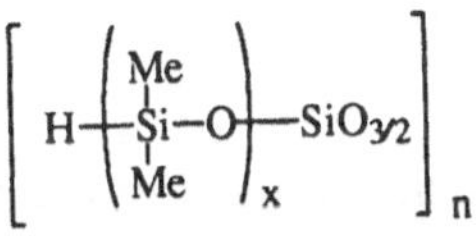

$$\left[H\!-\!\begin{pmatrix}Me\\|\\ \underset{|}{Si}\!-\!O\\|\\Me\end{pmatrix}_{\!x}\!-\!SiO_{y2}\right]_{n}$$

x = 0, 1, 2 ...

n = 6, 8, 10 ...

$$\left[H\!-\!\begin{pmatrix}Me\\|\\ \underset{|}{Si}\!-\!O\\|\\Me\end{pmatrix}_{\!n}\!-\!SiR_{x}\right]_{(4-x)}$$

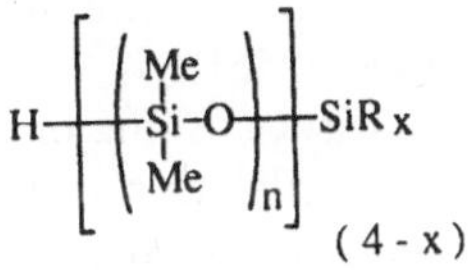

x = 0, 1

R = H, Alkyl, Aryl, Mesogen

groups the sequence of reactions to synthesize linear connected cycloaliphatic or aromatic ring mesogens with aid of the transition metal catalyzed Kumada/Negeshi cross coupling reaction [261, 262] has been selected (Scheme 60).

Of course the wealth of well known organic reactions such as esterification, etherification and others may be applied as well to the synthesis of mesogens. Monomeric, oligomeric and polymeric siloxanes can also be prepared by standard methods (see Sect. 3.1).

$$H_7C_3-\!\bigcirc\!-\!\bigcirc\!-OH \quad \xrightarrow{PPh_3 / Br_2} \quad H_7C_3-\!\bigcirc\!-\!\bigcirc\!-Br$$

$$\left[H_7C_3-\!\bigcirc\!-\!\bigcirc\!-Zn \right]_2 \quad \xleftarrow{ZnBr_2} \quad \left[H_7C_3-\!\bigcirc\!-\!\bigcirc\!-Li \right] \quad \Bigg\uparrow Li ,ZnBr_2 \tag{60}$$

Pd (O)   Br-◯-CN (with P and CN substituents)

$$H_7C_3-\!\bigcirc\!-\!\bigcirc\!-\!\bigcirc\!-CN \quad (F)$$

3.3.2.3
Linking Mesogens and Siloxane by Hydrosilylation

The connection of both components can be done via the hydrosilylation reaction. This reaction is an addition of a SiH bond to a double bond, catalyzed by Pt and other transition metal complexes or peroxides and is well known since the fifties [236]. Regioselective hydrosilylation may be obtained using Rh catalysts [264] e.g. $Rh(Ph_3P)_3$ Cl. The hydrosilylation was introduced to the field of lc siloxane chemistry by Finkelmann, Rehage and Kollmann in 1979 [265] and became a powerful tool on this field. Starting materials are Si-H functional siloxanes and mesogenic compounds with terminally unsaturated spacers. As catalyst mostly Pt-compounds are used such as $H_2PtCl_6 \cdot 6H_2O$, platinum(II)-olefin-complexes and platinum(0) complexes with phosphines [266].

Due to the importance of this reaction, it is necessary to discuss some of the peculiarities of the addition Si-H-groups to olefins. The hydrosilylation usually yields β-adducts so the silicon unit is connected to the ω-position of the olefin (Eq. 61).

In addition to the desired reaction several side reactions are observed. This is of minor significance in the case of monomers as they may be purified by distillation or recrystallization. In the case of polymers most of the side products are introduced to the final product without the opportunity of separating them from the regular structures. These side reactions have effects on structure and molecular weight of the synthesized lc polymer. The degrees of polymerization obtained by hydrosilylation reactions are relatively low. So 25 was the highest obtained DP of the main chain polymer resulting from the reaction of (4-vinylphenyl)dimethylsilane in the presence of Speier's catalyst [267].

If we consider main chain polymers, the degree of polymerization is reduced dramatically by isomerization reactions of ω-olefinic groups accompanied by lower reactivity or by their reduction to an alkane. When the formation of

$$\equiv Si-H \quad + \quad CH_2=CH-R \quad \xrightarrow{(\text{Cat.})} \quad \equiv Si-CH_2-CH_2-R \tag{61}$$

$$\beta \text{ - adduct}$$

Si—O—C-bonds is observed, the hydrolytic stability of the reaction product can be reduced.

In the case of side chain lc polymers the molecular weight itself is only marginally affected by these side reactions – if possible branching reactions are neglected – but the regular structure of the lc side chain polymer. Lc-polymers with hydrolytically unstable Si—O—C-bonds can be purified by chromatography on silica-columns due to the high affinity of resulting SiOH-groups to the silica surface.

Therefore some important side reactions, well known from the literature are listed below (Eqs. 62–67).

α - Addition

$$\equiv Si-H \quad + \quad CH_2=CH-R \quad \xrightarrow{(\text{Cat.})} \quad \equiv Si-CH(CH_3)-R \tag{62}$$

$$\alpha \text{ - adduct}$$

Isomerization

$$CH_2=CH-CH_2-R \quad \xrightarrow{(\text{Cat.})} \quad H_3C-CH=CH-R \tag{63}$$

Hydrogenation

$$\equiv Si-H \quad + \quad CH_2=CH-R \quad \xrightarrow{(\text{Cat.})} \quad H_3C-CH_2-R \tag{64}$$

Allylic compounds containing in the γ-position an atom with a free electron pair undergo a splitting reaction.

$$\equiv Si-H \quad + \quad CH_2=CH-CH_2-R \quad \xrightarrow{(\text{Cat.})} \quad \equiv Si-R \; + \; CH_2=CH-CH_3 \tag{65}$$

$$R = -OR', \; -Cl, \; -O_2CR, \; ....$$

A mechanism for the side reaction of allyl compounds was proposed by Speier et al. [268].

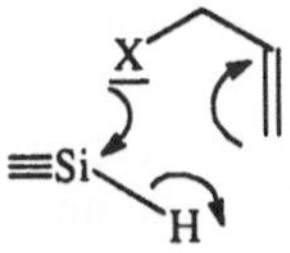

$$\tag{66}$$

$$X : -Cl, \; -OR'$$

In the case of phenylallyl ether the addition of H-Si(OEt)$_3$ yields beside of 70% of the desired product $\approx 25\%$ by-products [269], the addition to polymethyl-H-siloxane is reported free of isomerization [270].

The loss of significant amounts of propylene is one of the most important side reaction of allyloxy groups containing mesogens and can lead to a loss up to 30 % of the C3-spacer [271, 272]. A similar splitting reaction accompanied with branching reactions was reported with allyl carbamates [273].

Also chain branching of polydimethylsiloxane in the presence of H_2PtCl_6 is a possible side reaction [274].

$$
\begin{array}{cccccc}
& \overset{CH_3}{\underset{CH_3}{|}} & \overset{CH_3}{\underset{CH_3}{|}} & \overset{CH_3}{\underset{CH_3}{|}} & & \overset{CH_3}{\underset{CH_3}{|}} & \overset{CH_3}{\underset{CH_3}{|}} \\
---- & -\text{Si}-\text{O}-\text{Si}-\text{O}-\text{Si}- & ---- & + & ---- & -\text{Si}-\text{O}{\cdot}\text{Si}-\text{H}
\end{array}
$$

(67)

$$
\xrightarrow{H_2PtCl_6} \quad ---- -\overset{CH_3}{\underset{CH_3}{\text{Si}}}-\text{O}-\overset{CH_3}{\underset{O}{\text{Si}}}-\overset{CH_3}{\underset{CH_3}{\text{Si}}}- \quad ---- \quad + \quad CH_3-\overset{CH_3}{\underset{CH_3}{\text{Si}}}-\text{H}
$$
$$
CH_3-\text{Si}-CH_3
$$

Additionally reactions of the Si-H-bond with other functional groups can be observed. Normally Pt-catalysts favor the addition of the SiH-group to the ω-alkenes, while no addition to carbonyl moieties is observed. So it is possible to obtain functionalized (e.g. C=O groups) silicon containing liquid crystals.

Contrary, if sec. alcoholic OH groups are present, these groups can be involved to the hydrosilylation if the catalyst concentration is relatively high ($[H_2PtCl_6]/[SiH] > 10^{-4}$) [275], resulting in silyl ethers.

To control the progress of the hydrosilylation reaction it is recommended to observe the decrease of the SiH-concentration by ^{1}H-NMR at 4.7-4.8 ppm (or IR at 2, 160 cm^{-1}).

Beside the mentioned problems of the hydrosilylation reaction it is very difficult to remove the Pt-catalyst from the reaction product [276]. A general method to reduce the Pt-content of lc siloxanes does not exist. Several authors recommend a stoichiometric excess of 10% of the olefinic compound to get a total conversion of the Si-H groups [277, 278].

The polymers resulting from the hydrosilylation reaction contain remarkable amounts of mesogens not connected with the siloxane. These residues not only result from the applied excess of mesogens but also from side products of hydrogenation, isomerization, or splitting reactions. In the case of main chain systems this leads to a significant reduction of the DP. In the case of side chain systems these impurities may influence the phase behaviour and dielectric properties [279].

To reduce the amount of unreacted, hydrogenated or isomerized mesogens it is recommended to remove them by 8-10 precipitations. The amount of impurities can be reduced by this method below < 0.1% [280].

Other methods for purification are chromatographic procedures and the purification with supercritical carbon dioxide [281].

Although all these possible complications of the hydrosilylation reaction may influence the determination of phases, the phase transition temperatures or other physical properties found in the literature give a good survey of the phase behaviour LCP siloxanes synthesized via hydrosilylation reactions.

3.3.2.4
Coupling of Mesogens and Siloxane by Grignard Reactions

A second but less versatile method to connect siloxanes with mesogenic groups is the Grignard reaction (Eq. 68).

$$\equiv SiCl \ + \ XMg(CH_2)_nR \ \longrightarrow \ \equiv Si(CH_2)_nR$$

$$X : Cl, \ Br, \ Tos \tag{68}$$

An example for the application of the Grignard reaction is the synthesis of a siloxane containing a Schiff Base [282]. In this case the siloxane unit works as a side group (Scheme 69).

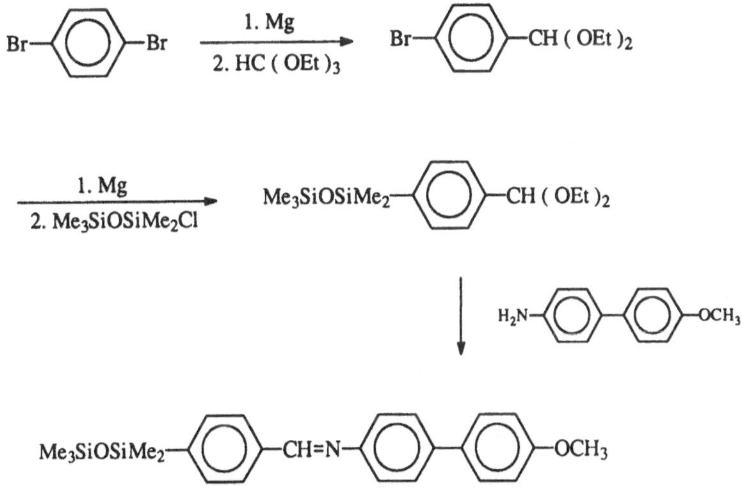

$$(69)$$

3.3.2.5
Coupling of Mesogens and Siloxane by Esterification

A third method of connecting the mesogenic groups to the siloxane is the esterification of siloxanes bearing OH functions (alcohols, phenols) with acid chlorides. This type of reaction was mainly applied to main chain polymers [283, 284] (Eq. 70).

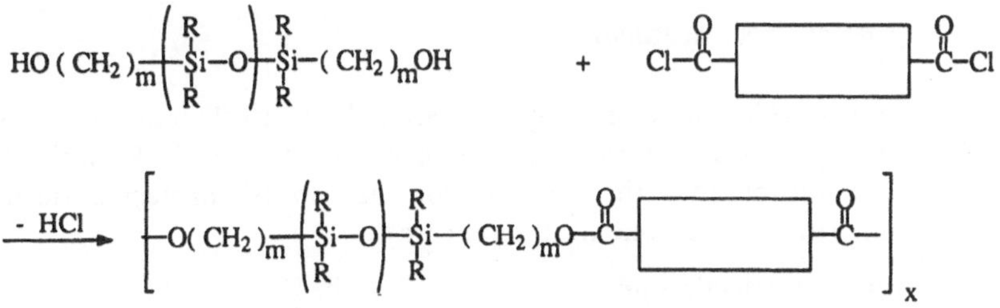

$$(70)$$

Also the siloxane component can be converted to an acid chloride while a silicon free diol is used as co-component for the formation of the polymer [285] (Scheme 71).

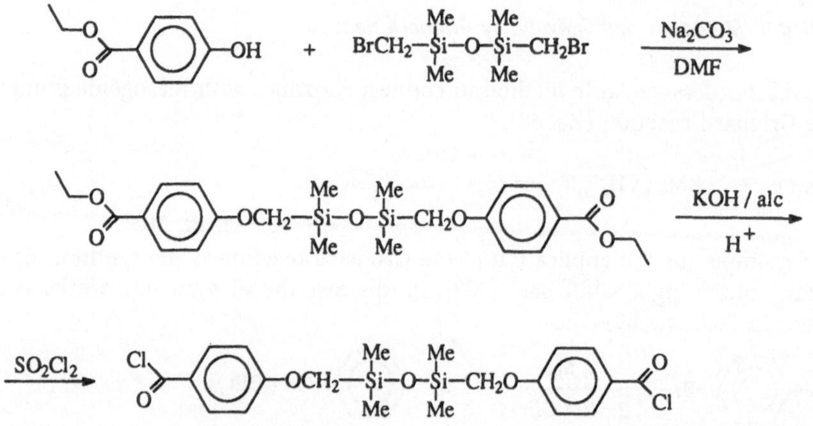

$$(71)$$

By a similar approach 4,4'-dihydroxy-1,3-bis (p-phenoxymethyl)-tetramethyl disiloxane (72) was obtained [285].

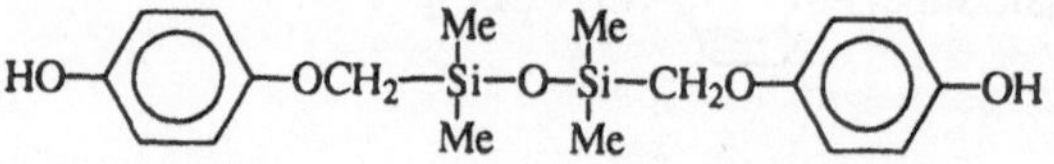

$$(72)$$

3.3.3
LC-Polysiloxanes Containing Mesogen-Modifications

Mostly it is sufficient to connect the mesogenic groups with the siloxane in a one step reaction. But sometimes it is necessary to modify the LCP after the coupling reaction. This procedure is requested if the hydrosilylation is hindered by certain functions of the mesogen – e.g. amino groups reduce the activity of the Pt-catalyst – or the extent of expected side reactions or by-products is not acceptable.

3.3.3.1
Esters by Polymer Analogue Reactions

The subsequent introduction of ester groups after the hydrosilylation is one of the widely used methods. The synthetic route has to provide LCPs with OH functions. That means that the OH-function has to be protected during hydrosilylation. This can be done in different ways.

a) Acetyl group as protection group

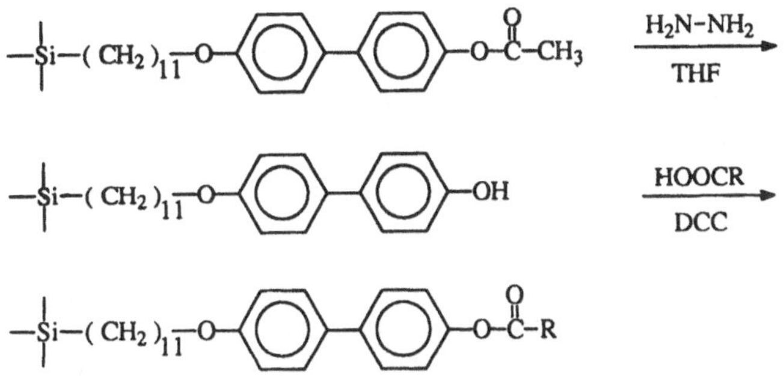

$$-\underset{|}{\overset{|}{Si}}-(CH_2)_{11}-O-\underset{\text{(biphenyl)}}{\bigcirc\!\!-\!\!\bigcirc}-O-\overset{O}{\overset{\|}{C}}-CH_3 \quad \xrightarrow[\text{THF}]{H_2N-NH_2}$$

$$-\underset{|}{\overset{|}{Si}}-(CH_2)_{11}-O-\bigcirc\!\!-\!\!\bigcirc-OH \quad \xrightarrow[\text{DCC}]{\text{HOOCR}} \tag{73}$$

$$-\underset{|}{\overset{|}{Si}}-(CH_2)_{11}-O-\bigcirc\!\!-\!\!\bigcirc-O-\overset{O}{\overset{\|}{C}}-R$$

The use of acetyl derivatives is explained by scheme 73 [286]:
The yield of the converted mesogenic esters varies between 79% and 89%.
b) Trimethylsilyl groups as protection groups
   The trimethylsilyl protecting group was employed to introduce methacrylic functions [287] (Eq. 74).

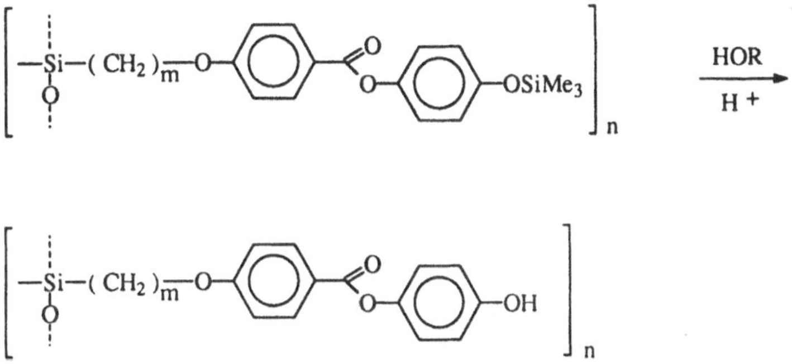

$$\left[-\underset{\overset{|}{O}}{\overset{|}{Si}}-(CH_2)_m-O-\bigcirc-\overset{O}{\overset{\|}{C}}-O-\bigcirc-OSiMe_3\right]_n \quad \xrightarrow[\text{H}+]{\text{HOR}}$$

$$\tag{74}$$

$$\left[-\underset{\overset{|}{O}}{\overset{|}{Si}}-(CH_2)_m-O-\bigcirc-\overset{O}{\overset{\|}{C}}-O-\bigcirc-OH\right]_n$$

Subsequent esterification by an acid chloride in the presence of an HCl acceptor or by an appropriate acid anhydride yields the desired ester.

3.3.3.2
Introduction of Amino Functions

Although the following example of a chloromethylation reaction of a siloxane polymer was not applied to a polymer exhibiting a lc phase it is interesting to see that reaction conditions can be employed, which normally affect the siloxane component dramatically (Scheme 75) [288].

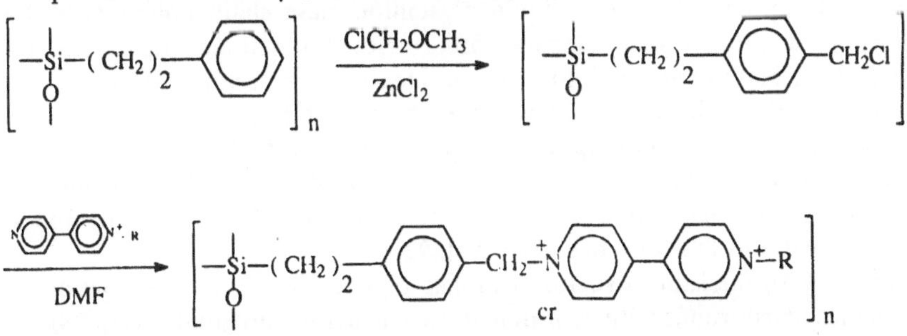

$$\left[-\underset{\overset{|}{O}}{\overset{|}{Si}}-(CH_2)_2-\bigcirc\right]_n \quad \xrightarrow[\text{ZnCl}_2]{\text{ClCH}_2\text{OCH}_3} \quad \left[-\underset{\overset{|}{O}}{\overset{|}{Si}}-(CH_2)_2-\bigcirc-CH_2Cl\right]_n$$

$$\tag{75}$$

$$\xrightarrow[\text{DMF}]{\overset{N}{\bigcirc}\!\!-\!\!\overset{N^+}{\bigcirc}\,R} \quad \left[-\underset{\overset{|}{O}}{\overset{|}{Si}}-(CH_2)_2-\bigcirc-CH_2-\overset{+}{N}\bigcirc\!\!-\!\!\bigcirc N^+-R\right]_n$$
$$\text{cr}$$

Up to a degree of 85% chloromethylation cross linking has not been observed. Chloromethylsiloxanes could also be used as starting materials for introducing amino functions like carbazole groups [289].

3.3.3.3
Introduction of Chemically Sensitive Groups

Siloxanes (cont. 11% spiropyrane groups) were prepared from MeH siloxane by initial treatment with 4-methoxyphenyl 4-(5-hexenyloxy) benzoate and the N-hydroxysuccinimide ester of 10-undecenoic acid and subsequent replacement of the succinimide ester group, forming an amide group [290].

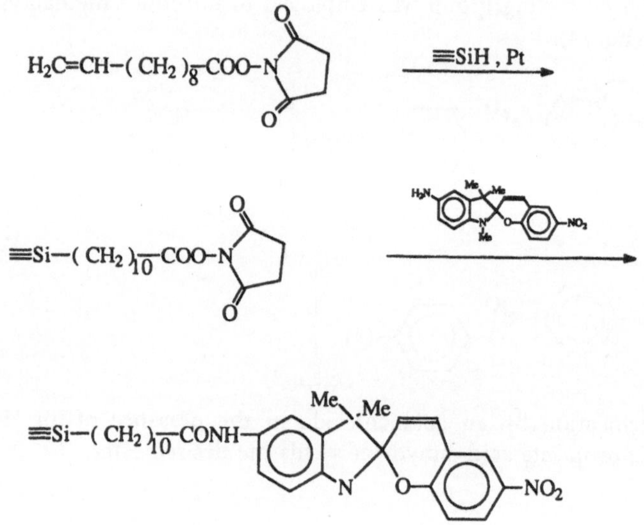

(76)

3.3.4
LC-Main Chain Polymers

The general approach to obtain liquid crystalline main chain polymers is the connection of bifunctional siloxane units with bifunctional mesogens by methods described previously [291]. Depending on the experimental conditions, the reaction can also lead to cyclic polymers by back biting.

Beside reactions for coupling of mesogenic groups with the siloxane blocks there exist some specific processes. So the reaction of an α, ω-diaminosiloxane with α, ω-dianhydrides can be used and was introduced for siloxane block copolymers without lc properties (Eq. 77) [292].

Products with the same structure can also be obtained by hydrosilylation of α, ω-dihydrogensiloxanes with α, ω-diolefines, e.g. α, ω-diallylimides (Eq. 78).

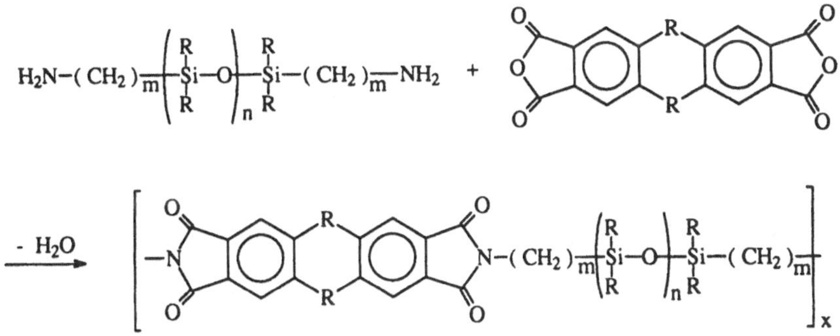

$$H_2N-(CH_2)_{\overline{m}} \left[\begin{matrix} R \\ Si-O \\ R \end{matrix} \right]_n \begin{matrix} R \\ Si-(CH_2)_{\overline{m}}NH_2 \\ R \end{matrix} \quad + \quad \text{(dianhydride)}$$

$$\text{(77)}$$

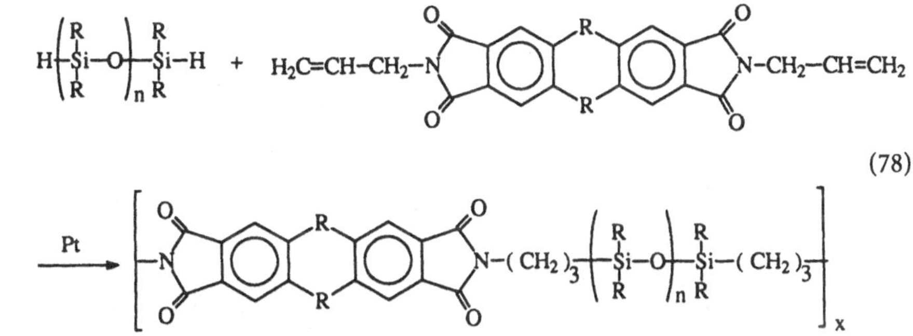

$$\xrightarrow{-H_2O} \left[\text{(diimide)} -N-(CH_2)_{\overline{m}} \left[\begin{matrix} R \\ Si-O \\ R \end{matrix} \right]_n \begin{matrix} R \\ Si-(CH_2)_{\overline{m}} \\ R \end{matrix} \right]_x$$

$$H \left[\begin{matrix} R \\ Si-O \\ R \end{matrix} \right]_n \begin{matrix} R \\ Si-H \\ R \end{matrix} \quad + \quad H_2C=CH-CH_2-N \text{(diimide)} N-CH_2-CH=CH_2$$

$$\text{(78)}$$

$$\xrightarrow{Pt} \left[-N \text{(diimide)} N-(CH_2)_3 \left[\begin{matrix} R \\ Si-O \\ R \end{matrix} \right]_n \begin{matrix} R \\ Si-(CH_2)_3 \\ R \end{matrix} \right]_x$$

In this case, side reactions are relatively less important [293]. The length of the siloxane block is determined before the connection with the organic co-component, but it is also possible to start with a disiloxane, and to elongate the siloxane block by an equilibration reaction, catalyzed by a sulfocation exchanger (Eq. 79) [294].

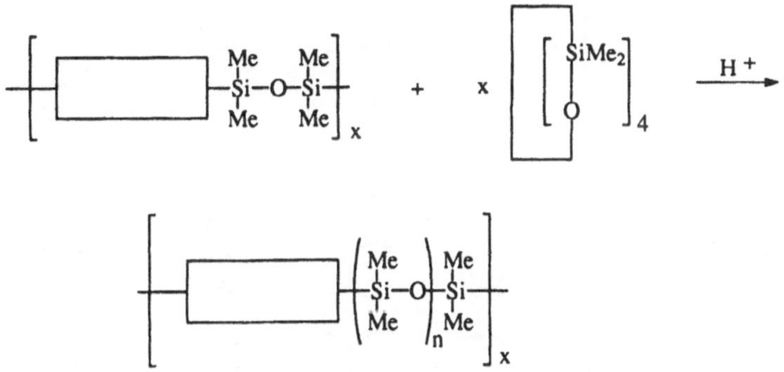

$$\left[\text{———} \begin{matrix} Me & Me \\ Si-O-Si \\ Me & Me \end{matrix} \right]_x \quad + \quad x \left[\begin{matrix} SiMe_2 \\ O \end{matrix} \right]_4 \quad \xrightarrow{H^+}$$

$$\text{(79)}$$

$$\left[\text{———} \left[\begin{matrix} Me \\ Si-O \\ Me \end{matrix} \right]_n \begin{matrix} Me \\ Si \\ Me \end{matrix} \right]_x$$

Several examples of main chain polymers of the type **80** are collected in Table 3.9 [295].

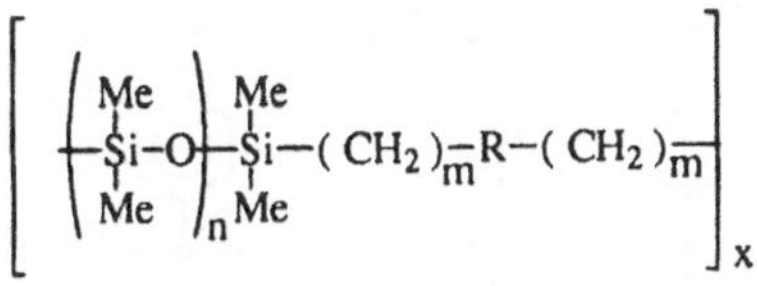

$$\left[\left[\begin{matrix} Me \\ Si-O \\ Me \end{matrix} \right]_n \begin{matrix} Me \\ Si-(CH_2)_{\overline{m}}R-(CH_2)_{\overline{m}} \\ Me \end{matrix} \right]_x$$

$$\text{(80)}$$

Table 3.9. Main-chain LC-polysiloxanes of the structure:

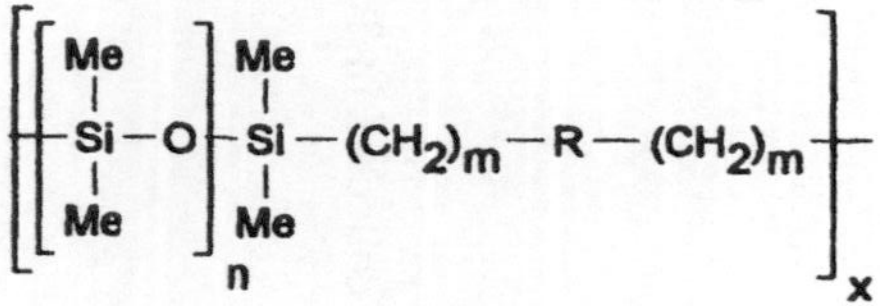

$$\left[\begin{array}{c}\text{Me} \\ \text{Si}-\text{O} \\ \text{Me}\end{array}\right]_n\left[\begin{array}{c}\text{Me} \\ \text{Si}-(CH_2)_m-R-(CH_2)_m \\ \text{Me}\end{array}\right]_x$$

R	m	n	Phase transition/°C	η_{inh} /dL·g^{-1}	$M\cdot10^{-3}$ /g·mol^{-1}	Ref.
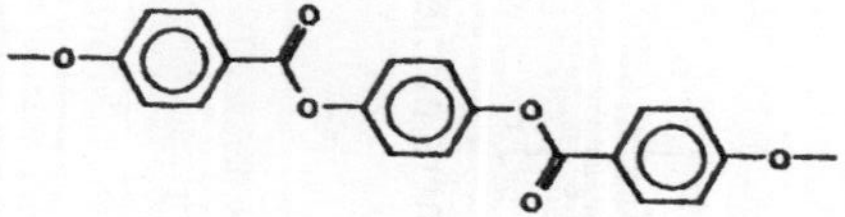 (structure 1)	3	2	K 35 LCx 130 I			[295]
	3	3	G 5 LCx 114 I			[295]
	3	5	G −13 LCx 94 I			[295]
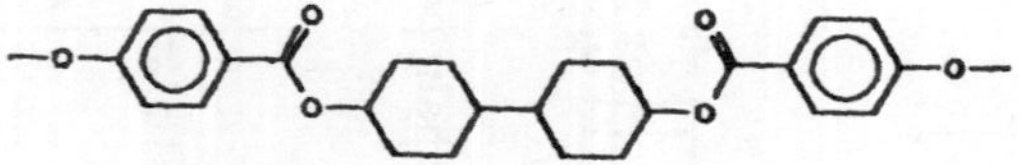 (structure 2)	3	2	G −14 LCx 166 I			[295]
	3	3	S 151 LCx 127 I			[295]
	3	4	G −47 LCx 121 I			[295]
	3	5	G −57 I			[295]
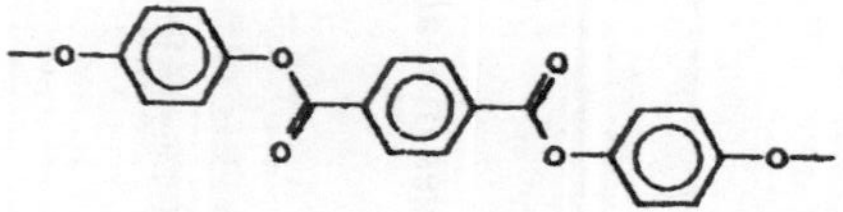 (structure 3)	1	1	K 199 N 208 I	0.161		[285]
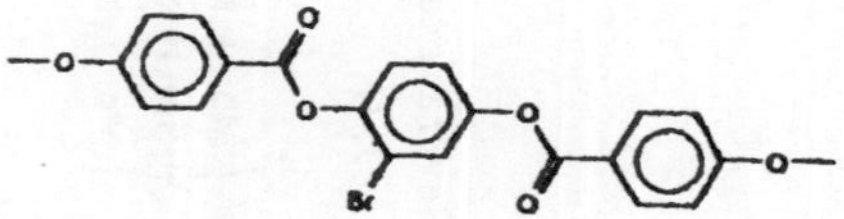 (structure 4)	1	1	K 43 N 77 I	0.48		[285]

n	m	G (°C)	K (°C)	S (°C)	N / LCx (°C)	I	M		Ref
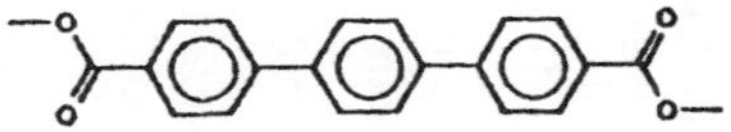 3	1		K 78	S 177		I	0.21	13	[283]
3	2		K 51	S1 126, S2 140		I	0.29	15	[283]
4	1		K 102	S 164		I	0.22	14	[283]
5	1		K 106	S 168		I	0.24	16	[283]
6	1		K 108	S1 160, S2 172		I	0.27	20	[283]
11	1		K 109	S1 134, S2 157		I	0.26	17	[283]
1	1	G 17			N 237	I	0.24	6	[294]
1	2	G −6			LCx 200	I	0.20	6.5	[294]
1	2.4	G −9			LCx 200	I	0.20	7	[294]
1	4.5	G −26			LCx 100	I	0.19	8	[294]
1	7	G −32			LCx 30	I	0.18	9.5	[294]
1	9	G −42			–	I	0.18	9.6	[294]
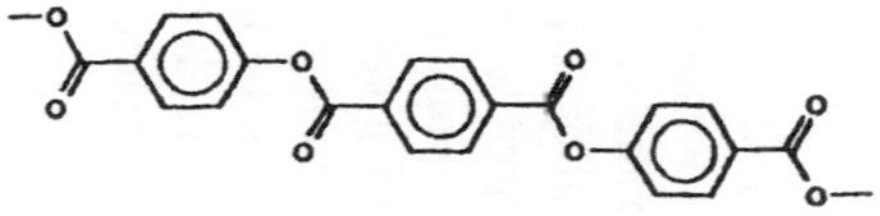 3	1		K 116	S 155	N 171	I	0.14	5	[283]
3	2		K 96	S1 110, S2 134		I	0.15	7	[283]
3	3		K 76	S1 80, S2 97		I	0.17	13	[283]
4	1		K 108	S1 138, S2 160		I	0.15	10	[283]
5	1		K 117	S 144		I	0.14	8	[283]
6	1		K 104	S 128		I	0.11	6	[283]
11	1		K 122	S 156		I	0.15	18	[283]

*M determined by different methods: Ref [283] by SEC, Ref [294] osmometrically, LCx = LC-phase not identified.

3.3.5
LC-Polysiloxanes with Pendant Mesogens

3.3.5.1
Linear Siloxanes with Mesogenic Side Groups

3.3.5.1.1 *Thermotropic Materials.* Most of the liquid crystalline siloxanes synthesized until now are based on a linear siloxane backbone and elongated or disc like mesogenic groups connected to the backbone by a spacer with 3–11 methylene groups. Additionally the mesogenic core and the methylene groups are connected by an ether bridge. The reason for the great number of compounds of this type is the simple synthesis via the hydrosilylation reaction. Most of these LCPs are based on a polymethyl-H-siloxane with a DP between 30 and 60, depending on the commercial source. Also disiloxane as the shortest backbone for liquid crystalline siloxanes was employed [296].

The thermal properties of linear lc siloxanes are not sensitive towards the molecular weight of the starting siloxane if the DP is higher than 10 (Fig. 3.10) [297, 298], so any phase transition temperatures can be compared from this point of view (Tab. 3.10). On the other hand some other experimental details may influence the thermal properties as mentioned before. Additionally methods of determination of the phase transitions such as PVT-measurements [299], optical methods, X-ray and DSC measurements (heating rate, heating up, cooling down, first run, second run) may give different results.

General relationships between structure and properties of liquid crystalline side chain polymers are reported by several authors [300–323].

A series of monomeric mesogens is listed in Table 3.11, which may be compared to the properties of corresponding lc siloxanes.

Apart from homopolymers interesting opportunities to modify the properties of lc siloxanes may be obtained by the synthesis of co-polymers (Tab. 3.12), terpolymers (Tab. 3.13) or polymers with more components.

The simplest case is a lc siloxane based on polymethyl-*H-co*-dimethylsiloxane so the resulting product contains mesogenic groups as well as non mesogenic methyl groups. The ratio of mesogenic to non mesogenic units can not varied arbitrarily because a phase separation occurs for high amounts of dimethyl-siloxy units, indicated by two glass transition temperatures [324–327].

Copolymers are of great interest as various functionalities in different fractions can be introduced to the lc siloxane system. So important properties e.g. glass transition, photo reactivity, chirality, dipole moment and radically polymerizable groups can be easily be tailored to specific applications.

Additionally functional groups of copolymers may be obtained by polymer analogue reactions. This method is of great value to modify materials with groups sensitive towards the hydrosilylation reaction.

Nonmesogenic silphenylene groups can be used as well [328].

A great variety of side chain polymers, both homo- and copolymers have been described e.g. with biphenylcarboxylate esters [329], α-methylstilbenes

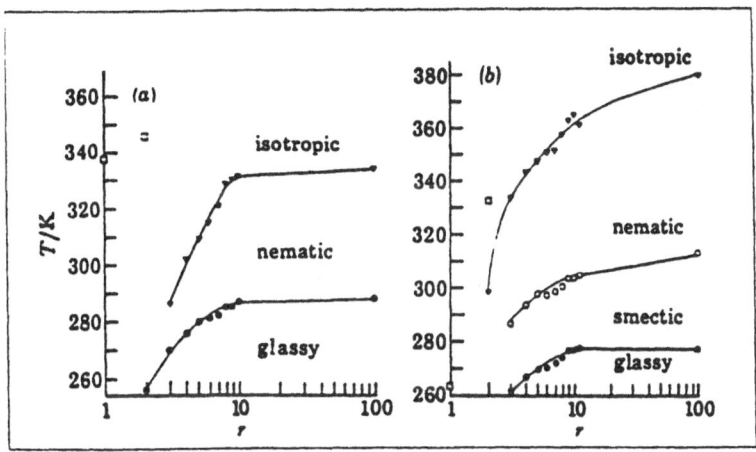

Fig. 3.10. Phase behaviour of mondisperse side chain siloxanes as function of DP (r). Length of the spacer: (a) m = 3; (b) m = 6 [297]

[330, 331], alkenyl substituted cholesterols and related steroids [332], other chiral substituents [333–336], disubstituted dioxanes [337], hemiphasmidic mesogens [338], macrocyclic polyethers [339], carbazolyl groups [340], with hydrogen bonds so the induction of mesogenity is possible [341], with a carbonate group in the spacer [342], with siloxane units in the side chain and with paired mesogens [343–346].

Laterally attached mesogenic groups exhibit some peculiarities with respect to the optical properties and the phase behaviour of LCPs (Tab. 3.14). The first examples were synthesized with polymethacrylate backbones [347]. In the siloxane series the first polymer with laterally attached mesogens with two aromatic rings as core of the mesogen did not exhibit a lc phase [348]. Other examples were published later used three ring systems as mesogens did show lc properties [349–352]. Also copolymers of side-on and side-end fixed mesogenic groups were synthesized [353, 354]. Systems with laterally attached mesogens show a tendency towards nematic order and reduced clearing temperature.

As a special type of siloxane, copolymers with cross shaped mesogens, with lc behaviour should be mentioned [348].

3.3.5.1.2 *Lyotropic Materials.* Introducing hydrophilic side groups yields amphiphilic materials [355, 356]. These materials exhibit mesophases in the presence of water. The chemistry is the same as described previously for

Table 3.10. Linear liquid crystalline side chain polysiloxanes: homopolymers

a   Biphenyl derivatives

Mesogenic group	m	Phase transition/°C				$\Delta H_{K/S/N\rightarrow i}$ J/g	DP	Ref.
—(CH₂)m—O—⬡—⬡—O—CH₃	3	G 40	C 125	SX 152	I	23	120	[307]
	4	G 32	C 123	SX 117	I	97	120	[307]
—(CH₂)m—O—⬡—⬡—CN	3	G 28.3		SX 132.5	I	4.18	50	[277]
	3			SX 169.5	I	–		[308]
	4	G 14.5			I		50	[309]
	5			SX 152	I	5.43	50	[277]
	5	G 16.0			I	–		[308]
	6	G 14		SX 166	I	–	50	[310]

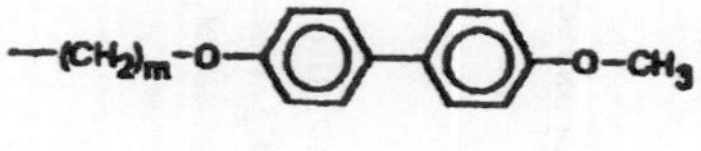

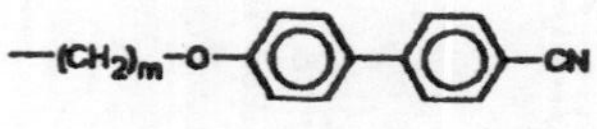

b   Azomethine derivatives

Mesogenic group	m	Phase transition/°C				$\Delta H_{K/S/N\rightarrow i}$ J/g	DP	Ref.
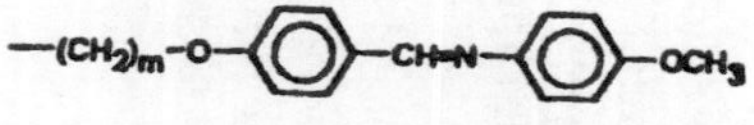 —(CH₂)m—O—⬡—CH=N—⬡—OCH₃		C 100		N 123	I		58	[311]
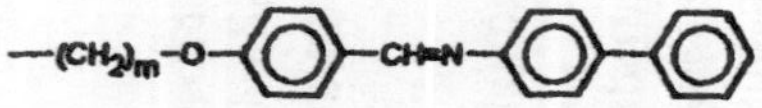 —(CH₂)m—O—⬡—CH=N—⬡—⬡		C 183	SX 262		I		58	[311]

c Alkoxyphenylbenzoates

Mesogenic group	m	Phase transition/°C				$\Delta H_{K/S/N \to I}$ J/g	DP	Ref.
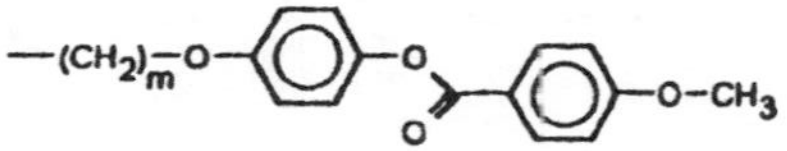	3	G 95	S_A 119	N 127	I		36	[315]
	3		S_A 74	–	I		35	[312]
	4		S_A 122	N 104	I		35	[312]
	5		S_A 109	–	I		35	[312]
	6		S_A 133	–	I		35	[312]
	10						35	[312]
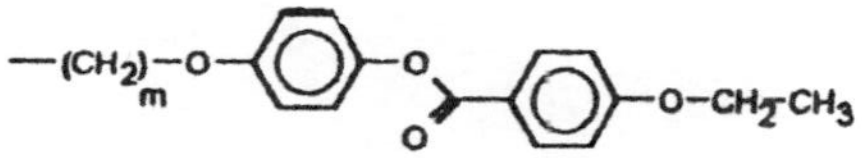	5		S_A 149		I		35	[312]
	3	G 46	S_E 67 $\quad S_A$ 169		I		36	[315]
	3	G 65	S_B 79 $\quad S_A$ 179		I			[334]
	5		S_A 163		I		35	[312]

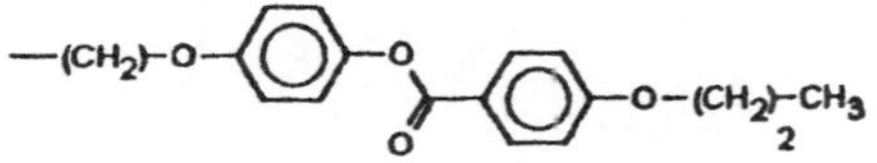

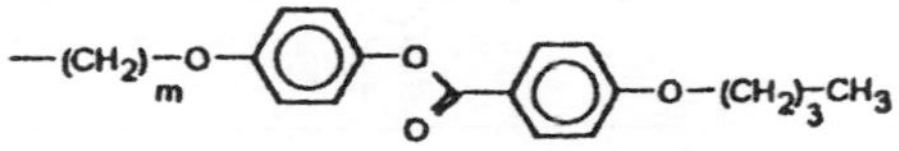

Table 3.10d. (*Contd.*)

Mesogenic group	m	Phase transition/°C (G)	(C)	(SX/Sc)	(N)	(I)	$\Delta H_{K/SN \to I}$ J/g	DP	Ref.
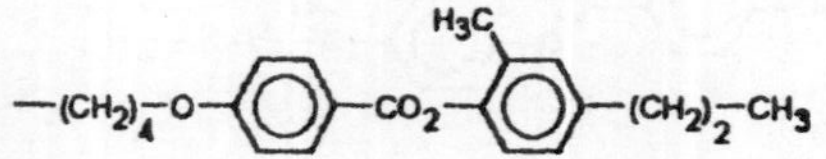 $-(CH_2)_4-O-\!\!\bigcirc\!\!-CO_2-\!\!\bigcirc\!(CH_3)\!-(CH_2)_2-CH_3$	4	G 15		SX 170		I		50	[310]
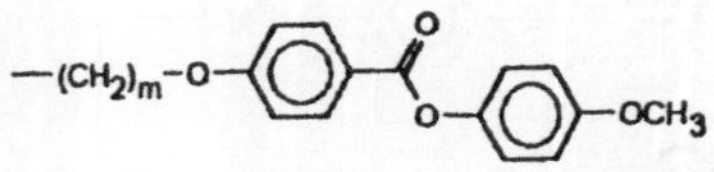 $-(CH_2)_m-O-\!\!\bigcirc\!\!-CO-O-\!\!\bigcirc\!\!-OCH_3$	3	G 15			N 61	I	2.2	120	[307]
	3	G 15			N 82	I	–		[322]
	3	G 30			N 87	I			[315]
	4	G 15			N 95	I			[317]
	5		C 87		N 115	I			[317]
	6	G 5	C		N 112	i			[317]
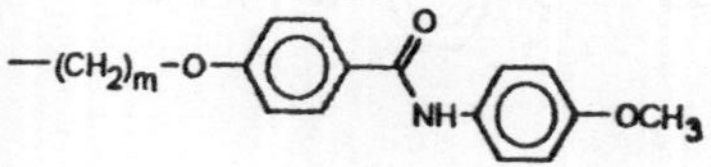 $-(CH_2)_m-O-\!\!\bigcirc\!\!-CO-NH-\!\!\bigcirc\!\!-OCH_3$	3		C > 350 dec.					58	[311]
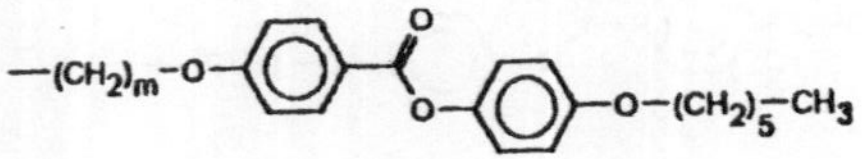 $-(CH_2)_m-O-\!\!\bigcirc\!\!-CO-O-\!\!\bigcirc\!\!-O-(CH_2)_5-CH_3$	3	G 15		SX 112	N	I	11.6	120	[307]
$-(CH_2)_m-O-\!\!\bigcirc\!\!-CO-O-\!\!\bigcirc\!\!-$	3	G 32		Sc? 95	N 112	I			[313]
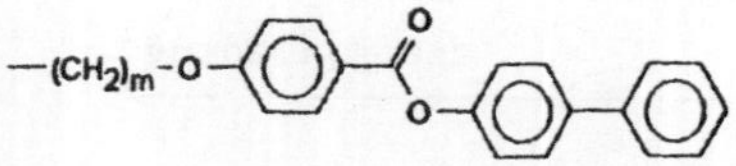 $-(CH_2)_m-O-\!\!\bigcirc\!\!-CO-O-\!\!\bigcirc\!\!\bigcirc$	3		C 95		N 125	I		58	[311]

Structure	n											Ref.
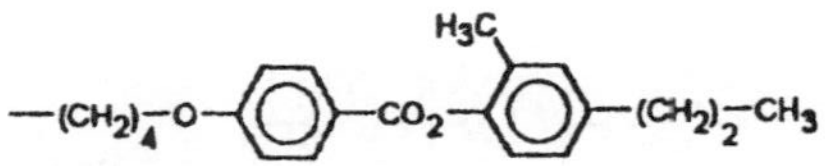 —(CH₂)₄—O—C₆H₄—CO₂—C₆H₃(CH₃)—(CH₂)₂—CH₃ (with H₃C)	4	G	15			SX	170			I	50	[310]
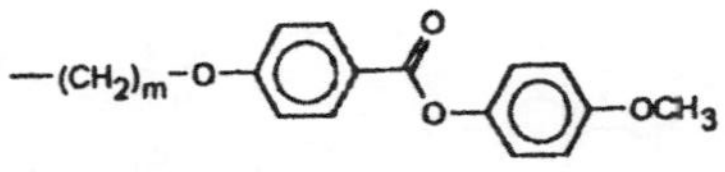 —(CH₂)ₘ—O—C₆H₄—C(=O)—O—C₆H₄—OCH₃	3	G	15			N	61	I	2.2	120	[307]
	3	G	15			N	82	I	–		[322]
	3	G	30			N	87	I			[315]
	4	G	15			N	95	I			[317]
	5	G		C	87	N	115	I			[317]
	6	G	5			N	112	i			[317]
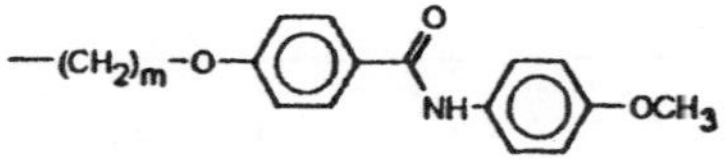 —(CH₂)ₘ—O—C₆H₄—C(=O)—NH—C₆H₄—OCH₃	3			C	>350 dec.	SX	46				58	[311]
—(CH₂)ₘ—O—C₆H₄—C(=O)—O—C₆H₄—O—(CH₂)₅—CH₃	3	G	15			SX	112					
						N	112	I	11.6	120	[307]
—(CH₂)ₘ—O—C₆H₄—C(=O)—O—(biphenyl)	3	G	32			Sc?	95					
						N	125	I			[313]
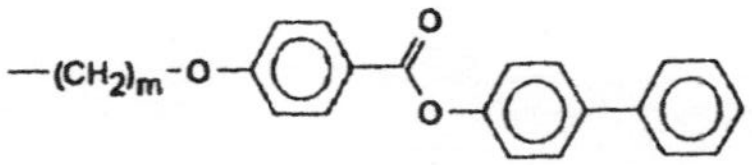 —(CH₂)ₘ—O—C₆H₄—C(=O)—O—(biphenyl)	3			C	95	N		I		58	[311]

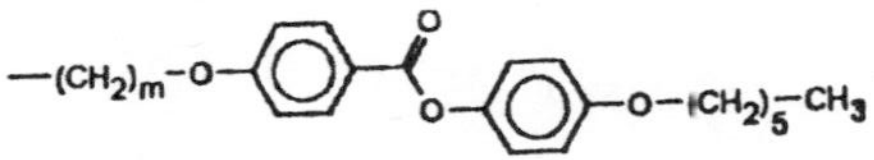

Table 3.10d. Derivatives of the 4-Hydroxybenzoic acid

Mesogenic group	m	Phase transition/ °C							$\Delta H_{K/SN \rightarrow I}$ J/g	DP	Ref.
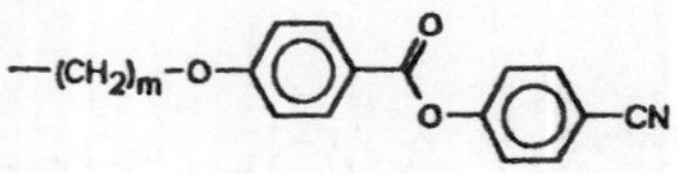 —(CH₂)ₘ—O—⟨⟩—C(=O)O—⟨⟩—CN	3	G	29.0			SX	161.5	I	2.01		[277]
	3	G	20.0			SX	61.0	I	1.9	120	[307]
	3	G	21.0			SX	130.0	I	1.76		[276]
	3	G	18.0			SX	133.0	I	1.59		[276]
	6	G	8	C	55	SX	181.5	I	3.18		[277]
	6	G	7	C	22	SX	148	I	3.68		[276]
	6			C	25	SX	153	I	3.22		[276]
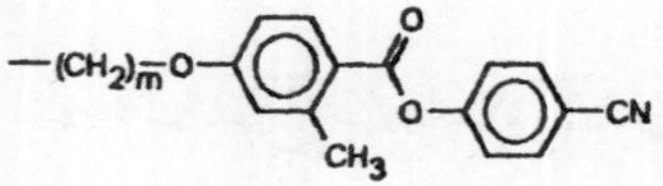 —(CH₂)ₘ—O—⟨⟩(CH₃)—C(=O)O—⟨⟩—CN	6	G	9			N	46			50	[310]

Table 3.10d. (Contd.)

Mesogenic group	m	Phase transition/ °C					ΔH$_{K/S/N \rightarrow I}$ J/g	DP	Ref.
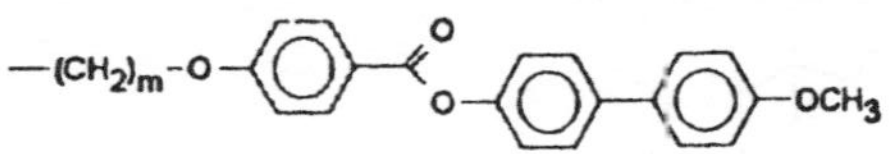 —(CH₂)m—O— ...—OCH₃	3	C	139	N	319	I		58	[311]
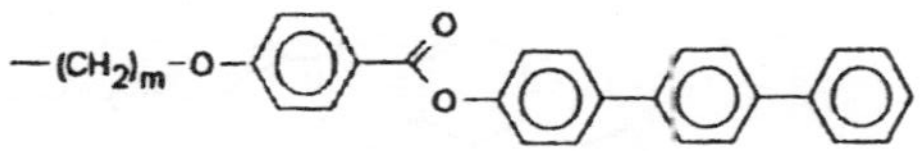 —(CH₂)m—O— ...	3	C	200	N	360 dec.	I		58	[316]
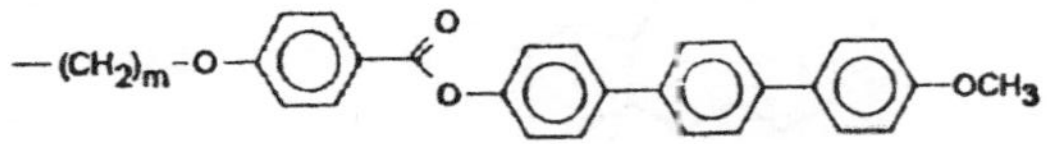 —(CH₂)m—O— ...—OCH₃	3	C	200	N	360 dec.	I		58	[316]

e Derivatives of the 4(4'-Hydroxy-phenyl-) benzoic acid

Mesogenic group	m	Phase transition/ °C					ΔH$_{K/S/N \rightarrow I}$ J/g	DP	Ref.
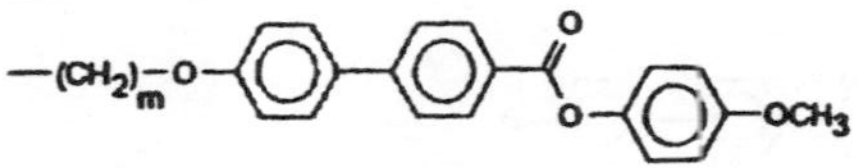 —(CH₂)m—O— ...—OCH₃	3	G	98	SX	258	N	I	35	[323]
	5	G	109	SX	295	N	I	35	[323]
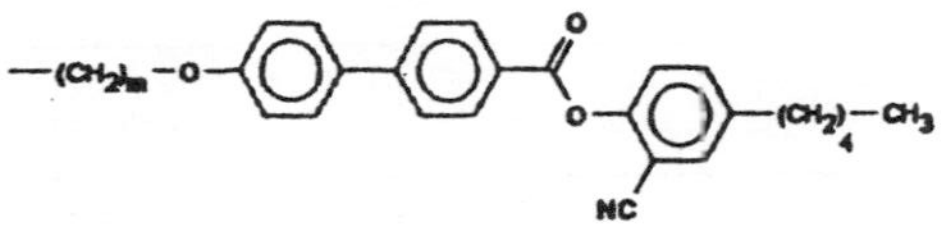 —(CH₂)m—O— ... NC, (CH₂)₄—CH₃	3	G	28	SX	112	N	I	35	[318]
	4	G	22	SX	95	N	I	35	[318]
	5	G	9	SX	116	N	I	35	[318]
	6	G	8	SX	95	N	i	35	[318]

Table 3.10. f (Contd.)

Mesogenic group	m	Phase transition/ °C				$\Delta H_{K/SN\to i}$ J/g	DP	Ref.
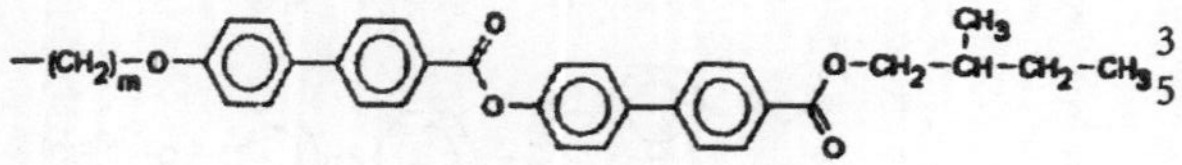 $-(CH_2)_m-O-$ [phenylene] $-C(O)O-$ [phenylene] $-O-CH_2-CH(CH_3)-CH_2-CH_3$	3	G 149	SX 254	N* 291	I			[323]
	5	G 114	SX 350 dec.		I			[323]
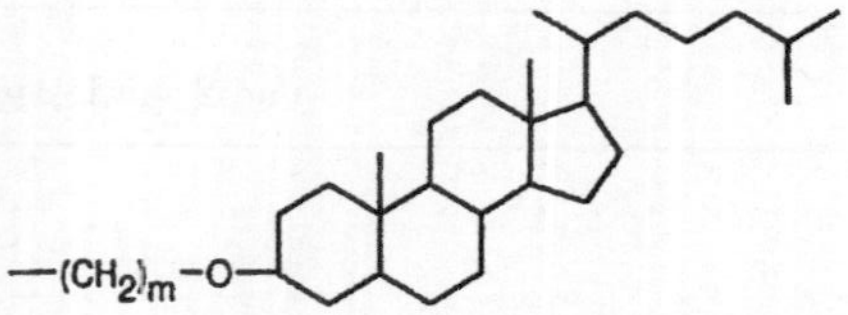 $-(CH_2)_m-O-$ [cholesteryl]	3	G 59	SX 174		I			[311]
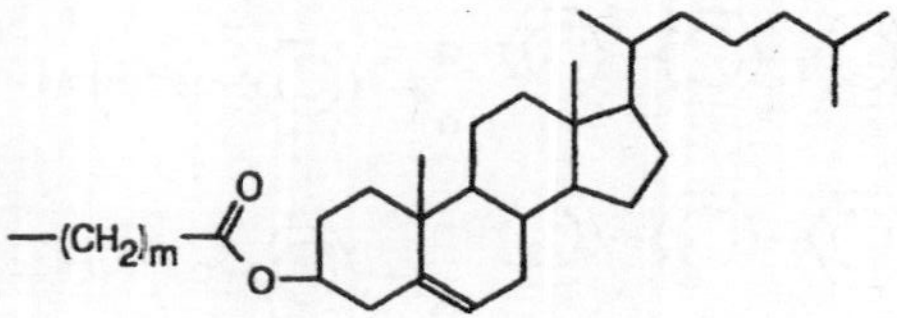 $-(CH_2)_m-O-C(O)O-$ [cholesteryl]	3	G 45	SX 115		I	2.7	120	[307]
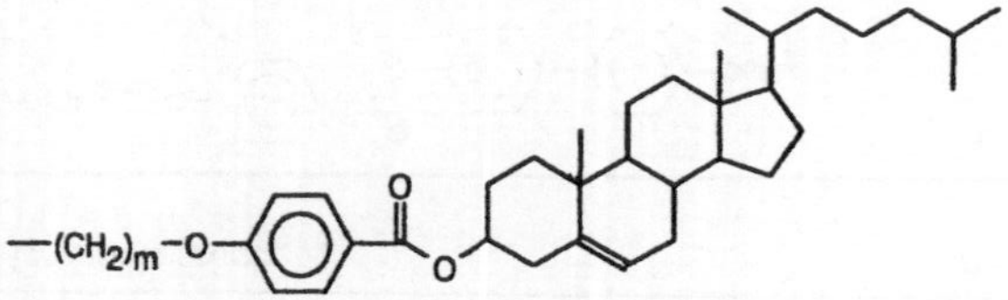 $-(CH_2)_m-O-$ [phenylene] $-C(O)O-$ [cholesteryl]	3	G 130	SX 287		I			[311]

g Discotics

Mesogenic group	m	Phase transition/ °C					$\Delta H_{K/SN \to i}$ J/g	DP	Ref.
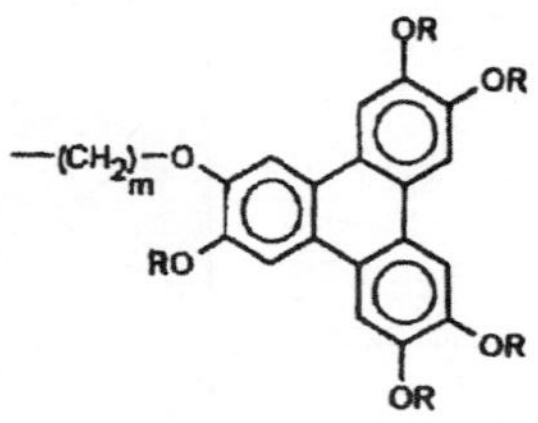 R: C_2H_5	11	G	−19	D	39	I		35	[321]

Table 3.11. Mesogens used for the synthesis of liquid crystalline polysiloxanes

a Biphenyl derivatives

Mesogenic group	m	Phase transition/ °C				Ref.
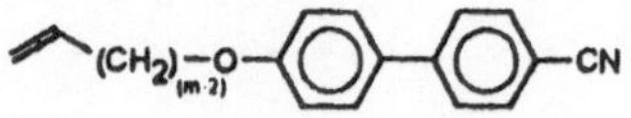 $CH_2{=}CH{-}(CH_2)_{(m-2)}{-}O{-}$⟨C6H4⟩⟨C6H4⟩${-}CN$	3	C	80	N	(79) I	[310]
	4	C	68	N	(36) I	[310]
	5	C	88	N	(71) I	[310]
	6	C	35	N	51 I	[310]
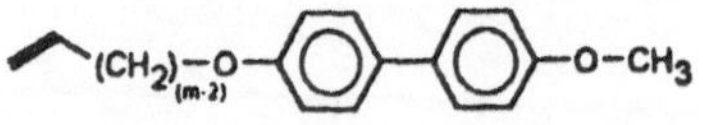 $CH_2{=}CH{-}(CH_2)_{(m-2)}{-}O{-}$⟨C6H4⟩⟨C6H4⟩${-}O{-}CH_3$	3	C	119		I	[307]
	4	C	116		I	[307]

b Azomethine derivatives

Mesogenic group	m	Phase transition/ °C				Ref.
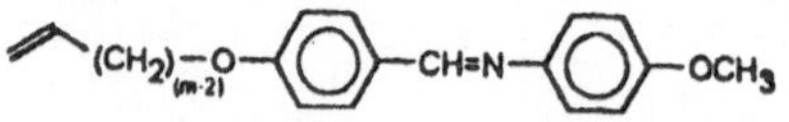 $(CH_2)_{(m-2)}{-}O{-}$⟨C6H4⟩${-}CH{=}N{-}$⟨C6H4⟩${-}OCH_3$	3	C	111		I	[311]
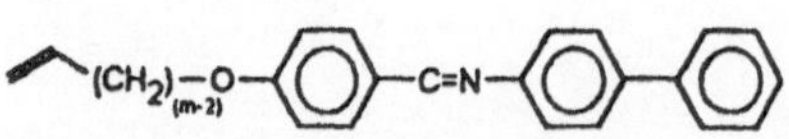 $(CH_2)_{(m-2)}{-}O{-}$⟨C6H4⟩${-}C{=}N{-}$⟨biphenyl⟩	3	C	152	N	166 I	[311]

c ω-Alkenyl derivatives of the 4-hydroxybenzoic acid

Mesogenic group	m	Phase transition/ °C				Ref.
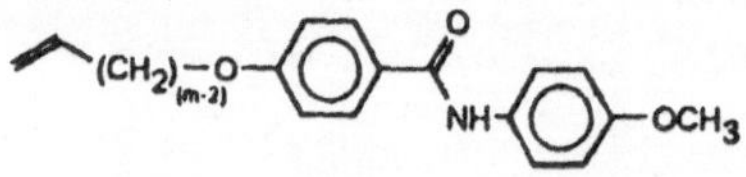 $(CH_2)_{(m-2)}{-}O{-}$⟨C6H4⟩${-}C({=}O){-}NH{-}$⟨C6H4⟩${-}OCH_3$	3	C	170		I	[311]
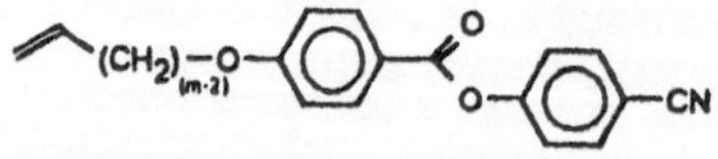 $(CH_2)_{(m-2)}{-}O{-}$⟨C6H4⟩${-}C({=}O){-}O{-}$⟨C6H4⟩${-}CN$	3	C	103		I	[307]
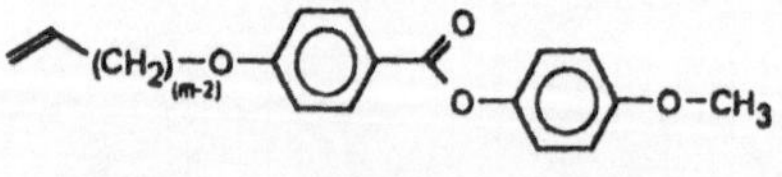 $(CH_2)_{(m-2)}{-}O{-}$⟨C6H4⟩${-}C({=}O){-}O{-}$⟨C6H4⟩${-}O{-}CH_3$	3	C	89		I	[307]
	4	C	87		I	[317]
	5	C	90		I	[317]
	6	C	63		I	[317]
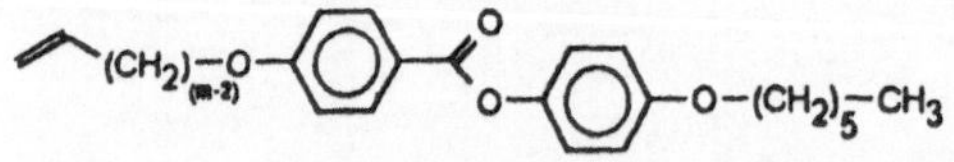 $(CH_2)_{(m-2)}{-}O{-}$⟨C6H4⟩${-}C({=}O){-}O{-}$⟨C6H4⟩${-}O{-}(CH_2)_5{-}CH_3$	3	C	61	N	77 I	[307]
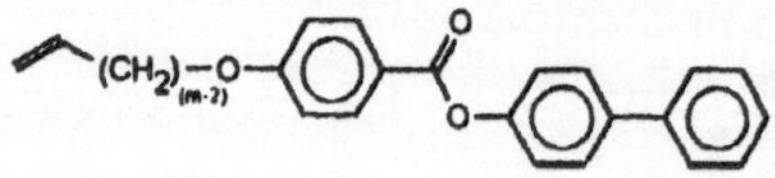 $(CH_2)_{(m-2)}{-}O{-}$⟨C6H4⟩${-}C({=}O){-}O{-}$⟨biphenyl⟩	3	C	139		I	[311]

Table 3.11 c (Contd.)

Mesogenic group	m	Phase transition/ °C					Ref.
(chemical structure) $(CH_2)_{(m-2)}$—O—⬡—C(=O)—O—⬡—⬡—⬡—OCH$_3$	3	C	147	N	249	I	[311]
	4	C	141	N	247	I	[311]
(chemical structure) $(CH_2)_{(m-2)}$—O—⬡—C(=O)—O—⬡—⬡—⬡—⬡	3	C	>350			I	[316]
(chemical structure) $(CH_2)_{(m-2)}$—O—⬡—C(=O)—O—⬡—⬡—⬡—⬡—OCH$_3$	3	C	>350			I	[316]

d ω-Alkenyl Derivatives of the 4(4'-Hydroxy-phenyl-) benzoic acid

Mesogenic group	m	Phase transition/ °C						Ref.
(chemical structure) $(CH_2)_{(m-2)}$—O—⬡—⬡—C(=O)—O—⬡—OCH$_3$	3	C	137			N	243	I [323]
	5	C	133	Sx	172	N	253	I [323]
(chemical structure) $(CH_2)_{(m-2)}$—O—⬡—⬡—C(=O)—O—⬡(NC)—$(CH_2)_4$—CH$_3$	3	C	84			N	157	I [318]
	4	C	76			N	126	I [318]
	5	C	70			N	141	I [318]
	6	C	71			N	123	I [318]
(chemical structure) $(CH_2)_{(m-2)}$—O—⬡—⬡—C(=O)—O—⬡—⬡—OCH$_3$	3	C	214			N	290	I [323]

e Chiral Mesogens

Mesogenic group	m	Phase transition/ °C						Ref.
(chemical structure) $(CH_2)_{(m-2)}$—O—⬡—C(=O)—O—⬡—C(=O)—O—CH$_2$—C*(CH$_3$)(H)—C$_2$H$_5$	3	C	49					I [319]
	4	C	43					I [319]
	5	C	63					I [319]
(chemical structure) $(CH_2)_{(m-2)}$—O—⬡—C(=O)—O—⬡—⬡—O—CH$_2$—CH(CH$_3$)—CH$_2$—CH$_3$	3	C	100	Sx	150	N*	188	I [323]
(chemical structure) $(CH_2)_{(m-2)}$—O—⬡—⬡—C(=O)—O—⬡—C(=O)—O—CH$_2$—CH(CH$_3$)—CH$_2$—CH$_3$	3	C	118	Sx	198	N*	213	I [323]
	5	C	105	Sx	198			I [323]
(chemical structure) $(CH_2)_{(m-2)}$—O—⬡—⬡—C(=O)—O—⬡—⬡—C(=O)—O—CH$_2$—CH(CH$_3$)—CH$_2$—CH$_3$	3	C	152	Sx	240	N*	278	I [323]
	5	C	135	Sx	295	N*	315	I [323]

Table 3.11 (*Contd.*)

f Steroid Derivatives

Mesogenic group	m	Phase transition/ °C				Ref.
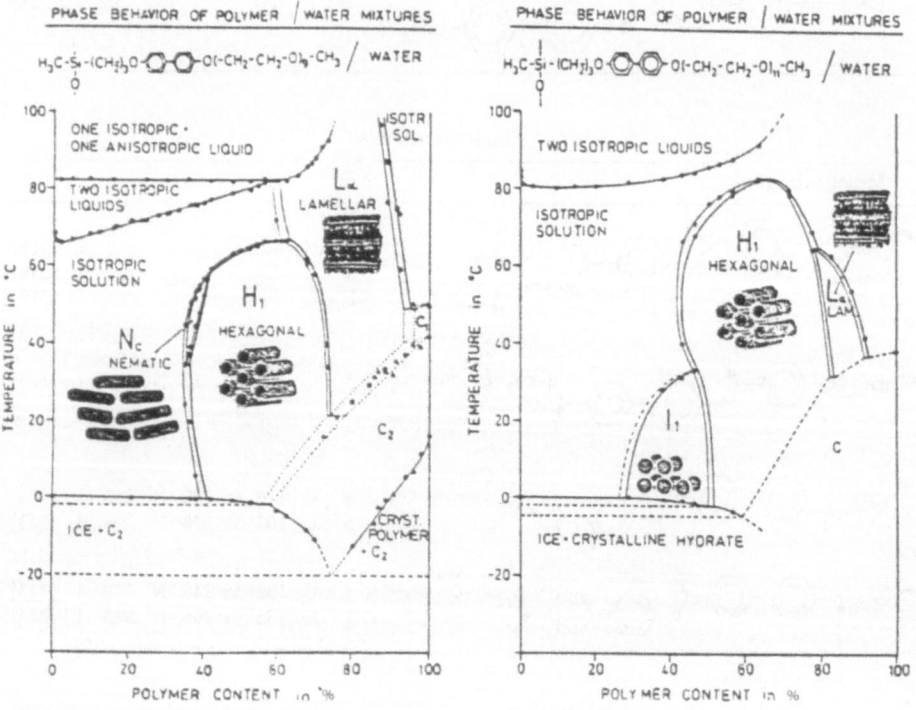	3	C	83	N*	98 I	[307]
	3	C	120	N*	243 I	[311]
	3	C	33 S 62		I	[311]

thermotropic liquid crystalline siloxanes. The following example may illustrate the general approach to synthesize these materials (Scheme 81).

$$H_2C=CH-(CH_2)_8-COCl \ + \ HO-(CH_2-CH_2-O)_4-CH_3 \ \xrightarrow{Pyr}$$

$$H_2C=CH-(CH_2)_8-CO_2-(CH_2-CH_2-O)_4-CH_3 \ \xrightarrow{\equiv SiH, Pt} \tag{81}$$

$$\equiv Si-(CH_2)-CO_2-(CH_2-CH_2-O)_4-CH_3$$

The phases which can be observed depend on the structure of the material and the concentration in the aqueous system, also on the degree of polymerization (Fig. 3.11).

Fig. 3.11. Binary phase diagrams of polymeric amphiphilic siloxanes [356]

Table 3.12. Linear liquid crystalline side chain polysiloxanes: copolymers

a   With Dimethylsiloxane

Mesogenic groups	m	Mol-%	Phase transition/°C		DP	Ref.
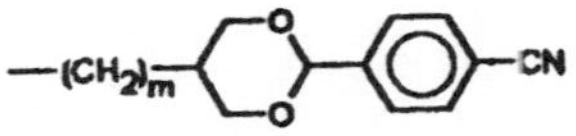 (1,3-dioxane–phenyl–CN), $-(CH_2)_m-$; $-CH_3$	11	100 / 0	G −10	S_A 135	I >27	[324]
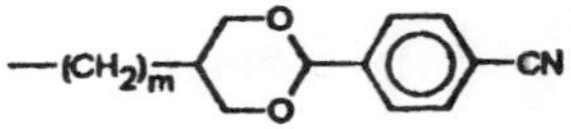 (1,3-dioxane–phenyl–CN), $-(CH_2)_m-$; $-CH_3$	11	33 / 67	G_1 −100 / G_2 −39	S_A 66	I >27	[324]
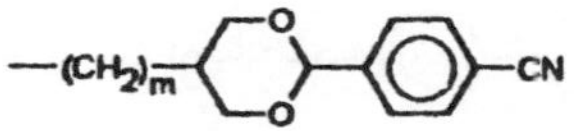 (1,3-dioxane–phenyl–CN), $-(CH_2)_m-$; $-CH_3$	11	17 / 83	G_1 −105 / G_2 −58	S_A −	I >27	[324]
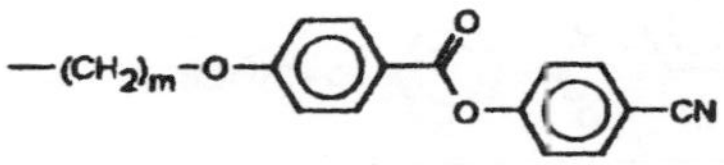 (phenyl–O–CO–phenyl–CN ester), $-(CH_2)_m-O-$; $-CH_3$	5	25 / 75	G −30	S 42	I	[295]

Table 3.12a. (Contd.)

Mesogenic groups	m	Mol-%	Phase transition/°C					DP	Ref
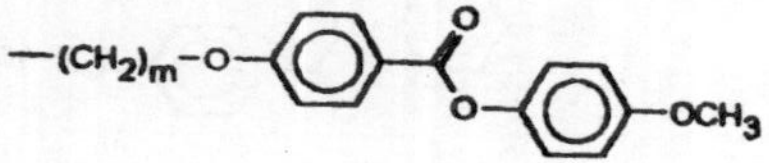	5	25	G	−30	S	23	I		[295]
		75							
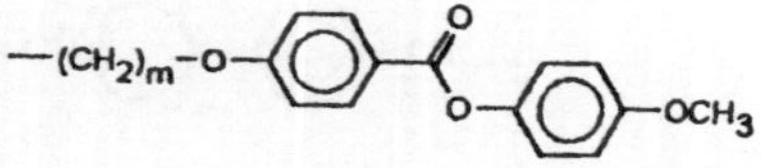	11	9.1	G₁	−114	Sx	−15	I		[295]
		90.9	G₂	−57					
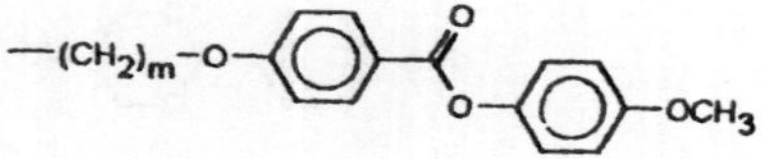	11	16.7	G	−46	Sx	11	LCx 64	I	[295]
		83.3							
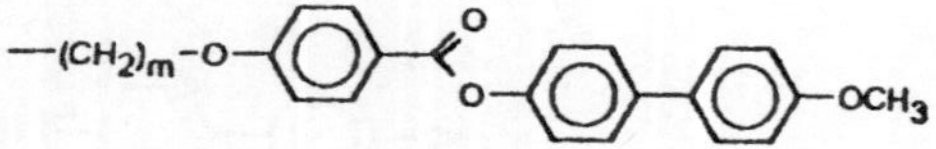	3	37	C	87	Sx	124	N 170	I	58 [316]
		63							

b With other mesogens

Mesogenic groups	m	Mol-%	G	Sx	N	I	DP	Ref.
$-(CH_2)_m-O$—C$_6$H$_4$—C$_6$H$_4$—CN	6	70	7	151		I		[310]
$-(CH_2)_m-O$—C$_6$H$_4$—CO_2—C$_6$H$_3$(CH_3)($(CH_2)_2$—CH_3)	4	30				I	>40	[280]
	6 / 4	50 / 50	4	86		I		[310]
	6 / 4	50 / 50	7	122		I		[310]
	6 / 4	32 / 68	4		53	I		[310]
$-(CH_2)_m-O$—C$_6$H$_4$—CO—O—C$_6$H$_4$—OCH_3	3 / 3	90 / 10	17		87	I	95	[413]
$-(CH_2)_m-O$—C$_6$H$_4$—CO—O—C$_6$H$_4$—Cl	4 / 4	90 / 10	16	82	99	I	95	[413]
	5 / 5	90 / 10	17	56	118	I	95	[412]
	6 / 3	92.5 / 7.5	7	60	108	I	95	[414]
	6 / 6	90 / 10	8	70	107	I	95	[413]
	6 / 6	85 / 15	9		107	I	95	[413]
	6 / 6	80 / 20	10	79	107	I	95	[413]

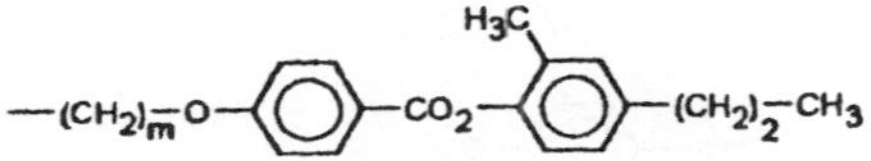

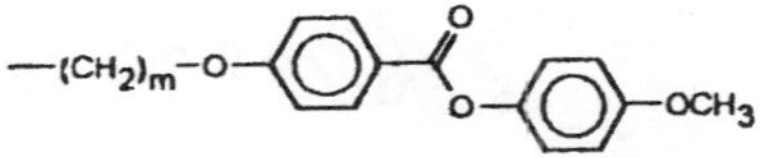

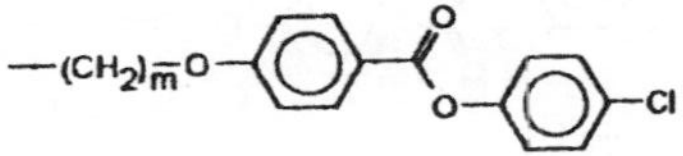

(Phase transition temperatures given in °C. G = glass, Sx = smectic X, N = nematic, I = isotropic.)

Table 3.12b. (*Contd.*)

Mesogenic groups	m	Mol-%	Phase transition/ °C							DP	Ref.
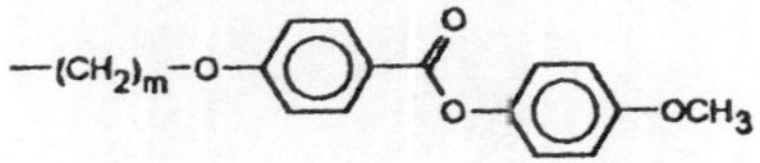 —(CH$_2$)$_m$—O—⟨C$_6$H$_4$⟩—COO—⟨C$_6$H$_4$⟩—OCH$_3$	6	92.5	G	7	Sx	70	N	113	I	50	[414]
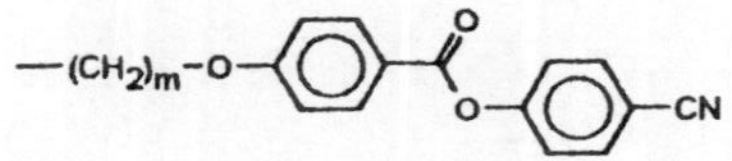 —(CH$_2$)$_m$—O—⟨C$_6$H$_4$⟩—COO—⟨C$_6$H$_4$⟩—CN	3	7.5									
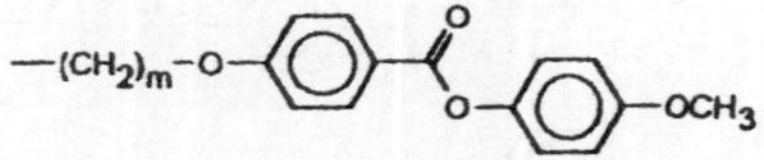 —(CH$_2$)$_m$—O—⟨C$_6$H$_4$⟩—COO—⟨C$_6$H$_4$⟩—OCH$_3$	6	90	G	6	Sx	53	N	102	I	95	[413]
—(CH$_2$)$_m$—O—⟨C$_6$H$_4$⟩—CO—⟨C$_6$H$_4$⟩—F	6	10									
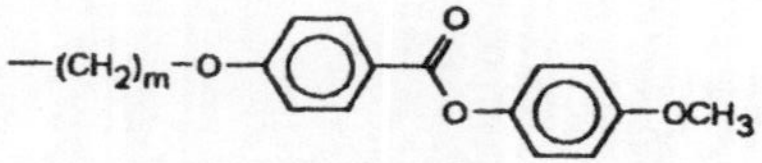 —(CH$_2$)$_m$—O—⟨C$_6$H$_4$⟩—COO—⟨C$_6$H$_4$⟩—OCH$_3$	6	90	G	7	Sx	89	N	117	I	95	[413]
—(CH$_2$)$_m$—O—⟨C$_6$H$_4$⟩—COO—⟨C$_6$H$_4$⟩—NO$_2$	6	10									
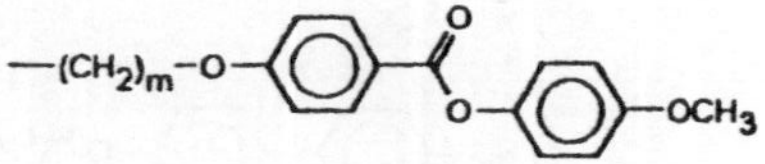 —(CH$_2$)$_m$—O—⟨C$_6$H$_4$⟩—COO—⟨C$_6$H$_4$⟩—O—⟨C$_6$H$_4$⟩—OCH$_3$	3	100	G	15			N	61	I	58	[311]
—(CH$_2$)$_m$—O—⟨C$_6$H$_4$⟩—COO—⟨C$_6$H$_4$⟩—⟨C$_6$H$_4$⟩—OCH$_3$	3	0									

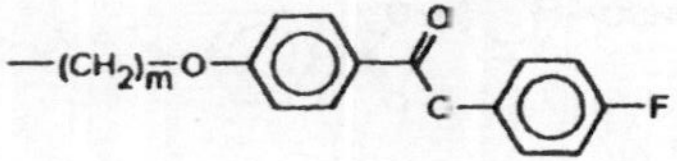

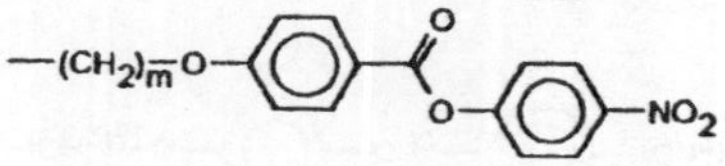

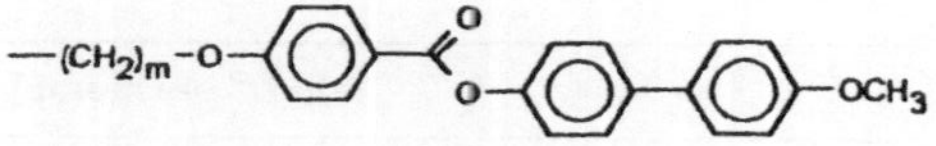

Structure 1: $-(CH_2)_m-O-C_6H_4-C_6H_4-COO-C_6H_4-OCH_3$

m	mol%	G	C	Sx	N	I	value	Ref
3	75 / 25		36		121	I	58	[311]
3	50 / 50		63		175	I	58	[311]
3	25 / 75		81		237	I	58	[311]
3	0 / 100		139		319	I	58	[311]
3	50	98		258		I	330	[415]

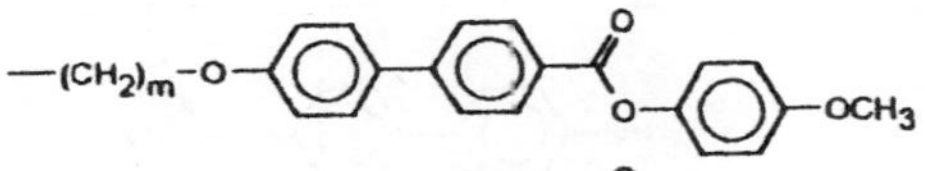

Structure 2: $-(CH_2)_m-O-C_6H_4-C_6H_4-COO-C_6H_4-OCH_3$

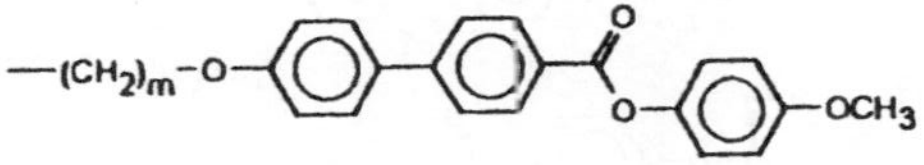

m	mol%	G	C	Sx	N	I	value	Ref
3	50	120		297		I	330	[415]
5	50		125	328		I	330	[415]

Table 3.12. (*Contd.*)

c With chiral compounds

Mesogenic groups	m	Mol-%	Phase transition/°C							DP	Ref.
(CN / CH₃ benzoate structure)	6	85	G	5	N*	31	I			50	[310]
(chiral ethyl-methyl biphenyl structure)	4	15									
(biphenyl OH + nitro benzoate chiral structure)	11	11	G	30	S_C*	156					[286]
	11	89			S_A	187	I				
(biphenyl OH + chloro benzoate chiral structure)	11	15	G	43	S_X	128					[286]
	11	85			S_C*	168	I				
(biphenyl OH + chloro chiral ester structure)	11	21			S_X	68					[286]
	11	79			S_C*	88					
					S_A	156	I				

Table 3.13. Linear liquid crystalline polysiloxanes: terpolymers

R	m	Mol%	Phase transition/°C							DP	Ref.
$-(CH_2)_m-O-$ [biphenyl]$-C(=O)O-$ [phenyl: NO_2, CH_3, OCH, C_6H_{13}]; $-(CH_2)_m-CH_3$; $-CH_3$	11	23	G	0			S_X 46 S_C^* 98		I		[416]
	7	4					S_A^* 144		I		
		73			C	83					
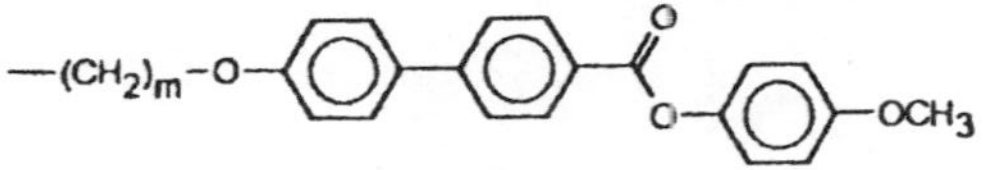 $-(CH_2)_m-O-$ [biphenyl]$-C(=O)O-$ [phenyl$-OCH_3$]; $-C(=O)O-$ [phenyl$-OCH_3$]; $-CH_3$	3	25					S 226		I	330	[415]
	3	25 / 50	G	98							
	3 / 5	25 / 25 / 50	G	120			S 230–260		I	330	[415]
	3 / 10	25 / 25 / 50	G	60			S 226		I	330	[415]
	5 / 5	25 / 25	G	103			S 288		I	330	[414]
$-(CH_2)_m-O-$ [biphenyl]$-CN$	10 / 10	25 / 25 / 50 / 50	G	92			S 246	N 286	I	330	[415]
	6										

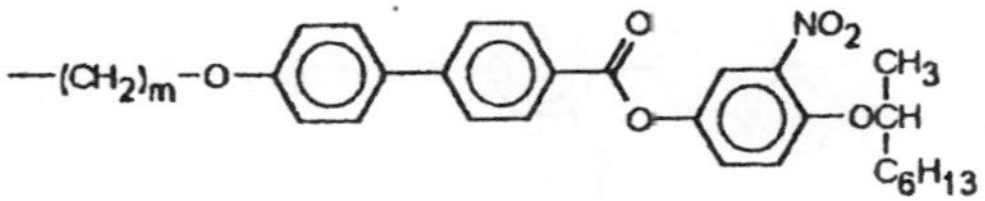

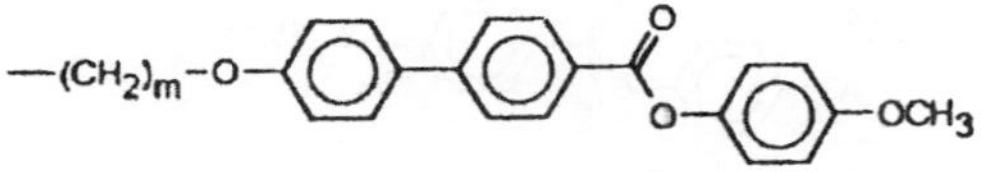

Table 3.13. (Contd.)

R	m	Mol%	Phase transition/°C			DP	Ref.
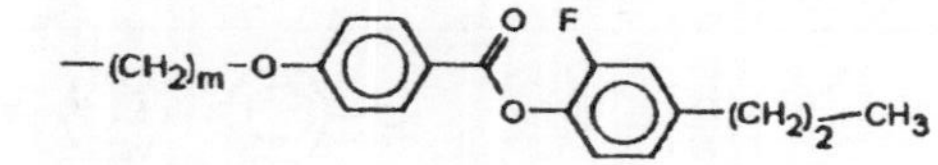 (structure: —(CH$_2$)$_m$—O— phenyl—C(=O)O— phenyl(F)—(CH$_2$)$_2$—CH$_3$)	4	45	G	Sx 122	I	50	[417]
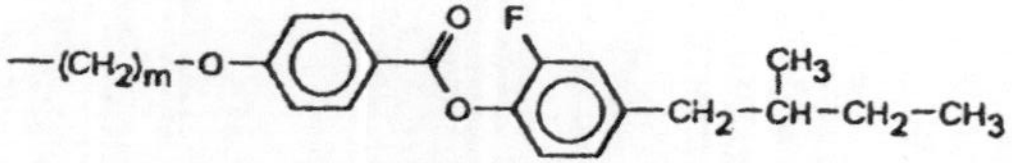 (structure: —(CH$_2$)$_m$—O— phenyl—C(=O)O— phenyl(F)—CH$_2$—CH(CH$_3$)—CH$_2$—CH$_3$)	4	5					

Table 3.14. LC side chain polysiloxanes with laterally attached mesogens

Mesogenic group	m	n	Phase transition/°C			DP	Ref.
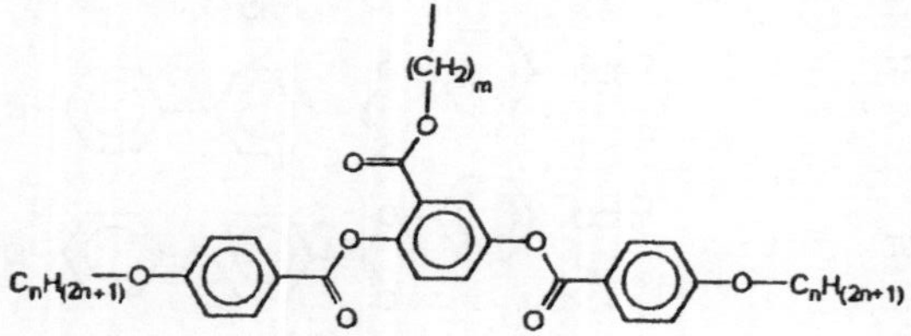 (structure: —(CH$_2$)$_m$— linked diester with C$_n$H$_{2n+1}$—O— and —O—C$_n$H$_{2n+1}$)	4	4	G	35	N 119 I	80	[349]
	4	6	G	23	N 99 I	80	[349]
	10	4	G	17	N 88 I	80	[349]
	10	6	G	13	N 91 I	80	[349]
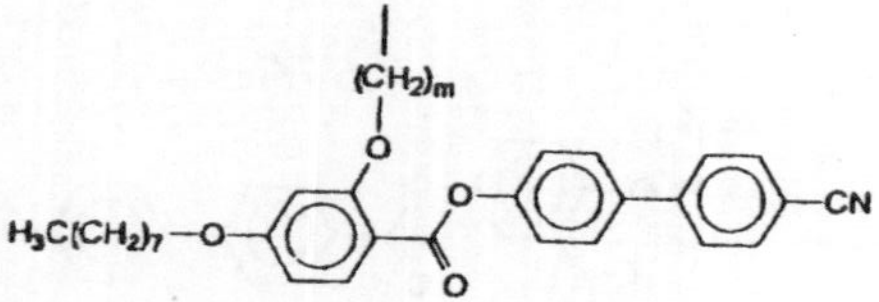 (structure: —(CH$_2$)$_m$—O— benzoate with CN-biphenyl and H$_3$C(CH$_2$)$_7$—O—)	5		G^a	15.7	N 68.6 I	46	[350]
	5		G^b	29.4	N 89.8 I	46	[350]
	6		G^a	11.8	N 67.7 I	46	[350]
	6		G^b	25.1	N 82.1 I	46	[350]

Used catalyst: ^{a}H$_2$PtCl$_6$
^{b}platinium divinyltetramethyldisiloxane complex

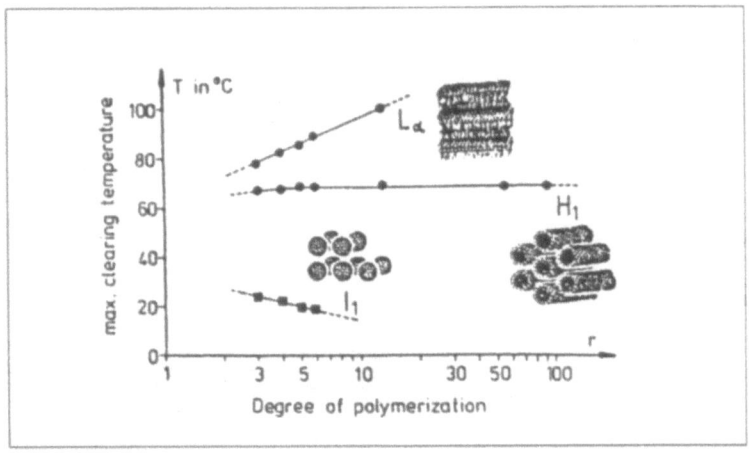

Fig. 3.12. Influence of the DP to the phase behaviour of lyotropic lc. siloxanes of the type

$$\left[\underset{\underset{\displaystyle |}{\overset{\displaystyle |}{\text{O}}}}{\text{Me}-\text{Si}}-(\text{CH}_2)_3-\text{O}-\text{O}-\text{O}-(\text{CH}_2-\text{CH}_2-\text{O})_9-\text{CH}_3 \right]_n$$

in the presence of water [356].

3.3.5.2
Cyclic Siloxanes with Mesogenic Side Groups (Tab. 3.15)

The synthesis of cyclic siloxanes with mesogenic side groups is similar to the linear species [357–359]. Some aspects have to be discussed with respect to the application properties of the materials. One of the most important points is the low definite degree of polymerization of the available cyclics. DPs from 4–7 are the most favoured species, however ring systems with DPs up to 24 have been studied already [360]. As cyclics are free of end groups the depression of T_G by trimethylsilyl groups is eliminated, but additional entropical effects play an important part [361]. So cyclics with DPs < 100 exhibit glass transition temperatures higher than the corresponding linear siloxanes [362]. This is also valid for cyclic lc siloxanes with DPs < 10. As a consequence of low DP the bulk viscosity is relatively low.

Statistical aspects are important in the case of co-oligomers with low DP. In long-chain co-polymers the statistical distribution yields inhomogeneities along the chain, but short-chain systems or cyclics with 4–10 siloxane units actually represent a collection of chemically different molecules.

The amount of the different species in the case of a copolymer is determined by a binominal distribution:

Assuming two mesogens M1 and M2 of the same reactivity attached to a cyclic siloxane with n siloxane units (Form. 82) where, k: is the number of mesogens M1 attached to the cyclic siloxane, n–k the number of mesogens M2, then the distribution is given by (Eq. 83). P: % of cyclic siloxane molecules substituted by k mesogenic groups M1. PM1: Mole-% of M1 in the mixture of both mesogens k: 0...n

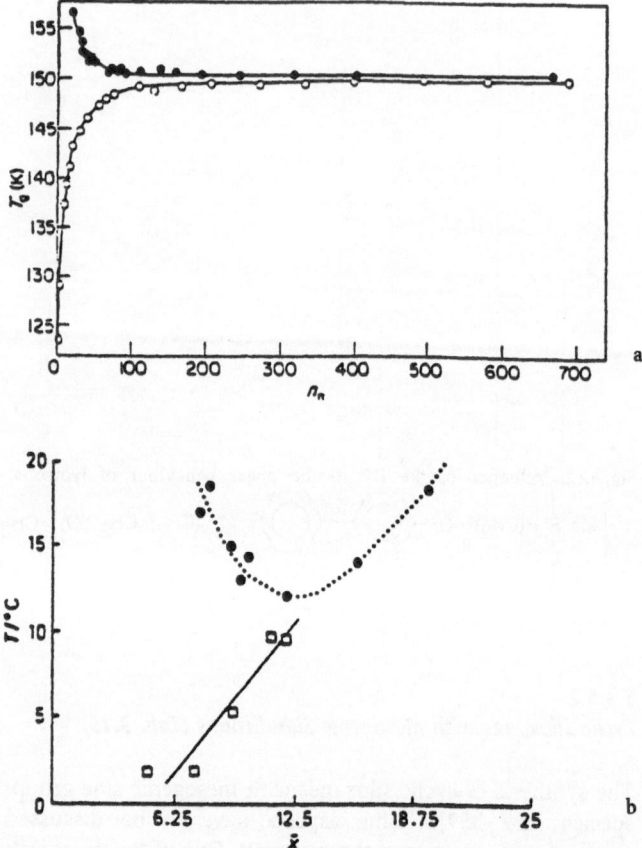

Fig 3.13. T_G as function of the DP   a) dimethylsiloxanes (○ linear, ● cyclic) [361]   b) liquid crystalline siloxanes containing the cyanobiphenyl mesogenic group (□ linear, ●cyclic) [360].

$$
\begin{bmatrix} Me-\underset{\underset{O}{|}}{Si}-M1 \end{bmatrix}_k
$$
$$
\begin{bmatrix} Me-\underset{\underset{O}{|}}{Si}-M2 \end{bmatrix}_{(n-k)}
$$

(82)

$$
p(n,k) = \frac{n!}{k!*(n-k)!} \; pM1^k *(1-pM1)^{(n-k)} *100
$$

(83)

Figure 3.14 shows that the hexamer corresponding to the overall composition of the co-oligomer mixture will be only found in a concentration of 31.25%.

Fig. 3.14. Product distribution of an hexameric copolymer consisting of two mesogens with the molar ratio 1:1

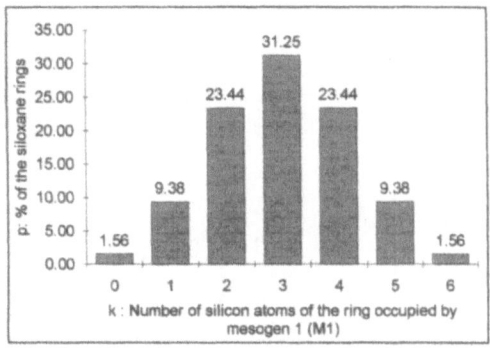

Fig. 3.15. Bundle structure of a cyclic liquid crystalline siloxane with 4 Si—O- units

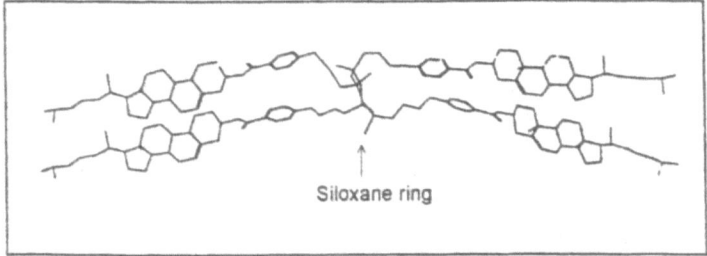

The phase behaviour of cyclic lc siloxanes is similar to the corresponding linear systems. With elongated mesogens calamitic structures were observed. Bundle or bowl like structures are the most probable ones [271–364].

Cyclic lc siloxanes also exhibit blue phases, which may be retained at RT [370].

3.3.5.3
Silsesquioxanes with Mesogenic Side Groups [363]

Liquid crystalline silsesquioxanes represent a special group of lc side-chain polymers with backbones ranging from centrosymmetric cage-like structures of siloxanes or heterosiloxanes to linear ladder structures. They may be viewed as the structural link between linear side-chain siloxanes and surface treated silica or quartz. Lc silsesquioxanes can be synthesized by combining mesogens with spherical silsesquioxanes bearing Si—H-groups. The procedure is, in general, the same as described in the previous chapters for linear and cyclic systems.

Most of the synthesized lc silsesquioxanes exhibit smectic phases (Tab. 3.16). Lateral connection of a mesogen results in a nematic phase. All these phases are optically uniaxially positive, i.e. calamitic structures are favoured.

Liquid crystalline silsesquioxanes can be considered as structural derivatives of cyclic lc siloxanes because they are double ring structures i.e. two siloxane

Table 3.15. Phase Transition Temperatures of Lc Cyclosiloxanes

a   Derivatives of the 4-Hydroxybenzoic acid

Mesogenic group	m	Phase transition/°C					DP	Ref.
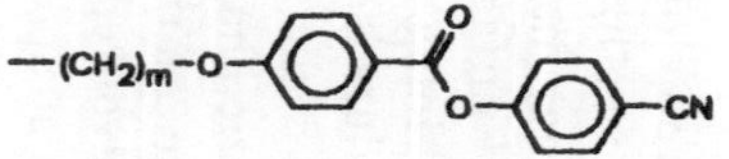	3	G	27		N	104	I	5	[418]
	3	G	31		N	122	I	5	[418]
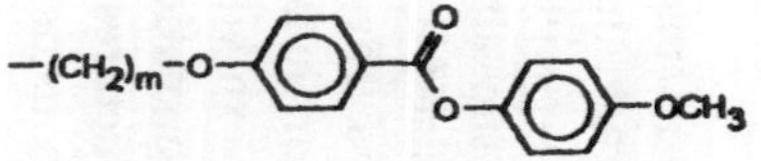	3	G	26		N	107	I	4	[271]
	3	G	21		N	82	I	5	[271]
	3	G	22		N	84	I	6	[271]
	3	G	23		N	83	I	7	[271]
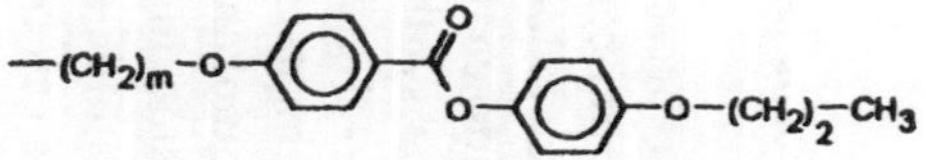	3	G	32		N	112	I	4	[271]
	3	G	30		N	94	I	6	[271]
	3	G	27		N	89	I	7	[271]
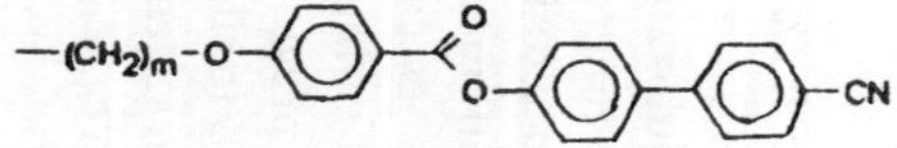	3	G	59	S_A	>300		I	4	[271]

Table 3.15. Phase Transition Temperatures of Lc Cyclosiloxanes

b Derivatives of Cholesterol

Mesogenic group	m	Phase transition/ °C	DP	Ref.
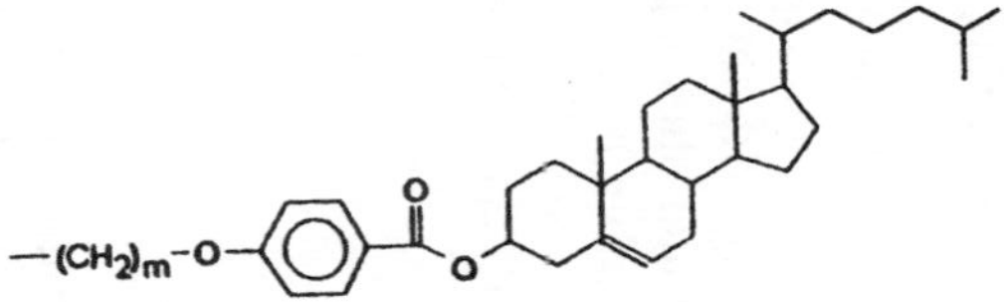	3	G 70 S_B 88 S_A 240 N^* 271 I	4	[271]
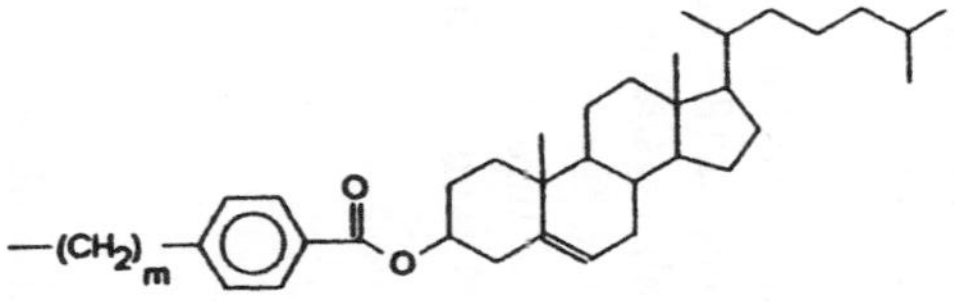	3	G 82 S_A 246 I		[271]

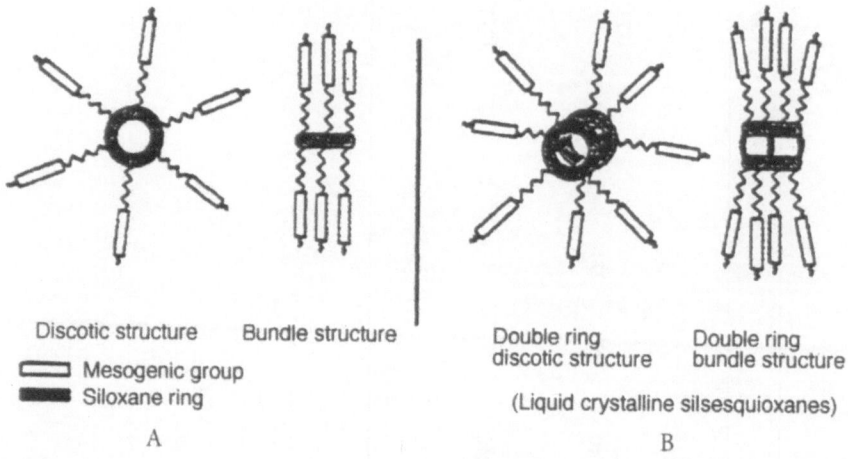

Fig. 3.16. Schematic structures of cyclic liquid-crystalline siloxanes A and liquid-crystalline silsesquioxanes B

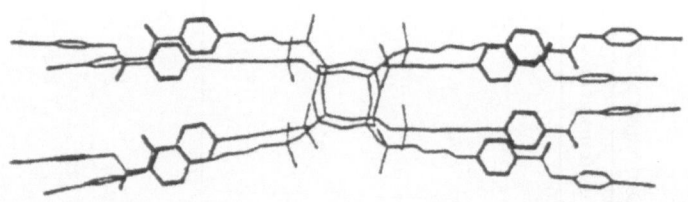

Fig. 3.17. Structure of a liquid crystalline silsesquioxane

rings connected by oxygen bridges (Fig. 3.16). In this case the ring backbone is extremely rigid and of well defined structure.

The most probable arrangement in solution is a star-like one, but in the liquid-crystalline state a calamitic bundle structure is discussed (Fig. 3.17).

The first liquid-crystalline silsesquioxanes $[R—SiO_{3/2}]_n$ with the structure **84** have been obtained by hydrolysis of alkoxysilanes under shearing conditions [371] where R is a mesogen connected to the silicon atom by a spacer group. These materials exhibit a cholesteric structure if

$R = —(CH_2)_3—O—C_6H_4—COO—C_6H_5$ and $—(CH_2)_3—O—C_6H_4COO$-Cholesteryl.

$$—O—\underset{\underset{O}{|}}{\overset{\overset{R}{|}}{Si}}—O— \tag{84}$$

During the process of hydrolysis at ambient humidity in the presence of dibutyltindiacetate and ageing at elevated temperatures the cholesteric phase is crosslinked so no clearing temperatures can be observed. The structures of the backbones of these silsesquioxanes have not been investigated in detail but it can be concluded from ^{29}Si-NMR-spectra that the Si-atoms have different chemical surroundings.

The synthesis of liquid-crystalline silsesquioxanes with well-defined backbones should therefore start from functionalized silsesquioxanes [372]. This method provides the opportunity to obtain a series of lc silsesquioxanes with different structures.

The structures are polyhedral frameworks with various degrees of symmetry [373, 374] with the silicon atoms at the corners and oxygen atoms between them (Fig. 3.18).

The silicon atoms can be substituted by hetero atoms [375] e.g. by Al, B, Fe. Moreover silsesquioxanes can exhibit ladder structures [376]. Therefore, silsesquioxanes represent a huge source of different backbones for lc systems.

Despite the attention directed to oligosilsesquioxanes for long years [377–386], the first oligomeric forms [387] of the basic members of this class (85)

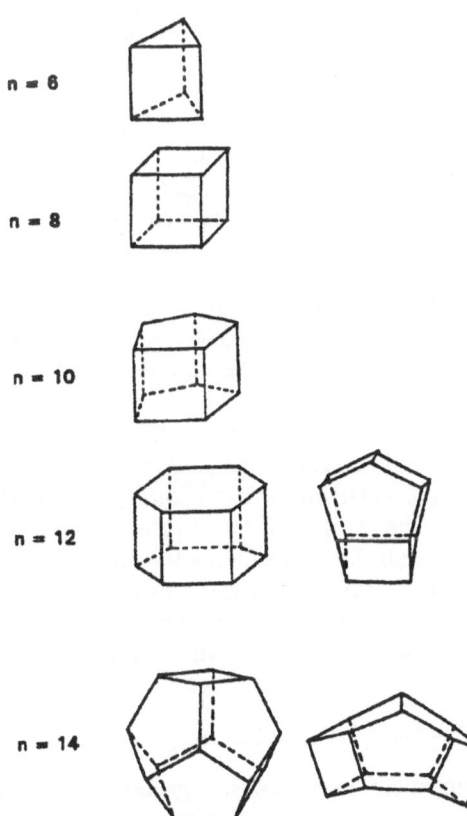

n = 6

n = 8

n = 10

n = 12

n = 14

Fig. 3.18. Schematic structures of silsesquioxanes. The corners of the polyhedral frameworks are silicon atoms, the oxygen atoms are between them [374, 377, 378, 379]

namely:

$$\left(HSiO_{3/2} \right)_n \qquad\qquad n = 6,8,10 \qquad\qquad (85)$$

were synthesized [388] in 1970 in moderate yields. Since, oligomeric silsesquioxanes and their Si—H containing derivatives are available in good quantities [389, 390], investigations of liquid crystalline polysiloxanes could be extended to silsesquioxanes. These molecules, provided with mesogenic groups, represent liquid crystalline side-chain polymers or oligomers of the general formula (86):

$$[RSiO_{3/2}]_n \quad n = 6, 8, 10, \ldots \qquad\qquad (86)$$

Lc silsesquioxanes with $n = 8$ and derivatives of the siloxy-substituted silsesquioxanes of the formula (87) have been synthesized.

$$[R—Si(CH_3)_2—O—SiO_{3/2}]_n \quad \text{with} \quad n = 6, 8, 10, \ldots \qquad (87)$$

The side-chain liquid crystalline silsesquioxanes prepared until now were synthesized by the hydrosilylation reaction employing di-cyclopentadienyl-platinumdichloride [391, 392] as catalyst.

It is recommended to use the mesogenic alkene in excess up to 100% to prevent dimerization or oligomerization of partially substituted backbones. Co-oligomers were prepared using the alkenes in the appropriate ratio to achieve the required statistical mix of side chains. The statistical considerations as mentioned before are also valid for the silsesquioxane chemistry. The hydrosilylation reaction is also accompanied by a loss up to 30% of the spacer when the propenyloxy spacer is used, as indicated by the evolution of propylene. The resulting Si—O—C bonds can be identified by ^{29}Si-NMR, but during the purification process the Si—O—C containing by-products can be removed. The highly symmetric structures of siloxy-substituted lc silsesquioxanes have been confirmed by two signals at 14.1 and -99.2 ppm in the ^{29}Si-NMR spectrum.

In spite of the spherical shape of the core (centrosymmetric with $n = 8$) silsesquioxanes with mesogenic side groups exhibit mesophases, mostly smectic phases. Substitution of the silsesquioxane core with leterally linked mesogens yields nematic materials. The phase transition temperatures of these compounds are summarized in Table 3.16.

Conoscopic polarizing microscopic observations of homeotropic oriented lc phases have shown, that all investigated liquid-crystalline silsesquioxanes – even the laterally substituted compounds – are monoaxially positive; thus, calamitic structures are favoured and discotic structures should have negative axiality [393].

Even the smallest centrosymmetric siloxane backbone of the structure also yields calamitic structures with liquid crystalline properties [394, 395] (Form. 88).

$$Si[—O—SiMe_2(CH_2)_{10}COOChol]_4 \qquad\qquad (88)$$

The arrangement of mesogenic groups with extended spacer units can be imagined as independent of geometry and symmetry of the backbone. So

Table 3.16. Phase Transitions of Liquid crystalline Silsesquioxanes

a   Silsesquioxanes of the type $(R\text{-}SiO_{3/2})_n$

Mesogenic group R	m	Phase transition/ °C	$\Delta H_{K/S/N\rightarrow I}$ J/g	DP	Ref.
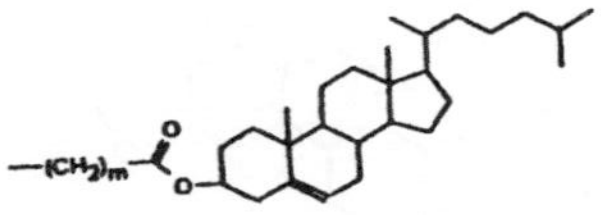	10	G (46) C 128 S$_A$ 157 I	4.3	8	[363]

b Silsesquioxanes of the type $(R-Si(CH_3)_2O-SiO_{3/2})_n$

Mesogenic group R	m	Phase transition/°C				$\Delta H_{K/S/N \to I}$ J/g	DP n	Ref.
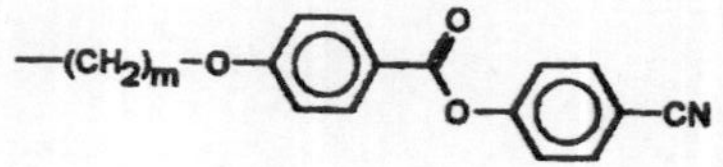 -(CH₂)m-O-⬡-O-C(=O)-⬡-CN	3	C 119	Sₐ 146		I	4.0	8	[363]
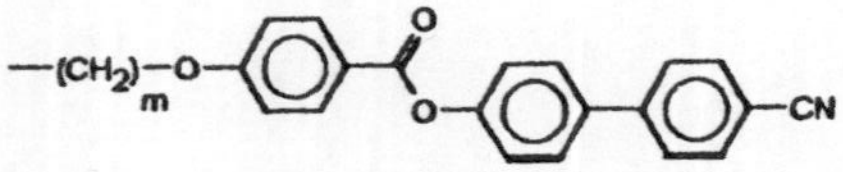 -(CH₂)m-O-⬡-O-C(=O)-⬡-⬡-CN	3	G 102	Sₓ 127	Sₐ >300	I		8	[363]
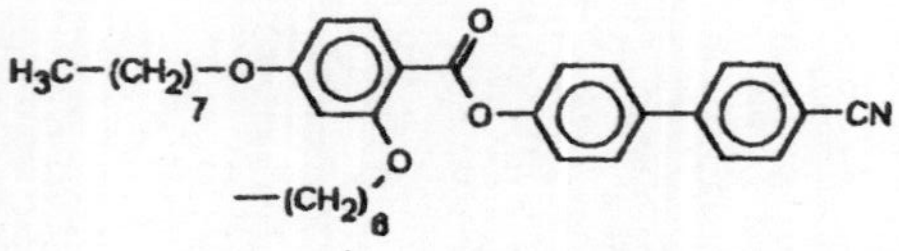 H₃C-(CH₂)₇-O-⬡-C(=O)-O, -(CH₂)₆-O-⬡-⬡-CN	6	G 25	N 44		I	1.3	8	[363]
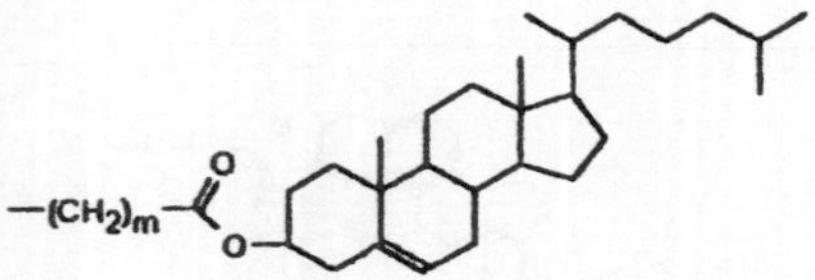 -(CH₂)m-O-C(=O)- cholesteryl	10	G 30	Sₓ 37	Sₐ 142	I	4.0	6	[363]
	10	C 108	Sₐ 160		I	5.0	8	[363]
	10	C 122	Sₐ 164		I	4.0	10	[363]
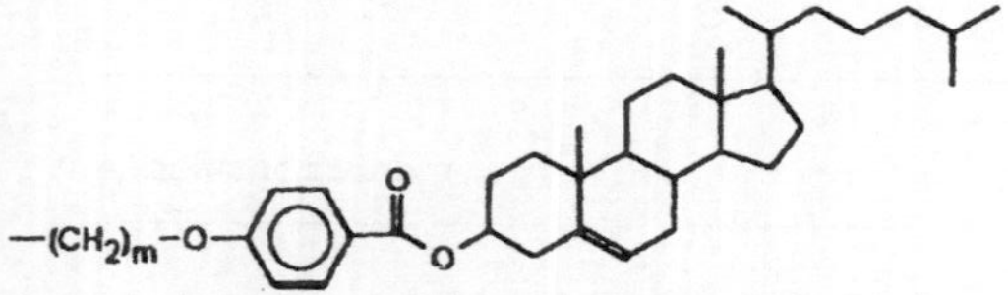 -(CH₂)m-O-⬡-C(=O)-O- cholestanyl	3	G (87) C 115	Sₐ 262		I	6.4	10	[363]

c Co-oligomers of the type [R—Si(CH₃)₂O—SiO3/2]ₙ

Mesogen R	m	Mol%	Phase transition/ °C						$\Delta H_{K/S/N \to I}$ J/g	DP n	Ref.
	10	50	G	35	S_X	63	S_A	148	I	5.0	8	[362]
—CH₃		50										
	10	50	G	17	S_X	60	S_A	107	I	5.0	8	[362]
		50										
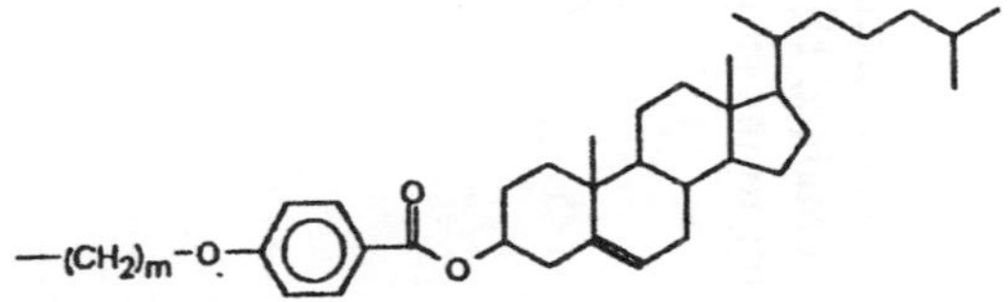	3	60			S_X	114						
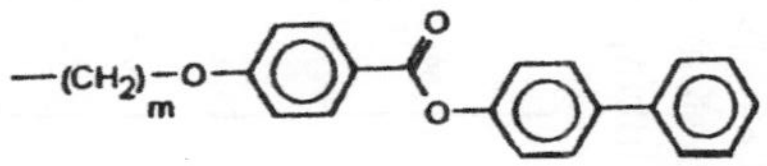	3	40					S_A	213	I	4.0	8	[362]

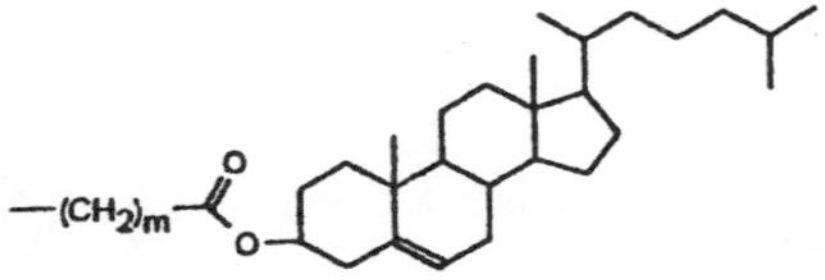

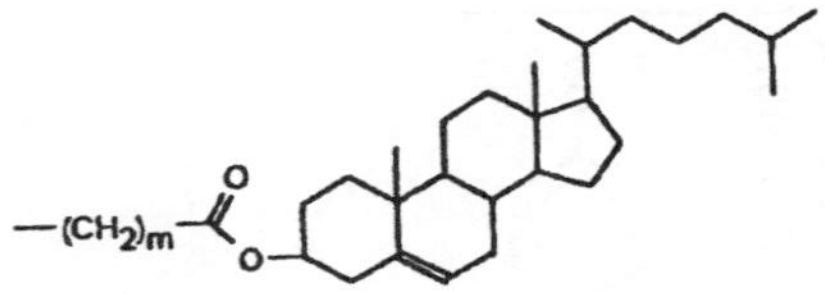

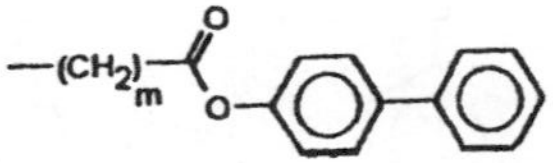

Fig. 3.19. Star like arrangement of the mesogenic groups of a liquid crystalline silsesquioxane in the isotropic phase or in solution

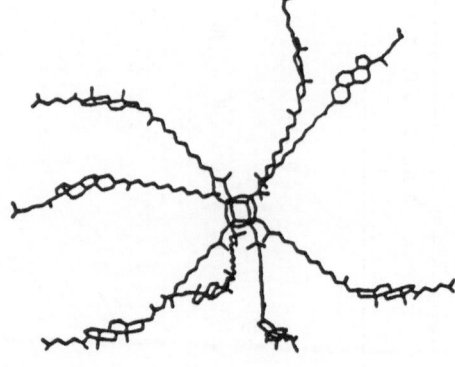

mesogens with appropriate spacer groups yield calamitic structures whether they are connected by a single atom, a flat disc like or a spherical backbone. This is in accordance with a model of calamitic mesogens weakly coupled to ring structures [396].

3.3.6
Liquid Crystalline Siloxane Networks

Lc-siloxane networks can be obtained from

a) Liquid crystalline siloxane main chain systems
b) Liquid crystalline siloxane side chain systems
c) Liquid crystalline – siloxane free – main chain systems

by crosslinking with a siloxane bridge or in case a) and b) with other – siloxane free – bridge units. Depending on T and crosslink density elastomers or resins are obtained.

3.3.6.1
Elastomers

Examples of elastomers dominated by a liquid crystalline siloxane side chain system are obtained by simultaneous addition of the mesogens and the cross-linking agent to the siloxane backbone [397–401] (Form. 89).

$$
\left[\text{Me}-\underset{\underset{O}{|}}{\overset{|}{\text{Si}}}-(\text{CH}_2)_{\overline{4}}\text{O}-\underset{}{\bigcirc}-\text{COO}-\underset{}{\bigcirc}-\text{OMe} \right]_x
$$

$$
\left[\text{Me}-\underset{\underset{O}{|}}{\overset{|}{\text{Si}}}-(\text{CH}_2)_{\overline{2}}\left(\underset{\underset{\text{Me}}{|}}{\overset{\overset{\text{Me}}{|}}{\text{Si}}}-\text{O} \right)_5 \right]_y
$$

(89)

These networks may be swollen by chiral compounds to form cholesteric elastomers. It is also possible to use siloxanes as a crosslinking agent [402, 403], Eq. (90).

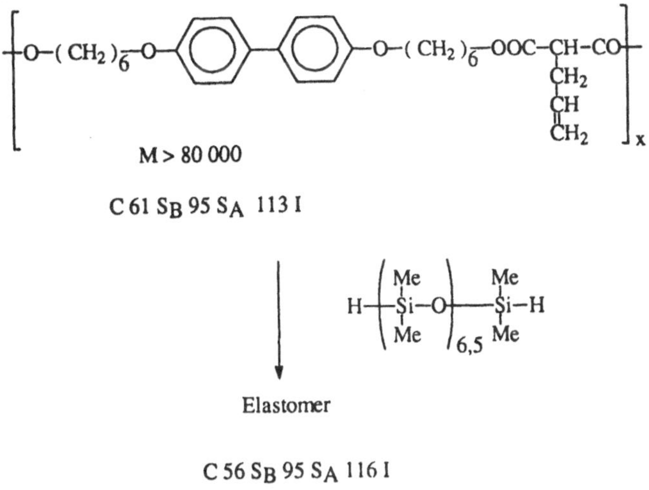

$$\left[O-(CH_2)_{\overline{6}}O-\bigcirc-\bigcirc-O-(CH_2)_{\overline{6}}-OOC-\underset{\underset{\underset{CH_2}{\overset{\Vert}{CH}}}{\overset{|}{CH_2}}}{CH}-CO \right]_x$$

$$M > 80\,000$$

$$C\ 61\ S_B\ 95\ S_A\ 113\ I$$

(90)

$$\downarrow \quad H-\left(\underset{Me}{\overset{Me}{Si}}-O\right)_{6,5}\underset{Me}{\overset{Me}{Si}}-H$$

Elastomer

$$C\ 56\ S_B\ 95\ S_A\ 116\ I$$

Also liquid crystalline main chain polymers with mesogenic side chains exhibiting ferroelectric behaviour were crosslinked with siloxanes [404–406].

The mesogens are usually oriented by stretching or compressing the liquid crystalline elastomer. After reducing the load the material relaxes again into the non oriented state. Nevertheless, it is possible to preserve the oriented state by a two step reaction [407], employing compounds with double bonds showing different reactivities towards the hydrosilylation reaction (Scheme 91).

3.3.6.2
Resins

For technical, temperature independent, non dynamic applications e.g. filters, polarizers and similar products highly crosslinked liquid crystals are required. To obtain liquid crystalline siloxanes which can be crosslinked by radical polymerization methacrylic groups were introduced according to a polymer analogue reaction (Scheme 92).

According to this scheme crosslinkable cholesteric liquid crystalline siloxanes were synthesized [408]. In general, any type and mixture of R may be choosen. Thus lc materials with tailored properties exhibiting S_A-, S_C-, S_C*-, N-phases can be obtained. The crosslinking reaction can be done thermally by addition of a radical initiator or photochemically with the aid of a photosensitizer.

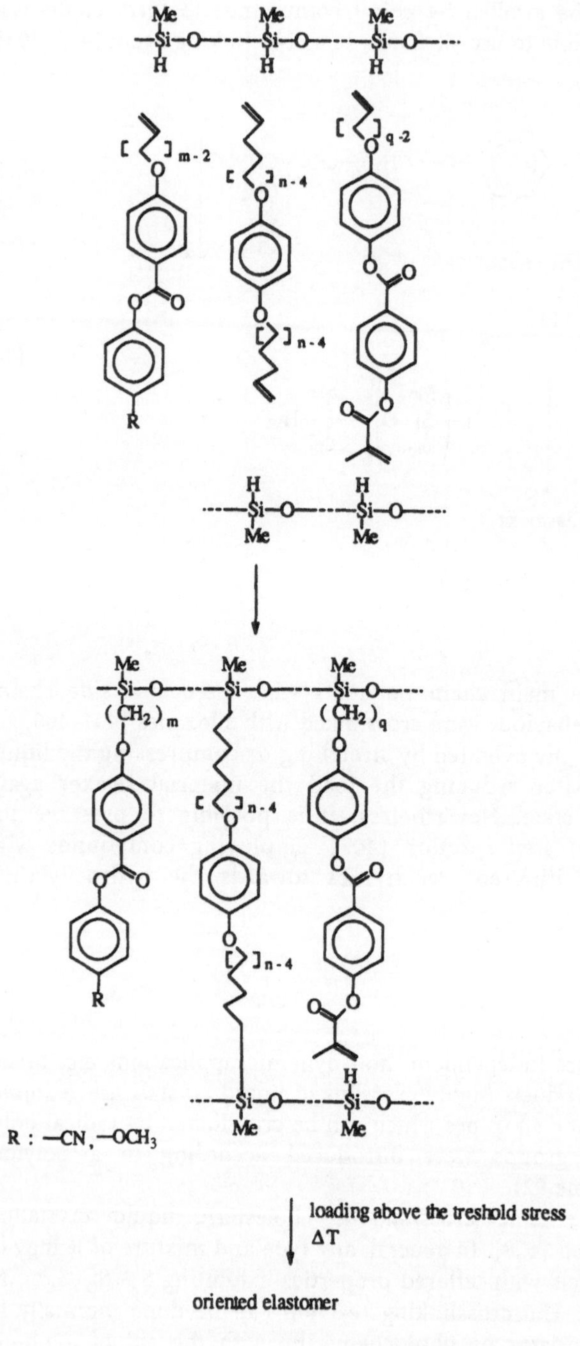

(91)

R : —CN, —OCH₃

loading above the treshold stress
ΔT

oriented elastomer

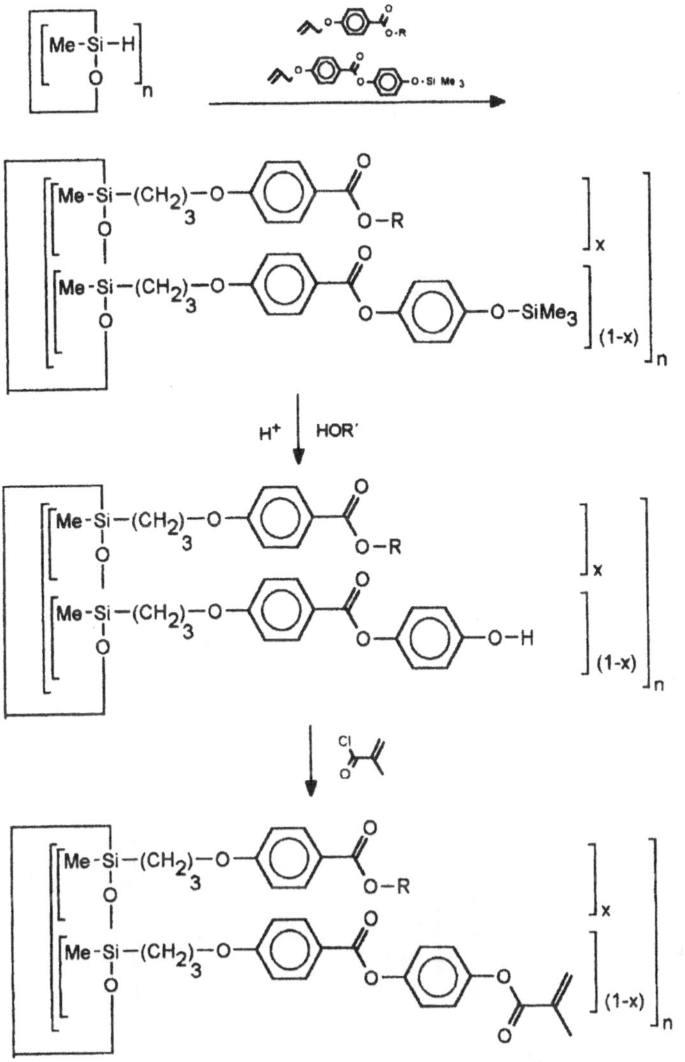

(92)

3.3.7
Complex Structures

A special type of liquid crystalline siloxanes are polycondensation products of dihydroxysiliconphthalocyanines [409–412]. The highest degrees of polymerization of up to 100 were obtained in tributylamine in the presence of catalysts e.g. CdCl$_2$, FeCl$_2$, CaCl$_2$ or via trifluoracetyl derivatives. To improve the solubility alkoxy substituted derivatives were synthesized (Eq. 93). The phase transitions of these polysiloxanes are listed in Table 3.17.

(93)

Table 3.17. Phase transitions of liquid crystalline Phthalo-cyaninatopolysiloxanes of structure

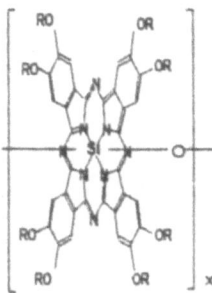

R: C_nH_{2n+1}

n	Phase transition/ °C						Ref.
2	C	–	D	–	I		[411]
4	C	111.4	D	–	I		
6	C	71.0	D	–	I		
8	C	59.2	D	84.0	I		
10	C	58.2	D	87.3	I		
12	C	58.8	D	93.6	I		
14	C	55.4	D	107.2	I		
16	C	56.2	D	89.3	I		
18	C	56.7	D	86.6	I		

3.4
References

1. Noll W (1968) Chemie und Technologie der Silicone, 2nd edn Verlag Chemie, Weinheim, FRG
2. Kaiser W, Riedle R (1982) In: Winnacker Küchler Chem Technologie, Vol 6, 4th edn 816
3. Deubzer B (1989) In: Silicone in Chemie und Technologie Vulkan-Verlag, p 99
4. Rochow EG (1987) Silicium und Silicone, Springer Verlag, Berlin Heidelberg New York
5. Wilcock DF, Patnode W (1946) J Am Chem Soc 68: 358
6. Kochs P (1987) In: Methoden der organischen Chemie, Makromolekulare Stoffe, Band E 20/III, 1782
7. Yang MH, Yang CK, Liaw WC (1990) CA 113: p 21275
8. Reedy JD, Furbee HD (1976) US 3. 983. 148
9. Harada K, Fukuhira M (1989) CA 110 213605c
10. Williams jr RE (1982) US 4. 412. 080
11. Yeboah YD (1982) US 4. 423. 240
12. Williams jr RE (1984) US 4. 447. 630
13. Lartigue-Peyrou F (1988) FR 2. 605. 634
14. Okawa N (1990) CA 112 218088n
15. Seyferth D, Prud'homme C, Wiseman GH (1983) Inorg Chem 22: 2163
16. Rutz W, Lange D, Popowski E, Kelling H (1986) Z Anorg Allg Chem 536: 197
17. Spörk H et al. (1977) US 4. 032. 557
18. Ciobanu A, Lachian N, Bostan M, Hamcive V (1979) CA 90 104619h
19. Paetzelt D, Ruehlmann K (1979) Plaste Kautsch 26: 606

20. Wick M, Kreis G, Kreuzer F-H (1982) In: Ullmanns Encyclopädie der technischen Chemie, Vol 21, Verlay Chemie, Weinheim, FRG
21. Tagiguchi TJ (1959) J Am Chem Soc 81: 2359
22. Yeboah YD, Venditti MP (1985) US 4. 497. 942
23. Hajjar AL (1983) DE-OS 3. 244. 500
24. Kaiser W, Riedle R (1982) In: Winnacker Küchler Chem Technologie Bd 64Aufl 830
25. Thomas DR, Francis J (1971) US 3. 626. 805
26. Zhikin DY, Ivanov VI, Chernyak VI (1983) CA 99 140010x
27. Nietzsche et al. (1973) DE 2148669
28. Kendrick TC, Parbhoo B White JW (1989) In: Patai S, Rappoport Z (eds) The chemistry of organic silicon compounds, New York, p 1302
29. Sauvet G, Lebrun JJ, Sigwalt P (Publ 1984) Cationic Polym Relat Processess, Proc Int Symp 6the 1983, 237
30. Chojnowski J, Rubinsztajn S, Wilcek L (1986) Actual Chim 3: 56
31. Tartakovskaya LM, Kopylov VM, Zhdanov AA (1984) CA 101 73186d
32. Wilczek L, Chojnowski J (1981) Macromolecules 14: 9
33. Sigwalt P, Stannet V (1990) Makromol Chem Makromol Symp p 217
34. Bryk MT, Skobets IE, Vasil'ev NG (1978) CA 89 215825 w
35. Voronkov MG, Mileshkevich VP, Yuzhelevshii YuA (1974) The siloxane bond p 203
36. Saam JC Zeigler JM, Gordon Fearon FW (eds) (1990) In: Silicon-based Polymer Science, American Chem Soc Washington p 71
37. Sigwalt P, Gobin G, Niwl P (1991) Makromol Chem Makromol Symp p 229
38. Sigwalt P, Moreau M, Masure M (1990) Polym Prepr (Am Chem Soc Div Polym Chem) 31 40
39. Chojnowski J, Scibiorek M (1976) Makromol Chem 177: 1413
40. Lebrun JJ, Sauvet GG, Sigwalt P (1984) In: Cationic polymerization and related processes Academic Press, New York p 237
41. Stark FO, Falender JR, Wright AP, Wilkinson G, Sone, FGA Pergamon, W (eds) (1982) In: Comprehensive Organometallic Chemistry, Vol 2 322
42. Gold'in GS et al. (1976) CA 84 60042j
43. Hyde JF (1949) UD 2. 490. 357
44. Warrick EL (1949) US 2. 634. 252
45. Hurd DT, Osthoff RC, Corrin MC (1954) J Am Chem Soc 76: 249
46. Grubb WT, Osthoff RC (1955) J Am Chem Soc 77: 1405
47. Chojnowski J, Mazurek M (1975) J Makromol Chem 176: 2999
48. Chojnowski J, Mazurek (1977) J Makromol Chem 177: 1005
49. Bargain M, Millet C (1976) DE 2724194
50. Hubert S, Hemery P, Boileau S (1986) Makromol Chem, Makromol Symp 6: 247
51. McGrath JE, Sormani PM et al. (1986) Makromol Chem, Makromol Symp 6: 67
52. Fleischmann G, Eck H, Wenzeler P (1989) EP 302. 492
53. Miyasaka T, Higuchi Y (1990) CA 112: 139948q
54. McGrath JE, Sormani PM et al. (1986) Polym Prepr (Am Chem Soc, Div Polym Chem) 27: 152
55. McGrath JE, Sormani PM et al. (1985) Polym Prepr (Am Chem Soc, Div Polym Chem) 26: 258
56. Sigwalt P, Deffieux A, Stannett VT, Naylor DM (1991) Polymer 32: 1084
57. Sigwalt P et al. (1986) J Polym Sci, Part C: Polym Lett 24: 319
58. Chawla AS, StPierre LE (1972) J Polym Sci 16: 1887
59. Mazurek M, Scibiorek M, Chojnowski J, Zavin B, Zholarov AA (1980) Eur Polym J 66: (1) 67
60. Voronkov MG, Borisov SN (1970) E Ya Lukevits p 9
61. Eaborn E (1960) Organosilicon Compounds p 89
62. Craig DP (1954) J Chem Soc p 332
63. Smith AL (1991) Chem Anal 112: 3
64. Barton ThJ, Boudjouk P, Zeigler GM, Gordon Fearon FW (eds) (1990) In: Silicon-based Polymer Science American Chem Soc, Washington p 3
65. Mark JE in ibid p 47

66. Bassindale AR, Taylor PG, Patai S, Rappoport Z (eds) (1989) In: The Chemistry of Organosilicon Compounds, New York, p 896
67. Burkhardt J (1989) In: Silicone Chemie und Technologie Vulkan-Verlag p 23
68. Grassi N, MacFarlane IG (1978) J Eur Polym 14: 875
69. Kendrick TC, Parbhoo B, White JW (Ed) by Patai S, Rappoport Z, (1989) In: The chemistry of organosilicon compounds New York Ch 21 p 1319
70. Hochgeschwender K (1989) In: Silicone Chemie und technologie Vulkan-Verlag p 39
71. Kochs P, Parlar H, Wrobel D, unpublished results: Photochemischer Abbau von Trimethylsilanol und Hexamethyldisiloxan in der Gasphase
72. Wacker AK-Fluids Technical Data Sheets
73. Owen MJ, Zeigler GM, Gordon Fearon FW, (1990) (eds) In: Silicon-based Polymer Science Amer Chem Soc, Washington p 705
74. Ott MA (1977) Paint Coat 67: (8) 31
75. Sanders AJ, Kasprzak K, Simoneau T (1981) Chem Times Trends 4: 61
76. Stark FO, Falender JR, Wright AP (1982) In: Comprehensive Organometallic Chemistry vol 2, p 345
77. Flaningham OL, Langley NR (1991) Chem Anal 112:, 135
78. Merker RL (1956) J Polym Sci 22: 353
79. Rasoul HAA, Zeigler GM, Gordon Fearon FW, (eds) (1990) In: Silicon-based Polymer Science, Amer Chem Soc Washington p 91
80. Wacker Phenyl-Methyl-Silicon Fluids Technical Data Sheets
81. Dvornic PR, Lenz RW (1990) High Temperature Siloxane Elastomers, Hüthig & Wepf Verlag, p 27
82. Hechtl W (1989) In: Silicone Chemie und Technologie Vulkan-Verlag, p 49
83. Wrobel D, In: ibid p 65
84. Wake WC (1987) Crit Rep Appl Chem 16: 89
85. Polmanteer KE (1987) In: Rubber Chemistry and Technology 61 470
86. Patzke J, Wolfarth E (1973) Chemiker-Zeitung 4 176
87. Weis JC (1984) (Ed) by Morrell SH In: Silicone Rubber, p 85
88. Gilson J-M, de Habimana, J DE-OS 4205201 27. 08. 92
89. Fletcher HJ, Hunter MJ (1949) J Am Chem Soc 71 2918
90. Noshay A, Matzner M, Williams TC (1973) Ind Eng Chem Prod Res Dev 12: 268
91. Bailey DL, DE 1018063, 23. 12. 54
92. Matzner M, Noshay A, Barclay R, US 3,701,815, 311072
93. Creamer CE, DE 1800968, 03. 12. 68
94. Wewers D (1989) In: Silicone Chemie und Technologie Vulkan Verlag p 81
95. Yilgör I, Riffle JS, Ward RS (1987) Polym Prep 28: 254
96. Braun F, Willner L, Kosfeld R (1989) J Organomet Chem 366: 53
97. Yilgör I, Riffle JS, McGrath JE (1985) ACS Symp Ser 282: 161
98. Boutevin B, Youssef B (1989) Macromol, Chem 190: 277
99. Barnes GH, US 3,398,174, 20. 08. 68
100. Speier JL, US 2,924,588, 09. 02. 60
101. Simmler W, US 3,481,962, 02. 12. 69
102. Kamei M, Ichinoe S, EP 385732, 07. 02. 90
103. Kojima K, Gore CR, Marvel CS (1966) J Polym Sci A-1, 4: 2325
104. Gornowicz GA, Lee C-I (1988) Polym Mater Sci Eng 59 1009
105. Ohyanagi M et al. (1985) Polym commun 26 249
106. Ohyanagi M, Kadayama Y, Kato T, Nenese H, Jkeda K, Sekine Y (1986) Makromol Chem Rapid Commun 7 153
107. Kumpf RJ, Gordon B (1989) Makromol Chem Rapid Commun 10: 617
108. Joyner MM (1986) Text Chem Color 18: 34
109. Maki H, Honguchi H, Sugar T, Vumon S (1971) CA 74: 32907 v
110. Bailey DL, US 2, 947, 771 02. 08. 60
111. Spinu M, McGrath JE (1991) J: Polym Sci; Part A: Polym Chem 29: 657
112. Au GV, Braunsberger K, Huhn K, Kovar I, DE-OS 3418358 21. 12. 85
113. Gamon N, Braunsberger K, DE-OS 3637837 19. 05. 88

114. Smith CB, Burrill PM, DE 294786 29. 06. 86
115. Helary G, Sauvet G (1992) Eur Polym J 28: 37
116. Ona I, Ozaki M, EP 350604 24. 05. 89
117. Elsbernd CS, Spinu M, McGrath JE (1990) Silicon Based Polymr Science: A comprehensive Resource, American Chemical Society p145
118. Sormani PM, Minton RJ, McGrath JE (1985) ACS Symp Ser 286: 147
119. Eckberg RP, Riding KD (1989) Polym Mater Sci Eng 60: 222
120. Crivello JV, Lee JL (1990) J Polym Sci; Part A: Polym Chem 28: 479
121. Guise GB, Jones FW (1977) Text Chem Color 9: 66
122. Plueddemann EP, Fanger G (1959) J Am Chem Soc 81: 1632
123. Plueddemann EP, US 3, 057, 901 09. 10. 62
124. Mukundan AL, Balasubranian K, Srinivasan KSV (1988) Polym Commun 29: 310
125. Desorcie JL, O Brien MJ, US 5, 010, 118 23. 04. 91
126. Stein J, Leonard TM, Desorcie JL, EP 404030 17. 12. 90
127. Eckberg RP, US 4, 987, 158 22. 01. 91
128. Crivello JV, Fan M (1991) J Polym Sci; Part A: Polym Chem 29: 1853
129. Riding KD, Farley DE, EP 449027 12. 03. 91
130. Hockemeyer F, John P, Preiner G, EP 169295 25. 07. 85
131. Preiner G, Hockemeyer F, Burger C, EP 283008 17. 03. 88
132. Varaprath PJ, Wright AP (1988) Polym Prepr Am Chem Soc, Div Polym Chem 29: 534
133. Canivenc E, Gay M, 281718 10. 08. 87
134. Motegi H, Sunaga T, Zenbayashi MOG, DE-OS 4010153 04. 10. 90
135. Jewel BS, Riffle JS, Allison D, McGrath JE (1985) Polymer Preprints, ACS Div P Chem 30: 295
136. Eckberg RP et al. (1989) Adhesives Age p 24
137. Eckberg RP, Riding KD (1990) ACS Symp Ser 417: 382
138. Crivello JV, Fan M, Bi D (1992) J Appl Polym Sci 44: 9
139. Okamoto Y, Grossan D, Ferrigno K, Nabos S (1988) Polym Sci, Technol p 201
140. Röltgen H (1986) Coating 11: 399
141. Müller U, Timpe H-J, Neuenfeld J (1991) Eur Polym J 27: 621
142. Martin ER, DE 2649854 12. 07. 84
143. Riess G, Hurtvez, G Bahadur P (1985) In: Encyclopedia of Polymer Science and Engineering 2nd ed John Wiley, New York Vol 2, p 324
144. Plumb JB, Atherton JH (ed) by Allport DC, James WH, (1973) In: Block Copolymers James Appl Sci Publ London, p 305
145. Noshay A, McGrath JE (1977) in block Copolymers, Overview and Critical Survey, Academic Press, New York, p 400
146. Bailey DL, O Connor FM, Brit Pat 880022 22. 05. 58
147. Köpnick H, Delfs D, Simmler W, DE 1108917 15. 06. 61
148. Bailey DL, O Connor FM, DE 1012602 25. 07. 57
149. Koepnick H, Delfs D, DE 1257433 28. 12. 67
150. Delaval JCA, Guinet PAE, Brit Pat 1060057 16. 07. 65
151. Beattie JH, Stuart RS, USPat 3541127 17. 11. 70
152. Plueddeman EP, DE 1445347 16. 09. 71
153. Haman H et al. (1979) Plaste Kautsch 26 (11) 619
154. Ritter J, Dubjaga JG, Komarowan AB (1979) Plaste Kautsch 26 (11) 624
155. Polyurethanes, Chemistry and Technology Part II eds Saunders JH, Frisch KH (1964) Interscience Publishers, John Wiley, New York, p 67
156. Becker W, Braun D, Hanser Verlag München (1973) Kunststoffhandbuch Polyurethane Neuherausgabe p101
157. Kira-ly Z, Vincent B (1992) Int 28 139
158. Blevins CH, Murphy CJ, Mlatoch PL, Greene CH, US Pat 5045571 03. 09. 91
159. Vaughn HA, Brit Pat 989379, 140465
160. Vaughn HA (1969) Am Chem Soc Div Org Coat Plast Chem pap 29 133
161. Noshay A, Matzner M, Williams TC (1973) Ind Eng Chem Prod Res Dev 12 (4) 268
162. Horlacher P et al. DE-OS 3924992 28. 07. 89

163. Serini V et al. DE-OS 3929401 05. 09. 89
164. Freitag D et al. EP 374635 18. 12. 89
165. Horlacher P, et al. EP 411370 14. 07. 90
166. Vaughn HA, US Pat 3419 634 31. 12. 68
167. Matssuwaka K, Inoue H (1990) J Polym Sci Polym Lett Edn 28 13
168. Riffle JS, McGrath JE, Freelin R6, Banthia AK (1981) J Makromol Sci Chem 15 (5) 967
169. Hawkins CM, Galluci RR, EP 402674 16. 06. 89
170. Maglia TL, LeGrand DG (1970) Polym Eng Sci 10 (6) 349
171. Estes GM, Cooper SL, Tobolsky AV (1970) J Macromol Sci Revs Macromol Chem C4 (2) 313
172. Kambour RP (ed) by Aggarwal (1970) In: Block Polymers, SL p 263
173. Riffle JS, McGrath JE, Yilgor I, Iran C, Wilker CL, Banthia Ak (1983) Polym Prepr 24 (1)
174. Vaughn HA, US Pat 3. 219. 634 und 3. 419. 635 31. 12. 68
175. Viventi RV, US Pat 3. 600. 288 17. 08. 71
176. Developments in Blockcopolymers (ed) by I Goodman (1982) Applied Science Publishers, p 164
177. Matzner M, Noshay A, McGrath JE (1973) Polym Prepr Am Chem Soc Div, Polym Chem 14 (1) 68
178. Yilgör I, McGrath JE (1988) in Adv Polym Sci 86 1
179. Madec PJ, Marechal EJ (1978) J Polym Sci Polym Chem Ed 16 3165
180. Shinoda H, Ohthaguro M, Iimuro S, EP 399827 24. 05. 90
181. Hilty TC, Revis A, Jones DP, EP 424000 08. 10. 90
182. Webster OW, US Pat 4. 417. 034 22. 11. 83
183. Farnharm WB, Sogah DY, US 4. 414. 372 08. 11. 83
184. Spinelli HJ, EP 420679 28. 09. 90
185. Spinelli HJ, EP 422805 28. 09. 90
186. Mikami R, Okawa T, Yoshitake M, EP 400613 30. 05. 90
187. Collyer AA et al. (1991) J Polym Sci, Part A 29: 193
188. Noshay A, Matzner M (1973) Angew Makromol Chem 37: 215
189. Sybert PD, US Pat 5. 011. 899 30. 04. 91
190. Chow SW, Byck JS, US Pat 3. 562. 353 09. 02. 71
191. Ai H, Ikeda A, Matsuoka Y, EP 177792 12. 0985
192. Chung JL, US Pat 4. 996. 278 26. 02. 91
193. Keohan FL, Swint SA, Buese MA (1991) J Polym Sci, Part A: Polym Chem 29 303 (1991)
194. Rich JD, 171288 and 030189, US Pat 4. 794. 153 and 4. 795. 680
195. Kreuzer F-H, Maier L, Wenski G, DE-OS 3. 925. 099 31. 01. 91
196. McGrath JE (1983) Pure Appl Chem 55: 1573
197. Müller KF, Harisiades P, EP 362145 25. 09. 89
198. Neale RS, Schilling CL, EP 405494 27. 06. 90
199. Sveda M, US Pat 2. 562. 000 24. 07. 51
200. Merker RL, Scott MJ (1964) J Polym Sci, Part A: Gen Pap 2: p 7 and 31
201. Williams, EA, Wengrovius JH et al. (1991) Macromolecules 24: (7) 1445
202. Webster OW, Shea KJ, Loy DA (1990) Polym Mater Sci Eng 63: 281
203. Noshay A, Matzner M, Williams TC (1973) Ind Eng Chem Prod Res Dev 12: 268
204. Morton M, Rembaum AA, Bostick (1964), J Appl Polym Sci 8: (6) 2707
205. Dean JW (1970) J Polym Sci, Part B: Polym Phys 8: (10) 677
206. Kumar RC, Andrus MH, Mazurek MH, 140890, EP 413550
207. Inoue H et al. (1988) J Polym Sci, Part A: Polym Chem 26: 1077
208. Clouet G, EP 421894 26. 09. 90
209. Mignani G, Meyrueix R, Prud homme C, WO 91/02018 and 91/02019 21. 02. 91
210. Kawakami Y, Karasawa H, Aoli T, Yamamura Y, Hisada H, Yamashita Y (1985) Polym J 17: 1159
211. Kawakami Y, et al. (1985) Polym Commun 26: 133
212. Pascault J-P, Camberlin Y (1986) Polym Commun 27: 230
213. Ibemesi J, et al. (1990) Mater Res Soc, Symp Proc 171: 105
214. Ibemesi J, Govzdic N, Vermin M, Lynch MY, Meier DJ (1985) ACS Polym Divis Polymer Preprints 26: 18

215. Bostick EE, Fessler WA, US Pat 3. 578. 726 11. 05. 71
216. Wu TS, US Pat 3. 325. 530 13. 06. 67
217. Dvornic PR (1992) Polymer Bull 28: 339
218. Bard JK, Burnier JS, EP 423
219. Leibfried RT, EP 423412, 17. 01. 90
220. Bortolin RB, Parbhoo B, DE-OS 4023931 07. 02. 91
221. Kozakai S, DE-OS 4022218 24. 01. 91
222. Guo X, Rempel CL (1990)ACS Polym Division Polym Prepr 31: 422
223. Nagasaki S, Kurosawa K, Suda M, Takahashi S, Tsuruta T (1990) Macromol Chem 191 2103
224. Plate´ NA (1990) J Membr Sci 52: 289
225. Masuda T, Isobe E, Higashimura T (1985) Macromolecules 18 841
226. Nagase Y, Veda T, Matsui K, Udukura M (1991) J Polym Sci Part B: Polym Phys 29: 171
227. Koßmehl G, Neumann W (1986) Makromol Chem 187: 1371
228. Robertson JR, et al.   EP 395583 18. 04. 90
229. Dvornic PR, Lenz RW (1990) In: "HighTemperature Siloxane Elastomers",Hüthig & Wepf Verlag p 136
230. Yilgör I, McGrath JE, Riffle JS, Wickes CL (1982) Polym Bull 8: 535
231. Yilgör I, McGrath JE, Wilkes GL, Tyogi G (1982) Polym Bull 8: 543
232. Koßmehl G, Neumann W (1986) Makromol Chem 187 1381
233. Saam JC, Ward AH, Fearon FW (1973) Advanc Chem Ser 129, 239
234. Saam JC, Ward AH, Fearon GF (1972) Polym Prepr Am Chem Soc Div Polym Chem 13: 524
235. Noll W (1968) Chemistry and Technology of Silicones, Academic, New York
236. Flory P (1984) Advances in Polymer Science: 53: 1
237. Makarova NN, Petrova IM, Godovsky YK, Lavruchin BD, Zhdanov AA (1983) Dokl Akad Nauk SSSR, 269: 1369
238. Polishchuk AP, Timofeeva TV, Makarova NN, Antipin MYu, Struchkov T (1991) Liquid Crystals, 9: 433
239. Kipping FS (1912) J Chem Soc, 101: 2108
240. Godovsky YK, Papkov VS (1989) Advances in Polymer Science, 88: 131
241. Godovsky YK, Papkov VS (1986) Makromol Chem, Macromol Symp, 4,71
242. Miller KJ, Grebowicz J, Wesson JP, Wunderlich B (1990) Macromolecules 23: 849
243. Pochan JM, Frois MF, Goedde AJ, Beatty CL, Pochan DF (1979) Liquid Crystals, the fourth state of the matter (Saeva ed FD), 41: Marcel Dekker
244. Beatty CL, Karasz FE (1975) J Pol Sci, Polym Phys Ed, 13: 971
245. Young CW, Servais PC, Currie CC, Hunter MJ (1948) J Amer Chem Soc, 70: 3758
246. Kögler G, Hasenhindl A, Möller M (1989) Macromolecules 22: 4190
247. Godowsky YK, Mamaeva II, Makarova NN, Papkov VS, Kuzmin NN (1985) Makromol Chem, Rapid Commun, 6: 797
248. Möller M, Siffrin S, Out R, Boileau S (1992) Amer Chem Soc Polym Prepr 33: 176
249. Godovsky YK (1992) Angew Markromol Chem, 202/203: 187
250. Kyes PH, Daniels WB (1975) J Chem Phys, 62: 2000
251. Volino F, Dianoux A-J (1978), Ann Phys 3: 151
252. Hyde JF, DeLong RC (1941) J Amer Chem Soc, 63: 1194
253. Burkhard CA (1945) J Amer Chem Soc, 67: 2173
254. Godowski YK, Makarova NN, Mamaeva II (1986) Makromol Chem, Rapid Commun, 7: 325
255. Godovsky YK, Makarova NN, Petrova IM, Zhdanov (1984) Makromol Chem Rapid Commun, 5: 427
256. Makarova NN, Godovsky YK, Kuzmin NN (1987) Makromol Chem, 188: 119
257. Makarova NN, Godovsky YK (1986) Vysokomol Soedin, B 28: 243
258. Brown JF (1963) J Polym Sci, C-1: 83
259. Tsvetkov VN, Andrianov KA, Makarova NN, Vitovskay GM, Rjumzev EI (1973) Europ Polym J, 9: 27
260. Demus D (1989) Liq Cryst, 5: 75

261. Hayashi T, Konishi M, Kobori Y, Kumada M, Higuchi T, K Hirotsu (1984) J Amer Chem Soc, 106: 158
262. Poetsch E (1988) Kontakte (Darmstadt) (2), 15
263. Speier JL, Webster JA, Barnes GH (1957) J Amer Chem Soc, 79: 974
264. Crivello JV, Bi D (1993) J Polym Sci, Part A, 31: 3121
265. Finkelmann H, Rehage G, Kollmann G, Wacker-Chemie (05111979) DOS 2, 993, 591
266. Ojima I, The hydrosilylation reaction in: The Chemistry of Organic Silicon Compounds, Ed S Patai, Z, Rappoport, 1479 (John Wiley & Sons 1989)
267. Itsuno S, Chao D, Ito K (1993) J Polym Sci: Part A, Polym Chem, 31: 287
268. Ryan JW, Menzie GK, Speier JL (1960) J Amer Chem Soc, 82: 3601
269. Belyakova ZV, Pomerantseva MG, Bykovchenko VG, Chernyshev EA (1976) Zh Obshch Khim, 46: 1034, CA 85: 108696
270. Torrès G, Madec P-J, Maréchal E (1989) Makromol Chem, 190: 203
271. Kreuzer F-H, Andrejewski D, Haas W, Häberle N, Riepl G, Spes P (1991) Mol Cryst Liq Cryst, 345
272. Yu JM, Teyssié D, Boileau S (1992) Polym Bull, 28: 435
273. Yu JM, Teyssié D, Boileau S (1993) J Polym Sci Part A, 31: 2373
274. He X, Lapp A, Herz J (1988) Makromol Chem, 189: 1061
275. Torrès G, Madec P-J, Maréchal E (1989) Makromol Chem, 190: 2789
276. Gray GW, Lacey D, Nestor G, White MS (1986) Makromol Chem, Rapid Commun, 7: 71
277. Gemmell PA, Gray GW, Lacey D (1985) Mol Cryst Liq Cryst, 122: 205
278. Apfel MA, Finkelmann H, Janini GM, La RJ, Lukmann BH, Price A, Roberts WL, Shaw TJ, Smith CA (1985) Anal Chem, 57: 651
279. Attard GS, Moura-Ramos JJ, Williams G (1987) Makromol Chem, 188: 2769
280. Nestor G, White MS, Michael S, Gray GW, Lacey D, David K, Toyne KJ (1987) Makromol Chem, 188: 2759
281. Krishnamurthy S, Chen SH (1989) Makromol Chem, 190: 1407
282. Young WR, Haller I, Green C (1971) Mol Cryst Liq Cryst, 13: 305
283. Braun F, Willner L, Heß M, Kosfeld R (1990) Makromol Chem, 191: 1775
284. Zuev VV, Smirnova GS, Tarasova IN, Skorokhodov SS (1989) Vysokomol Soedin, SerB 31: 784, CA112 (20): 180266d
285. Jo BW, Jin J-I, Lenz RW (1982) Eur Polym J, 18: 233
286. Kapitza H, Zentel R, Twieg RT, Nguyen C, Vallerien SU, Kremer F, Willson CG (1990) Adv Mater, 2: 539
287. Andrejewski D, Gohary M, Luckas HJ, Winkler R, Kreuzer F-H (1990) Consortium für elektrochemische Industrie GmbH, EP 358, 208 A2 Chem Abstr 133 68893a
288. Yamada K, Ueno Y, Ikeda K, Takamiya (1990) Makromol Chem, 191: 2871
289. Strohriegl P (1986) Makromol Chem, Rapid Commun, 7: 771
290. Cabrera I, Krongauz V, Ringsdorf H (1987) Angew Chem, 99: 1204
291. Aguilera C, Bartulin J, Hisgen B, Ringsdorf H (1983) Makromol Chem, 184: 253
292. Yilgör I, McGrath JE (1988) Adv in Polym Sci, 86: 1
293. Wenski, G Maier L, Kreuzer F-H (1993) Consortium für elektrochemische Industrie GmbH, DE 3, 925, 099 A1 (1991), Chem Abstr 114: 248026z
294. Zuev VV, Smirnova GS, Nikonorova NA, Borisova TI, Skorokhodov SS (1990) Makromol Chem, 191: 2865
295. Ringsdorf H, Schneller A (1981) Brit Polym J, 13: 43
296. Aquilera C, Bernal L (1984) Polymer Bull, 12: 383
297. Finkelmann H (1983) Phil Trans R Soc, Lond, A, 309: 105
298. Finkelmann H, Rehage G (1984) Advances in Pol Sci, 60/61: 99
299. Frenzel J, Rehage G (1983) Makromol Chem, 184: 1685
300. Rehage G (1984) Nachr Chem Tech Lab, 32: 287
301. Shibaev VP, Platé (1985) Pure & Appl Chem, 57: 1589
302. Zentel R, Wu J (1986) Makromol Chem, 187: 1727
303. Sigaud S, Achard MF, Hardouin F, Mauzac M, Richard H, Gasparoux H (1987) Macromolecules, 20: 578

304. Finkelmann H (1987) Angew Chem, 99: 840
305. Krücke B, Schlossarek M, Zaschke H (1988) Acta Polym, 39: 607
306. Dowell F (1989) Mater Res Soc Symp Proc, 134: 47
307. Finkelmann H, Rehage G (1980) Makromol Chem Rapid Commun, 1: 31
308. Ringsdorf H, Schneller A (1982) Makromol Chem Rapid Commun, 3: 557
309. Simon R, Coles HJ (1986) Liq Cryst, 1: 281
310. Gemmell PA, Gray GW, Lacey D (1983) Polym Preprints (Amer Chem Soc) 24: 253
311. Janini GM, Laub RJ, Shaw TJ (1985) Makromol Chem, Rapid Commun, 6: 57
312. Achard MF, Hardouin F, Sigaud G, Mauzac M (1986) Liq Cryst, 1: 203
313. Roth H, Krücke B (1986) Makromol Chem, 187: 2655
314. Krücke B, Zaschke H, Kostromin SG, Shibaev VP (1985) Acta Polym, 36: 639
315. Krücke B, Roth H, Rickmeyer T, Doberstein A (1988) Acta Polym, 39: 188
316. Apfel MA, Finkelmann H, Janini GM, Laub RJ, Lühmann B-H, Price A, Roberts WL, Shaw TJ, Smith CA (1985) Anal Chem 57: 651
317. Finkelmann H, Rehage G (1980) Makromol Chem, Rapid Commun, 1: 733
318. Surendranath V, Johnson DL (1989) Mol Cryst Liq Cryst, 167: 207
319. Finkelmann H, Rehage G (1982) Makromol Chem, Rapid Commun, 3: 859
320. Hsu C-S, Chu P-H (1990) J Amer Chem Soc, Polym, Prepr 31: 1, 490
321. Kreuder W, Ringsdorf H (1983) Makromol Chem, Rapid Commun, 4: 807
322. Stevens H, Rehage G, Finkelmann H (1984) Macromolecules, 17: 851
323. Jones BA, Bradshaw JS, Nishioka M, Lee ML (1984) J Org Chem, 49: 4947
324. Hsu CS, Percec V (1987) Makromol Chem Rapid Commun, 8: 331
325. Hsu CS, Percec V (1987) Polymer Bulletin, 17: 49
326. Hsu CS, Percec V (1987) Polymer Bulletin, 18: 91
327. Tinh NH, Achard MF, Hardouin F, Mauzac M, Richard H, Sigaud G (1990) Liq Cryst 7: 385
328. Itoh M, Lenz W (1992) J Polym Sci, Part A, 30: 303
329. Bradshaw JS, Schregenberger C, Chang KH-C, Markides KE, Lee ML (1986) J Chromatogr, 358: 95
330. Percec, V, Hsu CS, Tomazos D (1988) J Pol Sci, Part A, 26: 2047
331. Percec V, Tomazos D (1989) Macromolecules, 22: 1512
332. Adams NW, Bradshaw JS, Bayona J-M, Markides KE, Lee ME (1987) Mol Cryst Liq Cryst. (1987) 147: 43
333. Suzuki T, Okawa T, Ohnuma T, Sakon Y (1988) Makromol Chem Rapid Commun 9, 755
334. Aggarwal SK, Bradshaw JS, Eguchi M, Parry S, Rossiter BE, Markides KE, Lee ML (1987) Tetrahedron, 43: 451
335. Dumon M, Nguyen HT, Mauzac M, Destrade C, Gaparoux H (1991) Liq Cryst, 10: 475
336. Naciri J, Pfeiffer S, Shashidar R (1991) Liq Cryst, 10: 585
337. Hsu CS, Rodriguez-Prada JM, Pecec V (1987) J Polym Sci, Part A, 25: 2425
338. Percec V, Heck J (1991) Polymer Bull, 25: 431
339. Wen J-S, Hsiue G-H, Hsu C-S (1990) Makromol Chem, Rapid Commun 11: 151
340. Lux M, Strohriegl P, Höcker H (1987) Makromol Chem 188: 811
341. Kumar U, Fréchet JMJ (1992) Adv Mat 4: 665
342. Marignan Gde, Teyssié D, Boileau S, Malthête J, Noël C (1988) Polymer, 29: 1318
343. Engel M, Hisgen B, Keller R, Kreuder W, Reck B, Ringsdorf H, Schmidt H-W, Tschirner P (1985) Pure & Appl Chem, 57: 1009
344. Nagase Y, Y Takamura Y, Abe H, Ono K, Saito T, Akiyama E (1993) Makromol Chem, 194: 2517
345. Diele S, Hisgen B, Reck B, Ringsdorf H (1986) Makromol Chem, Rapid Commun 7: 267
346. Diele S, Oelsner S, Kuschel F, Hisgen B, Ringsdorf H, Zentel (1987) Makromol Chem 188: 1993
347. Hessel V, Finkelmann H (1985) Polym Bull, 14: 3751
348. Berg S, Krone V, Ringsdorf H (1986) Markromol Chem, Rapid Commun 7: 381
349. Keller P, Hardouin F, Mauzac M, Achard MF (1988) Mol Cryst Liq Cryst, 155: 171
350. Lee MSK, Gray GW, Lacey D, Toyne KJ (1989) Makromol Chem, Rapid Commun 10: 325
351. Gray GW, Hill JS (1990) Mol Cryst Liq Cryst, Lett, 7: 47
352. Gray GW, Hill JS, Lacey D (1991) Mol Cryst LiqCryst 197: 43

353. Gray GW, Hill JS, Lacey D (1989) Angew Chem, Adv Mater, 101: 1146
354. Achard MF, Leroux N, Hardouin F (1991) Liq Cryst, 10: 507
355. Finkelmann H, Lühmann B, Rehage G (1982) Colloid & Polymer Sci, 260: 56
356. Lühmann B, Finkelmann H (1987) Colloid & Polymer Sci, 265: 506
357. Kreuzer, F-H, Gawhary M, Winkler R, Finkelmann (1982) Consortium für elektro-chemische Industrie GmbH, EP 60 355 (1981), Chem Abstr 98: 10013u
358. Riepl G, Kreuzer F-H, Miller A (1989) EP 333 022 A2 Chem Abstr 112: 67360s
359. Spes P, Hessling M, Kreuzer F-H (1991) DE 394 0148 A1 (1991), Chem Abstr 115: 123530y
360. Richards RDC, Hawthorne WD, Hill JS, White MS, Lacey D, Semlyen JA, Gray GW, Kendrick TC (1990) J Chem Soc, Chem Commun, 95
361. Clarson SJ, Semlyen JA, Dodgson K (1985) Polymer, 26: 930
362. Clarson SJ, Semlyen JA, Dodgson K (1990) Am Chem Soc, Polym Prepr 31: 563
363. Kreuzer F-H, Maurer R, Spes P (1991) Makromol Chem Makromol Symp 50: 215
364. Bunning TJ, Klei HE, Samulski ET, Crane RL, Linville RJ (1991) Liq Cryst 10: 445
365. Pachter R, Bunning TJ, Adams WW (1991) Comp Polym Sci, 1: 179
366. Bunning TJ, McNamee SG, Klei HE, Samulski ET, Ober CK, Adams WW (1992) Amer Chem Soc, Polym Prepr 33: 315
367. Pachter R, Bunning TJ, Socci EP, Farmer BL, Crane RL, Adams WW (1992) Amer Chem Soc, Polym Prepr 33: 671
368. Socci EP, Farmer BL, Bunning TJ, Pachter R, Adams WW (1993) Liq Cryst, 13: 811
369. Pachter R, Bunning TJ, Crane RL, Adams WW (1993) Makromol Chem, Theory Simul 2: 337
370. Gilli JM, Kamaye´ M, Sixou P (1991) Mol Cryst Liq Cryst 199: 79
371. Finkelmann H, Rehage G, Kreuzer F-H (1982) Consortium für elektrochemische Industrie GmbH, DE 3119459 A1 (1982), Chem Abstr 98: 81903v
372. Fryer C, Weidner R, Zeller N, Kreuzer F-H, Spes P (1990) Poster-Presentation on the International Symposium on Organisilicon Chemistry, Edinburgh
373. Day VW, Klemperer WG, Mainz VV, Millar DM (1985) J Am Chem Soc 107: 8262
374. Agaskar PA, Day VW, Klemperer WG (1987) J Am Chem Soc 109: 5554
375. Feher FJ, Budzinowski TA, Weller KJ (1989) J Am Chem Soc 111: 7288
376. Brown JF, Vogt Jr. , LH, Katchman A, Eustance JW, Kiser KM, Krantz KW (1960) J Am Chem Soc 82: 6194
377. Barry AJ, Daudt WH, Domicone JJ, Gilkey JW (1955) J Am Chem Soc 77: 4248
378. Sprung MM, Guenther FO (1955) J Am Chem Soc, 77: 3990
379. Brown JF, Jr. , Vogt LH, Prescott PI (1964) J Am Chem Soc, 86: 1120
380. Vogt LH, Brown JF (1963) J Inorg Chem, 2: 189
381. Olsson K (1958) Ark Kemi 13: 367
382. Olsson K, Gronwall C (1961) Ark Kemi, 17: 529
383. Olsson K, Gronwall C (1964) Ark Kemi, 22: 237
384. Larsson K (1960) Ark Kemi, 16: 203
385. Larsson K (1960) Ark Kemi 16: 209
386. Wiberg E, Simmler W (1995) Z Anorg Allg Chem 282: 330
387. Müller R, Kohne R, Sliminski S (1959) J Prakt Chem 9: 71
388. Frye CL, Collins WT (1970) J Am Chem Soc 92: 5586
389. Hoebbel D, Pitsch I, Hiller W, Dathe S, Popowski E, Sonneck G, Reiher T, Janck H, Scheim U (1990) Akademie der Wissenschaften der DDR, EP-A 0348705 (1990), Chem Abstr 113: 125354b
390. Weidner R, Zeller N Deubzer B, Frey V (1990) Wacker-Chemie GmbH, EP-A 0367222 (1990), Chem Abstr 113: 116465m
391. Chatt J, Vallerino LM, Venanzi LM (1957) J Chem Soc (London) 2496
392. Clark HC, Manzer LE (1973) J Organomet Chem 59: 411
393. Gasparoux H (1984) J d Chim Phys 81: 795
394. Imai T, Koide N (1985) Toray Silicon Co, Ltd, EP 0163 495 A2 (1985), Chem Abstr 104: 139447k
395. Okawa T (1989) T Suzuki Toray Silicon Co, Ltd, EP 0338576 (1988), Chem Abstr 112: 159243j

396. Everitt DRR, Care CM, Wood RM (1987) Mol Cryst Liq Cryst 157: 55
397. Finkelmann H, Kock H-J, Rehage G (1981) Makromol Chem Rapid Commun 2: 317
398. Meier W, Finkelmann H (1991) MRS Bulletin 16: 29
399. Gleim W, Finkelmann H (1987) Makromol Chem, 188: 1489
400. Hammerschmidt K, Finkelmann H (1989) Makromol Chem, 190: 1089
401. Meier W, Finkelmann H (1990) Makromol Chem, Rapid Commun 11: 599
402. Zentel R, Reckert G (1986) Makromol Chem, 187: 1915
403. Hanus K-H, Pechhold W, Soergel F, Stoll B, Zentel R (1990) Colloid & Polym Sci, 268: 222
404. Zentel R, Reckert G, Bualek S, Kapitza H (1989) Makromol Chem, 190: 2869
405. Zentel R (1989) Angew Chem, Adv Mat, 101: 1437
406. Vallerien SU, Kremer F, Fischer EW, Kapitza H, Zentel R, Poths H (1990) Makromol Chem, Rapid Commun 11: 593
407. Küpfer J, Finkelmann H (1991) Makromol Chem, Rapid Commun 12: 717
408. Kreuzer F-H, Maurer R, Stohrer J (1992) Abstr of the 14th Int Liquid crystal conference, Pisa, 936
409. Joyner RD, Kenney ME (1962) Inorg Chem, 1: 717
410. Orthmann E, Wegner G (1986) Makromol Chem, Rapid Commun 7: 243
411. Sauer T, Wegner G (1989) Makromol Chem, Macromol Symp 24: 303
412. Exsted BJ, Urban MU (1990) Amer Chem Soc, Polym Prepr, 31: 663
413. Finkelmann H, Kiechle U, Rehage G (1983) Mol Cryst Liq Cryst, 94: 343
414. Haase W, Pranoto H (1985) Polym Sci Technol, 28: 313
415. Markides KE, Chang HC, Schregenberger CM, Tarbet BJ, Bradshaw JS, Lee MJ (1985) J High Res Chrom & Chrom Commun, 8: 516
416. Poths H, Andersson G, Skarp K, Zentel R (1992) Adv Mater, 4: 792
417. Coles HJ, Simon R (1985) Mol Cryst Liq Cryst, Letters, 1: 75
418. Jakob E, Kreuzer F-H, Häberle N, Haase W (1994) Adv Mat, 6: 150

CHAPTER 4

Poly(organosilanes), Poly(carbosilanes), Poly(organosilazanes)

P. Kochs

4.1
Poly(organosilanes)

Introduction

Attempts to completely replace the carbon in n-alkanes by the next element of group IV, silicon has been of special interest to practical as well as theoretical chemists ever since the work of Kipping [1]. Interest in the group of compounds, known as poly(organosilanes) **2b** has not diminished to the present day, especially since it is a well known fact that **2b** has different chemical and physical properties compared with n-alkanes. For example, the delocalization of electrons of the Si—Si-σ linkages in **2b** results in different UV absorption characteristics. The chemical reaction behaviour also differs markedly from that of n-alkanes, since the Si—Si linkage is unstable because of the unoccupied d-orbitals of the silicon atom. Hexamethyldisilane, for example, can easily be dissociated with iodine, with the formation of iodine trimethylsilane Eq. (1). This reaction is of industrial interest [13]:

$$(CH_3)_3Si - Si(CH_3)_3 \xrightarrow{I_2} 2 (CH_3)_3Si - I$$

$$\tag{1}$$

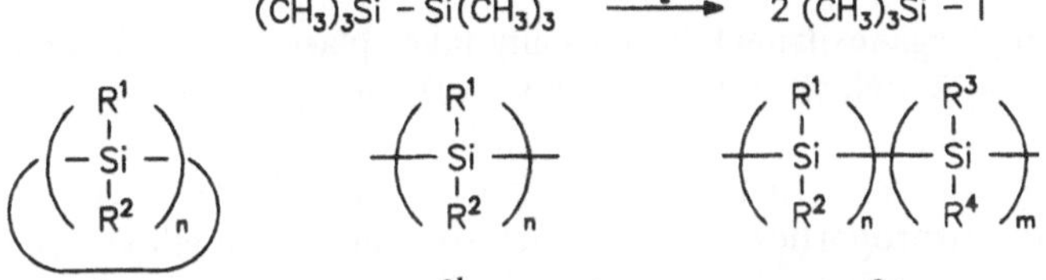

$$2a \qquad\qquad 2b \qquad\qquad 2c$$

The preparation and industrial use of poly(organosilanes) **2a** with alkyl and/or aryl radicals has experienced a revival, in view of the availability of alkyl and arylchlorosilanes as raw materials, whose large-scale production by the Rochow process [2] has decisively influenced the dynamic development of silicon chemistry. It is probable that, as far back as 1924, Kipping [1] had already discovered a poly(organosilane) with the probable structure **2b** (R1=R2=C$_6$H$_5$); which was prepared from dichlorodiphenylsilane **2d** (R1=R2=C$_6$H$_5$; R3=Cl) by the reductive coupling with sodium metal in toluene, but which could not be further identified because of its insolubility in organic solvents. Burkhard was able to obtain a better idea of the reaction of dichlorodimethylsilane **2d** (R1=R2=CH$_3$;

R3=Cl) with sodium metal in benzene. He was able to isolate a definite 6-ring **2a** (R1=R2=CH$_3$; $n = 6$), besides insoluble poly(dimethylsilane) **2b** (R1=R2 =CH$_3$) which decomposed, without melting, at temperatures above 250 °C [3]. This preparation of a low molecular weight ring raised hopes that it would be possible to prepare soluble, linear polymers–but these initially came to nothing because of further failures [4]. The preparation of soluble, linear polysilanes was first described by West et al. These were obtained by the co-condensation of dichlorodimethylsilane **2d** (R1=R2=CH$_3$; R3=Cl) and dichlorophenylmethyl-silane **2d** (R1=C$_6$H$_5$; R2=CH$_3$; R3=Cl) with sodium [5, 6, 7], producing poly(dimethylsilane-*co*-methylphenylsilane) **2c** (R1=R2 = CH$_3$; R3=C$_6$H$_5$),

$$
\begin{array}{ll}
R^2 - \overset{\displaystyle R^1}{\underset{\displaystyle R^3}{\underset{|}{\overset{|}{Si}}}} - Cl & \quad R^1, R^2 \;\; = \; \text{Alkyl, Aryl} \\
 & \quad R^3, R^4 \;\; = \; \text{Alkyl, Aryl, Cl}
\end{array}
\tag{2}
$$

2d

The most important research groups in the USA working on poly(organosilanes) **2b** and their derivatives are listed in Ref [8]. In Europe, there is the group headed by Hengge [9], in Germany the one led by Sartori [27] and in Japan the group led by Nagai [10].

Research on poly(organosilanes) experienced a further upswing when it became known that the conversion of **2b** (R1=R2=CH$_3$) into poly(carbosilanes), followed by thermal decomposition into β-silicon carbide gave easy access to ceramic products [11, 12]. Poly(organosilanes) **2b** and **c** as such deserve attention as auxiliaries in photolithography and as radical initiators in polymerization technology.

4.1.1
Preparation of the Raw Materials

The synthesis of poly(organosilanes) **2a** normally takes place via dichlorodiorganosilanes **2d** (R1=R2=alkyl and/or aryl; R3=Cl) through reductive coupling with an alkali metal such as sodium or lithium. To prepare crosslinked poly(organosilanes), one uses trichloromonoorganosilanes **2d** (R1=aryl/alkyl; R2=R3=Cl). The disproportion of disilanes is another method where poly(organosilanes) are formed.

Several methods for preparing **2b** have been described [28]:

a) the so-called "direct synthesis" which takes place via copperdoped silicon and methyl chloride according to Eq. (3)

$$
2\,CH_3Cl \;\; + \;\; Si \;\; \xrightarrow[\text{temp.}]{\;Cu\;} \;\; (CH_3)_2SiCl_2 \;\; + \;\; \text{further chlorosilanes}
\tag{3}
$$

A detailed description of the product composition of this so-called "Rochow synthesis" is given in (13). Dichlorodimethylsilane (**2d**, R1=R2=CH$_3$; R3=Cl) and trichloromethylsilane (**2b**, R1=CH$_3$; R2=R3=Cl) are now available in large-scale quantities.

b) the reaction of chlorosilanes with Grignard reagents [14] Eq. (4):

$$SiCl_4 \ + \ RMgX \ \xrightarrow{(C_2H_5)_2O} \ RSiCl_3 \ + \ R_2SiCl_2 \ + \ R_3SiCl \ + \ MgXCl \qquad (4)$$

The Grignard synthesis produces mixtures, which is why a cumbersome fractionation often has to be carried out [14]. Yields of dichlorodiorganosilane are usually unsatisfactory, but these can be improved by varying the Grignard reagent/silicon tetrachloride ratio [28]. For example, the reaction of trichloromethylsilane (2d, $R1=CH_3$; $R2=R3=Cl$) with excess (6–8 times more) of alkylmagnesium halide produces an approximately 85% yield of the mixed alkylmethyldichlorosilane [37] Eq. (5):

$$CH_3SiCl_3 \ + \ RMgX \ \longrightarrow \ R(CH_3)SiCl_2$$

$$R = CH_3(CH_2)_n- \ ; \ n = 5, 6, 7 \qquad (5)$$

c) the reaction of organo-H-silanes with olefins Eq. (6):

$$HSiCl_3 \ + \ H_2C=CH-R^4 \ \xrightarrow{Pt-Complex} \ R^4-CH_2-CH_2-SiCl_3 \qquad (6)$$

The reaction is preferably carried out using complex platinum compounds in solvents as catalysts. Apart from the complex platinum compounds, based on vinyl siloxane [15] commonly used nowadays, one can also use chloroplatinic acid [14]. The selectivity of the platinum compound used should be emphasised. Thus, the reaction of dichlorosilane 2d ($R1=R2=H$; $R3=Cl$) with $Pt(PPh_3)_4$ and α-olefins produces only the monosubstituted alkylated silanes Eq. (7), whereas chloroplatinic acid, H_2PtC_{16} forms the dichlorodialkylsilanes [24 b] Eq. (8):

$$H_2SiCl_2 \ \xrightarrow[Pt(PPh_3)_4]{RCH=CH_2} \ RCH_2-CH_2-Si(H)Cl_2 \qquad (7)$$

$$H_2SiCl_2 \ \xrightarrow[H_2PtCl_6]{RCH=CH_2} \ (RCH_2-CH_2)_2SiCl_2 \qquad (8)$$

It is interesting to note that the addition reaction of β-olefins and silanes containing Si—H, using platinum catalysts, produces n-alkyl silanes, this being analogous to the displacement of a double bond. Thus, the addition of n-pentene-2 to dichloromethylsilane or trichlorosilane in the presence of chloro-platinic acid H_2PtCl_6 results in the formation of the corresponding n-pentyl-silanes [33] 9a and 10a which can also be obtained by Si—H addition onto the isomeric α-olefin Eq. (9 and 10):

$$\begin{array}{c} CH_3CH=CH-CH_2-CH_3 \\ or \\ H_2C=CH-CH_2-CH_2-CH_3 \end{array} \ \xrightarrow[H_2PtCl_6]{HSiCl_3} \ \underset{\underset{Cl}{|}}{\overset{\overset{Cl}{|}}{CH_3Si}}-CH_2CH_2CH_2CH_2CH_3 \qquad (9)$$

$$\begin{array}{c} CH_3CH=CH-CH_2-CH_3 \\ or \\ H_2C=CH-CH_2-CH_2-CH_3 \end{array} \ \xrightarrow[H_2PtCl_6]{HSiCl_3} \ Cl-\underset{\underset{Cl}{|}}{\overset{\overset{Cl}{|}}{Si}}-CH_2CH_2CH_2CH_2CH_3 \qquad (10)$$

The yield of the platinum-catalysed addition reaction is favourably influenced by ultrasound [34a].

The use of peroxides and UV light as activators for the Si—H addition reaction with olefinic double bonds has been described in Ref [16].

Olefin addition to Si—H linkages can also take place without a catalyst although, in this case, the reaction must take place under pressure and at elevated temperatures [36].

d) the substituent exchange of R against R′, X (X=halogen) and hydrogen by co- and disproportioning in silanes [14] is of industrial importance.

The substitution reaction of aromatically bound chlorine, e.g. chlorobenzene through the silicon-hydrogen bond in trichlorosilane (2d, R1=H; R2=R3=Cl), with the formation of the industrially important trichlorophenylsilane (2d, R1=C_6H_5; R2=R3=Cl) is carried out on a large scale Eq. (11) [17a–b]:

$$\langle\bigcirc\rangle\!-\!Cl \xrightarrow[\text{temp. and/or Pd}]{\text{HSiCl}_3,\ -\text{HCl}} \langle\bigcirc\rangle\!-\!SiCl_3 \tag{11}$$

The reaction takes place purely thermally and under pressure, but can be carried out at normal pressure and elevated temperatures in the presence of palladium as catalyst. Yields are unsatisfactory in relation to the trichlorosilane used (maximum 38% [17a]), since the hydrogen chloride that is formed breaks down the C — Si bond.

Table 4.1 lists dichlorodiorganosilanes 2b which are important for preparing poly(organosilanes) having the structure 2a.

Table 4.1. Chlorosilanes (2d) used for Syntheses of Poly (organosilane)s

R^1	R^2	Synthetic Method [Reference]	b, p (°C)/mm Hg	Yield (%)
CH_3	CH_3	Grignard [18], Rochow	70/760	11[a], 85
C_2H_5	C_2H_5	Grignard [19], Rochow	129/744	22
$n\text{-}C_3H_7$	$n\text{-}C_3H_7$	Grignard, Rochow [20a, 20b]	172–173/730	10–15
$i\text{-}C_3H_7$	$i\text{-}C_3H_7$	Grignard [22]	67–69/11	11[b]
$n\text{-}C_4H_9$	$n\text{-}C_4H_9$	Grignard [21]	62–63.5/4.5	11[b]
$i\text{-}C_4H_9$	$i\text{-}C_4H_9$	Grignard [21]	93/16	20[b]
$n\text{-}C_5H_{11}$	$n\text{-}C_5H_{11}$	[23]		
$n\text{-}C_6H_{13}$	$n\text{-}C_6H_{13}$	Addition [24b]	150/4	
$n\text{-}C_7H_{15}$	$n\text{-}C_7H_{15}$	[25]		
CH_3	C_6H_{11}	Addition [33]	201–202/740	100
C_6H_5	CH_3	Grignard [26, 29] Substitution	82.5/13	73
C_6H_5	C_6H_5	Grignard [29], Rochow	123–126/2	77
CH_3	C_2H_5	Addition [30, 36]	100/744	–
CH_3	$n\text{-}C_3H_7$	Addition [31, 36]	123–124/747	72[c]
CH_3	$n\text{-}C_4H_9$	Addition [32, 36]	147.5–148/744	–
CH_3	$n\text{-}C_5H_{11}$	Addition [33]	169–170/740	94

Table 4.1. (Contd.)

CH$_3$	n-C$_6$H$_{13}$	Grignard [37]		85 [37]
		Addition [34]	192/743	95 [34a]
CH$_3$	n-C$_8$H$_{17}$	Grignard [37]	110–116/20	85 [37]
		Addition [36]		–
CH$_3$	n-C$_{12}$H$_{25}$	(35)		–
CH$_3$	n-C$_{18}$H$_{37}$	Addition [36]	200–210/6	–

[a] related to SiCl$_4$.
[b] related to Grignard reagents.
[c] noncatalyzed reactions.

Chlorosilanes are strongly corrosive and extreme care should therefore be taken when handling them, because of their irritant effect on the skin and mucous membranes.

4.1.2
Preparation of Poly(organosilanes)

4.1.2.1
Preparation by Reductive Coupling of Halogen Silanes with Alkali Metals

The most widely used method for the production of poly(organosilanes) **1b** is the reductive coupling of dichlorosilanes **2d** (R1=R2=alkyl and/or aryl; R3=Cl) with metallic sodium [3] Eq. (12):

$$n(CH_3)_2SiCl_2 \xrightarrow[-2n \ NaCl]{\substack{1) \ 2n \ Na/Benzene \\ 2) \ 98°C \\ 3) \ 115°C/10h}} \left(\begin{array}{c} CH_3 \\ | \\ Si \\ | \\ CH_3 \end{array} \right) \tag{12}$$

Copolymers can be obtained by the simultaneous use of different dichlorosilanes Eq. (13). West et al. have prepared poly(dimethylsilane-*co*-(methylphenylsilane)) as follows [38]:

$$m \ (C_6H_5)CH_3SiCl_2 \ + \ n(CH_3)_2SiCl_2 \xrightarrow[-2(m+n) \ NaCl]{+ \ 2(m+n) \ Na} \left(\begin{array}{cc} CH_3 & CH_3 \\ | & | \\ Si & Si \\ | & | \\ CH_3 & C_6H_5 \end{array} \right) \tag{13}$$

In this reaction, the chlorosilane mixture is added, drop by drop, to a dispersion of sodium in toluene. After treatment, the organic phase in tetrahydrofuran is subjected to fractionated precipitation with methanol and 2-propanol, resulting in the formation of the following oligomeric and polymeric poly(organosilanes) **2c** (R1=R2=R3=C$_3$; R4=C$_6$H$_5$) (Table 4.2):

Tables 4.3 and 4.4 list the properties structure of further poly(organosilanes) [83] with mixed alkyl/aryl radicals in **2b** Polymers with bimodal distribution are obtained, figures for MW referring to the higher peak of GPC analyses

Table 4.2. Fractionated precipitation of crude **2c** (R_1^1, R_1^2, R^3=(CH$_3$, R=C$_6$H$_5$) (Data adapted from [38])

Fraction	yield (%)[a]	m.p(°C)	Si—Me/Si—Ph	M_n	Si-Atoms
1	7.6	190–210	0.592	150 000	1550
2	2.6	155–190	0.608	80 000	830
3	1.9	176–185	0.617	34 000	350
4	1.9	124–135	0.567	17 000	175
5	2.5	135–155	1.358	31 000–18 000	370/210
6	10.2	82–100	1.067	12 000	135
7	4.9	115–135	1.083	27 000	310
8	6.2	106–116	0.950	27 000	300
9	6.7	80–95	1.008	10 000	110
10	3.6	69–76	1.008	6 400	75
11	5.1	63–83	1.008	6 500	70
12	10.2	55–62	0.942	3 500	40
13	22.9	oil	–	Oligomers	10

[a] The total yield of crude products (72%) are taken as 100%.

Table 4.3. Properties of poly(organosilanes) **26**. (Data adapted from [83])

Homopolymers wd	yield(%)	M_w^1	λ_{max}
R^1=CH$_3$; R^2=C$_3$H$_7$	32	640 000	306
R^1=CH$_3$; R^2=n-C$_4$H$_9$	34	110 000	304
R^1=CH$_3$; R^2=n-C$_6$H$_{13}$	11	520 000	306
R^1=CH$_3$; R^2=C$_{12}$H$_{25}$	9	480 000	309
R^1=CH$_3$; R^2=C$_6$H$_5$CH$_2$CH$_2$	35	290 000	303
R^1=CH$_3$; R^2=C$_6$H$_{11}$	25	800 000	326
R^1=CH$_3$; R^2=C$_6$H$_5$	55	190 000	335
R^1=CH$_3$; R^2=p-Tol	25	75 000	337
R^1=CH$_3$; R^2=p-Biphenyl	40	80 000	352
R^1=R^1=n-C$_4$H$_9$	12	1 800 000	314
R^1=R^1=n-C$_6$H$_{13}$	9	2 500 000	316

Table 4.4. Properties of copoly (organosilanes) **2c**. (Data adapted from [83])

		m/n	M_w	λ_{max}
R^1=R^2=CH$_3$	R^3=CH$_3$; R^4=n-C$_6$H$_{13}$	1.52	170 000	303
R^1=R^2=CH$_3$	R^3=CH$_3$; R^4—C$_6$H$_5$	1.51	900 000	330
R^1=R^2=CH$_3$	R^3=R^4=C$_6$H$_5$	1.13	350 000	351
R^1=CH$_3$; R^2=C$_6$H$_{11}$	R^3=CH$_3$; R^4=C$_6$H$_5$(CH$_2$)$_2$	1.49	150 000	310
R^1=CH$_3$; R^2=C$_6$H$_{11}$	R^3=CH$_3$; R^4=p-Tol	1.78	92 000	338
R^1=CH$_3$; R^2=C$_6$H$_5$(CH$_2$)$_2$	R^3=CH$_3$; R^4=C$_6$H$_5$	1.77	400 000	326

(polystyrene as internal standard). The bathochromic displacement of λ max is described in more detail in Sect. 4.4.2.

Reductive coupling has also been interpreted from a mechanistic point of view. West et al., for example, have postulated anions, radical anions, radicals and di-radicals as intermediate stages [39,40]. Zeigler demonstrated by interception reactions [41,42] that a radical intermediate stage occurs, at least at certain stages of the reaction. Here, solvent effects also play a part.

As Table 4.2 shows, a mixture with a wide molecular weight distribution is formed during reductive coupling. By simultaneously applying ultrasound to the reductive coupling reaction, it can be demonstrated that the formation of low-molecular weight components can be suppressed [43a–c] and only monomodal polymers are formed.

It is reported in the literatue that sonochemical homopolymerization under standard conditions (sodium metal, toluene as solvent, elevated temperature) takes place only in the presence of α-aryl substituents, for example 2d (R1= C_6H_5; R2=CH_3; R3=Cl). Dichlorodialkylsilanes (2d, e.g. R1=R2=C_6H_{11}; R3= Cl) will not enter the reaction under these conditions, but can be copolymerized with arylsilanes [43a]. Dichlorodialkylsilanes will, however, undergo homo-polymerization under modified experimental conditions, e.g. by using potassium and polar solvents [44].

In reductive coupling, solvent effects, temperature and ultrasonic activation determine the yield of polymer produced, as well as the mean molecular weight distribution.

Effect of solvent:

In 1984 West et al. showed that the addition of aprotic solvents such as diglyme or 1,2-dimethoxyethane to the toluene solution increases polymer yield under reflux conditions. The yield of high molecular weight compounds can also be increased under certain conditions [103]. This effect has been observed on a number of examples [104] (Table 4.5). Here, toluene-diglyme blends (with pure toluene as control), were used. R stands for the ratio of high to low molecular substances. The favourable effect of diglyme is particularly pronounced in the case of symmetrically substituted 2b whereas in the case of aromatically sub-stituted 2b the effect is not so obvious, unless a polar substituent is present in the aromatic nucleus.

Effect of reaction temperature:

The nature of the substituents in 2b influences the exothermic character of the reductive coupling Eq. (12). For example 2d (/R1=CH_3; R2=C_6H_5; R3=Cl) reacts strongly exothermically with a relatively high yield of bimodal polymer (Table 4.6), but produces a relatively low molecular weight. No marked induction period is evident, which is thought to be due to the large number of reactive centres [104]. Since the ratios in 2d (R1=R2=n-C_6H_{13}; R3=Cl) are reversed, and independent investigations [105] have shown that the induction period decreases with increasing surface area of the sodium particles, it can be assumed that the reaction temperature strongly influences the coupling reaction. This has

Table 4.5. Influence of toluene/diglyme ratio on the synthesis of polysilanes 2b. (Data adapted from [104])

Polymer 2b	vol% of toluene in the reaction mixture	yield (%)	$M_w \times 10^{-3}$	R
1 R^1=cyc.C_6H_{11}; R^2=CH_3	100	18	801, 4.5	8.7
2	90	35	1477, 24.8	0.12
3	75	32	23.1	–
4	25	33	16.5	–
5 R^1=n-$C_{12}H_{25}$; R^2=CH_3	100	8	1345, 9.4	2.73
6	70	33	476, 40.7	0.74
7 R^1=R^2=n-C_6H_{13}	100	5.9	1982, 1.2	3.12
8	70	37	2091, 24.4	1.27
9 R^1=R^2=n-$C_{10}H_{21}$	100	3	521, 14.1	5.2
10	70	34	570, 27.4	2.3
11 R^1=p-t-BuC_6H_4; R^2=CH_3	100	14	153	–
12	70	8.1	55.9, 0.64	9.1
13 R^1=p-$MeOC_6H_4$; R^2=CH_3	100	12	12.8	–
14	70	25	14.4	–
15 R^1=p-$MeOC_6H_4(CH_2)_3$; R^2=CH_3	100	16	333, 10.4	0.90
16	70	44	164, 12.0	0.91

been demonstrated in more detail by R. D. Miller [104], using as example the formation of 2b (R1=CH_3; R2=C_6H_5) in combination with solvent variations and the addition of complexing additives (Table 4.6).

The effects are very marked. Whereas, for example, bimodal distribution with 12% high molecular weight material results under reflux conditions, lowering the test temperature to 65 °C produces 10% monomodal, high molecular weight material. Complexing agents such as diglyme or 15-Crown-5 greatly increase yields, but markedly reduce the molecular weight.

In contrast to 2b with aromatic substituents, derivatives of 2b with two alkyl radicals cannot be activated at low temperatures without using additives. Yields of high molecular weight polymer, without additive, are, for example for 12-Crown-4 < 1% [104].

Activation with ultrasound:

Matyjaszewski and Kim have reported that ultrasound promotes the formation of high molecular weight, monomodal polymers in the condensation of 2d (R1=CH_3; R2=C_6H_5; R3=Cl) [106]. These authors and others [107] have found that 2d (R1=R2=n-C_6H_{13}; R3=Cl) in toluene, with the same activator, produces only negligible yields of high molecular weight product.

Matters take a different turn, however, if simultaneous use is made of complex formers, which have an activating effect on the reaction, (Table 4.7) [104]. In the case 2d (R1=CH_3; R2=C_6H_5; R3=Cl) the addition of diglyme or 15-Crown-5 improved yields of monomodal, high molecular weight material, although this had a relatively low M_w in relation to the inactivated reaction (Table 4.7). If

Table 4.6. Influence of temperature, solvents and complexing agents on the synthesis of polysilanes 2b ($R^1=CH_3$, $R^2=C_6H_5$)

Solvents	Temperature °C	Additive	yield (%)	$M_w \times 10^{-3}$	$M_n \times 10^{-3}$
1 Toluene	Reflux	–	25	383(12) [d]	267
				16(88)	8.1
2 Toluene/15% Heptane[a]	Reflux	–	9	1390(60) [d]	375
				10.5(40)	6.0
3 Toluene/15% Heptane[a]	Reflux	Diglyme 15%[b]	25	23.8	9.7
4 Toluene	65	–	10	1073	377
5 Toluene/15% Heptane[a]	65	–	9	1367	580
6 Toluene/70% Heptane[a]	65	–	3	1600(31) [d]	1300
				5.5(69)	4.5
7 Toluene/15% Heptane[a]	65	Diglyme 5%[b]	25	19.2	8.3
8 Toluene/15% Heptane[a]	65	Diglyme 15%[c]	28	14.2	6.7
9 Toluene/15% Heptane[a]	65	15-Crown-5[c] (5 mol%)	40	10.2	4.7

[a] vol.% Heptane.
[b] vol.% Diglyme.
[c] mol% 15-Crown-5, relative to sodium.
[d] GPC-Analyses.

Table 4.7. Influence of ultrasound and complexing agents are the formation of poly(organosilanes) 2b. (Data adapted from [104])

R^1_1	R^2_2	Additive	Tempera-ture(°C)		$M_w \times 10^{-3b}$	$M_n \times 10^{-3b}$
1 C_6H_5	CH_3	–	20	12	145.7(65%)	90.6
					9.7(35%)	6.2
2 C_6H_5	CH_3	Diglyme (15%)[c]	20	20	15.4	5.8
3 C_6H_5	CH_3	15-Crown-5 (5%)[d]	20	25	12.7	3.8
4 n-C_6H_{13}	CH_3	–	20	<1	1.0	0.83
5 n-C_6H_{13}	CH_3	Diglyme (15%)[c]	20	15	64.5	29.0
6 n-C_6H_{13}	CH_3	15-Crown-5 (5%)[d]	20	25	33.8	15.6
7 c-Hexyl	CH_3	–	20	<1	747.5	205.6
8 c-Hexyl	CH_3	Diglyme (15%)[c]	20	40	14.5	6.7
9 c-Hexyl	CH_3	Diglyme (15%)[d]	20	48	11.2	4.3
10 n-C_6H_{13}	n-C_6H_{13}	–	20	<1	1100(44%)	576
					6.8(56%)	4.4
11 n-C_6H_{13}	n-C_6H_{13}	Diglyme (15%)[c]	20	15	59.1	21.1
12 n-C_6H_{13}	n-C_6H_{13}	15-Crown-5 (5%)[d]	20	23	61.1	23.5
13 n-C_6H_{13}	n-C_6H_{13}	15-Crown-5 (5%)[d]	20	50	73.3	27.3
14 n-C_6H_{13}	n-C_6H_{13}	15-Crown-5 (5%)[d]	20	45	67.9	20.3
15 n-C_6H_{13}	n-C_6H_{13}	15-Crown-5 (5%)[d]	20	40	31.5	10.1
16 n-C_6H_{13}	n-C_6H_{13}	18-Crown-6 (5%)[d]	20	10	74.5	30.1

[a] added after dispersion of the sodium in toluene/heptane (volume ratio 85:15).
[b] from 6PC measurements.
[c] vol% in the toluene/heptane mixture (85:15).
[d] mol% of crown ether relative to sodium.

complexing agents are added, pure alkyl derivatives will rapidly polymerize, monomodal products being formed.

A combined radical ion/radical mechanism has been suggested for the copolymerization reaction [43a], Scheme (14).

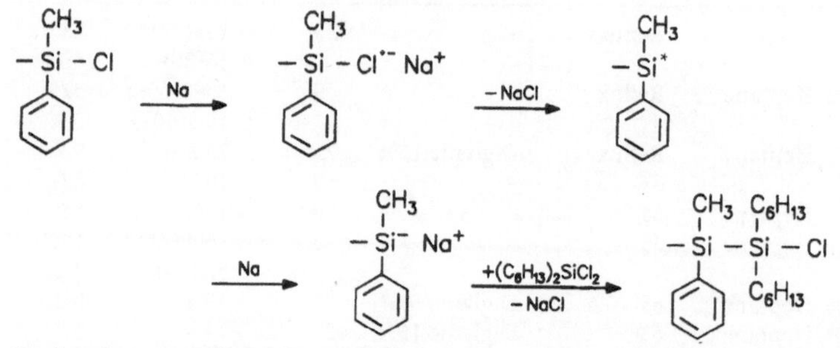

$$\begin{array}{c} CH_3 \\ | \\ -Si-Cl \\ \end{array} \xrightarrow{Na} \begin{array}{c} CH_3 \\ | \\ -Si-Cl^{-\cdot}\ Na^+ \\ \end{array} \xrightarrow[-NaCl]{} \begin{array}{c} CH_3 \\ | \\ -Si^\cdot \\ \end{array}$$

$$\xrightarrow{Na} \begin{array}{c} CH_3 \\ | \\ -Si^-\ Na^+ \\ \end{array} \xrightarrow[-NaCl]{+(C_6H_{13})_2SiCl_2} \begin{array}{c} CH_3\ \ C_6H_{13} \\ | \ \ \ \ | \\ -Si-Si-Cl \\ | \\ C_6H_{13} \\ \end{array} \tag{14}$$

Reductive coupling is noted for the many different results that can be achieved. Thus, **2d** (R1=R2=C_6H_5; R3=Cl) with lithium as reducing agent Eq. (15) produces practically only low molecular weight ring compounds [61]:

ring compounds[61]:

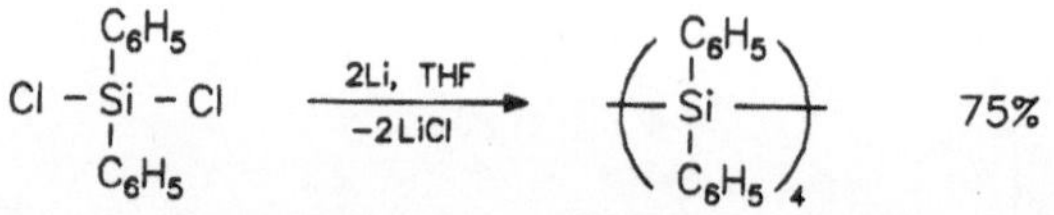

$$\begin{array}{c} C_6H_5 \\ | \\ Cl-Si-Cl \\ | \\ C_6H_5 \\ \end{array} \xrightarrow[-2LiCl]{2Li,\ THF} \left(\begin{array}{c} C_6H_5 \\ | \\ Si \\ | \\ C_6H_5 \\ \end{array}\right)_4 \quad 75\% \tag{15}$$

Greater excess of lithium shifts the equilibrium towards the 5-ring [61], which is more stable thermodynamically.

The low molecular weight, cyclic rings can be converted into open-chain, terminal α,ω-Cl oligosilanes, which can be condensed further (Scheme 16), enabling specific acyclic oligosilanes to be synthesized [62]:

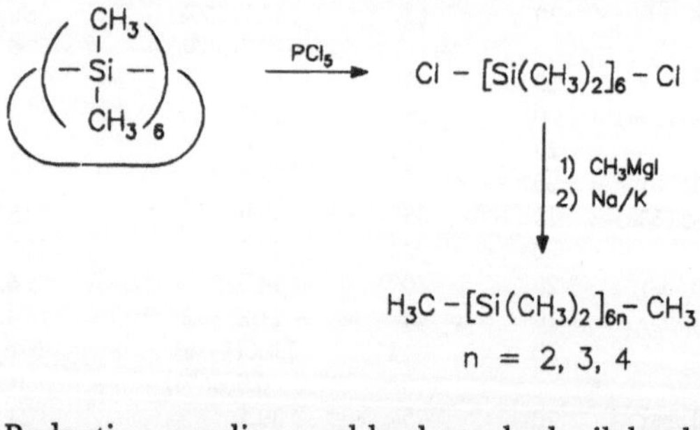

$$\left(\begin{array}{c} CH_3 \\ | \\ -Si- \\ | \\ CH_3 \\ \end{array}\right)_6 \xrightarrow{PCl_5} Cl-[Si(CH_3)_2]_6-Cl$$

$$\downarrow \begin{array}{l} 1)\ CH_3MgI \\ 2)\ Na/K \end{array} \tag{16}$$

$$H_3C-[Si(CH_3)_2]_{6n}-CH_3$$
$$n = 2,\ 3,\ 4$$

Reductive coupling enables branched, silyl-substituted poly(organosilanes) **17b** and **18b** (ladder polymers) to be prepared from the Cl-substituted disilanes **17a** and **18a** in accordance with Eqs. (17 and 18) [45]:

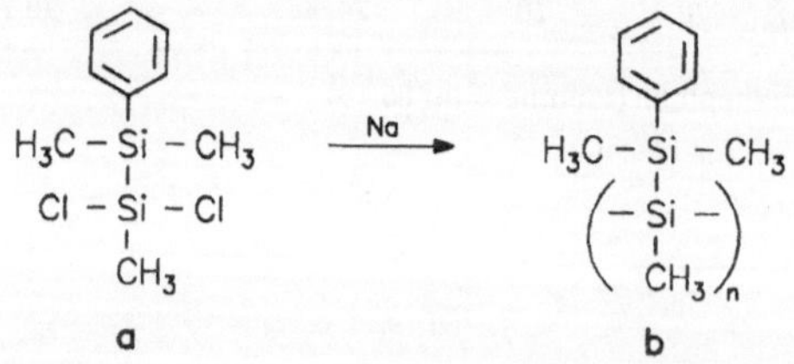

$$\begin{array}{c} H_3C-Si-CH_3 \\ | \\ Cl-Si-Cl \\ | \\ CH_3 \\ \end{array} \xrightarrow{Na} \begin{array}{c} H_3C-Si-CH_3 \\ \left(\begin{array}{c} -Si- \\ | \\ CH_3 \end{array}\right)_n \\ \end{array} \tag{17}$$

a   b

$$
\begin{array}{c}
\underset{\displaystyle \overset{\displaystyle \text{CH(CH}_3)_2}{|}}{\text{Cl} - \text{Si} - \text{Cl}} \\
\underset{\displaystyle \overset{}{|}}{\text{Cl} - \text{Si} - \text{Cl}} \\
\overset{|}{\text{CH(CH}_3)_2} \\
\mathbf{a}
\end{array}
\quad \xrightarrow{\text{Li}} \quad
\left(
\begin{array}{c}
\overset{\displaystyle \text{CH(CH}_3)_2}{-\,\text{Si}\,-} \\
-\,\text{Si}\,- \\
\underset{\displaystyle \text{CH(CH}_3)_2}{}
\end{array}
\right)_n
\qquad (18)
$$

b

4.1.2.2
Preparation of Poly(organosilanes) by Electrolysis of Halogen Silanes and Halogenized Disilanes

The electrochemical reduction of halogen silanes is a good method for synthesizing Si—Si linkages, although so far it has only been possible to prepare **2b** (R1=R2=CH_3; Eq. (19)), with relatively low molecular weights [46–48]. Under specific conditions, e.g. absence of solvent and addition of *tert*-butylammonium chloride compounds with molecular weights between 3000 and 6000 are formed [202]. The electrochemical reaction of **2d** (R1=R2=C_6H_5; R3=Cl) produces the cyclic tetramer **10** Eq. (20) [47], that of **2d** (R1=CH_3; R2=C_6H_5; R3=Cl) the halogenized dimer **21a** Eq. (21) [46]:

$$
n\,(CH_3)_2 SiCl_2 \; + \; 2n\,e^- \; \longrightarrow \; \left(\!\!\begin{array}{c} \overset{\displaystyle CH_3}{|} \\ Si \\ | \\ CH_3 \end{array}\!\!\right)_n \; + \; 2n\,Cl^- \qquad (19)
$$

$$
4\,(C_6H_5)_2 SiCl_2 \; + \; 8\,e^- \; \longrightarrow \;
\begin{array}{c}
H_5C_6 \quad\quad C_6H_5 \\
H_5C_6 \quad\quad C_6H_5 \\
H_5C_6 \quad\quad C_6H_5 \\
H_5C_6 \quad\quad C_6H_5 \\
\mathbf{a}
\end{array}
\; + \; 8\,Cl^- \qquad (20)
$$

$$
2\,C_6H_5 Si(CH_3)Cl_2 \; + \; 2\,e^- \; \longrightarrow \; (C_6H_5 Si(CH_3)Cl)_2 \; + \; 2\,Cl^- \qquad (21)
$$

a

Generally speaking, the results of electrochemical reduction very much depend on initial conditions. The materials used for the anodes and cathodes, for example, influence the reaction [48].

Aprotic solvents are indispensable for electrochemical reductions because of the susceptibility of the chlorosilanes used in these reactions to hydrolysis. These solvents must be capable of dissolving the silanes sufficiently and must not suffer reduction in the course of the reaction. They should, furthermore, dissolve the added electrolyte and have a high dielectric constant. Since reduction occurs via anionic and radical intermediate stages, it must be ensured that the solvent does not react with these [48].

Oligomeric silanes can be conveniently prepared by the electrolysis of monomeric and/or dimeric chlorosilanes, with copper as the anode material and platinum for the cathode, as well as 1, 2-dimethoxyethane as solvent. For example, the reduction of chloromethyldiphenylsilane **2d** (R1=R2=C_6H_5; R3= CH_3) produces high yields of the disilane **22a** [49] Eq. (22).

Under well defined reaction conditions using electrochemical techniques Biran et al. were able to produce oligomeric and polymeric silane compounds, partly in high yields [209]. Table 4.8 lists the results and experimental conditions:

Table 4.8. Electrodecmical reduction of various silanes. (Data adapted from [209])

Silane	Cosilane	Product	yield(%)
$2(CH_3)_3SiCl$	–	$(CH_3)_3Si\text{-}Si(CH_3)_3$	74
$(C_6H_5)_2(CH_3)SiCl$	$(CH_3)_3SiCl$	$(C_6H_5)_2(CH_3)Si\text{—}Si(CH_3)_3$	90
$(C_6H_5)_2(CH_3)SiCl_2$	$(CH_3)_3SiCl$	$(C_6H_5)(CH_3)SiCl\text{—}Si(CH_3)_3 +$	75
		$CH_3)_3Si\text{-}Si(CH_3)(C_6H_5)\text{-}Si(CH_3)_3$	25
$(CH_3)_2SiCl_2$	$(CH_3)_3SiCl$	$(CH_3)_3Si\text{-}Si(CH_3)_2\text{-}Si(CH_3)_3$	60
$n(CH_3)_2SiCl_2$	–	$-(Si(CH_3)_2)n\text{-}; n = 16-32;$	30
$n(CH_3)(C_6H_5)SiCl_2$	–	$-((CH_3)(C_6H_5)Si)_n\text{-}; M_n = 930, M_w = 3395$	75

The poly(organosilane) **2b** $(R_1=R_2=CH_3)$ produced from $(CH_3)_2SiCl_2$ is obtained in low yields only (30%), since the intermediate silyl anion attacks the solvent (THF), where ring opening reaction occurs.

It is also possible by this reaction procedure to produce low molecular weight cyclic carbosilanes, (e.g. **56** $(R=R'=CH_3)$) from $ClCH_2Si(CH_3)_2$, though the authors were not able to synthesize high molecular weight poly(carbosilanes) [209].

$$2\ (C_6H_5)_2(CH_3)SiCl \xrightarrow[-2Cl^-]{+2e^-} \underset{a}{(C_6H_5)_2(CH_3)Si\text{—}Si(CH_3)(C_6H_5)_2} \tag{22}$$

4.1.2.3
Preparation by dehydrogenating polymerization of hydrogen-containing mono and disilanes with metalocene catalysts

The dehydrogenating polymerization of hydrogen-containing monosilanes was first described by Aitken [50, 51]. Dimethyl titanocene and dimethyl zirconocene proved to be the most suitable catalyst systems. The reactivity of the hydrogen-containing original silane is governed by the number of hydrogen atoms present. The following reactivity scale applies as a rule of thumb:

$RSiH_3 \ll R2SiH_2 \ll R3SiH$
    R=alkyl/aryl

Triaryl/alkyl silanes R3SiH cannot react under the conditions of dehydrogenating polymerization.

The difference between hydrogen-containing disilanes and monosilanes is to be found in their reactivities – the latter have low reactivity whereas, for example the symmetrical 1, 1, 2, 2-tetramethyldisilane **23a** reacts spontaneously [52, 203] with the formation of hydrogen-containing poly(organosilane) **23b**:

$$\underset{a}{(CH_3)_2HSi\text{—}SiH(CH_3)_2} \xrightarrow{\overset{H_3C-Ti-CH_3}{}} \underset{b}{(CH_3)_2HSi\text{—}(Si(CH_3)_2)_n\text{—}SiH(CH_3)_2} \tag{23}$$

The number of Si atoms per polymer produced by the reaction depends upon reaction conditions. Whereas a maximum of five Si atoms are found if the reaction is carried out at room temperature, this is increased if the reaction takes place at elevated temperatures (Table 4.9) (52).

During the dehydrogenative polymerization of substituted trisilane- and tetrasilane-derivatives two mechanism are operating, which are discussed in the literature under the terms "σ-bond metathesis" and "β^*-bond elimination [206], while the dehydrogenative polymerization of certain monosilanes such as phenylsilane $C_6H_5SiH_3$ proceeds via an intermediate transitionmetal-silyl complex, in which repeated insertion of a silyl moiety occurs and thus Si—Si-bonds are being formed [207].

Table 4.9. Reaction products of $C_{P_2}T, Me_2$ catalyzed reactions of H-Containing disilanes. (Data adapted from [52])

Silane	Reactions at 25–35°C		Reactions at 130°C		Polymer
	X^a	X^b	X^a	X^b	
$(CH_3)_3SiSiH(CH_3)_2$	3, 4, 5, 6, 7	3, 4	3, 4, 5, 6, 7, 8, 9, 10	4, 5, 6	linear
$(CH_3)_2HSiSiH(CH_3)_2$	3, 4, 5, 6, 7	3, 4	3, 4, 5, 6, 7, 8, 9, 10	3, 4, 5, 6	linear
$(CH_3)H_2SiSiH_2(CH_3)$	3, 4, 5, 6, 7, 8				branched

^{a} number of Si-atoms in the crude reaction mixture.
^{b} number of Si-atoms of the main products.

Linear and branched, hydrogen-containing polysilanes are obtained, depending on the degree of substitution, which are accessible to further dehydrogenating polymerization. The polysilane **2b** (R1=CH_3; R2=H) is obtained through dehydrogenating polymerization of 1, 2-dimethyldisilane (see also Table 4.9). This is an orange-coloured polymer which is stable at room temperature, but self-igniting when exposed to air [52].

4.1.2.4
Preparation by Disproportionation of Chlorinated and Methoxylated Disilanes

The mixture of chloromethylsilanes such as *sym*-dimethyltetrachlorodisilane **24a** produced in the Rochow synthesis experiences disproportionation in the presence of catalysts, to form chlorinecontaining poly(organosilanes) **24b** Eq. (24) [53]:

$$(CH_3)_xSi_2Cl_{6-x} \xrightarrow{\;R_4PBr\;} [((CH_3)_2Si)(CH_3SiCl)_{1.6}(CH_3Si)_8] + \text{Monomer} \quad (24)$$

a   b

The reaction is speeded up by a number of catalysts, e.g. tertiary amines, although hexamethylphosphoric acid tris-amide seems to be the most effective [53].

The methoxylated disilanes **25a** which can be prepared from **24a** are the starting substances for the preparation of polysilanes with the structure **25b** some of which also contain silicon-bound methoxy groups Eq. (25). Alkoxide anions such as sodium methylate are used as catalysts:

$$(CH_3)_xSi_2(OCH_3)_{6-x} \xrightarrow{\text{NaOCH}_3} [((CH_3)_2Si)_x(CH_3SiOCH_3)_y(CH_3Si)_2] \tag{25}$$

a   b

Mixtures of **25a** are usually employed for the polymerization reaction which must take place in the presence of an inert gas and total absence of water, at elevated temperatures. The catalyst can be, for example, lithium methylate prepared from *n*-butyl lithium and absolute methanol [54]. The reaction is of industrial interest and the subject of various patents [55a–b]. Fibers and ceramic materials can be produced from type **25b** polymers by pyrolysis.

4.1.2.5
Preparation from Masked Disilenes by Anionic Polymerization

Si—Si double bonds exist only in sterically hindered substituents on the Si atom [56a–b]. Technically speaking, the polymerization of monomers containing Si—Si double bonds could be used to prepare poly(organosilanes) **2b** analogous to the polymerization of olefins. The use of masked and thus stabilized disilanes here represents an advance [57]. Polymers are obtained by anionic polymerization, of **26a** yields with up to 79%, depending on the reaction conditions Eq. (26), molecular weight distribution being narrower than for the method described in Sect. 4.1.2.1 [57].

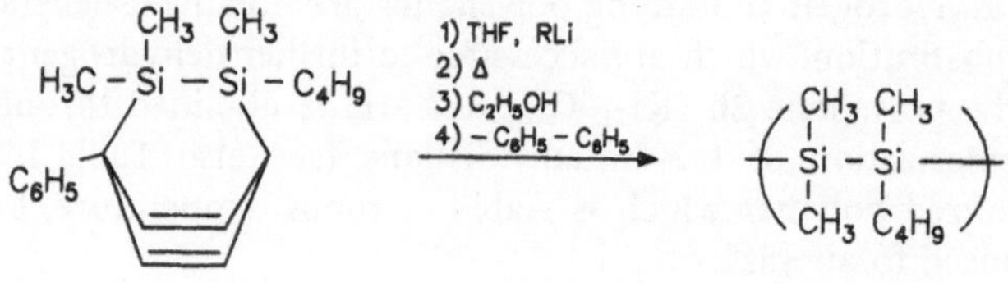

$$\tag{26}$$

It is assumed that the mechanism involves anions, so-called living polymers being formed as well as intermediate and final stages. The addition of methyl methacrylate, for example, produces copolymers [57].

4.1.2.6
Preparation from 1, 4-Disilacyclohexadiene Derivatives [58]

During attempts to isolate the monomeric, cyclic carbosilanes **27b** prepared by reacting acetylene and **27a** (R—OCH₃) as well as subsequent thermolysis Eq. (27) by distillation, polymer compounds are also obtained. The structure of **27b** has been confirmed by IR and NMR-spectroscopic tests [59].

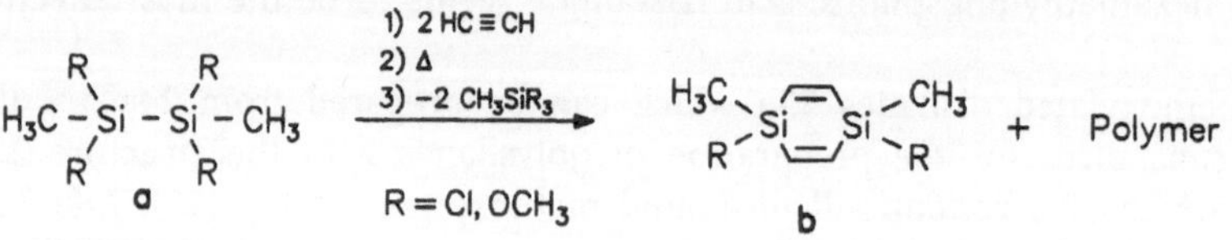

$$\tag{27}$$

Thermolysis of the polymeric residue after separation of the volatile reaction products yields 2b (R1=CH$_3$; R2=OCH$_3$) in the case of 27a (R=OCH$_3$) as starting product, or 2b (R1=CH$_3$; R2=Cl) with 27a (R=Cl) as the starting material. Poly(methylmethoxysilane) 2b (R1=CH$_3$; R2=OCH$_3$) is described as being very stable [58]. The structures were confirmed by IR, UV and NMR spectroscopic tests.

4.1.2.7
Preparation by Anionic and Cationic Ring Opening Polymerization of Cyclotetrasilanes [60]

It has long been known that cyclic poly(organosiloxanes) of type 28a can be converted into linear, high molecular weight poly(organosiloxanes) 28b by acid as well as alkaline catalysis, involving ring opening polymerization [108] Eq. (28).

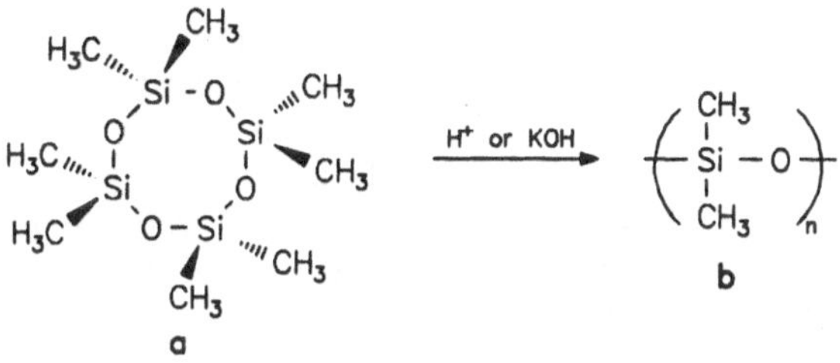

$$ \tag{28} $$

The obvious step was to transfer this reaction principle also to low molecular weight poly(organosilanes) with few chain segments 2b (for example n=4). Surprisingly, such reactions were first described in 1990 [60].

The cyclosilane 2a (R1=C$_6$H$_5$; R2=CH$_3$; n = 4), prepared in situ from 2a (R1=R2=C$_6$H$_5$; n = 4) via Triflat interchange Eq. (29), can be polymerized anionically, producing not only higher ring systems (n = 5,6) but also high polymer products with molecular weights of between 3000 and 20000:

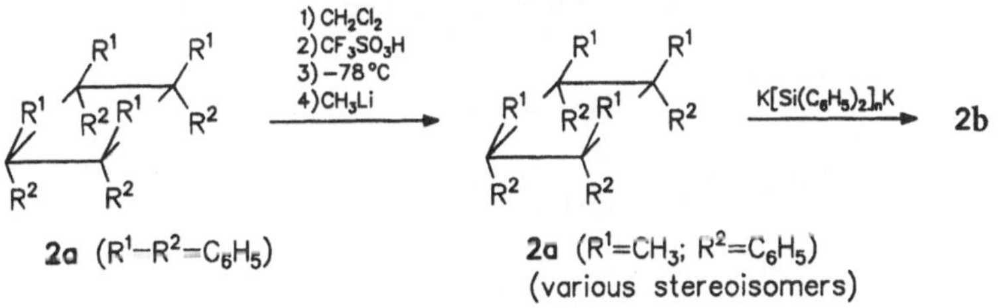

$$ \tag{29} $$

2a (R^1–R^2=C$_6$H$_5$)   **2a** (R^1=CH$_3$; R^2=C$_6$H$_5$)
(various stereoisomers)

The stereochemistry of the intermediate stage 1 (R1=CH$_3$; R2=CH$_6$H$_5$; n = 4) has been clarified. The "all *trans*" (=1,3-*cis*): 1,2-*cis*: 1,2,3-*cis* ratio is about 50:30:20 [60].

Anionic polymerization of the isolated stereoisomers and their polymerization leads one to expect products with controlled tacticity, eg. Eq. (30) with, possibly, different physical properties [60] and this is being investigated at the moment.

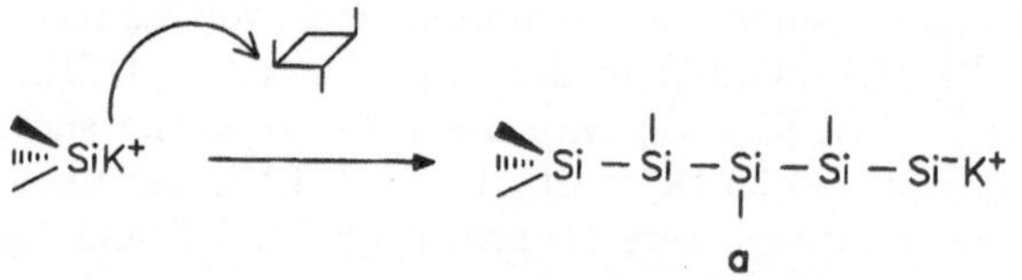

$$\overset{\displaystyle\curvearrowright\quad\square}{\overset{\blacktriangleright}{\diagdown}}SiK^+ \quad\longrightarrow\quad \overset{\blacktriangleright}{\diagdown}Si - \overset{|}{Si} - \overset{|}{Si} - \overset{|}{Si} - Si^-K^+ \qquad (30)$$

a

The cyclosilane 2a (R1=R2=OCH$_3$; $n = 4$) can be cationically polymerized with CF$_3$SO$_3$H, molecular weights of 44000 being obtained [60].

4.1.2.8
Assessment of the Methods of Preparation of Poly(organosilanes) Described in Sect. 4.1.2.1–4.1.2.7

High molecular weight poly(organosilanes) are preferred for synthesizing poly(carbosilanes) and the ß-silicon carbide made from them (see Sect. 1.5.), since only the high degree of crosslinkage will ensure adequate ceramic yields. Up to now, unfortunately, products with sufficiently high molecular weights can only be prepared by condensation through reductive coupling, analogous to the Wurtz synthesis (see Sect. 4.1.2.1). It would be preferable, also for safety reasons, to find a gentler method than one which uses sodium metal, but at the moment the only promising alternative would seem to be polymerization of hydrogen-containing silanes using transition metal catalysts [82] as well as the disproportion of disilanes [204].

4.1.3
Chemical Properties of Poly(organosilanes)

4.1.3.1
Ring Opening of Cyclic Poly(organosilanes)

Low molecular weight cyclic poly(organosilanes) add haloid acids, halogens and alkali metals with the formation of adducts 31a, b and c (Scheme 31).

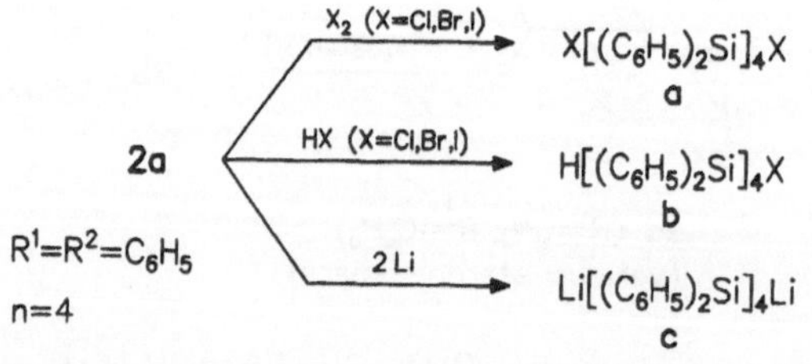

$$
\begin{array}{c}
\textbf{2a} \\
R^1=R^2=C_6H_5 \\
n=4
\end{array}
\begin{cases}
\xrightarrow{\;X_2\;(X=Cl,Br,I)\;} & X[(C_6H_5)_2Si]_4X \\
& \qquad\quad a \\
\xrightarrow{\;HX\;(X=Cl,Br,I)\;} & H[(C_6H_5)_2Si]_4X \\
& \qquad\quad b \\
\xrightarrow{\;2\,Li\;} & Li[(C_6H_5)_2Si]_4Li \\
& \qquad\quad c
\end{cases}
\qquad (31)
$$

forming open-chain α, ω-terminal, functional poly(organosilanes) which can be reacted further. This applies to phenylated as well as alkylated compounds [61] (see also Sect. 4.1.2.1).

The α, ω-chlorine-terminated derivative 32a for example, can be converted into the corresponding diol 32b by hydrolysis. This can then be converted into the cyclic oxo-compound 32c (Scheme 32):

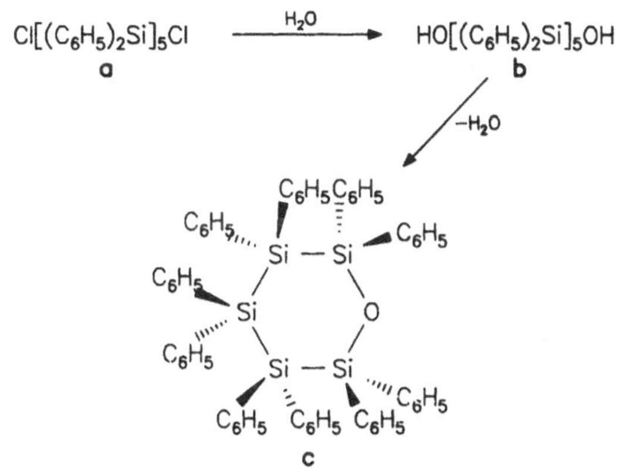

$$Cl[(C_6H_5)_2Si]_5Cl \xrightarrow{\quad H_2O \quad} HO[(C_6H_5)_2Si]_5OH$$

a   b

$$\searrow -H_2O$$

(32)

c

4.1.3.2
Cleavage of the Si—Si Bonds in Poly(organosilanes)

Strong nucleophilic agents such as trialkyl/arylsilyl anions cause the Si—Si bond in **2b** to dissociate, forming reactive silyl anions as reactive intermediate stages [63].

Poly(organosilanes) can be broken down with excess alkali metal to produce cyclic, low molecular weight oligomers [63]. The decomposition rate depends upon the substituents in **2b**, the type of alkali metal used, the solvent and temperature. The intermediates postulated are silyl anions and radicals. The additional use of ultrasound facilitates the selective cleavage of the Si—Si bonds in high molecular weight poly(organosilanes) **2b** [63], but the formation of low molecular weight, cyclic products is not evident.

Thermal degradation, at about 150 °C, in the presence of an inert gas such as nitrogen and/or argon, produces low molecular weight, cyclic silanes [63]. This thermal decomposition reaction can be carried with or without solvent.

Organic peroxides split the Si—Si bond, with the insertion of an oxygen atom Eq. (33), producing the cyclic siloxane **33a** [64].

Stretched cyclosilanes such as **2a** (R1=R2=CH$_3$; $n = 4$) *react spontaneously with atmospheric oxygen [61], sulfur and selenium [65], [66]*[1].

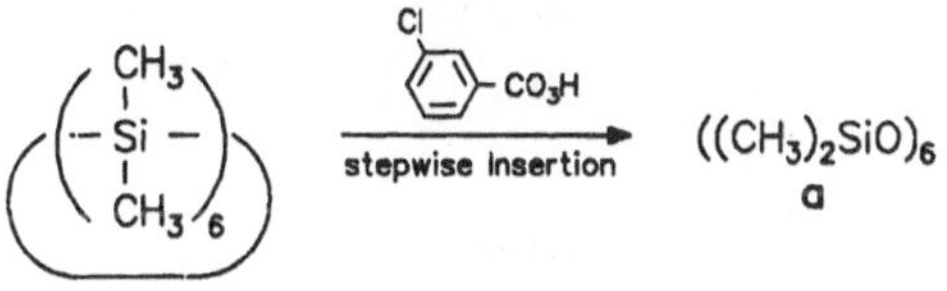

(33)

4.1.3.3
Transposition Reactions of Poly(organosilanes)

Linear, trimethylsilyl-terminal blocked poly(dimethylsilanes) of type **34a** are converted into the branched isomer **34b** in the presence of aluminium chloride which acts as a catalyst Eq. (34) [67]:

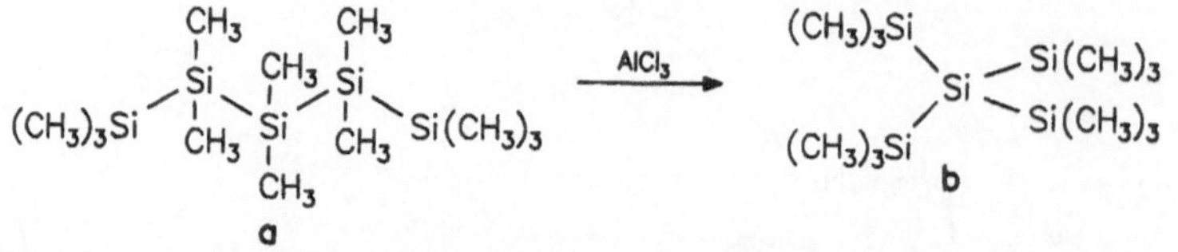

$$
\text{(structure a)} \xrightarrow{\text{AlCl}_3} \text{(structure b)} \tag{34}
$$

In the case of cyclic poly(dimethylsilanes) branching is accompanied by ring contraction Eq. (35). Here, higher substituted poly(dialkylsilanes) need not necessarily follow this reaction [68]:

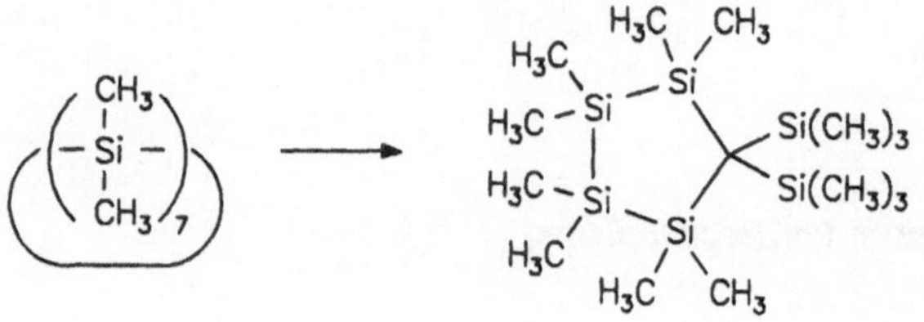

$$
\text{(structure)} \longrightarrow \text{(structure)} \tag{35}
$$

4.1.3.4
Interchange of Alkyl and Aryl groups in Poly(organosilanes)

The exchange of methyl groups in cyclic poly(dimethylsilanes), e.g. **36a** and **37a** can be achieved with hydrogen chloride/aluminium chloride and hydrogen-containing chlorosilanes Eq. (36–37), with the formation of the chlorinated derivatives **36b** and **37b** which can be further functionalized [61], since the newly formed Si—Cl bonds can react with nucleophilic agents:

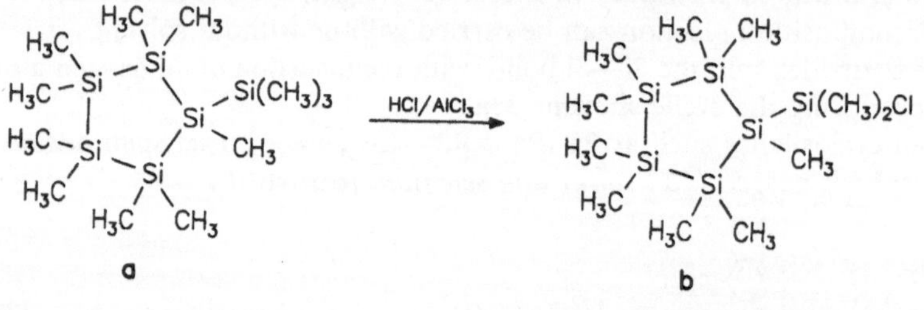

$$
\underset{\textbf{a}}{\text{(structure)}} \xrightarrow{\text{HCl/AlCl}_3} \underset{\textbf{b}}{\text{(structure)}} \tag{36}
$$

$$
\underset{\textbf{a}}{\text{(structure)}} \xrightarrow[\text{PtCl}_4]{\text{HSiCl}_3} \underset{\textbf{b}}{\text{(structure)}} \tag{37}
$$

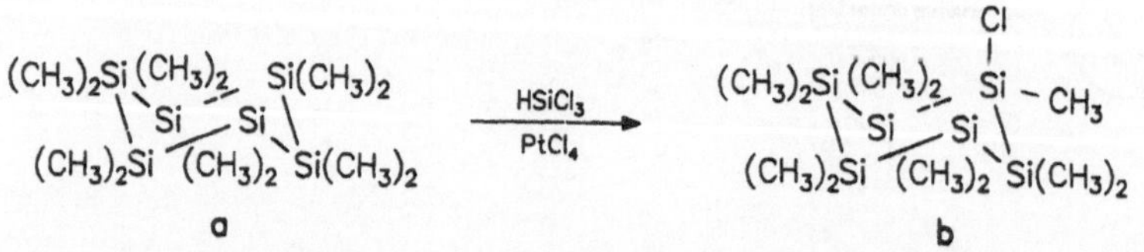

When poly(organosilanes) with aromatic radicals are reacted with trifluoro-methylsulfonic acid, benzene is formed Eq. (38), accompanied by a maximum of 90% exchange (in methylene chloride as solvent). The incorporation of triflate groups permits further functionalization of the poly(organosilane) chain, which also enables copolymers with interesting properties to be produced, since the triflate group itself is accessible to nucleophilic substitution [43a]:

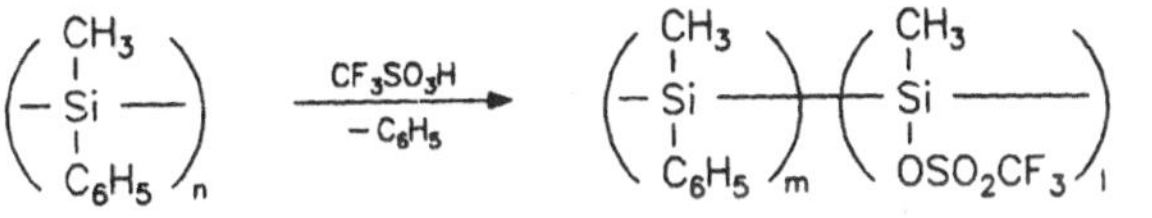

$$\left(-\underset{\underset{C_6H_5}{|}}{\overset{\overset{CH_3}{|}}{Si}}-\right)_n \quad \xrightarrow[-C_6H_5]{CF_3SO_3H} \quad \left(-\underset{\underset{C_6H_5}{|}}{\overset{\overset{CH_3}{|}}{Si}}-\right)_m \left(-\underset{\underset{OSO_2CF_3}{|}}{\overset{\overset{CH_3}{|}}{Si}}-\right)_l \tag{38}$$

$$n = m + l$$

4.1.3.5
Photolysis of Poly(organosilanes)

Cyclic, low molecular weight polydimethylsilanes such as **39a**, (e.g. **39d**) undergo ring contraction when irradiated with UV light, with the elimination of $(CH_3)2Si$: yielding **39b** or c. Here, the disilylene derivative produced in the intermediate stage can be identified by various reagents through interception reactions [69]. Open-chain structures **39d** are produced through the elimination of hydrogen from the solvent present (Scheme 39).

The spectroscopic identification of $(CH_3)2Si$: has been described in Refs. [70, 71],

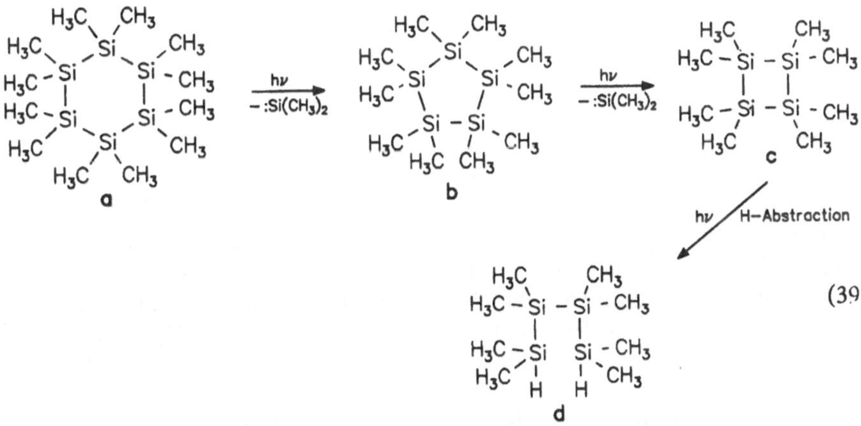

$$\tag{39}$$

High molecular weight poly(organosilanes) **2b**, too, are split up into smaller fragments when irradiated with light of a specified energy. Proof that silyl radicals play a part in this is the fact that the expected decomposition products are formed in the presence of halogen hydrocarbons (silyl radicals remove chlorine atoms from C—Cl-containing compounds). Thus, when **2b** (R1=cyclo-C_6H_{11}; R2=CH_3) is irradiated at 300 nm in the presence of carbon tetrachloride hexachloroethane is formed Eqs. (40, 41) [72, 73]:

$$SiR_2: + CCl_4 \longrightarrow ClSiR_2^. + CCl_3^. \tag{40}$$

$$2 CCl_3^. \longrightarrow Cl_3C-CCl_3 \quad 33\% \tag{41}$$

The same authors irradiated various poly(organosilanes) **2b** (substituents see Table 4.10), dissolved in triethylsilane at 254 nm. The following product distribution is evident (Table 4.10):

Table 4.10. Yields and product distribution of polysilane (**2b**) triethylsilane mixtures photolyzed at 254nm. (Data adapted from [73])

Product	$R^1=n$-C_4H_9 $R^2=n$-C_4H_9	$R^1=n$-C_6H_{13} $R^2=CH_3$	$R^1=$cyclo-C_6H_{11} $R^2=CH_3$
$(C_2H_5)_3SiR_1R_2H$	59	70	71
$HSiR_1R_2SiR_1R_2H$	11	11	14
$(C_2H_5)_3SiSiR_1R_2Si(C_2H_5)_3$	n.a.[a]	3	n.a.
$HSiR_1R_2OSiR_1R_2H$	n.a.[a]	n.a.[a]	2
$(C_2H_5)_3SiOSiR_1R_2SiR_1R_2H$	n.a.[a]	n.a.[a]	3

[a]not available.

In all cases the adduct obtained from triethylsilane with the dialkyl substituted silylene is the main product. Similar results have been observed with methanol and n-propyl alcohol in toluene as solvent [73].

The exact mechanism of the photolytic degradation is not yet known [74]. The above mentioned elimination of: $Si(CH_3)_2$ by photochemical decomposition of poly(dialkylsilanes), according to Eq. (42).

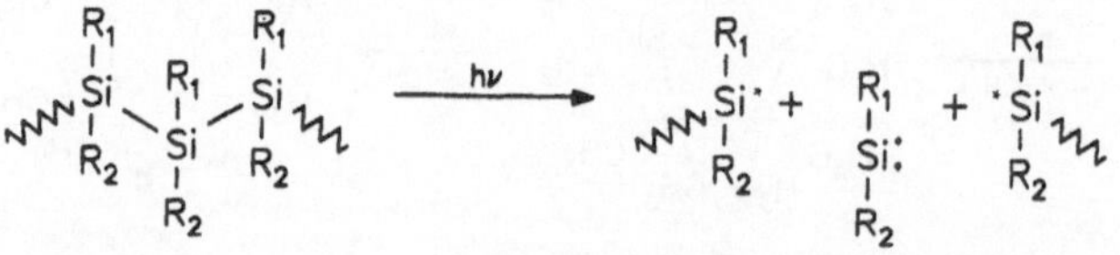

$$\tag{42}$$

is far more complex in character, as is evident from ESR-spectroscopic tests. Investigations of the hyper-fine structure of the spectra obtained give no indication of radicals resulting from the radical decomposition of the chain Eq. (42) [75]. In the case of poly(di-n-hexylsilane) 1 ($R1=R2=n$-C_6H_{13}) the proton dissociations are interpreted as indicating the existence of a polysilylated silyl radical [76].

The quantum yield of the photolysis of poly(dialkylsilanes) **2b** for chain cleavage in solution is high ($\beta=0.5-1.0$) and is consistent with the spectral bleaching described below [77] (see Sect. 4.1.4.3).

Apart from chain cleavage, crosslinkage to form polymers **43c** has been observed in the case of type **43a** aryl-substituted poly(organosilanes) [78-79], the following reaction scheme being suggested (Scheme 43):

4.1.3.6
Conversion of Poly(organosilanes) into Poly(carbosilanes)

When poly(organosilanes) are heated under pressure in an inert atmosphere, e.g. of argon, to 450–470 °C complex transposition reactions take place, with insertion of methylene groups Eq. (44) into the poly(organosilane) skeleton,

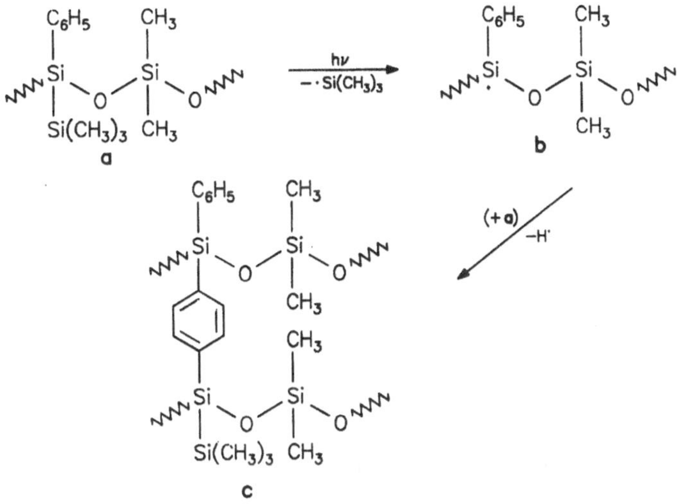

(43)

forming poly(carbosilanes) **44a**. The chemical and applicational aspects of the reaction have been described in detail by Yajima [80]:

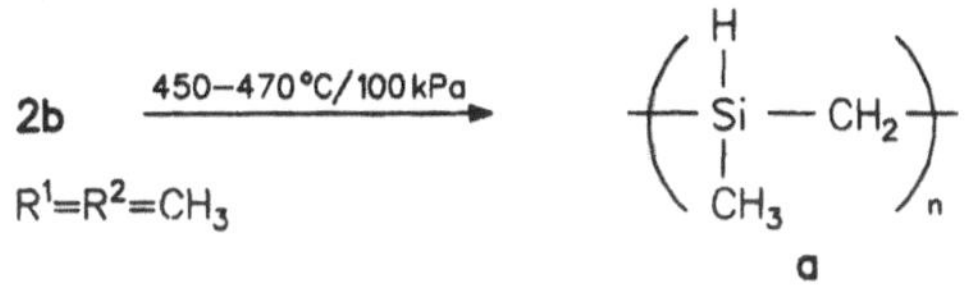

(44)

Poly(carbosilanes) **44a** can be dissolved in hexane and fractionated, important intermediates for the preparation of β-silicon carbide being produced after the removal of low molecular weight compounds (see also Sect. 4.2.1.1).

4.1.4
Physical Properties of Poly(organosilanes)

4.1.4.1
Molecular Weight Distributions

The physical properties of poly(organosilanes) depend on the nature of the substituents R1 and R2 in **2b**. Homopolymers **2b**(R1=R2=CH$_3$) are crystalline and insoluble. These properties change when the number of carbon atoms in the substituent molecule is increased. For example, **2b**(R1=CH$_3$; R2=n-C$_3$H$_7$) is already a soluble solid, whereas 1 (R1=CH$_3$; R2=n-C$_6$H$_{13}$) has elastomeric properties.

The molecular weight distributions of poly(organosilanes) depend on a number of experimental factors (see especially Sect. 4.1.2.1.) and are normally bimodal. Determinations of the numeric molecular weight M$_w$ by gel permeation chromatography (GPC), using an internal standard such as polystyrene, only yields the relative values for M$_w$. The actual values, determined by light scattering, are much higher [81], as shown in Table 4.11:

Table 4.11. Degree of polymerization(n) and 2 characteristic ratio of poly (organosilanes)s **2b** from light scattering measurements. (Data adapted from [81])

Polymer 1	Solvent	n	c_∞
R^1—Ch$_3$; R^2—n-C$_3$H$_7$	Tetrahydrofuran	2 500	19
R^1—CH$_3$; R^2—C$_6$H$_{11}$	Cyclohexane	20 000	14
R^1—CH$_3$; R^2—n-C$_6$H$_{13}$	Cyclohexane	37 000	21
R^1—CH$_3$; R^1—n-C$_6$H$_{13}$	Tetrahydrofuran	31 000	20
R^1—CH$_3$; R^2—C$_6$H$_5$	Tetrahydrofuran	400	$64 = 20$

The interpretation of the figures for C_∞ is that, at a working temperature of $-25\,°C$, the poly(organosilane) chains will be less flexible in solution than those of polyolefins.

The ladder polymer prepared from **18a** and 1,2-dichloro-1,1,2,2-tetraisopropyl-disilane by condensation with lithium (also see Eq. 18) has a complex structure. Structures of types **45a–c** ($R=$—CH(CH$_3$)$_2$) [45] have been identified:

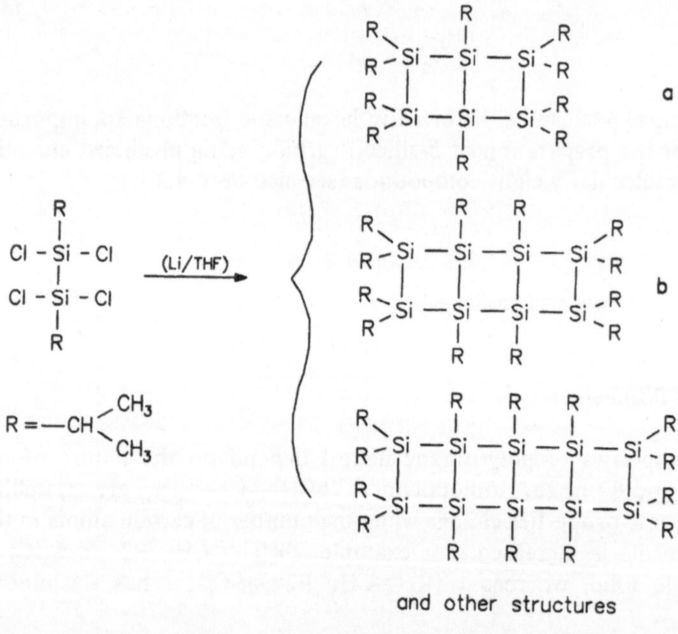

$$(45)$$

and other structures

Mean molecular weights of up to 25000 have been achieved through fractionated precipitation with solvents. The polymer blend **45a–c** is a soft, waxy, reddish-yellow polymer with an average molecular weight of 2000. Thermogravimetric analysis and the high ceramic yield indicate that no volatile silane compounds are set free when the substance is heated to $-800\,°C$. The following decomposition reaction has been suggested Eqs. (46–47) [45]:

$$\text{crude reaction mixture } \mathbf{45a-c} \quad \longrightarrow \quad CH_4 \;+\; 3/2\,H_2 \;+\; C \;+\; SiC \tag{46}$$

$$\text{crude reaction mixture } \mathbf{45a-c} \quad \longrightarrow \quad C_2H_4 \;+\; 3/2\,H_2 \;+\; SiC \tag{47}$$

The reaction Eqs. (46 and 47) are in agreement with the high ceramic yield [45].

4.1.4.2
Electron Spectra

Although a σ-linkage system is present in the poly(organosilane) chain of type **2b**, this group of compounds exhibits strong absorption bands in the UV region, whose maxima depend on the following factors:

a) Substitution at the Si atom: whereas alkyl-substituted, atactic and amorphous products absorb at 300–325 nm, products with sterically demanding groups are displaced towards longer waves [84, 85]. Poly(organosilanes) with aromatic-bound radicals likewise experience a shift of the absorption spectrum towards its red end, a phenomenon known as bathchromy [85]. This bathochromic shift occurs in the solid state as well as in solution.

b) Conformation: the thermochromic shift λmax of different poly-(organosilanes) is said to be due to a change in conformation within the Si—Si chain [86–90]. For example, the UV spectra of poly(di-n-hexylsilane) 1 $(R1=R2=n\text{-}C_6H_{13})$ and higher homologues exhibit a maximum λmax at 370–380 nm below 40 °C, which changes to 320–325 nm above this temperature [91–92]. Figure 4.1 shows the temperature-dependent UV spectrum of poly(methyl-n-hexysilane) 1 $(R1=CH_3;\ R2=n\text{-}C_6H_{13})$ in hexane. At room

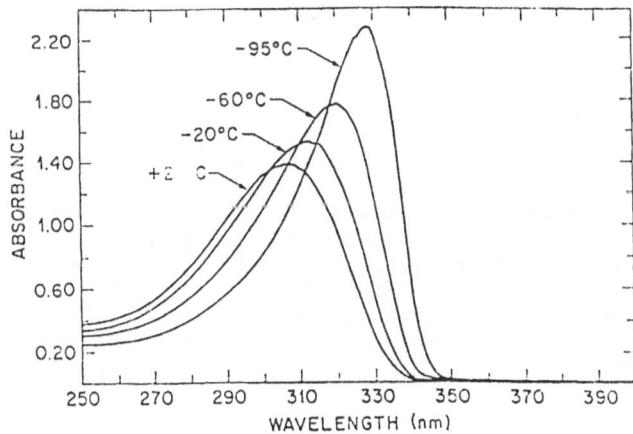

Fig. 4.1. UV absorption spectrum of poly(methyl-n-hexylsilane (0.004% solution in hexane). Reproduced from [93]

temperature λmax is 308 nm. When the solution is cooled, there is a continuous shift towards the red end of the spectrum and λmax reaches about 328 nm at $-95\,°C$.

An unusually low UV absorption maximum with λmax at 285 nm is shown by **1** (R1=$CH_2CH_2CF_3$; R2=CH_3) [133].

It has been suggested that the temperature-dependent displacement of λmax could be interpreted with a change of conformation from the preferred *trans*-configuration towards *gauche* in **2b** the energy difference between the two forms in **2b** (R1=R2=CH_3), calculated by the MNDO method, amounting to 1.4 kcal [94]. Sophisticated analysis shows further minima and maxima when the torsion angle is altered [94].

c) Molecular weight: As *n* in **2b** increases, λmax tends towards a maximum value (Fig. 4.2):

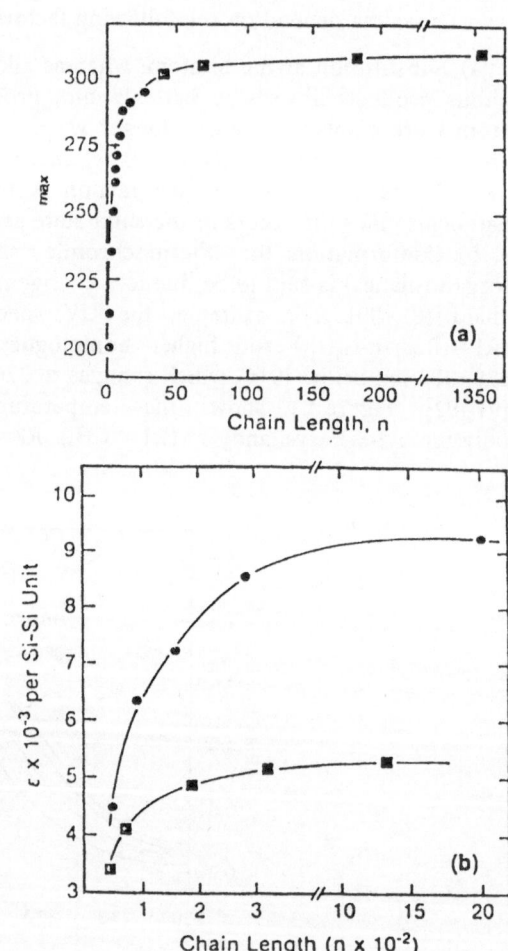

Fig. 4.2. a UV maxima, plotted against n in **2b** for poly(dialkylorganosilanes). **b** extinction coefficient ϵ per Si—Si unit as a function of chain length. Solid circles denote R1=R2=CH_3, solid squares R1=CH_3, R2=n-$C_{12}H_{25}$. Reproduced from [74]

4.1.4.3
Spectral Bleaching

In Sect. 1.3.5, it was shown, using linear and cyclic silanes as example, that by subjecting the material to irradiation by a specified amount of energy ηw can cause the elimination of SiR2 monomer units, this being linked with a change in molecular weight distribution [74]. This phenomenon is referred to as "spectral bleaching". Fig. 4.3 shows an example, for **2b** (R1=CH$_3$; R2=C$_{12}$H$_{24}$) [74].

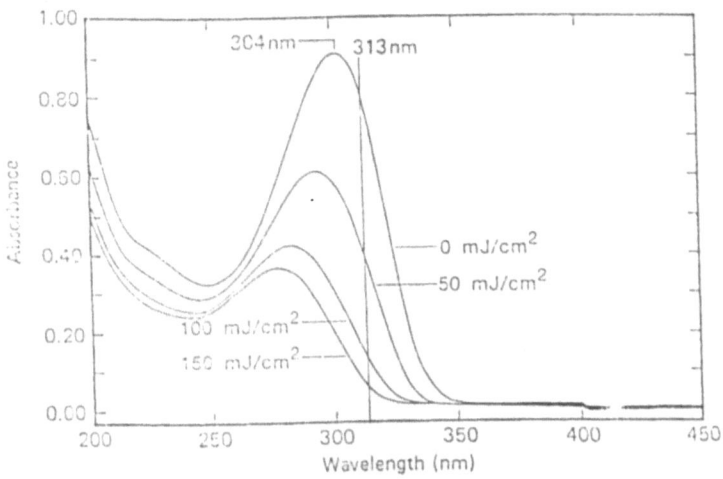

Fig. 4.3. Spectral bleaching through irradiation of **2b** (R1 = CH$_3$; R2 = C$_6$H$_{13}$; film) at 313 nm. Reproduced from [74]

The effect is observed not only in solution but also in films and this serves as the basis for applications of poly(organosilanes) in lithography [74]. For applications it is important whether they are carried out in the presence of atmospheric oxygen or in vacuum. Oxygen normally accelerates chain degradation [74], but other possible causes may be the effects of the substituent in **2b**, the initial molecular weight and the glass transition temperature T_g of the poly (organosilane). According to Ban and Sukegawa [100], the fragments of **2b** (R1=CH$_3$; R2=C$_6$H$_5$) and **2b** (R1=CH$_3$; R2=n-C$_3$H$_7$) are completely oxidized after irradiation at 254 nm in the presence of atmospheric oxygen, whilst this is only partly the case for irradiation at 330 nm.

The presence of additives such as, for example, 1,4-bis-(trichloromethyl) benzene accelerates the process of spectral bleaching in poly(organosilane) films [101]. The possible mechanism has been discussed by Miller [74].

By exchanging an alkyl radical in **2b** with organo-metallic ligands, e.g. **48a** it is possible to photochemically stabilise the poly(organosilane) skeleton [102]. Starting products such as **48c**, for example, can be obtained via salt elimination, followed by reductive coupling scheme (48):

$$(\eta^5\text{-}C_5H_5)Fe(CO)_2^- \; Na^+ \; + \; (CH_3)SiCl_3 \xrightarrow{\;-NaCl\;} (\eta^5\text{-}C_5H_5)Fe(CO)_2(CH_3)SiCl_2$$

a   b

$$+m \left(-\underset{\underset{C_6H_5}{|}}{\overset{\overset{CH_3}{|}}{Si}} - \right)$$

(Na, toluene)

(48)

$$[[(\eta^5\text{-}C_5H_5)Fe(CO)_2(CH_3)Si\,]_n[(C_6H_5)(CH_3)Si]_m]_x$$

c

Figure 4.4 shows the photolysis of two solutions, prepared from **2b** (R1=CH$_3$; R2=C$_6$H$_5$) and **48c** with a comparable molecular weight in tetrahydrofuran.

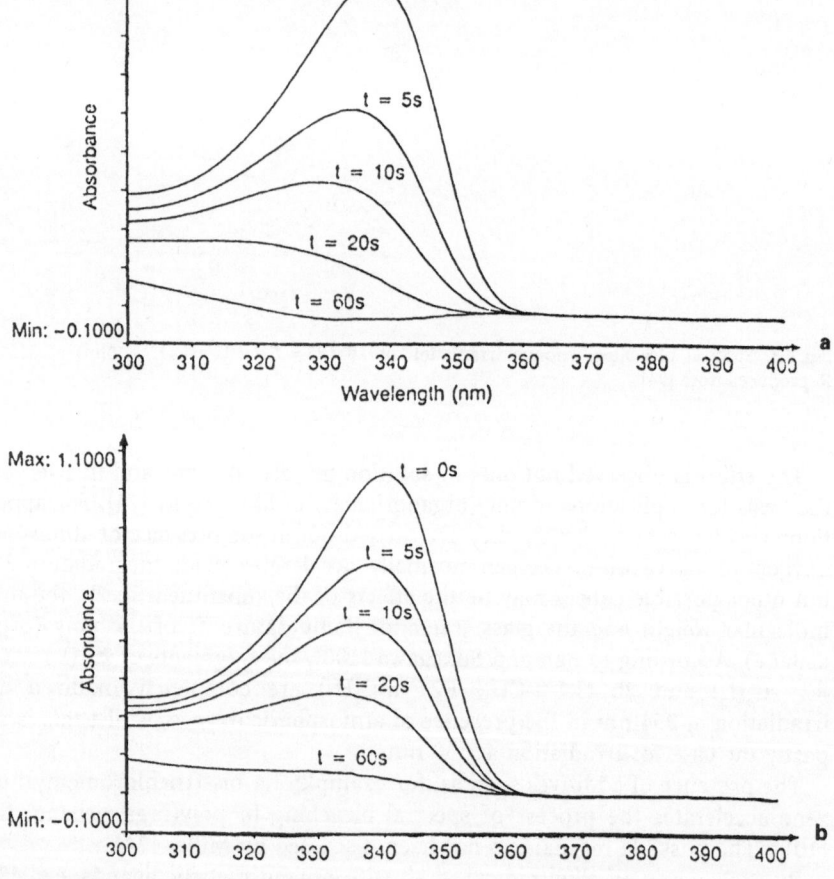

Fig. 4.4. Photolysis of **a** **2b** (R1=CH$_3$; R2=C$_6$H$_5$) and **b** **48c** in tetrahydrofuran at 270 nm; reproduced from [102]

4.1.4.4
Semi-Conductor Properties of Poly(organosilanes)

According to a suggestion by Matsumoto et al [95], poly(organosilanes) represent the segment of a chain within organic and inorganic silicon compounds such as disilanes, crystalline silicon, silicon-based polymer conductors, polysilane blends, silicon clusters and amorphous silicon, whose conductor characteristics can conveniently be described by the energy band model. Here, poly(organosilanes) take up a central position, with typical semi-conductor properties. The activation energy that must be expended in order to make electrons available from the valency band into the conductivity band, for the transport of electrical charges, amounts to about 4 eV for **2b** ($R1=R2=CH_3$; $n = 10$). Theoretical considerations indicate that the activation energy is influenced by substituents and the conformation of **2b** [95].

It can be demonstrated with the help of fluorescene analysis, that exciton migration in **2b** ($R1=CH_3$; $R2=n\text{-}C_3H_7$; measured in film form) is extremely high [96].

Characteristics of the electron transport mechanism in poly(methylphenyl-silane) **2b** ($R1=CH_3$; $R2=C_6H_5$) are qualitatively similar to those of poly(N-vinylcarbazole) [97].

It is hardly surprising that the conductivity of films based on **2b** ($R1=CH_3$; $R2=C_6H_5$) or **2c** ($R1=H$; $R2=CH_3$; $m = 0.35$; $R3=CH_3$; $R4=C_6H_5$; $n = 0.65$) can be increased by doping with ferric chloride or antimony pentafluoride [98].

4.1.4.5
Piezochromy of Poly(organosilanes)

Piezochromy (i.e. the dependence of the absorption maximum under pressure) was first observed in the case of **2b** ($R1=R2=n\text{-}C_6H_{13}$) [99]. The effect was confirmed on films made from **2b** ($R1=R2=n\text{-}C_5H_{11}$). The suggestion is that the change in wave length is due to a change in conformation of 7/3 helical to planar zig-zag. The latter conformation has a higher energy content and is not observed under normal conditons.

UV, FT-IR and Raman spectroscopy are used to ascertain whether piezochromy is present [99].

4.1.5
Applications of Poly(organosilanes)

The physical and chemical properties of poly(organosilanes) described in Sects. 4.1.3 and 4.1.4 make them predestined for a number of interesting applications. Attention is currently focusing on the thermal transposition of **2b** ($R1=R2=CH_3$) in **44a** and the subsequent conversion into β-silicon carbide fibers. This process is important from the industrial point of view.

4.1.5.1
Preparation of ß-Silicon Carbide from Poly(organosilanes)

The poly(organosilane) **44a** described in Sect.1.3.6 can be pyrolysed to form ß-silicon carbide, hydrogen and methane being liberated Eq. (49):

$$
\left[\begin{array}{c} CH_3 \\ | \\ -Si-CH_2- \\ | \\ C_6H_5 \end{array} \right] \xrightarrow[-H_2, -CH_4]{N_2 \text{ or Ar, } 1300\,°C} \quad \text{ß–SiC} \tag{49}
$$

Closer investigations have shown that the product formed is a β-SiC reinforced, amorphous silicon carbide matrix. The product is being marketed in Japan under the name of "Nicalon" and is used as a reinforcing fiber in ceramics [61]. Ichikawa et al. [116] have given a clear picture of the reaction Eq. (49). The precondition for a good ceramic yield is that no purely linear structures are present in **41**, but only crosslinked ones [118].

The intermediate stage of the poly(carbosilane) **44a** can be circumvented if the original polymer is substituted correspondingly. Thus, **2c** (R1=R2=R3=CH$_3$; R4= C$_6$H$_5$), when applied to substrates, irradiated by UV rays followed by pyrolysis, produces silicon carbide-based coatings [109]. The stage of a "polysilastyrene" **50a** will probably have been passed Eq. (50):

$$
\begin{array}{c}
\textbf{2c} \\
R^1=R^2=R^3=CH_3 \\
R^4=C_6H_5
\end{array}
\longrightarrow
\begin{array}{c}
 \\
(-\!Si-CH_2)_n- \\
| \\
CH_3 \\
\textbf{a}
\end{array}
\longrightarrow \quad \text{ß–SiC} \tag{50}
$$

4.1.5.2
Preparation of Photoresist Materials from Poly(organosilanes)

Spectral bleaching (lowering the molecular weight by UV irradiation), which has been described in Sect. 4.1.4.3, makes poly(organosilanes) interesting as photoresists.

Photoresists can continue to crosslink when exposed to light, and can thus become sparingly soluble (negative photoresists). Alternatively, light may cause them to break up into small pieces which can then be removed with solvent (positive photoresists). The latter include poly(organosilanes), provided no crosslinking agents are present.

Silicon surfaces can be selectively doped. For this, a mask made of SiO$_2$ or Si3N$_4$ must be applied to confine penetration of foreign atoms to the exact area to be doped. First of all, the entire silicon surface is oxidized. A photoresist coating of specified thickness is then applied on top of the oxide film. After exposure to light through a mask, and development to remove low molecular weight components from the poly(organosilane) and etching, to remove the SiO$_2$, the positive is obtained.

Because of their sensitivity to UV light, poly(organosilanes) can be used to increase information density of electronic circuits since this directly depends on the radiation wave length. At present, light in the visible range is mainly being used for this purpose ("high resolution photoresists" [110–112]).

4.1.5.3
Derivatives of 1 as Radical Photoinitiators

As mentioned in Sect.1.3.5, radical intermediate stages are produced during the photolysis of **2b**. These can initiate the polymerization of olefinic double bonds [113, 114].

One interesting property of poly(organosilanes) is that they are unaffected by oxygen. Because of their lower efficiency compared, for example, with benzoinmethyl ether, it may be that there is an intramolecular hydride transfer reaction Scheme (51):

$$\underset{CH_3}{\overset{C_6H_5}{\underset{|}{\overset{|}{\sim\!\!\sim Si}}}}\!\!-\!\!\underset{CH_3}{\overset{C_6H_5}{\underset{|}{\overset{|}{Si\sim\!\!\sim}}}} \longrightarrow \underset{CH_3}{\overset{C_6H_5}{\underset{|}{\overset{|}{\sim\!\!\sim Si^\bullet}}}} + \underset{CH_3}{\overset{C_6H_5}{\underset{|}{\overset{|}{\sim\!\!\sim Si^\bullet}}}} \longrightarrow \underset{CH_3}{\overset{C_6H_5}{\underset{|}{\overset{|}{\sim\!\!\sim Si}}}}\!\!-\!\!H + \underset{a}{\overset{C_6H_5}{\overset{|}{H_2C=Si\sim\!\!\sim}}} \qquad (51)$$

and that a "silaolefin" **51a** is formed Scheme (52):

$$\overset{C_6H_5}{\overset{|}{H_2C=Si\sim\!\!\sim}} \xrightarrow{O_2} \underset{O-O}{\overset{C_6H_5}{\underset{|}{\overset{|}{\sim\!\!\sim Si - CH_2}}}} \xrightarrow{-CH_2O} \overset{C_6H_5}{\underset{\gtrless}{\overset{|}{Si}}}=O \longrightarrow \underset{a}{\overset{C_6H_5}{\underset{\gtrless}{\overset{|}{Si}}}}-O- \qquad (52)$$

Reichmanis et al. [117] have given an overview of applications of poly (organosilanes) and other organo-silicon compounds in microlithography.

4.1.6
Toxicological Properties of Poly(organosilanes)

In contrast to type **52a** poly(organosiloxanes), the toxicological properties of poly(organosilanes) have not, so far, been described in the published literature. Whereas the former are considered to be physiologically compatible and non-toxic to man and the environment, this is not necessarily the case with **1**, because of its greater reactivity. The starting product of **1**, hexamethyldisilane, has been tested for its toxicological properties and found to be not acutely toxic when administered dermally and orally, although it does have a sensitising effect [115].

4.2
Poly(carbosilanes)

4.2.1
Preparation of Poly(carbosilanes)
with Alternating Silicon and Carbon Atoms

The thermolysis of **2b** to **44a** was described in Sects.4.1.3.6 and 4.1.5.1. The starting product of **44a** is poly(silaethylene) **53a** (R1=R2=R3=R4=H): Like the analogous poly(organosilanes), **53a** can contain linear, cyclic or polycyclic

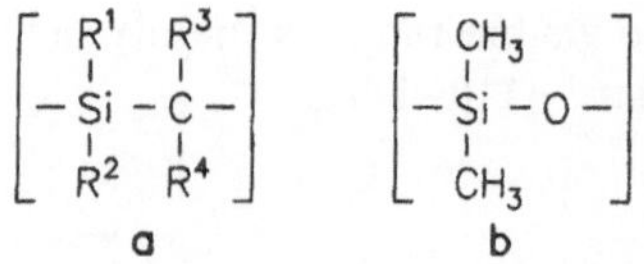

$$
\begin{bmatrix}
& R^1 & R^3 & \\
-& Si - C & -\\
& R^2 & R^4 &
\end{bmatrix}
\qquad
\begin{bmatrix}
& CH_3 & \\
-& Si - O & -\\
& CH_3 &
\end{bmatrix}
\tag{53}
$$

a   b

structures. Although poly(carbosilanes) **53a** (R1=R2=CH$_3$; R3=R4=H) are isoelectronic with the industrially important poly(organosiloxanes) **53b** (R1=R2=CH$_3$),

Surprisingly few investigations have so far been carried out on this group of compounds **53a** and little information is therefore available in the published literature.

Several methods of synthesizing **53a** are known and these are discussed below.

4.2.1.1
Thermolysis of Poly(organosilanes)

In the technical literature, the products obtained by thermolysis at 450–470 °C are said to have poly(carbosilane) structures Eq. (44), but more precise investigations of structure by IR and NMR spectroscopy have shown that the products obtained by thermolysis of **2b** possess a more complex structure than **44a**. Under drastic reaction conditions, crosslinkage and cyclization processes take place, the structures **54a** [120] and **54b** [119] being suggested (Scheme 54):

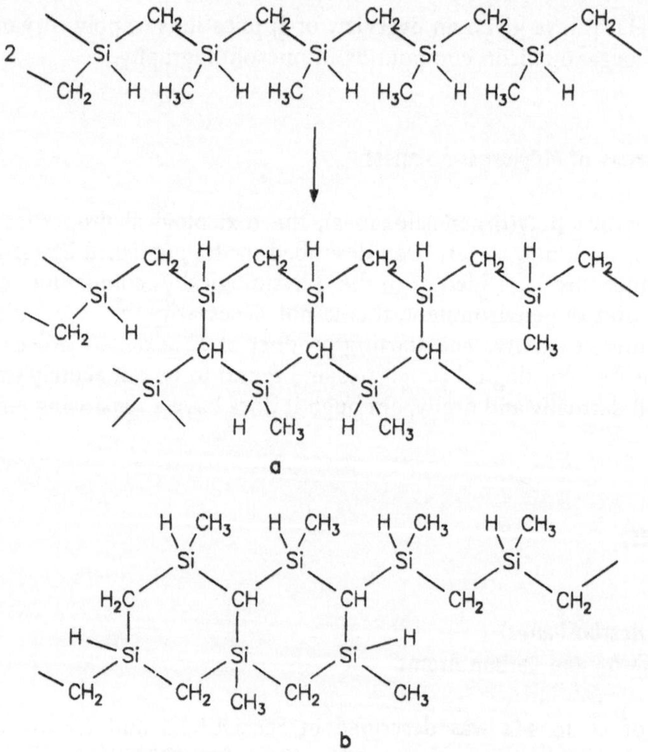

(54)

Ichikawa [121] has suggested a method of preparation for **53a** (R1=R2=CH$_3$; R3=R4=H) from **2b** (R1=R2=CH$_3$). After filtering off the insoluble residue and removing the low molecular substances, a solid with an MW of 2980 and an M$_n$ of 1470 remains behind.

4.2.1.2
Ring-Opening Polymerization of 1.3-Disilacyclobutane Derivatives

Low molecular weight poly(carbosilanes) can be obtained via organometallic, Grignard-type compounds and chlorosilane derivatives. According to Sommer et al., the simplest starting product 2,4-dimethyl-2,4-disilapentene (**55a** is prepared according to Eq. 55, [122]:

$$(CH_3)_3SiCH_2MgCl \ + \ (CH_3)_3SiCl \ \xrightarrow{-MgCl_2} \ (CH_3)_3SiCH_2Si(CH_3)_3 \qquad (55)$$
$$a$$

Besides magnesium, one can also use sodium or lithium as condensing agent [123], and it is also possible gradually synthesize higher homologues such as **56a** Eq. (56):

$$(CH_3)_3SiCl \ + \ ClCH_2Si(CH_3)_2OC_2H_5 \ \xrightarrow[-2NaCl]{+2Na} \ (CH_3)_3SiCH_2Si(CH_3)_2OC_2H_5 \ (56)$$
$$a$$

The gradual synthesis by the lithium method is suitable for preparing CH$_3$ ((CH$_3$)2SiCH$_2$)$_n$-H up to n=5 [124], but is unsuitable for preparing higher and high molecular weight polymers. It was Goodwin [125] who acheived the breakthrough by reacting (halomethyl) halosilanes **57a** with sodium and, after removing the volatile fraction up to 250 °C, obtained a polymer with an average molecular weight of 850 and a viscosity of 300–400 centistokes Eq. (57):

$$n \ XCH_2SiRR'-X \ \xrightarrow[-2NaCl]{+2Na} \ (RR'SiCH_2)_n \qquad (57)$$
$$a$$

The assumption that the reaction according to Eq. (57) takes place via the 1,3-disilacyclobutane derivative **58** is an obvious one and it is hardly surprising that this group of products has been investigated with special reference to its possible use for preparing linear poly(carbosilanes) **53a** [118].

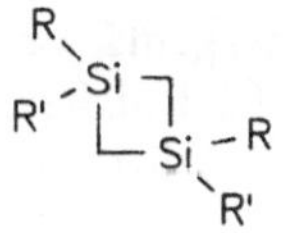

$$(58)$$

4.2.1.2.1 *Preparation of Raw Materials, Preparation of 1,3 disilacyclobutane derivatives from (halomethyl)halosilanes.* When (halomethyl) halosilanes, e.g. **57a** (X=Cl; R=R'=CH$_3$), are reacted with magnesium under certain conditons, the main products are derivatives of **58** (R=R'=CH$_3$). Cyclic Trimers **59a**, tetramers **59b** and the polymeric compounds **53a** (R1=R2=CH$_3$; R3=R4=H) [126] can also be found (Scheme 59):

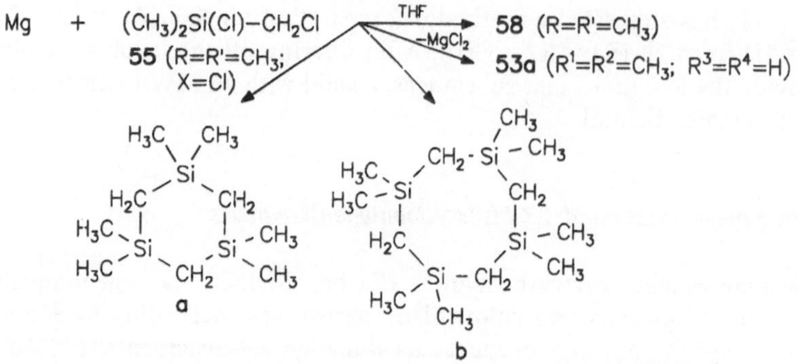

$$
Mg + (CH_3)_2Si(Cl)-CH_2Cl \xrightarrow[-MgCl_2]{THF} \begin{array}{l} 58 \ (R=R'=CH_3) \\ 53a \ (R^1=R^2=CH_3; \ R^3=R^4=H) \end{array}
$$

55 (R=R'=CH₃; X=Cl)

(59)

A detailed reaction scheme, with possible intermediate stages, has been given by Kriner [126].

The molar ratio of the reaction products 53a (R1=R2=CH₃; R3=R4=H) 58 (R=R'=CH₃)/59a/59b, greatly depends on experimental conditions. The end result is definitely affected by the type of solvent used, the halosilane and the magnesium metal. More detailed information on the subject is given in Table 4.12 [126], which shows that the best yields of 58 are obtained if the chlorosilane 57a is used the magnesium being added portionwise.

Table 4.12. Yield of 53a (R¹=R²=CH₃, R³=R⁴=H) 58 (R=R'= CH₃) 59a and 59b in relation to the procedure. Data adapted from [126]

	Silane added to Mg in Et₂O	Silane added to Mg in THF	Mg added to the Scheme in THF
53a	–	58	13.3
58	10	6.7	50
59a	–	1.4	18.6
59b	–	–	1.3

It is obvious that the addition of magnesium to the chlorosilane gives the best yields of 58 ("inverse addition"). Derivatives of 58 with functional groups such as chlorine are open to further reactions, e.g. with Grignard reagents, or to reduction with lithium aluminium hydride, so that a wide range of 1.3-disilacyclobutane derivatives 56 can be synthesized in this way. Table 4.13 lists some examples [126]:

Preparation of 1.3 disilacyclobutane derivatives from monosilacyclobutane derivatives by thermolysis. 1,3-disilacyclobutane derivatives (58 or 60b) are formed through pyrolysis at high temperatures in an enclosed system, from derivatives of monosilacyclobutane 60a Eq. (60):

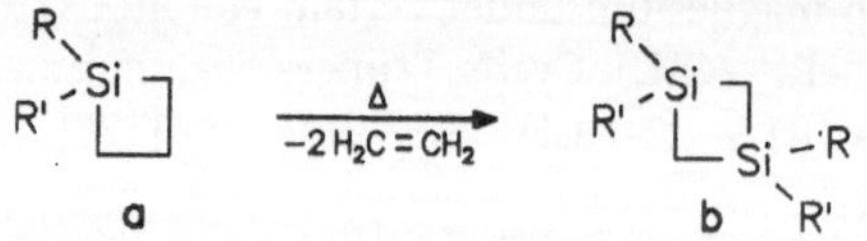

(60)

Table 4.13. Syntheses of various disilacyclobutanes (**58**) either from **57a** on by modification of **58**. Data adapted from [126]

Substituents of 58	Synthetic method	Yield (%)	b.p. (°C mm) m.p (°C)
R=R'=CH$_3$	from 57a by inverse addition	50	60 (100) −9 bis −8.5
R=Cl, R'=CH$_3$	from 57a by inverse addition	13.5	59 (20) −7 bis −6.6
R=OC$_2$H$_5$, R'= CH$_3$	from 57a by inverse addition	35.3	78–79 (16)
R=H; R'=CH$_3$	from 58 (R=Cl; R' =CH$_3$) by reduction with LiAlH$_4$	95.5	96 (760)
R=C$_6$H$_5$; R' =CH$_3$	from 58 (R=Cl; R' =CH$_3$ by reaction with C$_6$H$_5$MgCl	93.1 (crude)	99 (1.1)

In the case of **60a** (R=R'=Cl), Interrante has described an experimental set-up [127] which is based on work done by Nametkin et al. [128]. After pyrolysis at 840 °C and sublimation, 30% of a colorless solid are obtained whose structure **60a** (R=R'=Cl) is confirmed by spectroscopic data.

If the reaction is carried out in the liquid as well as in the gas phase, polymerization is invariably observed [129–131].

From the mechanistic point of view the reaction **60a**→ **60b** is interesting, insofar as it can be assumed that the reaction proceeds via silaethene **61**

$$R^1R^2Si{=}CR^3R^4 \tag{61}$$

Sommer et al. have shown [132] that this is indeed the case. The intermediate silaethanes **61** not only lead to **56** but can be shown to be adducts in the presence of various intercepting agents such as benzophenone.

Ring-Opening Polymerization of Monosila and 1.3-Disilacyclobutane Derivatives, using transition metal catalysts. Because of the stress conditions in **60b** and **60a**, it is possible to polymerize these ring systems with various compounds, e.g. phenyl lithium [134], as well as by heating [135]. Polymerization with small amounts of platinum (about 10–100 ppm) on activated charcoal as carrier material, or chloroplatinic acid, without closely identifying the polymers produced, was first described by Weyenberg and Nelson [136].

More precise experimental details have been given by Levin and Carmichael [137]. The polymerization of **58** (R=R'=CH$_3$) with H$_2$PtCl$_6$ × 6H$_2$O is carried out under various conditions (varying amounts of platinum and water, open/closed system) and essentially produces two polymers with an M$_n$ of 1.2×10^5 and 1.7×10^5 respectively. An ionic mechanism has been suggested for the formation of **53a** (R1=R2=CH$_3$; R3=R4=H) [138]. The characterization of the polymers obtained by GPC and thermogravimetric methods is described in some detail. The authors state that the poly(carbosilane) formed under the reaction conditions is more resistant to high temperatures than polydimethylsiloxanes **53b** and polyisobutylene.

The polymerization rate is influenced by the type of platinum catalyst used [139]. Thus, di-μ-chloro-dichloro(bis-cyclohexane)-diplatinum (II) $(C_6H_{10})2PtC_{14}$ has proved to be the most reactive and, in the case of **53a** $(R=R'=CH_3)$ initiates ring opening even at room temperature, molecular weights of up to 500 000 (ebullioscopic in toluene) are achieved. Triethylsilane $(C_2H_5)_3$ Si-H can be used to control the molecular weight Eq. (62). Here it is possible to obtain telomers with molecular weights of between 540 and 4800 [139].

$$\textbf{58 (60a)} \quad \xrightarrow{(C_2H_5)_3SiH} \quad (C_2H_5)_3Si(CH_2Si(CH_3)_2)_{2n}-H \tag{62}$$
$$(R=R'=CH_3) \qquad\qquad\qquad a$$

The parent substance of **53a** (R1=R2=R3=R4=H) has been prepared by L. Interrante as follows, giving a total yield of 25–52%[127,140] (Scheme 63)

$$\textbf{58 (60a)} \xrightarrow[\text{quarz tube}]{840\,°C} \textbf{58 (60a)} \xrightarrow[C_6H_6]{H_2PtCl_6} \textbf{53a}$$
$$(R=R'=Cl) \qquad\qquad (R=R'=Cl) \qquad\qquad (R^1=R^2=R^3=R^4=Cl)$$

$$\xrightarrow{\text{LiAlH}_4} \tag{63}$$

$$\textbf{53a} \nwarrow$$
$$(R^1=R^2=R^3=R^4=H)$$

The polymer, a clear, viscous liquid, is characterized by ^{13}C and ^{29}Si nuclear resonance spectroscopy and IR analysis, as well as a monomodal molecular weight distribution with $M_n = 12300$ and $M_W = 33000$ (GPC with polystyrene as internal standard). The substance has good heat resistance, up to 500 °C, without appreciable loss of weight. The ceramic yield after pyrolysis at 1000 °C is likewise good.

Thorough investigations of **53a** (R1=R2=CH_3; R3=R4=H) to characterize the material from the physico-chemical standpoint, have been carried out by Mark et al. [141].

Compounds other than platinum ones can also be used as catalysts for ring opening polymerization. Those suggested include $IRCl_6^{--}$, $RuCl_4^{--}$, $AuCl_4^{-}$, $PdCl_2$, RuI_3, $PdBr_2$, $AuCl_3$, $CuCl_2$, $CuCl$, π-crotyl nickel and chromium compounds [118].

4.2.1.2.2 *Ring-Opening Polymerization of Monosila and 1,3-Disilacyclobutane Derivatives by Heating.* Besides ring opening polymerization using heavy metal catalysts, the polymerization of **58** and **60a** at high temperatures is also worth mentioning. Since, in the case of **60a** the reaction is uneven in the gas phase, this method is not of great practical interest. In addition to the desired polymers, the 1,3-disilacyclobutane derivatives are also formed Eq. (60) [129, 130]. A silane (61) is possibly formed as reactive intermediate.

If the reaction is carried out in the liquid phase, the formation of polymeric material is evidently favored. The polymerization of **58** $(R=R'=CH_3)$, which proceeds smoothly to **53a** (R1=R2=CH_3; R3=R4=H) at 300 °C in the presence of an inert gas, has been studied and described [142].

The thermal polymerization of **58** (R=R′=C₆H₅) takes place at temperatures as low as 180–200 °C and produces a white, crystalline powder which is soluble in benzene and other normal solvents [143].

4.2.1.2.3 *Ring-Opening Polymerization of Monosila- and Disilacyclobutane Derivatives by Alkalis.* Russian authors have stated that alkalis such as potassium hydroxide, alkali metal silanols and alkyl lithium compounds are suitable for the ring opening polymerization of **58** and **60a** [143–145], the latter only being able to polymerize type **60a** compounds, whereas **58** does not react [144–146] Eq. (64). Here, according to the authors, there are steric reasons for the differences in the mode of reaction:

$$\textbf{60a} \xrightarrow{\text{R″Li}} \text{R″(RR′SiCH}_2\text{CH}_2\text{CH}_2\text{)}_n\text{Li} \qquad (64)$$

It is possible to polymerize the structure **65a** only via the monosilacyclobutane ring to form **65b** Eq (65) [144]:

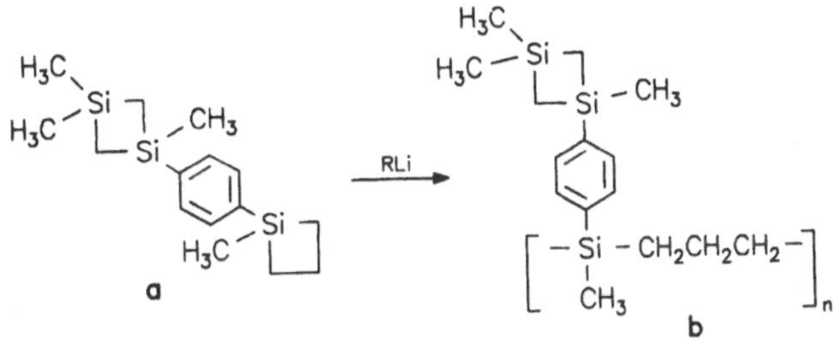

(65)

4.2.1.3
Preparation of Poly(carbosilanes) by Reacting Silicon Metal with Dichloro or Trichloromethane in the Presence of a Copper Catalyst

Patnode and Schiessler pointed out for the first time that, if the Rochow synthesis is carried out with methane derivatives having higher chlorine contents than monochloromethane, e.g. di or trichloromethane, the reaction will produce oligomeric and polymeric, chlorinated poly(carbosilanes) [147]. Fritz [148] clarified the very complex structures after the reduction of all chlorine-carbon linkages. He also carried out the reaction in a fludized bed reactor. Scheme 66 lists various structures (**66a–e**) which are obtained during the reaction of methylene chloride with copper-doped silicon metal. Here we are dealing mainly with derivatives of 1,3,5-trisilacyclohexane (**66a–d**) and open-chain poly(carbosilanes) **66e**.

Reaction with trichloromethane under the same conditions likewise results in the formation of derivatives of 1,3,5-trisilacyclohexane **67a** as well as branched (**67b–f**) and unbranched, open-chain poly(carbosilanes) (**67g**) after reduction of

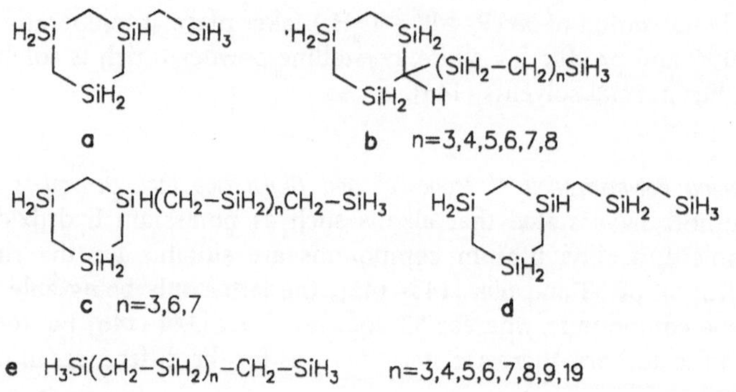

$$H_2Si\underset{SiH_2}{\overset{SiH\quad SiH_3}{\diagup\!\!\diagdown}}$$

$$\cdot H_2Si\overset{SiH_2}{\diagup\!\!\diagdown}\underset{SiH_2\quad H}{\diagdown}(SiH_2-CH_2)_nSiH_3$$

a

b   n=3,4,5,6,7,8

$$H_2Si\underset{SiH_2}{\overset{SiH(CH_2-SiH_2)_nCH_2-SiH_3}{\diagup\!\!\diagdown}}$$

$$H_2Si\underset{SiH_2}{\overset{SiH\quad SiH_2\quad SiH_3}{\diagup\!\!\diagdown}}$$

c   n=3,6,7

d

(66)

e $H_3Si(CH_2-SiH_2)_n-CH_2-SiH_3$ n=3,4,5,6,7,8,9,19

the chlorine-carbon bond [148] (Scheme 67)

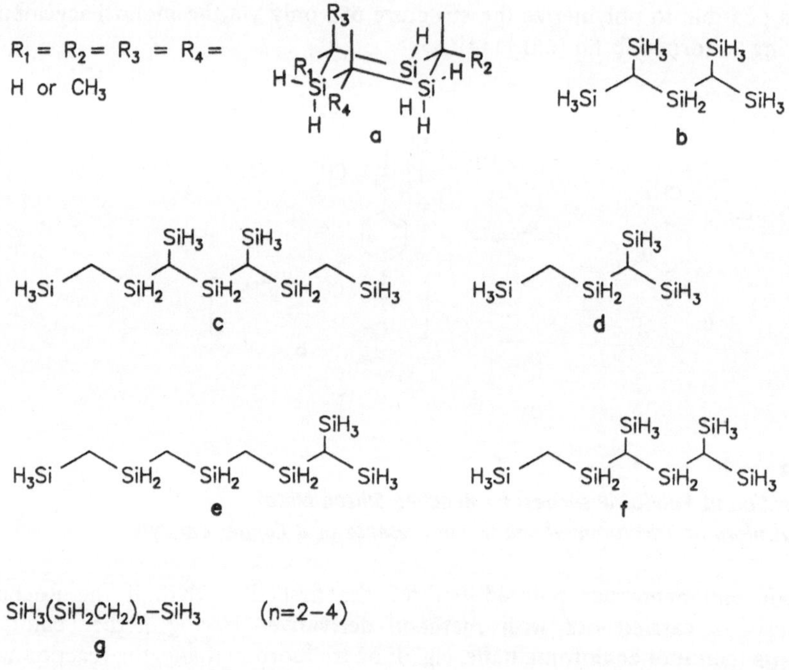

$R_1 = R_2 = R_3 = R_4 =$
H or CH$_3$

a

b

c

d

(67)

e

f

$SiH_3(SiH_2CH_2)_n-SiH_3$ (n=2–4)

g

4.2.1.4
Preparation of Poly(carbosilanes) by Pyrolysis of Tetramethylsilane

The pyrolysis of tetramethylsilane has been studied in great detail by Fritz [149, 150]. The reaction products formed are exclusively cyclic (**68a–e**) and polycyclic, saturated and unsaturated compounds, adamantane structures being in the foreground. Scheme 68 lists a few of the many different poly(carbosilanes) which are obtained by the pyrolysis of tetramethylsilane. All the structures are confirmed by modern chromatographic analysis and spectroscopic techniques [149, 150].

Pyrolysis has a certain economic interest since, at about $700\,^{\circ}C$, poly(carbosilanes) are formed which are soluble in aromatic and chlorinated hydrocarbons, from which ceramic SiC fibres can be drawn out [151]. The process can be perfected to a degree where low-molecular weight components and the unreacted starting compounds can be circulated, which greatly increases the total yield [151]. Pyrolysis of tetramethylsilane according to Fritz [149, 150]. Some typical structures are shown in Scheme (68).

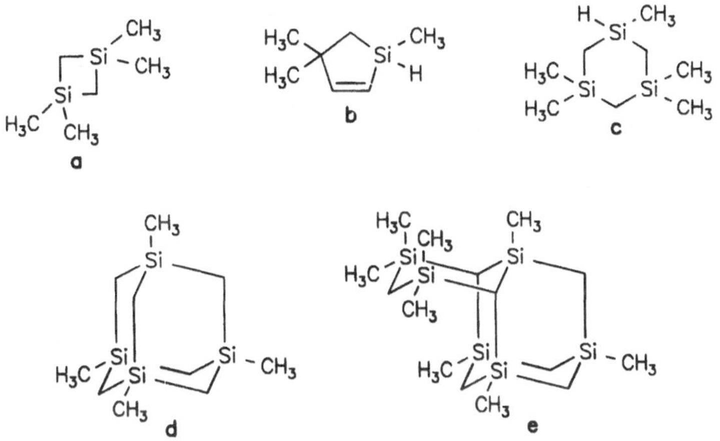

$$(68)$$

4.2.1.5
Preparation of poly(carbosilanes) by the Wurtz synthesis from dihalomethane and dihalosilane derivatives

The reaction discussed in Sect. 4.1.2.1, involving the reductive coupling of dihalosilanes similar to the Wurtz synthesis, can be expanded in the presence of dihalomethane derivatives, resulting in the formation of poly(carbosilane) structures (Eq. 69) [152]:

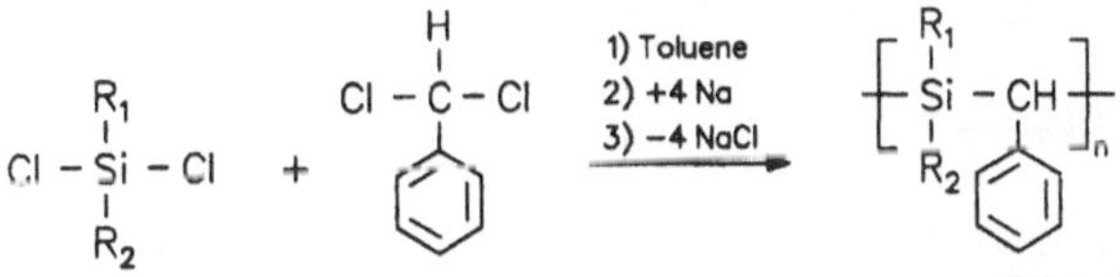

$$(69)$$

53a (R^3=H; R^4=C$_6$H$_5$)

As described in Sect. 4.1.2.1, the reaction takes place in toluene, the polymer being obtained through precipitation with ethanol or isopropanol. Analysis of nuclear resonance spectra shows that the silicon-carbon polymers produced do not consist of strictly alternating silicon and carbon atoms, but that silicon and carbon are statistically distributed along the chain [152]. The authors described a number of derivatives and the yields, the numerical average M_n and the ratio M_w/M_n. no further characterization of the polymers is given (Table 4.14).

Table 4.14. Syntheses of poly (carbosilane) **53a** by polycondensation of dichlorosilanes with phenyl-dichloromethane (molar ratio 1:1). Data adapted from [152]

Dichlorosilane	Reaction product **53a**	Yield (%)	$M_n \times 10^3$	M_w/M_n
$R_1=R_2=CH_3; R_3=Cl$	$R_1=R_2=CH_3; R_3=H;$ $R_4=C_6H_5$	51	1.6	3.0
$R_1=H; R_2=CH_3;$ $R_3=Cl$	$R_1=CH_3; R_2=R_3=H;$ $R_4=C_6H_5$	54	3.0	1.8
$R_1=H; R_2=CH_3;$ $R_3=Cl^*$	$R_1=CH_3; R_2=R_3=H;$ $R_4=C_6H_5$	15	3.1	2.3
$\left.\begin{array}{l}R_1=R_2=CH_3; R_3=Cl\\R_1=H; R_2=CH_3; R_3=Cl\end{array}\right]^+$	$R_1=CH_3; R_2=CH_3 \text{ (H)};$ $R_3=H; R_4=C_6H_5$	25	1.8	4.0

$^+$Molar ratio of the two.
*Molar Silanes 1:1 ratio of Silan/phenyldi-chlormethane 2:1.

P. Sartori et al. have described the direct synthesis of poly(diphenylcarbosilane) from methylene bromide and dichlorodiphenylsilane **2d** ($R1=R2=C_6H_5$; $R3=Cl$) in accordance with Eq. 70 [53].

$$n(C_6H_5)_2SiCl_2 \quad + \quad n\,CH_2Br_2 \xrightarrow[-2n\,NaCl,\,-2n\,NaBr]{+4n\,Na} \textbf{53a} \;\; (R^1,R^2=C_6H_5; R^3,R^4=H) \qquad (70)$$

Yields, elementary analyses, IR, ^{1}H-NMR, ^{13}C-NMR- and ^{29}Si-NMR spectroscopic data are discussed, and used to interpret the chemical structure. It is concluded that the indicated structure **53a** with alternating silicon and carbon units is strongly idealized. Detailed analysis indicates that, in addition, there are C_n- and Si_n- blocks, SiC_3-, SiC_4-, SiC_3C_n- and $SiC_2(C_n)_2$- structures as well as SiC_3CH_2Br-terminal groups.

The same author also examined copolymeric poly(carbosilanes) for their silicon content [154]. The reaction according to Eq. (71)

$$n(C_6H_5)_2SiCl_2 \quad + \quad x \cdot m(CH_3)_2SiCl_2 \quad + \quad x \cdot (n+m)\,CH_2Br_2 \xrightarrow{Na}$$

$$(((C_6H_5)_2Si)_n((CH_3)_2Si)_m(CH_2)_{n+m})_x \qquad\qquad (71)$$
$$\textbf{a}$$

with n/m = 9;4;2;1;0,5;0,25;1/9;0/1

produces poly(diphenyl-*co*-dimethyl) carbosilane **71a** whose molecular structure is described as mainly linear with some cyclic and macrocyclic components. As the methyl content increases, there will be branching and crosslinkage caused by reactions of the methyl group and resulting in decreasing solubility in organic solvents such as xylene and carbon tetrachloride. Since the reaction Eq. (71) is carried out in the presence of xylene, silylation is observed on the aromatic ring system, even though this is of minor importance. The composition is further more governed by the ratio of the two starting silanes. Phenyl-rich structures are marked by their high content of alternating carbosilane elements. Increasing the methyl content favours the formation of block polymers [154].

According to Sartori et al. poly(dimethyl-*co*-styrene) **72a** can be prepared by direct reaction of **2c** (R1=R2=CH_3; R3=Cl) with styrene in the presence of sodium metal in xylene [155] Eq. (72).

$$x \cdot n\,(CH_3)_2SiCl_2 \quad + \quad x\,C_8H_8 \quad \xrightarrow[-\text{NaCl}]{+\text{Na/Xylol}} \quad ((Si(CH_3)_2)_nC_8H_8)_x \tag{72}$$
$$\text{a}$$

Apart from the above mentioned spectroscopic data, the mean molecular weights, melting ranges and densities were determined to characterize the substances. Aromatic silylations in the styrene molecule were detected among other things.

4.2.1.6
Preparation of Poly (carbosilanes) by the Thermal Decomposition of Grignard Compounds

Interrante describes a highly branched poly (carbosilane) prepared by the thermal decomposition, in diethyl ether, of the Grignard compound formed from (chloromethyl) trichlorosilane and magnesium [156]. The poly (dichloro-carbosilane) which can be isolated and has the structure **53a** (R1=R2=Cl; R3=R4=H) may be shown, by ^{1}H, ^{13}C and ^{29}Si-spectroscopic analysis, to be branched and to have the following structural components: $SiCl_3$-, $-SiCl_2CH_2$, $=$ and $SiClCH_2 \equiv Si-CH_2^-$. Elementary analysis indicates the incorporation of ethyl groups through co-reacting solvent. According to the authors, the formation of poly(dichlorocarbosilane) **53a** (R1=R2=Cl; R3=R4=H) takes place according to the following reaction, magnesium chloride being successively separated (Scheme 73).

$$Cl_3SiCH_2Cl \xrightarrow{Mg/(C_2H_5)_2O} Cl_3SiCH_2MgCl \xrightarrow{-MgCl_2} Cl_3SiCH_2SiCl_2CH_2Cl$$

$$\xrightarrow[-n\,MgCl_2]{+n\,Cl_3SiCH_2MgCl}$$

$$(SiCl_2CH_2)_n$$
$$\textbf{53a}\ (R_1=R_2=Cl;\ R_3-R_4=H) \tag{73}$$

$$\downarrow LiAlH_4/(C_2H_5)_2O$$

$$(SiH_2CH_2)_n$$
$$\textbf{53a}\ (R_1=R_2=\acute{R}_3=R_4=H)$$

The dichloro derivative **53a** (R1=R2=R3=R4=H), which is reduced with lithium alanate, exhibits a broad molecular weight distribution, between 300 and

50 000 amu with a maximum between 300 and 3000 atomic weight units. The polydispersity of about 7 likewise indicates a broad molecular weight distribution [156].

4.2.2
Reactions of Poly(carbosilanes)

Like the chemical characterization of the poly(carbosilanes), which was described in Section 4.2.1, little work has been done on the chemical reactivity of this group of compounds. This is probably due to the fact that poly(carbosilanes) are essentially of industrial interest and greatest attention is paid to ceramic yields during the thermal decomposition of form silicon carbide. A number of reactions of **53a** (R1=R2=R3=R4=alkyl and/or aryl and/or hydrogen) have become known.

4.2.2.1
Metallization of Poly(Carbosilanes)

Seyferth and Lang investigated the metallization of **53a** (R1=R2=CH3; R3=R4=H) in rather more detail. The metallized products **74b** are prepared in situ only and immediately reacted with monochloroalkyl or monochloroalkenylsilanes forming substituted silyl derivatives at the carbon atom of the methylene group **74c**, Scheme (74) [157]:

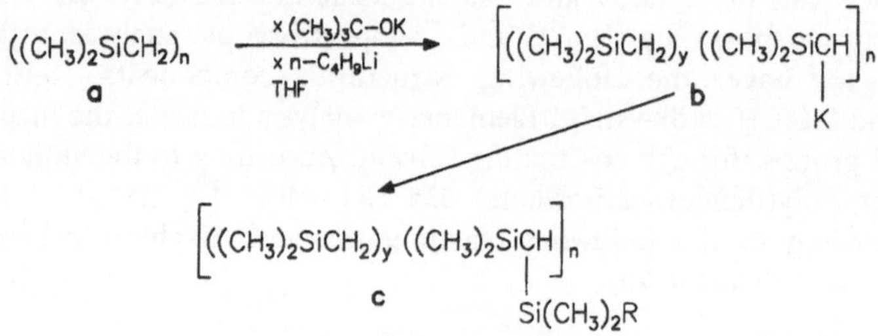

$((CH_3)_2SiCH_2)_n$
a
x $(CH_3)_3C-OK$
x $n-C_4H_9Li$
THF
$[((CH_3)_2SiCH_2)_y((CH_3)_2SiCH]_n$
b
K
$[((CH_3)_2SiCH_2)_y((CH_3)_2SiCH]_n$
c
$Si(CH_3)_2R$

(74)

The degree of metallization in **74b** is determined by the alkali/poly(carbosilane) ratio. At a ratio of 1, about every fourth carbon atom is metallized ($y=3$), as has been shown by ^{1}NMR spectroscopic tests. If the amount of alkali is increased, there will be chain cleavage in the starting material [157].

The silyl derivatives **74c** prepared by metallization can be subjected to a further crosslinking reaction, e.g. with azoisobutyronitrile and 1,1,3,3,5,5-trimethyltrivinyl cyclotrisilazane **75**, if R=vinyl or R=H in **74c**

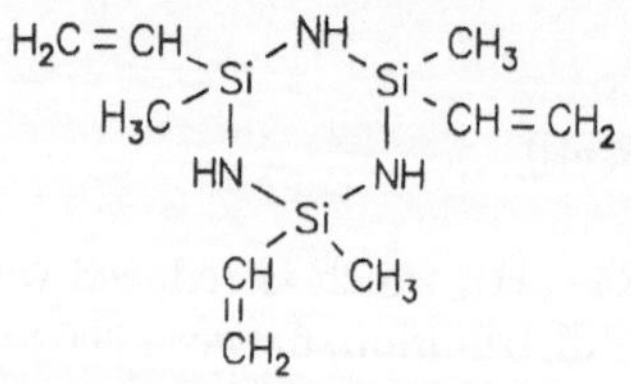

$H_2C=CH$
NH
CH_3
Si
Si
H_3C
$CH=CH_2$
HN
NH
Si
CH
CH_3
CH_2

(75)

The polymers obtained in this manner are characterized with regard to their ceramic yield [157].

4.2.2.2
Reaction of Poly(Carbosilanes) with Chlorotrimethylsilane/Aluminium Chloride

By reacting **53a** (R1=R2=CH_3; R3=R4=H) with chlorotrimethylsilane **2d** (R1=R2=R3=CH_3) in the presence of aluminium chloride as catalyst and successive reduction with lithiumaluminumhydride, Dunogues and Pillot have introduced silicon-chlorine and silicon-hydrogen bonds which are suitable for further reactions (Scheme 76) [158]. Their aim was to convert linear poly-(carbosilanes) **53a** into highly crosslinked products with high ceramic yields.

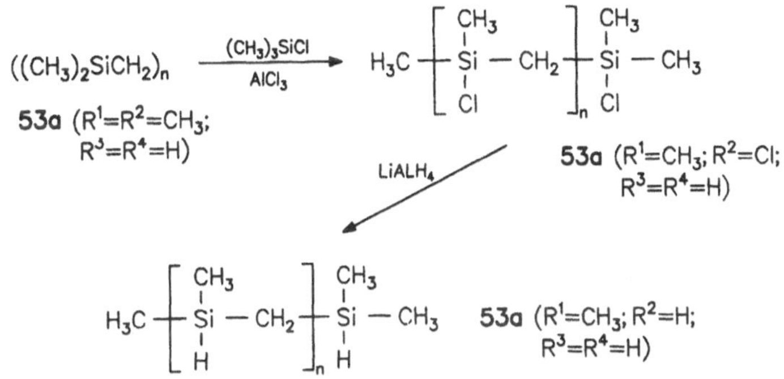

$$((CH_3)_2SiCH_2)_n \xrightarrow[AlCl_3]{(CH_3)_3SiCl} H_3C {-}\left[\begin{matrix} CH_3 \\ | \\ Si - CH_2 \\ | \\ Cl \end{matrix}\right]_n\!\!\!{-}\begin{matrix} CH_3 \\ | \\ Si - CH_3 \\ | \\ Cl \end{matrix}$$

53a (R^1=R^2=CH_3; R^3=R^4=H)

53a (R^1=CH_3; R^2=Cl; R^3=R^4=H)   (76)

LiALH$_4$

$$H_3C {-}\left[\begin{matrix} CH_3 \\ | \\ Si - CH_2 \\ | \\ H \end{matrix}\right]_n\!\!\!{-}\begin{matrix} CH_3 \\ | \\ Si - CH_3 \\ | \\ H \end{matrix}$$ **53a** (R^1=CH_3; R^2=H; R^3=R^4=H)

If, in reaction **53a** (R1=R2=CH_3; R3=R4=H) with an initial molecular weight (M_n) of 250000 is used, chain degradation will occur (in the case of **53a** (R1=CH_3; R2=H; R3=R4=H), M_n will be 2100 to 2600.

Derivatization of the chlorinated products can be achieved with sodium/toluene, potassium/tetrahydrofuran, methylamine/methylene chloride and ammonia/methylene chloride [158]. Crosslinked products are formed, which are soluble to insoluble in standard solvents and are characterized by spectroscopic data as well as molecular weight determination by gel permeation chromatography.

The same applies to hydrogen-containing derivatives, which can be branched by the addition of 1.3-butadiene or divinyl benzene, using Speier's catalyst (see also Eqs. (6) and (7) Table 4.15).

Condensation with potassium or sodium results in products with cyclic structural components which are formed through intermolecular and intramolecular reactions. Reaction with potassium appears to be more effective, since higher molecular weight polymers are formed. This is also evident in that a higher ceramic yield is obtained, which is consistent with the theory advocated by Schilling [159] that higher crosslinked poly(carbosilanes) give better ceramic yields.

An interesting variant of the halogenization reaction with chlorosilanes has been worked out by Sartori, using hydrogen chloride and hydrogen bromide. The poly(diphenylcarbosilane) discussed in Sect 4.2.1.5 Eq. (70) is completely halogenated by these two compounds in the presence of aluminium chloride as catalyst, benzene being liberated (Eq.(77) [160]

Table 4.15. Reactions of functionalized derivatives of **53a**. (data adapted from [158])

Starting Material **53a**, $R^1=CH_3, R^2=Cl$ $R^3=R^4=H$	Reaction product M_n	Polydispersity	%, 950 °C
Conversion with			
Sodium/Toluene	4250 + unsol. fractions	5.7	11.4
Potassium/Tetrahydrofuran	4930 + unsol. fractions	6.1	43.5
$(CH_3)_2NH/CH_2Cl_2$	2480	1.7	33.5
NH_3/CH_2Cl_2	4120	8.9	37.5
H_2O	2960	2.5	54.3
$(R^1=CH_3; R^2=H;$ $(R^3=R^4=H)$			
1.3 Butadiene	insoluble	insoluble	14
Divinylbenzene	insoluble	insoluble	24.7

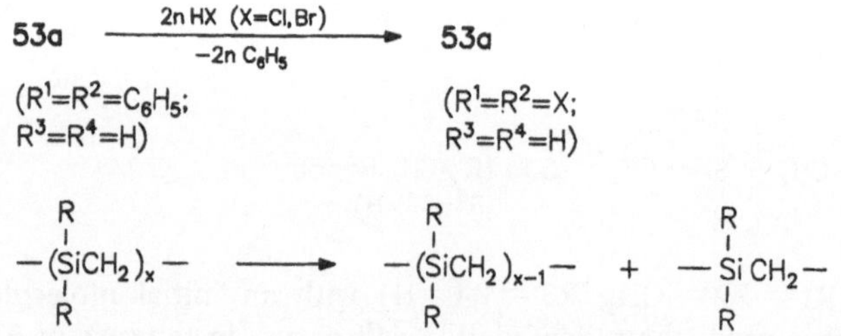

$$\textbf{53a} \quad \xrightarrow[-2n\,C_6H_5]{2n\,HX\ (X=Cl,Br)} \quad \textbf{53a}$$

$$(R^1=R^2=C_6H_5; \qquad\qquad (R^1=R^2=X;$$
$$R^3=R^4=H) \qquad\qquad R^3=R^4=H)$$

(77)

$$-\overset{\overset{\displaystyle R}{|}}{\underset{\underset{\displaystyle R}{|}}{(SiCH_2)}}_x- \quad\longrightarrow\quad -\overset{\overset{\displaystyle R}{|}}{\underset{\underset{\displaystyle R}{|}}{(SiCH_2)}}_{x-1}- \quad + \quad -\overset{\overset{\displaystyle R}{|}}{\underset{\underset{\displaystyle R}{|}}{Si}}\,CH_2-$$

53a (R1=R2=Cl; R3=R4=H) can be conveniently fluorinated by exchanging the chlorine atoms against fluorine ones, using lithium fluoride [168].

4.2.2.3
Thermal Decomposition of Poly(carbosilanes)

It has already been pointed out that poly(carbosilanes) can be used as starting materials for ceramic fibers (silicon carbide). Here, the degree of crosslinkage of the starting polymer before pyrolysis is of decisive importance. According to Seyferth, purely linear poly(carbosilanes) leave no residue after pyrolysis (no ceramic yield), but low molecular weight, volatile cyclic compounds. This is said to be due to a radical chain fission mechanism [118] (Scheme 78):

4.2.2.4
Further Reactions of Poly(carbosilanes)

53a (R1=R2=CH₃; R3=R4=H) has only little stability at temperatures of between 400 and 450°C in the presence of atmospheric oxygen, the reaction

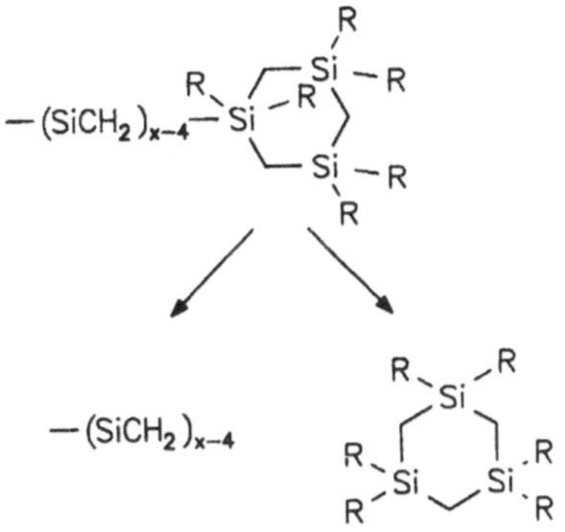

(78)

resulting in the formation of poly(dimethylsiloxane) **53b** (R1=R2=CH$_3$) with the insertion of oxygen [139].

Photochlorination of **53a** (R1=R2=CH$_3$; R3=R4=H) in carbon tetrachloride at 20 °C results in the chlorination of the methylene groups and reduction of the molecular weight, with the formation of ≡Si—Cl— and —SiCl$_3$ bounds [139].

4.2.3
Poly(carbosilanes) with Longer Carbon Chains

If the number of carbon atoms in the structural formula **53a**, is increased, compounds with alternating silicon atoms and carbon chains are obtained, with x > 1 carbon atom (Form. 79).

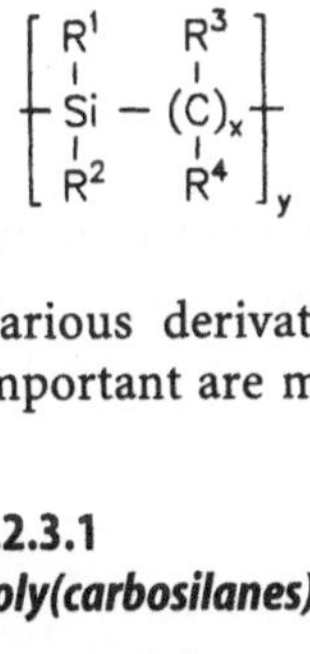

$$\left[\begin{array}{cc} R^1 & R^3 \\ | & | \\ Si & - (C)_x \\ | & | \\ R^2 & R^4 \end{array} \right]_y$$

(79)

Various derivatives of **79** have been described in the literature. The most important are mentioned below.

4.2.3.1
Poly(carbosilanes) with Two Carbon Atoms in the Chain

This group of compounds was first described by Curry [161]. They can be prepared by intermolecular Si—H addition to silanes which carry one hydrogen atom and one vinyl group on the same silicon atom **80a** in the presence of a platinum catalyst (finely divided platinum deposited on activated charcoal).

By 1960, ^{1}H-NMR spectroscopy had been perfected, so that the same author was able to clarify the complex reaction [162]. In the case of dimethylvinylsilane, for example **80a** (R1=R2=CH$_3$) the isomeric poly(carbosilane) **53a** (R1=R2=CH$_3$; R3=H; R4=CH$_3$) and the isomeric, dimeric five-ring systems **80c** (R1=R2=CH$_3$) and **80d** (R1=R2=CH$_3$) were isolated besides **79** (R1=R2=CH$_3$; R3=R4=H; x = 2).

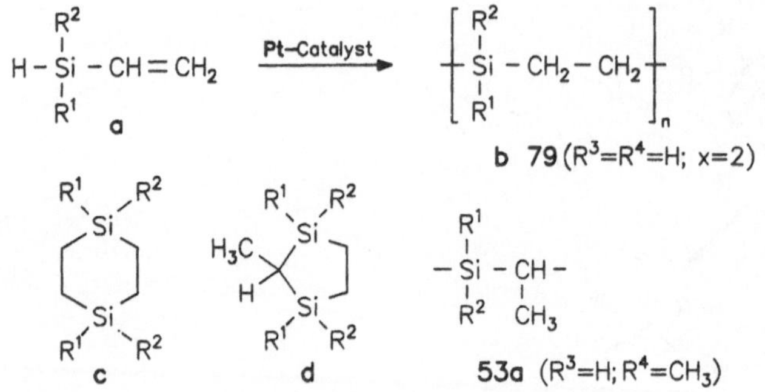

$$
\underset{a}{\underset{\underset{R^1}{|}}{\overset{\overset{R^2}{|}}{H-Si}} - CH=CH_2} \xrightarrow{\text{Pt–Catalyst}} \underset{\substack{b \ \ 79\,(R^3=R^4=H;\ x=2)}}{\left[\underset{R^1}{\overset{R^2}{|}}{\underset{|}{Si}} - CH_2 - CH_2 \right]_n}
$$

c

d

$$
-\underset{R^2}{\overset{R^1}{|}}{\underset{|}{Si}} - \underset{CH_3}{\overset{|}{CH}} -
$$

53a $(R^3=H; R^4=CH_3)$

(80)

If **80a** $(R1=R2=Cl)$ is polymerized in the presence of chloroplatinic acid, ß-hydrosilylation takes place exclusively. The molecular weight achieved depends largely on the solvent used, the highest values being obtained in monochlorobenzene, depending on the concentration ($Mn = 5500$; $MW = 17$ 200; 60% vol% solvent). Subsequent reduction of the Si—Cl bonds with lithium aluminium hydride results in the formation of the base polymer **79** $(R1=R2=R3=R4=H)$, the degree of polymerization being unchanged, so that no C—C— or Si—C bonds are split up Eq. (81) [163]:

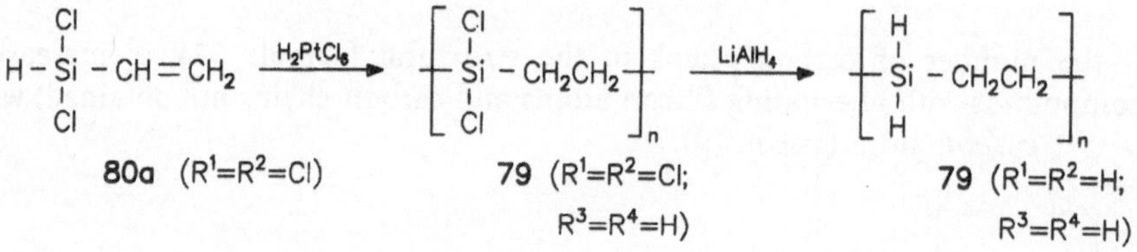

$$
\underset{\underset{Cl}{|}}{\overset{\overset{Cl}{|}}{H-Si}} - CH=CH_2 \xrightarrow{H_2PtCl_6} \left[\underset{Cl}{\overset{Cl}{|}}{\underset{|}{Si}} - CH_2CH_2 \right]_n \xrightarrow{LiAlH_4} \left[\underset{H}{\overset{H}{|}}{\underset{|}{Si}} - CH_2CH_2 \right]_n
$$

80a $(R^1=R^2=Cl)$ **79** $(R^1=R^2=Cl;$ **79** $(R^1=R^2=H;$

$R^3=R^4=H)$ $R^3=R^4=H)$

(81)

The ceramic yields of **79** $(R1=R2=R3=R4=H; x = 2)$, referred to in the literature as "poly(vinylsilane)"[163], are described as good [163].

4.2.3.2
Poly(carbosilanes) with Three Carbon Atoms in the Chain

These products are prepared by ring opening polymerization of mono-tsilacyclobutanes **60a**. Here it is possible to suppress the formation of 1,3-disilacyclobutanes under certain reaction conditions. This is best done at low temperatures in the presence of a platinum catalyst such as chloroplatinic acid H_2PtCl_6 [136] Eq. (82):

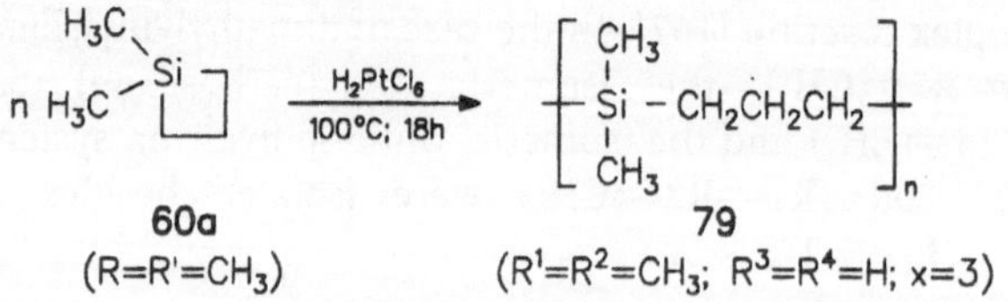

n **60a**

$(R=R'=CH_3)$

$$
\xrightarrow[100°C;\ 18h]{H_2PtCl_6} \left[\underset{CH_3}{\overset{CH_3}{|}}{\underset{|}{Si}} - CH_2CH_2CH_2 \right]_n
$$

79

$(R^1=R^2=CH_3;\ R^3=R^4=H;\ x=3)$

(82)

The resultant polymer is characterized by means of IR and NMR spectroscopy. A hydrogen-containing silane can be used additionally as chain regulator [136] (see also Eq. 62).

A variant, which likewise has three successive carbon atoms between the silicon atoms in the chain with a formal double bond, is the reaction product obtained by the thermal ring opening polymerization of 1,1-dimethyl-2,3-benzo-1-silacyclobutene 83a into the open chain 83b Eq. (83) [165]:

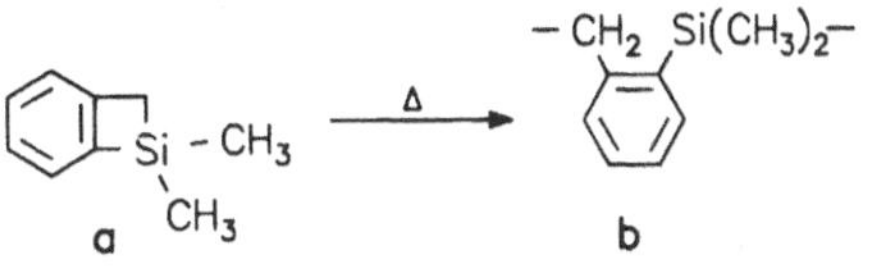

$$
\text{(83)}
$$

The polymerization reaction according to Eq. (83) can also be initiated by activators such as n-butyllithium or radical initiators. The mechanistic aspects are fully discussed in [165].

4.2.3.3
Poly(carbosilanes) with Four Carbon Atoms in the Chain

Ring opening polymerization of 1,1-substituted monosilacyclopentanes 84a in the presence of aluminium chloride results in the formation of the open-chain polymer 79 [164] Eq. (84):

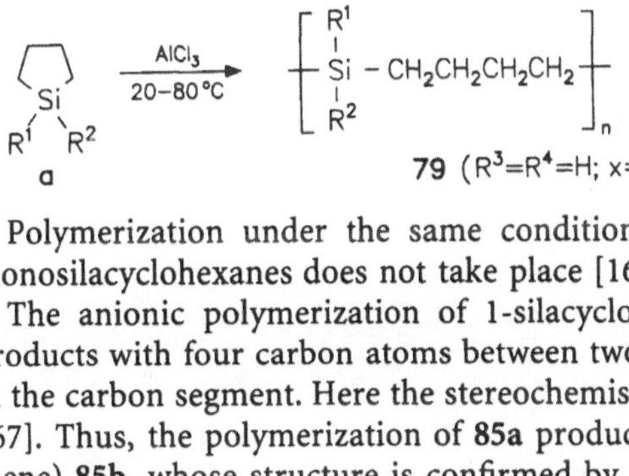

$$
\text{(84)}
$$

79 ($R^3=R^4=H$; x=4)

Polymerization under the same conditions of correspondingly substituted monosilacyclohexanes does not take place [164].

The anionic polymerization of 1-silacyclo-3-pentene derivatives 85a yields products with four carbon atoms between two silicon atoms and a double bond in the carbon segment. Here the stereochemistry has been studied in detail [166, 167]. Thus, the polymerization of 85a produces poly(1.1-dimethyl-sila-*cis*-pent-3-ene) 85b, whose structure is confirmed by ^{1}H-NMR spectroscopy. It is worth noting that the *cis* position of the two hydrogen atoms in the starting compound 85a remains intact [167] Eq. (85):

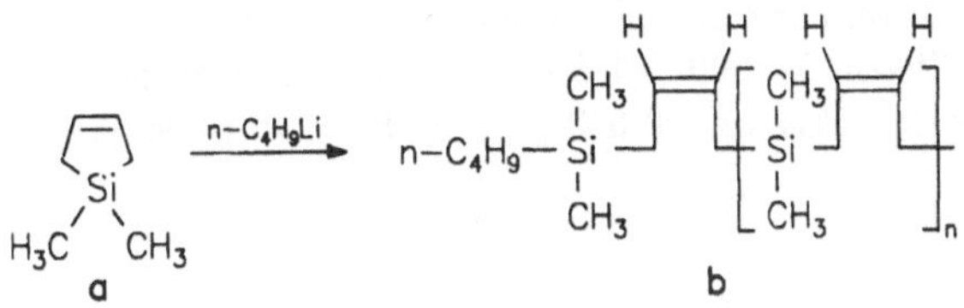

$$
\text{(85)}
$$

The ring opening polymerization with n-butyllithium results in cleavage of the Si—C bond in 85a and, depending on the solvent used, yields different molecular weights (THF/HMPA: MW = 120 000 and M_n = 30 400; THF/TMEDA: MW = 158 000 and M_n = 69 000) [167].

Under different reaction conditions the ring opening polymerization of **85a** can lead to the cleavage of the C — C double bond [169] (formation of **86a**),

$$85a \xrightarrow[\substack{(i-C_4H_9)_3Al \\ 25°C}]{WCl_6, Na_2O_2} \quad +CH=CH-CH_2-Si(CH_3)_2-CH_2+_n \tag{86}$$
$$\qquad\qquad\qquad\qquad\qquad\qquad a$$

4.2.4
Poly(carbosilanes) with Acetylene Groups in the Chain

Polymers with alternating silicon atoms and $-C{\equiv}C-$ groups (e.g. **87a**) are prepared from $BrMgC{\equiv}CMgBr$ [170] or $LiC{\equiv}CLi$ [171] and dialkyldichlorosilanes Eq. (87):

$$n\, R_1R_2SiCl_2 \;+\; n\, LiC{\equiv}CLi \xrightarrow{-2n\,LiCl} \;\; +R_1R_2Si\,C{\equiv}C+_n \tag{87}$$
$$\qquad\qquad\qquad\qquad\qquad\qquad\qquad\qquad a$$

According to the literature however, high molecular weights have not been achieved with this reaction.

By polymerizing diphenyldiethynylsilane (**88a**) in the presence of a catalyst such as $MoCl_5$ or WCl_6, a violet colored polymer (**88b**) is obtained whose structure appears to be complex when subjected to ^1H and ^{13}C-NMR spectroscopic tests (the final analyses had not yet been concluded at the time of publication [172]. Tentative interpretation indicates that the principal component is a methylenesilacyclobutene structure Eq. (88):

$$(C_6H_5)_2Si(C{\equiv}CH)_2 \xrightarrow{MoCl_5\ or\ WCl_6}$$
$$a$$

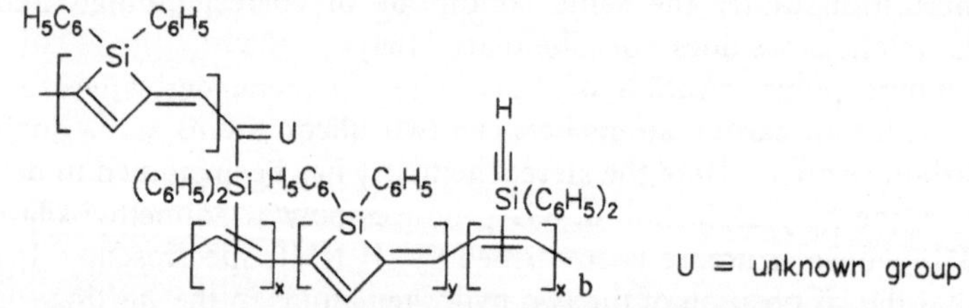

$$\tag{88}$$

U = unknown group

The molecular weights of **88b** that can be achieved depend on the catalyst system, solvent and temperature. For example, WCl_6/toluene at 25 °C produces a violet substance with a MW of 34 000, whilst $MoCl_5$/benzene at 60 °C will produce a violet polymer with a MW of 100 000 [172] (GPC with polystyrene as internal standard).

4.2.5
Silane Dendrimers

The preparation of silane dendrimers produces interesting polycarbosilane-structures. According to van der Made and van Leeuwen [205] the synthesis of silane dendrimers starts with the exhaustive allylation of silicon tetrachloride with allylmagnesium bromide to produce tetraallylsilane **89a** which is hydrosilylated with trichlorosilane to give **89b**. If all the SiCl$_3$-groups in **89b** are reacted with allylmagnesium bromide, a dendrimer with 12 allyl end-groups **89c** is formed, which can be converted according to the same reaction scheme to yield a product containing 972 allyl end-groups Scheme (89):

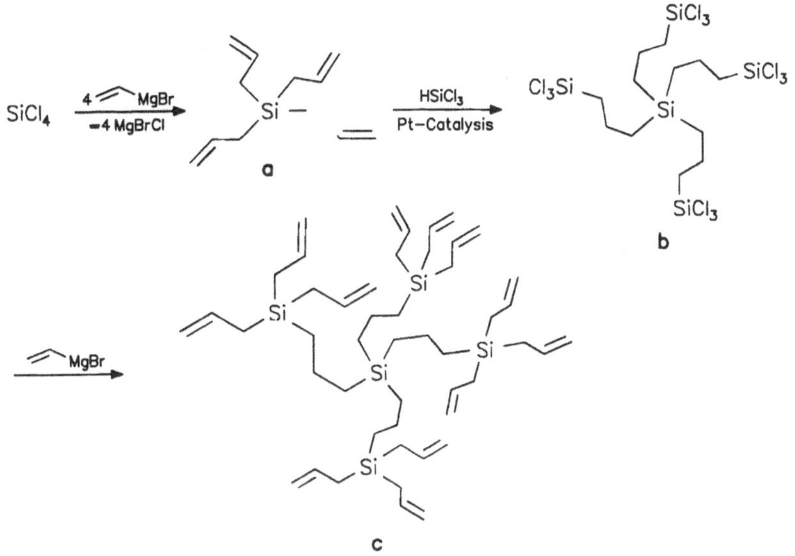

(89)

The dendrimer **89c** has a molecular mass of 73912 g/mol, the formula reads as $C_{4368}H_{7764}Si_{485}$ (white solid). Each of the two steps (alkenylation and hydrosilylation) provides almost quantitative yields and after purification, the neat silane dendrimer, whose structure is corroborated by ^{1}H-, ^{13}C- and ^{29}Si nuclear magnetic resonance spectra as well as elemental analyses. All structures such as **89c** (wax-like consistency) are soluble in common solvents such as hexane, diethylether, chloroform, acetonitrile, ethyl acetate and dimethylformamide.

4.2.6
Poly(carbosilanes) with Phenyl Groups in the Chain

It is obvious to see whether it is possible to insert activated ring compounds such as quinones into the Si—Si-bond in **2b**. This is carried out successfully indeed with p-benzoquinone or 9,10-phenanthraquinone according to Tanaka's work with the aid of a catalyst based on palladium [210] Eq. (90)

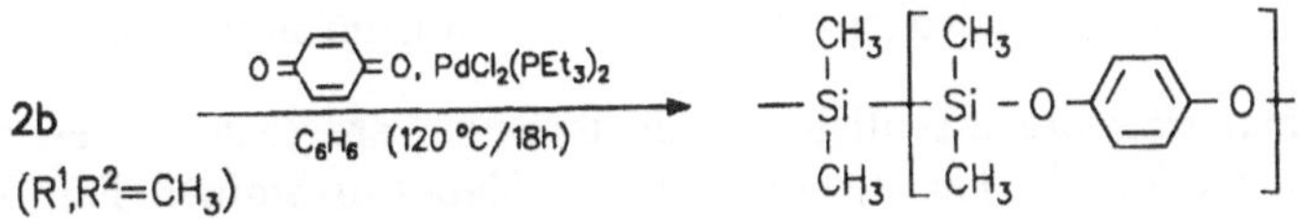

(90)

The molecular weight of the resulting polymer, poly((p-phenylendioxy) dimethylsilylene) (yield 85%) is $MW = 1.5 \times 10^4$.

In other poly(organosilanes) being already modified with ethylene-bridging or oxygen-atoms in the chain 9,10-phenanthrylenedioxy-units can be introduced according to the same procedure [210].

4.3
Poly(organosilazanes)

If one replaces the oxygen atom in poly(organosiloxanes), e.g. 28b by the iso-steric-NH group, one obtains poly(organosilazanes) 91.

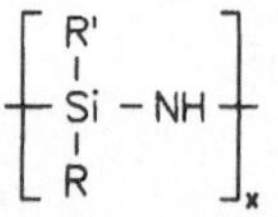

$$\left[\begin{array}{c} R' \\ | \\ Si - NH \\ | \\ R \end{array} \right]_x \qquad (91)$$

Because of their high reactivity with water and protic solvents, ammonia is liberated in accordance with Eq. (92).

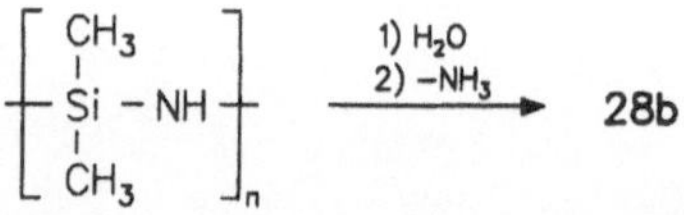

$$\left[\begin{array}{c} CH_3 \\ | \\ Si - NH \\ | \\ CH_3 \end{array} \right]_n \xrightarrow[\text{2) } -NH_3]{\text{1) } H_2O} \quad 28b \qquad (92)$$

Poly(organosilazanes) 91 are difficult to handle and therefore have not achieved the same degree of importance from the preparative and industrial points of view as the isoelectronic poly(organosiloxanes) 28b which are far less chemically reactive. This is also the reason for the existence of little published information on these compounds.

Here we intend to deal mainly with more recent developments, since a comprehensive survey dealing with poly(organosilazanes) up to 1967 has already been published [194].

Poly(organosilazanes) have recently achieved a certain importance as the preliminary stage in the preparation of silicon nitride Si_3N_4 and silicon carbonitride which is formed by subjecting 91 to high temperature pyrolysis [173].

Poly(organosilazanes) 91 can be prepared by several different methods, the most important being

a) the ammonolysis of Chloroorganosilanes
b) the ring opening polymerization of low-molecular weight, cyclic poly(organosilazanes)
c) the thermolysis of silylamines
d) the equilibration reaction of silicon compounds containing chlorine and nitrogen
e) the deprotonization of hydrogen-containing poly(organosilazanes).

The starting materials are essentially products which are easily accessible preparatively and technically, e.g. dichlorodiorganosilanes or amines. For this reason we shall only deal with cylic poly(diorganosilazanes) with a definite ring size when discussing the preparation of the starting products.

4.3.1

**Preparation of Cyclic, Low-Molecular Weight Poly(organosilazanes)
as Starting Material for the Preparation
of Polymeric Poly(organosilazanes) by the Ammonolysis
of Dichlorodiorganosilanes**

The preparation of poly(organosilazanes) by the ammonolysis of hydrogen-containing silanes such as dichlorosilane **2d** ($R_1=R_2=H; R_3=Cl$) was first described by Stock and Somiesky [174]. Since **2d** ($R_1=R_2=CH_3; R_3=Cl$) is now freely available industrially, this is now preferred for the reaction with ammonia in accordance with Eq (93).

$$x\ (CH_3)_2SiCl_2\ +\ 3x\ NH_3\ \xrightarrow{-2x\ NH_4Cl}\ \mathbf{96a} \atop (R=R'=CH_3) \tag{93}$$

The reaction normally takes place with the formation of the stress-free trimer (**96a**; $R=R'=CH_3$) and tetramer (**96b**; $R=R'=CH_3$).

Good instructions are available for the preparation of **96a** and **96b** since they are preferred for further investigations on account of their good availability. The method described by Wannagat is worth mentioning [175], since it produces a total yield of silazane mixture **96a/b** of 83.5%, calculated on the amount of chlorosilane used. It is, however, essential in all cases to separate higher oligomers (91, $x > 4$ and < 10) by fractionation.

W. Fink [177] has provided detailed descriptions of the preparation of further derivatives of **96a/b** with other radicals R/R'. Yields, after deducting the ammonium salt. (Eq. 93) are almost quantitative.

Whether the trimeric and tetrameric cyclic silazanes or polysilazanes are formed, depends on the substituents on the silicon as well as on the nitrogen atom. If one starts out from dichlorosilane ($2; R_1=R_2=H; R_3=Cl$) one obtains only polymeric silazanes [174]. 2 ($R_1=CH_3; R_2=H; R_3=Cl$) produces cyclic compounds of the type **96a/b** and polymers [178], whilst bulky groups such as, for example, α-naphthyl [179] or tertiary butyl [180] favor the formation of the silane diamine derivative, Eq. (94):

$$(CH_3)_2SiCl_2\ +\ 2\ RNH_2\ \xrightarrow{-2\ HCl}\ (CH_3)_2Si(NHR)_2 \tag{94}$$
$$R=\alpha\text{-naphthyl, -tert butyl}$$

The stretched four-ring system **96c** has also been suggested as the starting material for preparing poly(organosilazanes) by ring opening polymerization [181]. A comprehensive survey of the preparation of **96c** has been given by Fink [177]. One of the most important methods is the reaction of double-metallised bis-methylamino dimethylsilane with dichlorodimethyl silane ($2; R_1=R_2=CH_3; R_3=Cl$), which generally produces good yields, Eq. (95).

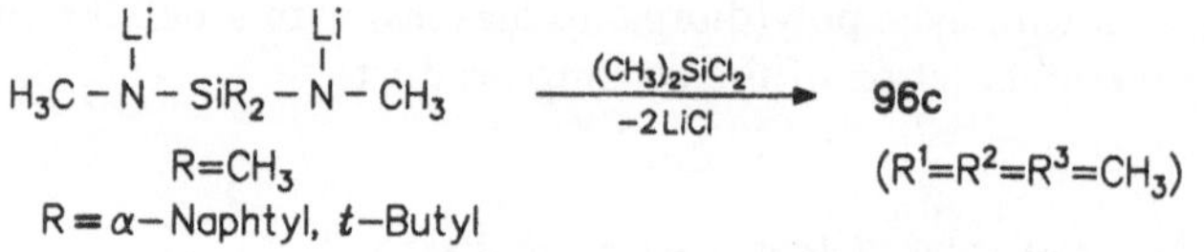

$$\underset{\substack{R=CH_3 \\ R=\alpha-\text{Naphtyl, } t-\text{Butyl}}}{\overset{\substack{Li \qquad Li \\ | \qquad | \\ H_3C-N-SiR_2-N-CH_3}}{}} \xrightarrow[-2\,LiCl]{(CH_3)_2SiCl_2} \underset{(R^1=R^2=R^3=CH_3)}{\mathbf{96c}} \tag{95}$$

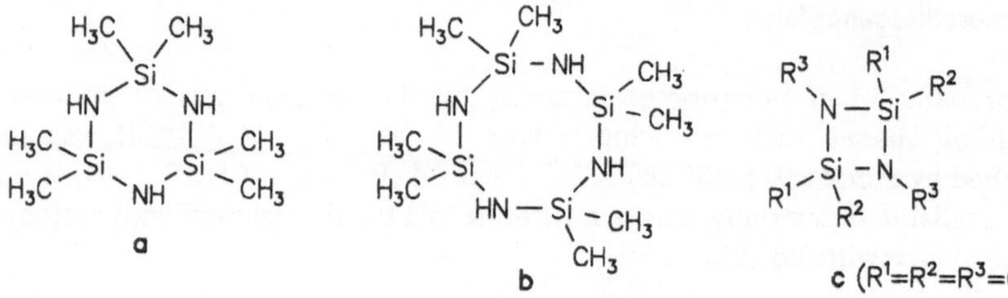

$$\tag{96}$$

4.3.2
Ring-Opening Polymerization of Cyclic Poly(organosilazanes)

4.3.2.1
Ring-Opening Polymerization of Cyclic Poly(organosilazanes)
with Ammonium Salts and Acids

Krüger and Rochow have described the ring opening polymerization of **96a** and **96b** by means of ammonium salts such as NH_4I, NH_4Br, NH_4Cl and amidosulphonic acid [176], which in all cases produces crosslinked poly(organosilazanes) **97a** with the formation of ammonia Eq. (97).

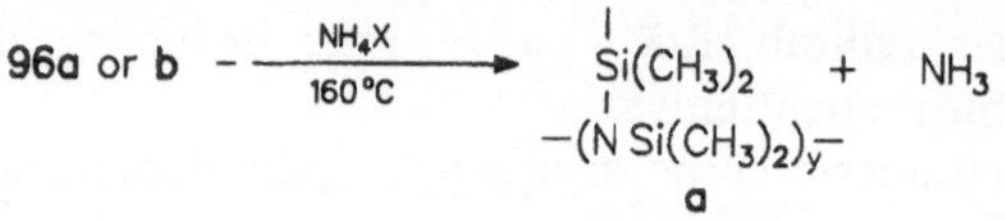

$$\mathbf{96a \text{ or } b} \xrightarrow[160\,°C]{NH_4X} \underset{\mathbf{a}}{\overset{\substack{| \\ Si(CH_3)_2 \\ | \\ -(N\,Si(CH_3)_2)_{\overline{y}}}}{}} + NH_3 \tag{97}$$

The compounds, which initially still contain a little chlorine (originating from the ammonium halide) are converted into chlorine-free products by treating them with ammonia. The type of catalyst influences the reaction conditions and polymer yields (Table 16).

Whereas sulphamic acid produces oil-like products, ammonium salts generate wax-like polymers.

The molecular weights and analytical data of the poly(organosilazanes) prepared in this way (Table 4.16) do not exhibit major differences and are determined using the cryoscopic method in benzene, the molecular weight being around 10000. Infrared and ^{1}H-NMR spectroscopic tests indicate the presence of two different structural elements **98a** (cyclic) and **98b** (linear) [176].

An interesting variant of ammonolysis has been described by Arai et al., involving the reaction of the pyridine adduct of dichlorosilane with ammonia **98a** [186], pyridine being used as solvent, Eq. (98). The perhydridosilazane **98b** obtained is a polydisperse solid which dissolves in chloroform. Its average

Table 4.16. *Yields of branched poly(organosilazane)s by reactions of **96a/b** mixtures with various catalyts

Catalyst	Catalyst concentration (%)	yield of polymers (%)
NH_4Cl	5	11
NH_4Br	5	57
NH_4Br	1	67–73
NH_4Br	10	61
NH_4Br	45	49
NH_4J	5	69

*data adapted from [176].

$$SiH_2Cl_2 * 2\,Pyr \xrightarrow[80\,°C;\,77\%]{+2NH_3,\ -2C_5H_5N,\ -2NH_4Cl} \mathbf{96a\text{--}c} \qquad (98)$$

$$(R^1=R^2=R^3=H)$$

a

b

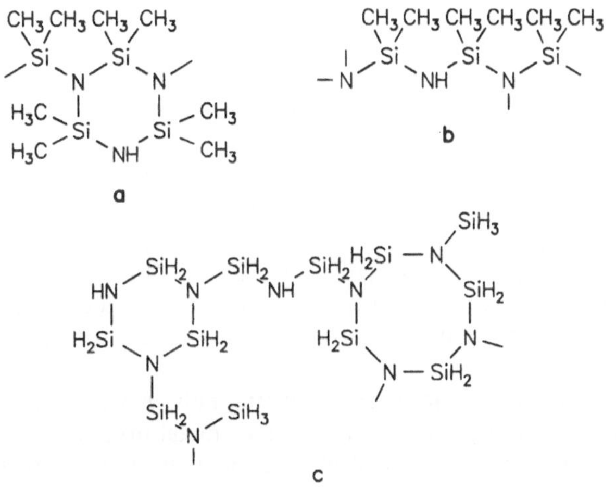

(99)

c

molecular weight is 1300, as determined by gel permeation chromatography. ^1H and ^{29}Si nuclear resonance spectroscopic tests indicate the structure **99c** with linear and cyclic constituents.

4.3.2.2
Ring-Opening Polymerization of Cyclic Poly(organosilazanes) Using Organo-Alkali Compounds

The ring opening polymerization of **96c** was first investigated by Seyferth, molecular weights (M_n) being ≤3000. Soum analysed the reaction under different conditions, e.g. examining the effect of substituents [181]. The surprising observation is that **96a/b** does not react, **96c** only if there are no bulky radicals such as *tert*-butyl on the nitrogen atom. The methyl group reacts most readily, the ethyl group less so, molecular weights, M_n, of ≤2000 being obtained in the case of **96c** ($R_1=R_2=CH_3$; $R_3=C_2H_5$). In the case of **96c** ($R_1=R_2=R_3=CH_3$) molecular weights of up to 16000 are obtained, depending on the type of initiator used (Table 17).

Table 4.17. Syntheses of poly(organosilazane)s by ring-opening polymerization of **96c** in THF at 25 °C. (Data adapted from [181])

Initiator (I)	[M]/(I)	Reaction time (h)	M_n (y/mol)
Methyllithium	10	8	4400
t-Butyllithium	10	8	4200
Phenyllithium	10	8	4200
Sodiumnaphtyl	200	2	16000
Sodium-α-methylstyrol	10	4	4000

The authors characterize the reaction as kinetically controlled (effect of substituents; stretched ring systems react rapidly compared with **96 a/b**); steric hindrance at the nitrogen atom in **96c**).

If one partly replaces the methyl groups in **96c** by vinyl substituents (**96c**; ($R_1=R_3=CH_3$; $R_2=-CH=CH_2$), high polymerisation rates at temperatures of -40 °C and molecular weights of up to 100000, with narrow distribution, are observed [181]. From this one can conclude that initiation and chain growth proceed rapidly, possibly via nitrogen anions as active centers. The polymers obtained are described as colorless compounds, soluble in tetrahydrofuran and chloroform, such solutions being relatively stable towards air. The $^1H-$, $^{13}C-$, ^{29}Si and ^{15}N nuclear resonance spectra are simply constructed, the authors coming to the conclusion that the structures are essentially linear, with few, if any, branching points [181].

Attempts have also been made to initiate ring opening polymerization using common alkalis such as potassium hydroxide in place of metal organyls [187]. The reaction proceeds irregularly, especially at elevated temperatures, with the development of the hydrocarbons which form the basis of **91** as well as with the formation of ammonia and brittle solids which are assumed to be highly crosslinked poly(organosilazanes).

The thermal decompositions of **96a/b**, in the absence of catalyst, leads to the formation of **91** with an average molecular weight of 1200 Dalton and predominantly linear structural constituents [188], ammonia being liberated (see also Sect. 4.3.3).

4.3.2.3
Ring-Opening Polymerization of Cyclic Poly(organosilazanes) Using Heavy Metal Catalysts

In 1983, Zoeckler and Laine reported for the first time the ring opening of **96b** with carbonyls of the VIIIth side group, e.g. triruthenium dodecacarbonyl and hexaruthenium hexadecacarbonyl in the presence of hexamethyldisilazane as chain stopper [182], Eq. (100).

$$96b \xrightarrow[\text{Ru}_3(\text{CO})_{12};\ 135\,°C]{(CH_3)_3NHSiNHSi(CH_3)_3} 96$$

$$(R=R'=CH_3); \quad x=1-12$$

$$96b \xrightleftharpoons[-M]{MH_2} H \dashv Si(CH_3)_2NH \vdash_{\overline{4}} H \tag{100}$$

$$96b \xrightleftharpoons{MH_2} HM \dashv Si(CH_3)_2NH \vdash_{\overline{4}} H$$

The molecular weights obtained are comparatively modest, although they can be increased by modifying the reaction in the presence of hydrogen or metal hydrides [183]. Open-chain silazanes with a terminal amino and a SiH group are postulated as the intermediate stages, which can be condensed further [183], Scheme (101):

$$H \dashv Si(CH_3)_2NH \vdash_{\overline{4}} H \quad + \quad HM \dashv Si(CH_3)_2NH \vdash_{\overline{4}} H \xrightleftharpoons{-MH_2}$$

$$(M = \text{transition metal}) \qquad\qquad H \dashv Si(CH_3)_2NH \vdash_{\overline{8}} H \tag{101}$$

Since all reaction stages in Eq. (101) are reversible the molecular weights are not very high according to the authors.

The postulated mechanism is supported by the fact that the symmetrical tetramethyl disilazane 102a reacts with ammonia, with the formation of polymeric, open-chain silazanes, if the reaction is catalyzed with $Ru_3 (CO)_{12}$ [182] Eq. (102).

$$\underset{a}{HSi(CH_3)_2NHSi(CH_3)_2H} \xrightarrow[\text{Ru}_3(\text{CO})_{12}/60°C]{NH_3,\ -H_2} \underset{b}{H \dashv Si(CH_3)_2NH \vdash_x H} \tag{102}$$

Here, molecular weights of 1000–2000 Dalton are obtained, these depending on the reaction time and the partial pressure of ammonia. There has been no further characterization by the authors [183] of the polymers thus obtained.

The reaction Eq. (100) has also been described by Blum and Laine (184) under modified conditions, platinum-coated activated charcoal being used as catalyst besides the above described, ruthenium-based catalyst besides the above described, ruthenium-based catalyst systems. Without further characterization, a maximum average molecular weight of approx. 1000 Dalton is obtained.

A detailed method of preparation, using different hydrogen-containing silanes $(H_2Si(C_2H_5)_2, H_3SiC_6H_5,$ and $H_3Si-n-C_6H_{13})$ is described in [183], although there is no accurate characterization of the polymers obtained.

4.3.3
Deaminization/Condensation Reaction Through the Thermolysis of Silylamines

The thermolysis of silylamines having the structures 108 and 109 has been investigated by Verbeek, the starting materials being alkyltriamino-and dialkyl-diaminosilane compounds which can be thermolysed at temperatures of up to 800 °C [189] Eqs. (103, 104):

$$R_2Si(NHCH_3)_2 \xrightarrow{\Delta\ 200-800\,°C} CH_3NH_2 \ + \ \mathbf{96} \tag{103}$$
a

$$RSi(NHCH_3)_3 \xrightarrow{\Delta\ 520\,°C/3h} CH_3NH_2 \ + \ \mathbf{96} \tag{104}$$
b

where $R = CH_3$ or C_6H_5. Since the products formed at the high pyrolysis temperatures according to Eq. (103) – methylamine and cyclic trisilazane – can be reversibly decomposed back into the starting compound 103a, a better polymer yield is achieved.

Reaction (104) has been more closely studied by Penn et. al. [189,190], polymers with a molecular weight of 4000 Dalton being obtained, whose structures are state by Wynne [191] to be as follows 105:

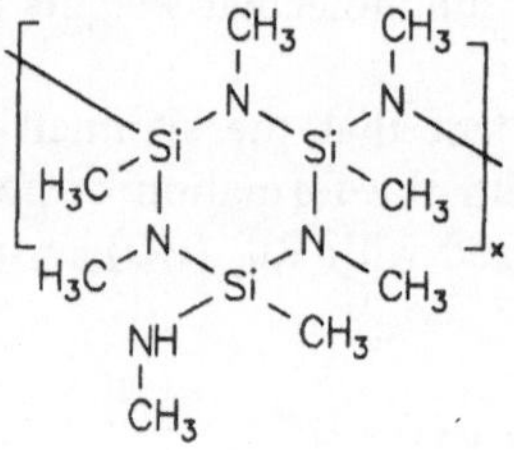

$$\tag{105}$$

The pyrolysis behavior of the polymers 105 form Si_3N_4 and silicon carbonitride was investigated and it was found that the deaminization-condensation reaction has not yet been understood from the mechanistic point of view [192].

Polymerization of tris-(N-methylamino)-methyl silane, $CH_3—Si(NH—CH_3)_3$ at 520 °C for up to 4.4 hours results in the formation of transparent, reddish-brown products which are soluble in methylene chloride and chloroform. The maximum molecular weights obtained are 4200 Dalton [201].

4.3.4
Preparation of Poly(organosilazanes) by the Equilibration Reaction of ≡Si—Cl— with Compounds Containing ≡Si—N=

The reaction between hexamethyldisilazane and trichlorosilane – both of which are available industrially – is complex [143]. Here, it is notable that there is no silylation reaction at the NH—group of hexamethyldisilazane Eq. (106):

$$(CH_3)_3Si - NH - Si(CH_3)_3 \quad \xrightarrow[\quad\quad]{HSiCl_3; \; -HCl} \not\rightarrow \quad (CH_3)_3Si - \overset{\displaystyle SiHCl_2}{\underset{\displaystyle |}{N}} - Si(CH_3)_3 \qquad (106)$$

Instead, there is nucleophilic substitution of Cl in $HSiCl_3$ by nitrogen, with the formation of **107a**, Eq. (107), chlorotrimethylsilane being eliminated at the same time. The reaction can be continued in the case of **107a** through further-substitution of **a**, Eqs. (108) and (109), but the last step (109) is reversible.

$$(CH_3)_3Si - NH - Si(CH_3)_3 \quad \xrightarrow[-(CH_3)_3SiCl]{HSiCl_3} \quad \underset{a}{(CH_3)_3Si - NH - SiHCl_2} \qquad (107)$$

$$(CH_3)_3Si - NH - SiHCl_2 \quad \xrightarrow[-(CH_3)_3SiCl]{(CH_3)_3SiNHSi(CH_3)_3} \quad \underset{a}{((CH_3)_3Si - NH)_2 - SiHCl} \qquad (108)$$

$$\underset{a}{((CH_3)_3Si - NH)_2 - SiHCl} \quad \xrightleftharpoons[-(CH_3)_3SiCl]{(CH_3)_3SiNHSi(CH_3)_3} \quad \underset{b}{((CH_3)_3Si - NH)_3 - SiH} \qquad (109)$$

As more **109a** is formed, more chlorine-containing tetrasilazane (**110a**) is also produced. The favourable conditions for this are thought to be due mainly to steric considerations [193] Eq. (110)

$$\underset{\mathbf{109a}}{((CH_3)_3Si - NH)_2 - SiHCl} \quad \xrightarrow{-(CH_3)_3SiNHSi(CH_3)_3} \quad \underset{a}{(CH_3)_3SiNHSiH(Cl)NHSiH(Cl)NHSi(CH_3)_3} \qquad (110)$$

The occurrence is superimposed by another reaction sequence during which a *N*-silylated tetrasilazane species **111a** is formed from **109a**. The hydrogen chloride produced is capable of decomposing the silazane used, Eqs. (111, 112).

$$\underset{\mathbf{109a}}{((CH_3)_3Si - NH)_2 - SiHCl} \quad \xrightarrow{-(CH_3)_3SiNHSi(CH_3)_3} \quad \underset{a}{(CH_3)_3SiNHSiH(Cl)NH\overset{\displaystyle Si(CH_3)_3}{\underset{\displaystyle |}{Si}}H(NHSi(CH_3)_3)_2} \qquad (111)$$

$$(CH_3)_3Si - NH - Si(CH_3)_3 \quad \xrightarrow{+3\,HCl} \quad NH_4Cl \; + \; 2\,(CH_3)_3SiCl \qquad (112)$$

110a and **111a** form the basis for further cyclization and branching reactions, although it has not, so far, proved possible to analytically identify individual species in the finished reaction product. The mixture may be characterized by means of elementary analysis, ^1H—NMR, IR and gel permeation chromatography [193]. The molecular weight M_n according to the above mentioned method of preparation is ~ 3500 Dalton (MW ~ 15000 Dalton), the average empirical formula of the polymer being given as $-[HSi\,(NH)_{1.5}]_x[HSiNH\,(NHSi(CH_3)_3]_y-$.

4.3.5
Preparation of Poly (organosilazanes) by Polycondensation of ≡SiH— and =N—H-Containing Oligomeric Poly(organosilazanes)

The following reaction principle is based on work done by Seyferth [1965] which combines the ammonolysis described in Sect. 4.3.1 with a reaction described as dehydrocyclodimerization [196].

A prerequirement for this reaction is the presence of at least one hydrogen atom of the silicon as well as at the nitrogen atom of the silazane linkage Eq. (113).

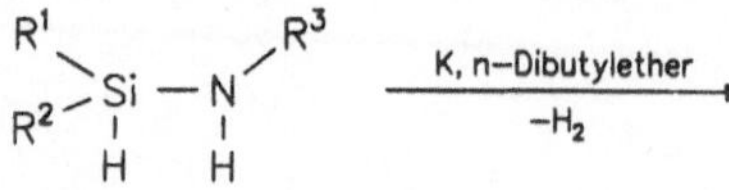

$$\begin{array}{c} R^1 \\ R^2 \end{array}\!\!\!> \!\!\! \underset{\underset{H}{|}}{Si}\!-\!\underset{\underset{H}{|}}{N}\!\!\!<\!\!\!^{R^3} \quad \xrightarrow[-H_2]{K,\ n-Dibutylether} \quad 96c \tag{113}$$

A Si=N double bond (114b) is released from the amide ion 114a as a possible intermediate stage, with the elimination of potassium hydride, Eq. (114). Afterwards cyclodimerization into 96c occurs Eq. (115).

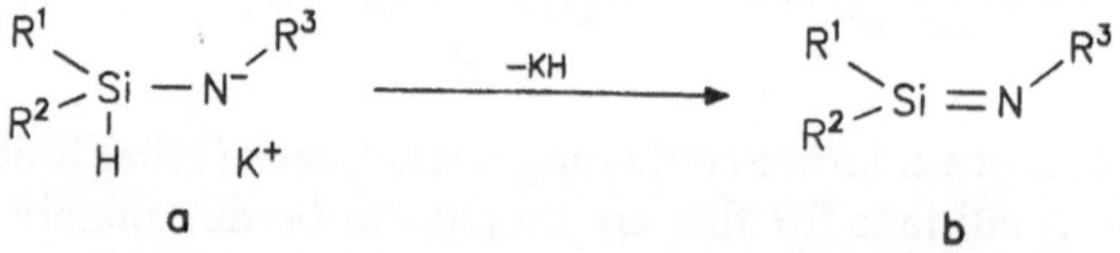

$$\begin{array}{c} R^1 \\ R^2 \end{array}\!\!\!> \!\!\! \underset{\underset{H}{|}}{Si}\!-\!\underset{\underset{K^+}{}}{N}\!\!\!<\!\!\!^{R^3} \quad \xrightarrow{-KH} \quad \begin{array}{c} R^1 \\ R^2 \end{array}\!\!\!> \!\!\! Si\!=\!N\!\!\!<\!\!\!^{R^3} \tag{114}$$

$$\qquad\qquad a \qquad\qquad\qquad\qquad\qquad\qquad b$$

$$2\ \begin{array}{c} R^1 \\ R^2 \end{array}\!\!\!> \!\!\! Si\!=\!N\!\!\!<\!\!\!^{R^3} \quad \longrightarrow \quad 96c \tag{115}$$

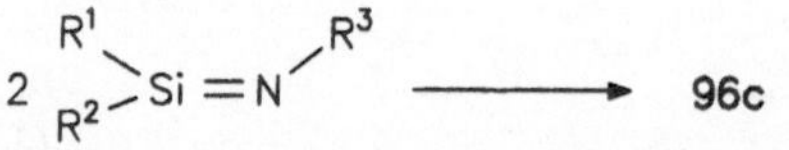

Ammonolysis of 2 (R_1=CH$_3$; R_2=H; R_3=Cl) produces a mixture of cyclic and, possibly, linear oligomers [197] with the spectroscopically checked structure 91 (R=CH$_3$; R'=H) with x = 4.7 − 5.4 (cryoscopic method of determination using benzene as solvent), as a colorless oil which is affected by moisture. If alkalis are added which are strong enough to deprotonize the NH-function, the tetramer 116a present in the mixture, in a suitable solvent, first produces the dimer 116b, hydrogen being formed. This dimer then condenses further to form lamellar structures Eq. (116).

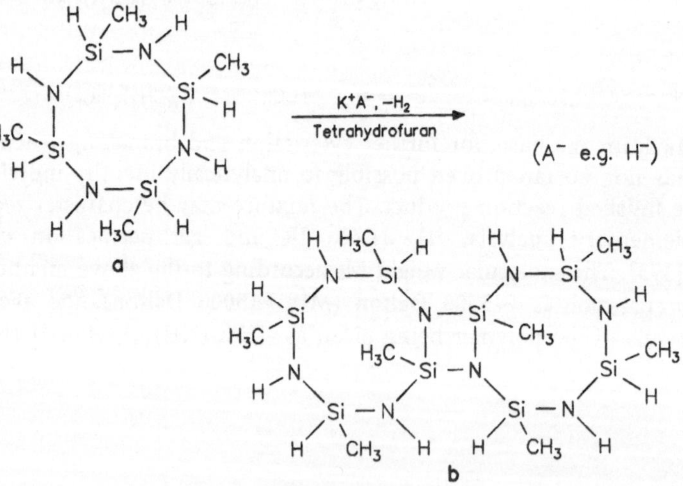

$$\xrightarrow[Tetrahydrofuran]{K^+A^-,\ -H_2} \qquad (A^-\ e.g.\ H^-) \tag{116}$$

After all the hydrogen has been liberated, a "living polymer" having the general structure $[CH_3SiHNH)_a(CH_3SiN)_b(CH_3SiHNK)_c]_n$ is formed, which reacts with electrophilic reagents such as iodine trimethylsilane to form the corresponding trimethylsilyl derivative. The poly(organosilazane) is obtained in the form of a white solid having a molecular weight of 1180, which is soluble in hexane, benzene, diethyl ether and tetrahydrofuran [195]. ^{1}H-NMR analysis gives us the empirical formula $(CH_3SiHNH)_{0.39}(CH_3SiN)_{0.57}(CH_3SiHNCH_3)_{0.04}$. This compound is a suitable preliminary stage for the pyrolysis and preparation of Si_3N_4 [195] and silicon carbonitride Eq. (117):

$$2\ (CH_3SiHNH)(CH_3SiN) \xrightarrow[83\%]{\Delta\ 1000°C,\ N_2} Si_3N_4 + SiC + C + 2\ CH_4 + 4\ H_2 \qquad (117)$$

4.3.6
Chemical Properties of Poly (organosilazanes)

The Si — N bond in low molecular weight poly (organosilazanes) is unstable and will break down because of solvolysis, especially in the presence of acids. During this reaction siloxanes are normally formed and the amine on which poly (organosilazane) is based [176]. High molecular weight poly (organosilazane) are unaffected by pure water and alkalis, a fact which is probably due to the hydrophobic character of the polymers [176].

Oxidation with oxygen or benzoyl peroxide results in elastomerlike products, but further information is not given in Ref. [176].

The higher the molecular weight, the more resistant will be the product to hydrolysis and oxidation [195].

Solvolysis of low molecular weight poly (organosilazanes), e.g. alcohols, leads to the formation of silyl derivatives. This reaction is activated by catalysts such as ammonium chloride [13].

Cyclic poly (organosilazanes) can be dissociated by alcohol and Lewis acids (e.g. Al_2Cl_6) [175], derivatives 118a and 118b of hexamethyl disilazane being formed Scheme (118).

$$96b \begin{cases} \xrightarrow{ROH} NH(Si(CH_3)_2OR)_2 & a \\ \xrightarrow{Al_2Cl_6} NH(Si(CH_3)_2Cl)_2 & b \end{cases} \qquad (118)$$

One important property of poly (organosilazanes) is their thermal decomposition into silicon nitride of silicon carbonitride, a reaction which also has industrial importance. The properties of the pyrolysis product depend on a number of factors which form the subject of numerous publications and patents [198]. Until recently, ceramic materials were made using powder coating technology. This technique cannot, however, be used to make coatings, to impregnate porous products and to produce ceramic multifilaments. The so-called precursor technique uses poly(organosilazanes) with an average molecular weight of around 1600–3000 g/mol (osmometric determination). A product with the formula 119, for example, dissolved in tetrahydrofuran, produces a clear melt after removal of the solvent, which will produce a compound with the empirical formula $Si_{1.0}N_{1.3}C_{1.6}$ if pyrolyzed at temperatures above 1400 °C in an atmosphere of nitrogen. Partial crystallization of α-silicon nitride and hexagonal silicon carbide will occur above 1500 °C [198, 199].

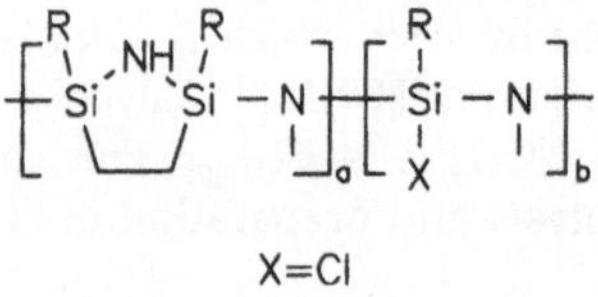

$$(119)$$

X=Cl

4.4
Poly(carbosilazanes)

Poly(carbosilazanes)**120** are formed by the insertion of one or more carbon atoms between the silazane linkage in **105**.

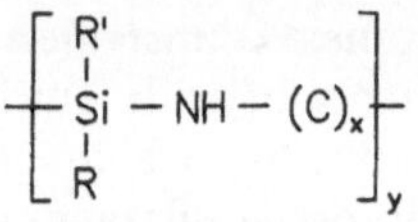

$$(120)$$

C_x can also, for example, be an aromatic ring system (**121b**). The reaction of 1.3-diphenyl-2.2.4.4-tetramethylcyclodisilazane **121a** with aromatic diamines is one way of preparing such products [200] Eq. (121).

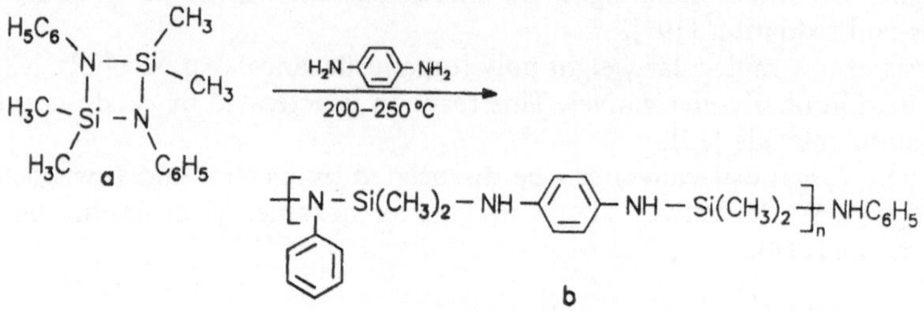

$$(121)$$

After removal of volatile products by distillation, a brown resin with a wide melting range remains behind. The molecular weights obtained are relatively low, the maximum being 1240 [200].

4.5
References

1. Kipping FS (1924) J Chem Soc 125: 2291
2. Rochow EC, Pat US (1941) General Electric, 2 380: 995
3. Burkhard CA (1949) J Amer Chem Soc 71: 963
4. Wesson JP and Williams TC (1981) J Polym Sci Polym Chem Ed 19: 65
5. Helmer B and West R (1982) J Organomet Chem 21: 236
6. Mazdyasni KS, West R and David LD (1978) J Am Ceram Soc 61: 504

7. West R, David LD, Djurovich PI, Stearley KI, Srinivasan KSV and Yu, H (1981) J Amer Chem Soc 103: 7352
8. Zeigler M and Gordon Fearon (eds) (1990) Silicon-based polymer science, A comprehensive resource. Advances in Chemistry Series Nr. 224, American Chemical Society, Washington DC, 285
9. Hengge E (1974) Topics Current Chem 51: 1
10. Watanabe H, and Nagai Y (1985) In: H. Sakurai (ed) Organosilicon and bioorganosilicon chemistry. Ellis Horwood, London, Chap 9 and references therein
11. a) Yajima S, Hayashi J and Omori M (1975) Chem Lett 931
    b) Yajima S, Okamura K and Hayashi J (1975) Chem Lett 1209
12. a) Yajima S, Hasegawa Y, Hayashi Y and Iimura M (1978) J Mat Sci 13: 2569
    b) Hagasewa Y, Iimura M and Yajima S (1980) ibid. 15: 720
13. Kochs P (1989) Chemiker-Ztg 113: 225
14. Noll W (1968) Chemie und Technologie der Silicone. Verlag Chemie, Weinheim 38 ff
15. Karstedt BD (1966), DE 1 668 159 General Electric
16. Pietrusza EW, Sommer LH, and Whitmore FC (1948) J Amer Chem Soc 70: 484
17. a) Barry AJ, De Pree L, Gilkey JW and Hook DE (1947) J Amer Chem Soc 69: 2916
    b) Gilman WF and Miller LS, (1951) J Amer Chem Soc 73: 2367
18. Gilliam WF, Liebhafsky HA and Winslow AF (1941) J Amer Chem Soc 63: 801
19. a) Curran C, Witucki RM and McCusker PA (1950) J Amer Chem Soc 72: 4471
    b) Kohlschütter HW, Scheele W and Jackel G (1949) ref. in W. Kohlschütter, Fortschr. Chem Forsch 1: 1
20. a) Smetankina NP and Nikishin, GJ (1955) J. Gen Chem USSR 25: 2305
    b) Joklik J and Bazant V (1965) Collection Czech. Chem Commun 30(9): 2928
21. Hurd CD and Yarnall WA (1949) J Amer Chem Soc 71 755
22. Petrov AD, Cernysev and Cernysev EA (1952) Ber Akad Wiss UdSSR 86(2): 737
23. Polysilanes on the basis of the silane in question are described in the literature, though no precise preparation procedure for the starting material is available
24. a) Palmer KW and Kipping FS (1930) J Chem Soc (London), 1020
    b) Koga I, Terni Y and Ohgushi M (1977) U.S. 4.309.558 Chisso Corp.
25. Polysilanes on the basis of the silane in question are described in the literature, though no precise preparation procedure is available
26. Hyde JF and de Long RC (1941) J Amer Chem Soc 63: 1194
27. Adrian Th, Habel W and Sartori P (1991) Chemiker-ztg. 115: 1
28. A comprehensive evaluation of the preparation of organic halosilanes can be found in Pawlenko S (1980) Methoden der Organischen Chemie (Houben-Weyl), 4th edn. 96 ff., Georg Thieme Stuttgart New York
29. Rosenberg SD, Wallburn JJ and Ramsden HE (1957) J Org Chem 22: 1606
30. a) Balis EW, Liebhafsky HA und Getz DH (1949) Ind Engng Chem 41: 1459
    b) Ponamorenko VA, Sokolov BA and Petrov AD (1957) Izvest Akad. Nauk SSSR., Otdel Khim Nauk 1956: 628. Cited in Chem Abstr 51: 1027. $CH_3Si(H)Cl_2$ reacts with vinyl chloride under the catalysis of platinized carbon to a mixture of CH_3SiCl_3, $CH_3Si(CH_2CHCl-CH_3)Cl_2$, starting material and $CH_3(C_2H_5)SiCl_2$, which is isolated by fractioned distillation.
    Barry AJ, DePree L, Gilkey JW and Hook DE (1947) J Amer Chem Soc 69: 2916
31. a) Petrov AD, Ponomarenko VA, Cherkaev VG and Zadorozhnii NA (1959)U.S.S.R. 114 156. Cited in Chem Abstr 53: 14003
    b) Barry AJ, DePree L, Gilkey JW and Hook DE (1947) J Amer Chem Soc 69: 2916
32. a) Nesmezanov AN, Freidlina RKh and Chukovskaya EC (1957) Tetrahedron 1: 248. Cited in Chem Abstr 52: 295 (1958). Thermal telomerization of $CH_3(H)SiCl_2$ with ethylene is effected under elevated temperature and pressure, whereby $CH_3(n-C_4H_9)SiCl_2$ is isolated among other homologues by fractioned distillation.
    b) Barry AJ, DePree L, Gilkey JW and Hook DE (1947) J Amer Chem Soc 69: 2916
33. Speier JL, Webster JA, und Barnes GH (1957) J Amer Chem Soc 79: 974
34. a) Pat US (1983) 4 447 633, North Dakota State University (inv.: Ph.R. Boudjouk)
    b) Barry AJ, DePree L, Gilkey JW and Hook DE (1947) J Amer Chem. Soc. 69: 2916

35. Polysilanes on the bases of the silane in question are described in the literature, though no precise preparation procedure for the starting material is available
36. Barry AJ, DePree L, Gilkey JW and Hook DE (1947) J Amer Chem Soc 69: 2916
37. Adrianov, KA, Izmailov BA, Minyaeva NP, Tsvetkova NV and Prozumentova EA (1976) J Gen Chem USSR 46: 2255
38. West R, David LD, Djurovich PI, Yu H and Sinclair R (1983) Ceramic Bull. 62: 899
39. West R (1986) J Organomet Chem 300: 327
40. Zhang Y-H. and West R (1984) J Polym Sci Polym Chem Ed. 22: 225
41. Zeigler JM (1986) Polym. Prepr. (Am Chem Soc Div Polym Chem) 27: 109
42. Zeigler JM (1987) ibid. 28: 424
43. a) Matyjaszewski K, Hrkach J, Hwan-Kyu K and Ruehl K in Silicon-Based Polymer Science, A Comprehensive Resource, Edited by John M. Zeigler and Gordon Fearon FW, Advances in Chemistry Series Nr. 224, American Chemical Society, Washington DC 1990, S. 285 ff
    b) Matyjaszewski K, Chen YL und Kim HK (1988) in Inorganic and Organometallic Polymers; Zeldin M, Wynne KJ, and Allcock HR Editors; ACS Symposium Series 360; American Chemical Society; Washington DC, page 78
    c) Matyzjaszewski K and Chen YL, (1988) J Organomet Chem 340: 7
44. Kim HK (1990) Polym Prepr (1990) Am Chem Soc Div Polym Chem 31 (2): 278
45. Nagai Y, Watanabe H, Matsumoto H, Naoi Y and Sotou N (1990) Zeigler JM, Gordon Fearon FW (eds) In: Silicon-based-polymer science, A comprehensive resource. Advances in Chemistry Series Nr. 224, American Chemical Society, Washington DC, 505
46. Doctoral Thesis by R. Firgo, Technical University Graz (Austria), 1980
47. Litscher G (1977) Doctoral thesis Technical University Graz (Austria)
48. Kalchauer W (1986) Doctoral thesis Technical University Graz (Austria)
49. Kunai A, Kawakami T, Toyoda E and Ishikawa M (1991) Organometallics 10: 2001
50. Aitken CT, Harrod JF and Samuel E (1985) J Organomet Chem. 279: C11
51. Aitken CT, Barry JP, Gauvin F, Harrod JF, Malek A and Rousseau D (1989) Organometallics 8: 1732
52. Hengge E, Weinberger M and Jammegg Ch (1991) J Organometallics Chem 410: C1-C4
53. Calas, R, Dunogues J, Deleris G and Duffault N (1982) J Organometallc Chem 225: 117
54. Burns GT (1991) OL DE 39 24 232 A 1, Dow Corning Corp.
55. a) Baney RH, Pat US (1985) Dow Corning Corp 4, 534: 948
55. b) Baney RH (1982) OL DE 30 41 762 A 1, Dow Corning Corp
56. a) West R et al. (1984) J Organometal Chem 3: 793
56. b) West R (1987) Angew Chem 99: 1231
57. Sakamoto K, Obata K, Hirata H, Nakajima M and Sakurai H (1989) J Amer Chem Soc 111: 7641
58. Rhein RA (1990) In: John Zeigler M and Gordon Fearon FW (eds) Silicon-Based-Polymer Science, A Comprehensive Resource, Advances in Chemistry Series Nr. 224, American Chemical Society, Washington DC p 309
59. DE-B-1 768 896 (1971) Dow Corning Corp. (inv.: Atwell WH)
60. Gupta Y and Matyjaszewsky K (1990) ACS Polym Prepr 31 (1): 46
61. West R (1989) In: Patai S, Rappoport Z (eds) The chemistry of organic silicon compounds Part 2. Wiley J, Chichester 1208
62. Boberski WG and Alfred AL (1974) J Organometal Chem 88, C 27
63. Kim HK, Uchida H and Matyjaszewskyi (1990) Polym Prepr (Am Chem Soc Div Poly Chem) 31(2): 276
64. Alnaimi IS and Weber WP (1983) Organometallics 2: 903
65. a) Weidenbruch M and Schaefer A (1984) J Organometal Chem 269: 231
    b) Hengge E and Schuster HG (1982) J Organometal Chem 231: C17
66. a) Wojnowska M, Wojnowski W and West R (1980) J Organometal Chem 199: C1
    b) Carlson CW and West R (1983) Organometallics 2: 1798
67. Kumada M (1975) J Organomet Chem 100: 127
68. Blinka TA and West R (1986) Organometallics 5: 128, ibid 5, 133
69. Ishikawa M and Kumada M (1972) J Organometal 42: 325

70. Drahnak TJ, Michl J and West R (1979) J Amer Chem Soc, 101: 5427
71. Gaspar PP, Holter D, Konieczny C and Corey JY (1987) Acc Chem Res 20, 329 and references cited therein
72. Wilt JW (1983) in "Reactive Intermediates", Editor Abramovitch, R.A. Plenum, New York, Chapter 3
73. Trefonas III PT, West R and Miller RD (1985) J Amer Chem Soc 107: 2737
74. Miller RD (1990) In: John Zeigler M and Gordon Fearon FW (eds) Silicon-based-polymere science, A comprehensive Resource, Advances in Chemistry series Nr. 224, American Chemical Society, Washington DC, p 413
75. Michl J, Downing JW, Karatsu T, Klingensmith KA, Wallraff GM and Miller RD (1988) in "Inorganic and Organometallic Polymers", Editors: Zeldin M, Wynne KJ and Allcock HR, ACS Symposium series 360; American Chemical Society; Washington, DC Chapter 4
76. McKinley AJ, Karatsu T, Wallraff GM, Miller RD, Sooriyakumaran R, and Michl J (1988) J Organometallics, 7: 2569
77. Miller RD, Guillet JE, and Moore J (1988) J. Polym Prepr 29: 552
78. Trefonas III P, Miller RD and West R (1981) J Polym Sci Polym Lett Ed 21: 823
79. West R, Zhang X-H, Djurovich PI and Stüger H (1986) in "Science of Ceramic Chemical Processing", Editors Hench LL and Ulrich R D; Wiley, New York; Chapter 36: page 337
80. Yajima S (1983) Am Ceram Soc Bull 62: 893
81. Cotts PM, Miller RD, Trefonas III PT, West R and Fickes G (1987) Macromolecules 20: 1046
82. Aitken CT, Harrod JF and Samuel E (1985) J Organometal Chem 279: C11;
    Aitken CT, Harrod JF, Samuel E (1986) J Amer Chem Soc 108: 4059;
    Aitken CT, Harrod JF, Samuel E (1986) Can J Chem 64: 1677
83. West R in "The Chemistry of Functional Groups", Series Editor Patai S, "The Chemistry of Organic Silicon Compounds", Editors Patai S and Rappoport Z, An Interscience Publication, J Wiley, Chichester, 1989, S 1225
84. Zeigler JM, Harrah LA and Johnson AW (1985) Proc SPIE 539: 166
85. Trefonas III PT, West R, Miller RD and Hofer D (1983) J Polym Sci Polym Lett Ed 21: 823
86. West RJ (1986) J Organometal Chem 300: 327 and references cited therein
87. Harrah LA und Zeigler JM (1987) Macromolecules 20: 2039
88. a) Zeigler JM (1989) Synthetic Metals 28: C 581
    b) Miller RD and Michl J (1989) Chem Rev 89: 1359
89. Miller RD, Hofer D, Rabolt J, Sooriyakumaran R, Willson CG, Fickes GN, Guillet JE and Moore J in "Polymers for High Technology: Electronics and Photonics"; ACS Symp Ser 346; Editor Bowden MJ and Turner SR, American Chemical Society, Washington DC, 1987, page 170
90. Farmer BL, Rabolt JF and Miller RD (1987) Macromolecules 20: 1169
91. a) Rabolt JR, Hofer D, Miller RD and Fickes GN (1986) Macromolecules 19: 611
    b) Miller RD and Sooriyakumaran R (1988) ibid 21: 3120
92. Kuzmany H, Rabolt JF, Farmer BL and Miller RD (1986) J Chem Phys 85: 7413
93. Harrah LA und Zeigler JM in "Photophysics of Polymers", Editors Hoyle CE and Torkelson JM; ACS Symposium Series 358; American Chemical Society, Washington DC, 1987, page 482–498
94. Welsh WJ (1990) Polym Prepr (Am Chem Soc Div Polym Chem) 31(2), 232
95. Matsumoto N, Takeda K, Teamea H and Fujino M In: Silicon-Based-Polymer Science, A Comprehensive Resource, Edited by John Zeigler M and Gordon Fearon FW, Advances in Chemistry Series Nr 224, American Chemical Society, Washington DC 1990, page 515
96. Kepler RG and Zeigler JM In: John Zeigler M and Gordon Fearon FW (eds) Silicon-Based-Polymer Science, A Comprehensive Resource, Advances in Chemistry Series Nr 224, American Chemical Society, Washington DC 1990, page 459
97. Abkowitz MA, Stolka M, Weagley RJ, McGrane KM and Knier FE ibid, page 467
98. Lee YL et al. (1991) Polym Prepr Am Chem Soc Div Polym Chem 32(3), 228–229
99. Rabolt JF (1990) Polym Prepr, Am Chem Soc Div Polym Chem 31(2), 262
100. Ban H and Sukegawa K (1987) J Appl Polym Sci 33: 2787

101. Miller RD, Hofer D, McKean DR, Wilson CG, West R and Trefonas III PT in "Materials for Microlithography"; Editors: Thompson LF, Wilson CG and Frechet JM; ACS Symposium Series 266, American Chemical Society, Washington, DC, 1984, page 294
102. Pannell KH, Rozell JM Jr, Vincenti S In: Silicon-Based-Polymer Science, A Comprehensive Resource, Edited by John Zeigler M and Gordon Fearon FW, Advances in Chemistry Series Nr 224, American Chemical Society, Washington DC 1990, page 329
103. Miller RD, Hofger D, McKean DR, Wilson CG, West R and Trefonas III P in "Materialsfor Microlithography", ACS Symposium Series 266; Editors: Thompson LF, Wilson CG and Frechet TMJ; American Chemical Society, Washington DC, 1984, Chapter 14
104. Miller RD, Thompson D, Sooriyakumaran R and Fickes GN (1991) in Journal of Polymer Science, Part A: Polymer Chemistry, Vol 29: page 813–824
105. Worsfold DJ in "Inorganic and Organometallic Polymers", ACS Symposium Series Nr. 360; Editors: Zeldin M, Wynne KJ and Allcock HR; American chemical Society, Washington DC, 1988, Chapter 9
106. Kim HK and Matyjaszewsky K (1988) J Amer Chem Soc 110: 3321
107. Miller RD, Hofer D, Rabolt J, Sooriyakumaran R, Wilson CG, Fickes GN, Guillet JE and Moore J in "Soluble Polysilanes in Photolithography", ACS Symposium Series 346; Editors: Bowden MJ and Turner SR; American Chemical Society, Washington DC, 1987, Chapter 15
108. Kochs P in "Methoden der Organischen Chemie", Makromolekulare Stoffe, Band E20, page 2219–2236 (Houben-Weyl), Georg Thieme, Stuttgart
109. West R in "Ultrastructure Processing of Ceramics, Glasses and Composites" (Editors: Hench LL and Ulrich RD, Chapter 19, part 3, Wiley, New York, 1984
110. Miller RD, Hofer D, McKean DR, Wilson CG, West R and Trefonas P III in "Materialsfor Microlithography" (Editors: Thompson L, Wilson CG und Frechet JMJ), ACS Symp Ser 266, American Chemical Society., Washington DC, 1984, page 293–310
111. Zeigler JM, Harrah LA and Johnson AW (1985) SPIE Vol. 539, Advances in Resist Technology and Processing, 166
112. Miller RD, Rabolt JF, Sooriyakumaran R, Fleming W, Fickes GN, Farmer BL and Kuzmay H in "Inorganicand Organometallic Polymers" (Editors; Zeldin M, Wynne KJ and Allock HR), ACS Symp Ser 360, American Chemical Society, Washington, DC, 1988 (Chapter 4)
113. Wolff A and West R (1987) Appl Organomet Chem, 1: 7
114. West R, Wolff AR and Peterson DJ (1986) J Radiat Curing 13: 35
115. Unpublished results of WACKER CHEMIE GmbH, Munich (Germany) in cooperation with HAZLETON, Lyon (France)
116. Ichikawa H, Machino F, Teranishi H and Ishikawa T in Silicon-Based-Polymer Science, A Comprehensive Resource, Edited by John M Zeigler and Gordon Fearon FW, Advances in Chemistry Series Nr. 224, American Chemical Society, Washington DC 1990, S. 619
117. Reichmanis E, Novembre AE, Tarascon RG, Shugard A and Thompson LF in Silicon-Based-Polymer Science, A Comprehensive Resource, Ediited by John M Zeigler and Gordon Fearon FW, Advances in Chemistry Series Nr 224, American Chemical Society, Washington DC 1990, S 265
118. Seyferth D (1988) Am Chem Soc Symp Ser 360: 21
119. Hasegawa Y and Okamura K (1986) J Mater Sci 21: 321
120. Yajima S (1983) Am Ceram Soc Bull 62: 893
121. Ichikawa H, Machino F, Teranishi H and Ishikawa T in: Silicon-Based-Polymer Science, A Comprehensive Resource, Edited by John M Zeigler and Gordon Fearon FW, Advances in Chemistry Series Nr 224, American Chemical Society, Washington DC 1990, S 619
122. a) Sommer LH, Goldberg GM, Gold J and Whitmore FC (1947) J Amer Chem Soc 69: 980
     b) Bluestein B (1948) J Amer Chem Soc 70: 3068
123. Goodwin Jr JT, Baldwin WE and McGregor RR (1947) J Amer Chem Soc 69: 2247
124. Sommer LH, Mitch FA and Goldberg GM (1949) J Amer Chem Soc 71: 2746
125. Pat US 2 483 972 (1949), Dow Corning (inv.: Goodwin JT)
126. Kriner WA (1964) J Org Chem 29: 1601
127. Jung WH and Interrante L (1991) Polym Prepr 32(3), 588

128. Nametkin NS, Gusel'nikov LE, Vdovin WM, Grinberg PL, Zavyalov VI and Oppengeim VD (1966) Dokl Akad Nauk SSSR 171(3), 630
129. Damrauer R (1972) Organometal Chem Rev A: 8: 67
130. Auner N and Grobe J (1980) J Organomet Chem 188: 151
131. a) Auner N, Grobe J (1980) J Organomet Chem 197: 13
     b) Auner N, Grobe J (1980) J Organomet Chem 197: 147
     c) Auner N and Grobe J (1982) Z anorg allg Chem 485: 53
132. Golino CM, Bush RD and Sommer LH (1975) J Amer Chem Soc 97: 25
133. Fujino M, Hisaki T, Fujiki M and Matsumoto N (1992) Macromolecules 25(3), 1079
134. Baum GA, ph.D Thesis, The Pennsylvania State University, 1955
135. Vdovin VM, Pushchevaya JS and Petrov AD (1961) Dokl Akad Nauk SSSR 141: 843
     Nametkin NS and Vdovin VM (1963) J Polym Sci C4: 1043
136. Weyenberg DR and Nelson LE (1965) J Org Chem 30: 2618
137. Levin G and Carmichael JB (1968) J Polym Sci Part A-1: Vol 6, 1
138. Kriner WA (1966) J Polym Sci., Part A-1: Vol 4, 444
139. Bamford WR, Lovie JC and Watt JAC (1966) J Chem Soc C: 1137
140. Wu H and Interrante LV (1992) Macromolecules 25(6): 1940
141. a) Mark JE and Ko JH (1975) Macromolecules 8: 874
     b) Llorente A, Mark JE und Saiz E (1983) J Polym Sci., Polymer Phys Ed., 21: 1173
     c) Galiatsatos V and Mark JE (1987) Polymer Preprints 28: 258
142. Pat US 2 850 514 (1958), E.I. du Pont de Nemours (inv.: Knoth WH, Jr.)
143. Nametkin NS, Vdovin VM and Zavyalov VI (1965) Dokl Akad Nauk SSSR 162: 824
144. Review of ring opening reactions of silacyclobutanes: Nametkin NS and Vdovin VM (1974)Izv Akad Nauk SSSR: Ser Khim 1153
145. Nametkin NS, Vdovin VM and Zavyalov VI (1964) Izv Akad Nauk SSSR: Ser Khim 203
146. Seyferth D and Mercer J, not published, 1976, cited in reference 118
147. Pat US 2 381 000 und 2 381 002 (1945), General Electric (inv.: Patnode WI and Schiessler RW)
148. Fritz G and Worsching A (1984) Z anorg allg Chem 512: 103
149. Fritz G and Worns K-P (1984) Z anorg allg Chem 512: 103
150. Fritz G, Grobe J and Kummer D (1965) Advan Inorg Chem Radiochem 7: 349
151. DOL 2 236 078 (1974), Bayer AG (inv.: Verbeek M and Winter G)
152. Carlsson DJ (1990) Polym Prepr (Am Chem Soc Div Polym Chem.) 31(2), 268
153. van Aefferden B, Habel W and Sartori P (1990) Chemiker-Zeitung, 114 Jahrgang Nr 12, page 367
154. ibid., 115 Jahrgang (1991) Nr 10, page 309
155. ibid., 114 Jahrgang (1990) Nr 10, page 277
156. Whitmarsh C and Interrante LV (1991) Organometallics 10(5), 1336–1344
157. Seyferth D and Lang H (1991) Organometallics 10(3), 551–558
158. Dunogues J et al (1991) Chem Mater 3(2), 348–355
159. a) Schilling Jr CL, Wesson JP and Williams T (1983) Ceram Bull 62: 912
     b) Schilling Jr CL (1983) ONR Technical Report, No AD-A141546
160. Habel W, Mayer L and Sartori P (1991) Chemiker-Zeitung, 115 Jahrgang Nr 11, page 301
161. Curry JW (1956) J Amer Chem Soc 78: 1686
162. Curry JW (1961) J Org Chem 26: 1308
163. Corrin JP, Boury B, Leclerg D, Mutin PH, Planeix LU, Vioux A (1991) Organometallics 10(5), 1457
164. a) Vdovin VM, Pushchevaya KS, Belikova NA, Sultanov R, Plate AF and Petrov AD (1961) Dokl Akad Nauk SSSR 136
     b) Nametkin NS, Vdovin VM, Pushchevaya KS and Zavyalov V (1965) Izv Akad Nauk SSSR: Ser Khim 1453
165. a) Nametkin NS, Vdovin VM, Finkelshtein E Sh, Yatsenko MS and Ushakov NV (1969) Vysokomolekul Soedin B 11: 207
     b) Salamone JC and Fitch WL (1971) J Polymer Sci.: Part A 9: 1741
166. Horvath RF and Chan TH (1987) J Org Chem 52: 4489

167. Zhang X, Zhou Q, Weber WP, Horvath RF, Chan T-H and Manuel G In: Silicon-Based-Polymer Science, A Comprehensive Resource, Edited by John M Zeigler and Gordon Fearon FW, Advances in Chemistry Series Nr 224, American Chemical Society, Washington DC 1990, S 679
168. Habel W, Mayer L and Sartori P (1992) J prakt Chem 334: 327
169. Lammens H, Sartori G, Siffert J and Sprecher N (1971) J Polmer Sci Part B, Polymer Lett 9: 341
170. Luneva LK, Sladkov AM and Korshak VV (1967) Vysokomolekul Soedin A9, Nr 4: 910
171. Seyferth D, Third International Conference on Ultrastructure Processing of Ceramics, Glasses and Composites, San Diego, February 23–27, 1987
172. Barton Th, Maghsoodi SI, Pang Y (1991) Macromolecules 24(6), 1257
173. Laine RM, Blum YD, Chow A, Hamlin R, Schwartz KB and Rowcliffe DJ (1987) Polym Prepr (Amer Chem Soc.: Div Polym Chem.) 28: 393
174. Stock A and Somiesky K (1921) Ber Dtsch Chem Ges 54: 740
175. Wannagat U, Bogusch E and Geymayer P (1964) Monatshefte für Chemie 95(3), 52
176. Krüger CR and Rochow EG (1964) J Polymer Sci.: Part A, Vol 2, 3179
177. Fink W (1966) Angew Chem 78: 803
178. Brewer SD and Haber CP (1948) J Amer Chem Soc 70: 3888
179. Chugunov VS (1953) J allg Chemie (russ.) 23: 777 (1954); Chem Abstr 48: 4461
180. Sommer LH and Tyler LY (1954) J Amer Chem Soc 76: 1030
181. Duguet E, Schappacher M and Soum A (1992) Macromolecules 25(19), 4835
182. Zoeckler MT and Laine RM (1983) J Org Chem 48: 2539
183. Blum YD, Laine RM, Schwartz KB, Rowcliffe DJ, Bening RC and Cotts DB (1986) Mater Res Soc Symp Proc 73: 389
184. Blum YD and Laine RM (1986) Organometallics 5: 2081
185. Blum YD, Schwartz KB and Laine RM (1989) J Mater Sci 24: 1707
186. Arai M, Sakurada S, Isoda T and Tomizawa T (1987) Polym Prep (Amer Chem Soc., Div Polym Chem) 28: 407
187. a) Zhdanov AA, Kotrelev GV, Kazakova VV and Redkozubova YeP (1985) Polym Sci USSR 27: 1593
    b) Adrianov KA, Ismailov BA, Konov AM and Kotrelev GV (1965) J Organomet Chem 3: 129
    c) Adrianov KA, Kotrelev GV, Kamaritski BA, Unitski IH and Sidorova NI (1969) J Organomet Chem 16: 51
188. Redl G and Rochow EG (1964) Angew Chem 76: 650
189. Penn BG, Daniels JG, Ledbetter III FE and Clemons JM (1986) Poly Eng and Sci 26: 1191 ibid (1982) J Appl Polm Sci 27: 3751
190. Penn BG, Ledbetter III FE and Clemons JM (1984) Ind Eng Chem Process Res Div 23: 217
191. a) Wynne KJ and Rice RW (1984) Ann Rev Mater Sci 14: 297
    b) Wills RR, Mark RA and Mukherjee SA (1983) Am Ceram Soc Bull 62: 904
    Rice RR (1983) Cer Bull 62: 889
192. Laine, RM, Blum YD, Tse D and Glaser R (1988) Inorganic and Organometallic Polymers, ACS Symposium Series 360: Chapter 10, 124
193. Legrow GE, Lim Th.F, Lipowitz J and Reaoch RS (1988) Mat Res Soc Symp Proc 73: 553
194. Aylett BJ (1968) J Organomet Chem Rev 3: 151
195. Seyferth D In: John Zeigler M and Gordon Fearon FW (eds) Silicon-Based-Polymer Science, A Comprehensive Resource, Advances in Chemistry Series Nr 224, American Chemical Society, Washington DC 1990, S 565
196. Neth Appl 6 507 996 (1965), Monsanto Company; Chem Abstr 64: 19677 (1964)
197. Brewer SD and Haber CP (1948) J Amer Chem Soc 70: 3888
198. Peuckert M, Vaahs T and Brück M (1990) Adv Mater 2(9), 398
199. Datasheets of Hoechst AG, Frankfurt "Keramik aus Polymeren, Polysilazan VT 50 und ET 70", Frankfurt (1991)
200. Kulpinski J and Lasocki Z (1992) J Polym Sci.: Part A 30(3), 461
201. Penn BG, Ledbetter III FE, Clemons JM and Daniels JG (1982) J appl Poly Sci 27: 3751

202. Umezawa M, Ichikawa H, Ichikawa T and Nonaka T (1992) Denki Kagaku oyobi Kogyo Butsuri Kagaku 60(8): 743
203. Hengge E, Jammegg CH, Kalchauer W and Weinberger M (1992) Wacker Chemie GmbH Deutsche Offenlegungsschrift 41 10 917 A1
204. Kalchauer W, Pachaly B, Geisberger G and Rösch L (1992) Z anorg allg Chem 618: 148
205. van der Made AW and van Leeuwen PWNM (1992) J Chem Soc Chem Comm 1400
206. Hengge E and Weinberger M (1993) J Organomet Chem 443: 167
207. Harrod JF (1988) ACS Symp Ser 360: 89; Harrod JF (1988) NATO ASI Ser Ser E 141: 103
208. Kim HK and Matyjazewski K (1993) J Polym Sci, Polym Chem Part A 31(2), 399
209. Biran C, Bordeau M and Leger MP (1992) NATO ASI Ser E 206: 79
210. Yamashita H, Prabhakar Reddy N and Tanaka M (1993) Macromolecules 26: 2143

CHAPTER 5

Polycondensation of Silylated Monomers

H. R. Kricheldorf

5.1
Introduction

Silylated monomers useful for polycondensations are normally compounds with one or two trimethylsilyl groups attached to heteroatoms and replacing protons. The monomers most widely used for polycondensations in the form of trimethylsilyl derivatives are:

> aliphatic and aromatic diamines,aminophenols,aliphatic or aromatic amino acids, diphenols, bisthiophenols or hydroxythiophenols, aromatic hydroxy acids, aliphatic and aromatic dicarboxylic acids.

The advantage of using silylated monomers instead of the protonated precursors is fourfold. First, polycondensation of silylated monomers with halogen containing electrophiles (e.g. acid chlorides or difluoroaromatics) avoids the liberation of free hydrochloride or hydrofluoric acid, and thus, acid catalyzed side reactions. Second, the liberated trimethylsilyl derivatives, such as fluoro or chlorotrimethylsilane, trimethylsilylacetate, or hexamethyldisiloxane are highly volatile and easy to remove from the reaction mixture. Third; silylation improves the solubility of monomers in a given reaction medium or reduces the melting point. This latter aspect is, for instance, of interest for bulk condensations of high melting aromatic or heterocyclic dicarboxylic acids (T_m usually > 300 °C). Fourth, silylation enhances the volatility of monomers due to the elimination of H-bonds. The combination of silylation and distillation is an effective method of purification for many monomers. Aromatic diamines, aminophenols and hydroquinones, which are often contaminated with oxidation products or inorganic salts are typical substrates for such a purification procedure.

In almost all polycondensations silylated monomers play typically the role of the nucleophile. This situation is partially a consequence of the electropositive character of silicon which enhances the nucleophilicity of heteroatoms attached to silicon. Only one class of polycondensation involving a silylated electrophilic group (silylated carboxylic acids) seems to be known. In this connection it is noteworthy that the silylation of heteroatoms may have a significant influence on their reactivity. Silyl groups may play the role of protecting groups for O- or S-functionalities because they prevent deprotonation, and thus, the formation of nucleophilic anions. On the other hand, silylation of poorly nucleophilic aromatic amines, ureas, or lactams may enhance their nucleophilicity. In addition to avoiding deprotonation of the parent X-H group the influence of the silyl-group is based on two somewhat opposite interactions with the attached heteroatom:

A) the positive inductive effect via the σ-bond,
B) the electron-withdrawing effect of a $p\pi$-$d\pi$-bond

In addition of polycondensations of silylated monomers the present review will also discuss a few condensations in which silicon compounds play the role of catalysts or condensing reagents. In such cases silylated monomers are inter-mediately formed. Furthermore, a few polyadditions of silylated monomers are reported. Silylation procedures and properties of silylating reagents are discussed in Appendix B.

From the historical point of view it should be mentioned that the first poly-condensations of silylated monomers were those of N, N'-bissilylated piperazines with α, α'-dichlorxylylene (J.F. Klebe 1964) [1] the synthesis of polyimides from N,N'-bissilylated diamines (Boldebuch and Klebe 1967) [2] and various polycondensation of silylated bisthiocarboxylic acids (Kricheldorf and Leppert 1972) [3]. The first polyadditions of silylated monomers were described by Klebe in 1964 [4] who prepared polyureas from N,N'-bissilylated diamines and diisocyanates. Several reviews concentrating on selected aspects of the chemistry of silylated monomers have been published since 1964 [5–11]. A comprehensive review on polycondensation and polyaddition of silylated monomers was pub-blished by Katsarava and Vygodskii [12].

5.2
Polyureas and Related Polymers

The first synthesis of polyureas from silylated monomers was described by Klebe[13]. He reacted 2,4-toluylene diisocyanate with a variety of silylated diamines in refluxing dry toluene or THF, Eqs. (1, 2). The silylated polyureas formed in this way were soluble in the reaction media in contrast to nonsilylated polyureas. Klebe isolated the silylated polyureas, for instance, in the form of films. High molecular weights ($M_w = 120\ 000$) determined by light scattering were reported, but it is not clear, if association was completely avoided.

The N—Si-bonds of silylated polyureas are sensitive to hydrolysis, and exposure to moist air or precipitation into methanol removes the trimethylsilyl groups. Sentsova et al. [14] utilized this approach for the synthesis of silicon free

$$(1)$$

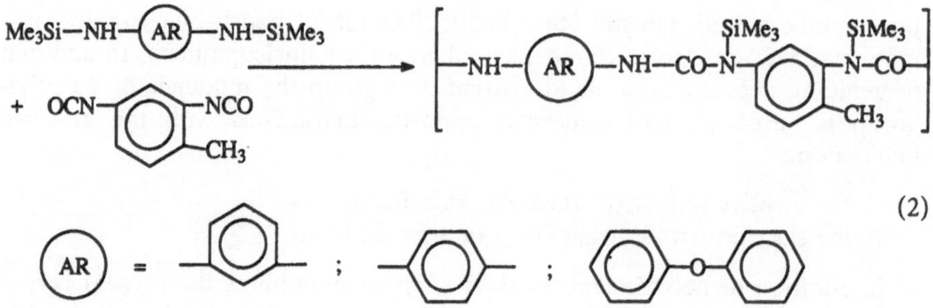

$Me_3Si-NH-(AR)-NH-SiMe_3$

$+ \quad OCN-\langle\ \rangle-NCO$ (with CH_3)

$\longrightarrow$

$\left[-NH-(AR)-NH-CO-N(SiMe_3)-\langle\ \rangle-N(SiMe_3)-CO- \right]$ (with CH_3)

(2)

$(AR) = -\langle\ \rangle- \quad ; \quad -\langle\ \rangle- \quad ; \quad \langle\ \rangle-O-\langle\ \rangle-$

polyureas from natural amino acids, such as ornithine, lysine or cystine. For this purpose not only the trimethylsilyl derivatives of these diamino acids but also their diisocyanates were used as starting materials Eq.[3]. Later Imai and coworkers [15] produced a variety of aromatic polyureas such as Form. 4 by the same method, and studied the influence of the reaction medium on the molecular weights. However, a strong solvent effect was not found.

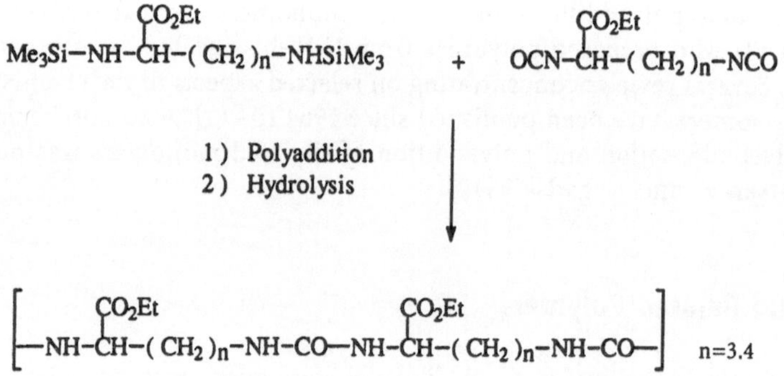

$$Me_3Si-NH-\underset{\underset{CO_2Et}{|}}{CH}-(CH_2)_n-NHSiMe_3 \quad + \quad OCN-\underset{\underset{CO_2Et}{|}}{CH}-(CH_2)_n-NCO$$

1) Polyaddition
2) Hydrolysis

(3)

$$\left[-NH-\underset{\underset{CO_2Et}{|}}{CH}-(CH_2)_n-NH-CO-NH-\underset{\underset{CO_2Et}{|}}{CH}-(CH_2)_n-NH-CO- \right] \quad n=3.4$$

$$\left[-NH-\langle\ \rangle-O-\langle\ \rangle-NH-CO-NH-\langle\ \rangle-N-CO- \right]$$ (with CH_3)

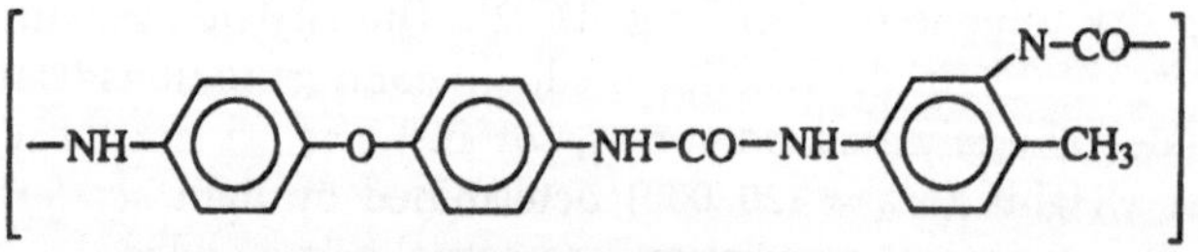

(4)

A different approach for the synthesis of polyhexamethylene urea was reported by Di Salvo [16]. 1,6-Diaminohexane was heated with trimethylsilylisocyanate in a sealed tube Eq. (5). Furthermore, the bisurea of 1,6-diaminohexane was heated with hexamethyldisilazane. In both cases a complex polycondensation and degradation process occured and the polyureas were isolated in moderate yields and low molecular weights ($M_n < 2500$). Various aliphatic polyureas were prepared in higher yields (70 – 96 %) by Katsavara et al. [18] by polycondensation

$$NH_2-(CH_2)_6-NH_2 \quad \xrightarrow{(Me_3SiNCO)} \quad \left[-NH-(CH_2)_6-NH-CO- \right]$$ (5)

of silylated diamines with bis-(4-nitrophenyl)carbonate, Eq. (6). The ethylester of silylated l-lysine was also successfully condensed by this method. An alternative approach starts with the synthesis of a bis(4-nitrophenylcarbamate) of 1,6-diaminohexane followed by its polycondensation with silylated aliphatic diamines (Eqs. 7, 8). Silylated diamines as reaction partners of activated carbonates and carbamates (or activated esters, see below) have the advantage that the generation of free acidic phenols is avoided, which can protonate or form complexes with the diamine.

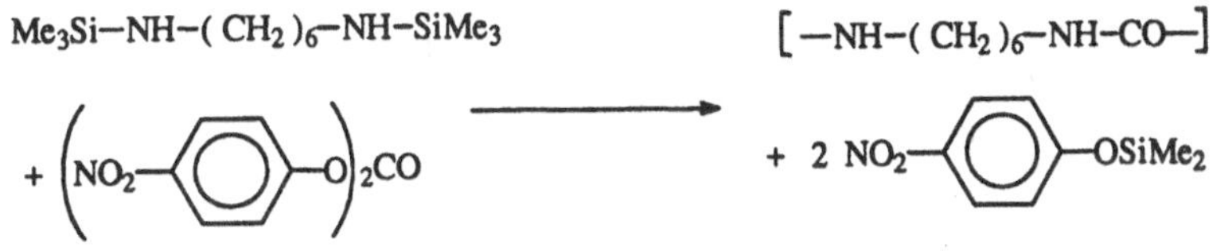

$$Me_3Si-NH-(CH_2)_6-NH-SiMe_3 \longrightarrow \left[-NH-(CH_2)_6-NH-CO-\right]$$

$$+ \left(NO_2-\bigcirc-O\right)_2CO \qquad + 2\ NO_2-\bigcirc-OSiMe_2 \tag{6}$$

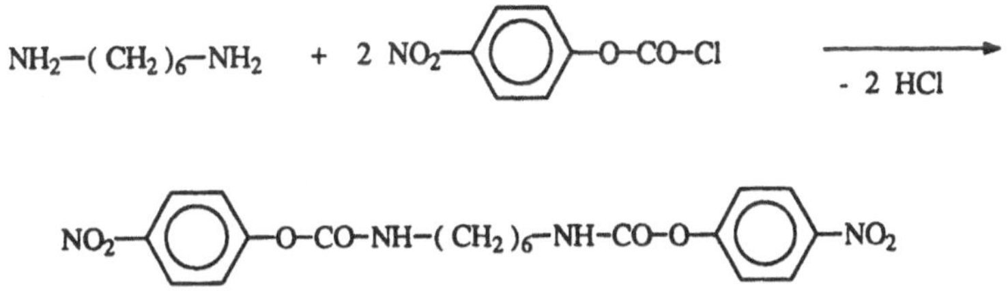

$$NH_2-(CH_2)_6-NH_2 \quad + \quad 2\ NO_2-\bigcirc-O-CO-Cl \xrightarrow[-2\ HCl]{} \tag{7}$$

$$NO_2-\bigcirc-O-CO-NH-(CH_2)_6-NH-CO-O-\bigcirc-NO_2$$

$$Me_3SiNH-\overset{\overset{\displaystyle CO_2CH_3}{|}}{CH}-(CH_2)_4-NHSiMe_3 \quad \Big\downarrow \quad -2\ NO_2-\bigcirc-O-SiMe_3 \tag{8}$$

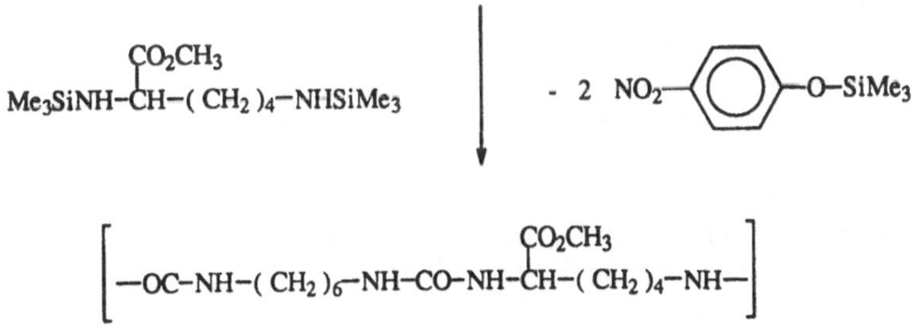

$$\left[-OC-NH-(CH_2)_6-NH-CO-NH-\overset{\overset{\displaystyle CO_2CH_3}{|}}{CH}-(CH_2)_4-NH-\right]$$

Polyureas derived from 1,4-dianilino-benzene were prepared by Oishi et al. [19] from its N,N'-bischloroformyl derivative and silylated secondary diamines, such as piperazine, 4,4'-diaminodiphenylether or 1,6-dianilinobenzene Eq. (9). Heating of the reactants in sulfolane up to temperatures of 220 °C was required for a complete conversion. In this way polyureas entirely devoid of H-bonds were prepared. High yields but only low or moderate inherent viscosities ($\eta_{in} < 0.35$

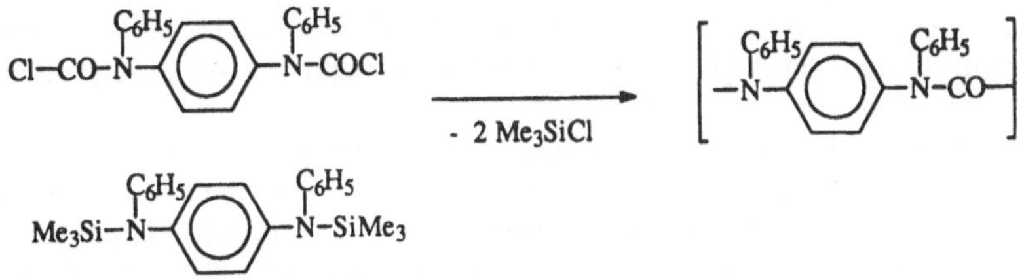

$$Cl-CO-\overset{\overset{\displaystyle C_6H_5}{|}}{N}-\bigcirc-\overset{\overset{\displaystyle C_6H_5}{|}}{N}-COCl \xrightarrow[-2\ Me_3SiCl]{} \left[-\overset{\overset{\displaystyle C_6H_5}{|}}{N}-\bigcirc-\overset{\overset{\displaystyle C_6H_5}{|}}{N}-CO-\right]$$

$$Me_3Si-\overset{\overset{\displaystyle C_6H_5}{|}}{N}-\bigcirc-\overset{\overset{\displaystyle C_6H_5}{|}}{N}-SiMe_3 \tag{9}$$

dl/g in NMP or H_2SO_4) were obtained. The fully aromatic polymer of Eq. (9) possesses the expected high thermostability [19].

Another group of thermostable polymers containing a cyclic urea in the backbone was prepared by Kricheldorf and Stöber[20] from N,N'-bistrimethylsilyl imidazolidone, Eq. (10). Whereas the silylation of primary amines does not

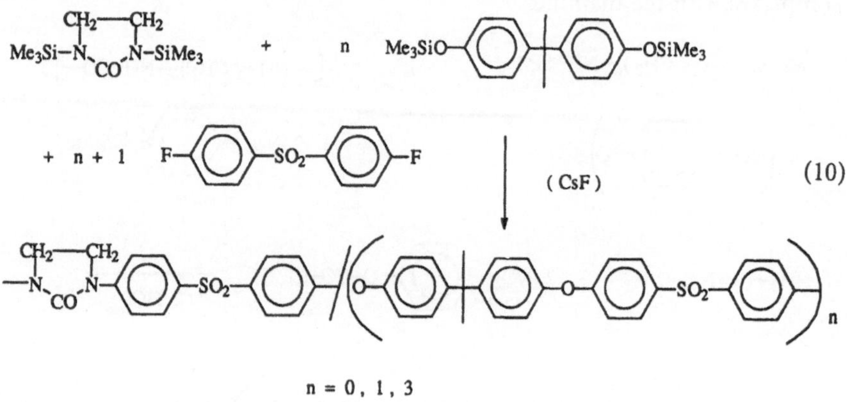

$$(10)$$

$$n = 0, 1, 3$$

significantly modify their nucleophilicity, the silylation of imidazolidone raises its reactivity by several orders of magnitude. Therefore polycondensation with activated difluoroaromatics proved to the successfull in the presence of CsF as catalyst. However, activated dichloroaromatics, such as 4,4'-dichlorodiphenylsulfone, were not electrophilic enough for a polycondensation. The product obtained from silylated imidazolidone and 4,4'-difluorodiphenylsulfone is an insoluble and infusible material (Eq. 10, $n = 0$). However, cocondensation with silylated bisphenol-A yielded soluble and meltable poly(ether-sulfone)s (Eq. 10, $n = 1.3$). The incorporation of the imidazolidone raises the T_g compared to the

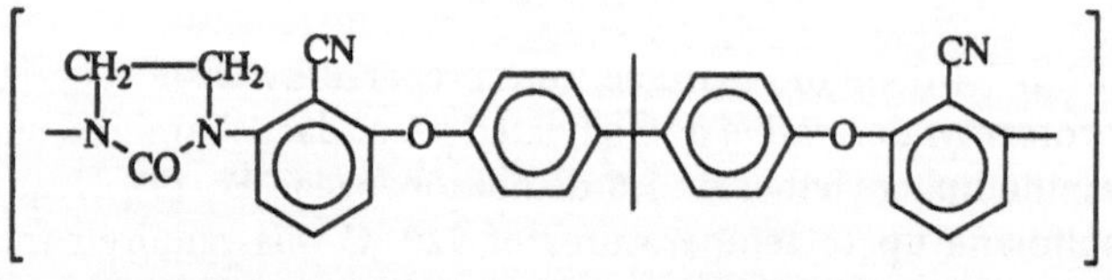

$$(11)$$

homopoly(ether-sulfone). Further poly(imidazolidone-ether)s were synthesized from 2,6-difluorobenzonitrile (Form. 11) or 2,6-difluoropyridine. High yields and moderate inherent viscosities were found in all cases. The N,N'-bissilylated benzoimidazolidone proved to be not reactive enough for such polycondensations.

In addition to the quasi normal polyureas discussed above several classes of so called poly(N-acylurea)s were synthesized in various ways. For instance, polycondensation of N,N''-bistrimethylsilyl imidazolidone with aliphatic dicarboxylic acid chlorides yields poly(N-acylimidazolidone)s as a new class of highly crystalline materials, Eq. (12). This polycondensation requires chloride ions as catalysts and gives high yields and moderate molecular weights. Remarkable are the violent reactions observed when succinylchloride or oxalylchloride were used as reaction partners. In these cases mainly volatile reaction products were formed. Cocondensation of silylated imidazolidone with silylated diphenols and dicarboxylic acid dichorides yields polyesters containing N-acylimidazolidone units, Eq. (13) [20].

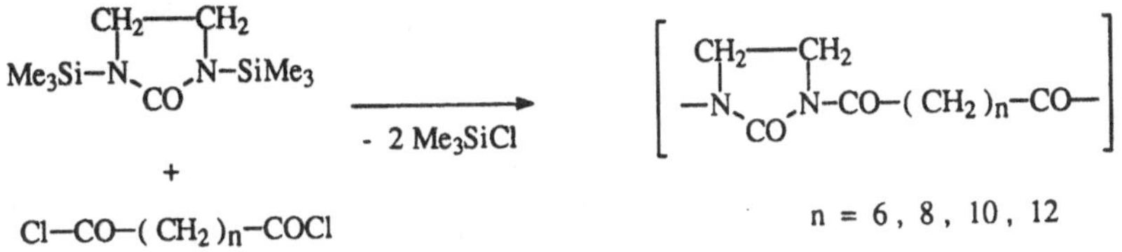

$$n = 6, 8, 10, 12 \tag{12}$$

$$(Cl^{\ominus}) \tag{13}$$

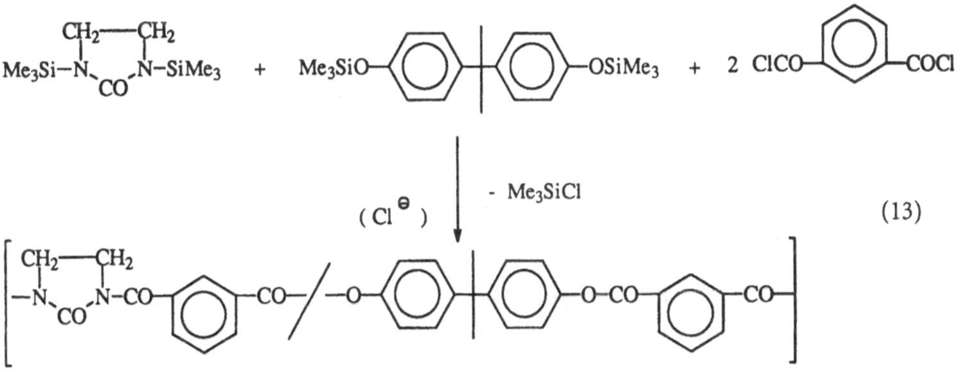

A special class of N-acylated polyureas are poly(allophanate)s. Polyurethanes and polyureas containing allophanate groups as branching points are known for decades. In these cases the allophanate groups results from addition of urethane groups onto isocyanates. Linear polyallophanates, such as those of Eq. (14), have never been synthesized before. Kricheldorf et al. [21] obtained these polyallophanates by polyaddition of silylated diamines onto butane-1,4-bis-(isocyanatoformate). In analogy to Klebe's synthesis of polyureas soluble silylated polymers were formed in the first stage and precipitation of the reaction mixture into methanol yields the desilylated polymers. High yields and moderate viscosities were found in all cases. Like polyurethanes,

$$Me_3Si-NH-(CH_2)_n-NH-SiMe_3 \quad + \quad OCN-CO-O-(CH_2)_4-O-CO-NCO$$

1) Polyaddition
2) Methanolysis

$$\tag{14}$$

$$[-NH-(CH_2)_n-NH-CO-NH-CO-O-(CH_2)_4-O-CO-NH-CO-]$$

polyallophanates are thermaly unstable above 200 °C and reversible melting and crystallization is not feasible, when the melting points are high.

Finally the synthesis of linear polybiurets by Kricheldorf et al. [22–24] should be mentioned. Trisubstituted biuret groups are usually byproducts of the synthesis of polyurethanes and polyureas resulting from the addition of urea groups onto isocyanates [25]. Linear polybiurets were prepared for the first time by a mixed polycondensation/polyaddition of silylated diamines with chloroformylisocyanate, Eq. (15). The silylated polymers formed in the first stage loose their silyl groups in contact with moisture or methanol Eq. (16). When aromatic diamines are used as starting materials the biurets precipitate in an

$$Me_3Si-NH-(CH_2)_n-NH-SiMe_3 \quad \xrightarrow{- Me_3SiCl} \quad \left[\begin{array}{c} SiMe_3 \\ | \\ -N-(CH_2)_n-NH-CO-NH-CO- \end{array} \right] \quad (15)$$

$$+ \ ClCO-NCO$$

$$\downarrow \ CH_3OH \quad - \ CH_3OSiMe_3$$

$$\left[-NH-(CH_2)_n-NH-CO-NH-CO- \right] \quad (16)$$

oligomeric form from inert reaction media such as CH_2Cl_2 or $CHCl_3$. However, polar solvents such as acetonitrile, DMF or DMSO cannot be used, because they react with chloroformylisocyanate. Linear polybiurets are slowly crystallizing materials with unexpectedly low crystallinities and melting points. Despite intensive spectroscopic studies, their structure-property relationships are not fully understood [22].

Multiblock polybiurets containing polyether segments (Form. 17) can be prepared in "one-pot"-procedure by cocondensation of poly(ethylene oxide)s containing two silylated amino endgroups and other silylated diamines [23]. The

$$\left[-NH-\overset{CH_3}{\underset{|}{CH}}-CH_2-(O-CH_2-CH_2)_n-O-CH_2-\overset{CH_3}{\underset{|}{CH}}-NH-CO-(-NH-CO-NH-(A)-NH-CO)_m NH-CO-)_n \right]$$

$$(A) = -(CH_2)_6- \ ; \ -\text{⟨O⟩}-CH_2-\text{⟨O⟩}- \quad (17)$$

average lengths of the hard segments vary with the molar ratio of silylated diamine and poly(ethylene-oxide). Polybiurets are more sensitive to hydrolytic degradation than polyureas [23]. When a natural diamine such as the ethylester of l-lysine is used as building block [23], segmented polybiurets might be useful as biode gradable materials for pharmaceutical applications. A second approach which allows the synthesis of segmented polybiurets quite conveniently consists of a stepwise cocondensation of poly(ethylene glycol)s with chloroformylisocyanate and with silylated diamines, Eq. (18) [24]. The resulting multiblock copolymers differ from the former class (Form. 17) in that the ether and biuret blocks are linked by allophanate groups which are more sensitive to hydrolysis than biuret groups. This second approach is also more versatile, because it allows the synthesis of various segmented polybiurets from oligomeric or polymeric diols (e.g. THF-diol).

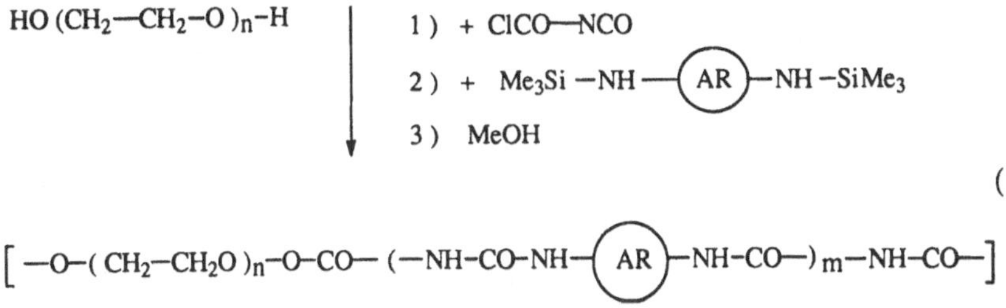

$$HO\,(CH_2{-}CH_2{-}O\,)_n{-}H$$

1) + ClCO—NCO

2) + Me$_3$Si —NH—(AR)—NH –SiMe$_3$

3) MeOH

$$(18)$$

$$\left[\,{-}O{-}(\,CH_2{-}CH_2O\,)_n{-}O{-}CO{-}\,({-}NH{-}CO{-}NH{-}\!\!\left(\,AR\,\right)\!\!{-}NH{-}CO{-})_m{-}NH{-}CO{-}\right]$$

Not only polyureas of any kind but also aliphatic polyurethanes can be prepared from silylated aliphatic diamines [25]. Polycondensation of N,N'-bistrimethylsilyl-1,6-diaminohexane with the bischloroformate of 1,3-butanediol at 20 °C yielded a polyurethane in film-forming quality Eq. (19) [26]. Yet, this approach is not recommended for less reactive aromatic diamines, because at

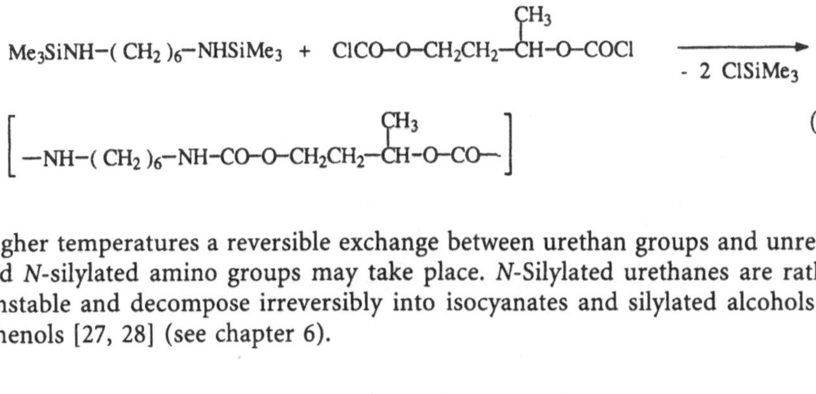

$$Me_3SiNH{-}(\,CH_2\,)_6{-}NHSiMe_3 \;+\; ClCO{-}O{-}CH_2CH_2{-}\overset{\overset{\displaystyle CH_3}{|}}{CH}{-}O{-}COCl \xrightarrow{}$$

$$-\,2\ ClSiMe_3$$

$$\left[\,{-}NH{-}(\,CH_2\,)_6{-}NH{-}CO{-}O{-}CH_2CH_2{-}\overset{\overset{\displaystyle CH_3}{|}}{CH}{-}O{-}CO{-}\right] \qquad (19)$$

higher temperatures a reversible exchange between urethan groups and unreacted N-silylated amino groups may take place. N-Silylated urethanes are rather unstable and decompose irreversibly into isocyanates and silylated alcohols or phenols [27, 28] (see chapter 6).

5.3
Polyamides

Polyamides, in particular aliphatic polyamides, can be prepared in high yields and with high molecular weights by a variety of methods [29]. Therefore, it does not make sense to use silylated aliphatic diamines for the synthesis of normal nylons. However, in the case of diamino acid alkyl esters, (e.g. derivatives of lysine or ornithine) storage and handling is plagued by their instability resulting from a high tendency of intramolecular aminolysis (i.e. formation of lactams). In this case silylation of the aliphatic amino groups reduces their nucleophilicity and stabilizes the monomers. Using the N,N'-bistrimethylsilyl derivatives of l-lysine methyl or ethyl esters, Katsarava et al. [30] were able to prepare numerous optically active polyamides. The highest yields (mostly $>90\%$) and inherent viscosities (η_{inh} up to 1.0 dl/g) were obtained when dicarboxylic acid chlorides were used as reaction partners Eq. (20). Apparently due to steric hindrance the ethylester of lysine gave considerably lower yields and molecular weights.

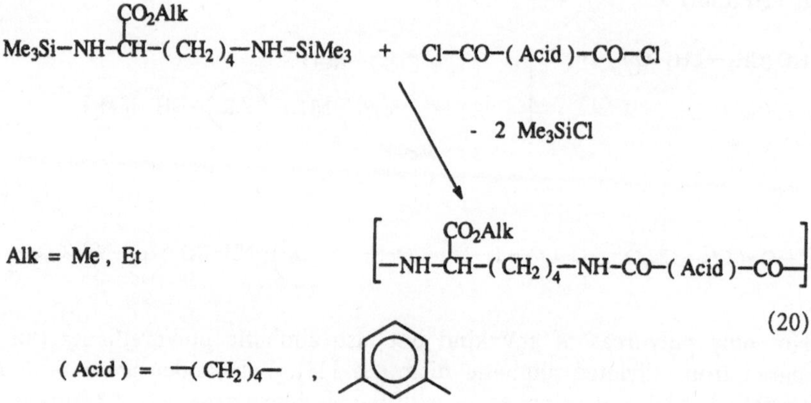

$$\text{Me}_3\text{Si}-\text{NH}-\overset{\overset{\displaystyle CO_2Alk}{|}}{\text{CH}}-(\text{CH}_2)_4-\text{NH}-\text{SiMe}_3 \quad + \quad \text{Cl}-\text{CO}-(\text{Acid})-\text{CO}-\text{Cl}$$

$$\searrow \quad -2\ \text{Me}_3\text{SiCl}$$

Alk = Me , Et

$$\left[-\text{NH}-\overset{\overset{\displaystyle CO_2Alk}{|}}{\text{CH}}-(\text{CH}_2)_4-\text{NH}-\text{CO}-(\text{Acid})-\text{CO}-\right]$$

$$(20)$$

$$(\text{Acid}) = -(\text{CH}_2)_4- \quad , \quad \bigcirc$$

In addition to acid chlorides a broad variety of activated esters of succinic, adipic, isophthalic and terephthalic acid were used as reaction partners Eq. (21) [30–32]. The yields and viscosities obtained from these less reactive dicarboxylic acid derivatives were lower. Nonetheless, this approach may be advantageous in the case of succinic acid because polycondensations of succinyl chloride with

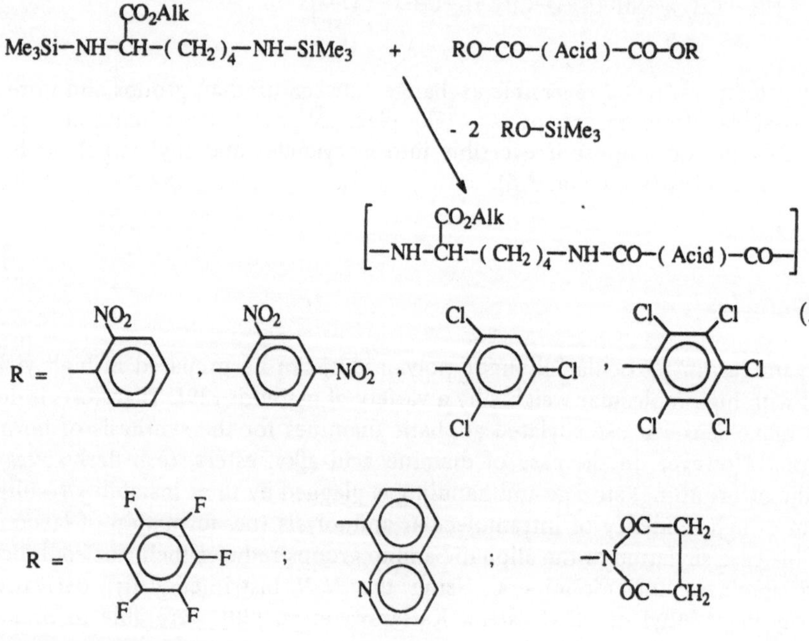

$$\text{Me}_3\text{Si}-\text{NH}-\overset{\overset{\displaystyle CO_2Alk}{|}}{\text{CH}}-(\text{CH}_2)_4-\text{NH}-\text{SiMe}_3 \quad + \quad \text{RO}-\text{CO}-(\text{Acid})-\text{CO}-\text{OR}$$

$$\searrow \quad -2\ \text{RO}-\text{SiMe}_3$$

$$\left[-\text{NH}-\overset{\overset{\displaystyle CO_2Alk}{|}}{\text{CH}}-(\text{CH}_2)_4-\text{NH}-\text{CO}-(\text{Acid})-\text{CO}-\right]$$

$$(21)$$

$$R = \text{(various aryl substituents)}$$

$$\text{Cl-CO-CH}_2\text{-CH}_2\text{-CO-NH-Lys} \quad \xrightarrow[\text{- HCl}]{} \quad \begin{matrix} \text{CH}_2\text{-OC} \\ | \\ \text{CH}_2\text{-OC} \end{matrix} \hspace{-0.3em} \rangle\hspace{-0.3em}\text{N-Lys} \qquad (22)$$

diamines bear a high risk of cyclization Eq. (22), which is completely suppressed, when activated esters are used. The influence of the reaction medium on the polycondensation of silylated methyl lysinate and bis(4-nitrophenyl)adipate is illustrated by the data listed in Table 5.1.

The synthesis of high molecular weight polyaramides is usually conducted in such a way that the free aromatic diamine is polycondensed with dicarboxylic acid chloride in an amide-type solvent, such as N,N-dimethylacetamide or N-methylpyrolidone. This method is plagued by two problems. First traces of moisture hydrolyze the acid chloride groups. Second, the HCl liberated in the course of the polycondensation protonates the amino groups, and albeit the solvent acts as HCl acceptor, the protonation slows down the polycondensation process. The advantage of silylated aromatic diamines is now twofold. First, their rapid hydrolysis dries up the reaction mixture if moisture is present at all. Second, the formation of HCl is avoided. Furthermore, in the case of less nucleophilic amines (bearing electrons withdrawing substituents) the silylation may activate the nucleophilicity slightly (see sect. 5.2). The easy reaction of silylated (aniline-type) aromatic amines with acid chlorides was firstly reported by Kricheldorf [33] and later by Bowser et al. [34].

The first example of a successful polycondensation of silylated aromatic diamines was the synthesis of polyamic acids from pyromellitic anhydride described by Klebe in 1964 [1]. This reaction leads to polyimides and is discussed in Sect. 5.4 in more detail. The usefulness of silylated aromatic diamines in polycondensations with dicarboxylic acid chlorides was first studied by Kaneda et al. [35, 36]. 3,8-Diaminophenanthridinone and terephthaloylchloride served as starting materials and LiCl, $CaCl_2$ and/or chlorotrimethylsilane were added to the reaction mixture (Form. **23**). In this way the silylated diamine was generated and

Table 5.1 Influence of the reaction medium on the polycondensation of O-methyl-N,N'-bistri methylsilyl-L-lysine with bis-(4-nitro-phenyl)adipate at 25 °C.

Reaction medium	Yield (%)	m.p.[a] (°C)	h_{inh}/c^b (dl/g)	$[\alpha]_D^c$
Hexamethylphosphorustriamide	71	115–118	0.52	−17
N-Methylpyrolidone	71	112–115	0.48	−16
Dimethylacetamide	79	115–120	0.45	−17
Acetonitrile	84	140–145	1.00	−16
1,2-Dichloroethane	73	120–125	0.78	−15
Benzene	49	120–125	0.80	−14
1,4-Dioxane	51	115–120	0.25	−15

[a] determined by optical microscopy
[b] measured in tetrachloroethane/phenol (weight ratio 3:1)
[c] measured in dimethylsulfoxide: c = 20 g/l

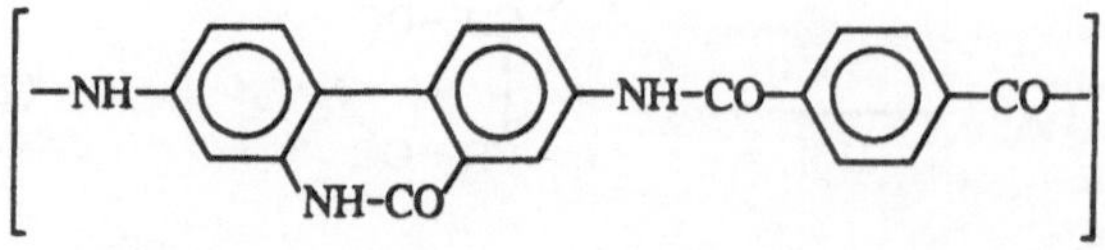

$$\left[-NH-\underset{\underset{NH-CO}{\diagdown}}{\bigcirc}-\bigcirc-NH-CO-\bigcirc-CO- \right] \qquad (23)$$

condensed in situ, and the highest viscosities were indeed obtained from reaction mixtures containing the silylated diamine (Table 5.2). This approach is well suited, when pure diamines are available or when the boiling point of the silylated diamine is too high for distillation in vacuo. The alternative approach consists of the distillation and isolation of bissilylated diamines prior to polycondensation. Its advantage is an intensive purification of the diamine, but it is not applicable to diamines containing three or more aromatic or heterocyclic rings due to the lack of volatility.

This second approach was elaborated and explored in much detail by Imai and coworkers [37–44]. When polycondensation of N, N'-bistrimethylsilyl-p-phenylenediamine and terephthaloylchloride were compared with analogous polycondensations of the free p-phenylene diamine a much faster reaction was found for the silylated diamine along with, higher molecular weights. The highest inherent viscosities determined for the Kevlar resulting from free p-phenylene diamine only reached a value around 3.5 dl/g. However, commercial Kevlar also prepared from free p-phenylenediamine may have inherent viscosities around 6.0 dl/g when measured in conc. H_2SO_4. Thus, the maximum viscosities obtained by the "silyl method" are not much higher than those of the commercial polyamide. Nonetheless, the usefulness of the "silyl method" was recently confirmed by two other groups.

Table 5.2 Influence of LiCl, $CaCl_2$ and Me_3SiCl on the inherent viscosity of the polyamide prepared from 3, 8-diaminophenanthridinone and terephthaloylchloride.

Solvent[a]	LiCl				Me_3SiCl		
	1	2	3	4[b]	1	2	3[c]
HMPA-NMP (1:1, V/V)	3.71	3.74	3.79	–	6.13	6.68	–
NMP-DMAc (1:1, V/V)	0.83	2.62	3.91	4.01	0.74	–	1.84
Solvent	$CaCl_2$	3 LiCl + 2Me_3SiCl	$CaCl_2$ + 2Me_3SiCl	$CaCl_2$ + 4Me_3SiCl	$CaCl_2$ + 4Me_3SiCl	3 LiCl	
HMPA + NMP (1:1, V/V)	–	4.04	–	–	–		
NMP-DMAc (1:1, V/V)	4.48	4.75	3.84	4.67	5.43		

[a] HMPA = Hexamethylphosphorus triamide; NMP = N-Methylpyrolidone; DMAc = N,N-Dimethylacetamide
[b] molar ratio: LiCl/diamine
[c] molar ratio: Me_3SiCl/diamine

Hatke et al. [45] studied the synthesis of polyaramides based on 2,2'-dimethylbenzidine and terephthalic acid or aryl substituted terephthalic acids (Forms. 24). In the case of terephthalic acid and phenylterephthalic acid the polycondensation of the free diamine was compared with that of the silylated diamine. The data of Table 5.3 clearly demonstrate the advantage of the silyl

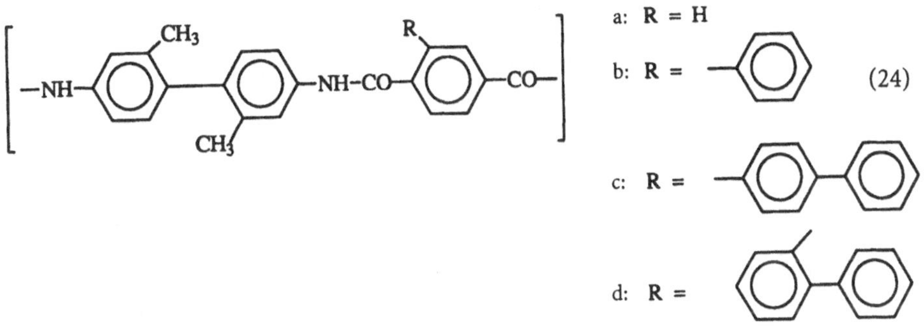

a: R = H

b: R = (phenyl)

c: R = (biphenyl)

d: R = (biphenyl substituted)

(24)

method. Kricheldorf and Schmidt [46] compared the polycondensation of phenoxy terephthalic acid and p-phenylene diamine under various reaction conditions (Table 5.4). Again the silylated diamine gave the highest viscosities. Remarkable is the finding that the free diamines gives better results in the absence of tert.-amines. Furthermore, the polycondensation of phenoxy terephthalic acid with triphenylphosphite and pyridine ("Higashi method" [47]) proved to be inferior to polycondensations of phenoxyterephthaloylchloride with either free or silylated p-phenylene diamine.

The relatively high volatility of silylated aromatic diamines allowed Takahashi et al. [43] to prepare ultrathin films (thickness around 0.5 µm) on an aluminum substrate. The success of this method is based on the simultaneous evaporation (and deposition) of terephthaloyl chloride and bistrimethylsilyl diamines. The thermal and electrical properties of these films were studied in some detail. Less successful proved to be the polycondensation of silylated

Table 5.3 Polycondensation of silylated 2,2'-dimethylbenzidine (A) or free 2,2'-dimethylbenzidine (B) with terephthaloylchloride or phenylterephthaloylchloride.

Synthetic[a] method	Dicarboxylic acid	Yield in %	η_{inh}^{c} (dl/g)
A	Terephthalic a.	99	4.01
B[b]	Terephthalic a.	94	1.49
A	Phenylterephth a.	99	2.52
B[b]	Phenylterephth.a.	89	0.96

[a] All polycondensations were conducted in NMP-containing 3 weight % of LiCl at $-20\,°C$
[b] triethylamine was added as HCl acceptor
[c] measured at 25 °C with c = 5 g/l in DMAc + LiCl (4 % /v)

Table 5.4 Yield and inherent viscosities of polyaramides prepared from 1, 4-diaminobenzene and phenoxyterephthalic acid under various reaction conditions.

Reaction partners	Temp.(°C)	Yield (%)	η_{inh}^a (dL/g)
1, 4-diaminobenzene, phenoxyterephthaloyl chloride	− 15	99	3.13
	0	99	3.20
1, 4-diaminobenzene, phenoxyterephthaloyl chloride, pyridine	− 15	99	2.34
	0	99	2.30
1, 4-diaminobenzene, phenoxyterephthaloyl chloride, N, N-dimethylaniline	− 10	99	2.22
1, 4-diaminobenzene, phenoxyterephthaloyl chloride, triethylamine	− 15	98	0.65
N, N' -bis (trimethylsilyl)-1, 4-diaminobenzene, phenoxyterephthaloyl chloride	− 10	99	3.08
	− 15	99	5.13
	− 18	99	6.13
1, 4-diaminobenzene, phenoxyterephthalic acid, triphenyl phosphite, pyridine	120	99	2.46

[a] Measured at 25 °C with c = 0.1 g/dL in concentrated H_2SO_4

4,4'-diaminodiphenylether and diphenylisophthalate [42]. Cesium fluoride was used as catalyst and both reaction time and reaction medium were varied over a broad range. However, the inherent viscosities never exceeded 0.43 dl/g (in DMAC at 30 °C).

Among the numerous polyaramides prepared by Imai and coworkers, fluorinated polyaramides are worth mentioning. They were either prepared by polycondensation of tetrafluoroisophthaloylchloride **25a** [41] and tetrafluoroterephthaloylchloride **25b** [41] with various silylated diamines or by polycondensation of silylated tetrafluoro-1,3-diaminobenzene **25c** [44]. It was shown that the thermal stability of the fluorinated polyaramides is lower than that of the analogous fluorine free polyaramides [39].

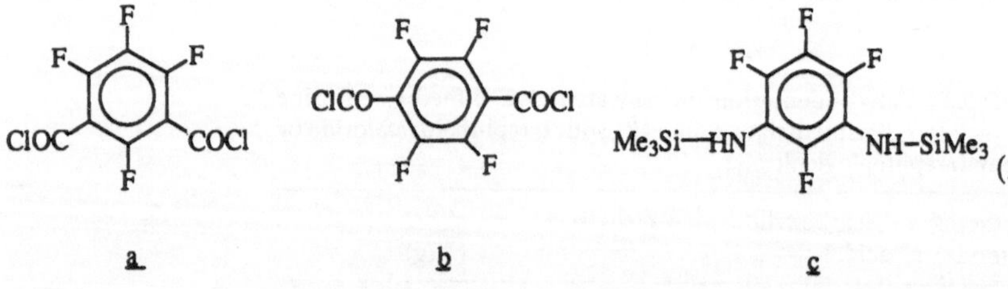

(25)

a   b   c

As demonstrated by Oishi et al. [39, 40] the silylation is particularly useful in the case of sterically hindered diamines of low nucleophilicity, such as (26a−c). Due to the steric hindrance reaction temperatures in the range of 80−200 °C are now required for high conversions.

An additional advantage of the silyl method becomes evident, when deuterated polyaramides should be synthesized. The absence of acidic protons

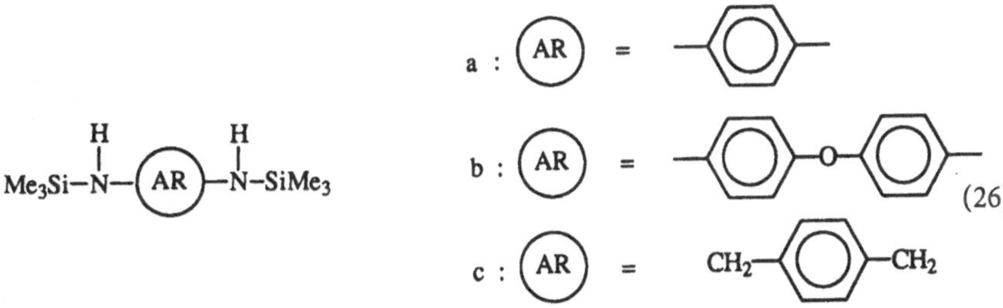

a : (AR) = —◯—

Me₃Si–N—(AR)—N–SiMe₃

(H above N's)

b : (AR) = —◯—O—◯—

c : (AR) = CH₂—◯—CH₂

(26)

prevents an undesired H/D-exchange. The selectively deuterated polymers (27a,b and 28a,b) were prepared in this way [48]. As reported by Kricheldorf and Schmidt [48] the ^{2}H NMR spectroscopy of these polyaramides gave the unexpected result that the H-bond forming polymers (28a and b) contain more mobile phenyl and phenylene groups than the N-substituted analogs (27a and b). Further structure-property relationships of particular interest is the good

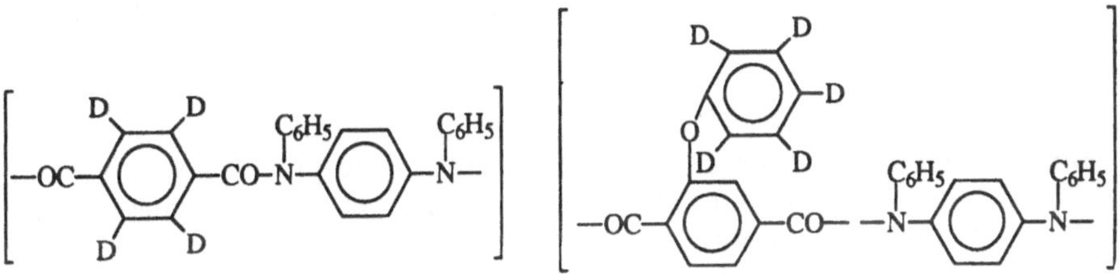

a

b (27)

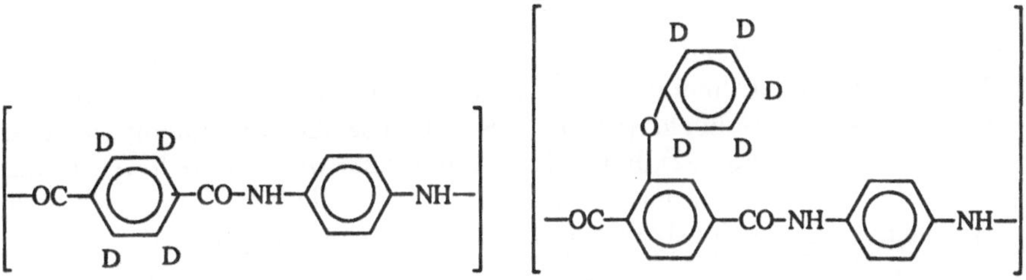

a

b (28)

solubility of several substituted rigid rod polyaramides such as (24a−d or 29a,b) in nonacidic solvents. In some cases lyotropic solutions were found, whereas the few meltable Kevlar-type polyaramides reported sofar [49] do not form a liquid crystalline melt.

Concerning the synthesis of polyamides, it should be noted that four research groups reported on the silicon promoted polycondensation of 4-aminobenzoic acid. Kozyukov et al. [50] mentioned shortly the polycondensation of N,O-bistrimethylsilyl-4-aminobenzoic acid in NMP under the catalytic influence of large amounts of Lewis acids Eq. (30). Unfortunately the resulting

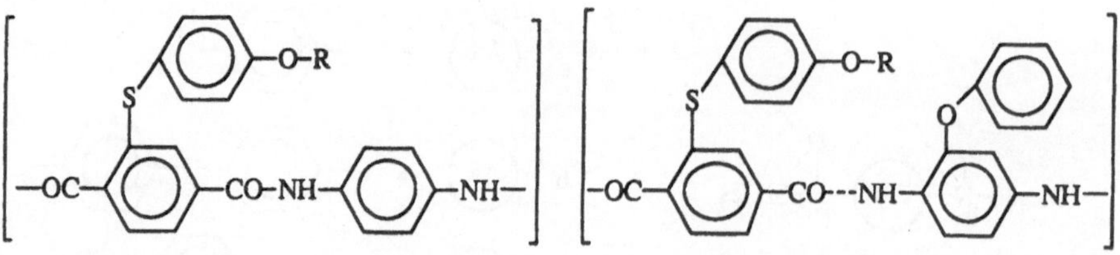

a   b   (29)

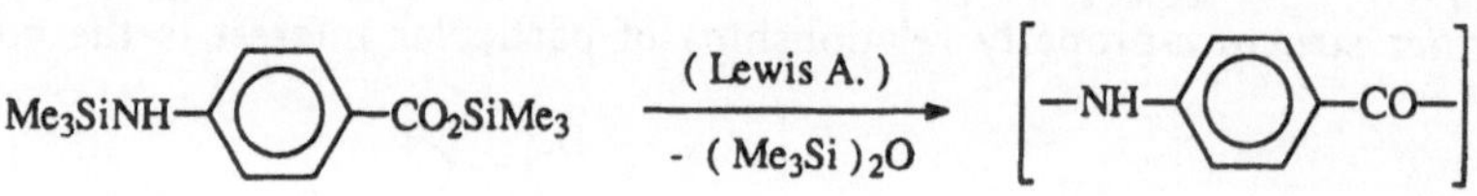

$$R = \text{—⟨SO}_2\text{—⟩} \quad ; \quad -(CH_2)_3-O-\text{⟨⟩}-CO-\text{⟨⟩}$$

$$Me_3SiNH-\text{⟨⟩}-CO_2SiMe_3 \xrightarrow[-(Me_3Si)_2O]{(Lewis A.)} \left[-NH-\text{⟨⟩}-CO-\right] \quad (30)$$

poly(4-benzamide) was not characterized at all and no precise procedure was given. When Kricheldorf and Löhden [51] attempted to reproduce these results using ZnCl$_2$ or AlCl$_3$ as catalyst, only dark colored impure oligoamides or tars were obtained. However, in inert solvents such as m-terphenyl, diphenylether or Marlotherm-S yellowish, highly crystalline poly(benzamide)s of low molecular weight ($\eta_{inh} \leqslant 0.1$ dl/g in H$_2$SO$_4$) were obtained. Strohriegel and Heitz described polycondensations of 4-aminobenzoic acid or N-methyl-4-aminobenzoic acid upon heating with SiCl$_4$ in pyridine [52, 53] Eq. (31). High yields and M$_n$'s up to 11 000 g/mol were obtained. However, a problem is the purification of the polyamides from the liberated SiO$_2$. In order to get rid of this purification problem. Ackar et Golioglu [54] proposed to use dichlorodimethylsilane as condensing reagent, Eq. (32), but yield and molecular weight were lower. In fact none of these methods enabled the synthesis of a really high molecular weight poly(4-benzamide) obviously due to the insolubility of this polymer.

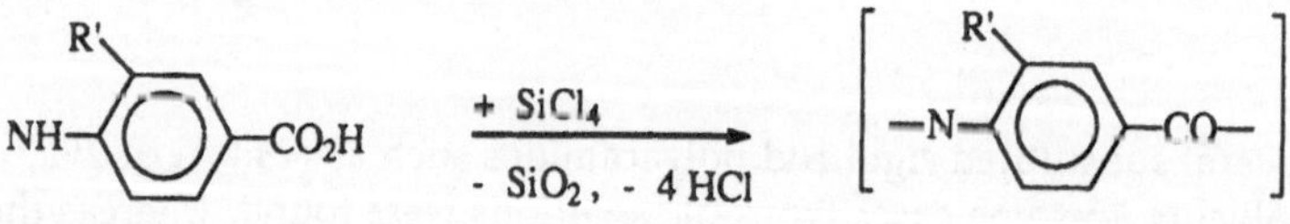

$$\text{NH}-\text{⟨⟩}-CO_2H \xrightarrow[-SiO_2, -4HCl]{+SiCl_4} \left[-N-\text{⟨⟩}-CO-\right] \quad (31)$$

R = H, CH$_3$

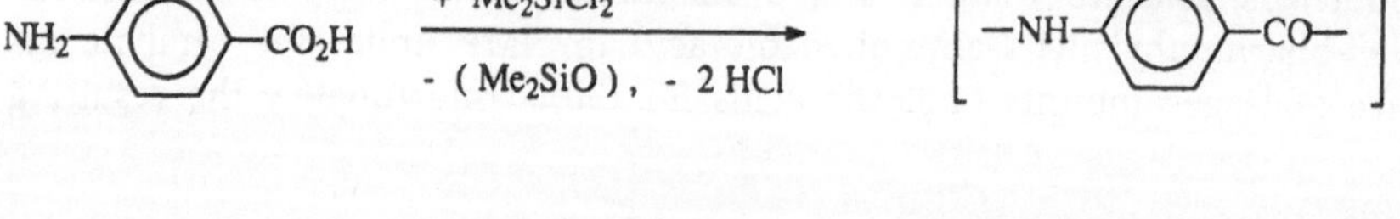

$$\text{NH}_2-\text{⟨⟩}-CO_2H \xrightarrow[-(Me_2SiO), -2HCl]{+Me_2SiCl_2} \left[-NH-\text{⟨⟩}-CO-\right] \quad (32)$$

Finally, two recent reports on the synthesis of chiral polyamides should be mentioned. Rodriguez-Galán et al. [55] studied the polycondensation of methylene-protected l-tartaric acid bispentachlorophenylester with N, N'-bistrimethylsilyl-1,9-diaminononane or -1,2-diaminododecane, Eq. (33). Surprising is the strong influence of the solvent on the molecular weights as illustrated in Table 5.5. Katsarava et al. [56] investigated the polyaddition of 2,2'-p-phenylenebis(Δ^2-5-oxazolone)s. When 1,6-diaminohexane and N, N'-bistrimethylsilyl-1,6-diaminohexane were compared as reaction partners significantly higher molecular weights were obtained with the free diamine. This negative result is

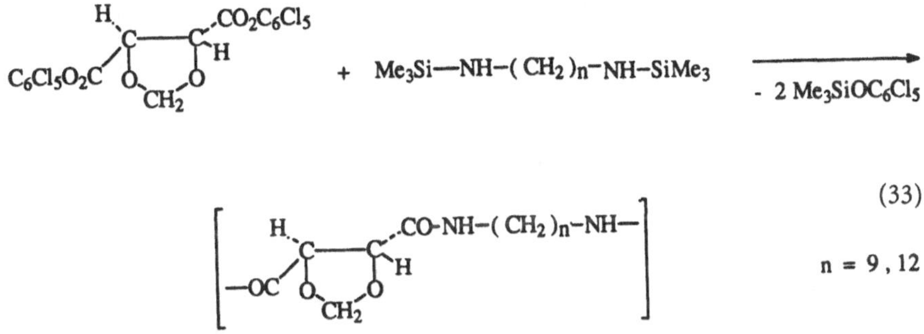

$$ (33) $$

$$ n = 9, 12 $$

Table 5.5 Solution polycondensation of bispentachlorophenyl 1- 2, 3-methylene-L-tartrate with silylated 1, 9-diaminononane or 1, 12-diaminododecane.

Solvent[a]	Silylated diamine of	Yield (%)	$[\eta]$[b] (dL/g)	$[\alpha]_D^{20}$[c]
Dimethylsulf-Oxide	1.9-Nonane	86	0.30	− 12.1
	1.12-Dodecane	77	0.40	− 9.60
Hexamethyl-phosphortriamide	1.9-Nonane	79	0.78	− 13.29
	1.12-Dodecane	84	0.50	− 9.76
N-Methylpyrro-lidone-2	1.9-Nonane	78	0.75	− 12.40
	1.12-Donane	82	0.89	− 9.41
Chloroform	1.9-Nonane	91	0.61	− 12.20
	1.12-Dodecane	79	1.33	− 9.59

[a] Monomer conc. 2.6 mol/l; reaction time 48 h (Nonane) or 72 h (Dodecane) at 25 °C
[b] Intrinsic viscosity measured in dichloroacetic acid at 25 °C
[c] Molar optical rotation measured in CHCl, at c = 1 g/dL

presumably a consequence of the fact that the silylated amide groups formed upon the additional step are energetically less favorable than the silylated amino groups of the starting material. This is a typical difference between a polyaddition and polycondensations of silylated diamines. Nonetheless, polyadditions of silylated lysine esters are still attractive Eq. (34), because the free lysine esters are not stable. The resulting polyamides, Eq. (35) are of interest as

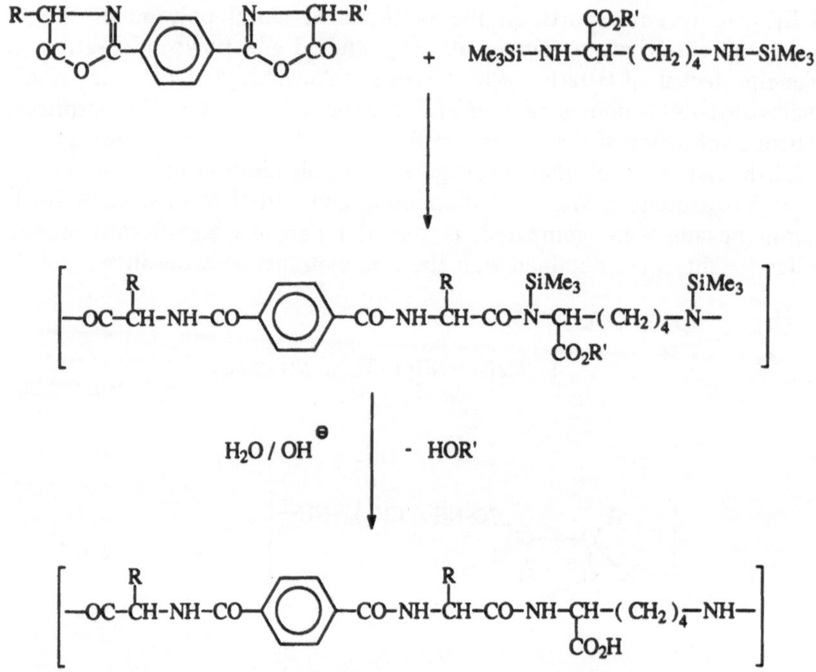

$$R-CH-N \underset{OC-O-C}{\overset{\quad}{\bigcirc}} C-O-CO \quad N-CH-R' \qquad + \quad Me_3Si-NH-\overset{CO_2R'}{\underset{}{CH}}-(CH_2)_4-NH-SiMe_3$$

(34)

$$\left[-OC-\overset{R}{\underset{}{CH}}-NH-CO-\bigcirc-CO-NH-\overset{R}{\underset{}{CH}}-CO-N-\overset{SiMe_3}{\underset{CO_2R'}{CH}}-(CH_2)_4-\overset{SiMe_3}{\underset{}{N}}- \right]$$

$$H_2O / OH^{\ominus} \quad \bigg| \quad - HOR'$$

(35)

$$\left[-OC-\overset{R}{\underset{}{CH}}-NH-CO-\bigcirc-CO-NH-\overset{R}{\underset{}{CH}}-CO-NH-\overset{}{\underset{CO_2H}{CH}}-(CH_2)_4-NH- \right]$$

biodegradable, materials, because they contain peptide bonds sensitive to enzymatic cleavage.

5.4
Polyimides

Polyimides have been known for several decades and have found wide-spread technical application as thermostable duromers, engineering plastics, thermosetting resins or photosensitive lacqers. A broad variety of chemical structures and synthetic methods has been developed over the past three decades [57]. The synthesis of polyimides from N, N'-bissilylated diamines and pyromellitic dianhydride (PMDA) or benzophenone-3,3',4,4'-tetracarboxylic anhydride was first described by Boldenbuck and Klebe in a patent claim [58]. The condensation of silylated diamine with a dianhydride leads, at low temperatures, to the formation of a polyamic acid with silylated carboxyl groups, Eq. (36). The silylated polyamic acids have the advantage to be soluble in various organic solvents, e.g. THF, so that casting of films or spinning of fibers is easily feasible. Furthermore, nearly quantitative cyclization was reported at the relatively low temperature of 150 °C, Eq. (37) [58]. Due to these advantages Greber et al. [59, 60] studied silylation and processing of poly(amic acid)s prepared in a conventional

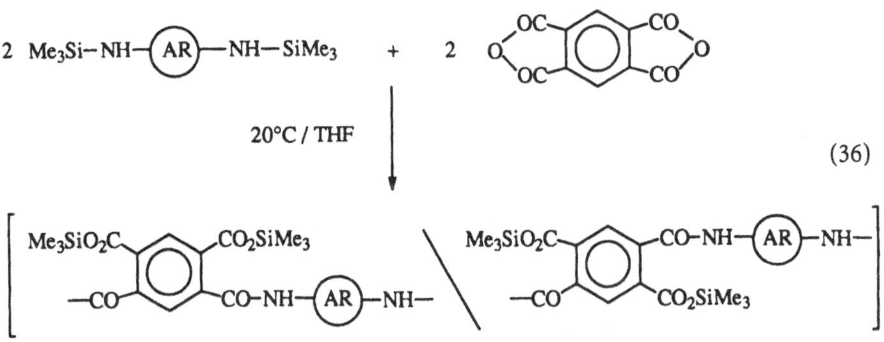

$$2 \ Me_3Si-NH-\!\!\left(AR\right)\!\!-NH-SiMe_3 \quad + \quad 2 \quad \text{[anhydride structure]}$$

20°C / THF

(36)

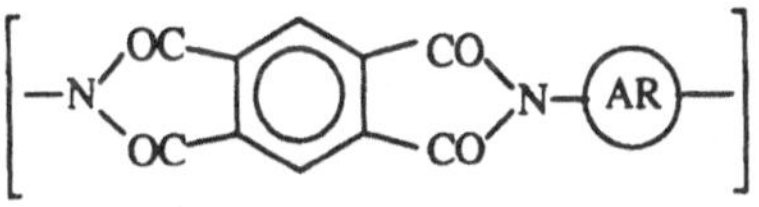

$$\left[\text{Me}_3\text{SiO}_2\text{C}-\!\!\!\!\text{CO}_2\text{SiMe}_3 \quad \text{and} \quad \text{Me}_3\text{SiO}_2\text{C}-\!\!\!\!\text{CO-NH-}\!\!\left(AR\right)\!\!-\text{NH-} \right]$$

$\geq$ 150°C   - 2 Me$_3$SiOH

$$\left[-N\text{OC}\!\!-\!\!\text{CO}\!\!-\!\!N\!\!-\!\!\left(AR\right)\!\!- \right]$$

(37)

$$\left(AR\right) = \text{[aromatic structures]}$$

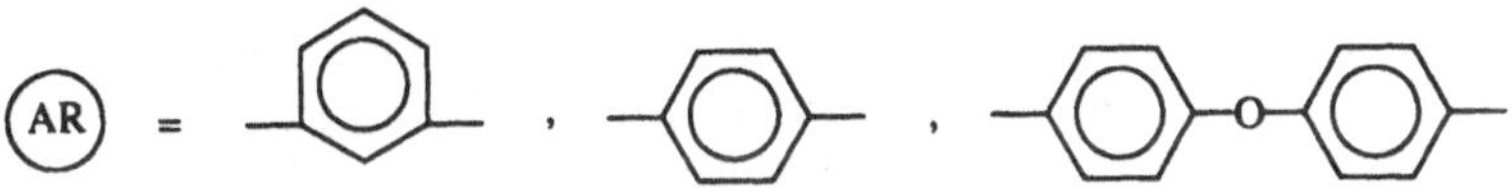

manner from nonsilylated diamines. With bistrimethylsilylacetamide a quantitative silylation of various polyamic acids in polar solvents was conducted, Eq. (38) and after processing cyclization was completed at 220–250 °C.

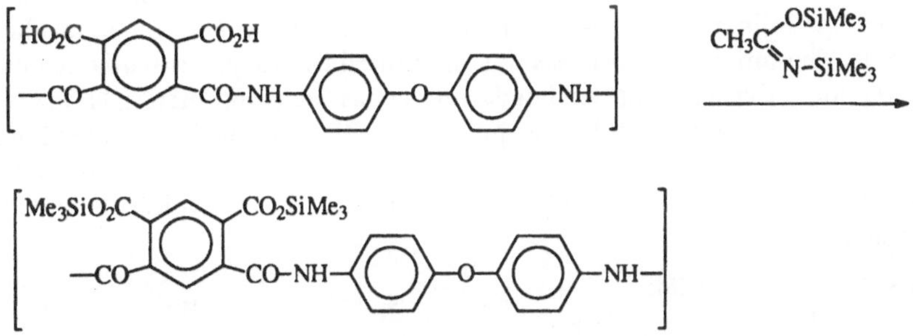

$$\left[\text{HO}_2\text{C}-\!\!\!\!\text{CO}_2\text{H} \quad \text{CO-NH-}\!\!\!\!-\text{O}-\!\!\!\!-\text{NH-} \right] \xrightarrow{\text{CH}_3\text{C}\begin{smallmatrix}\text{OSiMe}_3\\\text{N-SiMe}_3\end{smallmatrix}}$$

(38)

$$\left[\text{Me}_3\text{SiO}_2\text{C}-\!\!\!\!\text{CO}_2\text{SiMe}_3 \quad \text{CO-NH-}\!\!\!\!-\text{O}-\!\!\!\!-\text{NH-} \right]$$

Further studies of polycondensations with silylated and free diamines were reported by Korshak et al. [61]. A comparative study conducted with diphenylether tetracarboxylic anhydride and 1,6-diaminohexane or N, N'-bistrimethylsilyl-1,6-diaminohexane, Eq. (39), revealed that the silylated diamine yields a higher reaction rate and higher molecular weights of the silylated polyamic acid. However, when a "one-pot synthesis" of copolyamides (Form. 40) was conducted in nitrobenzene at 200 °C the free diamines and the silylated diamines gave nearly identical results.

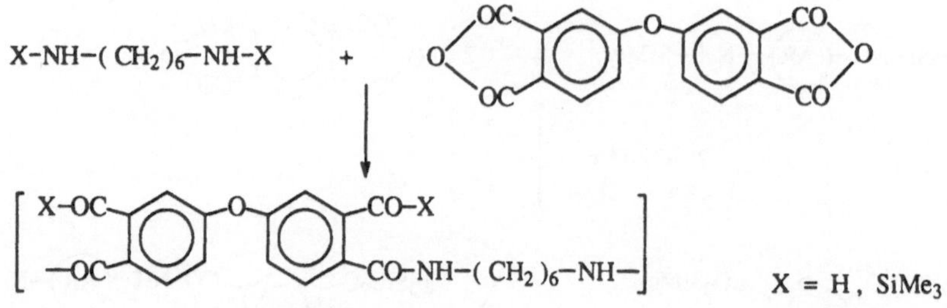

$$\text{X–NH–(CH}_2)_6\text{–NH–X} \quad + \quad \text{[structure]}$$

(39)

$$\left[\text{structure} \right] \quad \text{X = H, SiMe}_3$$

$$\text{structure}$$

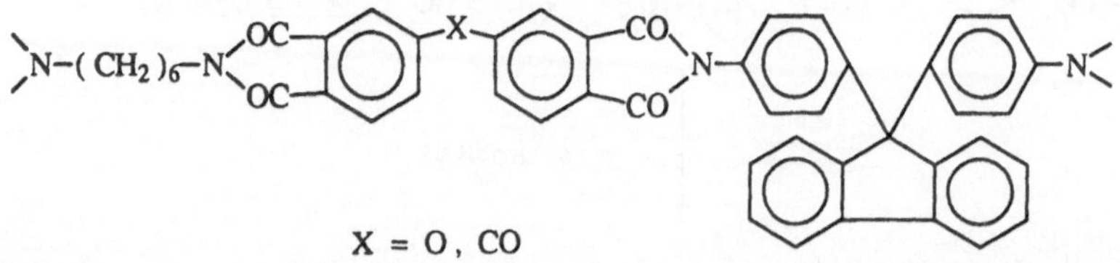

(40)

X = O, CO

A comparative study of silylated and nonsilylated 4,4′-diaminodiphenyl ether with PMDA as reaction parter was also conducted by Oishi et al. [62]. They found that the cyclization of the silylated polyamic acid starts at a somewhat lower temperature (approx. 130°C). When compared under identical conditions the silylated polyamic acid cyclized faster by a factor 2 than the free polyamic acid. The influence of various solvents used in the preparation of the silylated polyamic acid on the molecular weight was studied. The best results were obtained with N,N-dimethylacetamide. Furthermore, model reactions with phthalic anhydride and silylated aniline were reported, but such model reactions were described by Pratt [63] much earlier.

Simultaneous evaporation of silylated 4,4′-diaminodiphenylether and PMDA in a high vacuum and condensation of the vapors on an aluminum substracted was used by Iojima et al. [64] to prepare ultrathin films of an amorphous polyimide. The imidization process was monitored by IR-spectroscopy. Another characteristic advantage of silylated amines as starting materials for the preparation of polyimides and other polymers is the intermediate protection of an OH-group by silylation. The synthesis of polyimides with a pendant phenol group from silylated 2,4-diaminophenol, Eqs. (41 and 42) is a typical example for this strategy.

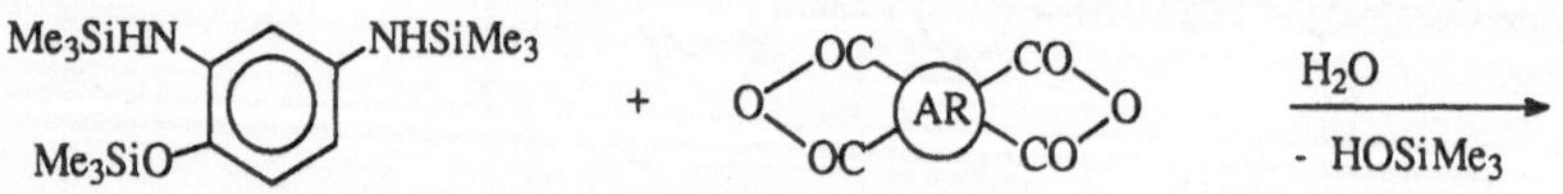

$$\text{Me}_3\text{SiHN}\text{–}\underset{\text{Me}_3\text{SiO}}{\bigcirc}\text{–NHSiMe}_3 \quad + \quad \text{structure} \quad \xrightarrow[\text{– HOSiMe}_3]{\text{H}_2\text{O}}$$

(41)

5.5
Polybenzoxazoles

The benzoxazole ring may be prepared from a silylated o-aminophenol and carboxylic acid chloride in a two-step reaction. The first step consists of a highly selection acylation of the silylated amino group, Eq. (43), at low or moderate

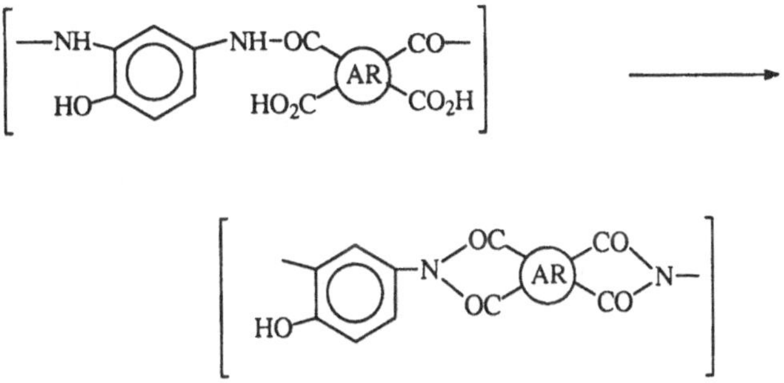

$$(42)$$

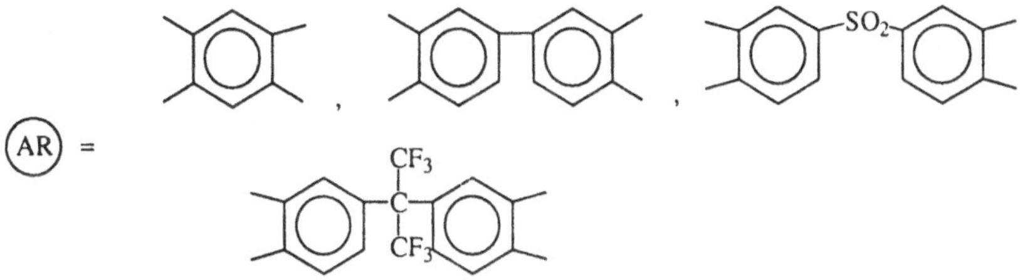

temperatures ($< 100 \,°C$). Under these conditions the siloxy group does not react with carboxylic acid chloride (in the absence of chloride or fluoride ions) and thus, side reactions (e.g. of the OH-groups) leading to crosslinks in the case of tetrafunctinal monomers are avoided. The second step consists of the cyclization of the siloxyamide intermediate at temperatures above $250 \,°C$ Eq. (44). For

$$(43)$$

$$(44)$$

Y, Z = functional groups

complete cyclization temperatures around $300 \,°C$ are recommendable and even $350 \,°C$ may be advisable in some cases. Such high temperatures may be used because fully aromatic polybenzoxazoles usually have the properties of (and are designed as) thermostable engineering plastics.

Korshak and coworkers [66–69] were the first to study the syntheses of polybenzoxazoles from silylated diamino diphenols. The silylated monomers they used (Forms **45a–c** and **46a, b**) were obtained by silylation with

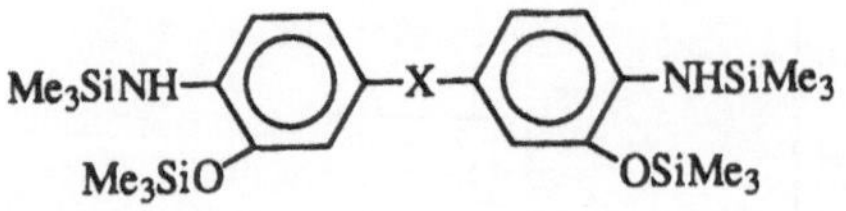

$$a : X = O$$
$$b : X = CH_2 \quad (45)$$
$$c : X = CMe_2$$

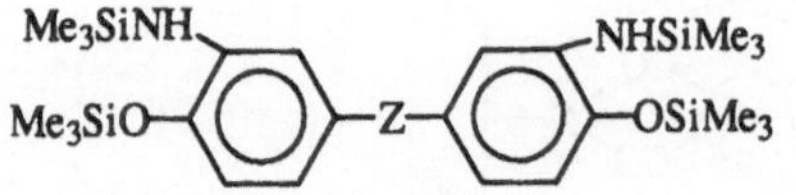

$$a : Z = CMe_3$$
$$(46)$$
$$b : Z =$$

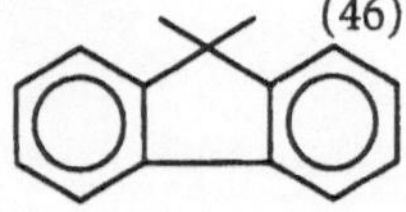

N-trimethylsilyl diethylamine [66,67]. The polycondensation of the silylated aminophenols was conducted at 20–25 °C with isophthaloylchloride, terephthaloylchloride *ortho*-or *para*-carboranedicarboxylic acid chlorides and with the acid chlorides **47a** or **b**. The cyclization of the initially formed polyamides was completed at a final reaction temperature of 250 °C [68, 69].

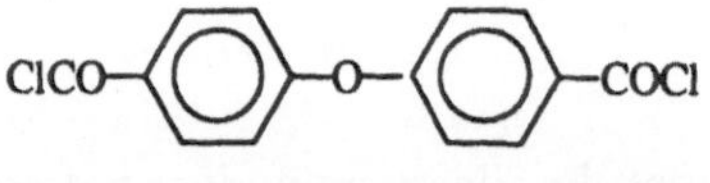

$$(47)$$

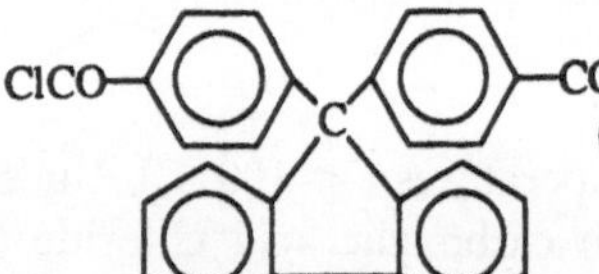

a   b

More recently Maruyama et al. [70] reported an analogous synthesis of several polybenzoxazoles. In addition to the silylated diaminodiphenol **45c** the monomers **48a** and **b** were used in combination with isophthaloyl chloride, terephthaloylchloride or dicarboxylic acid chlorides such as **47a** or **49a** and **49b**. In

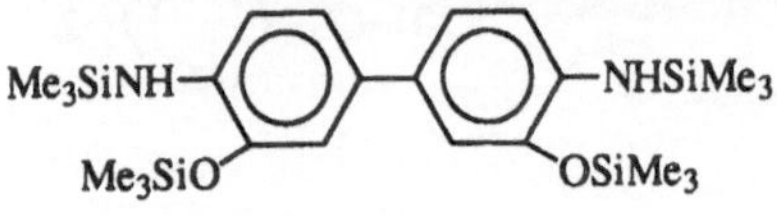

$$(48)$$

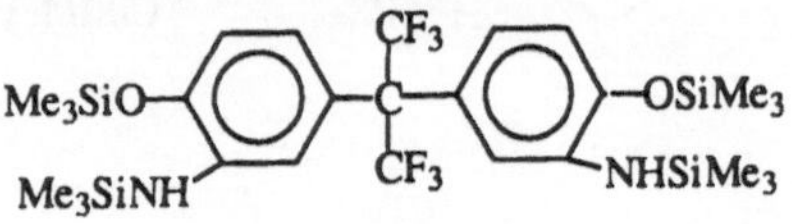

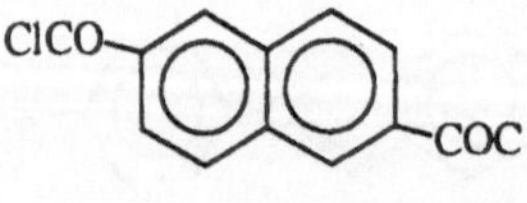

a 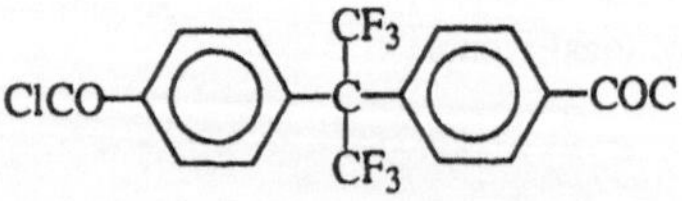   b

$$(49)$$

a   b

some cases the siloxygroups of the initially formed polyamides were hydrolyzed and the resulting polyamides with free phenolic groups were then cyclized in the form of films at 250 °C. Imai and coworkers [71,72] also synthesized a new class of poly(benzoxazole-amide)s, Eqs.(50 and 51). Silylated 2,4-diamino-phenols served as starting material and in addition to isophthaloylchloride or

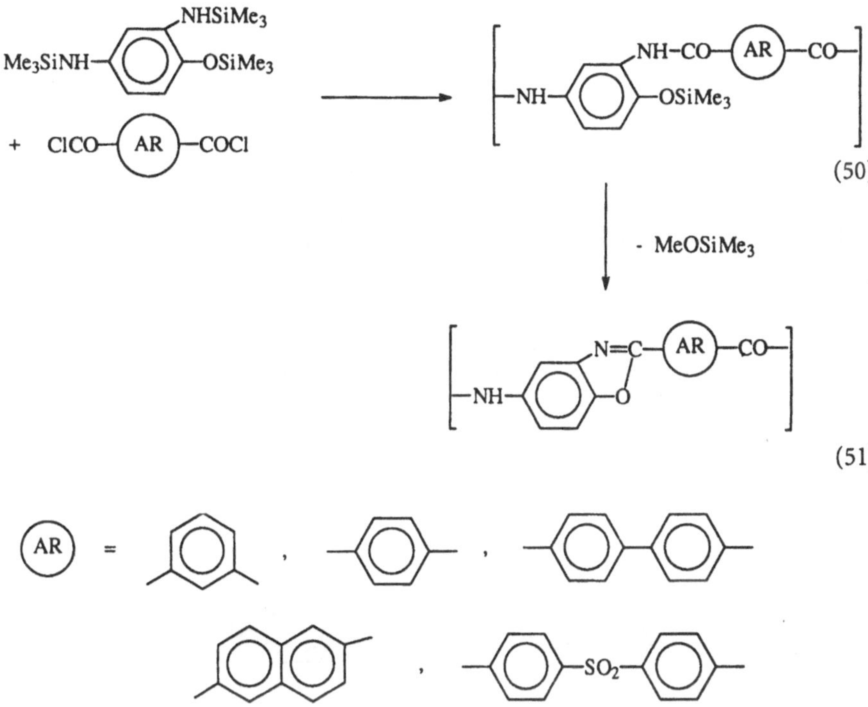

(50)

$$- \text{MeOSiMe}_3$$

(51)

terephthaloylchloride, the acid chlorides **49a** and **52a, b** were used as reaction partners. In some cases materials with high inherent viscosities were obtained and their mechanical properties were studied, because such poly(benzoxazole amide)s may be of interest as high strength fibers.

(52)

a   b

Poly(benzobisoxazole)s and poly(benzobisthiazole)s are of particular interest because they possess an intrinsically stiff and linear main chain, which is also fully conjungated. However, these polymers are infusible and only dissolve in conc. sulfuric acid. Therefore, it is important to synthesize substituted analogs which are meltable and fusible. Unfortunately the most common and technically applied synthetic procedure involves heating of the starting materials in poly-phosphoric acid to 200 °C or higher temperatures [73, 74]. Under these harsh conditions most aromatic or aliphatic substituents of terephthalic acid are somehow damaged. As reported by Kricheldorf et al. [75–78] the polycondensa-tion of silylated monomers such as **53a** or **b** with substituted terephthaloylchlor-ides proceeds under milder conditions due to the absence of acidic tons and allowed the synthesis of numerous substituted poly(benzobisoxazole)s and poly(benzobisthiazole)s.

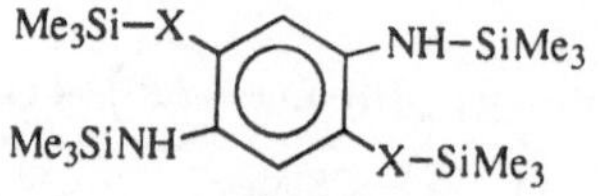

$$\text{Me}_3\text{Si–X}\overset{}{\underset{\text{Me}_3\text{SiNH}}{\bigcirc}}\overset{\text{NH–SiMe}_3}{\underset{\text{X–SiMe}_3}{}}$$

$$\underline{a} : X = O$$

$$\underline{b} : X = S$$
(53)

Detailed studies with low molecular weight model compounds revealed that complete cyclization of poly(benzoxazole)s requires temperatures up to 350 °C [75, 76] whereas 250 °C is high enough for a complete cyclization in the case of poly(benzobisthiazole)s [76, 77]. In addition to homopolymers with one aromatic substitutent (eg. 54a–d) copolymers with two (e.g. 55) or three different aromatic

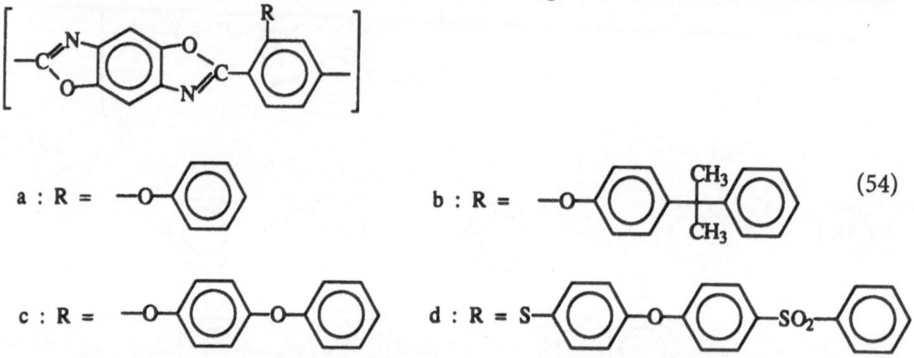

$$a : R = \quad\quad b : R = \quad\quad (54)$$

$$c : R = \quad\quad d : R =$$

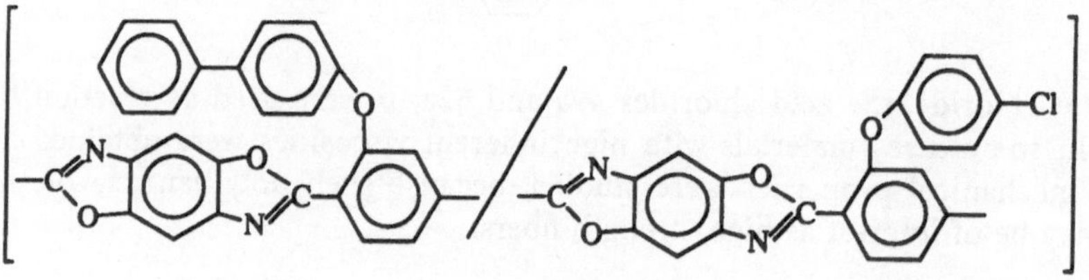

(55)

side chains were prepared. In the case of poly(benzobisthiazole)s also comonomers having aromatic and aliphatic substituents were combined (Form. 56). When the formation of poly(benzobisoxazole)s from alkoxy or alkylthio substituted terephthalic acids was compared (Forms. 57 and 58) a more rapid cyclization was observed for the alkylthiosubstituents 57b and 58b possibly due to a steric effect [78]. Furthermore, substituted poly(biphenylbisoxazole)s 59 should be mentioned [79]. The silyl method has the additional advantage that copolymers of bisoxazol and aramide units can be prepared in an "one-pot

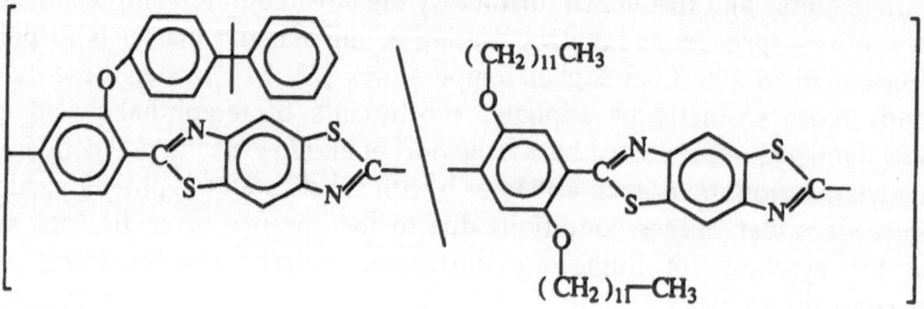

(56)

procedure" by cocondensation of monomers **53a** or **b** with silylated *p*-phenylene diamine (Forms. **60** and **61**).

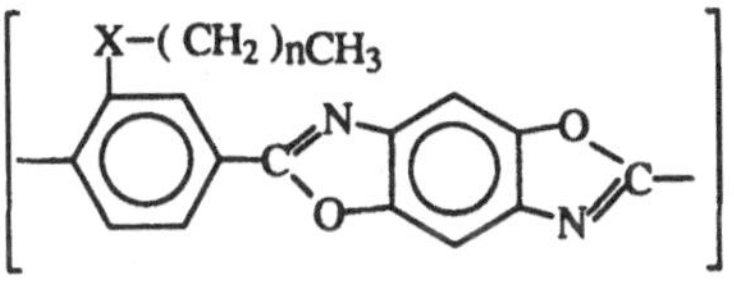

a : X = O

b : X = S

(57)

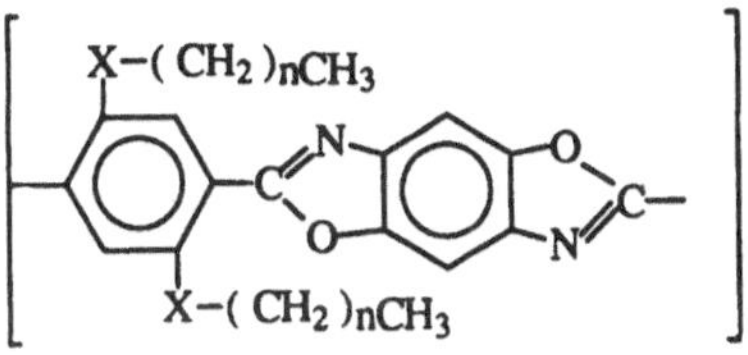

a : X = O

b : X = S

(58)

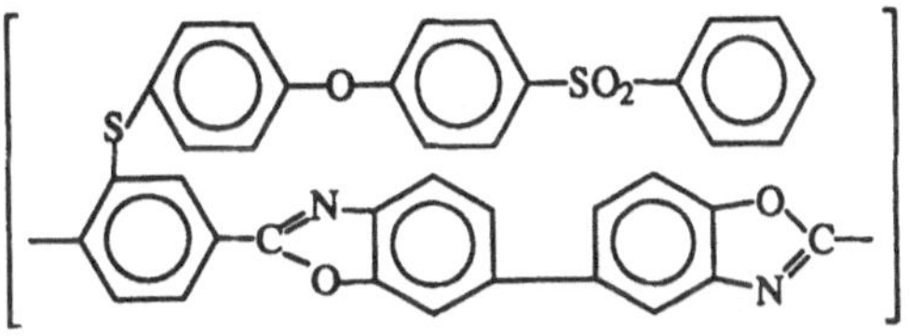

(59)

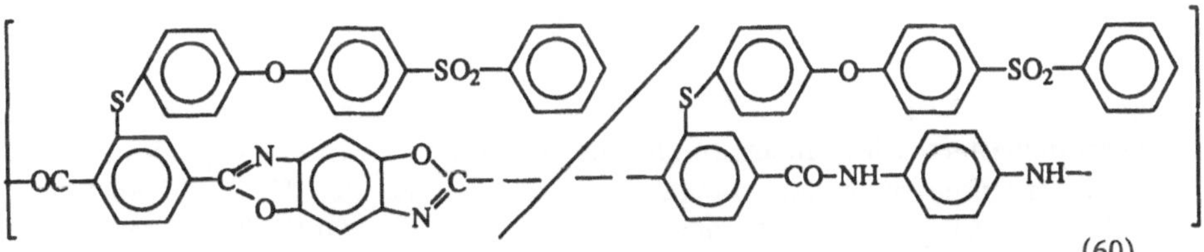

(60)

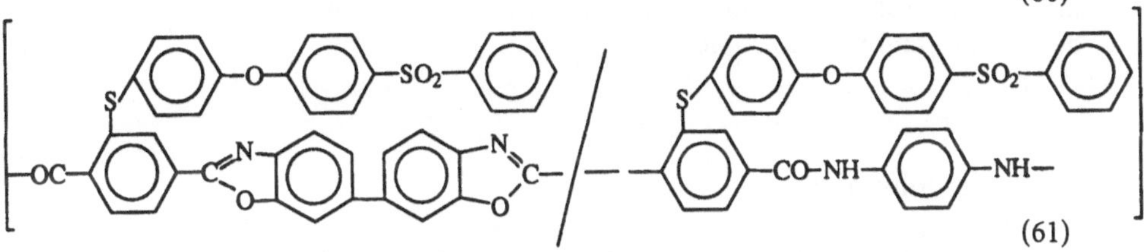

(61)

Kricheldorf and Thomsen [80] also reported the synthesis of poly(benzoxazole-ester)s in an "one-pot procedure". Trisilylated 3-amino-4 hydroxybenzoic served as starting material. It was acylated by various aromatic acetoxy acid chlorides at temperatures below 100 °C, Eq. (62) and cyclization was achieved by heating the resulting amide to 350 °C, Eq. (63). At this temperature the rapid cyclization is followed by slower polycondensation, Eq. (64), which represents a special example of the "silylacetate method" discussed in section 5.7. Due to the numerous steps and some side reactions involved in this "one-pot procedure" the molecular weights of the resulting polyesters are only moderate or low but the yields are high and allow a reasonable preparation of pure benzoxazole monomers by hydrolysis of the crude polyesters. The 2-(hydroxyaryl)

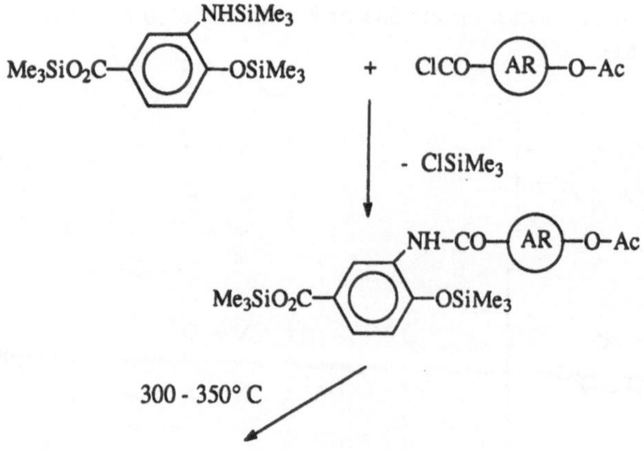

(62)

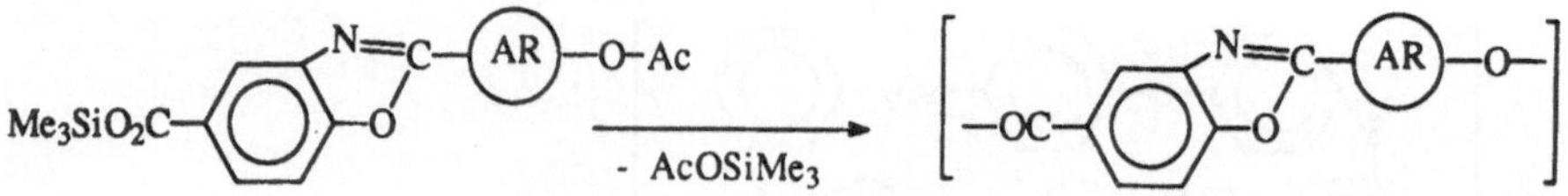

(63)

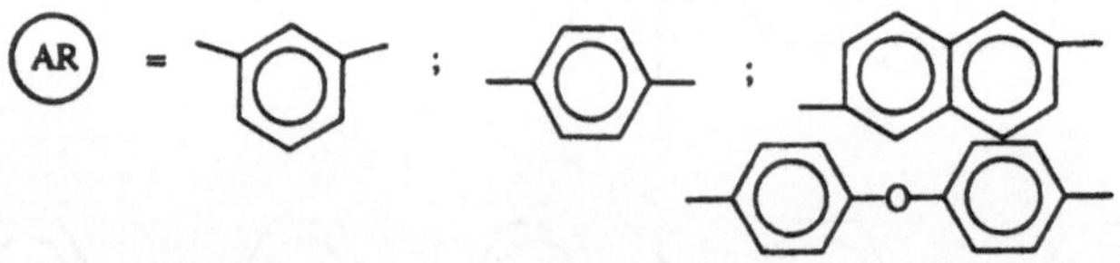

(64)

benzoxazole-5-carboxylic acids thus obtained, may then be used to prepare homopolyesters with higher molecular weights or various copolyesters by cocondensation with other aromatic hydroxy acids [80–82]. The homopolyesters possess an exceptionally high degree of crystallinity (> 90 %). They are neither meltable nor soluble in common solvents, whereas some of the copolyesters may be meltable and thermotropic.

5.6
Aromatic Polyethers

Aromatic polyethers, such as poly(phenylene oxide), poly(ether-sulfone)s or poly(ether-ketone)s are widely used as thermostable engineering plastics either in a pure form or as components of blends and composites. A further important application is that as membranes for reverse osmosis or gas-separation. Regardless of the final application a variation of their properties by a variation of the chemical structure is of great interest. The polycondensation of silylated diphenols with activated difluoro- or dichloroaromatics is an approach which allows the synthesis of a broad variety of aromatic polyethers, and in most cases their synthesis may be conducted in an "one pot procedure". However, the silylation of phenolic OH-groups provides an excellent protection against any kind of electrophilic attack, and heating of a silylated diphenol with a pure difluoro- or dichloroaromatic does not result in any polycondensation even at temperatures up to 350 °C. Therefore an activation of the siloxy groups is required to enable a satisfactory polycondensation.

As illustrated by the data listed in Table 5.6 fluoride ions and, in particular, cesium fluoride, are the most effective catalysts for polycondensations of

Table 5.6 Yields and properties of poly(ether-sulfone)s prepared from bistrimethylsilyl bisphonol-A and 4,4'-difluorodiphenylsulfone.

No	Catalyst	Temp.[a] (°C)	Time (h)	Yield (%)	η_{inh}^{b} (dl/g)	M_W (GPC)[c]
1	none	360	4.0	0	–	–
2	LiF, LiCl CaF$_2$, BaF$_2$, PbF$_2$	300	1.0	0	–	–
3	KF	270	1.0	95	0.1	–
4	KF	270, 320	0.5, 0.5	97	0.3	15,000
5	KF	270, 320	0.5, 2.5	97	0.4	28,000
6	KF + Crown-8	270, 320	0.5, 2.5	98	0.5	25,000
7	CsF	180, 320	0.5, 2.5	98	1.4	200 000
8	CsF	180, 320	0.5, 2.0	91	1.7	300 000
9	CsF	180, 340	0.5, 2.5	97	1.5	100 000

[a] initial and final reaction temperatures
[b] measured with c = 2 g/l at 20 °C in tetrachloroethane
[c] from GPC measurements in tetrahydrofuran calibrated with commerical polystyrene standards

silylated diphenols and difluoro aromatics. This finding first reported by Kricheldorf and Bier [83] is not surprising because the fluoro-silicon bond is the strongest of all Si-bonds and because fluoride ions are strong nucleophiles in the absence of H-bonds. Cesium as counterion is advantageous because CsF shows a relatively high solubility in nonprotic reaction media. Syntheses of low molecular weight arylethers may be promoted by tetraalkylammonium fluorides at temperatures below 100 °C [84], but tetraalkylammonium ions are not thermostable enough to catalyze polycondensations of silylated diphenols in bulk. The reaction mechanism proposed by Kricheldorf and Bier, Eqs. (65, 66) [83] assumes a nucleophilic attack of the fluoride ion on the silicon via its free d-orbitals. The generation of a phenoxide ion, Eq. (65), albeit at low concentration, results in a net gain of nucleophilicity and enables the reaction with the fluoroaromatic, Eq. (66). Hedrich et al. [85] have suggested that a pentacoordinated silicon

$$R-O-SiMe_3 + CsF \quad \rightleftharpoons \quad R-O^{\ominus} \, Cs^{\oplus} + F-SiMe_3 \qquad (65)$$

$$R-O^{\ominus} \, Cs^{\oplus} + R'-F \quad \rightleftharpoons \quad R-O-R' + CsF \qquad (66)$$

complex is the active species, Eq. [67]. Their argument against free phenoxide ions is the absence of ether exchange reactions, Eq. [68] which may occur at temperatures above 120 °C [86–89]. However, it must be taken into account that the concentration of phenoxide ions generated from silylated phenols is several orders of magnitude lower than in the case of the standard method using

$$
\begin{array}{c}
\text{Me} \\
\text{Me}{-}\overset{|}{\underset{|}{\text{Si}}}{-}\text{O}{-}\text{R} \quad + \quad \text{R'}{-}\text{F} \quad \longrightarrow \quad \text{R}{-}\text{O}{-}\text{R'} \quad + \quad \text{F}{-}\text{SiMe}_3 \\
\overset{\ominus}{\text{F}}\diagup\text{Me}
\end{array}
$$

$$\text{R , R`} = \text{different aromatic residues}$$

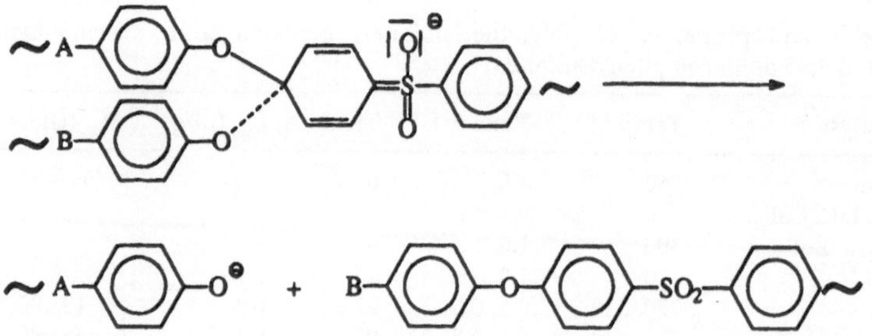

(67)

(68)

preformed potassium salts of diphenols [86, 87]. Another argument against a predominant role of the pentacoordinated silicon complex is the strong catalytic effect of chloride ions in polycondensations of silylated diphenols with various acid chlorides (Sect. 5.7 and 5.8). Furthermore, it was found that the nucleophilicity of the phenol is decisive for the reaction rate indicating that the nucleophilic attack onto the fluoroaromatic is the rate-determining reaction step. For instance the silyl derivatives of the acidic 4,4'-dihydro-xydiphenylsulfone or of the sterically hindered tetramethylbisphenol-A react much slower than bisphenol-A or phenylhydroquinone [83, 90]. Nonetheless, it must be emphasized that the reaction mechanism is not fully elucidated yet.

The polycondensation of silylated diphenols with activated difluoroaromatics catalyzed by CsF has the advantage that it may be conducted in bulk. No expensive reaction medium is required and the volatile fluorotrimethylsilane is the only byproduct, Eq. (69). Hence purification of the polyethers from inorganic

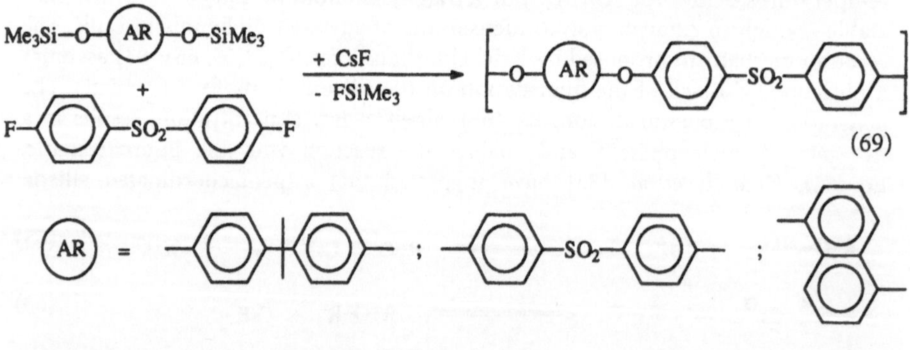

(69)

salts is not required. As reported by Kricheldorf et al. [83, 91] GPC and light-scattering measurements agree in that poly(ether-sulfone)s with weight average molecular weights up to 300 000 may be obtained in this way (the VPO data in Ref. 83 are a mistake).

This synthetic method (denoted method I) allows the preparation of different polyethers by variation of the electrophilic reaction partner. With

2,6-difluorobenzophenone polyethers containing pending ketogroups were synthesized for the first time (Form. **70**) [91]. These polyethers are amorphous

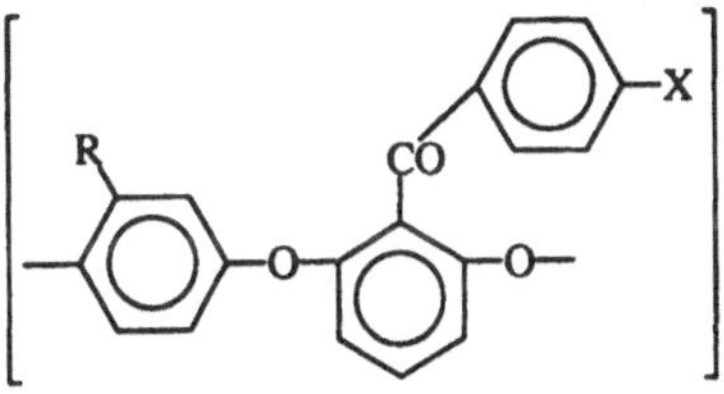

$$R = H, CH_3, C_6H_5$$

$$X = H, CH_3, OCH_3, SCH_3 \qquad (70)$$

and possess a high thermostability [92]. Furthermore, variation of X in Form. **70** allows one to introduce functional groups into the polyether. High molecular weight poly(benzonitrile ether)s were also obtained by bulk polycondensations of 2,6-difluorobenzonitrile (Forms. **71a, b**) and 2,4-difluorobenzonitrile (Form. **72a**). The 2,4-difluorobenzonitrile proved to be the less reactive isomer [93]. Furthermore, numerous poly(pyridine ether)s (Form. **73**) were prepared starting from 2,6-difluoropyridine [94]. However, these polycondensations also revealed a typical short-coming of "silyl-method I". The high volatility of some difluoro-aromatics makes it difficult to maintain the stoichiometry over the whole course of the polycondensation at least in small scale experiments.

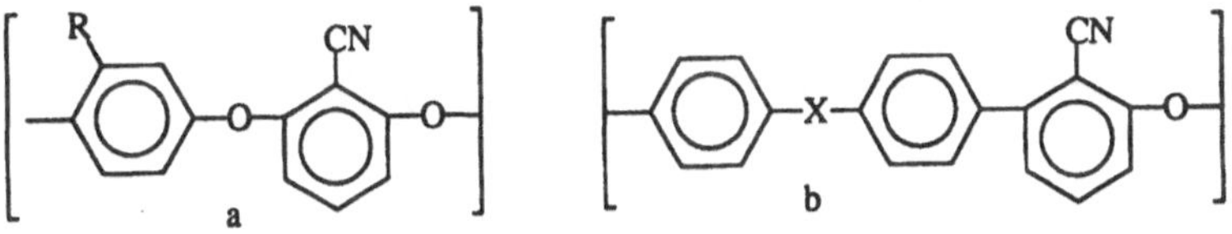

R = H, CH$_3$ $X = $ nil, O, SO$_2$, CMe$_2$, C–C$_6$H$_{10}$

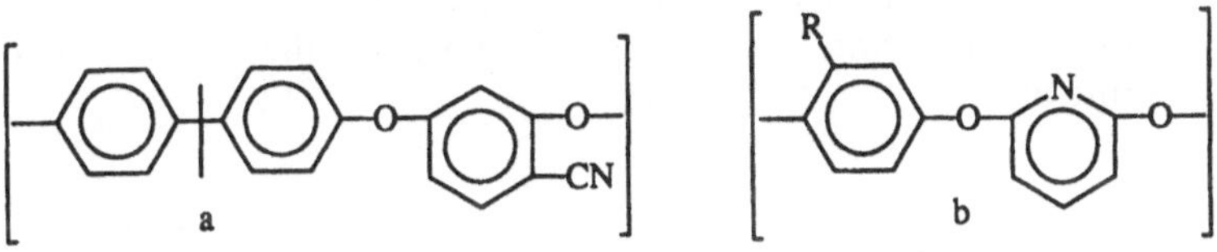

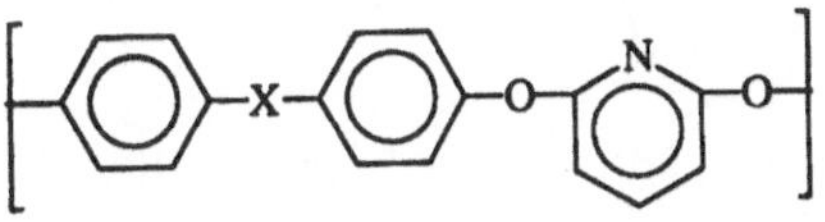

$$X = \text{nil}, O, SO_2$$

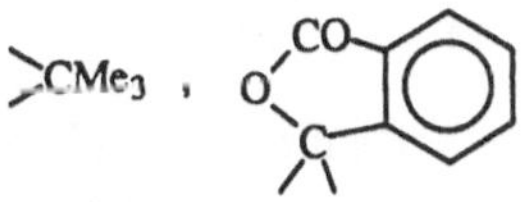

The synthesis of poly(ether-ketone)s from 4,4′-difluorobenzophenone revealed another problem of polycondensations in bulk. Poly(ether-ketone)s may be rapidly crystallizing materials and their melting temperatures may be as high as 320–420 °C [87]. Thus, the final reaction temperatures should be higher than 320 °C, but at such high reaction temperatures crosslinking takes place. Hence, the "silyl-method I" is not suited for the preparation of crystalline polyethers with T_m's above 320 °C. However, a broad variety of amorphous poly(ether-ketone)s such as **74a** and **b** or poly(ether-ketone)s with a lower melting temperature (e.g. Form. **75**) were synthesized [90, 95]. The amorphous character of **74a**,

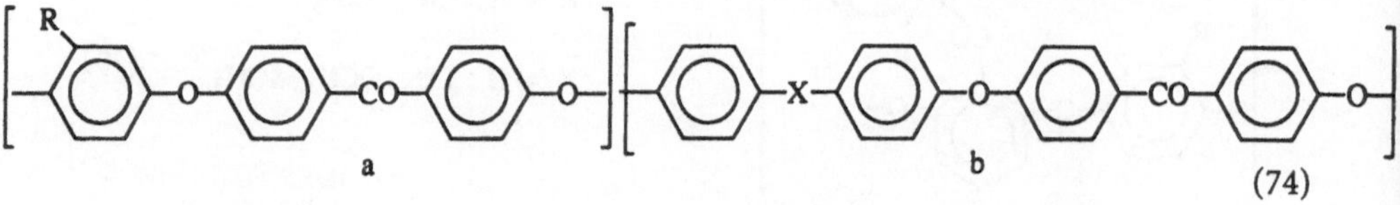

$$R = CH_3, C_6H_5 \qquad\qquad X = S, SO_2, CMe_2, C-C_6H_{10}$$

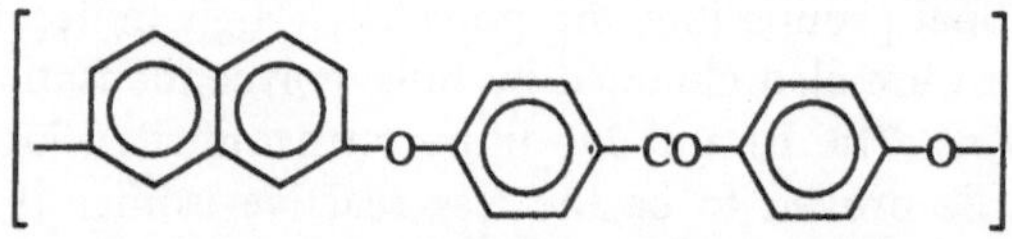

(74)

(75)

and **b** indicates that even small substituents attached to the poly(ether-ketone) backbone efficiently suppress the crystallization. Polycondensations of silylated methyl- or phenylhydroquinone with difluoroketones of increasing blocklength (Form. **76**) allowed Kricheldorf et al. [95] to study the substituent effect with

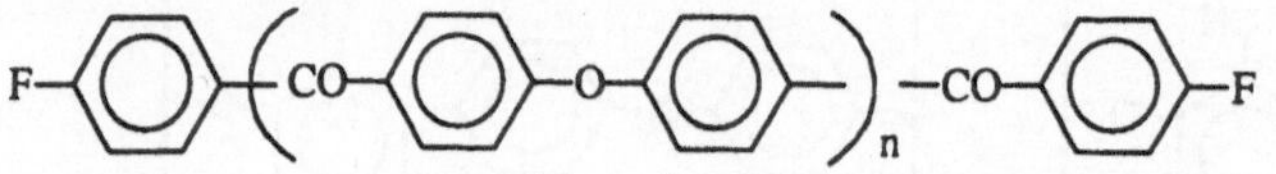

$$n = 0, 1, 2, 3$$

(76)

systematic "dilution" along the backbone. Furthermore, a group of poly(ether-ketone)s with a regular sequence of aliphatic spacers and aromatic blocks was synthesized and characterized [96]. These poly(ether-ketone)s (Form. **77**) are crystalline with melting temperatures below 300 °C, but they do not form a liquid crystalline phase.

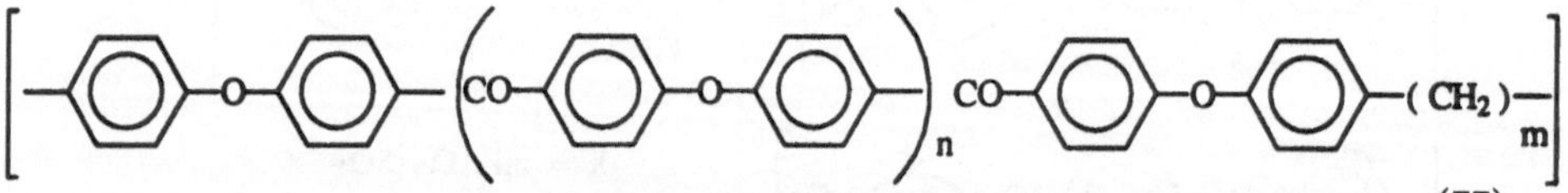

(77)

$$n = 1, 2, 3 \qquad\qquad m = 6, 10, 14$$

Poly(ether-sulfone)s and poly(ether-ketone)s can also be prepared from monomers containing two different functional groups, Eqs. (78) and (79) [97]. In contrast to the highly crystalline and high melting poly(ether-ketone) cocondensation of both monomers yields an amorphous poly(ether-ketone-sulfone). Cocondensation with 4,4'-difluorodiphenylsulfone yields oligomers with polymers

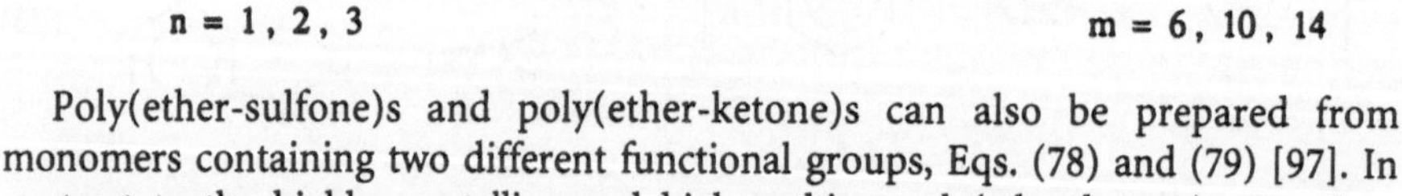

(78)

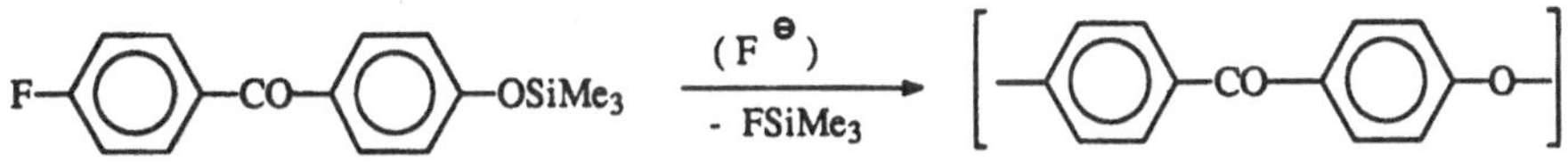

$$F-\bigcirc-CO-\bigcirc-OSiMe_3 \xrightarrow[- FSiMe_3]{(F^\ominus)} \left[-\bigcirc-CO-\bigcirc-O- \right]$$

$$(79)$$

with two reactive fluoroendgroups, Eq. (80). When a silylated bisphenol is used as comonomer in small quantities, oligomers or polymers with two tri-methylsiloxy endgroups are formed, Eq. (81) [97]. Such oligomers or polymers with two well defined endgroups are useful building blocks for the synthesis of A-B-A triblockcopolymers. Furthermore, star-shaped polyethers can be prepared by cocondensations of 4-fluoro-4′-trimethylsiloxy diphenylsulfone with tri- and multifunctional monomers such as **82a** or **82b**.

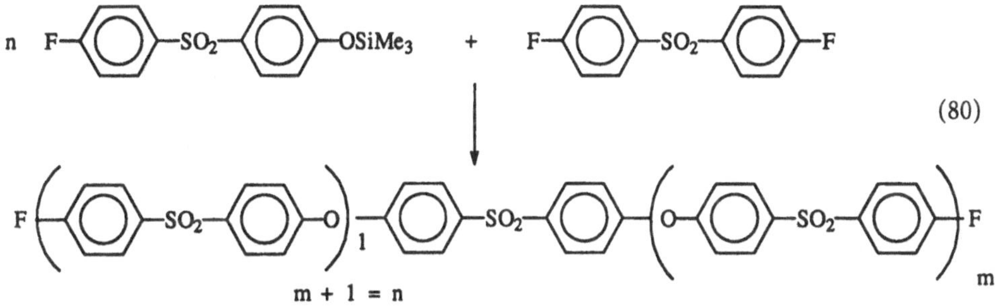

$$n \ F-\bigcirc-SO_2-\bigcirc-OSiMe_3 \ + \ F-\bigcirc-SO_2-\bigcirc-F$$

$$(80)$$

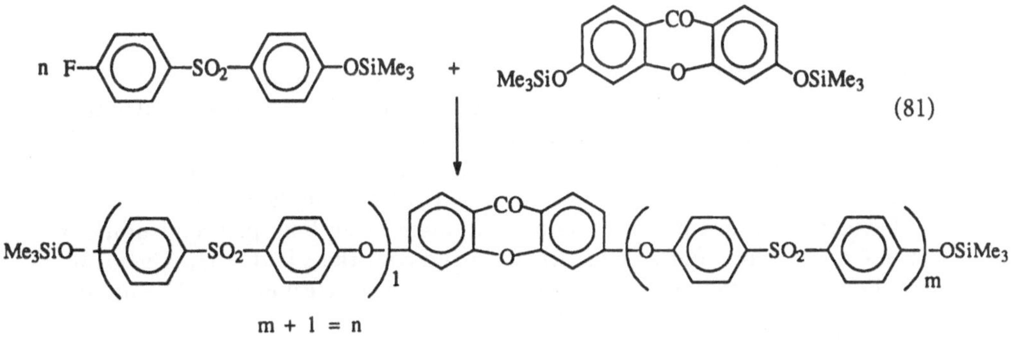

$$F\left(\bigcirc-SO_2-\bigcirc-O\right)_1-\bigcirc-SO_2-\bigcirc-\left(O-\bigcirc-SO_2-\bigcirc\right)_m F$$

$$m + 1 = n$$

$$n \ F-\bigcirc-SO_2-\bigcirc-OSiMe_3 \ + \ Me_3SiO-\bigcirc\!\!\!\bigcirc\!\!\!\bigcirc-OSiMe_3$$

$$(81)$$

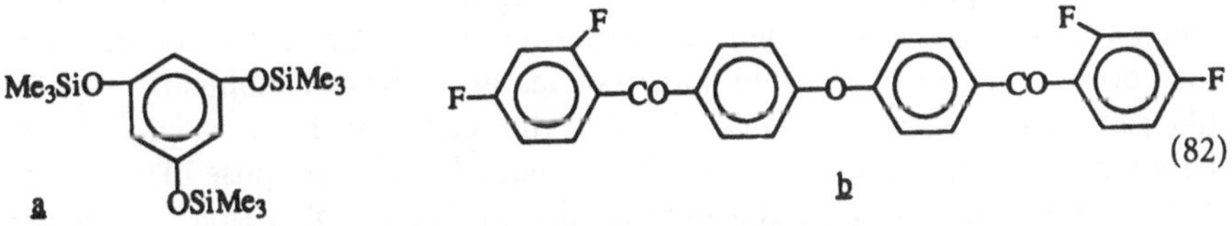

$$Me_3SiO-\left(\bigcirc-SO_2-\bigcirc-O\right)_1-\bigcirc\!\!\!\bigcirc\!\!\!\bigcirc-\left(O-\bigcirc-SO_2-\bigcirc\right)_m-OSiMe_3$$

$$m + 1 = n$$

$$Me_3SiO-\bigcirc-OSiMe_3$$
$$\underset{OSiMe_3}{}$$

a

$$F-\bigcirc\!\!^{F}-CO-\bigcirc-O-\bigcirc-CO-\bigcirc\!\!^{F}-F$$

$$(82)$$

b

A typical advantage of "silyl method I" is its potential for the synthesis of copolyethers, because either mixtures of different silylated diphenols or mixtures of different difluoroaromatics may be used in this process. For instance, mixtures of 2,6-difluorobenzophenones and 2,6-difluorobenzonitrile or 2,6-difluoropyridine yield polyethers with various functional groups in the backbone [98]. Poly(ether-sulfone)s containing pyridine units are of particular interest, because alkylation of the pyridine ring yields a new class of polyelectrolytes (Form. 83) which may be useful as membranes for gas-separation [99].

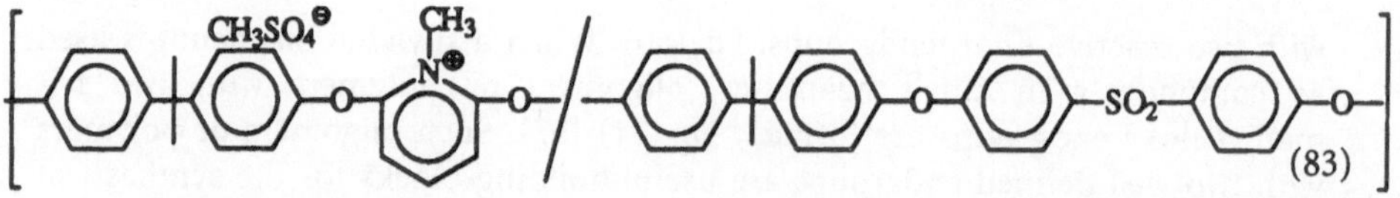

$$(83)$$

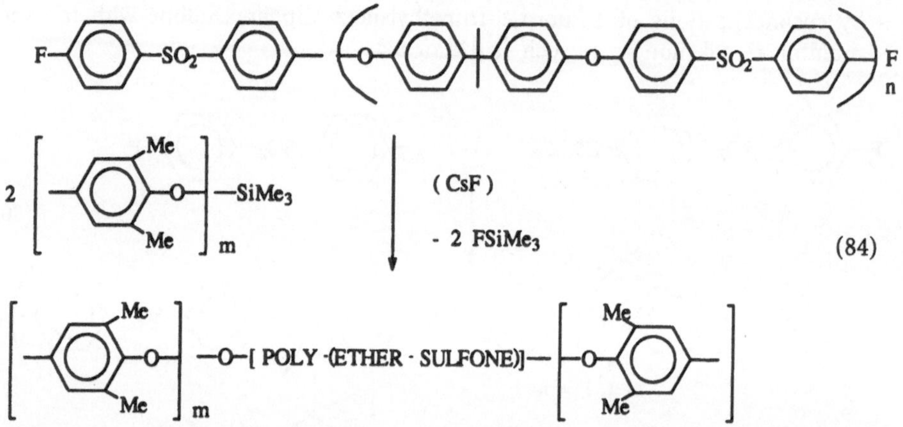

$$(84)$$

Another advantage of polycondensations based on CsF activated silylated diphenols was utilized by Hedrich et al. [85, 100]. Working in dry NMP as reaction medium they synthesized a series of A-B-A-triblockcopoly(ether-sulfone)s. Silylated poly(phenylene oxide) and fluoroterminated oligo(ether-sulfone)s were used as reaction partners, Eq. (84). Under the given reaction conditions the ether exchange reaction, Eq. (68) was entirely avoided.

Further variations of the polyether backbone using bulk polycondensations via method I were reported by Kricheldorf and coworkers [101–103]. The incorporation of ester groups into aromatic polyethers is of great interest, because the resulting poly(ether-ester)s allow transesterification with various aromatic polyesters [104]. The resulting multiblock copoly(ether-ester)s may possess useful mechanical properties. Two methods were reported for the synthesis of polyethers containing a variable percentage of ester groups. One method is based on the cocondensation of silylated hydroxyacids and silylated diphenols, Eq. (85) [83, 102]. Interestingly silylated dicarboxylic acids, such as isophthalic or terephthalic acid, are not nucleophilic enough for this purpose. The second method is a twostep process starting with the synthesis of oligoesters terminated by siloxy groups. Such oligoesters result from a large excess of the silylated diphenol. The mixture of oligoesters and excess diphenol is then polycondensed with the appropriate amount of a difluoroaromatic, Eqs. (86) and (87). It is important that the first step is catalyzed by a trace of chloride ions and the second step by a larger molar amount of fluoride ions [101] (relative to chloride ions). Even poly(ether-ester)s containing imide groups can be prepared in this way (Form. 88).

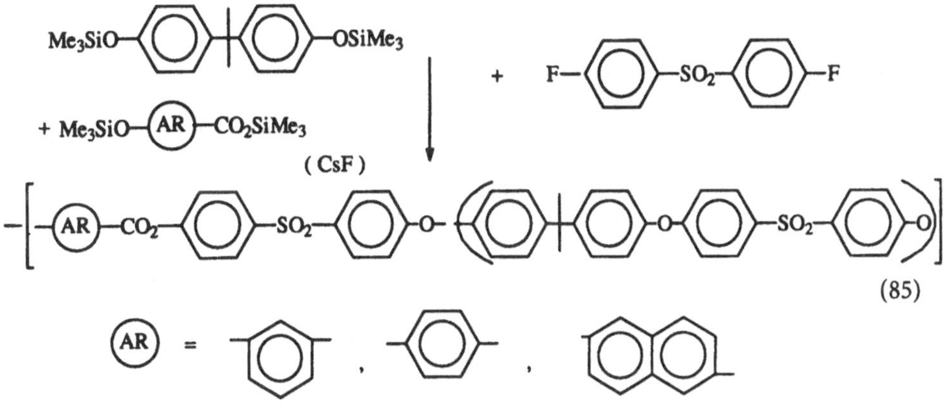

$$Me_3SiO-\!\!\bigcirc\!\!-\!\!\bigcirc\!\!-OSiMe_3 \quad + \quad F-\!\!\bigcirc\!\!-SO_2-\!\!\bigcirc\!\!-F$$

$$+ \; Me_3SiO-(AR)-CO_2SiMe_3$$

$$(\,CsF\,)$$

$$-\!\!\left[(AR)-CO_2-\!\!\bigcirc\!\!-SO_2-\!\!\bigcirc\!\!-O-\!\!\left(\!\!\bigcirc\!\!-\!\!\bigcirc\!\!-O-\!\!\bigcirc\!\!-SO_2-\!\!\bigcirc\!\!-O\right) \right]$$

$$(85)$$

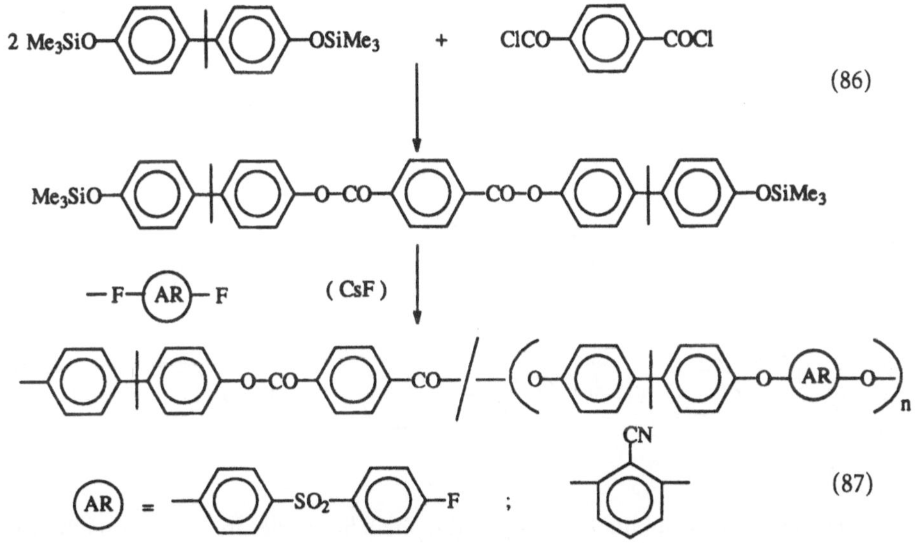

$$(AR) = \bigcirc\!\!- \;,\; -\!\!\bigcirc\!\!- \;,\; \bigcirc\!\!\bigcirc$$

$$2\; Me_3SiO-\!\!\bigcirc\!\!-\!\!\bigcirc\!\!-OSiMe_3 \quad + \quad ClCO-\!\!\bigcirc\!\!-COCl \qquad (86)$$

$$Me_3SiO-\!\!\bigcirc\!\!-\!\!\bigcirc\!\!-O-CO-\!\!\bigcirc\!\!-CO-O-\!\!\bigcirc\!\!-\!\!\bigcirc\!\!-OSiMe_3$$

$$-F-(AR)-F \qquad (\,CsF\,)$$

$$-\!\!\bigcirc\!\!-\!\!\bigcirc\!\!-O-CO-\!\!\bigcirc\!\!-CO-\!\!\!/\!\!-\left(O-\!\!\bigcirc\!\!-\!\!\bigcirc\!\!-O-(AR)-O\right)_n$$

$$(AR) = -\!\!\bigcirc\!\!-SO_2-\!\!\bigcirc\!\!-F \quad ; \quad \overset{CN}{\underset{}{\bigcirc}} \qquad (87)$$

$$\left[-\!\!\bigcirc\!\!-\!\!\bigcirc\!\!-O-CO-\!\!\overset{CO}{\underset{CO}{\bigcirc}}\!\!N-\!\!\bigcirc\!\!-CO-\left(O-\!\!\bigcirc\!\!-\!\!\bigcirc\!\!-O-(AR)-O\right) \right]$$

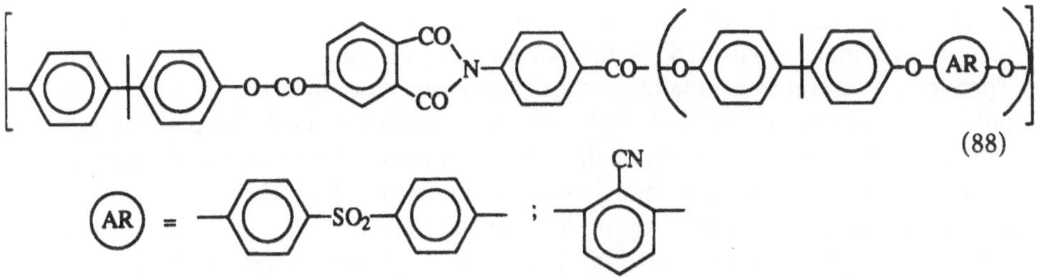

$$(88)$$

$$(AR) = -\!\!\bigcirc\!\!-SO_2-\!\!\bigcirc\!\!- \quad ; \quad \overset{CN}{\underset{}{\bigcirc}}$$

Polyethers containing amide groups were obtained by the following reaction sequence [102]. A silylated diphenol was synthesized by selective acylation of a silylated aminophenol with a dicarboxylic acid chloride, Eq. (89). The resulting "amide diphenol" mixed with another silylated diphenol was then subjected to a polycondensation with a reactive difluoroaromatic, Eq. (90). This two step synthesis may be conducted as an "one-pot procedure". Apparently due to side reactions of the amide groups only moderate inherent viscosities were obtained ($\eta^{inh} \leqslant 0,5$ dl/g in CH_2Cl_2/TFA). Higher molecular weights were found, when a N-substituted 4-aminophenol served as starting material (Form. **91**), but in this case the entire reaction sequence can not be conducted in an "one-pot procedure" [103].

$$2 \ Me_3SiO-\langle\bigcirc\rangle-NH-SiMe_3 \ + \ ClOC-\langle\bigcirc\rangle-COCl \longrightarrow \tag{89}$$

$$Me_3SiO-\langle\bigcirc\rangle-NH-CO-\langle\bigcirc\rangle-CO-NH-\langle\bigcirc\rangle-OSiMe_3$$

$$+ \ Me_3SiO-(AR)-OSiMe_3$$

$$+ \ F-\langle\bigcirc\rangle-SO_2-\langle\bigcirc\rangle-F \tag{90}$$

$$\left[\langle\bigcirc\rangle-NH-CO-\langle\bigcirc\rangle-CO-NH-\langle\bigcirc\rangle-O-\langle\bigcirc\rangle-SO_2-\langle\bigcirc\rangle-O-(AR)-O-\langle\bigcirc\rangle-SO_2-\langle\bigcirc\rangle-O\right]$$

$$\left[\langle\bigcirc\rangle-\underset{Me}{N}-CO-\langle\bigcirc\rangle-CO-\underset{Me}{N}-\langle\bigcirc\rangle-O-\left(\langle\bigcirc\rangle-\langle\bigcirc\rangle\right)-O-\langle\bigcirc\rangle-SO_2-\langle\bigcirc\rangle-O\right) \tag{91}$$

The lower nucleophilic reactivity of the phenolic species in the "silyl method I" which helps to reduce ether exchange reactions has, on the other hand, the disadvantage that the relatively expensive difluoroaromatics are necessary as reaction partners. Activated dichloroaromatics are not electrophilic enough. Therefore, a careful screening of reaction conditions that might enable satisfactory polycondensations of silylated diphenols with activated dichloroaromatics was conducted [91]. A combination of dry NMP and equimolar amounts of K_2CO_3 as catalyst and $Cl^{\ominus}$ acceptor was found to give optimum results [91] (silyl method II). Poly(ether-sulfone)s with M_W's up to 200 000 were obtained, Eq. (92). Under these reaction conditions difluoroaromatics do not yield higher molecular weights.

The disadvantage of a relatively expensive reaction medium is counterbalanced by several advantages. For instance, reaction partners with different volatility do neither affect the stoichiometry and the molecular weights, nor the molar composition of copolymers. Furthermore, dichloroaromatics can be used where the corresponding difluoroderivatives are not commercial. For this reasons, polyethers containing pyrazine (Form. **93a**) or pyridazine units

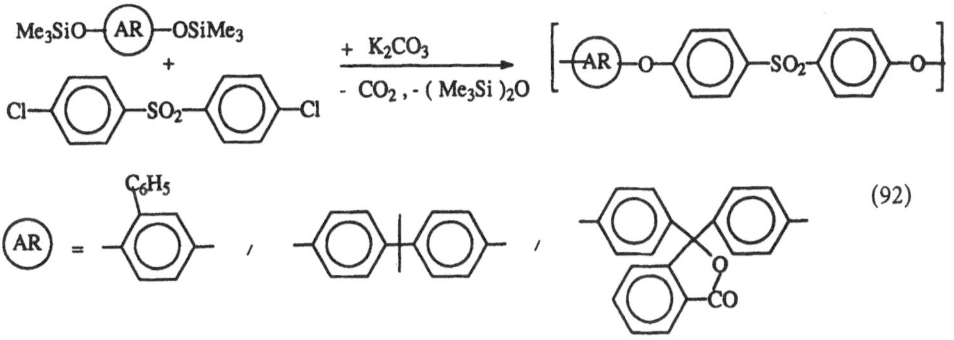

$$(92)$$

(Form. **93b**) were for the first time synthesized via the "silyl method II" [91, 105]. Aromatic polyether containing nitrogen heterocycles are of interest because alkylation yields novel chemically and thermally stable ionomers, such as **94** [105].

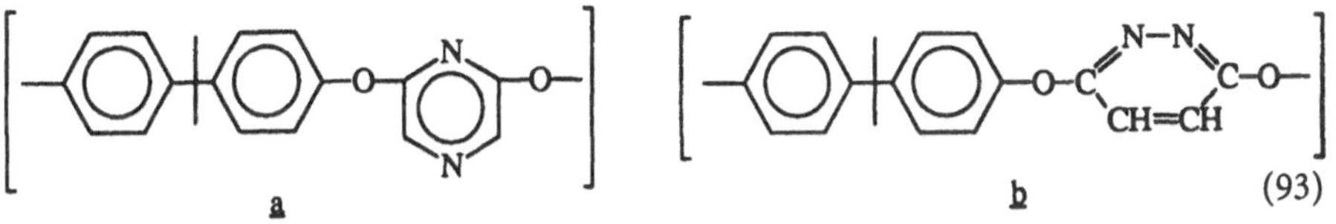

$$(93)$$

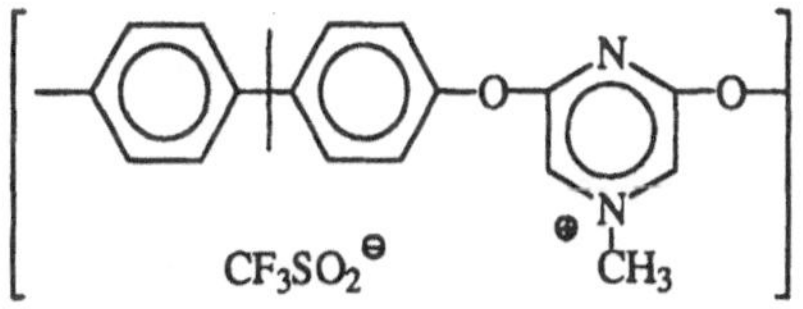

$$(94)$$

The screening of the reaction condition suitable for the "silyl method II" revealed that despite careful purification of solvent and monomers a slight excess of the dichloroaromatic is necessary to obtain high molecular weights. This observation suggests that a side reaction occurs which converts chlorosubstituents into nucleophilic groups which participitate in the polycondensation process. A hypothetical mechanism assumes that trimethylsilyloxide anions are formed in low concentrations which react with activated chloroaromatics by formation of silylated phenol groups, Eq. (95). The in-situ generated silylated phenols in turn participate in the polycondensation process.

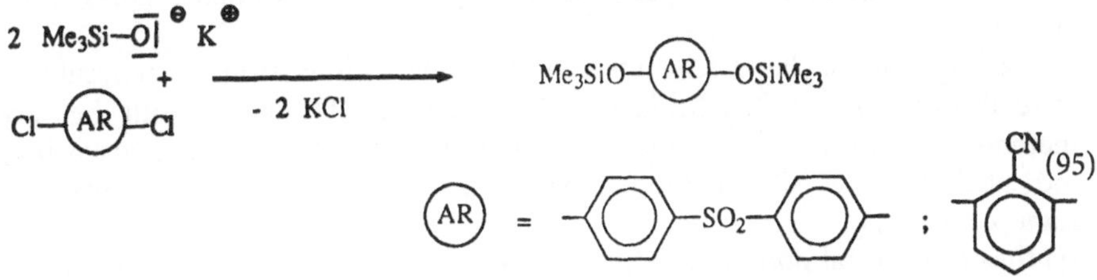

$$(95)$$

Based on this hypothesis a new method for the synthesis of homopolyethers from difluoro- or dichloroaromatics and K_2CO_3 was designed. By systematic variation of the reaction conditions it was found [106] that chlorotriphenylsilane or triphenyl silanol are suitable catalysts, Eq. (96). High molecular weight

$$F-\langle\bigcirc\rangle-SO_2-\langle\bigcirc\rangle-F \xrightarrow{[\ (C_6H_5)_3SiCl\ ]} \left[-\langle\bigcirc\rangle-SO_2-\langle\bigcirc\rangle-O- \right]$$

$$+ \ K_2CO_3 \qquad\qquad\qquad\qquad\qquad\qquad\qquad\qquad\qquad (96)$$

poly(ether sulfone)s were obtained from 4,4′-difluorodiphenylsulfone but only low molecular weights from 4,4′-dichlorodiphenylsulfone. The most interesting aspect of this "silyl method III" is the role of the silyl group, which reacts as a phase transfer catalyst with simultaneous modification of the transfered anion from carbonate to oxide.

The successful synthesis of polyethers from silylated phenols prompted several authors to study the synthesis of aromatic polysulfides from silylated thiophenols. Kricheldorf and Jahnke [107] isolated poly(ether-sulfide)s of moderate molecular weights, when silylated 4-mercaptophenol was polycondensed with 2,6-dichloropyridine or 3,6-dichloropyridazine (Forms. 97 and 98). The moderate

$$\left[-\langle\bigcirc\rangle-S-\langle\bigcirc_N\rangle-O- \right] \qquad \left[-\langle\bigcirc\rangle-S-C{\overset{N-N}{\underset{CH=CH}{\diagup\diagdown}}}C-O- \right]$$

$$\text{a} \qquad\qquad\qquad\qquad\qquad\qquad\qquad \text{b} \qquad (97)$$

$$\left[\langle\bigcirc\rangle-S-\langle AR\rangle-S- \right] \qquad\qquad\qquad\qquad\qquad (98)$$

$$\langle AR\rangle \ = \ -\langle\bigcirc\rangle-SO_2-\langle\bigcirc\rangle- \quad , \quad \overset{CN}{-\langle\bigcirc\rangle-} \quad , \quad -\langle\bigcirc_N\rangle-$$

molecular weights and the elemental analyses suggest that the sulfur causes side reactions which are probably radical in nature. Reaction steps involving radical cations are well known to occur in the synthesis of aromatic polysulfides [108, 109].

Starting from silylated 1,3-dimercaptobenzene various aromatic polysulfides were also prepared by Kricheldorf and Jahnke [110] (Form. 98). Both the "silyl methods I" and II" were used. A comparison of the inherent viscosities indicated that the "silyl method II" yields the higher molecular weights. The same conclusion was reached by Imai and coworkers [111] who studied the synthesis of several polysulfides (Forms. 99a and b) either by polycondensations in bulk (using CsF as catalyst) or in solution (in the presence of K_2CO_3). Again

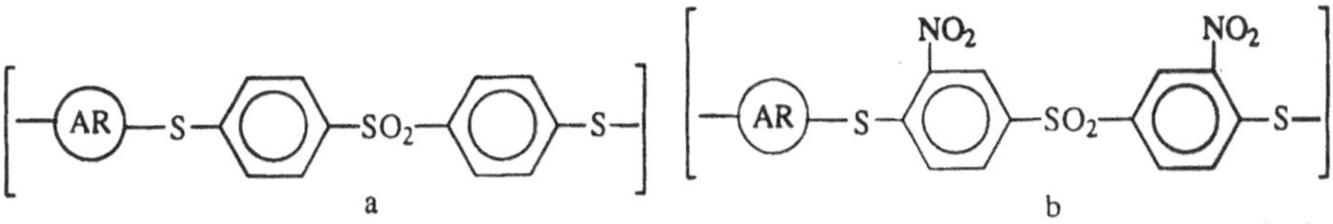

$$(99)$$

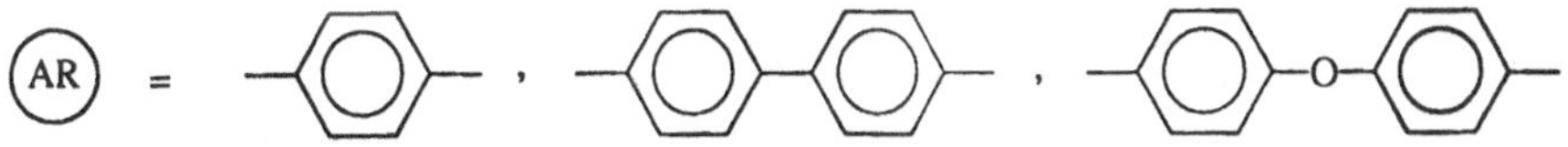

the low or moderate inherent viscosities suggest that the silylated mercapto groups cause side reactions. In other words the synthesis of aromatic polysulfides was not as successful as that of analogous polyethers.

5.7
Polyesters

The synthesis of polyesters from silylated monomers may be conducted in three ways. The oldest and probably most versatile method is based on the condensation of a silylated phenol group with an carboxylic acid chloride under elimination of chloro trimethylsilane, Eq. (100). This condensation reaction requires

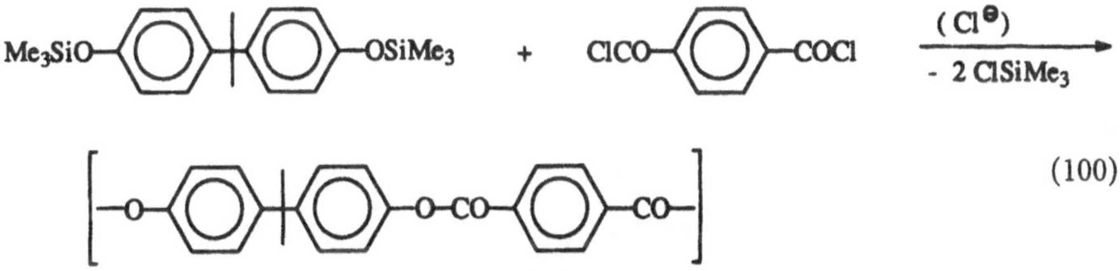

$$(100)$$

catalysis by an anion, in particular chloride ions. The role of the catalyst presumably consists of an attack on the silyl group via its free d-orbitals thereby generating a phenoxide anion, Eq. (101). This desilylation equilibrium is

$$(101)$$

certainly shifted to the left side, but the phenoxide ion is several orders of magnitude more nucleophilic than the siloxy group, and the net result is a gain in nucleophilicty. Besides chloride or fluoride ions, Lewis acids may also serve as catalysts activating the acid chloride. However, Lewis acids may cause side reactions, such as the Fries-rearrangement at higher temperatures, and thus, are less favorable as catalysts than chloride ions.

When pure silylated diphenols (e.g. resulting from the silylation with hexamethyldisilazane) are heated with a pure aromatic dicarboxylic acid chloride no reaction takes place even at 300 °C. Addition of a catalytic amount of chloride anions starts the polycondensation at temperatures above 100 °C. For preparative purposes the initial reaction temperature usually falls into the range of 120–150 °C. The final reaction temperature depends on the melting temperature of the resulting polyesters, but should not exceed 320 °C. This silyl method (I) has no particular advantage over the chemical polycondensation methods for normal polyarylates based on isophthalic or terephthalic acid [112]. In the case of the more reactive aliphatic dicarboxylic acid chlorides the initial reaction temperature may be as low as 100 °C and even lower. Such low reaction temperatures may be advantageous compared to the high temperatures required for the polycondensations of the free dicarboxylic acids which can undergo side reactions, such as decarboxylation, or formation of cyclic anhydrides above 200 °C.

The silyl method I has also proven to be favorable for polycondensation of aryloxy and arylthioterephthalic acids. When the free acids are heated with acetylated diphenols self catalylzed intramolecular acylation causes frequent termination steps [113], Eq. (102). The polycondensation of the corresponding substituted terephthaloyl chlorides, Eq. (103), can be started at lower tempera-

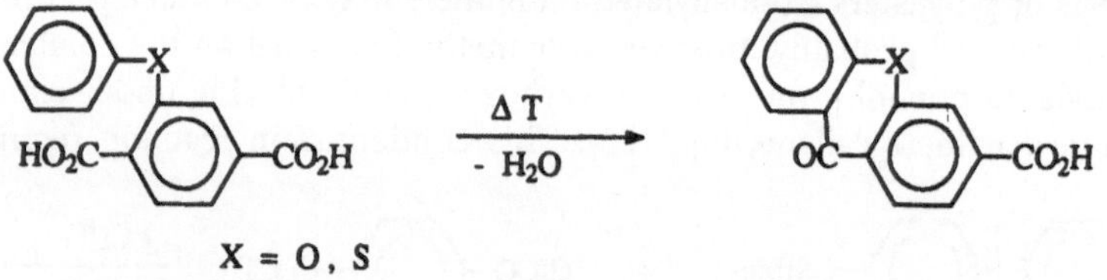

$$X = O, S \tag{102}$$

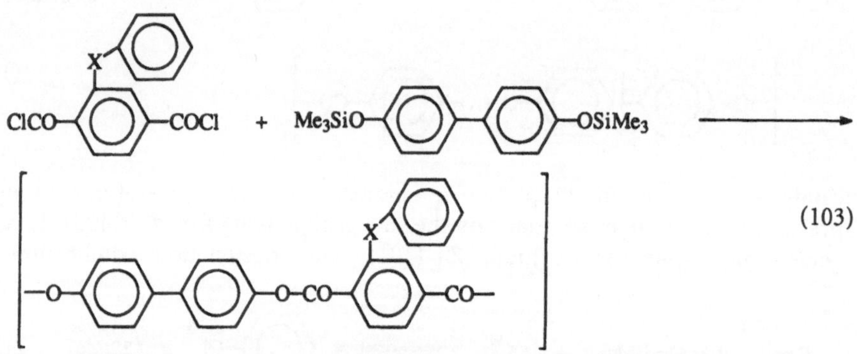

$$(103)$$

tures, involves fewer side reactions and gives higher molecular weights [114–117]. A comparison of three different synthetic methods is discussed in ref. 113. Homo- and copolyesters derived from aryloxy- or aryl-thioterephthalic acid are of particular interest because most of them form a liquid-crystalline (mainly nematic) melt. The silyl method I is also well suited for the preparation of polyesters from 2,5-dialkoxyterephthalic acids (e.g. **104**), which can again form liquid crystalline phases. Numerous examples of such LC-polyesters were synthesized and characterized by Kricheldorf and coworkers [115–117].

The absence of acidic protons is a significant advantage of the silyl method, when deuterated polyesters such as **105** [118, 119] should be synthesized. Selective

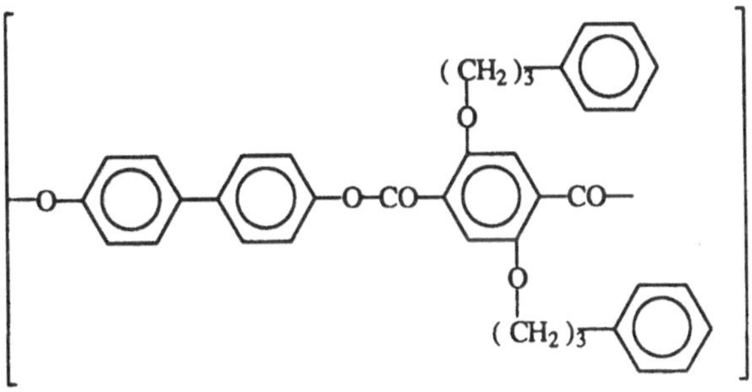

(104)

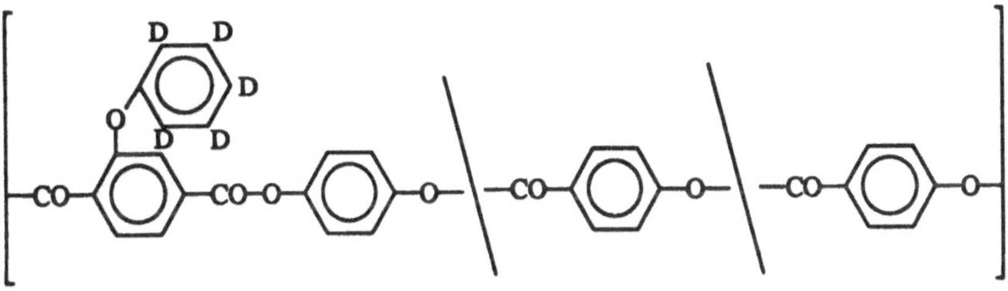

(105)

deuteration offers the opportunity to study the chain dynamics by ^{2}H NMR spectroscopy over a broad temperature range. Selectively deuterated liquid-crystalline or isotropic polyesters are under investigation by Kricheldorf et al. [120]. The silyl method I is also useful for the synthesis of polyesters from aromatic hydroxy acids. The trimethylsiloxy chlorides can easily be prepared by treatment of silylated hydroxy acids with thinoyl chloride Eq. (106) [120]. Also

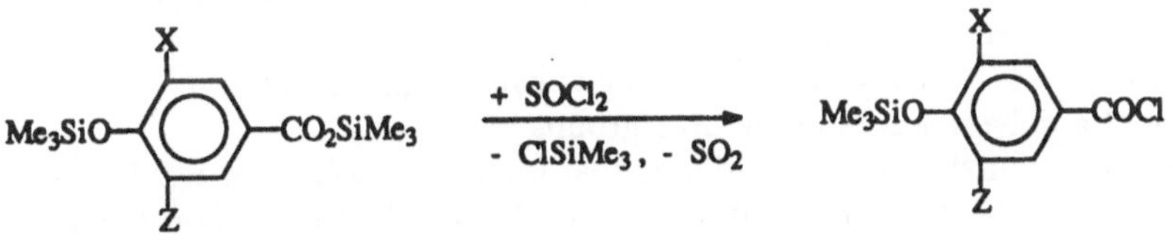

$$X = H, H, Cl, CH_3 \qquad\qquad Z = Cl, OCH_3, CH_3$$

(106)

this chlorination is catalyzed by chloride ions. The monomers thus obtained can be purified by distillation in vacuo if chloride ions are absent. Therefore, it is recommendable to use pyridine or triethylamine hydrochloride as catalysts for the chlorination. They can be removed from the reaction mixture by sublimation.

The chloride ion-catalyzed polycondensation yields highly crystalline infusible polyesters, when 4-hydroxybenzoic acids serve as starting materials, Eq. (107) [120]. The absence of acidic protons is here a significant advantage for the synthesis of branched polyesters from trifunctional monomers such as 3,5-dihydroxybenzoic acid. The cocondensation of 3-trimethylsiloxybenzoylchloride and 3,5-bistrimethylsilylbenzoylchloride, Eq. (108), yielded for the first time a series of more or less branched, soluble copolyesters as demonstrated by

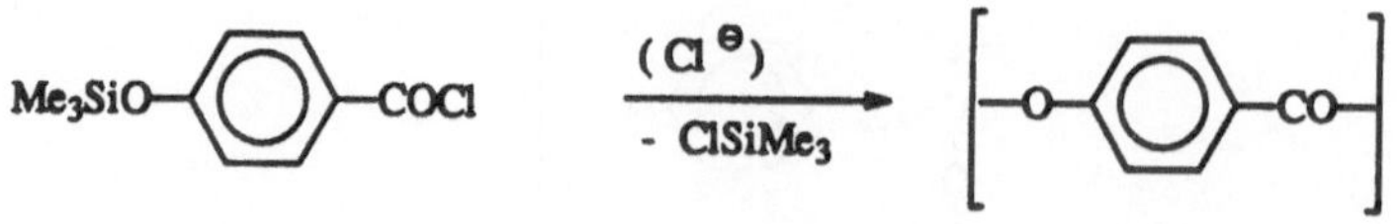

$$(107)$$

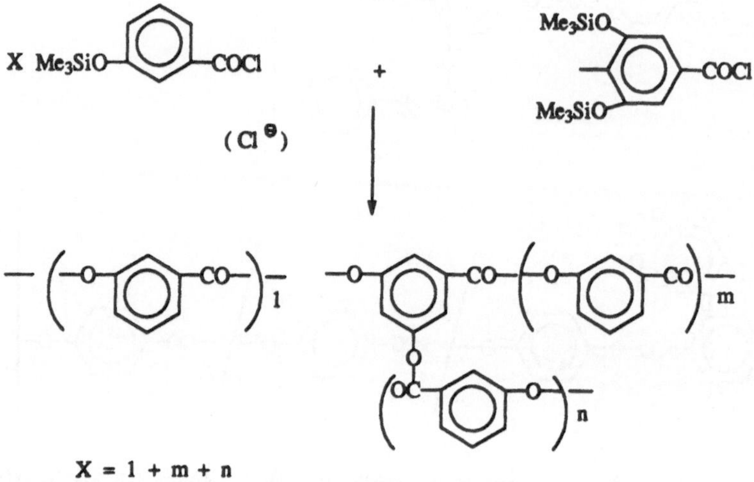

$$(108)$$

$$X = 1 + m + n$$

Kricheldorf and Schwarz [122]. Tréchet and coworkers [123] used the same approach later for the polycondensation of 3,5-bistrimethylsiloxybenzoylchloride. A sufficiently low reaction temperatures a highly branched non-crosslinked polyester (Form. **109**) was isolated. Furthermore, the polycondensation of silylated diphenols is particularly suited for the synthesis of copolyesters and modified polyesters. Polyesters containing phosphate or phosphonate group are discussed in Sect. 5.9, polyester containing anhydride groups in Sect. 5.8 and an example of a polyester containing urea groups in Sect. 5.2.

Furthermore, it should be mentioned that numerous poly(ester-imide)s were prepared by Kricheldorf and coworkers [124, 125] via the silylmethod I as illustrated by Form. **110**. Such poly(ester-imide)s may be thermotropic, or isotropic.

A second approach based on the condensation of silylated carboxyl groups with acetylated phenol groups was developed by Kricheldorf and coworkers [126–129]. As examplified for the synthesis of poly(4-oxybenzoate), Eq. (111), such a polycondensation is not trivial because it was also found that fully silylated 4-hydroxybenzoic acid does not react even at 400 °C, Eq. (112), and acetoxybenzoic acid alkyl or phenyl esters do not react either, Eq. (113) [126]. This "silyl method II" has again the advantage to be a relatively clean reaction which yields poly(4-oxybenzoate) single crystals of high perfection [126].

The "silyl method II" also allows the polycondensation of silylated dicarboxylic acids, Eq. (114) [127]. Temperatures around 300 °C are required and these polycondensations are best conducted in bulk, if the resulting polyesters are meltable. As illustrated by the data of Table 5.7, titanium tetrapropoxide is useful as catalyst [127]. This method is advantageous for the polycondensation of high melting monomers, because silylation lowers the melting points of the

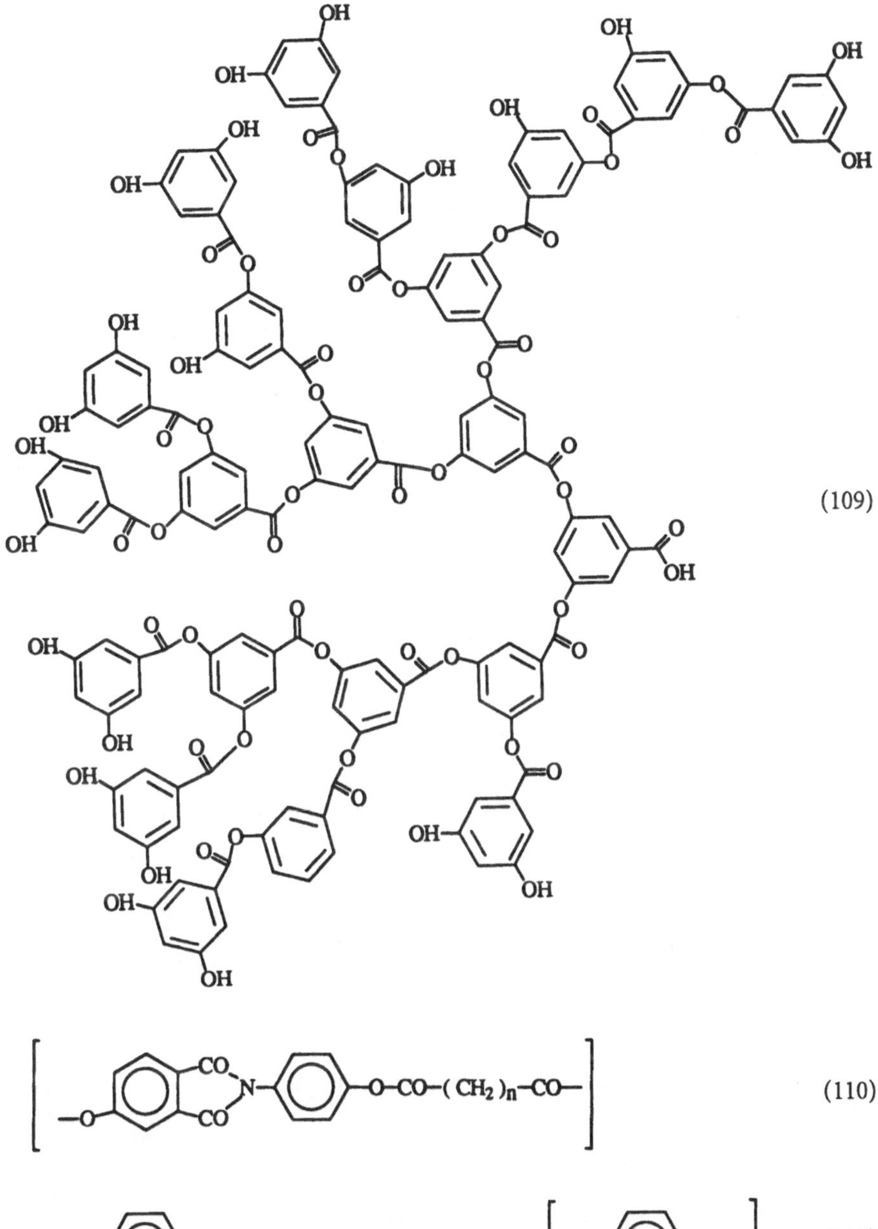

(109)

(110)

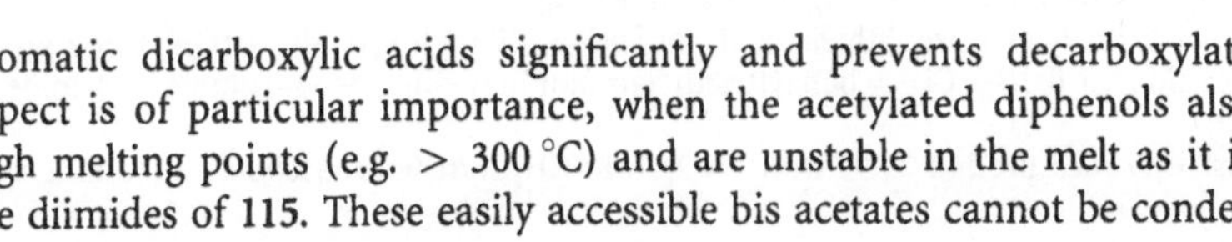

(111)

aromatic dicarboxylic acids significantly and prevents decarboxylation. This aspect is of particular importance, when the acetylated diphenols also possess high melting points (e.g. > 300 °C) and are unstable in the melt as it is true for the diimides of 115. These easily accessible bis acetates cannot be condensed with

$$\text{Me}_3\text{SiO}-\text{C}_6\text{H}_4-\text{CO}_2\text{SiMe}_3 \quad \xcancel{\longrightarrow} \quad \left[-\text{O}-\text{C}_6\text{H}_4-\text{CO}-\right]$$

$$\text{AcO}-\text{C}_6\text{H}_4-\text{CO}_2\text{R}$$

(112)

$$R = C_2H_5 , C_6H_5 \tag{113}$$

$$\text{AcO}-\text{C}_6\text{H}_4-\text{C(CH}_3)_2-\text{C}_6\text{H}_4-\text{OAc} \quad + \quad \text{Me}_3\text{SiO}_2\text{C}-\text{C}_6\text{H}_4-\text{CO}_2\text{SiMe}_3$$

Ti (OPr)$_4$
300°C

$\downarrow$ − 2 AcOSiMe$_3$

(114)

$$\left[-\text{O}-\text{C}_6\text{H}_4-\text{C(CH}_3)_2-\text{C}_6\text{H}_4-\text{O}-\text{CO}-\text{C}_6\text{H}_4-\text{CO}-\right]$$

Table 5.7 Reaction conditions and results of polycondensations conducted with silylated isophthalic acid and acetylated bisphenol A in bulk.

No.	Temp in °C	Time[a] in h	Catalyst partners (mole-ratio)[b]	Yield in %	η_{inh}[c] dL/g
1	300	5 + 2, 0 vac	–	0	–
2	300	4 + 1, 5 vac	MgO (1:10^{-3})	86	0, 36
3	300	4 + 1, 0 vac	Bu$_2$SnO (1:10^{-3})	oligom.	–
4	300	5 + 1, 0 vac	Sn(II) (1:10^{-3}) 2-ethylhexanoate	85	0, 21
5	300	4 + 1, 5 vac	Ti(OiPr)$_4$ (1:10^{-3})	89	0, 31
6	300	4 + 1, 0 vac	Ti(OiPr)$_4$ (2, 5:10^{-3})	91	0, 33
7	300	4 + 1, 0 vac	Ti(OiPr)$_4$ (5:10^{-3})	95	0, 41
8	300	4 + 1, 0 vac	Ti(OiPr)$_4$ (1:10^{-2})	95	0, 30
9	350	4 + 1, 5 vac	–	76	0, 48
10	350	4 + 1, 0 vac	Et$_3$BzlN$^+$Cl$^-$ (1:10^{-3})	crossl.	–
11	350	4 + 1, 0 vac	4-Toluols (1:10^{-3})	74	0, 52
12	350	4 + 1, 0 vac	MgO (1:10^{-3})	85	0, 48
13	350	4 + 1, 0 vac	Ti(OiPr)$_4$ (1:10^{-3})	81	0, 73
14	350	4 + 0, 5 vac	Ti(OiPr)$_4$ (5:10^{-3})	crossl.	–

Kricheldorf HR (1972) Lbbers D, Makromol Chem Rapid Commun 12 691 (1991)
[a] After heating for 4 or 5 h under nitrogen vacuum was applied.
[b] Mole ratio catalyst/silylated isophthalic acid; Bu$_2$SnO = dibutyltin oxide, Ti(OiPr)$_4$ = titanium (IV) isopropoxide; Et$_3$BzlN$^+$ Cl$^-$ = benzyltriethylammonium chloride; 4-Toluols = 4-toluenesulfonic acid.
[c] Inherent viscosity measured at 25 °C with $c = 2$g/L in CH$_2$Cl$_2$/trifluoroacetic acid (volume ratio 4:1).

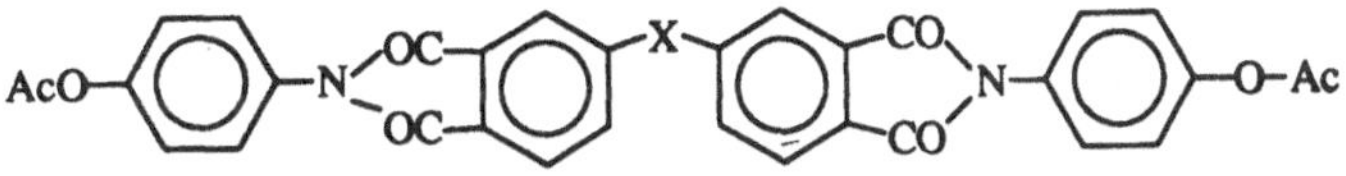

$$\text{AcO} - \text{C}_6\text{H}_4 - \text{N} \begin{smallmatrix} \text{OC} \\ \text{OC} \end{smallmatrix} - \text{C}_6\text{H}_3 - \text{X} - \text{C}_6\text{H}_3 \begin{smallmatrix} \text{CO} \\ \text{CO} \end{smallmatrix} \text{N} - \text{C}_6\text{H}_4 - \text{O} - \text{Ac}$$

(115)

a : X = O, b : X = CO, c : X = SO$_2$

free aromatic dicarboxylic acids, because initial reaction temperatures around 350 °C are required, where all monomers begin to decompose. However, with silylated dicarboxylic acids reaction temperatures below 300 °C may be used and numerous thermotropic poly(ester-imide)s such as that of 116 could be prepared

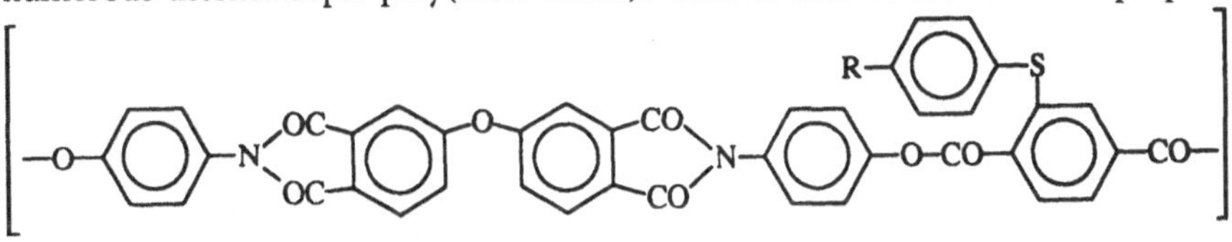

(116)

in this way [128, 129]. The silyl method II is also successful for the synthesis of hyper branched polyesters from 5-acetoxyisophthalic acid [130] or 3,5-bisacetoxybenzoic acid [131]. The silylated carboxyl group reduces the risk of proton catalysed crosslinking, Eq. (117). Despite the lower reactivity the degree of branching of the resulting dendridic polyester (Forms. 117 and 118) was as high as that of the polyester prepared from 3,5-bistrimethylsiloxybenzoylchloride (Form. 109). Silyl methods seem to be particularly useful for the synthesis of star-shaped and dendridic polycondensates.

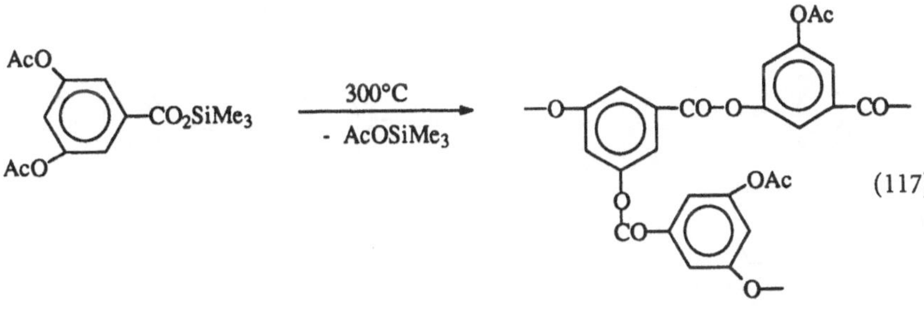

$$\xrightarrow[\text{- AcOSiMe}_3]{300°\text{C}}$$

(117)

An interesting variant of the silyl method II is based on a highly selective acylation silylated amino acids at the amino group. At moderate temperatures (< 100 °C) and in the absence of chloride ions the silylated carboxyle group remains inert. If an acetoxy benzoyl chloride is used as reaction partner this acylation yields a new "amidemonomer"in situ, Eq. (119) which upon heating to 300–400 °C will undergo polycondensation, Eq. (120) [132]. With para-substituted monomers this reaction sequence is best conducted as an "one-pot procedure" in a high boiling inert reaction medium and yields a highly crystalline poly-(estere-amide) with a melting point above 500 °C (with decomposition). Semicrystalline or amorphous poly(ester-amide)s can also be prepared from dicarboxylic dichlorides and diphenol acetates as illustrated by Eqs. (121) and

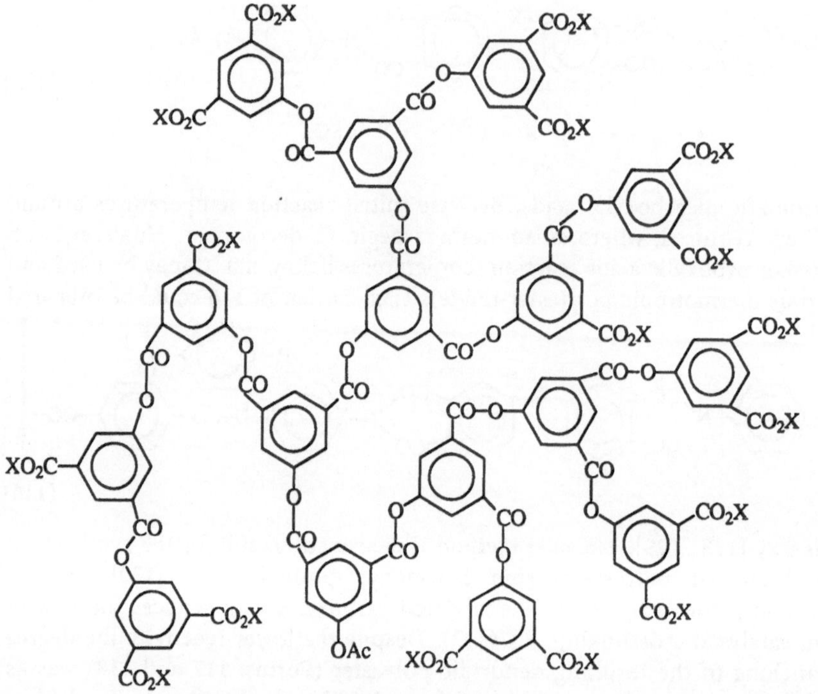

$$(118)$$

$$2a: X = H , \quad 2b: X = SiMe_3 , \quad 2c: X = C_2H_5$$

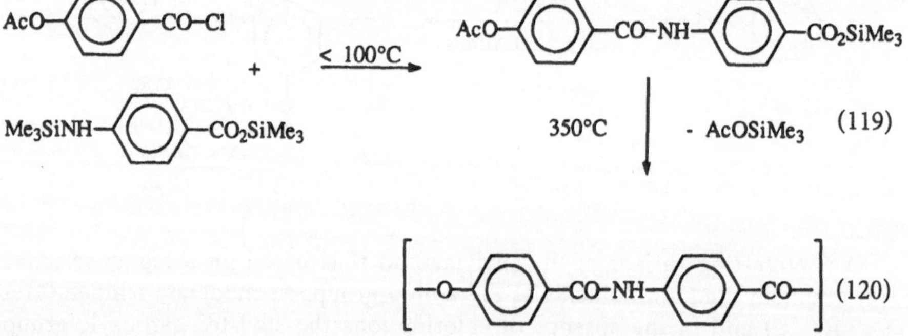

$$(119)$$

$$(120)$$

(122) [133]. Thus, this approach opens an access to a broad range of novel poly(ester-amide)s.

The third silyl method is a more exotic polycondensation yielding poly(alkylthio-ester)s. Silylated thiodicarboxylic acids serve as nucleophilic reaction partners of reactive aliphatic bromides, Eq. (123). The resulting polythioesters are of minor importance. However, this polycondensation method reported by Kricheldorf in 1972 [134] represents the first example of a polyester synthesis based on a silylated monomer.

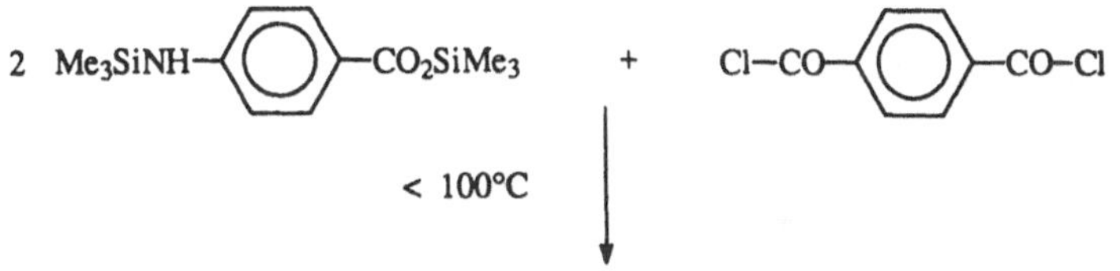

2 Me₃SiNH—⬡—CO₂SiMe₃   +   Cl—CO—⬡—CO—Cl

$$2\ Me_3SiNH\!-\!\!\bigcirc\!\!-\!CO_2SiMe_3\ +\ Cl\!-\!CO\!-\!\!\bigcirc\!\!-\!CO\!-\!Cl$$

< 100°C

(121)

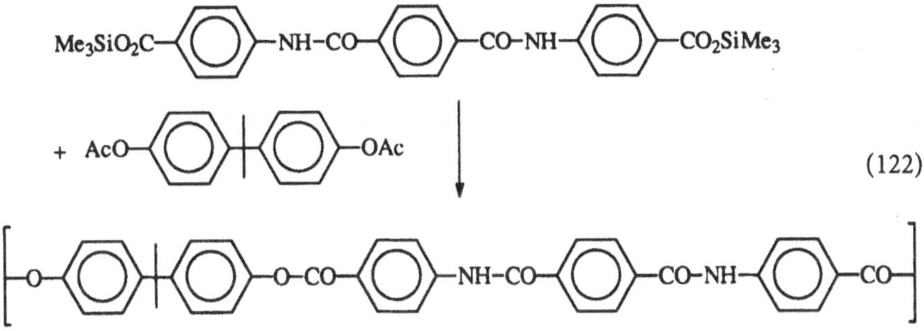

$$Me_3SiO_2C\!-\!\!\bigcirc\!\!-\!NH\!-\!CO\!-\!\!\bigcirc\!\!-\!CO\!-\!NH\!-\!\!\bigcirc\!\!-\!CO_2SiMe_3$$

+ AcO—⬡—|—⬡—OAc

(122)

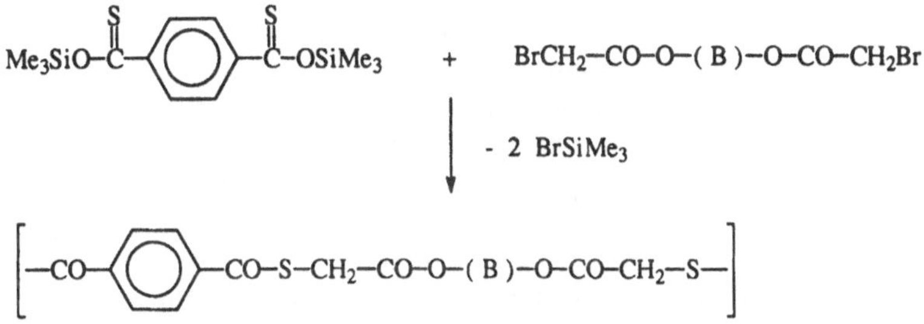

$$\left[\!-O\!-\!\!\bigcirc\!\!-\!|\!-\!\!\bigcirc\!\!-\!O\!-\!CO\!-\!\!\bigcirc\!\!-\!NH\!-\!CO\!-\!\!\bigcirc\!\!-\!CO\!-\!NH\!-\!\!\bigcirc\!\!-\!CO\!-\!\right]$$

$$Me_3SiO-\overset{S}{\overset{\|}{C}}\!-\!\!\bigcirc\!\!-\!\overset{S}{\overset{\|}{C}}\!-\!OSiMe_3\quad +\quad BrCH_2\!-\!CO\!-\!O\!-\!(\,B\,)\!-\!O\!-\!CO\!-\!CH_2Br$$

- 2 BrSiMe₃

(123)

$$\left[\!-CO\!-\!\!\bigcirc\!\!-\!CO\!-\!S\!-\!CH_2\!-\!CO\!-\!O\!-\!(\,B\,)\!-\!O\!-\!CO\!-\!CH_2\!-\!S\!-\!\right]$$

5.8
Polyanhydrides

Silylated carboxyl groups are nucleophilic enough to react with strong electrophiles such as phosphorus tribromide, thionylchloride or carboxylic acid chlorides. As demonstrated by Kricheldorf such condensations allow the syntheses of carboxylic acid bromides [135], acid chlorides [136] and anhydrides [137] without formation of free HCl or HBr which may cause side reactions to a considerable extent. The synthesis of aminoacid N-carboxyanhydrides according to Eq. (124) may be considered as a first model reaction for the synthesis of anhydride groups from silylated carboxylic acids [135, 137].

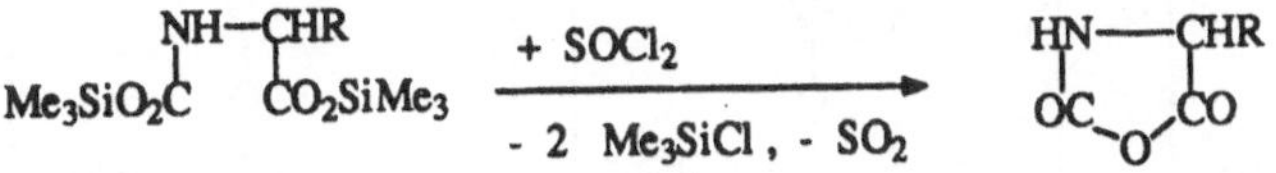

$$Me_3SiO_2C\overset{NH-CHR}{\underset{}{|\quad\ |}}CO_2SiMe_3\ \xrightarrow[-\ 2\ Me_3SiCl\,,\ -\ SO_2]{+\ SOCl_2}\ \overset{HN--CHR}{\underset{OC\diagdown_O\diagup CO}{|\qquad\ |}}$$

(124)

A polycondensation based on this approach was described at first by Kricheldorf et al. (136) for the synthesis of poly(acylsulfide)s. Silylated bisthiocarboxylic acids were polycondensed with the corresponding dicarboxylic acid chlorides, Eq. (125) under mild conditions. As demonstrated by Gupta [138]

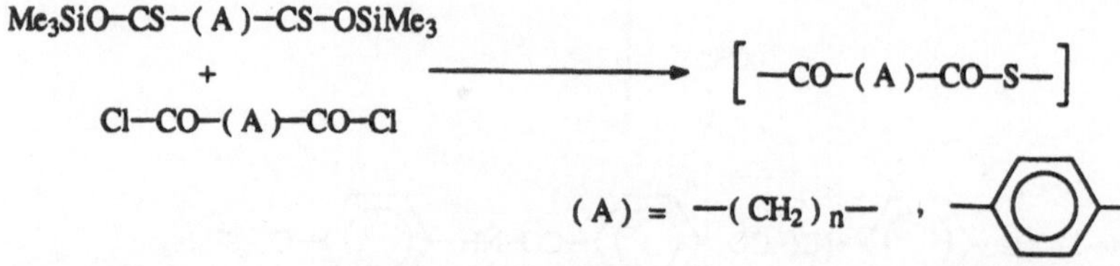

$$
\begin{array}{c}
\text{Me}_3\text{SiO–CS–(A)–CS–OSiMe}_3 \\
+ \\
\text{Cl–CO–(A)–CO–Cl}
\end{array}
\longrightarrow
\left[\text{–CO–(A)–CO–S–} \right]
\tag{125}
$$

$$ (A) = \text{–(CH}_2)_n\text{–} \quad , \quad \text{–}\langle\!\bigcirc\!\rangle\text{–} $$

and Kricheldorf et al. [139] silylated dicarboxylic acids undergo an analogous polycondensation with dicarboxylic acid chlorides yielding highly crystalline aliphatic polyanhydrides, Eq. (126). Since the nucleophilicity of the carboxyl

$$
\begin{array}{c}
\text{Me}_3\text{SiO}_2\text{C–(CH}_2)_n\text{–CO}_2\text{SiMe}_3 \\
+ \\
\text{Cl–CO–(CH}_2)_n\text{–COCl}
\end{array}
\xrightarrow[\text{- 2 Me}_3\text{SiCl}]{(\text{Cl}^\ominus)}
\left[\text{–CO–(CH}_2)_n\text{–CO–O–} \right]
\tag{126}
$$

groups is significantly lower than that of thiocarboxyl groups, chloride ions are required as catalysts, Eq. (126). In solution poly(anhydride)s are sensitive to hydrolysis and alcoholysis. Therefore their purification by reprecipitation is difficult to achieve without degradation. In this connection the silyl method, Eq. (126) has the advantage of yielding relatively pure polyanhydrides as evidenced by the elemental analyses of the crude reaction products (Table 5.8), because the volatile chlorotrimethylsilane is the only byproduct. A variant of the standard procedure is the polycondensation of silylated dicarboxylic acids with stoichiometric amounts of thionylchloride, Eq. (127) [139]. Unfortunately this

$$
\text{Me}_3\text{SiO}_2\text{C–(CH}_2)_n\text{–CO}_2\text{SiMe}_3
\xrightarrow[\substack{\text{- SO}_2 \\ \text{- 2 Me}_3\text{SiCl}}]{+\ \text{SOCl}_2}
\left[\text{–CO–(CH}_2)_n\text{–CO–O–} \right]
\tag{127}
$$

method has two short-comings. First, due to the volatility of thionylchloride it is difficult to maintain the stoichiometry over the whole course of the polycondensation. Second, for unknown reasons aliphatic monomers tend to yield crosslinked products [140]. However, in this way soluble and meltable polyanhydrides of substituted terephthalic acids were obtained (Form. 128a–c).

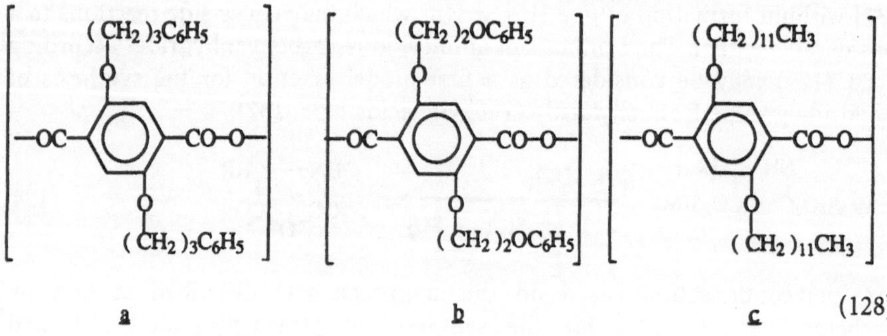

$$ \text{a} \qquad\qquad \text{b} \qquad\qquad \text{c} \tag{128} $$

Table 5.8 Yields and properties of polyanhydrides prepared by polycondensation of silylated dicarboxylic acids and dicarboxylic acid dichlorides.

Chemical Structure	Yield[b] (%)	η_{inh}[c] (dl/g)	Tm[d] (°C)	Elem. Form. (Form. weight)		Elemental Analyses C	H
[-OC)-(CH$_2$)$_6$-CO-O-]	98	0.43	80	C$_8$H$_{12}$O$_3$ (156.1)	Calcd. Found	61.56 61.55	7.69 7.74
[-OC)-(CH$_2$)$_8$-CO-O-]	99	0.31	92	C$_{10}$H$_{16}$O$_3$ (184.1)	Calcd. Found	65.24 64.86	8.69 8.77
[-OC)-(CH$_2$)$_{10}$-CO-O-]	99	0.32	97	C$_{12}$H$_{20}$O$_3$ (212.1)	Calcd. Found	67.94 68.10	9.43 9.53
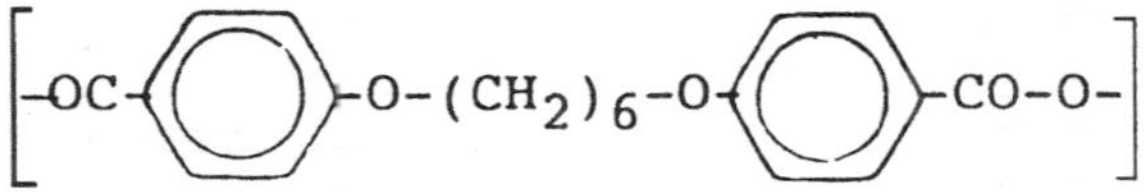 [-OC-⬡-O-(CH$_2$)$_6$-O-⬡-CO-O-]	99	0.23	145	C$_{20}$H$_{20}$O$_5$ (340.2)	Calcd. Found	70.60 69.33	5.88 6.09
-OC-(CH$_2$)$_8$-CO-O- / -OC-⬡-CO-O-	87	0.30	—	C$_{18}$H$_{20}$O$_6$ (332.2)	Calcd. Found	65.08 63.58	6.02 6.44
-OC-(CH$_2$)$_8$-CO-O- / -CO-⬡-O-(CH$_2$)$_6$-O-⬡-OC-O-	98	0.33	72	C$_{30}$H$_{36}$O$_8$ (524.3)	Calcd. Found	68.72 68.98	6.87 6.88

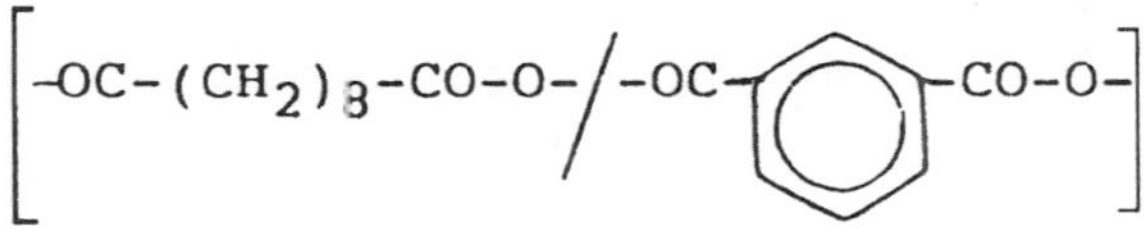

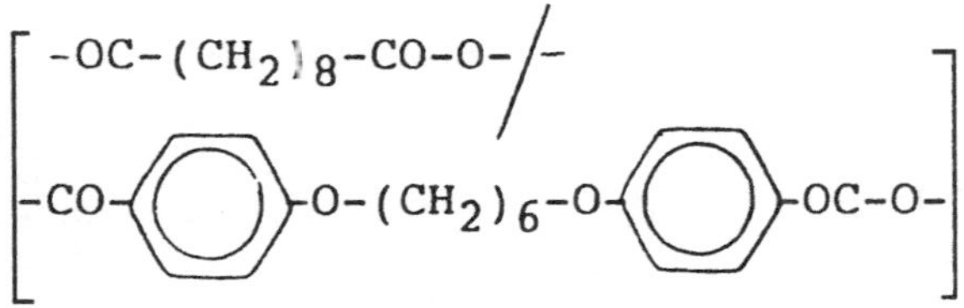

[a] Adapted from Ref. (139)
[b] Crude reaction products
[c] Measured at 35°C with c = 2 g/l in CHCl$_3$
[d] From DSC measurements with a heating rate of 20°C/min

Kricheldorf and Lübbers [139] studied the sequences of copolyanhydrides by ^{13}C NMR-spectroscopy. It was found that regardless of the reaction conditions a rapid acyl exchange takes place, and random sequences are the result, when the silylester of one dicarboxylic acid is condensed with the dichloride of another dicarboxylic acid (Fig. 5.1).

As demonstrated by Kricheldorf et al. [141–144] the synthesis of polyanhydrides by the silyl method is a versatile approach which does not only enable us to prepare homo- and copolyanhydrides, but also to prepare various poly(ester-anhydride)s. Poly(ester-anhydride)s are characterized by a backbone made up of ester and anhydride groups in a fixed or variable molar ratio. Aromatic poly(ester-anhydride)s are of interest for several reasons. First, aromatic polyesters ("polyarylates") are thermostable engineering plastics. The incorporation of anhydride groups allows one to vary the rate of biodegradation by hydrolysis. Second, the anhydride groups enable a selective modification of chemical structure and properties including crosslinking with difunctional or multi-functional reaction partners. Third, polyesters may form liquid crystalline phases, and the influence of the non-linear anhydride groups on the stability of such mesophases is at least of academic interest.

Thermotropic poly(ester-anhydride)s were obtained by Kricheldorf and Lübbers [141], when terephthaloylchloride was polycondensed with silylated 4'-hydroxybenzoic acid, Eq. (129) 6-hydroxy-2-naphthoic acid or 4'-hydroxybiphenyl-4-carboxylic acid (Forms. **130** and **131**). Amorphous isotropic poly(ester-anhydride)s were the result, when substituted 4-hydroxybenzoic acids were used as comonomers, Eq. (129), X, Z = OCH$_3$, Cl). Poly(ester-anhydride)s with a variable molar fraction of anhydride groups were prepared by cocondensations of terephthaloylchloride with a mixture of a silylated diphenol and a silylated hydroxy acid (Eq. 132, Forms. **133–135**). The combinations of

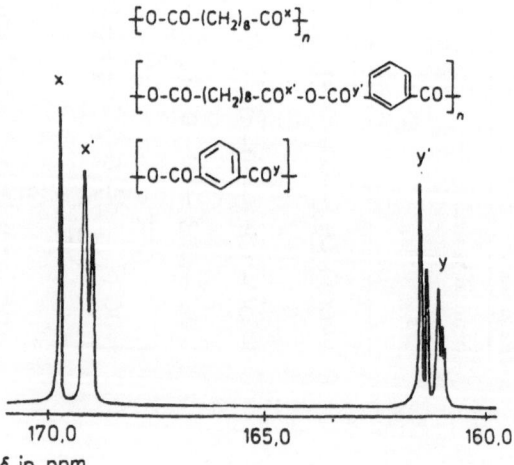

Fig 5.1 75,4MHz^{13}C NMR spectrum of a copolyanhydride prepared from silylated sebacic acid and isophthaloyl chloride (measured in CDCl$_3$ with internal TMS).

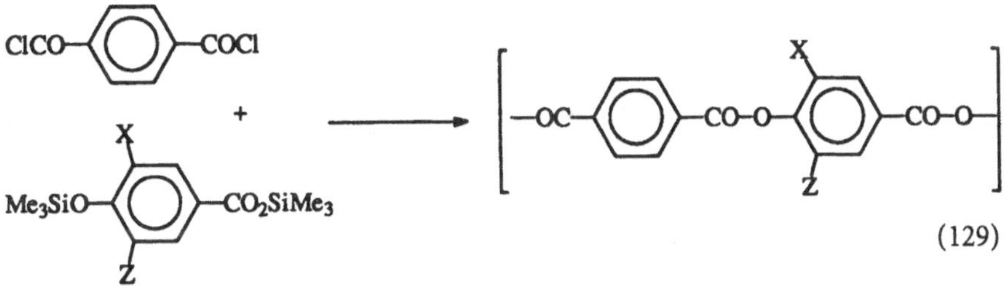

ClCO—⬡—COCl

+

Me₃SiO—⬡(X, Z)—CO₂SiMe₃   →   [—OC—⬡—CO-O—⬡(X,Z)—CO-O—]   (129)

X =	H	H	H	Cl	OCH₃
Z =	H	Cl	OCH₃	Cl	OCH₃

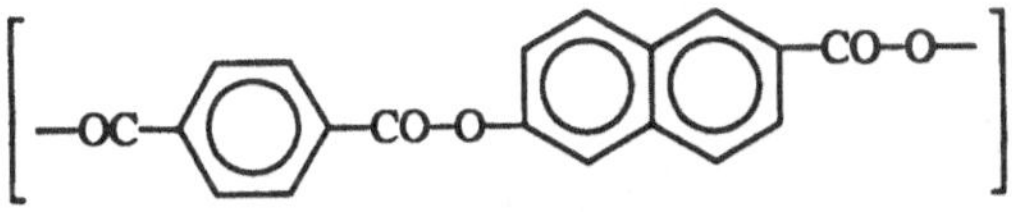

[—OC—⬡—CO-O—⬡⬡—CO-O—]   (130)

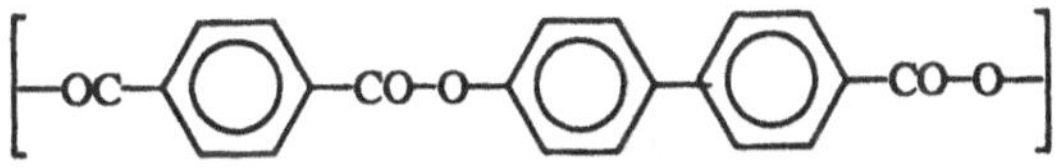

[—OC—⬡—CO-O—⬡—⬡—CO-O—]   (131)

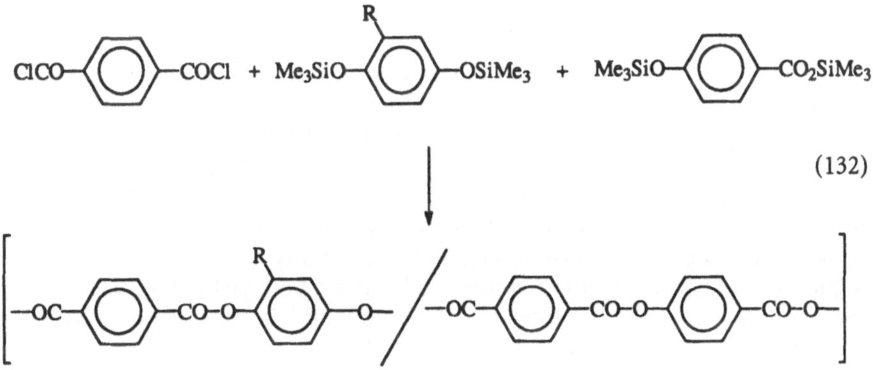

ClCO—⬡—COCl + Me₃SiO—⬡(R)—OSiMe₃ + Me₃SiO—⬡—CO₂SiMe₃

↓   (132)

[—OC—⬡—CO-O—⬡(R)—O— / —OC—⬡—CO-O—⬡—CO-O—]

R = CH₃ , C₆H₅

unsubstituted terephthalic acid with hydroquinone and 4-hydroxybenzoic acid are neither meltable nor soluble. With substituted hydroquinones or substituted 4-hydroxybenzoic acids thermotropic poly(ester-anhydride)s were obtained, (Forms. 132, 133). 2,5-Bisalkoxyterephthalic acids (Form. **134**) or bisphenol-A, (Form. 135), as comonomer mainly yield amorphous and isotropic poly(ester-anhydride)s.

All these polycondensations were conducted with chloride ions as catalyst. High yields were obtained in all cases, but the inherent viscosities scatter over a broad range depending on the comonomers. The formation of anhydride groups was documented by IR spectra [139, 141]. ^{13}C-NMR spectra suggest that little

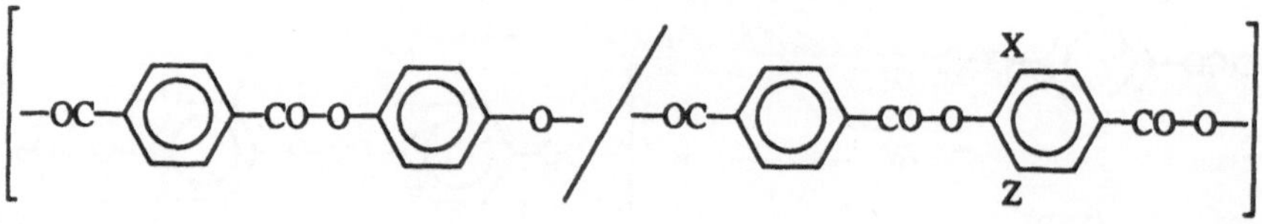

(133)

X , Z = H, OCH$_3$

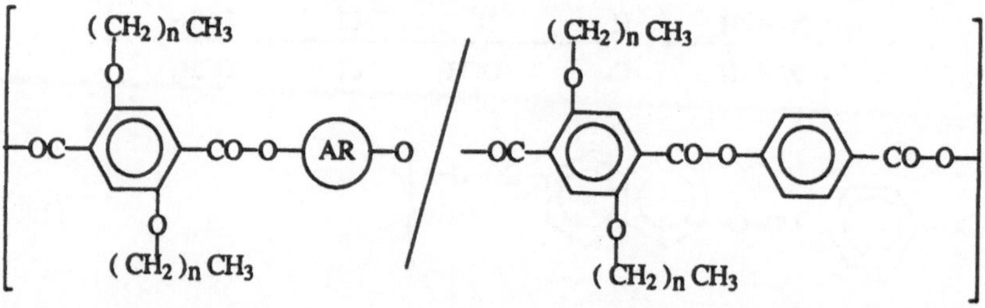

(134)

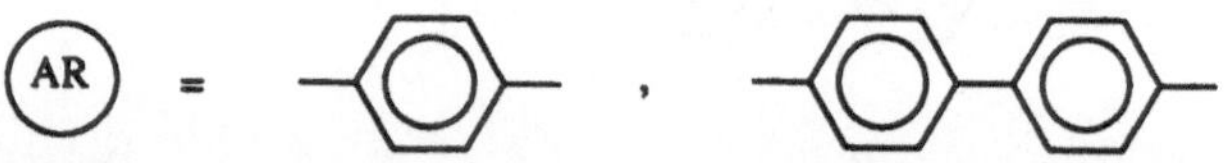

$$AR = \text{—⟨O⟩—} , \text{—⟨O⟩—⟨O⟩—}$$

transesterification occured despite rapid acyl exchange between anhydride groups [139, 143]. For comparison several poly(ester-anhydride)s were prepared by the only conventional method, namely polycondensation of free dicarboxylic acids and hydroxyacids by means of acetic anhydride [144,145]. Both WAXS powder patterns (Fig. 5.2) and DSC measurements indicate that the reaction products of the "acetic anhydride method" differ from those of the "silyl method" by a higher degree of crystallinity, lower solubility and blockness of their sequences [142, 146]. In other words the "silyl method" yields the more homogeneous and easier processable products. Finally, it should be noted that poly(ester-anhydride)s prepared by polycondensation of terephthaloylchloride with mixtures of silylated bisphenol-A and isophthalic acid (Form. 136) are insoluble in common inert solvents and more difficult to process and to

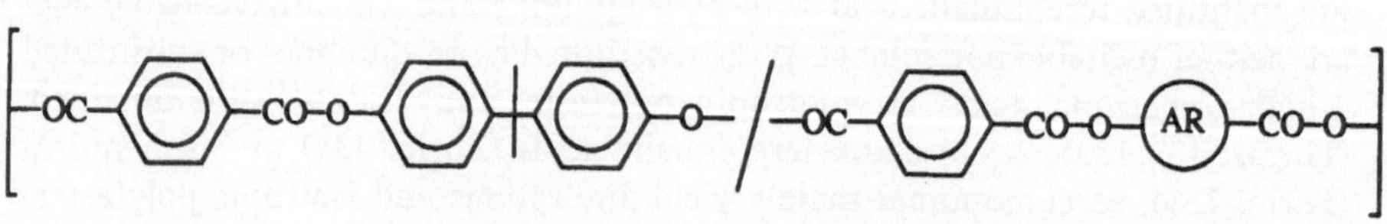

(135)

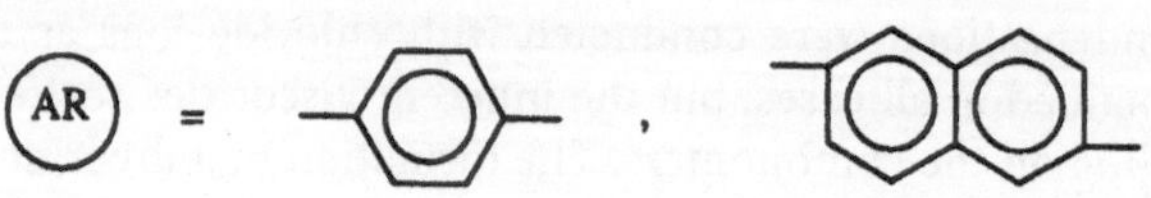

$$AR = \text{—⟨O⟩—} , \text{⟨OO⟩—}$$

Fig 5.2 WAXS powder patterns of poly(ester-anhydride)s: A) Polycondensation of terephthaloylchloride and bistrimethylzilyl 4-oxbenzoate, B) polycondensation of terephthalic acid 4-hydroxybenzoic acid and acetic anhydrid, C) polycondensation of mixed terephthalic-acetic anhydride with 4-hydroxybenzonic acid.

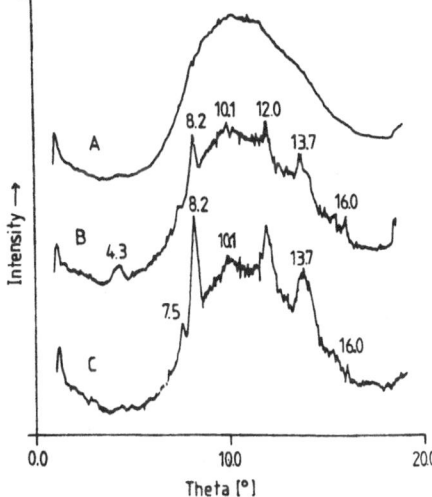

characterize than analogous poly(ester-anhydride)s based on hydroxy acids (Form. **135**) [147].

(136)

5.9
Phosphorus-Containing Polymers

Low molecular weight trisarylphosphates (e.g. trikresylphosphates) are widely used as flame-extinguishing additives of commercial engineering plastics. Low molecular weight additives have the general disadvantage that they slowly evaporate from the matrix and they may be washed out, when the matrix comes into contact with solvents. Incorporation of phosphate and phosphonate groups into polymer chains via covalent bond avoids these problems. Therefore numerous research groups studied the synthesis of phosphorous containinng polymers [147, 148].

The successful synthesis of aromatic polyesters by acylation of silylated diphenols (Sect. 5.7) prompted Kricheldorf and coworkers to study the polycondensation of silylated diphenols with aryldichlorophosphoridates

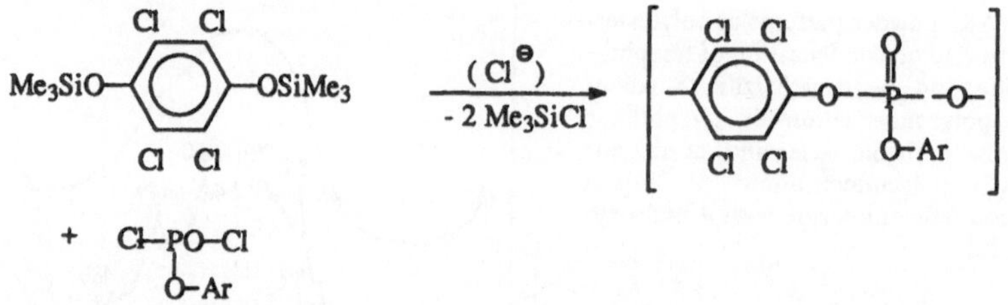

(137)

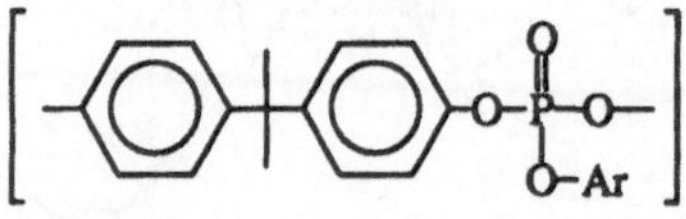

(138)

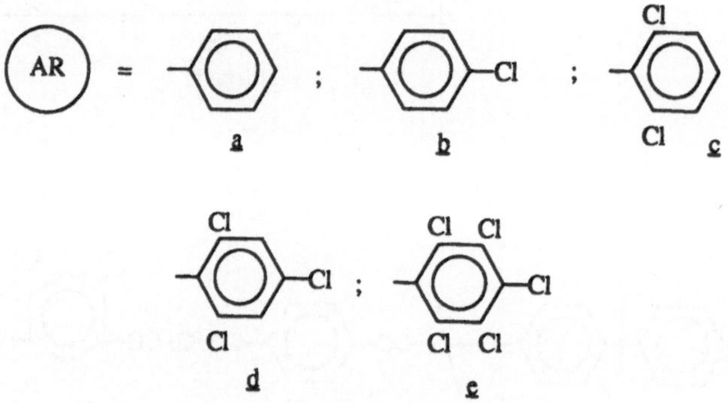

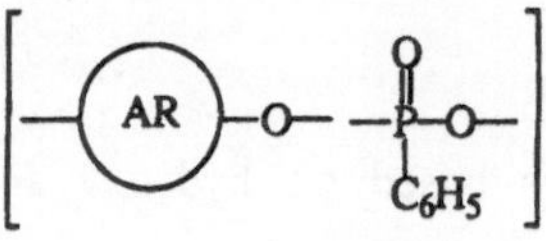

(139)

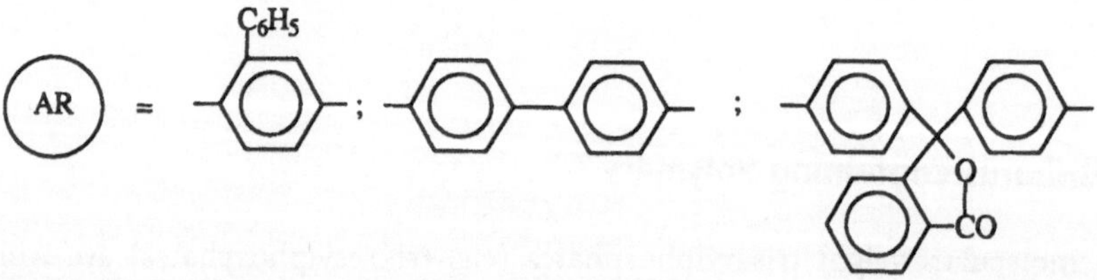

[149, 150]. In analogy to the synthesis of polyarylates chloride ions are required as catalyst. Silylated tetrachlorohydroquinone, Eq. (137) and bisphenol-A (Form. 138) served as starting material. The aryl residues of the phosphate group was varied in both series from unsubstituted phenyl to pentachlorophenyl (a–e in Forms. 137 and 138). The polyphosphate (137)e is remarkable because it represents the rare case of an absolutely H-free polymer. Despite high yields only low or moderate molecular weights (M_n < 15 000 g/mol) were obtained. However, polyphosphates of tetra-chlorohydroquinone are even more difficult to prepare by any other method, due to its acidity and steric hindrance.

In addition to polyphosphates several polyphosphonates were synthesized from commercial phenylphosphonic dichloride (Form. **139**) [151]. A variety of silylated diphenols including tetrasubstituted derivatives of bisphenol-A (Forms. **139** and **140**) were used as reaction partners. In analogy to the synthesis of polyphosphates high yields but only low or moderate viscosities were found. Not surprisingly the lowest viscosities were determined for the polyphosphonates of the sterically

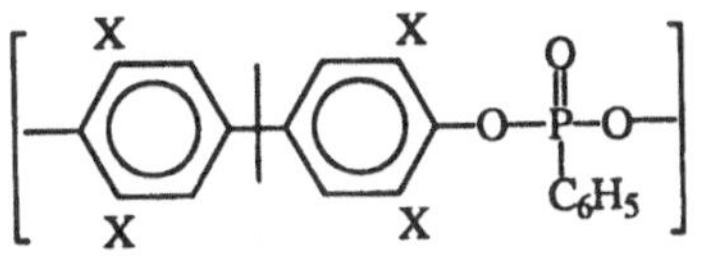

X = H, CH₃, Cl, Br   (140)

hindered tetra-substituted bisphenols. All polyphosphonates like all polyphosphates are amorphous materials with glass-transition temperatures up to 220 °C.

Whereas the silyl method is not particularly advantageous for the synthesis of homopolyphosphates or -polyphosphonates it is favorable for the preparation of copolyesters [152]. This aspect was illustrated by the synthesis of two series of copolyesters based on bisphenol-A and terephthalic acid (Form. **141**). In one series the terephthaloyl units was gradually replaced by phenylphosphonic acid

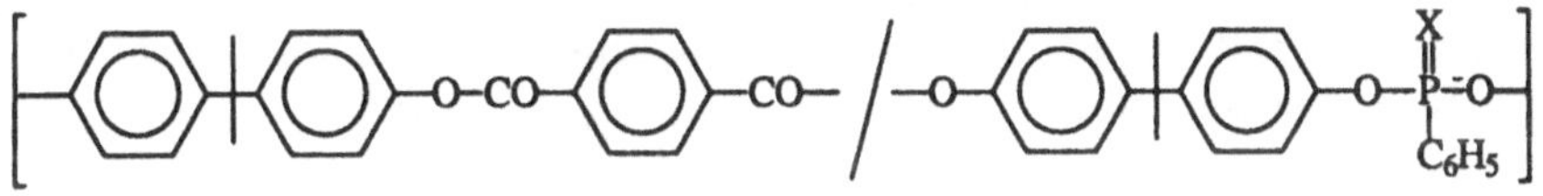

X = O, S   (141)

(10–50 mol %, Form. **141**, X = O), whereas in the second series phenylthiophosphonic was incorporated (Form. **141**, X = S). High yields and fairly high viscosities were obtained in most cases. The reaction conditions and some results are compiled in Tab. 1 of ref. [152]. A comparison with other synthetic methods clearly evidenced the advantage of the "silyl approach" [153]. Furthermore, the amorphous character and the good solubilities of all copolyesters suggests a random sequence, because long blocks of bisphenol-A and terephthalic acid tend to crystallize with the consequence of a poor solubility in most common solvents.

Kricheldorf et al. [150, 154] also studied the incorporation of phosphate and phosphonate groups into various liquid-crystalline polyesters. A first series of copolyesters was prepared from *trans*-1,4-cyclohexane dicarboxylic acid and substituted hydroquinones (Form. **142**). However these copolyesters turned out to be semicrystalline and at least biphasic obviously due to blocky sequences.

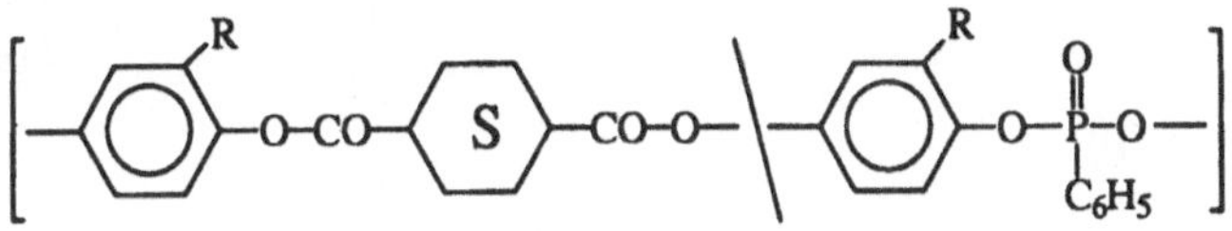

R = CH₃, Cl   (142)

Their thermal properties and phase transition are difficult to interprete, and thus, were not published in detail [153]. In contrast, amorphous and homogeneous copolyesters were obtained from silylated phenylhydroquinone and

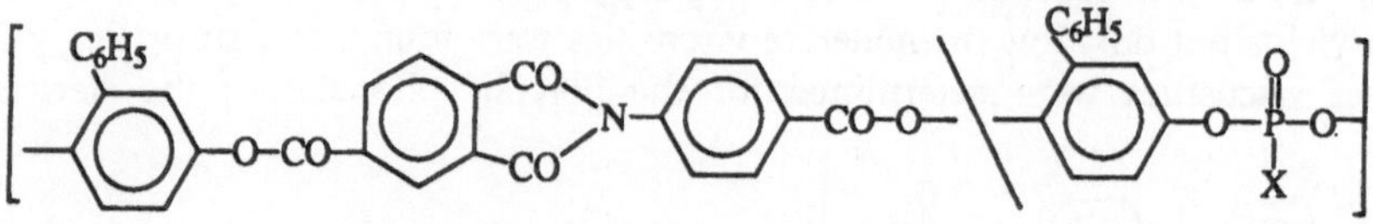

$$X = CH_3, \ C_6H_5, \ O\text{-}C_6H_5$$

(143)

mixtures of acid chlorides of N-(4-carboxyphenyl)trimellitimide and various phosphoric or phosphoric acids (Form. 143). Both glass-transition temperature and isotropization temperature decrease continuously with increasing molar fraction of phosphate or phosphonate groups, and thus, suggest the existence of more or less random sequences (Fig. 5.1). Furthermore, it was found that the nature of the residue attached to the P-atom (X in Form. 143) does not have a significant influence on the synthesis or properties of these copolyesters. However, the sp^3-hybridized phosphorus is unfavorable for the stability of the nematic phase. With more than 40 mol % of phosphate or phosphonate groups the mesophase disappears.

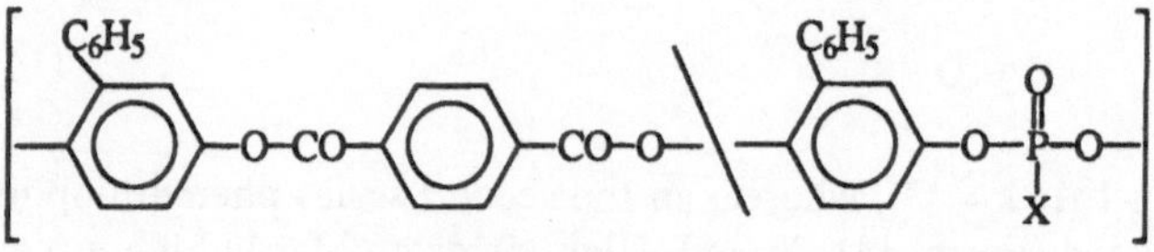

$$X = CH_3, \ C_6H_5, \ OC_6H_5, \ \text{-}O\text{-}$$

(144)

A similar series of amorphous and homogeneous copolyesters was prepared from silylated phenylhydroquinone and terephthaloylchloride (Form. 144) [154]. Again the properties do not depend on the nature of X, but more than 20 mol % of phosphate and phosphonate groups suffice to suppress the nematic phase. Also the copolyesters 145 and 146 were isolated as amorphous and homogeneous

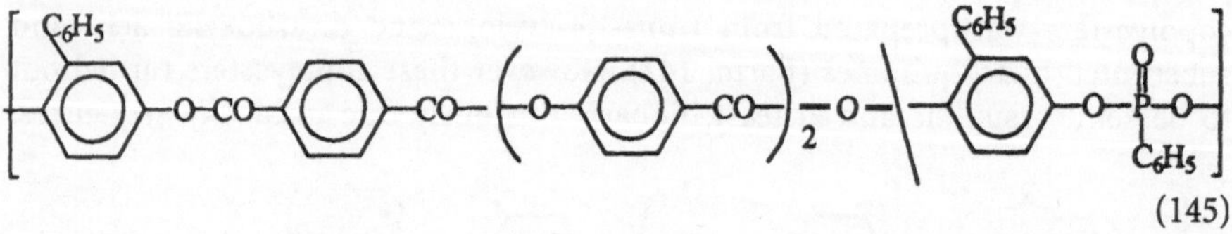

(145)

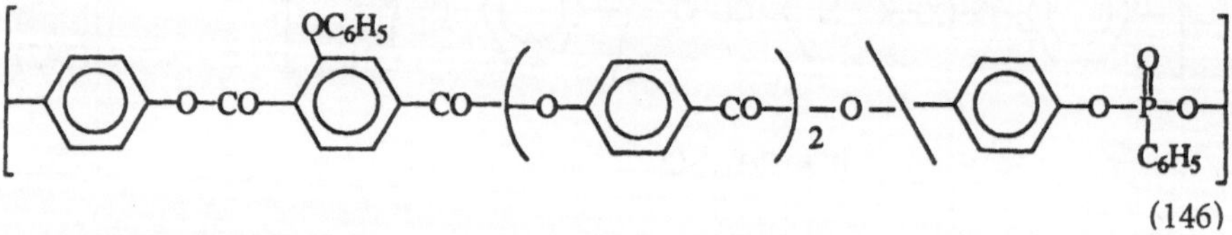

(146)

materials [154]. Due to the longer mesogenic repeating units of the parent homopolyesters the influence of the phenylphosphonate groups is weaker, and more than 70 mol % are required to eliminate the mesophase. Altogether these results demonstrate that the silyl method is well suited to prepare a broad variety of P-containing copolyester in high yields and with high molecular weights.

A completely different kind of silylated monomers and polycondensations process was first reported by Flindt and Rose [155]. They were able to prepare poly(bis(trifluoro-ethoxy)phosphazene from tris(trifluoroethoxy)-N-trimethyl-

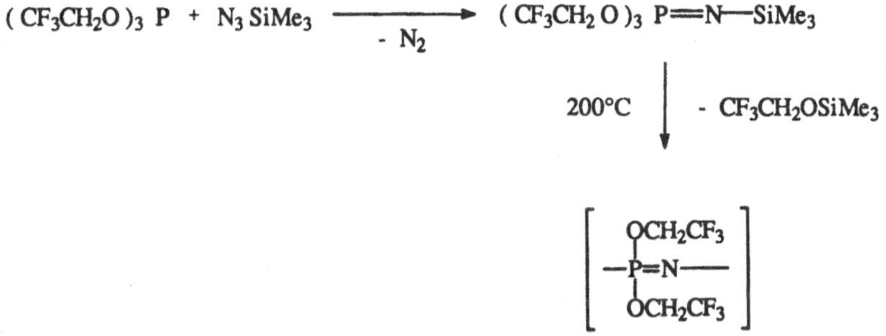

$$(CF_3CH_2O)_3\, P\ +\ N_3\, SiMe_3 \xrightarrow[-\ N_2]{} (CF_3CH_2\, O)_3\, P{=\!\!=}N{-\!\!-}SiMe_3 \tag{147}$$

$$200°C \quad \Big\downarrow \quad -\ CF_3CH_2OSiMe_3$$

$$\left[\begin{array}{c} OCH_2CF_3 \\ -P{=}N{-\!\!-} \\ OCH_2CF_3 \end{array} \right] \tag{148}$$

silylphosphinimine, which is easily accessible from tris(trifluoroethoxy) phosphite and trimethylsilylazide, Eqs. (147 and 148). Polycondensation conducted in bulk at 200 °C gave molecular weights around 10 000, and polycondensations in solution at 60 °C gave M_n's around 4000 and 6000. This working field was further developed by Wisian-Neilson and Neilson [156–166]. First attempts to prepare polyphosphazenes from fluoro- or bromo-N-trimethyl-

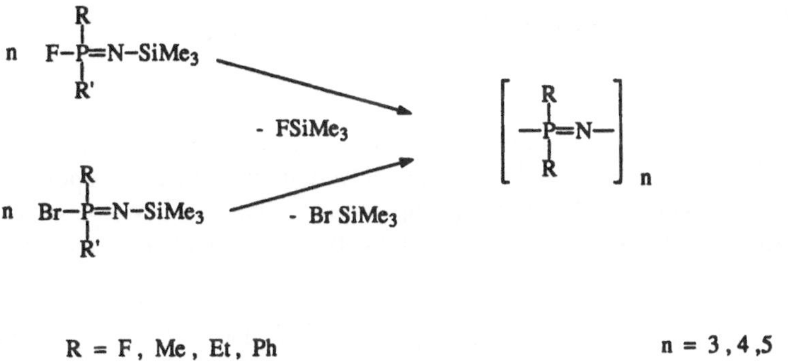

$$n\quad F{-}\underset{\underset{R'}{|}}{\overset{\overset{R}{|}}{P}}{=}N{-}SiMe_3 \tag{149}$$

$$-\ FSiMe_3$$

$$\left[\begin{array}{c} R \\ -P{=}N{-} \\ R \end{array} \right]_n$$

$$n\quad Br{-}\underset{\underset{R'}{|}}{\overset{\overset{R}{|}}{P}}{=}N{-}SiMe_3 \qquad -\ Br\, SiMe_3 \tag{150}$$

$$R = F,\ Me,\ Et,\ Ph \qquad\qquad n = 3,4,5$$

silylphosphinimines only yielded cyclophosphazenes with the cyclotetramer as the prevailing species, Eqs. (149 and 150) [156–159]. However, the nucleophilic substitution of the bromide residue with the trifluoroethoxy group yielded a series of N-trimethylsilyl-P-(trifluoro-ethoxy)phosphoranimines, Eq. (151)

$$Br-\underset{\underset{R'}{|}}{\overset{\overset{R}{|}}{P}}=N-SiMe_3 \xrightarrow[- Et_3N \cdot HBr]{CF_3CH_2OH/NEt_3} CF_3CH_2O-\underset{\underset{R'}{|}}{\overset{\overset{R}{|}}{P}}=N-SiMe_3 \quad (151)$$

$$RCH_2O-\underset{\underset{R'}{|}}{\overset{\overset{R}{|}}{P}}=N-SiMe_3 \xrightarrow[- CF_3CH_2OSiMe_3]{} \left[-\underset{\underset{R'}{|}}{\overset{\overset{R}{|}}{P}}=N- \right] \quad (152)$$

R = Me, Et, nPr, Me, Me, Me, CH₂Phe

R' = Me, Et, nPr, Et, n-Pr, Phe, Phe

[160, 161] which proved to be useful as monomers for the synthesis of numerous polyphosphazenes, Eq. (152) [156, 158, 159, 162–165].

These N-trimethylsilyl phosphoranimines were usually polycondensed in bulk by heating at 160–220 °C for 2–12 days in sealed evacuated glass or steel vessels. High yields (> 90 %) and high molecular weights (M_w = 50 000–250 000) were obtained. This new approach to the synthesis of polyphosphazene is remarkable for the following reasons. First, in this way dialkyl or alkyl-aryl polyphosphazenes were prepared for the first time. Second, this approach is highly versatile and not only allows the synthesis of numerous homopolymers but also the synthesis of copolymers from mixtures of two different phosphoranimines [156]. Third, the polymerization mechanism is complex and involves an interesting selectivity concerning the substituents attached to the P-atom. For instance, diphenylpolyphosphazenes 153a or polyphosphazenes

$$\left[-\underset{\underset{C_6H_5}{|}}{\overset{\overset{C_6H_5}{|}}{P}}=N- \right] \qquad \left[-\underset{\underset{CH_2SiMe_2R}{|}}{\overset{\overset{Me}{|}}{P}}=N- \right] \qquad \begin{array}{l} R = Me, Phe, CH=CH_2 \\ \\ H, (CH_2)_3 CN \end{array} \quad (153)$$

$$\underline{a} \qquad\qquad\qquad \underline{b}$$

with a silicon in β-position 153b were not accessible. In contrast a phosphine group in β-position does not hinder the polycondensation, and copolyphosphazenes with pending phosphine groups were isolated. They are of great interest as polymeric ligands of heavy metal complexes, Eq. (154) [156, 166].

$$\left[\underset{\underset{P-(Phe)_2}{\overset{|}{CH_2}}}{\overset{\overset{Me}{|}}{P}}=N- \Big/ \underset{\underset{Me}{|}}{\overset{\overset{Me}{|}}{P}}=N- \right] \xrightarrow{Fe_2(CO)_9} \left[\underset{\underset{\underset{Fe(CO)_4}{\vdots}}{\overset{|}{P-(Phe)_2}}}{\underset{\overset{|}{CH_2}}{\overset{\overset{Me}{|}}{P}}}=N- \Big/ \underset{\underset{Me}{|}}{\overset{\overset{Me}{|}}{P}}=N- \right] \quad (154)$$

$$\underline{a} \qquad\qquad\qquad\qquad\qquad \underline{b}$$

The alkyl and aryl substituted polyphosphazenes are noteworthy for their high thermostability with inital decomposition temperatures (in N_2) around 350–400 °C. Homopolymers may be semicrystalline, whereas all comopolymers

$$
\begin{array}{cc}
\overset{\displaystyle Me}{\underset{\displaystyle (CH_2)_n}{CF_3CH_2O-P=N-SiMe_3}} & \overset{\displaystyle Phe}{\underset{\displaystyle (CH_2)_n}{CF_3CH_2O-P=N-SiMe_3}} \\
CH=CH_2 & CH=CH_2
\end{array} \tag{155}
$$

$$ \underline{a} \qquad\qquad \underline{b} $$

$$ n = o, 1, 2 $$

are amorphous elastomeric materials. Incorporation of comonomers with unsaturated side chains (Forms. 155a, b) allow controlled crosslinking. The crosslinking reaction usually occurs spontaneously in the course of the polycondensation process. Further chemical modifications of the previously mentioned polyphosphazenes were reported [156].

The reaction mechanism of the polymerization of N-silylphosphoranimines has not been fully elucidated yet. However, all results agree in that a formation and ring-opening polymerization does not take place. Flindt and Rose [155]

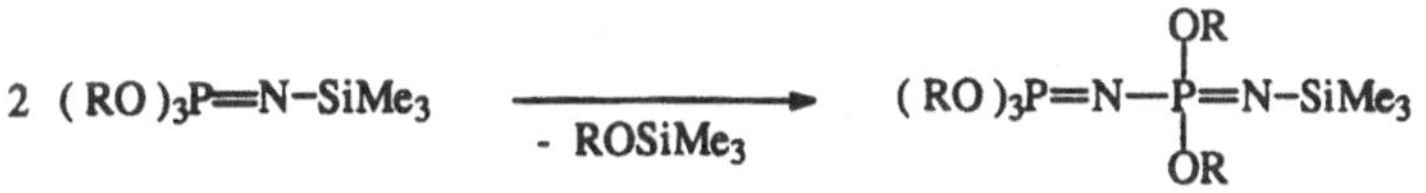

$$
2\ (RO)_3P=N-SiMe_3 \xrightarrow[-\ ROSiMe_3]{} (RO)_3P=N-\overset{\displaystyle OR}{\underset{\displaystyle OR}{P}}=N-SiMe_3 \tag{156}
$$

proposed a simple condensation mechanism according to the step growth kinetics and (hypothetically) formulated a dimer of structure 156 as reaction intermediate. In this connection it is of intertest to compare the thermal stability and reactivity of phosphoranimines $R_2XP = N—SiMe_3$ with regard to the leaving group X. The thermal stability increases in the order

$$ X = Br < F < CF_3CH_2O < PheO < Akoxy. $$

Alkoxysubstituted N-silylphosphoranimines are too stable to yield polyphosphazenes, the halogen substituted mainly yield cyclophosphazenes. Thus the optimum leaving groups for the polymerization are the trifluoroethoxy and phenoxy groups, Eq. (157) (the preparative potential of aryloxy-substituted monomers had not been explored at the time this review was written).

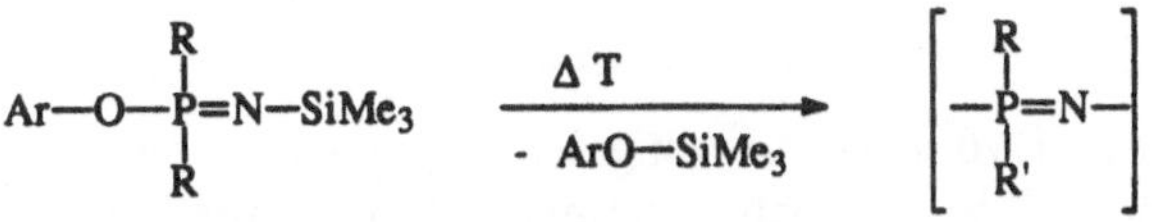

$$
Ar-O-\overset{\displaystyle R}{\underset{\displaystyle R}{P}}=N-SiMe_3 \xrightarrow[-\ ArO-SiMe_3]{\Delta T} \left[-\overset{\displaystyle R}{\underset{\displaystyle R'}{P}}=N-\right] \tag{157}
$$

$$ Ar = Aryl ; \quad R, R' = Alkyl\ or\ Aryl $$

Wisian-Neilson and Neilson reported [163, 165] that high molecular weight polyphosphazenes are formed in the beginning of the polymerization process at low conversion. This observation, later confirmed by Matyjazewski and coworkers [167, 168] is a convincing argument against a normal step growth process. Based on the working hypothesis of a chain growth mechanism, Matyjazewski and coworkers [167–170] conduct a systematic screening of potential catalysts. Whereas cationic and radical catalysts were not effective, anionic initiators, in particular the fluoride ion, proved to be active. The reaction sequence of Eqs. (158)–(160) was proposed as polymerization mechanisms. The authors also pointed out that at high conversions condensation steps may contribute to the chain growth, Eq. (161). This mechanism is certainly a simplication, which cannot explain all experimental findings, but it is a useful basis for further studies.

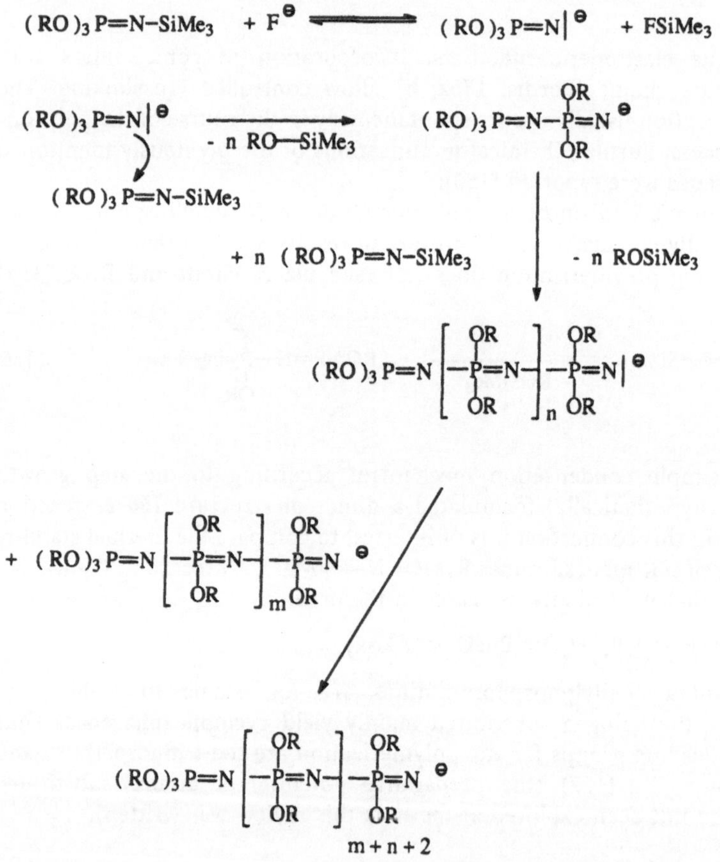

$$(RO)_3 P{=}N{-}SiMe_3 \; + F^{\ominus} \; \rightleftharpoons \; (RO)_3 P{=}N|^{\ominus} \; + FSiMe_3 \tag{158}$$

$$(RO)_3 P{=}N|^{\ominus} \quad \xrightarrow[- \, n \, RO{-}SiMe_3]{} \quad (RO)_3 P{=}N{-}\overset{\displaystyle OR}{\underset{\displaystyle OR}{P}}{=}N^{\ominus} \tag{159}$$
$$(RO)_3 P{=}N{-}SiMe_3$$

$$+ n \; (RO)_3 P{=}N{-}SiMe_3 \qquad\qquad - n \; ROSiMe_3$$

$$(RO)_3 P{=}N \left[-\overset{\displaystyle OR}{\underset{\displaystyle OR}{P}}{=}N- \right]_n \overset{\displaystyle OR}{\underset{\displaystyle OR}{P}}{=}N|^{\ominus} \tag{160}$$

$$+ \; (RO)_3 P{=}N \left[-\overset{\displaystyle OR}{\underset{\displaystyle OR}{P}}{=}N- \right]_m \overset{\displaystyle OR}{\underset{\displaystyle OR}{P}}{=}N^{\ominus}$$

$$(RO)_3 P{=}N \left[-\overset{\displaystyle OR}{\underset{\displaystyle OR}{P}}{=}N- \right] \overset{\displaystyle OR}{\underset{\displaystyle OR}{P}}{=}N^{\ominus} \atop {\scriptstyle m+n+2} \tag{161}$$

When Matyjazewski et al. [170, 171] reinvestigated the preparative route described by Flindt and Rose [155], namely treatment of trifluoroethoxyphosphines with trimethylsilylazide, they detected an additional polycondensation mechanism. Part of the phosphine exchanges an azide group with the trimethylsilylazide, Eq. (162) and undergoes a selfpolycondensation at relatively low temperatures (down to 70 °C) (Eq. 163). This hypothesis is supported by earlier

$$\text{Phe}(RO)_2P \ + \ N_3\text{—SiMe}_3 \quad \longrightarrow \quad \text{Phe}(RO)P\text{—}N_3 \ + \ RO\text{—SiMe}_3$$

$$\Delta T \ \Big\downarrow \ -N_2 \qquad\qquad (162)$$

$$\left[\begin{array}{c} \text{Phe} \\ | \\ -P{=}N- \\ | \\ OR \end{array} \right] \qquad\qquad (163)$$

work of Paciorek et al. [172] and Tesi et al. [173] who synthesized and polycondensed several phosphineazides via chlorophosphine Eqs. (164 and 165). All these results indicate that the chemistry of N-silylphosphoranimines is highly interesting, but deserves further intensive studies to explore all important preparative and mechanistic aspects.

$$R_2P\text{—Cl} \ + \ N_3SiMe_3 \quad \longrightarrow \quad R_2P\text{—}N_3 \ + \ Cl\text{—SiMe}_3$$

$$\Delta T \ \Big\downarrow \ -N_2 \qquad\qquad (164)$$

$$\left[\begin{array}{c} R \\ | \\ -P{=}N- \\ | \\ R \end{array} \right] \qquad\qquad (165)$$

5.10
Miscellaneous Polymers

This section summarizes polycondensations and model reactions that do not belong together but neither do they fit into the other sections. The first examples concern the synthesis of polyamines and these examples based on the N,N'-bissilylated piperazines are also the oldest examples of polycondensations utilizing silylated monomers as reaction partners [1].

Klebe studied systematicaly condensation reactions of N,N'-bissilylated piperazines (Forms. 166a–c and 167a–c) with benzylchloride as a model reaction for their polycondensation with α, α'-dichloroxylene, Eq. (168). The model reactions revealed that these alkylations are catalyzed by traces of ammonium chloride or sulfate and by the presence of sulfolane. In contrast to sulfolane dimethylsulfoxide causes side reactions. The polycondensations were conducted in bulk with or without addition of ammonium chloride. When N,N'-bistrimethylsilyl piperazine was used as monomer the volatile chlorotrimethylsilane escaped rapidly from the reaction mixture and the

$$\text{(166)}$$

$$\underline{a}: R = H, \quad \underline{b}: R = CH_3 \qquad \underline{c}$$

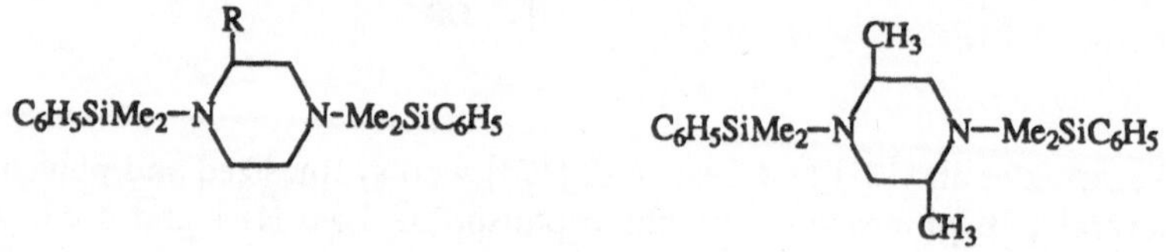

$$\text{(167)}$$

$$\underline{a}: R = H, \quad \underline{b}: R = CH_3 \qquad \underline{c}$$

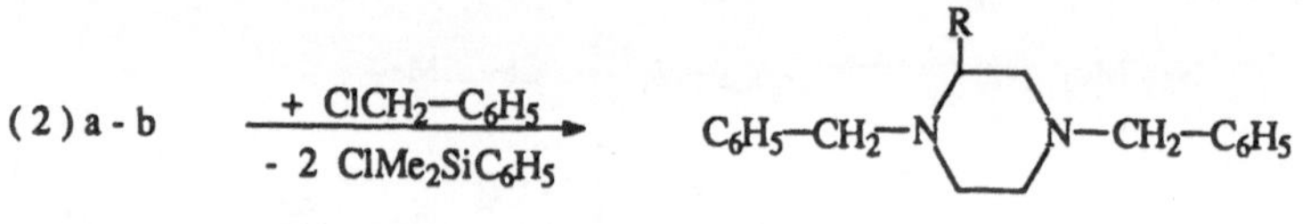

$$\text{(168)}$$

polyamine solidified in the oligomeric state without reaching higher molecular weights. Liquid homogeneous reaction mixtures were obtained, when the dimethylphenylsilyl derivatives of methyl or 2,5-dimethylpiperazine 167b,c were used as monomers. The polyamine chains remained dissolved in the liberated chlorodimethylphenylsilane and the reaction temperatures could be raised to 200 °C. In this way the polyamines 169b and c were isolated with number average molecular weights around 40 000 g/mol and 15 000 g/mol respectively. Their protonation, quarternization and mechanical properties were also studied [1].

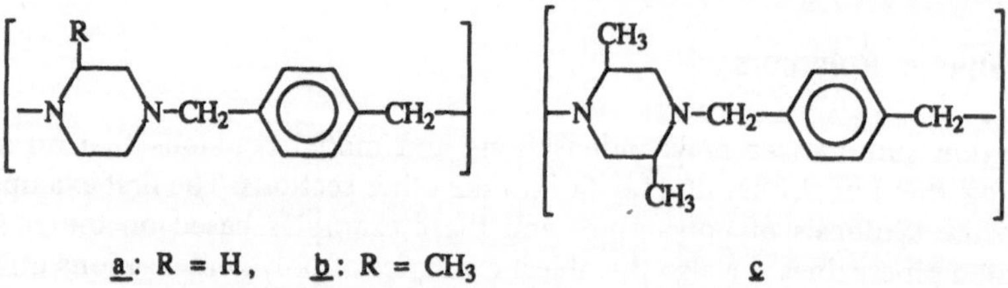

$$\text{(169)}$$

$$\underline{a}: R = H, \quad \underline{b}: R = CH_3 \qquad \underline{c}$$

Imai and coworkers [174] synthesized numerous aromatic polyamines starting from the bistrimethylsilylderivatives of m-phenylene diamine, p-phenylene diamine, 4,4'-diaminodiphenyl methane or 4,4'-diamino diphenyl ether. Their polycondensations with 4,4'-difluorodiphenylsulfone yielded the polysulfones 170a–d. With 4,4'-difluorobenzophenone the polyketones 171a–d were

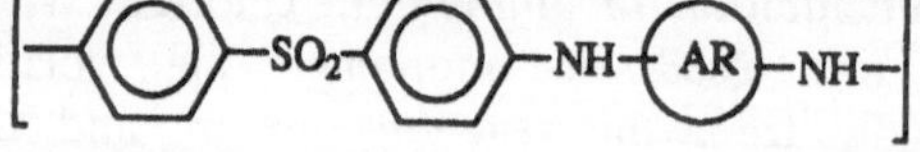

$$\underline{a} - \underline{d} \qquad \text{(170)}$$

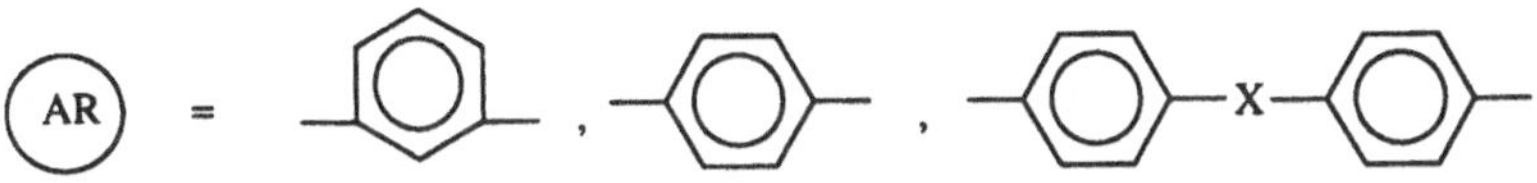

$$\boxed{AR} = \quad , \quad , \quad \underset{a}{} \qquad \underset{b}{} \qquad \underset{c\ :\ X\ =\ O}{} $$

$$\underline{d}\ :\ X\ =\ CH_2$$

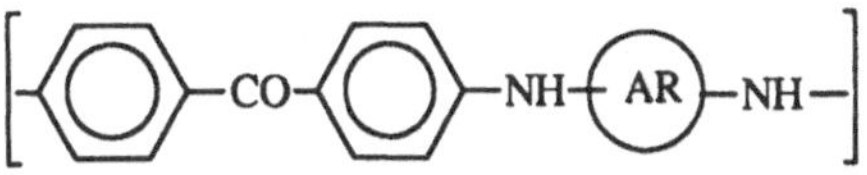

$$\left[\text{—}\bigcirc\text{—CO—}\bigcirc\text{—NH—}\boxed{AR}\text{—NH—} \right] \qquad \underline{a} \text{ - } \underline{d} \tag{171}$$

obtained. All polycondensations were conducted in DMSO with KF as catalysts in analogy to polycondensations of silylated diphenols (Sect.5.6). Inherent viscosities in the range of 0,3–0,66 dl/g (measured in NMP) were reported. The dichloro aromatics were not reactive enough and only yielded oligomers. Oligomers or black tars were also obtained, when silylated aromatic diamines were condensed with 4,4'-difluorodiphenylsulfone in bulk at temperatures in the range of 200–300 °C [175]. Thus, a polar solvent and lower temperatures seem to be necessary for a successful synthesis of aromatic polyamines.

In this connection it should be mentioned that silylated aliphatic amines are alkylated by epoxides, Eq. (172). Albeit, modell reactions with monofunctional

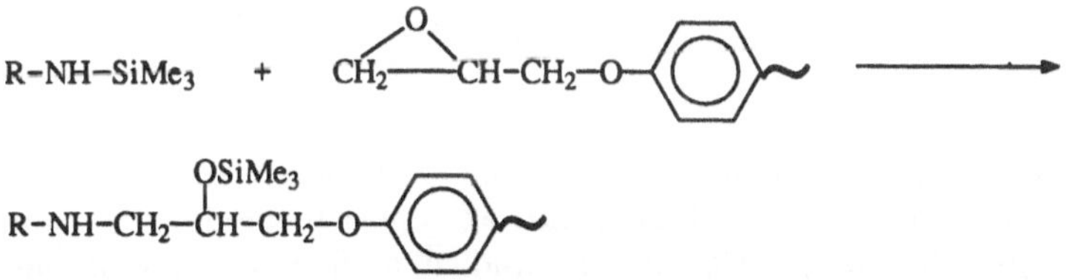

$$\tag{172}$$

reaction partners were successful [176] the synthesis of high molecular weight linear poly(alkyl-arylamine)s via this approach has not been reported yet. Presumably high conversion entail branching and crosslinking, because the secondary amino group formed by the first alkylation step is nucleophilic enough to attack another epoxide. Consequently silylated amines have been proposed as crosslinking additive for epoxide resins [177, 178].

As reported by Kurosaki and coworkers [179, 180] silylated aromatic diamines also proved to be useful for the synthesis of aromatic polyazomethines. The synthesis of polyazomethines is in most cases plagued by the early precipitation of oligomers from the liquid reaction medium or by an early solidification of the reaction mixture, when the polycondensation is conducted in bulk. The poly-condensation of N,N'-bissilylated diamines with the bisethylacetale of tereph-thalaldehyde at 30 °C in NMP yields a soluble precursor 173 which allows casting of films. The formation of the insoluble polyazomethine is then achieved at temperatures up to 200°C Eq. (174).

N,N'-Tetrasilylated aromatic diamines can be used in the same way, but their synthesis is more difficult and more expensive, Eqs. (175 and 176).

$$
\begin{array}{c}
\text{Me}_3\text{Si–NH–}\underset{}{\bigcirc}\text{–NH–SiMe}_3 \\
+ \\
(\text{EtO})_2\text{CH–}\underset{}{\bigcirc}\text{–CH(OEt)}_2
\end{array}
\quad\xrightarrow{\ 30°\text{C}\ }\quad
\left[\ \underset{\overset{|}{\text{N}}}{\overset{\text{X}}{}}\underset{}{\bigcirc}\underset{\overset{|}{\text{N}}}{\overset{\text{X}}{}}\underset{\overset{|}{\text{C}}}{\overset{\text{OEt}}{}}\underset{}{\bigcirc}\underset{\overset{|}{\text{C}}}{\overset{\text{OEt}}{}}\ \right]
\tag{173}
$$

$$200°\text{C} \quad\Big|\quad - \text{XOEt}$$

$$
\text{X} = \text{H, SiMe}_3 \qquad
\left[\ =\text{N–}\underset{}{\bigcirc}\text{–N}=\text{C–}\underset{}{\bigcirc}\text{–C}=\ \right]
\tag{174}
$$

$$100°\text{C} \quad\Big\uparrow\quad - \text{EtOSiMe}_3$$

$$\tag{175}$$

$$
\begin{array}{c}
(\text{Me}_3\text{Si})_2\text{–N–}\underset{}{\bigcirc}\text{–N–(SiMe}_3)_2 \\
(\text{EtO})_2\text{CH–}\underset{}{\bigcirc}\text{–CH(EtO)}_2
\end{array}
\quad\xrightarrow{\ 30°\text{C}\ }\quad
\left[\ \underset{\underset{\text{SiMe}_3}{|}}{\overset{|}{\text{N}}}\underset{}{\bigcirc}\underset{\underset{\text{SiMe}_3}{|}}{\overset{|}{\text{N–CH}}}\underset{}{\overset{\text{OEt}}{}}\underset{}{\bigcirc}\overset{\text{OEt}}{\text{CH}}\ \right]
\tag{176}
$$

Polymers containing (phenylthio)phenyldithiazyl groups were prepared by Chien and Ramakrishnan [181,182] from an anusual monomer, namely bistrimethylsulfur diimide. These authors studied at first a series of model reactions using *para*-substituted sulfenylchlorides as reaction partners, Eq. (177). The success of these model reactions encouraged the synthesis of

$$\text{Me}_3\text{Si–N}=\text{S}=\text{N–SiMe}_3$$

$$
+ \ \text{Cl–S–}\underset{}{\bigcirc}\text{–X}
\quad\xrightarrow{\ -2\ \text{Me}_3\text{SiCl}\ }\quad
\text{X–}\underset{}{\bigcirc}\text{–S–N}=\text{S}=\text{N–S–}\underset{}{\bigcirc}\text{–X}
\tag{177}
$$

$$\text{X} = \text{NO}_2,\ \text{OMe},\ \text{Me}$$

corresponding polymers from difunctional sulfenyl chlorides, Eq. (178). High yields and satisfactory elemental analyses were reported, but no information about viscosities or molecular weights. After doping with bromine vapour the electric conductivity of these materials was investigated, and conductivity values up to 10^{-3} Siemens were found [182].

As demonstrated in Sect. 5.6 and 5.7 silylated diphenols react easily with dicarboxylic acid dichlorides or activated dihalogenoaromatics. These successful

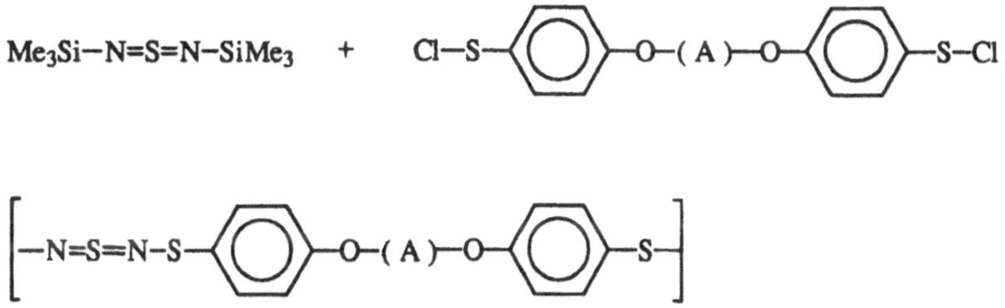

$$Me_3Si-N=S=N-SiMe_3 \quad + \quad Cl-S-\!\!\!\left\langle\bigcirc\right\rangle\!\!\!-O-(\,A\,)-O-\!\!\!\left\langle\bigcirc\right\rangle\!\!\!-S-Cl$$

$$\left[-N=S=N-S-\!\!\!\left\langle\bigcirc\right\rangle\!\!\!-O-(\,A\,)-O-\!\!\!\left\langle\bigcirc\right\rangle\!\!\!-S-\right] \tag{178}$$

$$(\,A\,) = (\,CH_2\,)_{10}\,; \quad -\!\!(\,CH_2CH_2-O\,)_n\,CH_2CH_2 \qquad n = 1\,,\,2$$

polycondensations prompted the author of this chapter to study polycondensations with a broader variety of reactive dichlorocompounds, such as
dichlorodiphenylsilane, dichlorodibutylstannane or phenyl boronic dichloride
[175]. In all cases triethylamine-HCl or benzyltriethylammonium chloride were
used as catalysts. Despite large quantities of these catalysts no substantial polycondensation took place between silylated bisphenol-A and dibutyldichlorostan-
nane. With diphenyldichlorosilane as reaction partner a sluggish polycondensation was achieved, but large amounts of a catalyst were required. At the
end of the polycondensation the catalyst had completely sublimed from the
reaction mixture as evidenced by ^1H NMR spectroscopic analysis of the remaining product. Nonetheless only polymers (Forms. **179a** and **b**) with low inherent
viscosities were obtained ($\eta_{inh} < 0.2$ dl/g in $CHCl_3$).

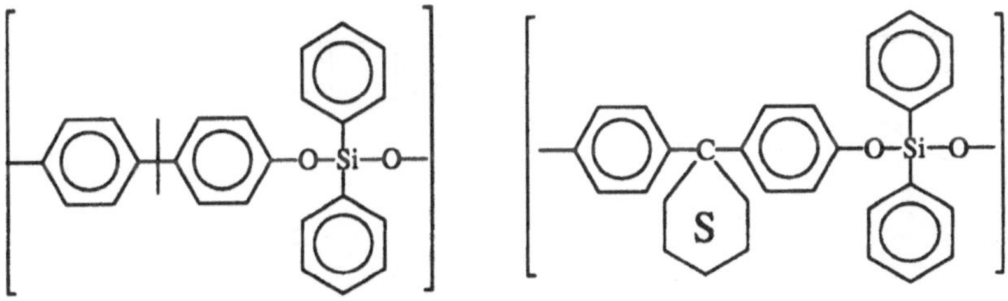

$$\tag{179}$$

a   b

Rapid polycondensations even below 150 °C were observed with phenyl-
boronic dichloride as comonomer. In addition to a homopolymer **180** two 1:1
copolyesters with isophthaloylchloride **181a** or phenylphosphonic dichloride
181b were prepared. However, only low inherent viscosities (< 0.15 dl/g in
CHCl$_3$ at 25 °C) were found in all three cases. Due to this unsatisfactory results
these polycondensations were not published in detail.

An electrochemical polycondensation of silylated thiophenes was reported by
Lemaire et al. [183] and later by Matsuda et al. [184, 185]. 2,5-Bistrimethylsilyl-
thiophene was electrochemically oxidized and polycondensed in nitrobenzene
containing Et_4NBF_4, Et_4NClO_4 or Et_4NPF_6 as electrolytes, Eq. (182). Films
with a thickness of 16.30 μm were obtained with conductivities up to
50 S/cm. Furthermore bis(2-thienyl)tetramethyldisilane **183a** and bis(2-thienyl)

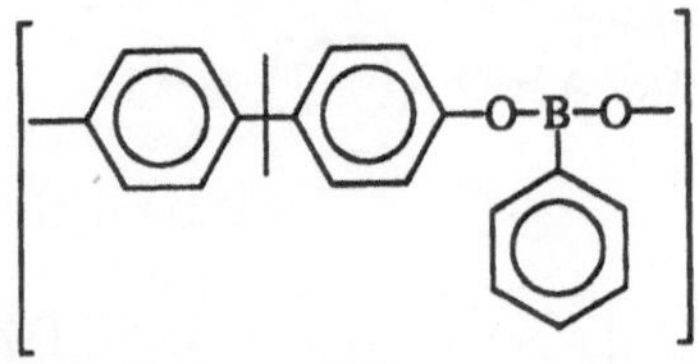

(180)

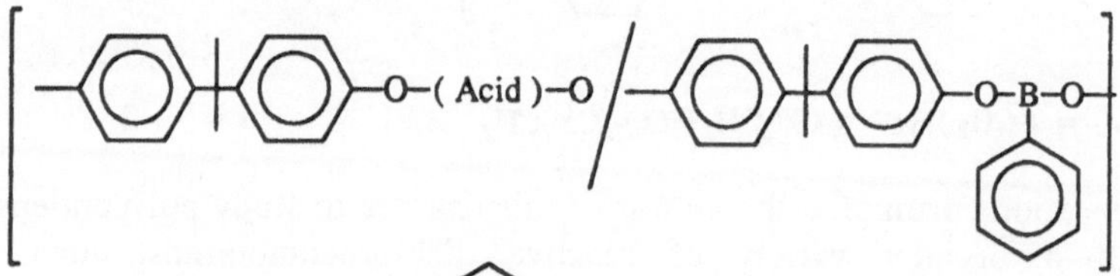

(181)

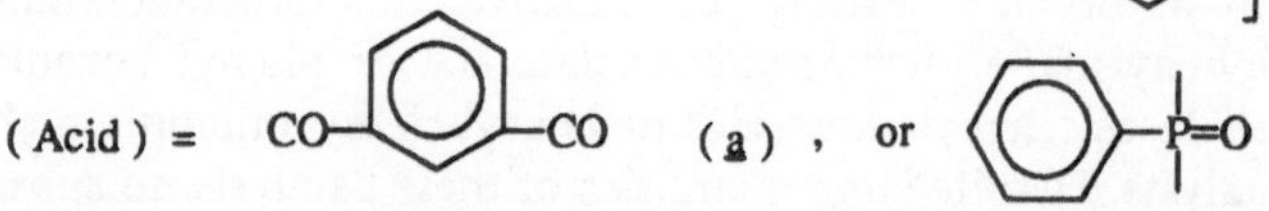

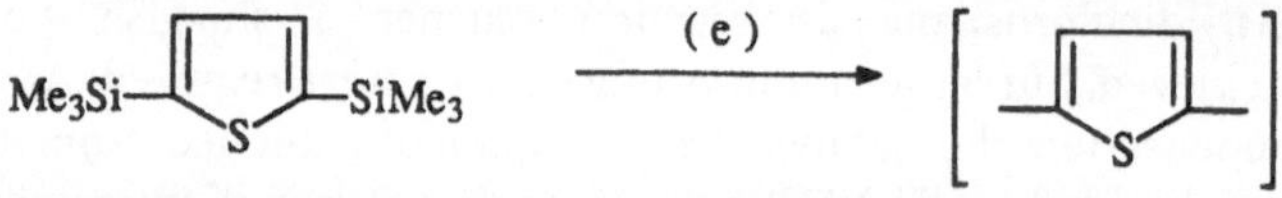

(Acid) = CO—⟨ ⟩—CO (**a**) , or ⟨ ⟩—P=O (**b**)

Me₃Si—⟨S⟩—SiMe₃ $\xrightarrow{(e)}$ [—⟨S⟩—] (182)

diphenylsilane **183b** were polycondensed under the same conditions and films with a conductivity up to 100 S/cm were isolated [185]. In all cases an almost complete desilylation was found.

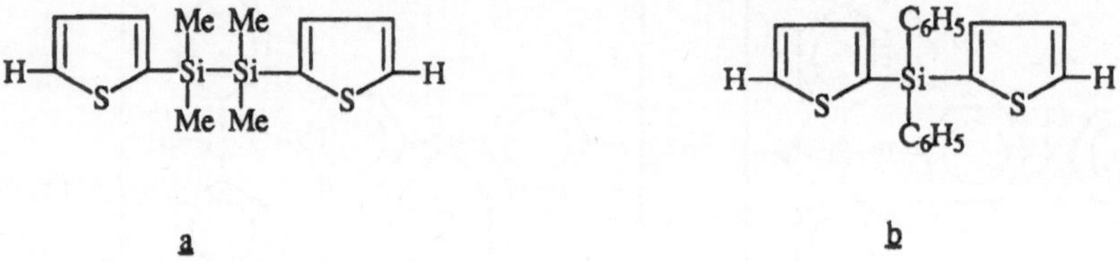

(183)

a   **b**

5.11
References

1. Klerbe JF (1964) J Polym Sci Part A 2: 2673
2. Boldebuch EM, Klerbe JF (1967) US.Pat. 3.303.157 (1967) to General Electric; C.A. 66 P 96 125 f
3. Kricheldorf HR, Leppert E (1972) Makromol Chem 158: 223
4. Klerbe JF (1964) J Polym Sci Part B 2: 1079
5. Klerbe JF (1972) Adv Org Chem 8: 97
6. Imai Y, Oishi Y (1989) Progr Polym Sci 14: 173
7. Rusanov AL (1990) Usp Khim 59 1492 (1990) CA 113: 232093n
8. Imai Y (1992) Makromol Chem, Macromol Symp 54/55: 389
9. Kricheldorf HR (1992) Makromol Chem, Macromol Symp 54/55: 365
10. Kricheldorf HR (1991) In: Sivaram S (ed.) Polymer sci contemporary themes vol. 1: Tata McGraw Hill, New Delhi, p 49

11. Imai Y (1991) In: Sivaram S (ed) Polymer Sci Contemporary Themes Vol. 1: Tata McGraw Hill Co., Ltd. New Delhi p. 3
12. Katsarava RD, Vygodskii YS (1992) Uspekhi Khimii 61: 1142 (1992) Russian Chem Reviews 61: 629
13. Klerbe JF (1964) J Polym Sci Letters Ed. (Part B 2) 1079
14. Sentsova TN, Butaeva VI, Davidovich YA, Rogozhin SN, Korshak VV (1977) Dokl Acad Sci 232: 335
15. Oishi Y, Padmanaban M, Kakimoto M, Imai Y (1992) J Polym Sci Part A, Polym Chem 30: 1363
16. DiSalvo AL (1974) J Polym Sci Polym Letters Ed 12: 641
17. Katsarava RD, Kartvelishvilii TM, Davidovich Y.A, Zaalishvili MM, Rogozhin SV(1983) Dokl Adad Nauk Sci SSR 266 363 C A 98: 35055x
18. Katsarava RD, Kartvelishvili TM, Zaalishvili MM (1984) Soobshch Akad Nauk Gruz SSR 113 533 C A 101: 171861g (1984)
19. Oishi Y, Padmanaban M, Kakimoto M, Imai Y (1987) J Polym Sci Part A, Polymer Chemistry 25: 3387
20. Kricheldorf HR, Stöber O (1992) Eur Polym J 28: 1377
21. Kricheldorf HR, Meier-Haack J (1991) Eur Polym J 27: 1039
22. Kricheldorf HR, Meier-Haack J (1992) Makromol Chem 193: 2631
23. Kricheldorf HR, Meier-Haack J (1993) Eur Polym J 29: 559
24. Kricheldorf HR, Meier-Haack J (1993) J Polym Sci 31: 1327
25. Eisenbach CD, Nefzger H (1991) In: Kricheldorf HR (ed.) Handbook of Polymer Syntheses Marcel Dekker Inc New York Chapter 12
26. Katsarava RD, Kartvelishvili TM (1986) Vysokomol Soedin Ser B 28: 377
27. Greber G, Kricheldorf HR (1968) Angew Chem 80: 1028
28. Kricheldorf HR (1970) Synthesis 649
29. Sekiguchi H, Coutin B (1992) In: Kricheldorf HR, (ed.) Handbook of Polymer Synthesis Marcel Dekker Inc, New York, Chapter 14
30. Katsarava RD, Kharadze DP, Japaridze NS, Omiadze TN, Avalishvili LM, and Zaalishvili MM (1985) Makromol Chem 186: 939
31. Katsarava RD, Kunchulia DP, Avalishvili LM, Zaalishvili MM (1979) Vysokomol Soedin Ser B 21 643 (1979) CA 92: 76962k
32. Katsarava RD, Kharadze DP, Avalishvili LM, Zaalishvili MM (1984) Makromol Chem, Rapid Commun 5: 585
33. Kricheldorf HR (1970) Chem Ber 103: 3353
34. Bowser JR, Williams PJ, Kurz K (1983) J Org Chem 48: ...
35. Kaneda T, Ishikawa S, Daimon H, Katsura T, Ueda M, Oda K, Hario M (1982) Makromol Chem 183: 417
36. Kaneda T, Ishikawa S, Daimon H, Katsura T, Ueda M, Oda K, Hario M (1982) Makromol Chem 183: 433
37. Oishi Y, Kakimoto M, Imai Y (1987) Macromolecules 20: 703
38. Oishi Y, Kakimoto M, Imai Y (1988) Macromolecules 21: 547
39. Kakimoto M, Oishi Y, Imai Y (1985) Makromol Chem, Rapid Commun 6: 557
40. Oishi Y, Kakimoto M, Imai Y (1987) J Polym Sci Part A, Polym Chem 25: 2493
41. Oishi Y, Harada S, Kakimoto M, Imai Y (1989) J Polym Sci Part A, Polym Chem 27: 3393
42. Oishi Y, Tanaka M, Kakimoto M, Imai Y (1991) Makromol Rapid Commun 12: 465
43. Takahashi Y, Jijima M, Oishi Y, Kakimoto M, Imai Y (1991) Macromolecules 24: 3543
44. Oishi Y, Harada S, Kakimoto M, Immai Y (1992) J Polym Sci Part A, Polym Chem 30: 1203
45. Hatke W, Schmidt HW, Heitz W (1991) J Polym Sci Part A, Polym Chem 29: 1387
46. Kricheldorf HR, Schmidt B (1992) B-rger R, Macromolecules 25: 5465
47. Yamazaki N, Matsumoto M, Higashi F (1975) J Polym Sci, Polym Chem Ed 13: 1373
48. Kricheldorf HR, Schmidt B (1992) Macromolecules 25: 6090
49. Kricheldorf HR, Schmidt B (1992) Macromolecules 25: 5471
50. Kozyukov VP, Mironova NV, Mironov VF, Obshchei Khim Z (1978) 48: 2541
51. Kricheldorf HR, Löhden G (unpublished results)
52. Strohriegel P, Heitz W (1985) Makromol Chem, Rapid Commun 6: 111

352   Chapter 5: Polycondensation of Silylated Monomers

53. Strohriegel P, Heitz W (1986) Makromol Chem, Rapid Commun 7: 513
54. Akar A, Galioglu O (1988) Macromol Chem, Rapid Commun 9: 12
55. Rodriguez-Galán A, Bou JJ, Muñoz-Guerra S (1992) J Polym Sci Part A, Polym Chem 30: 713
56. Katsarava R, Kharadze D, Kirmelashvili L, Medzmariashvili N, Goguadze T, Tsilanadze G (1993) Makromol Chem 194: 143
57. de Abajo J (1991) In Kricheldorf HR, (ed), "Handbook of Polymer Synthesis" Marcel Dekker, New York Part B, Chapter 15
58. Boldebuck EM, Klerbe JF (1967) US Pat 3.303.157 to General Electric CA 66: P96125f
59. Greber G, Darms R (1972) Ger Offen 2.206.359 to Ciba-Geigy AG, CA 77: P165517b
60. Greber G, Greber U, Kuhn M, Lohmann D, Gati S (1972) Ger Offen 2.206.379 to Ciba-Geigy AB, CA 77: P165520x
61. Korshak VV, Vinogradova SV, Vygodskii YS, Nagiev ZM, Urman YG, Alekseeva SG, Slonium IY (1983) Makromol Chem 184: 235
62. Oishi Y, Kakimoto M, Imai Y (1991) Macromolecules 24: 3475
63. Pratt J, Thames SF (1973) Synthesis 223
64. Iojima M, Takakhashi Y, Oishi Y, Kakimoto M, Imai Y (1991) J Polym Sci Part A, Polym Chem 29: 1717
65. Oishi Y, Shivasari M, Kakimoto M, Imai Y (1993) J Polym Sci Part A, Polym Chem 31: 293
66. Korshak VV, Bondarevskiy GS, Valetskiy PM, Kalachev AI, Rogoshin SV, Davidovin A, Vinogradova SV, Tseitlin GM (1975) USSR Pat 464590 (1975) CA 83: 59023b
67. Bondarevskiy GS, Kalachev AI, Valetskiy PM, Davidovich YA, Tseitlin GM (1976) Izv Acad Nauk SSSR, Ser Khim 920 CA 85: 63120q
68. Vinogradova SV, Korshak VV, Bondarevskiy GS, Valetskiy PM, Kalachev AI, Rogoshin SV, Davidovin A, Tseitlin GM, Stanko VI (1975) USSR Pat. 477177 CA
69. Semenov VI, Tseitlin GM, Valetskie PM, Vinogradova SV, Korshak VV, Sokolov LB, Zhinkin DY (1979) USSR Pat 663 699 (1979) CA 91: 40349f
70. Maruyama Y, Oishi Y, Kakimoto M, Imai Y (1988) Macromolecules 21: 2305
71. Imai Y, Kakimoto M, Oishi Y, Tanaka Y (1990) Jap Pat 01.292,034 (1989) to Cosmo Sekiyu KK, CA 112: 199388s
72. Tanaka Y, Oishi Y, Kakimoto M, Imai Y (1991) J Polym Sci Part A, Polym Chem 29: 1941
73. Wolfe JF, Arnold FE (1981) Macromolecules 14: 909
74. Wolfe JF (1987) Encycl Polym Sci 11: 601
75. Kricheldorf HR, Engelhardt J (1989) Makromol Chem 190: 2939
76. Kricheldorf HR, Engelhardt J, Pakull R, Eckhardt R, Leyrer V (1992) U.S.Pat. 5.134.219
77. Kricheldorf HR, Engelhardt J (1990) Makromol Chem 191: 2017
78. Kricheldorf HR, Domschke A (1994) Polymer 35: 199
79. Kricheldorf HR, Bürger R (1994) New Polym Mater 4: 119
80. Kricheldorf HR, Thomsen S (1991) J Polym Sci Part A, Polym Chem 29: 1751
81. Kricheldorf HR, Thomsen S (1993) Macromolecules 26: 6628
82. Kricheldorf HR, Thomsen S (1992) Makromol Chem 193: 2467
83. Kricheldorf HR, Bier G (1983) J Polym Sci, Polym Chem Ed 21: 2283
84. Saunders DG (1988) Synthesis 377
85. Hedrick JL, Brown HR, Hofer DC, Johson RD (1989) Macromolecules 22: 2048
86. Johnsen RN, Farnham AG, Clendinning RA, Hate WF, Merriam CN (1967) J Polym Sci, Polym Chem Ed 5: 2375
87. Attwood TE, Dawson PD, Freeman JC, Hoy LR, Rose JB, Staniland PA (1981) Polymer 22: 1976
88. Williams FJ, Donahue PE (1977) J Org J Chem 42: 3414
89. Williams FJ, Relles HM, Donahue PE, Manello JS (1977) J Org Chem 42: 3419
90. Kricheldorf HR, Bier G (1984) Polymers 25: 1151
91. Kricheldorf HR, Jahnake P (1990) Makromol Chem 191: 2027
92. Kricheldorf HR, Delius U (1989) Makromol Chem, Rapid Commun 10: 41
93. Kricheldorf HR, Meier J, Schwarz G (1987) Makromol Chem, Rapid Commun 8: 529
94. Kricheldorf HR, Schwarz G (1988) J Erxleben Makromol Chem 189: 2255
95. Kricheldorf HR, Delius U, Tönnes K-U (1988) New Polym Mater 1: 127

96. Kricheldorf HR, Delius U (1989) Macromolecules 22: 517
97. Kricheldorf HR, Adebahr T (1993) Makromol Chem 194: 2103
98. Kricheldorf HR, Delius U (1989) Makromol Chem 190: 1277
99. Kricheldorf HR, Jahnke P, Scharnagl N (1992) Macromolecules 25: 1382
100. Hedrick JL (1992) Polym Bull 27: 665
101. Kricheldorf HR, (1991) Berghahn M, Makromol Chem Rapid Commun 12: 529
102. Kricheldorf HR, Jürgens C (1993) Eur Polym J 29: 903
103. Kricheldorf HR, Schmidt B, Delius U (1990) Eur Polym J 26: 791
104. Kumpf RJ, Nerger D, Lantmann C, Pielartcik H, Wehrmann R (1991) ACS Polym Prepr 32: 280
105. Kricheldorf HR, Jahnke P (1992) J Polym Sci Part A, Polym Chem 30: 1299
106. Kricheldorf HR, Jahnke P (1991) Makromol Chem, Rapid Commun 12: 331
107. Kricheldorf HR, Jahnke P (1992) Polym Bull 28: 411
108. Lenz RW, Carrington WK (1959) J Polym Sci 41: 333
109. Koch W, Heitz W (1983) Makromol Chem 184: 779
110. Kricheldorf HR, Jahnke P (1991) Polym Bull 27: 135
111. Hara A, Oishi Y, Kakimoto M, Imai Y (1991) J Polym Sci Part A, Polym Chem 29: 1933
112. Kricheldorf HR, Schwarz G. (1979) Polym Bull 1: 383
113. Kricheldorf HR, Schwarz G, Ruhser F (1988) J Polym Sci Part A, Polym Chem 26: 1621
114. Kricheldorf HR, Beuermann I, Schwarz G (1989) Makromol Chem, Rapid Commun 10: 211
115. Kricheldorf HR, Engelhardt J (1990) J Polym Sci Part A, Polym Chem 28: 2335
116. Kricheldorf HR, Weegen-Schulz B, Engelhardt J (1991) Makromol Chem 192: 631
117. Kricheldorf HR, Engelhardt J, Weegen-Schulz B (1991) Makromol Chem 192: 645
118. Kricheldorf HR (1991) ACS Polym Prepr 32: 395
119. Kricheldorf HR, Gronski W, vom Stein T (manuscript in preparation)
120. Schwarz G, Alberts H, Kricheldorf HR (1981) Liebigs Ann Chem 1257
121. Kricheldorf HR, Schwarz G (1983) Makromol Chem 184: 475
122. Kricheldorf HR, Zher Zang Q, Schwarz G (1982) Polymer 23: 1821
123. Hawker CJ, Lee R, Fréchet JM (1991) J Am Chem Soc 113: 4583
124. de Abajo J, dela Campa J, Kricheldorf HR, Schwarz G (1990) Makromol Chem 191: 537
125. de Abajo J, de la Campa J, Kricheldorf HR, Schwarz G (1972) Eur Polym J 28: 261
126. Kricheldorf HR, Schwarz G, Ruhser F (1991) Macromolecules 24: 3485
127. Kricheldorf HR, Lübbers D (1991) Makromol Chem Rapid Commun 12: 691
128. Kricheldorf HR, Linzer V, Bruhn C, J Macromol Sci Pure and Appl Chem (in press) (LC-Polyimides 11)
129. Kricheldorf HR, Linzer V, de Abajo J, de la Campa J, J Macromol Sci Pure and Appl Chem in press (LC-Poly- imides 12)
130. Kricheldorf HR, Stöber O (1994) Makromol Chem Rapid Commun 15: 87
131. Kricheldorf HR, Stöber O (1995) Macromolecules 28: 2118
132. Kricheldorf HR, Löden G, Wilson DJ (1994) Macromolecules 27: 1669
133. Kricheldorf HR, Stöber O (Manuscript in preparation)
134. Kricheldorf HR, Leppert E (1972) Makromol Chem 158: 223
135. Kricheldorf HR (1971) Chem Ber 104: 3146
136. Kricheldorf HR (1972) Leppert E, Makromol Chem 158: 223
137. Kricheldorf HR (1971) Chem Ber 104: 87
138. Gupta B, US Pat 4.868.265 (1989) CA 112: 217746p (1990)
139. Kricheldorf HR, Lübbers D (1990) Makromol Chem, Rapid Commun 11: 83
140. Kricheldorf HR, Lübbers D (1990) Makromol Chem, Rapid Commun 11: 261
141. Kricheldorf HR, Lübbers D (1990) Makromol Chem, Rapid Commun 11: 303
142. Kricheldorf HR, Lübbers D (1991) Eur Polym J 12: 1397
143. Kricheldorf HR, Lübbers D (1992) Macromolecules 25: 1377
144. Kricheldorf HR, Lübbers D (1992) Eur Polym J 28: 887
145. Kricheldorf HR, Schwarz G (1992) In: Kricheldorf HR (ed) "Handbook of Polymer Synthesis" Marcel Dekker Inc New York, chapter 27, p. 1668
146. Griffin BP, McDonald WA (1982) Eur Pat 55527 to ICI plc CA 106: 67869r (1987)
147. Kricheldorf HR, Jürgens C (1994) Eur Polym J 30: 281

148. Penczek S, Lapienis G, In: Kricheldorf HR (ed) "Handbook of Polymer Syntheses"Marcel Dekker, New York 1992, Part B, p. 1077
149. Kricheldorf HR, Koziel HJ (1986) Macromol Sci A 23: 1337
150. Kricheldorf HR, Hüner R (1992) J Polym Sci Part A, Polym Chem 30: 337
151. Kricheldorf HR, Koziel H, Witek E (1988) Makromol Chem, Rapid Commun 9: 217
152. Kricheldorf HR, Koziel H (1988) New Polymer Mater 1: 143
153. Koziel H, Thesis, University of Hamburg
154. Kricheldorf HR, Lübbers D (1993) Polymer 34: 1515
155. Flindt EP, Rose H (1977) Z Anorg Allg Chem 428: 204
156. Wisian-Neilson P, Neilson RH (1988) Chem Reviews 88: 541
157. Wisian-Neilson P, Neilson RH, Cowley AH (1970) Inorg Chem 16: 1460
158. Wisian-Neilson P, Neilson RH (1980) J Am Chem Soc 102: 2848
159. Neilson RH, Wisian-Neilson P (1981) J Macromol Sci Chem A 16: 425
160. Neilson RH, Wisian-Neilson P (1982) Inorg Chem 21: 3568
161. Wisian-Neilson P, Neilson RH (1980) Inorg Chem 19: 1975
162. Neilson RH, Wisian-Neilson P, Morton DW, O'Neal HR (1981) ACS Symp.Ser. 171: 239
163. Neilson RH, Hani R, Wisian-Neilson P, Meister JJ, Roy AK, Hagnauer GL (1987) Macromolecules 20: 910
164. Neilson RH, Ford RR, Hani R, Roy AK, Scheide GM, Wettermark UG, Wisian-Neilson P (1987) ACS Polym Prepr 28: 442
165. Neilson RH, Ford RR, Roy AK, Scheide GM, Wettermark UG, Wisian-Neilson P (1988) ACS Symp Ser No. 360: 283
166. Roy AK, Hani R, Neilson RH, Wisian-Neilson P (1987) Organo-metallics 6: 378
167. Montague RA, Matyjazewski K (1990) J Am Chem Soc 112: 6721
168. Matyjazewski K, Dauth J. Montague R, Reddick C, White M (1991) ACS Polym Prepr 32: 305
169. Matyjazewski K, Montague R, Dauth J, Nuyken O (1992) J Polym Sci Part A, Polym Chem 30: 813
170. Matyjazewski K, Cypryk M, Dauth J, Nuyken O, White M (1992) Makromol Chem, Macromol Symp 54/55: 13
171. Matyjazewsksi K, Lindenberg MS, White ML (1991) ACS Polym Prepr. 3 1096
172. Paciorek KC, Kratzer R (1964) Inorg Chem 3: 594
173. Tesi G, Haber CP, Douglas CM (1960) Proc Chem Soc 219
174. Imai Y, Oishi Y (1989) Progr Polym Sci 14: 173
175. Kricheldorf HR, Stöber O (unpublished results)
176. Komarova LI, Salazkin SN, Vygodskii YS (1990) Vysokomol Soedin Ser. A 42 1571
177. Shulz IH, Zike CG (1963) US Pat 3.072.594 to CA 58: 6988 (1963)
178. Khrustaleva EM, Golubko GE, Zhinkina DE (1970) Plastmassy 1: 12
179. Mueto T, Miyazawa M, Matsumoto T (1987) Polym Prepr Jpn 36: 324
180. Muneto T, Miyazawa M, Matsumoto T, Kurosaki J, Jap Pat 0169.631 (1989) to Kuraray Co Ltd., CA 111: 154649r
181. Ramakrishnan S, Chien JW (1987) J Polym Sci Part A, Polym Chem 25: 1433
182. Chien JW, Ramakrishnan S (1988) Macromolecules 21: 2007
183. Lemaire M, Buchner W, Garreau R, Hou HA, Guy A, Roncali J (1990) J Electroanal Chem 281: 293
184. Matsuda H, Tanaka S, Kaeriyama K (1990) J Polym Sci Part A, Polym Chem 28: 1831
185. Matsuda H, Taniki Y, Kaeriyama K (1992) J Polym Sci Part A, Polym Chem 30: 1667

CHAPTER 6

Miscellaneous Applications of Silicon Reagents

H. R. Kricheldorf

6.1
Difunctional Isocyanates and Isothiocyanates

6.1.1
Introduction

Difunctional aliphatic or aromatic isocyanates and isothiocyanates are useful and versatile reagents for polyadditions and polycondensations. Furthermore, polymerizable, cyclic N-carboxy anhydrides can be obtained from α,β- and γ, isocyanato or isothiocyanatocarboxylic acids. Silicon reagents, in particular trimethylsilylazide and trimethylchlorosilane, have proved to be highly useful for the synthesis of various difunctional isocyanates and isothiocyanates. The silicon reagents were used in two quite different ways. Either the isocyanate and isothiocyanatogroup were prepared by means of a silicon reagent or the second functional group was introduced or stabilized ("protected") by a silyl group. When this second functional group is set free it can react with the isocyanate or isothiocyanate group either by formation of polymerizable heterocycles or by polycondensation.

6.1.2
Difunctional Isocyanates via Trimethylsilylazide

An useful reagent for the synthesis of isocyanate groups from acid chlorides is trimethylsilylazide [1-3]. This reagent is easily inflammable but not explosive and easy to synthesize from sodium azide and chlorotrimethylsilane [2, 4-6]. Its reaction with acid chlorides is catalyzed by pyridine, probably by the intermediate formation of an acylpyridinium ion, Eqs. (1-3). The acylation of trimethylsilylazide is also catalyzed by chloride ions, by analogy with most reactions of

$$R\text{--CO--Cl} + \text{py} \rightleftharpoons R\text{--CO--py}^{\oplus} + \text{Cl}^{\ominus} \quad (1)$$

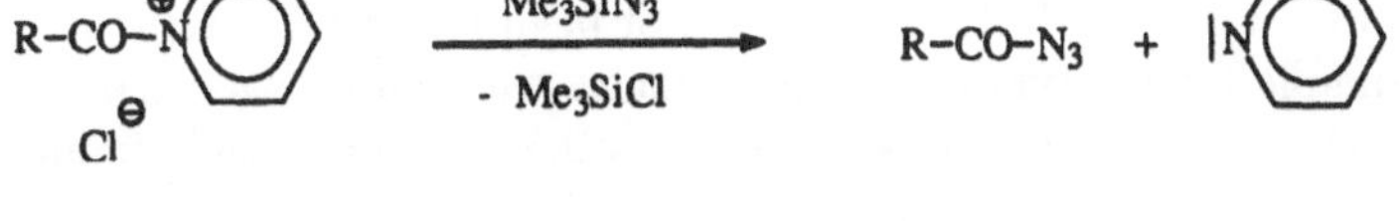

$$R\text{--CO--py}^{\oplus}\ \text{Cl}^{\ominus} \xrightarrow[\text{- Me}_3\text{SiCl}]{\text{Me}_3\text{SiN}_3} R\text{--CO--N}_3 + \text{py} \quad (2)$$

silylated nucleophiles, and in this case the intermediate formation of azide

$$\downarrow \tag{3}$$

$$R\text{--}N\text{=}C\text{=}O \;+\; N_2$$

anions is the most likely reaction mechanism, Eqs. (4, 5).

$$Me_3Si\text{--}N_3 \;+\; Cl^{\ominus} \;\rightleftharpoons\; Me_3SiCl \;+\; N_3^{\ominus} \tag{4}$$

$$R\text{--}CO\text{--}Cl \;+\; N_3^{\ominus} \;\longrightarrow\; R\text{--}CO\text{--}N_3 \;+\; Cl^{\ominus} \tag{5}$$

The Curtius method via trimethylsilylazide is advantageous over the classical methods based on metal azides or hydrazine + nitrous acid, when difunctional isocyanates with an additional electrophilic group are to be synthesized. A typical example is the synthesis of ω-bromoalkylene isocyanates, Eqs. (6–8) [7], because the bromoalkylgroup can react with both, metalazides and hydrazine. The synthesis of ω-bromoalkylene isocyanates is also an interesting example, how various silyl reagents can be combined in a multistep synthesis. Lactones are cleaved by bromotrimethylsilane [7] (and still better by iodotrimethylsilane [8, 9] to yield trimethylsilyl ω-bromoalkanoates, Eq. (6), which are then reacted

$$\underset{O\text{------}CO}{\overset{(CH_2)_n}{\frown}} \;+\; Me_3SiBr \;\longrightarrow\; Br\text{--}(CH_2)_n\text{--}CO\text{--}OSiMe_3 \tag{6}$$

$$+ \; SOCl_2$$

$$-\; Me_3SiCl \tag{7}$$

$$Br\text{--}(CH_2)_n\text{--}CO\text{--}Cl \;\xrightarrow[-\;Me_3SiCl]{+\;Me_3SiN_3}\; Br\text{--}(CH_2)_n\text{--}CO\text{--}N_3 \tag{8}$$

$$\downarrow$$

$$Br\text{--}(CH_2)_n\text{--}NCO \tag{9}$$

with thionyl chloride, Eq. (7). The silylester group is presumably the only kind of ester groups which can be transformed into acid chlorides (or acid bromides) without intermediate saponification. Furthermore, the formation of free HCl (or HBr) is avoided. The usefulness of this method is further discussed below (Sect. 6.1.3). Yield and properties of ω-bromoalkylene isocyanates are listed in Table 6.1. Polycondensations or other applications of these compounds for synthesis and modification of polymers have not been reported so far.

Another example for the synthesis of isocyanates with additional electrophilic group is the preparation of ω-isothiocyanato isocyanates from ω-isothiocyanato-

Table 6.1 Yields and properties of trimethylsilyl ω-bromocarboxylates, ω-bromoalkanoyl chlorides and ω-bromoalkyl isocyanates.

Functional ω-bromoalkyl compounds	Yield (%)	B.p. (°C/mbar)	η^{20}D
$Br-CH_2-CH_2-CO_2SiMe_3$	79	75 – 76/12	1.4453
$Br-(CH_2)_3-CO_2SiMe_3$	86	87 – 89/11	1.4489
		45 – 47/01	
$Br-(CH_2)_4-CO_2SiMe_3$	90	54 – 56/0.13	1.4522
$Br-(CH_2)_5-CO_2SiMe_3$	96	75 – 77/0.07	1.4541
$Br-(CH_2)_2-CO-Cl$	92	49 – 51/12	1.4939
$Br-(CH_2)_3-CO-Cl$	95	70 – 71/12	1.4896
$Br-(CH_2)_4-CO-Cl$	87	43 – 45/0.2	1.4913
$Br-(CH_2)_5-CO-Cl$	98	60 – 62/0.3	1.4877
$Br-(CH_2)_2-NCO$	82	44 – 46/12	1.4838
$Br-(CH_2)_3-NCO$	87	65 – 66/12	1.4840
$Br-(CH_2)_4-NCO$	91	84 – 86/12	1.4842
$Br-(CH_2)_5-NCO$	91	56 – 58/01	1.4861

carboxylic acid chlorides, Eq. (10). These acid chlorides can be prepared from the corresponding trimethylsilyl-ω-isothiocyanatocarboxylates and thionylchloride [10] or from N-trimethylsilyl lactames and thiophosgene, Eqs. (11–13) [11]. The

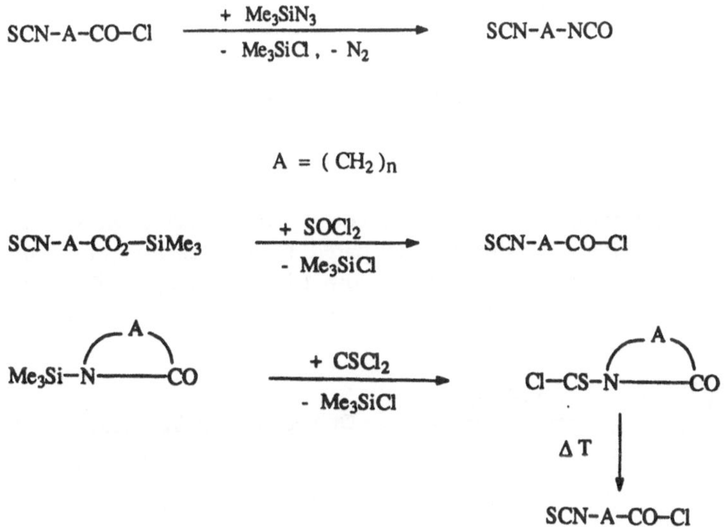

$$\text{SCN-A-CO-Cl} \xrightarrow[\text{- Me}_3\text{SiCl, - N}_2]{\text{+ Me}_3\text{SiN}_3} \text{SCN-A-NCO} \tag{10}$$

$$A = (CH_2)_n$$

$$\text{SCN-A-CO}_2\text{-SiMe}_3 \xrightarrow[\text{- Me}_3\text{SiCl}]{\text{+ SOCl}_2} \text{SCN-A-CO-Cl} \tag{11}$$

$$\text{Me}_3\text{Si-N}\overset{\frown{A}}{\text{---}}\text{CO} \xrightarrow[\text{- Me}_3\text{SiCl}]{\text{+ CSCl}_2} \text{Cl-CS-N}\overset{\frown{A}}{\text{---}}\text{CO} \tag{12}$$

$$\Delta T \downarrow$$

$$\text{SCN-A-CO-Cl} \tag{13}$$

obvious synthesis of polyureas or polyurethanes with thioureas or thiourethane groups from ω-isothioisocyanato-isocyanates and diamines or diols, Eqs. (14 and 15), has never been reported so far. Yet ω-isothiocyanato-isocyanates have proved to be useful spacers for binding of various substrates to polymer chains. The isocyanate group is much more electrophilic than the isothiocyanate group and allows a selective reaction with OH-groups. The isothiocyanate group can then be

$$\text{SCN-A-NCO} \xrightarrow{\text{+ NH}_2\text{-B-NH}_2} \text{(NH-CS-NH-A-NH-CO-NH-B--)} \qquad (14)$$

$$\text{SCN-A-NCO} \xrightarrow{\text{+ HO-B-OH}} \text{(-CS-NH-A-NH-CO-O-B-O-)} \qquad (15)$$

added to amino groups. Examples for this approach is the binding of nucleosides to polylysine, Eq. (16) [12] or the binding of enzymes to polyvinyl alcohol [13].

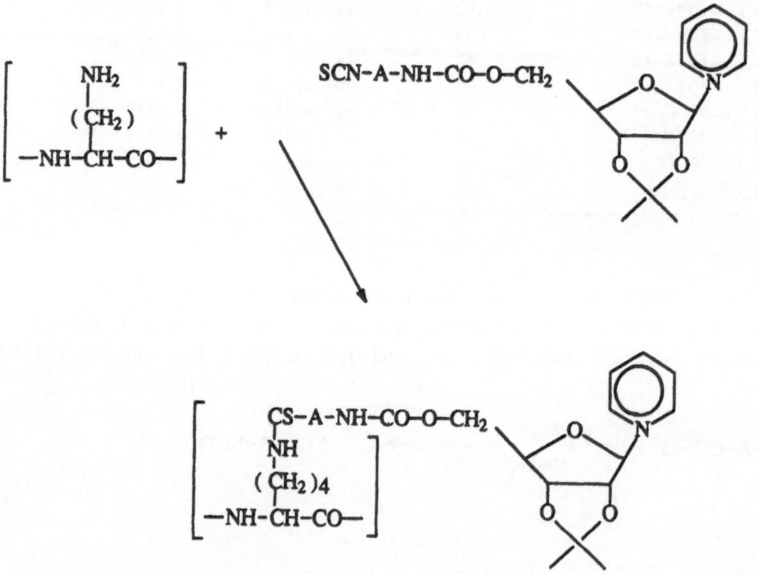

(16)

Furthermore, the synthesis of ω-isocyanatosulfonylchlorides is worth noting. 2-Isocyanatoethanesulfonylchloride was prepared by the reaction of sulfopropionic anhydride with trimethylsilylazide followed by chlorination of the resulting silylester with phosphorus pentachloride, Eqs. (17,18) [14]. An attempt to

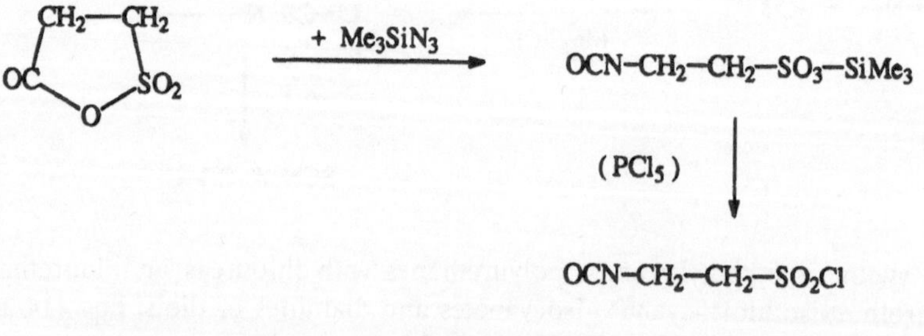

$$\xrightarrow{\text{+ Me}_3\text{SiN}_3} \text{OCN-CH}_2\text{-CH}_2\text{-SO}_3\text{-SiMe}_3 \qquad (17)$$

$$(\text{PCl}_5) \downarrow$$

(18)

$$\text{OCN-CH}_2\text{-CH}_2\text{-SO}_2\text{Cl}$$

prepare 3-isocyanatopropane sulfonylchloride from 3-chlorocarbonyl propane sulfonylchloride and trimethylsilylalzide, Eq. (19) was not quite successful. The desired product was spectroscopically identified, but not isolated [15].

$$\text{Cl—CO—(CH}_2\text{)}_3\text{—SO}_2\text{Cl} \quad \xrightarrow[\text{- N}_2,\ \text{- Me}_3\text{SiCl}]{\text{+ Me}_3\text{SiN}_3} \quad \text{OCN—(CH}_2\text{)}_3\text{—SO}_2\text{Cl} \qquad (19)$$

Finally, the synthesis of trimethylsiloxyarylisocyanates (Forms. **20, 21**) should be mentioned. As illustrated by Eq. (20) the corresponding trimethylsiloxyaryl

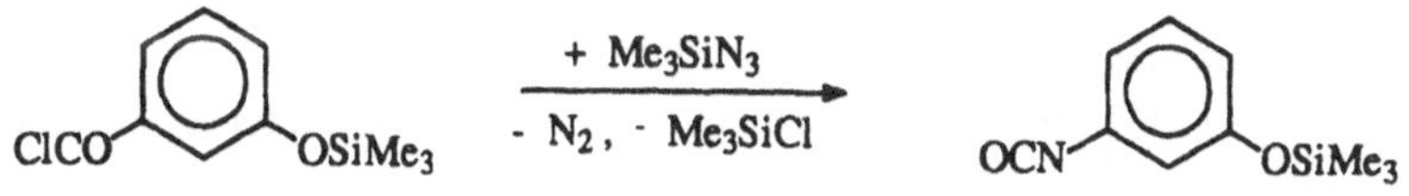

$$(20)$$

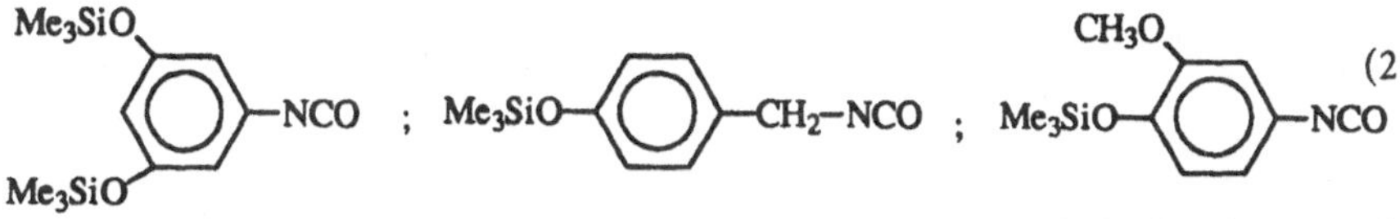

$$(21)$$

carboxylic acid chlorides are required as starting materials [16]. They can be prepared by treatment of silylated hydroxyacids with thionylchloride. An alternative approach reported for the synthesis of 4-trimethylsiloxyphenyliso-cyanate consists in the acylation of N,O-bistrimethylsilyl-4-aminophenol with phenylchloroformate [17]. As discussed in the next section the intermediately formed N-silylated urethanes decomposes spontaneously, Eq. (22). A third

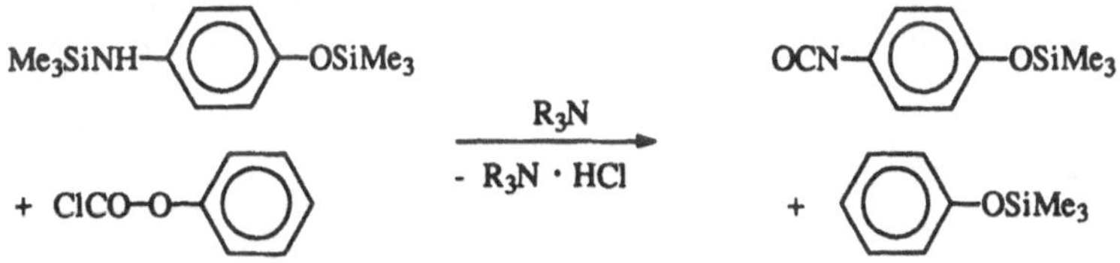

$$(22)$$

approach giving high yields of trimethylsiloxy alkyl- or arylisocyanates consists of the reaction between silylated hydroxyalkylamines or aminophenols and 4,4′-diisocyanato diphenyl methane [18]. The trimethylsiloxygroup can be acylated by functional acid chlorides, Eq. (23) [16]. Further examples of this reaction and its application for the synthesis of polyurethanes is discussed in the next section.

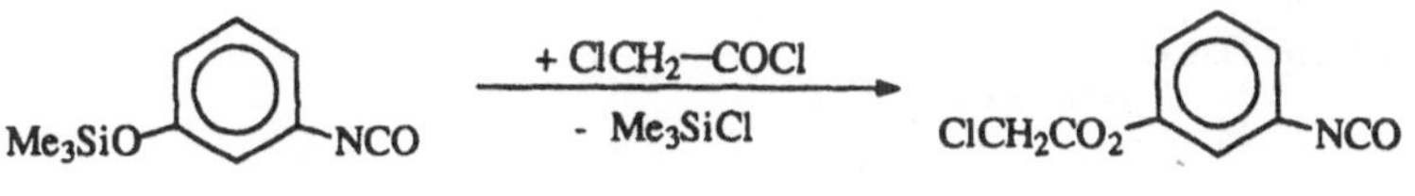

$$(23)$$

6.1.3
Synthesis and Reactions of Isocyanato- and Isothiocyanato Carboxylic Acid Chlorides

Several aliphatic and aromatic isocyanatocarboxylic acid chlorides were synthesized by direct phosgenation of ω-aminocarboxylic acids [19, 20]. An alternative and versatile approach is the reaction of trimethylsilyl-ω-isocyanato carboxylate with thionylchloride, Eq. (24) [21]. Two methods were developed for

$$XCN-A-CO_2-SiMe_3 \xrightarrow[-\ Me_3SiCl\ ,\ -\ SO_2]{SOCl_2} XCN-A-CO-Cl \qquad (24)$$

$$X = O, S$$

the synthesis of trimethylsilyl-ω-isocyanatocarboxylates. Firstly, cyclic anhydrides can be reacted with trimethylsilylazide (catalyzed by pyridin) so that the intermediately formed acyl azides are directly converted to isocyanate groups, Eqs. (25, 26) [22–26]. The advantage of this method is the one-step character.

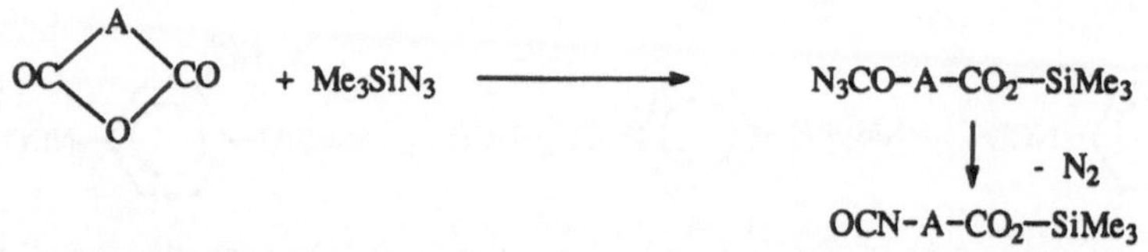

$$N_3CO-A-CO_2-SiMe_3 \qquad (25)$$

$$\downarrow -N_2$$

$$OCN-A-CO_2-SiMe_3 \qquad (26)$$

Characteristic short-comings is the limitation to symmetrically substituted, aliphatic anhydrides. The second method is a two-step procedure. The first step consist in the synthesis of N-phenoxycarbonyl amino acids, which are best, prepared from silylated amino acid, Eq. (27, 28) [27–29]. The silylation of the urethane group with trimethylchlorosilane and triethylamine leads to a spontaneous and irreversible cleavage of the urethane group, Eq. (29, 30). In this con-

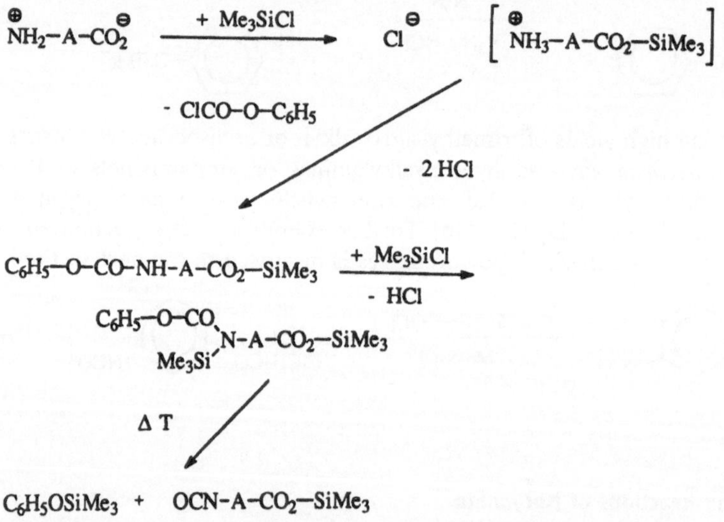

$$\overset{\oplus}{N}H_2-A-CO_2^{\ominus} \xrightarrow{+\ Me_3SiCl} Cl^{\ominus} \left[\overset{\oplus}{N}H_3-A-CO_2-SiMe_3 \right] \qquad (27)$$

$$- ClCO-O-C_6H_5$$

$$- 2\,HCl \qquad (28)$$

$$C_6H_5-O-CO-NH-A-CO_2-SiMe_3 \xrightarrow[-\ HCl]{+\ Me_3SiCl}$$

$$\underset{Me_3Si}{\overset{C_6H_5-O-CO}{\diagdown}} N-A-CO_2-SiMe_3 \qquad (29)$$

$$\Delta\,T$$

$$\qquad (30)$$

$$C_6H_5OSiMe_3 + OCN-A-CO_2-SiMe_3$$

nection it should be mentioned that the formation of isocyanates by N-silylation of urethane and order carbamic acid derivatives is a synthetic method of much broader scope and not limited to the preparation of silylated ω-isocyanato-carboxylic acids [30, 31]. However, it is worth noting that N-silylation and

$$Me_3Si-NH-A-CO_2-SiMe_3 \xrightarrow{(CS_2)} SCN-A-CO_2-SiMe_3 \qquad (31)$$

elimination of O-alkyl urethanes is more difficult and involves more side reactions that silylation of O-aryl urethanes [29]. Yields and properties of trimethylsilyl-ω-isocyanato carboxylates and acid chlorides derived from them are listed in Table 6.2.

The trimethylsilyl-ω-isothiocyanatocarboxylates required for the synthesis of ω-isothiocyanatocarboxylic acid chlorides, Eq. (11) can be prepared in various ways. Moderate yield are obtained by the reaction of CS2 with N,O-bistrimethylsilyl-ω-amino acids [32], Eq. (31). Better yields can be obtained by acylation of silylated amino acids, Eq. (27), with arylchlorothioformates or alkyl dithiochloroformates (by analogy with Eqs. 27, 28) and subsequent silylation of the thiourethane groups, Eqs. (32, 33) [33].

$$R-X-CS-NH-A-CO_2-SiMe_3 \xrightarrow[- \text{ HCl}]{+ \text{ Me}_3\text{SiCl}} \begin{matrix} R-X-CS \\ Me_3Si \end{matrix} N-A-CO_2-SiMe_3 \quad (32)$$

$$R = \text{Alkyl , Aryl}$$
$$X = \text{O , S}$$

$$R-X-SiMe_3 \;+\; SCN-A-CO_2-SiMe_3 \qquad (33)$$

However, the chlorothioformates required for this method are not commercial and require thiophosgen for their preparation. More convenient and less expensive, is the synthesis of crystalline N-(ethoxycarbonyldithiocarbonyl) amino acids (DTE-amino acids) [34, 35] and their silylation with trimethylchlorosilane. Their decomposition during silylation is spontaneous and quantitative, Eq. (34) [33].

$$C_2H_5-O-CO-S-CS-NH-A-CO_2H \xrightarrow[- \text{ 2 HCl}]{+ \text{ 2 Me}_3\text{SiCl}}$$

$$(34)$$

$$C_2H_5-O-SiMe_3 \;+\; COS \;+\; SCN-A-CO_2-SiMe_3$$

The carbonylchloride group is significantly more electrophilic than the isocyanato or isothiocyanato group. Therefore, a variety of difunctional iso(thio)cyanates can be prepared by selective modification of the acid chloride group (Table 6.3 and 6.4). Particularly successful proved to be the condensation of iso(thio)cyanatocarboxylic acid chlorides with silylated nucleophiles. The oldest example of such a reaction seems to be the acylation of hexamethyldisilthiane which may take two different courses depending on the molar ratios of the reactants, Eqs. (35, 36) [33].

$$XCN-A-CO-Cl \;+\; (Me_3Si)_2S - Cl\,SiMe_3 \longrightarrow XCN-A-CS-O-SiMe_3 \qquad (35)$$

More important for the synthesis of polymers is the acylation of N-trimethylsilyl lactams, Eq. (37) [36, 37]. The resulting N-(ω-isocyanatoacyl) lactams or the corresponding isothiocyanates can be reacted with diamines to yield poly(amide-ureas)s or poly(amide-thiourea)s. However, a clean reaction proceeding

Table 6.2 Yields and properties of silylated ω-isocyanato carboxylic acids.

Formula	Yield (%)	b.p. (°C/mbar)	$n^{20}D$ (m.p.)	Ref.
$OCN-(CH_2)_2-CO_2SiMe_3$	75–80	56–57/1	1.4308	28
$OCN-(CH_2)_3-CO_2SiMe_3$	–	41–43/0.1	1.4348	23
$OCN-(CH_2)_4-CO_2SiMe_3$	64	75–77/0.5	1.4350	46
$OCN-(CH_2)_5-CO_2SiMe_3$	–	69–71/0.01	1.4384	28
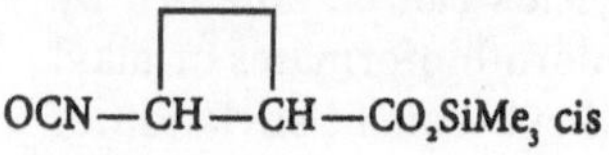 (cyclobutane) $OCN-CH-CH-CO_2SiMe_3$ cis	80–85	60–64/0.1	1.4490	24
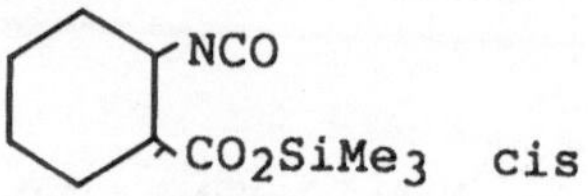 (cyclohexane) NCO / CO_2SiMe_3 cis	80–85	83–85/0.4	1.4581	21
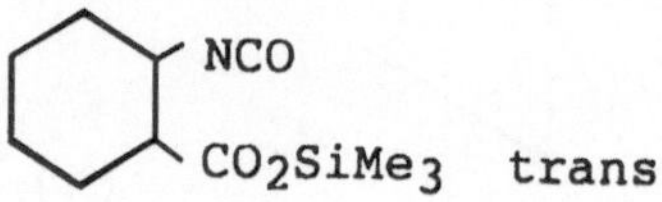 (cyclohexane) NCO / CO_2SiMe_3 trans	77	70–72/0.1	1.4536	24
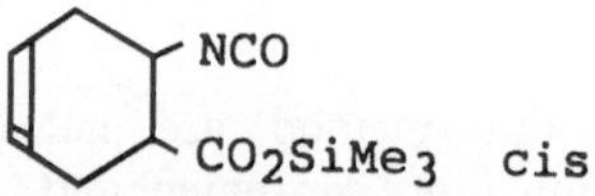 (cyclohexene) NCO / CO_2SiMe_3 cis	80–85	82–84/0.4	1.4666	21
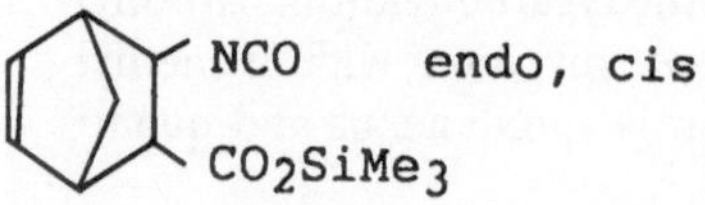 (norbornene) NCO endo, cis / CO_2SiMe_3	80	90–92/0.05	1.4825	24
$OCN-CH_2-CMe_2-CH_2CO_2SiMe_3$		54–56/0.1	1.4362	23
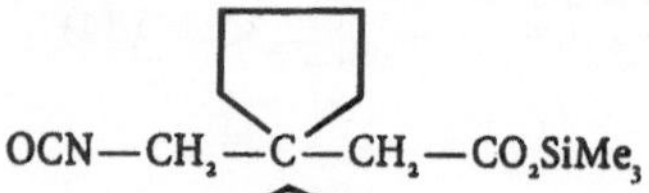 $OCN-CH_2-C$(cyclobutane)$-CH_2-CO_2SiMe_3$		74–76/0.06	1.4574	23
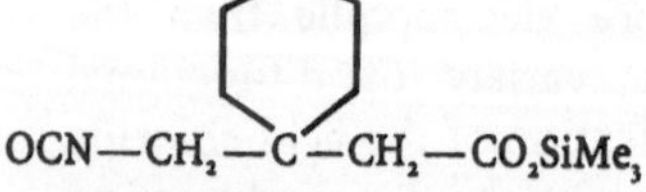 $OCN-CH_2-C$(cyclohexane)$-CH_2-CO_2SiMe_3$		100–102/0.02	1.4646	23
$OCN-CH_2-O-CH_2-CO_2SiMe_3$		72–74/0.1	1.4352	23
$OCN-CH_2-S-CH_2-CO_2SiMe_3$		68–70/0.1	1.4720	23
$OCN-CHMe-S-CHMe-CO_2SiMe_3$		88–90/0.01	1.4585	23
$OCN-CH_2-S-CMe_2-CO_2SiMe_3$		76–78/0.07	1.4640	23
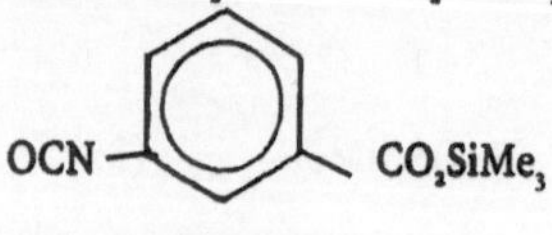 $OCN-$(benzene, meta)$-CO_2SiMe_3$		95–97/0.1	(26–28°C)	28
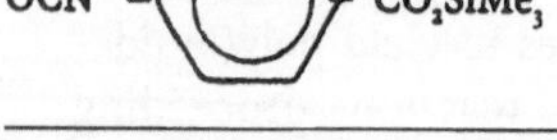 $OCN-$(cyclohexane)$-CO_2SiMe_3$		94–96/0.1	(48–50°C)	28

Formula	Yield (%)	b.p. (°C/mbar)	n^{20}D (m.p.)	Ref.
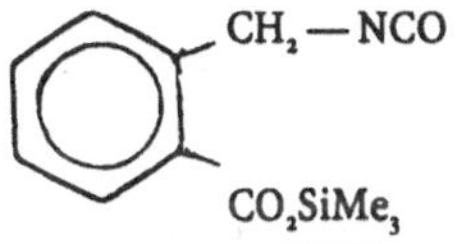 [structure: benzene ring with CH$_2$—NCO and CO$_2$SiMe$_3$ substituents]		106 – 108/0.1	1.5028	23
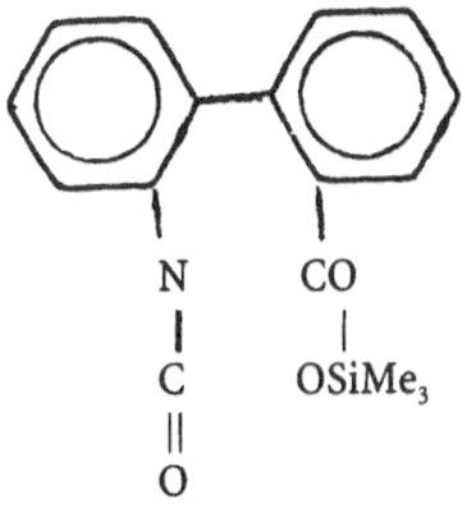 [structure: biphenyl with N–C=O and CO–OSiMe$_3$ groups]	81	114 – 116/0.001	–	46
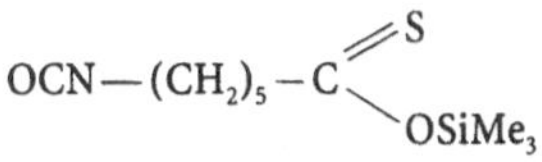 $OCN-(CH_2)_5-C{\overset{\displaystyle S}{\underset{\displaystyle OSiMe_3}{\Big<}}}$				20

$$2\ XCN\text{-}A\text{-}CO\text{-}Cl\ +\ (Me_3Si)_2S\ \longrightarrow\ (XCN\text{-}A\text{-}CO)_2S$$

$$(36)$$

$$X = O, S$$

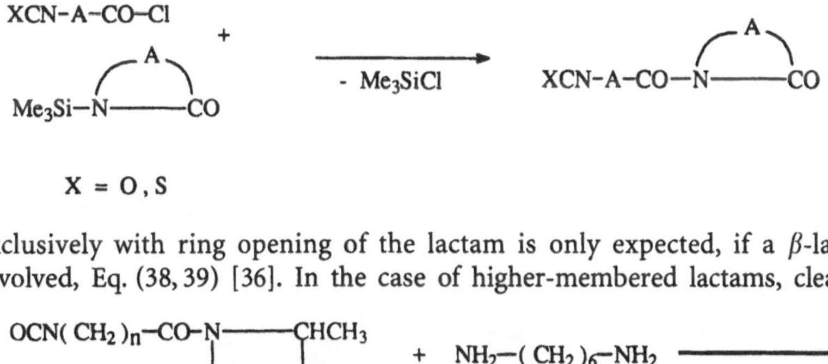

$$XCN\text{-}A\text{-}CO\text{-}Cl$$
$$+$$
[structure: $Me_3Si\text{-}N\underset{CO}{\overset{A}{\frown}}$] $\xrightarrow{\ -\ Me_3SiCl\ }$ $XCN\text{-}A\text{-}CO\text{-}N\underset{CO}{\overset{A}{\frown}}$

$$(37)$$

$$X = O, S$$

exclusively with ring opening of the lactam is only expected, if a β-lactam is involved, Eq. (38, 39) [36]. In the case of higher-membered lactams, cleavage at

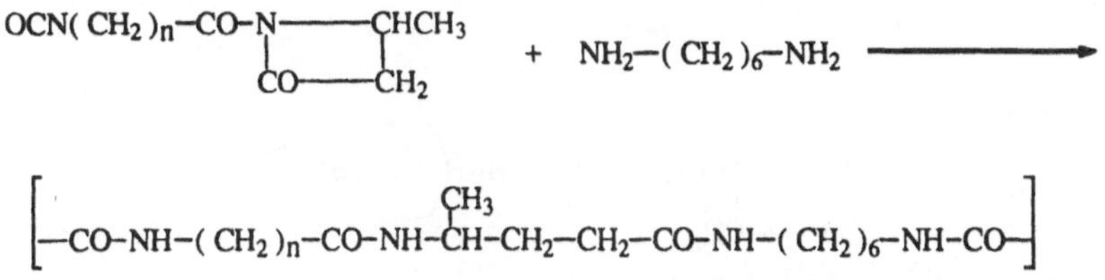

$$OCN(CH_2)_n\text{-}CO\text{-}N\underset{CO\quad\ CH_2}{\overset{\quad CHCH_3}{\big\lceil\quad\big\rceil}}\ +\ NH_2\text{-}(CH_2)_6\text{-}NH_2\ \longrightarrow$$

$$(38)$$

$$\left[-CO\text{-}NH\text{-}(CH_2)_n\text{-}CO\text{-}NH\text{-}\overset{\displaystyle CH_3}{\underset{}{CH}}\text{-}CH_2\text{-}CH_2\text{-}CO\text{-}NH\text{-}(CH_2)_6\text{-}NH\text{-}CO- \right]$$

the exocyclic carbonyl group with liberation of lactams is a competing reaction, Eq. (40). However, detailed studies of such a polycondensation have not yet been published. The reaction of N-(iso-isocyanatoacyl)-β-lactams with ω-amino alcohols may yield polymers with regular sequence of urea, amide and ester groups, Eq. (41) [36]. N-(ω-isocyanatoacyl) lactams react with alcohols under

Table 6.3 Yields and properties of various difunctional isocyanates.

Formula	Yield (%)	b.p. (°C/mbar)	n^{20}D (m.p.)	Ref.
OCN—(CH$_2$)$_2$—CO—Cl	85–95	76–78/12	1.4890	20
OCN—(CH$_2$)$_3$—CO—Cl	85–95	54–56/0.3	1.4690	20
OCN—(CH$_2$)$_4$—CO—Cl				
OCN—(CH$_2$)$_5$—CO—Cl	85–95	66–68/0.01	1.4625	20
OCN—C$_6$H$_4$—CO—Cl meta	85–95	74–76/0.1	1.5765	20
OCN—C$_6$H$_4$—CO—Cl para				

Formula	Yield (%)	b.p. (°C/mbar)	n^{20}D (m.p.)	Ref.
Me—CH—CH$_2$ | | OCN—(CH$_2$)$_2$—CO—N—CO	70–80	82–84/0.001	1.4896	35
Me—CH—CH$_2$ | | OCN—(CH$_2$)$_5$—CO—N—CO	70–80	124–128/0.001	1.4854	35
OCN—(CH$_2$)$_2$—CO—N(—(CH$_2$)$_3$—)CO	93	–	1.5018	36
OCN—(CH$_2$)$_3$—CO—N(—(CH$_2$)$_3$—)CO	72	–	1.4983	36
OCN—(CH$_2$)$_2$—CO—N(—(CH$_2$)$_4$—)CO	76	–	1.5040	36
OCN—(CH$_2$)$_2$—CO—N(—(CH$_2$)$_5$—)CO	73	–	1.5032	36
OCN—(CH$_2$)$_3$—CO—N(—(CH$_2$)$_5$—)CO	67	–	1.5002	36

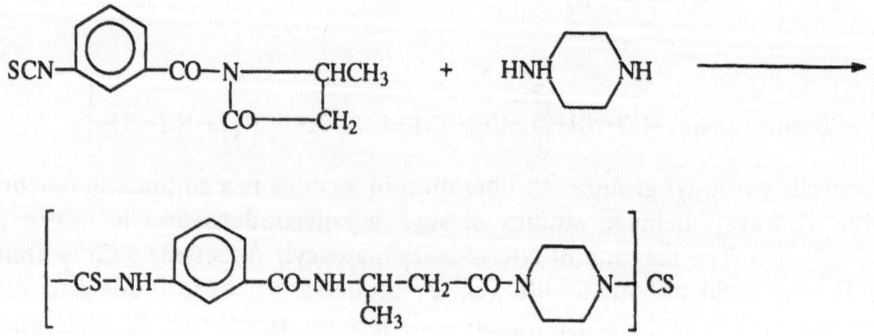

$$(39)$$

Table 6.4   Yield and properties of difunctional isothiocyanates.

Formula	Yield (%)	b.p. (°C/mbar)	$n^{20}D$ (m.p.)	Ref.
SCN—CH$_2$—CO$_2$SiMe$_3$	70—76	70—72/2	1.4880	10
Me | SCN—CH—CO$_2$SiMe$_3$ D, L	75—81	55—57/0.05	1.4772	10
CH—Me$_2$ | SCN—CH—CO$_2$SiMe$_3$ D, L	78—90	83—85/0.3	1.4778	10
CHMe$_2$ | CH$_2$ | SCN—CH—CO$_2$SiMe$_3$ D, L	67—80	75—78/0.001	1.4748	10
Phe | CH$_2$ | SCN—CH—CO$_2$SiMe$_3$ D, L	78—85	122—125/0.001	1.5300	10
SCN—(CH$_2$)$_2$—CO$_2$SiMe$_3$	70—82	86—88/1	1.4846	10
SCN—(CH$_2$)$_3$—CO$_2$SiMe$_3$	82—83	90—93/0.2	1.4770	10
SCN—(CH$_2$)$_5$—CO$_2$SiMe$_3$	74—84	95—89/0.001	1.4832	10
SCN—C$_6$H$_4$—CO$_2$SiMe$_3$ meta	83—86	107—110/0.001	(36—38 °C)	10
SCN—C$_6$H$_4$—CO$_2$SiMe$_3$ para	70—78	—	(106—108 °C)	10
SCN—(CH$_2$)$_2$—COCl	85—95	57—59/008	1.5516	10
SCN—(CH$_2$)$_3$—CO—Cl	95—95	75—77/0.001	1.5444	10
SCN—(CH$_2$)$_3$—CO—Cl	72	98—100/0.01	1.5343	11
SCN—(CH$_2$)$_5$—CO—Cl	85—95	92—94/0.001	1.5275	10
	69	—	1.5188	11
SCN—C$_6$H$_4$—CO—Cl meta	92	90—92/0.01	1.6675	20
(SCN—C$_6$H$_4$—CO)$_2$S meta	91	—	(103—105 °C)	20
MeCH—CH$_2$ | | SCN—(CH$_2$)$_2$—CO—N—CO	70—80	105—107/0.001	1.5458	35
MeCH—CH$_2$ | | SCN—(CH$_2$)$_5$—CO—N—CO	70—80	140—143/0.001	1.5286	35
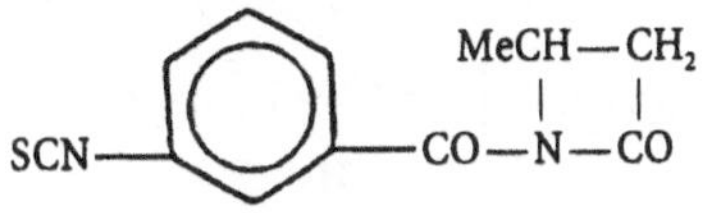 SCN—⬡—CO—N—CO with MeCH—CH$_2$ bridge	70—80	155—160/0.001	72—74 °C	35
(CH$_2$)$_3$ SCN—(CH$_2$)$_2$—CO—N——CO	77	—	54—56 °C	36

Table 6.4   (*Contd.*)

Formula	Yield (%)	b.p. (°C/mbar)	n²⁰D (m.p.)	Ref.
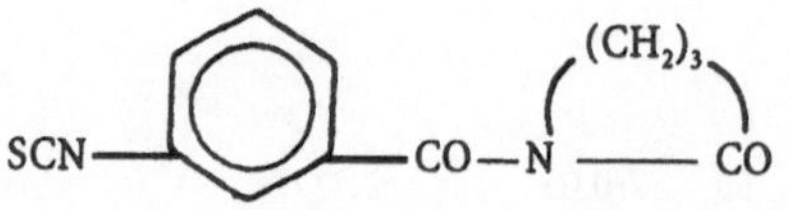 SCN—C₆H₄—CO—N⟨(CH₂)₃⟩CO	55	–	90–92 °C	36
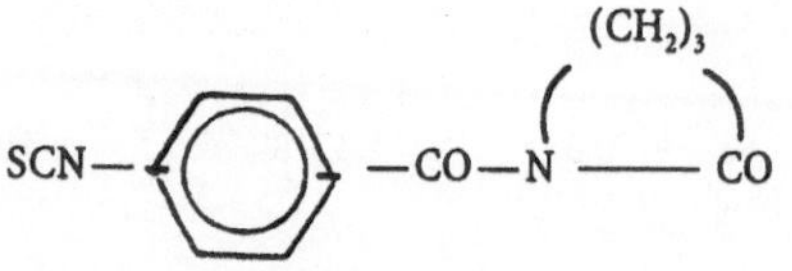 SCN—C₆H₁₀—CO—N⟨(CH₂)₃⟩CO	89	–	115–117 °C	36

$$R\text{-CO-N}\langle\text{(CH}_2)_n\rangle\text{CO} \;+\; NH_2\text{-R'} \longrightarrow R\text{-CO-NH-R'} \;+\; HN\langle\text{(CH}_2)_n\rangle\text{CO} \tag{40}$$

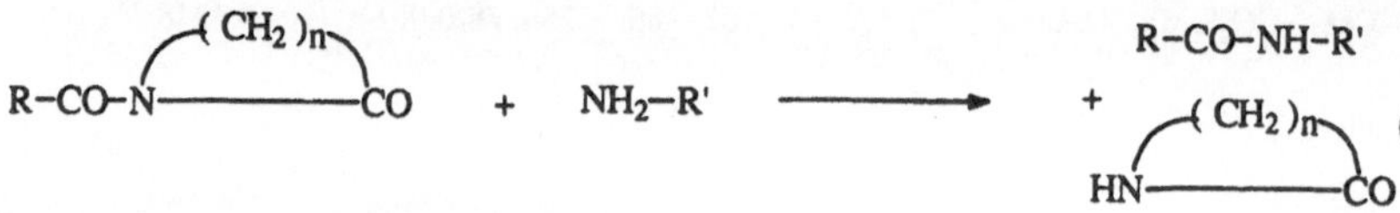

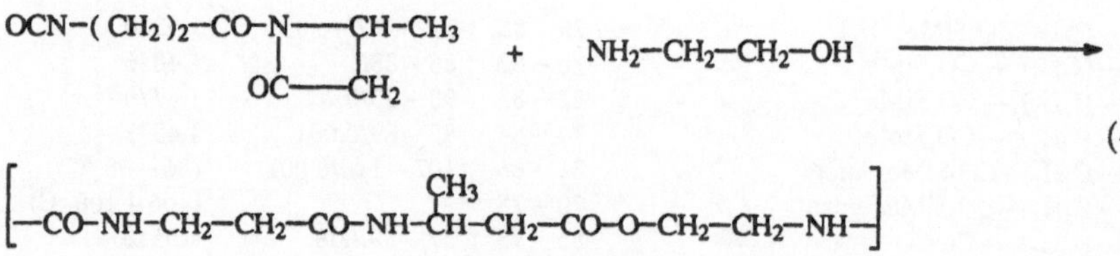

$$OCN\text{-(CH}_2)_2\text{-CO-N}\langle\text{CH-CH}_3 / \text{OC-CH}_2\rangle \;+\; NH_2\text{-CH}_2\text{-CH}_2\text{-OH} \longrightarrow \tag{41}$$

$$\left[\text{-CO-NH-CH}_2\text{-CH}_2\text{-CO-NH-}\overset{\text{CH}_3}{\text{CH}}\text{-CH}_2\text{-CO-O-CH}_2\text{-CH}_2\text{-NH-} \right]$$

mild conditions exclusively under formation of urethane groups, Eq. (42). When suitable alcohols are used such as *tert*-butanol, acidic cleavage of the urethane groups yields *N*-(ω-aminoacyl) lactams, Eq. (43) [38]. When the amino group is set free, alternating polyamides may be formed in the case of β-lactams, but this has not been reported so far, Eq. (44).

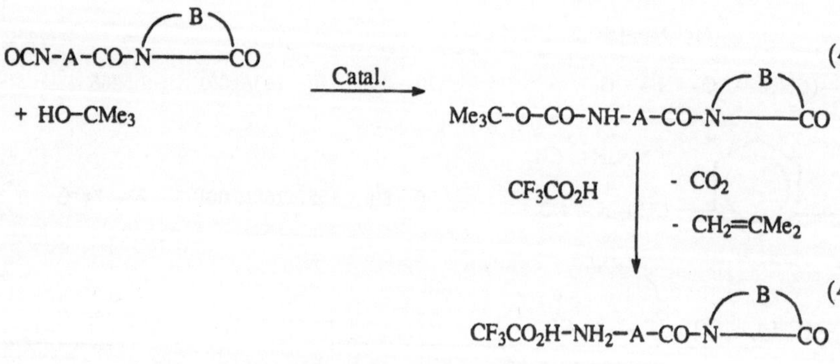

$$OCN\text{-A-CO-N}\langle B\rangle\text{CO} \;+\; HO\text{-CMe}_3 \xrightarrow{\text{Catal.}} Me_3C\text{-O-CO-NH-A-CO-N}\langle B\rangle\text{CO} \tag{42}$$

$$\downarrow CF_3CO_2H \quad \begin{array}{l} - CO_2 \\ - CH_2=CMe_2 \end{array}$$

$$CF_3CO_2H\text{-NH}_2\text{-A-CO-N}\langle B\rangle\text{CO} \tag{43}$$

$$\downarrow \quad \text{- CF}_3\text{CO}_2\text{H}$$

$$(44)$$

$$\left[-\text{NH-A-CO-NH-B-CO}-\right]$$

$$\begin{array}{l} \text{OCN}-(\text{CH}_2)_n-\text{CO-Cl} \\ \qquad\qquad\qquad\qquad + \\ \text{Me}_3\text{Si}-\text{NH}-(\text{A})-\text{CO-Cl} \end{array} \xrightarrow[\text{- Me}_3\text{SiCl}]{} \text{OCN}-(\text{CH}_2)_n-\text{CO-NH}-(\text{A})-\text{CO}_2\text{SiMe}_3$$

$$(45)$$

ω-Isocyanatoacylchlorides react easily with silylated amino acids, Eq. (45), [39, 40] and the resulting silylesters can be hydrolyzed to N-(ω-isocyanatoacyl) amino acids, Eq. (46). Yet their polycondensation is more or less affected by side reactions, such as formation of cyclic and linear ureas (see next section). More satisfactory is the synthesis of N, N'-bis(ω-isocyanatoacyl) pierazines and their polyaddition to polyureas, Eq. (47) or polyurethanes, (Form. 48) [41].

$$+ \text{H}_2\text{O} \quad\Bigg|\quad \text{- Me}_3\text{SiOH}$$

$$(46)$$

$$\text{OCN}-(\text{CH}_2)_n-\text{CO-NH}-(\text{A})-\text{CO}_2\text{H}$$

$$\text{OCN}-(\text{CH}_2)_n-\text{CO-N}\bigcirc\text{N-CO}-(\text{CH}_2)_n-\text{NCO} \quad + \quad \text{HN}\bigcirc\text{NH} \longrightarrow$$

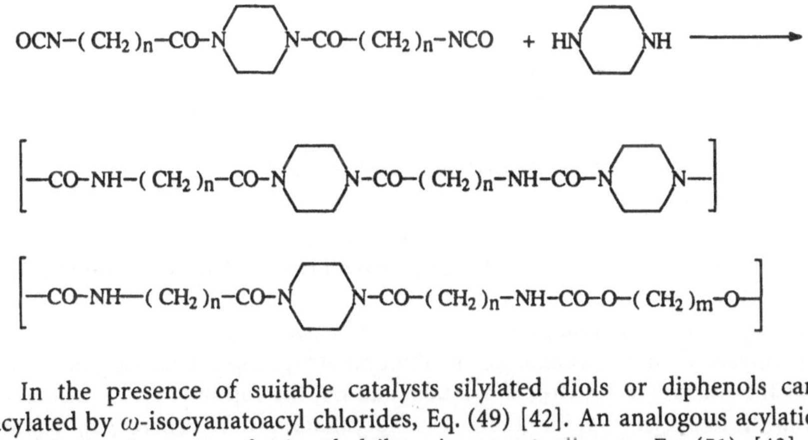

$$(47)$$

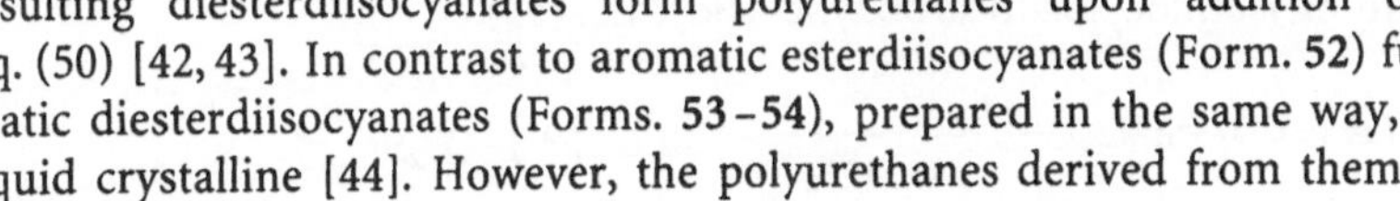

$$\left[-\text{CO-NH}-(\text{CH}_2)_n-\text{CO-N}\bigcirc\text{N-CO}-(\text{CH}_2)_n-\text{NH-CO-N}\bigcirc\text{N}-\right]$$

$$\left[-\text{CO-NH}-(\text{CH}_2)_n-\text{CO-N}\bigcirc\text{N-CO}-(\text{CH}_2)_n-\text{NH-CO-O}-(\text{CH}_2)_m-\text{O}-\right] \quad (48)$$

In the presence of suitable catalysts silylated diols or diphenols can be acylated by ω-isocyanatoacyl chlorides, Eq. (49) [42]. An analogous acylation is feasible in the case of trimethylsiloxy-isocyanatoalkanes, Eq. (51) [43]. The resulting diesterdiisocyanates form polyurethanes upon addition of diols, Eq. (50) [42, 43]. In contrast to aromatic esterdiisocyanates (Form. 52) fully aromatic diesterdiisocyanates (Forms. 53–54), prepared in the same way, may be liquid crystalline [44]. However, the polyurethanes derived from them are not thermostable enough to form a stable liquid-crystalline melt.

$$2\ \text{OCN}-(\text{CH}_2)_n-\text{CO-Cl} \quad + \quad \text{Me}_3\text{Si}-\text{O}-(\text{A})-\text{O-SiMe}_3 \xrightarrow[\text{- 2 Me}_3\text{SiCl}]{} \quad (49)$$

$$OCN-(CH_2)_n-CO-O-(A)-O-CO-(CH_2)_n-NCO$$

$$\downarrow \quad + \quad HO-(B)-OH \tag{50}$$

$$[-OC-NH-(CH_2)_n-CO-O-(A)-O-CO-(CH_2)_n-NH-CO-O-(B)-O-]$$

$$OCN-(A)-OSiMe_3 \quad + \quad Cl-CO-(CH_2)_n-NCO \xrightarrow{\quad} $$
$$- Me_3SiCl \tag{51}$$

$$OCN-(A)-O-CO-(CH_2)_n-NCO$$

OCN—⬡—O-CO—⬡—NCO $\qquad$ (52)

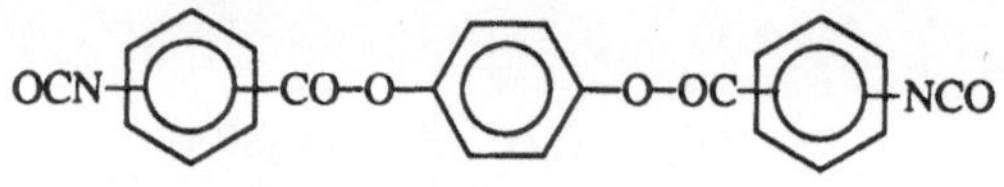

OCN—⬡—CO-O—⬡—O-OC—⬡—NCO $\qquad$ (53)

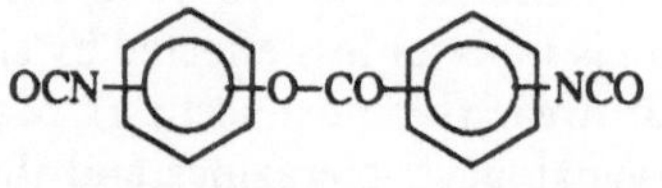

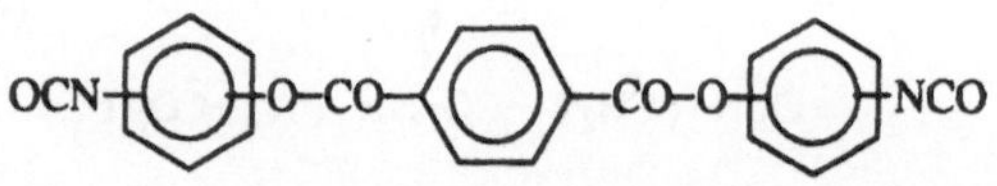

OCN—⬡—O-CO—⬡—CO-O—⬡—NCO $\qquad$ (54)

6.2
Syntheses of *N*-Carboxyanhydrides, Polypeptides and Polyamides

The trimethylsilyl isocyanato- and isothiocyanatocarboxylates mentioned above (Eqs. 26, 30–33) are useful reaction intermediates not only for the preparation of the corresponding acid chlorides. The silylated carboxylgroup is more sensitive to hydrolysis than isocyanate or isothiocyanate groups, because the water attacks the silicon (via the free *d*-orbitals) and not the carbonyl group, which is, of course, less electrophilic than the isocyanate group. Therefore, careful hydrolysis of the silylester group with neutral or slightly acidic water yields free isocyanato or isothiocyanatocarboxylic acids, Eq. (55).

$$XCN-(A)-CO_2-SiMe_3 \xrightarrow[- Me_3SiOH]{+ H_2O} XCN-(A)-CO_2H \tag{55}$$

$$X = O, S$$

The isocyanatocarboxylic acids are not stable above $0\,°C$ and, depending on their structure, either undergo cyclization, Eqs. (56, 57) [23, 24, 25] or polyaddi-

$$OCN-(CHR)_n-CO_2H \longrightarrow \text{[structure: HN——(CHR)$_n$ ring with OC, CO, O]} \tag{56}$$

$$n = 1, 2, 3$$

$$\text{[structure: benzene with CO$_2$H and NCO]} \longrightarrow \text{[structure: benzene fused ring with CO-O, CO, NH]} \tag{57}$$

tion, Eqs. (58, 59) [46, 47]. The oligomeric poly (N-carboxyanhydride)s which precipitate from the reaction mixture spontaneously decarboxylate in the course

$$OCN-(CH_2)_n-CO_2H \longrightarrow [-OC-NH-(CH_2)_n-CO-O-] \tag{58}$$

$$n = 4, 5$$

of several hours or days (depending on temperature and structure) yielding oligoamides, Eq. (60) [46, 47].

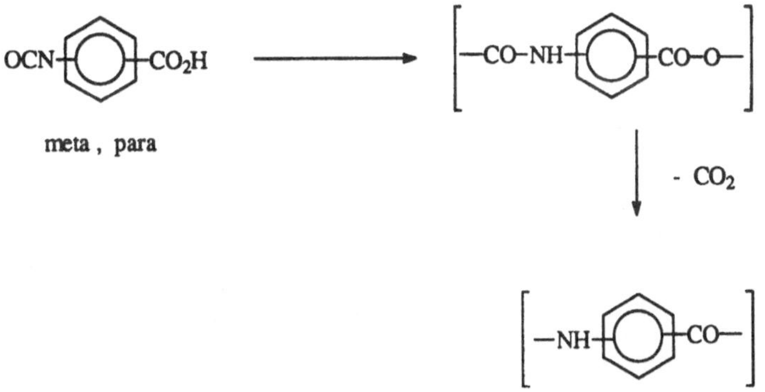

$$OCN-\text{[benzene]}-CO_2H \longrightarrow [-CO-NH-\text{[benzene]}-CO-O-] \tag{59}$$

meta, para

$$\downarrow -CO_2$$

$$[-NH-\text{[benzene]}-CO-] \tag{60}$$

In the case of cyclic aliphatic N-carboxyanhydrides polymerization can be initiated by basic catalysts, such as tertiary amines, or by nucleophiles, such as water or primary amines, Eq. (60) [48]. The five-membered rings yield polypeptides with moderate or high molecular weights, Eq. (61) [48–50]. The six-

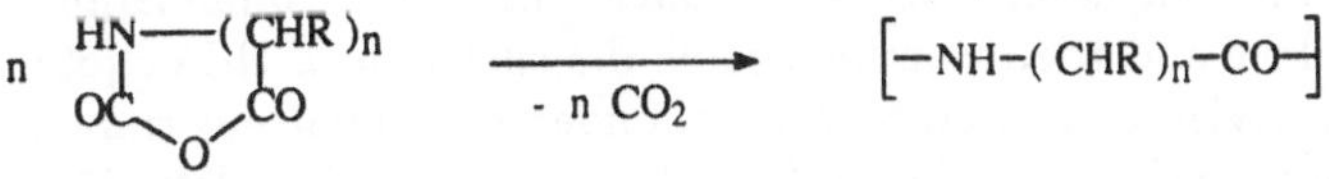

$$n \, \text{[structure: HN——(CHR)$_n$ ring with OC, CO, O]} \xrightarrow{-n\,CO_2} [-NH-(CHR)_n-CO-] \tag{61}$$

$$n = 1, 2, 3$$

membered rings (n = 2) yield low molecular weight polyamides (Nylon-3 type) (DP ⩽ 60) regardless of reaction conditions [51, 52] and the seven-membered rings (n = 3) only yield oligoamides [23].

Because the five-membered cyclic N-carboxyanhydrides (NCAs; oxazolidine-2,5-diones) are an important class of monomers for the preparation of polypeptides the synthesis, properties and polymerization was the object of numerous studies. Since up-to-date reviews of this topic are available [48–50] only syntheses based on silicon reagents need to be mentioned in this work. Phosgenation of N, O-bissilylated amino acids is a smooth reaction which leads to the desired NCA's in high yields and without significant side reactions in contrast to the phosgenation of free amino acids, Eq. (62) [53]. A particular advantage of

$$\text{Me}_3\text{Si}-\text{NH}-\text{CHR}-\text{CO}_2-\text{SiMe}_3 \xrightarrow[\ +\ \text{COCl}_2\ ]{} \begin{array}{c} \text{HN}\underline{\quad\quad}\text{CHR} \\ \text{OC} \qquad \text{CO} \\ \diagdown\text{O}\diagup \end{array} \tag{62}$$

this method is that silylation of trifunctional amino acids, such as serine, threonine, or tyrosine provides a protection of the side chain which stable enough during phosgenation, but easy to remove after the synthesis of an oligo- or polypeptide [53–55]. However, the silylated amino acids are extremely sensitive to hydrolysis and difficult to distill or to store. Treatment of silylated amino acids with carbondioxide followed by thionylchloride, Eqs. (63–65) is

$$\text{Me}_3\text{Si}-\text{NH}-\text{CHR}-\text{CO}_2-\text{SiMe}_3 \xrightarrow[]{\ +\ \text{CO}_2\ } \text{Me}_3\text{SiOCO}-\text{NH}-\text{CHR}-\text{CO}_2-\text{SiMe}_3 \tag{63}$$

$$\text{R}'-\text{O}-\text{CO}-\text{NH}-\text{CHR}-\text{CO}_2-\text{SiMe}_3 \xrightarrow[-\ \text{Me}_3\text{SiCl}]{\ +\ \text{SOCl}_2\ } \text{R}'-\text{O}-\text{CO}-\text{NH}-\text{CHR}-\text{CO}-\text{Cl} \tag{64}$$

$$\Big\downarrow {\scriptstyle -\ \text{R}'-\text{Cl}}$$

$$\begin{array}{c} \text{HN}\underline{\quad\quad}\text{CHR} \\ \text{OC} \qquad \text{CO} \\ \diagdown\text{O}\diagup \end{array} \tag{65}$$

another approach without significant advantages [56]. However, treatment of silylated N-methoxycarbonyl- or N-benzyloxycarbonyl amino acids with SOCl$_2$ or PBr$_3$ is a useful method for the preparation of NCA's Eqs. (64, 65, $\text{R}' = \text{Me}$, Bzl) because the formation of HCl or HBr which can cleave the NCA's, is avoided [57]. This method seems to be particularily useful for the synthesis of larger quantities of NCA's ($\geqslant 0{,}5$ mol) [48, 50].

Non-nucleophilic silylating reagents such as the combination of chlorotrimethylsilane and triethylamine enable the silylation of NCA's [58]. N-Silylated six- and seven-membered NCA rings undergo spontaneous ring-opening yielding quantitatively trimethylsilyl β- or γ-isocyanato carboxylates [23, 24]. In the case of five-membered NCA's an equilibrium of both isomers is established close to a molar ratio of 1:1 Eqs. (66 and Fig. 6.1) [58]. However, this

$$\begin{array}{c} \text{Me}_3\text{Si}-\text{N}\underline{\quad\quad}\text{CHR} \\ \text{OC} \qquad \text{CO} \\ \diagdown\text{O}\diagup \end{array} \xrightarrow[\quad]{\ \Delta\text{T}\ } \begin{array}{c} \text{N}-\text{CHR}-\text{CO}_2-\text{SiMe}_3 \\ \| \\ \text{C} \\ \| \\ \text{O} \end{array} \tag{66}$$

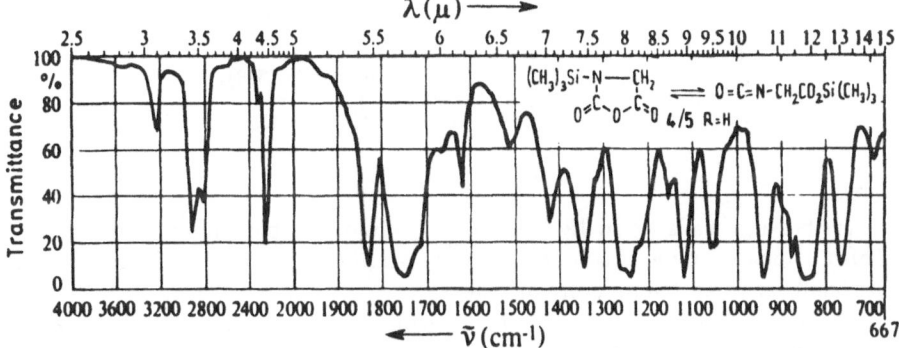

Fig. 6.1 IR-spectrum of the reaction mixture obtained immediately after silylation of glycine-*N*-carboxyanhydride with chlorotrimethylsilane and triethylamine "Isocyanate band" at 4.5 μm and "anhydridebands" at 5.6 and 6.3 μm

mixture is not stable because two different chain growth reactions may occur simultaneously and independently, Eqs. (67, 68). The result are copolymers con-

$$(67)$$

$$(68)$$

taining short blocks of nylon-2 (peptide) and nylon-1 (polyisocyanate) repeating units [58].

ω-Isothiocyanatocarboxylic acids liberated from the corresponding silylesters by hydrolysis with neutral water, Eq. (55) can undergo spontaneous cylization when the distance between both functional groups is short (i.e. α-, or β-position, Eq. (69) [59–62]. Nonetheless isothiocyanatocarboxylic acids differ from iso-

$$SCN-(CHR)_n-CO_2H \quad\rightleftharpoons\quad \qquad (69)$$

cyanatocarboxylic acids in several aspects. First, γ-isothiocyanato (and higher ω-isothiocyanato) carboxylic acids do not cyclize to an appreciable extent [59]. Second, the cyclization of α- and β-isothiocyanatocarboxylic acid in solution is

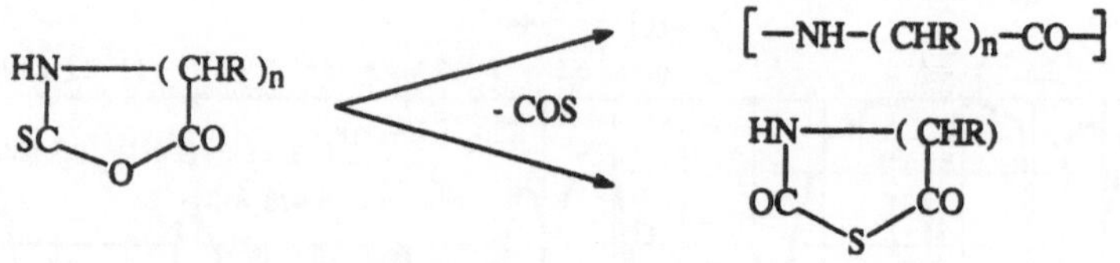

$$\begin{array}{c} HN\text{——}(CHR)_n \\ SC \diagdown_{O}\diagup CO \end{array} \diagup^{\nearrow \qquad [-NH-(CHR)_n-CO-]} \quad (70)$$

$$- COS \searrow \begin{array}{c} HN\text{——}(CHR) \\ OC\diagdown_{S}\diagup CO \end{array} \quad (71)$$

reversible and leads to an IR-spectroscopically detectable equilibrium [34, 60]. Third, aliphatic and aromatic isothiocyanatocarboxylic acids that do not cyclize are stable at room temperature or below.

The cyclization products, 2-thioxo-oxazolidones-5 and 2-thioxodihydro-1,3-oxazinones-6, are not stable at room temperature and polymerize, Eq. (70). Because of side reactions, such as isomerization, Eq. (71), the molecular weights of the resulting polypeptided and polyamides are not high [34, 61].

Aliphatic and aromatic isothiocyanatocarboxylic acids, even though stable at room temperature, undergo polycondensation at, higher temperatures yielding polyamides along with carbon oxysulfides, Eq. (72). This process is catalyzed by

$$SCN-(A)-CO_2H \xrightarrow[-COS]{\Delta \, T \, or \, NR_3} -[-NH-(A)-CO-]-$$

$$(72)$$

$$(A) = (CH_2)_m;$$

bases, in particular by tertiary amines [34]. This process is of interest for the synthesis of polyamides with an alternating sequence of aliphatic and/or aromatic amino acids, Eqs. (73-75). The synthesis of the corresponding monomers

$$\overset{\ominus}{Cl}\left[\overset{\oplus}{NH_3}-(A)-CO_2SiMe_3\right] \xrightarrow[-\,2\,HCl]{} SCN-(A)-CO-NH-(B)-CO_2SiMe_3$$

$$+ \quad SCN-(B)-CO-Cl \hspace{5.5cm} (73)$$

$$\Big\downarrow$$

$$SCN-(A)-CO-NH-(B)-CO_2H$$

$$\Delta T \Big\downarrow \;\; - COS \hspace{4cm} (74)$$

$$\Big\downarrow$$

$$[-NH-(A)-CO-NH-(B)-CO-] \quad (75)$$

starts with the conversion of ω-amino acids to their silylester hydrochlorides [62] followed by acylation with ω-isothiocyanatocarboxylic acid chlorides, Eq. (74) [63-65]. Both reactions and the hydrolysis of the silylester group may be conducted as an "one-pot-procedure" [63-65]. In addition to polyamides of ω-amino acids, alternating polyamides of sulfanilic acid (Form. 76) [66] and

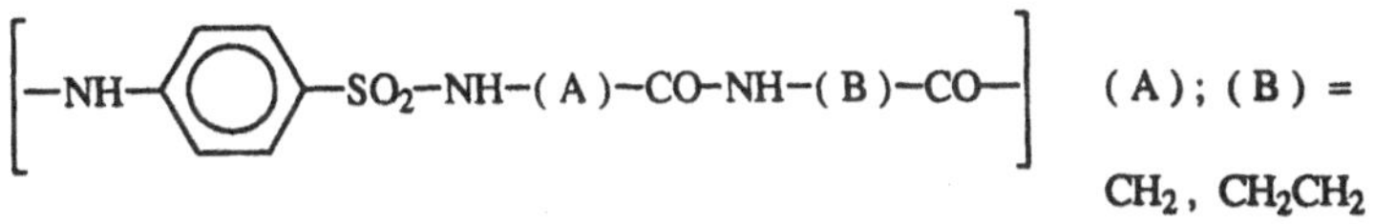

$$\left[-NH-\!\!\left\langle\!\!\bigcirc\!\!\right\rangle\!\!-SO_2-NH-(A)-CO-NH-(B)-CO- \right] \quad (A);\,(B) = \tag{76}$$

$$CH_2,\ CH_2CH_2$$

polydepsipeptides of hydroxy- or mercaptoacetic acid (Forms. 77 and 78) were synthesized in this way [67]. These results demonstrate the main advantage of

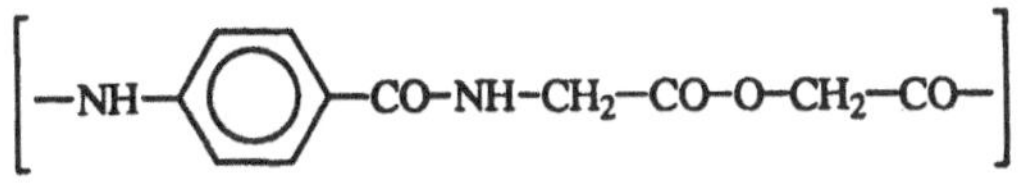

$$\left[-NH-\!\!\left\langle\!\!\bigcirc\!\!\right\rangle\!\!-CO-NH-CH_2-CO-O-CH_2-CO- \right] \tag{77}$$

$$\left[-NH-\!\!\left\langle\!\!\bigcirc\!\!\right\rangle\!\!-CO-O-CH_2-CO-NH-CH_2-CO- \right] \tag{78}$$

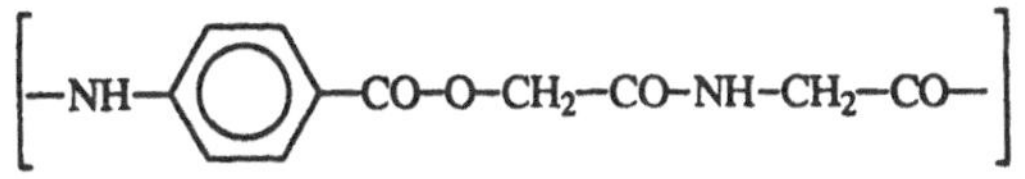

this strategy, namely its versatility. A characteristic short-coming are the relatively low molecular weights (< 10000 g/mol). The existence of side reactions causing termination steps is indicated by the presence of sulfur (typically $0,5$–$2,0$ weight %). A side reaction which is typical for α- and β-isothiocyanatoacyl amides is cyclization yielding thiohydantoins or 2-thioxoperhydro-1,3-diazine-6-ones (Eq. 79, X = S). The tendency of cyclization is even more pronounced in the case of α- or β-isocyanatoacyl amides (Eq. 79, X = O) [68,69]. However, cyclization may be avoided, when the isothiocyanato acyl group is attached to an imino acids, such as sarcosine, (Eq. 80) [70] or

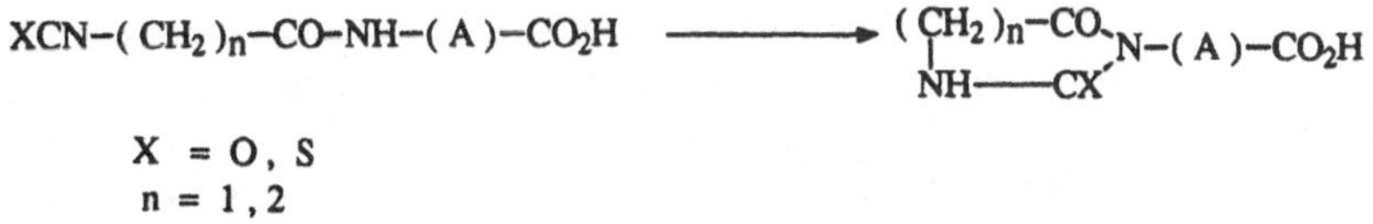

$$XCN-(CH_2)_n-CO-NH-(A)-CO_2H \longrightarrow \begin{array}{c}(CH_2)_n-CO\\ | \quad\quad\quad\ \\ NH\!-\!\!-CX\end{array}\!\!N-(A)-CO_2H \tag{79}$$

$$X = O,\ S$$
$$n = 1,2$$

$$SCN-(CH_2)_2\!-\!CO-\overset{\overset{\displaystyle CH_3}{|}}{N}-CH_2-CO-NH-CH_2-CO_2H \xrightarrow{\ -\,COS\ } \tag{80}$$

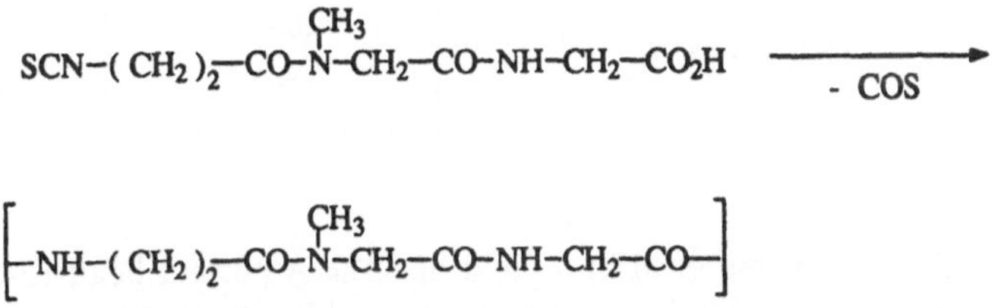

$$\left[-NH-(CH_2)_2\!-\!CO-\overset{\overset{\displaystyle CH_3}{|}}{N}-CH_2-CO-NH-CH_2-CO- \right]$$

proline. The alternating polypeptides and polyamides have served as useful models for studies of thermal degradation [71,72] or for ^{13}C- and ^{15}N NMR sequence analyses [63–65,73–75].

6.3
Silicon in Ring-Opening Polymerization

The role of silicon containing compounds in the field of ring-opening polymerization is twofold. On the one hand, numerous silicon-containing cyclic monomers were synthesized and polymerized, on the other hand, silicon-containing compounds play an increasing role as initiators.

6.3.1
Silicon Containing Monomers

Almost all Si-containing cyclic monomers belong to the following five classes of compounds.

1) Cyclic siloxanes including heterocycles with at least one Si—O—Si group.
2) Cyclosilazanes
3) Cyclic olefins
4) Silacycloalkanes
5) Heterocycles, containing Si—Si bonds.

Syntheses and polymerization of cyclic siloxanes were described in Chapter 3, cyclosilazanes in Chapter 4 and heterocycles containing Si—S bond were not studied in detail. Therefore this section will concentrate on monomers of classes 3–5.

Cyclic olefins containing silicon may be subdivided into two quite different groups:

I) Cyclic olefins with attached silyl groups (e.g. Formulas 81a, 82a, 83a–c, 87a,b, 89a–c, 91a,b).
II) Heterocyclic olefins with one or two Si-atoms as ring members (e.g. Formulas 95a–c, 96a–d, 98a–c, 99a,b).

Monomers of group I and their metathesis ring-opening polymerization (ROP) will be discussed first. Whereas, most Si-substituted cycloolefins are norbornene derivatives, only one smaller ring, namely 1-trimethylsilyl cyclobutene was synthesizede and polymerized, Eq. (81) [76]. Caseys catalyst yielded after 48 h at

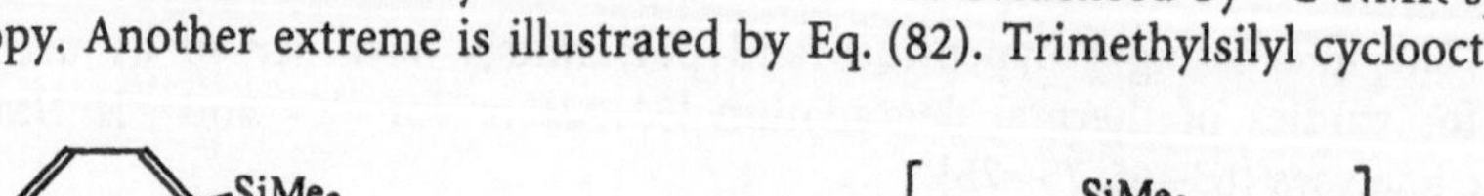

$$\text{(81)}$$

39 °C a high molecular weight material in a yield of 80%. A perfect head-to-tail structure and exclusively *cis* double bonds was evidenced by ^{13}C-NMR spectroscopy. Another extreme is illustrated by Eq. (82). Trimethylsilyl cyclooctatetrene

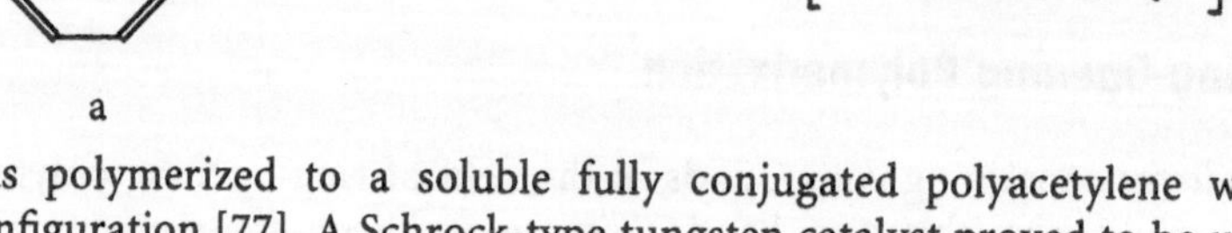

$$\text{(82)}$$

was polymerized to a soluble fully conjugated polyacetylene with all-*trans* configuration [77]. A Schrock-type tungsten catalyst proved to be useful for this purpose. The poly(TMS-acetylene) was characterized by UV-spectroscopy and DSC measurements.

Functionalized norbornenes, such as **83a–c** and **84a** were mainly used as

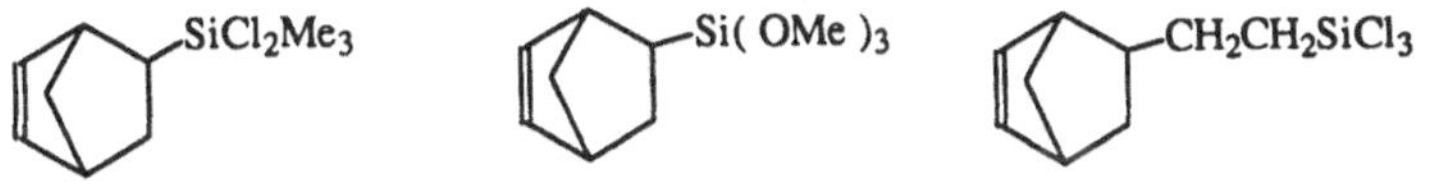

$$\text{(83)}$$

comonomers in combination with cyclopentene or 1, 5-cyclooctadiene [78–80]. The copolymerizations were usually conducted with WCl_6 based catalysts (Eq. 84). The purpose of such copolymerizations is the introduction of functional groups which are sensitive to hydrolysis and allow controlled crosslinking Eqs. (85, 86).

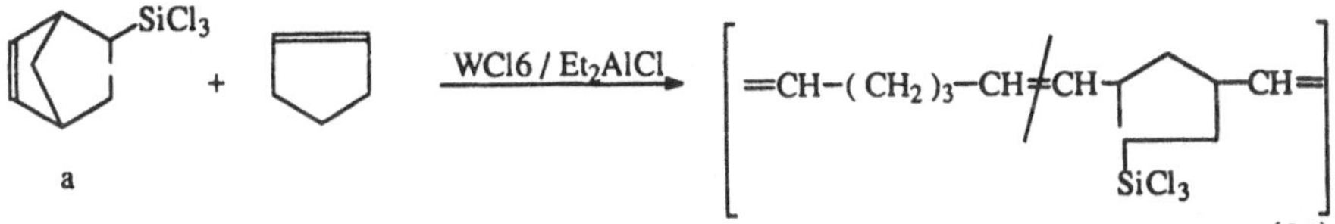

$$\text{(84)}$$

$$\text{(85)}$$

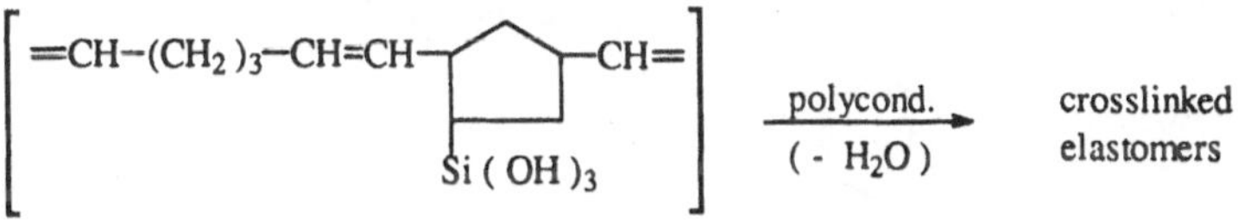

$$\text{(86)}$$

This approach enables the synthesis of colorless elastomers with a variable, easy to control number of crosslinks. In addition to norbornene derivatives larger cycles **87a** and **b** were synthesized and proposed for the same application.

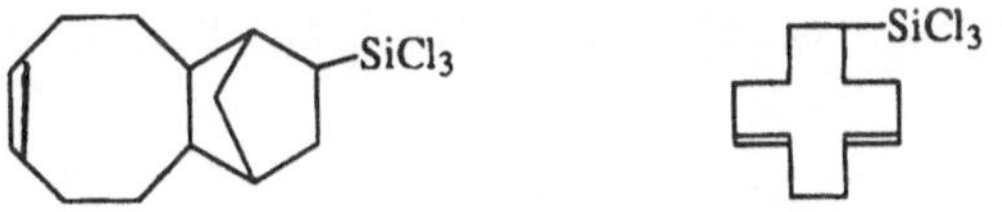

$$\text{(87)}$$

Another approach leading to elastomers via hydrolysis of di- or trifunctional silyl groups was described by the same research group. Cycloolefins were polymerized by metathesis ROP in the presence of the vinylsilanes **88a–c** [80].

$$CH_2{=}CH{-}SiCl_3 \qquad CH_2{=}CH{-}SiCl_2Me \qquad CH_2{=}CH{-}Si(OMe)_3$$

$$\text{(88)}$$

a   b   c

These vinylsilanes react as "chain stoppers" and yield multifunctional endgroups (see also the last reaction of this section).

Substituted norbornenes **89a–c** were also studied by a russian research group

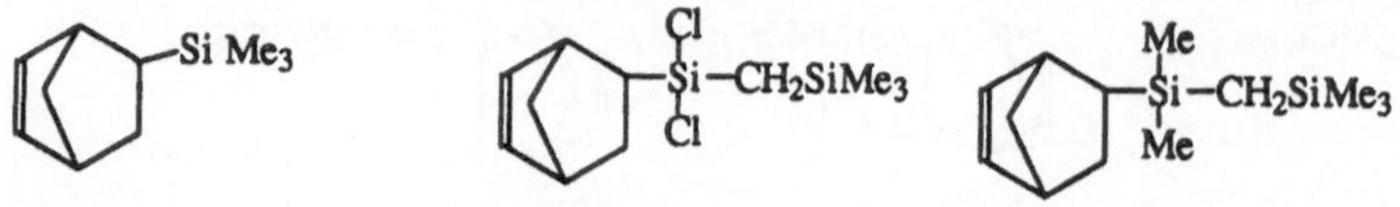

$$(89)$$

[81]. These monomers like other Si-substituted norbornenes were synthesized by a "Diels-Alder type" addition of vinylsilanes onto cyclopentadiene, Eq. (90).

$$(90)$$

$$X = Cl \, , Me \, , OMe$$

The polymerizations of monomers **89a** and **c** were conducted either with a homogeneous catalysts based on WCl_6 or with heterogeneous catalysts based on Re_2O_7/Al_2O_3. Higher yields and viscosities were obtained by homogeneous catalysts. The resulting polymers were characterized by IR-, ^1H and ^{13}C NMR spectroscopy. Membranes were cast from toluene solutions and various properties such as the permeability and diffusion coefficients of numerous gases were evaluated.

Two Si-containing norbornadienes **91a, b** were reported to polymerize exclus-

$$(91)$$

ively at the unsubstituted double bond, when exposed to WCl_6/Me_4Sn in toluene [82, 83]. The structural analyses by ^1H and ^{13}C NMR spectroscopy revealed that all four combination of *cis* and *trans* double bonds (**92a, b** and **93a, b**) were

$$(92)$$

$$(93)$$

(94)

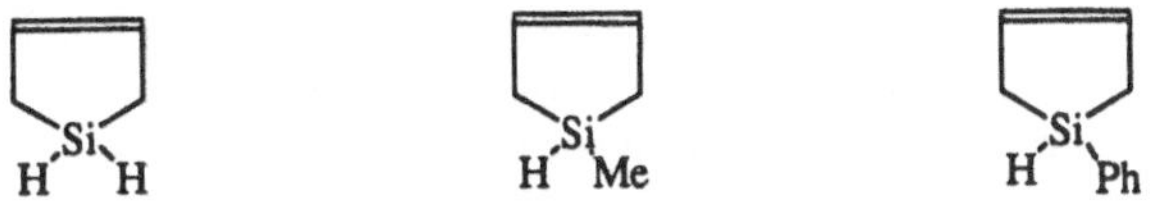

(95)

formed at nearly equal rates. The exotic monomer **94a** was polymerized either in bulk or in toluene solution by means of a "Schrock-catalyst" based on molybdenum [84]. In solution a M_w of 13200 was obtained, but a MW of 78200 in bulk.

Cyclic olefins containing a Si-atom as ring member (e.g. **95a–c**, **96a–d**, **98a–c** and **99a**) [85–93] may be polymerized by metathesis catalysts under cleavage of

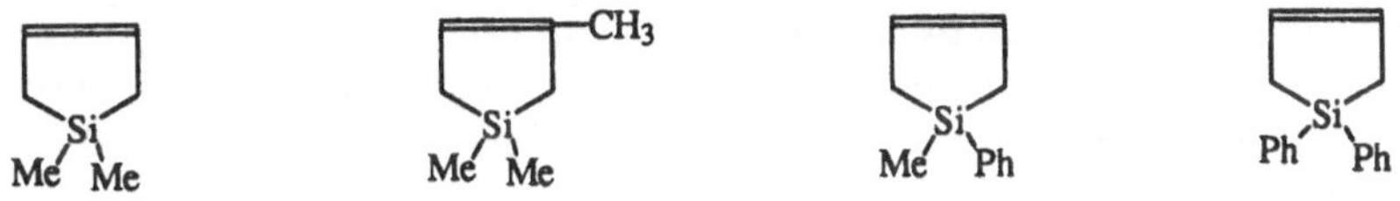

(96)

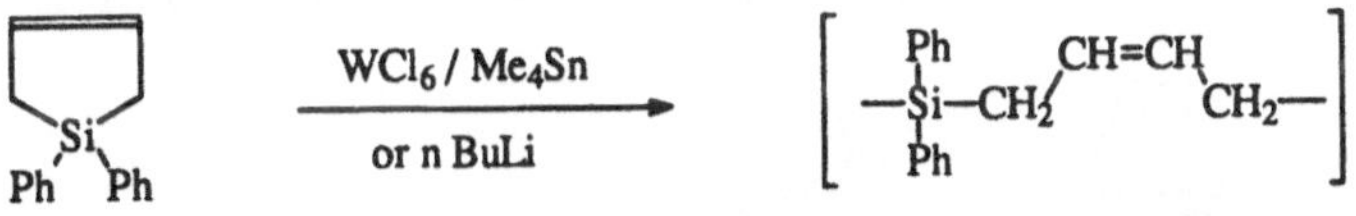

(97)

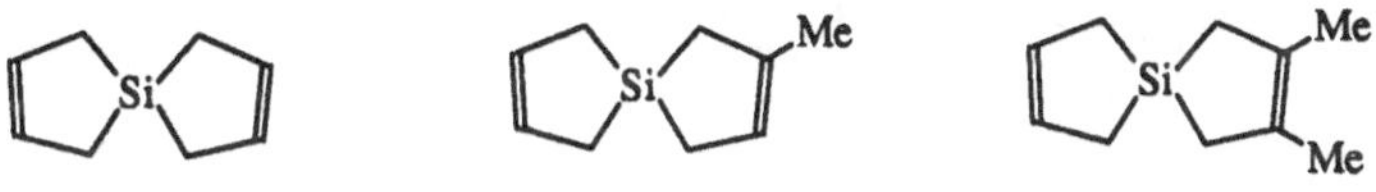

(98)

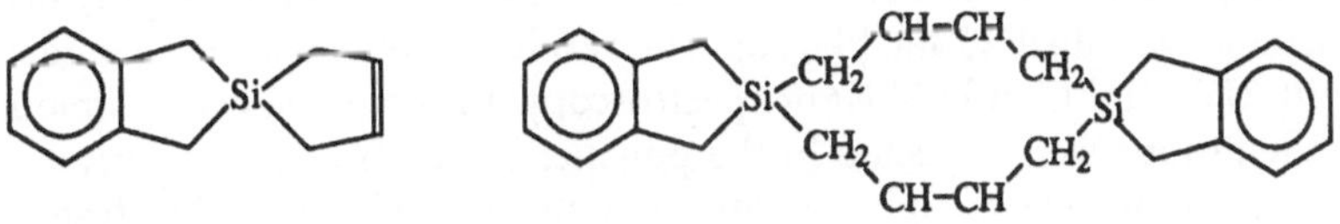

(99)

the double bond or by anionic initiators (e.g. *n*-BuLi) under cleavage of the Si—C bond. A comparison of both methods was only reported for monomer **96d**, Eq. (97) [91]. Due to considerable side reactions in the course of the anionic polymerization, higher yields and molecular weights (Mw ~ 436000, Mw ~ 143000) were obtained with the WCl_6/Me_4Sn initiator [91]. Another interesting difference between both mechanisms concerns the stereochemistry of the double bonds in the polymer backbone. No change of the cis structure is expected for the anionic polymerization, whereas the metathesis polymerizations may involve cis/trans isomerization. However, for the metathesis ROP's of monomers 94c and 96d, an almost perfect all-*cis* structure was reported [85, 91]. The anionic polymerizations of monomers 95a–c with either MeLi or n-BuLi (+ HMPT) in THF at low temperatures have in common that only low molecular weight polymers were obtained. Endgroup analyses were conducted by means of ^1H, ^{13}C- and ^{19}Si NMR spectroscopy [87, 89, 92]. Relative low molecular weights ($M_n \leqslant 5000$ g/mol) were also found, when 96d was polymerized with *n*-BuLi as initiator. In addition to end groups the existence of branching was discussed [91].

The poly(carbosilane)s resulting from anionic polymerizations of monomer 96b were analyzed in much detail with regard to their microstructure. It was found that both head-to-tail and head-to-head sequences were formed, and that the head-to-tail sequences are favored at lower reactions temperatures [88]. Furthermore, anionic copolymerizations of monomers 96a and 96d were studied [90]. Despite low polymerization temperatures ($-50\,°C$) random copolymers were formed as evidenced by ^1H, ^{13}C and ^{29}Si NMR spectroscopy.

Anionic polymerizations of the spiromonomers 98a–c with *n*-BuLi/HMPT in THF yielded complex mixtures of cyclic oligomers and polymers [93] which were analyzed by HPLC, GC/MS and NMR spectroscopic methods. The yield of cyclic dimer (99)b was particularly high, when the benzospiromonomer 99a was subjected to anionic polymerization [94].

In the case of the 2,3-benzo-cyclosilapentenes 100a–c metathesis ROP is not

(100)

a   b   c

possible, and the initiation with *n*-BuLi/HMPT in THF is the only polymerization method described so far [95, 96]. The resulting polymers were characterized by IR, UV, ^1H, ^{13}C and ^{29}Si NMR spectroscopy. In some cases small amounts of siloxane groups were detected in the polymer backbones. Particularly interesting is the influence of substituents on both T_g and T_m as illustrated in connection with Formulas 101a–c. Weber and coworkers also studied the thermal degradation of poly(carbosilane)s [86–96] and found that the benzosubstituted polymers possess the highest thermostabilities [95, 96].

Synthesis and polymerization of silacyclobutanes and silacyclobutenes has found increasing interest over a period of thirty years [97–104]. Anionic polymerizations initiated with alkyl lithium compounds seem to be best suited for

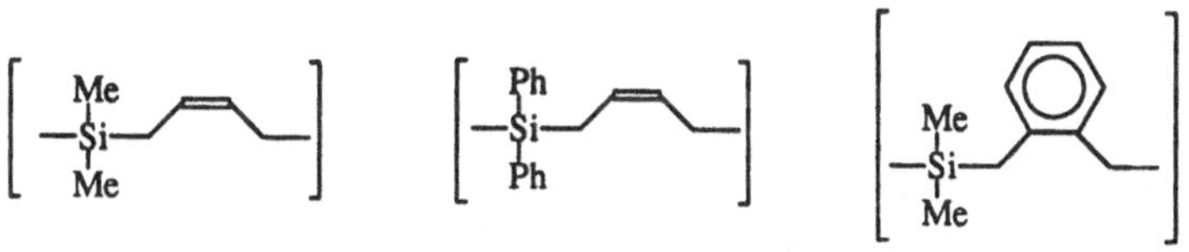

$$\left[\begin{array}{c} Me \\ -Si- \\ Me \end{array} \diagup\hspace{-0.3em}\diagup\hspace{-0.3em}\diagdown \right] \qquad \left[\begin{array}{c} Ph \\ -Si- \\ Ph \end{array} \diagup\hspace{-0.3em}\diagup\hspace{-0.3em}\diagdown \right] \qquad \left[\begin{array}{c} Me \\ -Si- \\ Me \end{array} \right]$$

(101)

a   b   c

$$T_g = -64\,°C \qquad T_g = -18\,°C \qquad T_g = 75\,°C$$
$$T_m = 25\,°C \qquad T_m = 130\text{-}165\,°C \qquad T_m = 163\text{-}165\,°C$$

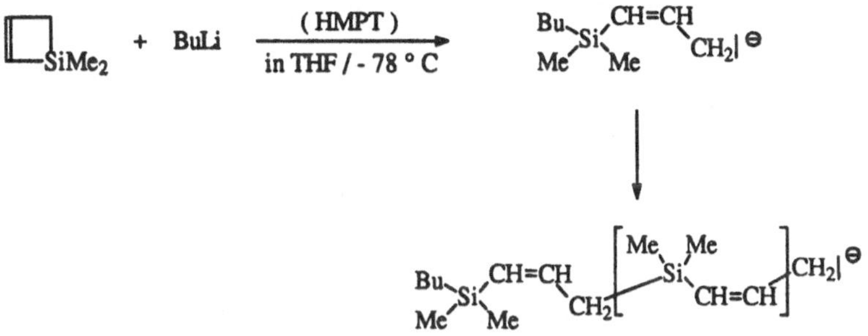

$$\square_{SiMe_2} + BuLi \xrightarrow[\text{in THF}/-78\,°C]{(HMPT)} \begin{array}{c} Bu \\ \diagdown \\ Me \end{array}\!\!Si\!\!\begin{array}{c} CH=CH \\ \diagup \\ Me \end{array}\!\!CH_2|^{\ominus}$$

(102)

$$\begin{array}{c} Bu \\ \diagdown \\ Me \end{array}\!\!Si\!\!\begin{array}{c} CH=CH \\ \diagup \\ Me \end{array}\!\!CH_2\!\!\left[\begin{array}{c} Me\quad Me \\ \diagdown Si \diagup \\ CH=CH \end{array} \right]\!\!CH_2|^{\ominus}$$

(103)

$$\square_{SiMe_2} + BuLi \longrightarrow \begin{array}{c} Bu-CH- \\ {}_{\ominus}CH-SiMe_2 \end{array}$$

(104)

the polymerization of these heterocycles. All research group agree in that the nucleophile attacks the silicon atom via its free d-orbitals, thereby generating a carbanion as the active chain ends. This mechanism is favored by the ring-opening of the highly strained cyclobutane ring, Eqs. (102, 103) [100]. It is thus more favorable than the anionic polymerization of the C=C double bond of silacyclobutenes which could involve an anion stabilized by delocalization via the Si-d-orbitals Eq. (104). In the case of 2,3-benzosilacyclobutanes 105a–c

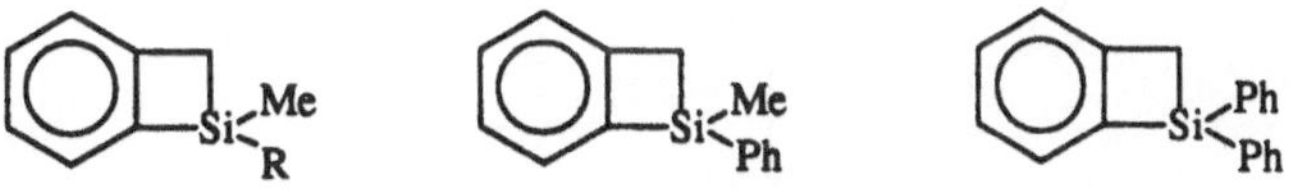

(105)

a   b   c

[103] or for the saturated silacyclobutanes 106a–d [97, 98, 102] an alternative to the formation of a carbanion by ring-opening (analogous to Eqs. 102 and 103)

$$\square_{Si}\!\!<\!\!\begin{array}{c} Me \\ Me \end{array} \qquad \square_{Si}\!\!<\!\!\begin{array}{c} H \\ R \end{array} \qquad \begin{array}{l} b\,:\,R = H \\ c\,:\,R = Me \\ d\,:\,R = Ph \end{array}$$

(106)

a

does not exist. Surprisingly the hydrid of **106b–d** does not cause significant side reactions. The ^{13}C and ^{29}Si NMR spectra of the poly(carbosilane)s derived from **106d** were reported to exhibit tacticity effects [102]. A high selectivity of the anionic ROP of silacyclobutanes is also demonstrated by the finding that the three vinyl substituted monomers **107a–c** yield non-crosslinked polymers under

$$
\underset{R}{\boxed{}}\!\!\!\text{Si–CH=CH}_2 \quad \xrightarrow[\text{(THF , - 78 °C)}]{\text{n-BuLi / HMPT}} \quad \left[\begin{array}{c} R \\ \overset{|}{\underset{\underset{CH_2}{\overset{|}{CH}}}{Si}}{-}CH_2\overset{CH_2}{\diagup}CH_2' \end{array} \right]
\tag{107}
$$

a : R = Me,   b : R = Ph,   c : R = CH=CH$_2$

suitable reaction conditions [103]. However, crosslinked gels are formed under harsher conditions [99].

Whereas all the monomers depicted in Eqs. (102)–(107) were polymerized by anionic initiators, the silacyclobutanes **108a–c** were reported [104] to polymer-

$$
\underset{\text{Z–SiMe}_3}{\overset{\text{Me}}{\boxed{}\!\!\!\text{Si}}}
\qquad
\begin{array}{l}
\text{a : Z = } -\text{CH}_2- \\
\text{b : Z = } -\text{CH}_2-\text{CH}_2-\text{CH}_2- \\
\text{c : Z = } -\bigcirc-
\end{array}
\tag{108}
$$

ize thermally or by action of a palladium based catalyst. Soluble, high molecular weight polymers were obtained under both conditions.

A particularly interesting class of monomers are the disilacyclobutanes. The chlorosubstituted monomers are, of course, not suited for anionic polymeriz-ation, but allow polymerizations catalyzed by platinum complexes Eqs.

$$
\underset{Cl}{\overset{Cl}{\diagdown}}\!\!Si\!\!\underset{CH_2}{\overset{CH_2}{\diagdown\diagup}}\!\!Si\!\!\underset{Cl}{\overset{Cl}{\diagup}} \quad \xrightarrow{\text{H}_2\text{PtCl}_6} \quad \left[\overset{Cl}{\underset{Cl}{\overset{|}{\underset{|}{Si}}}}{-}CH_2- \right]
\tag{109}
$$

$$
\text{(LiAlH}_4\text{)} \quad \Big| \quad \text{- 2 H}
$$

$$
\left[-\text{SiH}_2\text{–CH}_2- \right]
\tag{110}
$$

(109, 111) [105, 106]. After reduction with LiAlH$_4$ linear polycarbosilanes were obtained, Eqs. (110, 112) which may be considered as the Si-analogs of polyethy-lene and polypropylene.

These poly(carbosilane)s which were isolated in the form of colorless oils with M_n's above 10 000 g/mol are not only of theoretical interest. Like other aliphatic polycarbosilanes prepared from silacyclobutanes their thermostability is low, and their pyrolysis yields thermostable SiC ceramics [100–103, 106].

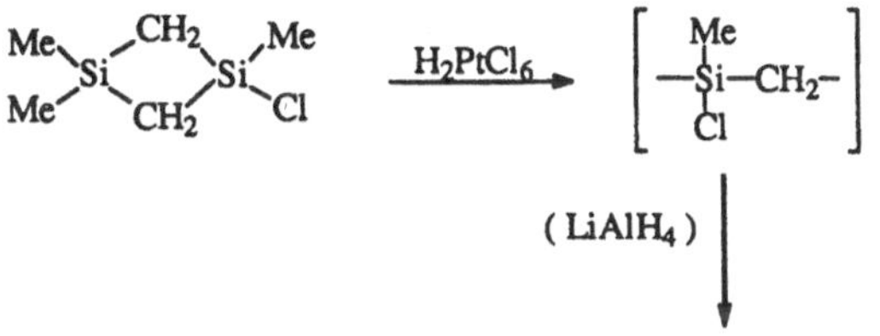

$$\underset{\text{Me}}{\overset{\text{Me}}{\text{Si}}}\underset{\text{CH}_2}{\overset{\text{CH}_2}{}}\underset{\text{Cl}}{\overset{\text{Me}}{\text{Si}}} \xrightarrow{\text{H}_2\text{PtCl}_6} \left[-\underset{\text{Cl}}{\overset{\text{Me}}{\text{Si}}}-\text{CH}_2- \right]$$

(111)

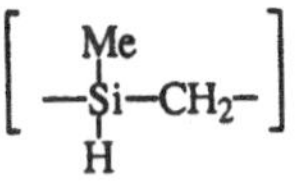

$$(\text{LiAlH}_4) \downarrow$$

$$\left[-\underset{\text{H}}{\overset{\text{Me}}{\text{Si}}}-\text{CH}_2- \right]$$ (112)

Interesting examples of cyclic carbosilanes with Si—Si bonds are the monomers **113a–d**. These 1,2,5,6-tetrasila-3,7-diynes polymerize anionically when treated with n-BuLi in THF at 20 °C yielding linear, high molecular weight polymers. Their degradation by BuLi or UV light and their electric conductivity (after doping with SbF_5) were studied [107]. Further information on cyclic disilanes is available from Chapter 4.

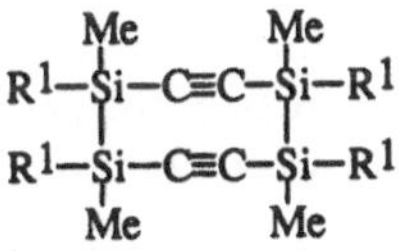

$$\begin{array}{c} \text{Me} \quad\quad \text{Me} \\ \text{R}^1\!-\!\underset{|}{\overset{|}{\text{Si}}}\!-\!\text{C}\!\equiv\!\text{C}\!-\!\underset{|}{\overset{|}{\text{Si}}}\!-\!\text{R}^1 \\ \text{R}^1\!-\!\underset{\text{Me}}{\overset{|}{\text{Si}}}\!-\!\text{C}\!\equiv\!\text{C}\!-\!\underset{\text{Me}}{\overset{|}{\text{Si}}}\!-\!\text{R}^1 \end{array}$$

a : R = Et   c : R = hexyl
b : R = Bu   d : R = Ph

(113)

Finally, a variety of heterocycles bearing Si-containing substituents should be mentioned. Biodegradable polyesters with a silyl-protected hydroxygroup were prepared by copolymerization of δ-valerolactone or ε-caprolactone with a t.Butyldimethylsiloxy substituted ε-caprolactone (Formula **116b** [108]. Desilylation with aqueous acetic acid yielded copolyesters with pending OH groups which allow controlled crosslinking with diisocyanates. The silyl protection was not only required for a clear polymerization process, but also for the synthesis of the monomer itself, because hydroxy caprolactone isomerizes to the thermodynamically stable hydroxyethylbutyrolactone, Eqs. (114–116). Another example for the synthesis of a biodegradable polymer with a Si-protected hydroxy group is the cationic polymerization of an 1-6-anhydro-D-glucose derivative outlined in Eq. (117) [109].

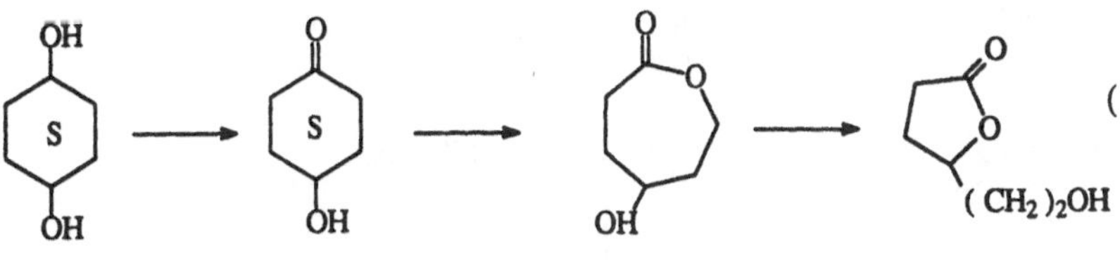

(114)

$$\downarrow$$

(115)

Me_3CSiMe_2O—[structure with S, =O] → Me_3CSiMe_2O—[lactone structure with =O]   (116)

    a   b

[sugar structure: CH_2—O, O, R^1O, OR^2, OR^2] $\xrightarrow[-60\,°C]{PF_3}$ [sugar structure: —O—CH_2, O, R^1O, OR^2, O—, OR^2]   (117)

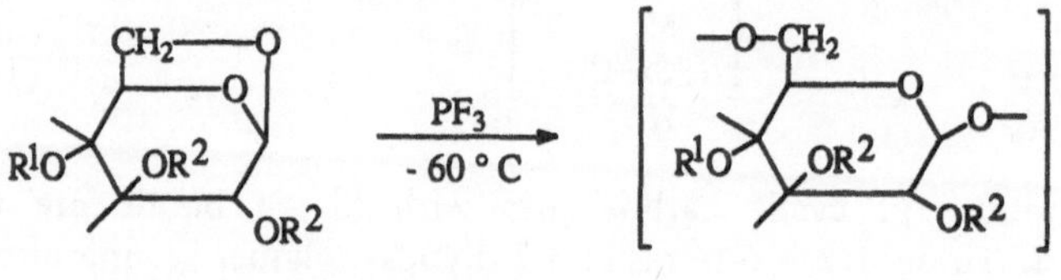

    a   b

R^1 = benzyl-, R^2 = Me_3CSiMe_2

Allcock and coworkers [110–112] studied synthesis and thermal polymerization (210–250 °C) of numerous cyclophosphazenes bearing one Si-containing substituent (Forms. 118a–c, 119a–c, 120a, b and 121a–g). Whereas in the case of 118a, 119a and 120a high molecular weights ($M_W \sim 10^5$–10^6) were obtained, all

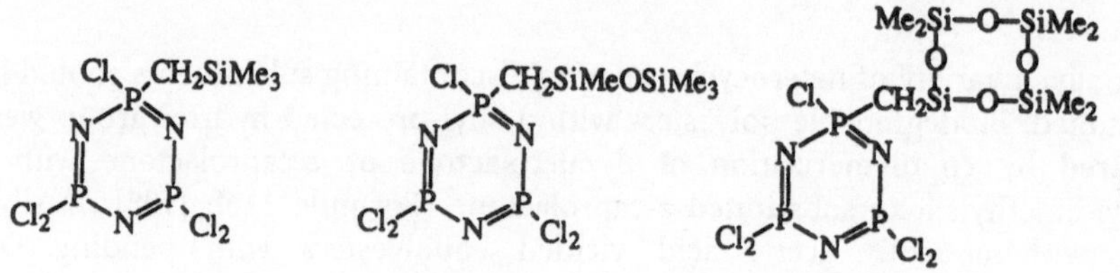

[Structure 118a: Cl, CH_2SiMe_3, P-N-P-N ring, Cl_2P, PCl_2]

[Structure 118b: Cl, $CH_2SiMeOSiMe_3$, ring, Cl_2P, PCl_2]

[Structure 118c: Cl, CH_2Si with Me_2Si—O—$SiMe_2$, O, O, Si—O—$SiMe_2$, ring, Cl_2P, PCl_2]   (118)

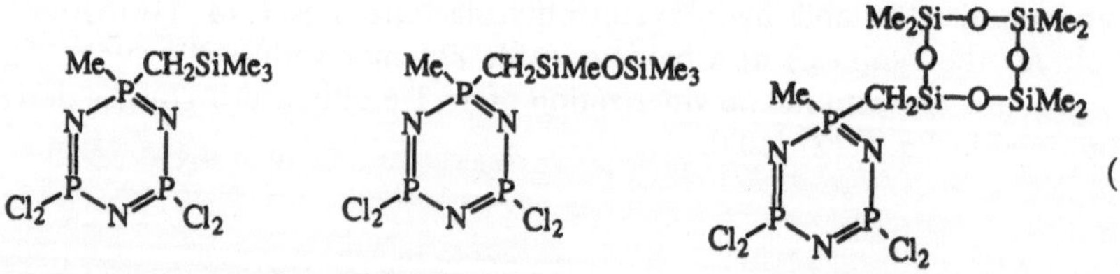

    a   b   c

[Structure 119a: Me, CH_2SiMe_3, ring, Cl_2P, PCl_2]

[Structure 119b: Me, $CH_2SiMeOSiMe_3$, ring, Cl_2P, PCl_2]

[Structure 119c: Me, CH_2Si with Me_2Si—O—$SiMe_2$, O, O, Si—O—$SiMe_2$, ring, Cl_2P, PCl_2]   (119)

    a   b   c

[Structure 120a: Me_3SiCH_2, CH_2SiMe_3, ring, Cl_2P, PCl_2]

[Structure 120b: $Me_3SiOSiCH_2$, $CH_2SiOSiMe_3$, ring, Cl_2P, PCl_2]   (120)

    a   b

monomers containing siloxane groups failed to yield polyphosphazenes. Side reactions such as cleavage of Si—O and Si—Cl bonds by chloride ions were discussed. In the case of monomers 121a–g the influence of the alkyl group R was investigated. Furthermore a comparison with monomers 122a–g was conducted and showed that the silyl group of monomers 121a–g is more favorable for the thermal polymerization than the methyl group. However, unfavorable are in both series 121 and 122 bulky substituents such as those of 121e,f and 122e,f. The isopropyl group 121c and 122c has an intermediate

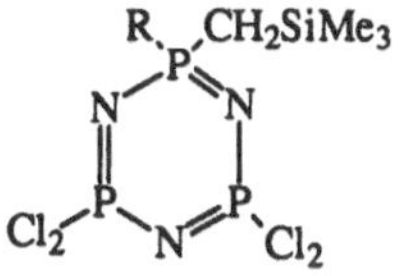

$$
\begin{array}{ccc}
& & a : R = Me \qquad e : R = t\text{-Bu} \\
& & b : R = Et \qquad f : R = neo\ C_5H_{11} \\
& & c : R = i\text{-Pr} \qquad g : R = Ph \\
& & d : R = n\text{-Bu}
\end{array}
\qquad (121)
$$

$$
\begin{array}{ccc}
& & a : R = Me \qquad e : R = t\text{-Bu} \\
& & b : R = Et \qquad f : R = neo\ C_5H_{11} \\
& & c : R = i\text{-Pr} \qquad g : R = Ph \\
& & d : R = n\text{-Bu}
\end{array}
\qquad (122)
$$

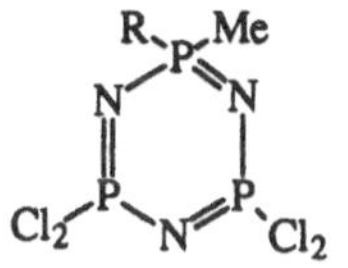

effect. All polyphosphazenes were subjected to nucleophilic substitution with the sodium salt of trifluoroethanol, Eq. (123). The resulting chlorine-free polyphosphazenes are of interest as oil and water resistant elastomers for higher temperatures. All attempts to polymerize the siloxane ring of monomers 118c or 119c failed, Eq. (124) [111].

$$
\left[\left(\begin{array}{c} Cl \\ -P=N- \\ Cl \end{array}\right)_2 -\left(\begin{array}{c} R \\ -P=N- \\ CH_2SiMe_3 \end{array}\right)\right] \longrightarrow \left[\left(\begin{array}{c} OCH_2CF_3 \\ -P=N- \\ OCH_2CF_3 \end{array}\right)_2 -\left(\begin{array}{c} R \\ -P=N- \\ CH_2SiMe_3 \end{array}\right)\right] \qquad (123)
$$

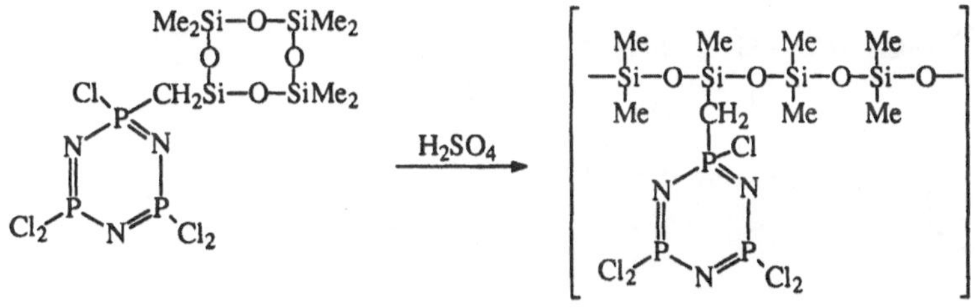

$$
\xrightarrow{\ H_2SO_4\ } \qquad (124)
$$

(118c)

Last but not least several epoxides with pending silyl-groups should be mentioned 125a, b, 127c, 128a, b, 129 [113–115]. Whereas no information is available on the polymerization of 125b the optical active enantiomer of 125a and the racemate were polymerized with a variety of catalysts [113]. These catalysts include KOH or KOt. Bu (in bulk or in DMSO), Et$_3$Al + H$_2$O and BF$_3$ (at −78 °C). The chemical structure and stereochemistry of oligomers and polymers was analyzed. It was found that the base catalyzed polymerizations involve

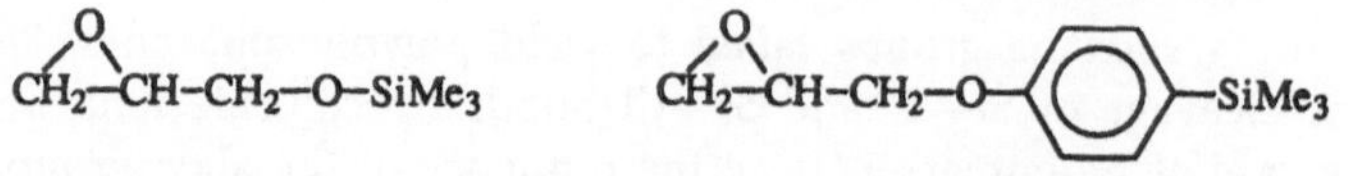

$$\begin{array}{cc} \text{CH}_2\text{—CH–CH}_2\text{–O–SiMe}_3 & \text{CH}_2\text{—CH–CH}_2\text{–O–}\!\!\bigcirc\!\!\text{–SiMe}_3 \end{array} \qquad (125)$$

a   b

a rearrangement so that mainly a polyoxetane backbone **126a** is formed instead of the expected polethyleneoxide structure **126b**.

Multifunctional epoxides **127c**, **128a,b**, **129** were prepared by platinum catalyzed hydrosilylation of vinylcyclohexene oxide [115]. The electron beam initiated polymerization of these monomers yielded crosslinked diomers. Only weak doses (1–3 mrad) of irradiation were required which makes this process attractive for commercial applications.

$$(126)$$

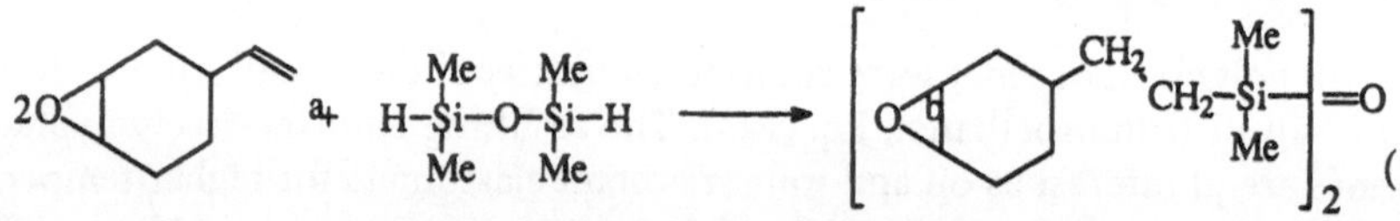

a   b

$$(127)$$

a   b   c

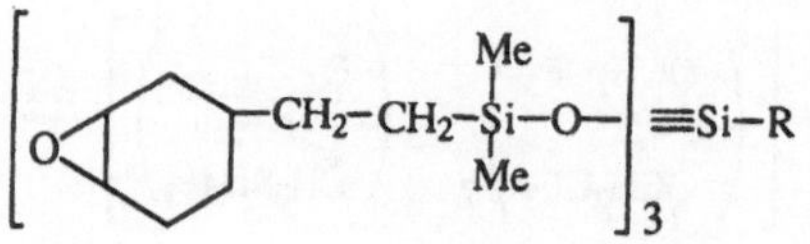

a : R = Me
b : R = Ph   (128)

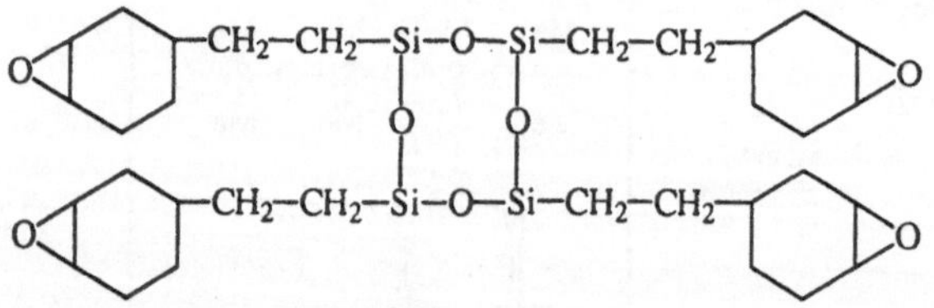

$$(129)$$

6.3.2
Silicon-Containing Initiators

Silicon-containing initiators are playing a permanently increasing role in polymer chemistry. Those components used for the initiation of vinyl monomers were described in chapter 1 (Sect. 1.5). The Si-containing initiators used for ROP may be subdivided into the following four groups according to the different polymerization mechanisms:

A) Initiators for metathesis ROP
B) Initiators for anionic polymerizations
C) Initiators for cationic polymerizations
D) Initiators of insertion mechanisms

Various silanes were extensively used as modifying agents of Ziegler-Natta type catalysts (Sect. 1.5). The modification of metathesis catalysts with silanes has not been of much interest and the results do indeed not justify a more intensive exploration of such an application. Apparently the first study of a silane modified metathesis catalyst was published by Günther et al. [116] who reacted methylcyclosiloxanes $(Me_3SiH-O)_n$ ($n = 4$ or 8), with WCl_6 prior to the polymerization of cyclopentene.

Another research group used the combination of tetra(n-propyl)silane and WCl_6 for the some monomer [117]. Prior to their publication a Russian research group had demonstrated [118] that various n-propyl silanes, such as $(nPr)_4Si$, $(nPr)_3SiPh$ or $(nPr)_3SiMe_2$, may be useful for the modification of WCl_6 as initiators in polymerizations of 1,5-cyclooctadiene. Also Et_3SiH or Ph_3SiOH served as coinitiators of WCl_6, but without detailed analysis [119, 120]. Furthermore, silica was proposed as an insoluble support of Mo or tungsten carbene complexes, but again without detailed study of its usefulness in polymerization processes [121, 122].

The most interesting Si-containing coinitiators proposed for metathesis ROP are the difunctional and tetrafunctional norbornenes depicted in Eq. (130) and Form. (131) [123]. After activation with titanacyclobutanes these coinitiators

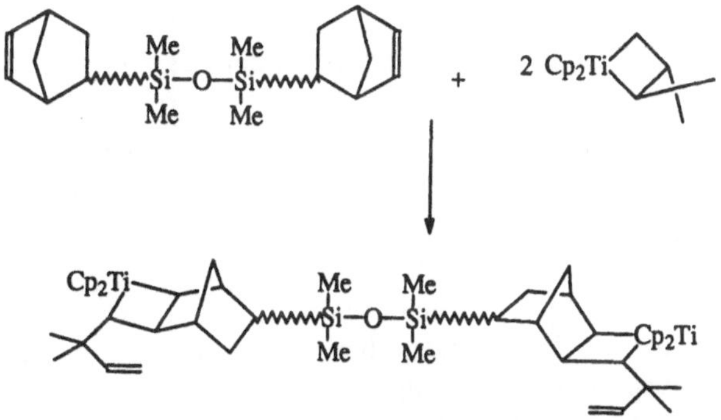

$$(130)$$

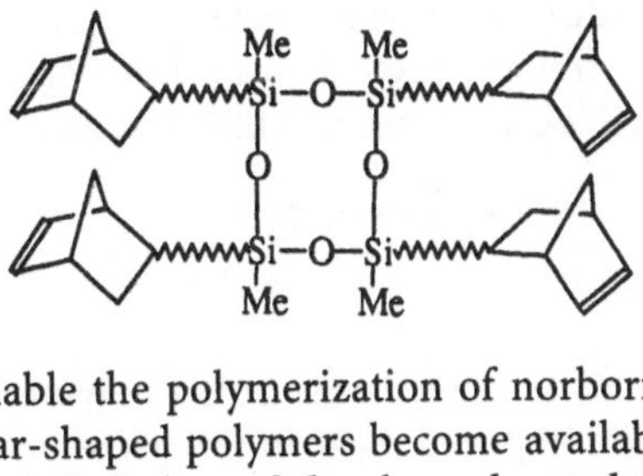

$$(131)$$

enable the polymerization of norbornene in such a way that A-B-A triblock or star-shaped polymers become available.

Only little work has been devoted to Si-containing anionic initiators [124, 125]. Phenoxytrimethylsilane in combination with CsF was used as initiator for the polymerization of epoxides, Eqs. (132, 133) [124]. The easy desilylation of silylated phenols by fluoride or chloride ions (Eq. 132) was known before from numerous

$$\text{Ph–O–SiMe}_3 + \text{CsF} \rightleftharpoons \text{Ph–O}^{\ominus}\ \text{Cs}^{\oplus} + \text{FSiMe}_3 \tag{132}$$

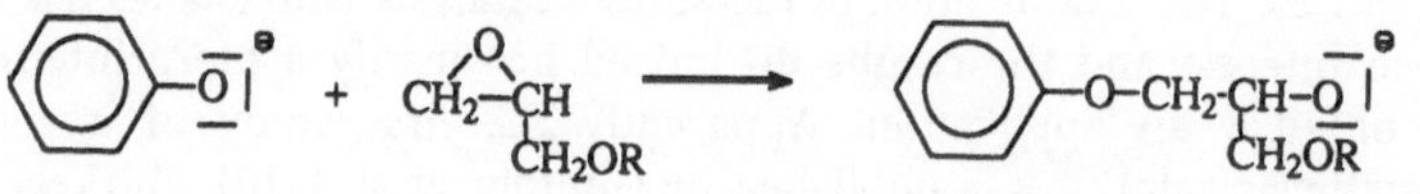

$$(133)$$

polycondensations of diphenols (Chapter 5). A potential advantage of phenol silylether/CsF initiators are the narrow MWD's of the resulting polyethers [124].

A two block polymer was obtained when a polymethacrylate with silylketene acetal endgroup (resulting from a GTP process, see Chapter 2) was used as initiator for a cyclocarbonate, Eq. (134) [125]. Again a fluoride containing anion

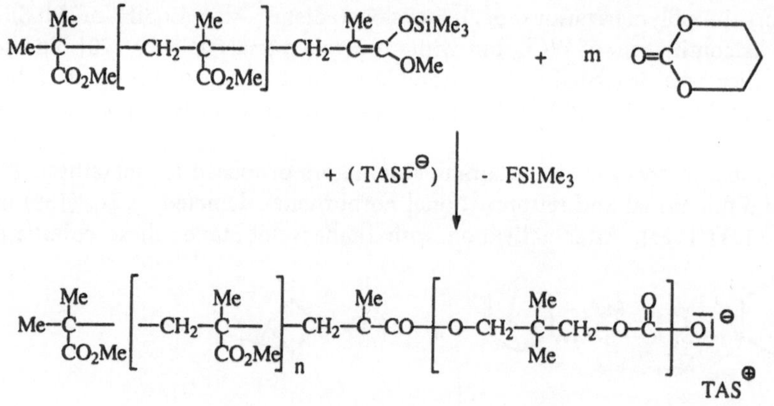

$$+ (\text{TASF}^{\ominus}) \quad \big| \quad - \text{FSiMe}_3$$

$$(134)$$

$$\text{TAS} = (\text{Me}_2\text{N})_3\,\text{S}^{\oplus}$$

served as activating species of the siloxy endgroup. However, it should be mentioned that the structure of the blockcopolymer depicted in Ref. 125 is unlikely and not proven by spectroscopic methods.

Three research groups [126–129] studied the usefulness of trimethylsilyl esters of strong acids 135a–c as cationic initiators for various heterocycles. 135a which is the ester of an acid of moderate strength proved to be not reaction enough to initiate 1,3-deoxepane, 1,3-dioxolane or trioxane. In contrast 135b

$$\text{Me}_3\text{SiO–}\overset{\overset{\displaystyle O}{\|}}{\text{P}}\text{–Ph}_2 \qquad \text{Me}_3\text{SiO–SO}_2\text{Me} \qquad \text{Me}_3\text{SiO–SO}_2\text{CF}_3$$

$$(135)$$

a   b   c

which is the ester of a stronger acid was found to initiate at least the polymerization of 1,3-dioxepane [126]. This initiation happened even in the presence of 2,6-bis-tert-butyl pyridine which acts as proton trap, and thus, prevents the competition of traces of free methane sulfonic acid. TMS triflate 132c seemingly initiated the cationic polymerizations of δ-valerolactone and ε-caprolactone [127]. Unfortunately these experiments were conducted in the absence of a proton trap, and thus, it is not clear to what extent the results are influenced by traces of triflic acid. However, the high initiating power of TMS-triflate was documented by the cationic polymerizations of THF and 2-methyloxazoline which proceed even in the presence of 2,6-bis-tert-butyl pyridine [128, 129].

The reaction of THF with **135c** was studied in much detail and both the complexation and the reaction sequence outlined in Scheme (136) was

$$
\text{CF}_3\text{SO}_3\text{SiMe}_3\text{-O} \rightleftharpoons \text{CF}_3\text{SO}_3^{\ominus} \; + \; \text{Me}_3\text{Si-O}^{\oplus}
$$

+ THF   + THF

(136)

$$
\text{Me}_3\text{SiO-}(\text{CH}_2)_4\text{-O}^{\oplus} \rightleftharpoons \text{Me}_3\text{SiO-}(\text{CH}_2)_4\text{-O-}(\text{CH}_2)_4\text{-OSO}_2\text{CF}_3
$$

$$
\text{CF}_3\text{SO}_3^{\ominus}
$$

evidenced by ^{1}H NMR spectroscopy. An interesting preparative aspect of polymerizations with silyltriflates is cationic grafting of THF onto a polysilane backbone, Eqs. (137, 138) [128]. In the case of 2-methyloxazoline two competing

$$
\left[\begin{array}{cc} \text{Me} & \text{Me} \\ -\text{Si-} & -\text{Si-} \\ \text{Ph} & \text{Ph} \end{array}\right] \xrightarrow[\; -\text{PhH}\;]{+ \text{CF}_3\text{SO}_3\text{H}} \left[\begin{array}{cc} \text{Me} & \text{Me} \\ -\text{Si-} & -\text{Si-} \\ \text{Ph} & \text{O} \\ & \text{SO}_2\text{CF}_3 \end{array}\right]
$$

(137)

+ n THF

$$
\left[\begin{array}{cc} \text{Me} & \text{Me} \\ -\text{Si-} & -\text{Si-} \\ \text{Ph} & \text{O} \\ & [(\text{CH}_2)_4\text{-O-}]_n\text{-SO}_2\text{CF}_3 \end{array}\right]
$$

(138)

initiation steps were detected by ^{1}H NMR-spectroscopy. In addition to the attack of **135c** onto the nitrogen, Eq. (139), the thermodynamically more favorable

$$
\text{CF}_3\text{SO}_3\text{SiMe}_3 \; + \; \underset{\underset{\text{Me}}{\text{C}}}{\text{N}}\text{O} \rightleftharpoons \text{CF}_3\text{SO}_3^{\ominus} \; + \; \text{Me}_3\text{Si-N}\underset{\underset{\text{Me}}{\text{C}}}{}\text{O}
$$

(139)

silylation of the oxygen was observed, Eq. (140). However both initiation reactions finally result in the same propagation mechanism [129]. A cationic mechanism was also reported for the polymerization of substituted oxatanes, **141b**,

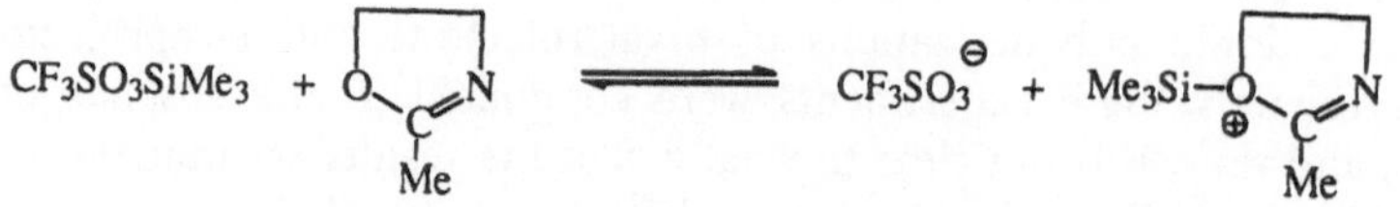

$$CF_3SO_3SiMe_3 \; + \; \underset{Me}{O{-}C{=}N} \; \rightleftharpoons \; CF_3SO_3^{\ominus} \; + \; \underset{Me}{Me_3Si{-}\overset{\oplus}{O}{-}C{=}N} \qquad (140)$$

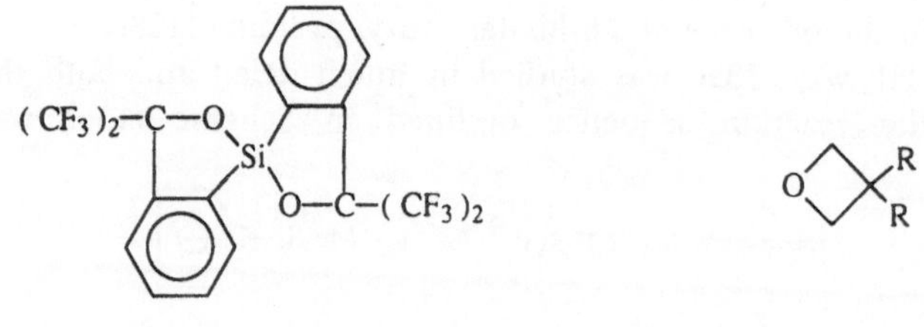

a   b

$$(141)$$

initiated with the spiro(benzoxasilole), **141a**. The polymerizations were conducted in the presence and absence of water or alcohols as coinitiators, but also in the presence of 2,6-bis-*tert*-butyl pyridine [130].

A new and highly efficient class of initiators capable of polymerizing various kinds of heterocycles is formed by the reaction of silicon hydrides with group III metal complexes. A first report, describing the polymerization of THF initiated by a silyl cobalt carbonyl complex dates back to 1970 [137]. More detailed studies of SiH/Co, SiH/Rh and SiH/Pt complexes and their catalytic activities were recently conducted by Crivello and coworkers [132–136]. Their results suggest, that the reaction of silicon hydrides with $Co_2(CO)_8$ or $Co_3(CO)_{12}$ yields silyl cobalt tetracarbonyl as the active initiator, Eqs. (142,143). Several arguments in favor of a cationic polymerization of epoxides, Eqs. (144,145) and

$$Co_2(CO)_8 \; + \; R_3SiH \; \longrightarrow \; HCo(CO)_4 \; + \; R_3SiCo(CO)_4 \qquad (142)$$

$$R_3SiH \; + \; HCo(CO)_4 \; \xrightarrow{-H_2} \; R_3SiCo(CO)_4 \qquad (143)$$

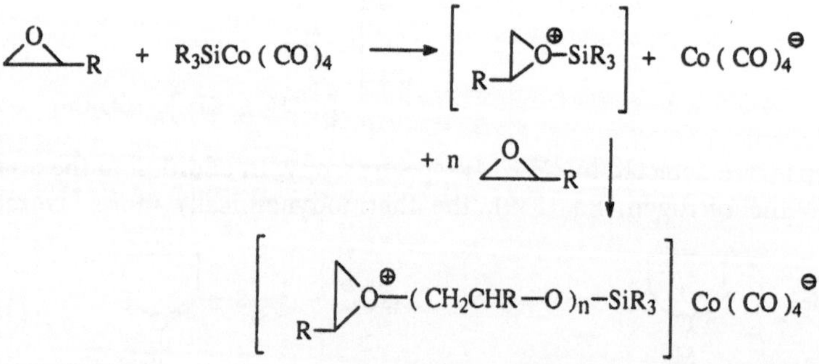

$$\underset{R}{\triangle}\!\!-O \; + \; R_3SiCo(CO)_4 \; \longrightarrow \; \left[\underset{R}{\triangle}\overset{\oplus}{O}{-}SiR_3 \right] + \; Co(CO)_4^{\ominus} \qquad (144)$$

$$+ \, n \; \underset{R}{\triangle}\!\!-O \; \Big\downarrow$$

$$\left[\underset{R}{\triangle}\overset{\oplus}{O}{-}(CH_2CHR{-}O)_n{-}SiR_3 \right] Co(CO)_4^{\ominus} \qquad (145)$$

other cyclic ethers involving silyl groups were presented. Firstly, the strong acid $HCo(CO)_4$ proved to be inactive as initiator. Secondly, Lewis bases such as phosphines and amines act as inhibitors. Thirdly, THF can be polymerized with the silyl-cobalt silyl-rhodium or silyl-platinum catalysts, albeit this monomer is not prone to anionic or radical ROP. Analogous reaction mechanisms were formulated for the rhodium catalysts, Eq. (146) [133, 134] and for the platinum

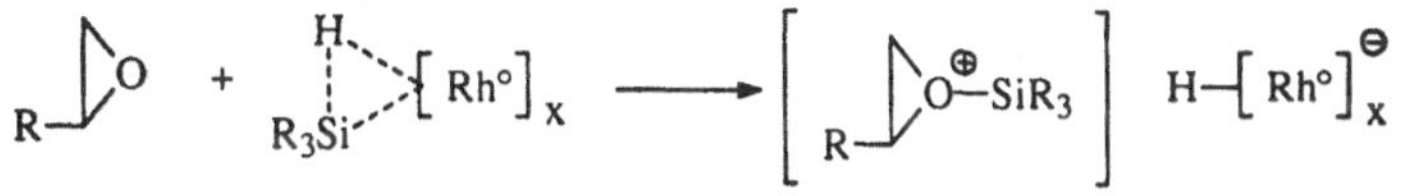

$$\text{R} \overset{\text{O}}{\underset{}{\triangleright}} \;+\; \underset{R_3Si}{\overset{H}{\triangleright}}[\text{Rh}^\circ]_x \;\longrightarrow\; \left[\text{R} \overset{\oplus}{\underset{}{\triangleright}} O\text{—SiR}_3 \right] \; H\text{—}[\text{Rh}^\circ]_x^{\ominus} \tag{146}$$

complexes [135, 136]. In addition to epoxides and THF, oxetanes and thiiranes were polymerized. Molecular weights in the range of 10^4–10^5 were obtained in most cases. An interesting preparative application is the grafting of epoxides on polysiloxanes, Eq. (147). When this manuscript was finished these studies were still in their exploratory phase.

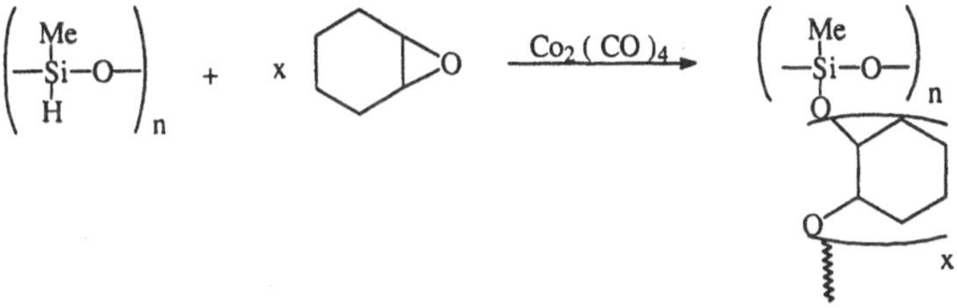

$$\left(\begin{matrix} \text{Me} \\ -\text{Si}-\text{O}- \\ \text{H} \end{matrix} \right)_n \;+\; x \; \overset{}{\underset{}{\bigcirc}}\!\!\!\triangleright O \;\xrightarrow{\text{Co}_2(\text{CO})_4}\; \left(\begin{matrix} \text{Me} \\ -\text{Si}-\text{O}- \\ \text{O} \end{matrix} \right)_n \tag{147}$$

Another research group [137] reported on the polymerization of cyclohexene oxide initiated by a combination of o-nitrobenzyl-triphenylsilyl ether and aluminum acetylacetonate activated by UV-light (365 nm). The authors asume that the photolytic cleavage of the 2-nitrobenzyl ether generates triphenylsilanol according to Eq. (148). The silanol should in turn form Al—O—Si bonds (Form. 149) which was considered to be the active site of the initiator.

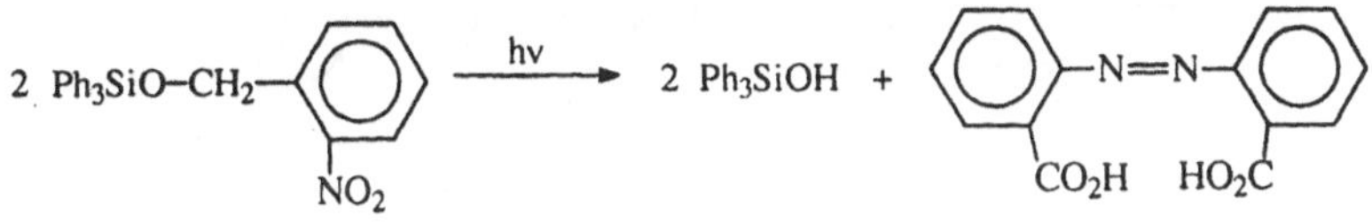

$$2 \; \text{Ph}_3\text{SiO-CH}_2\text{—}\overset{}{\underset{\text{NO}_2}{\bigcirc}} \;\xrightarrow{h\nu}\; 2 \; \text{Ph}_3\text{SiOH} \;+\; \overset{}{\underset{\text{CO}_2\text{H}}{\bigcirc}}\text{—N}{=}\text{N—}\overset{}{\underset{\text{HO}_2\text{C}}{\bigcirc}} \tag{148}$$

$$\begin{matrix} \text{Me}{\searrow} \\ \text{C}{=}\text{O}\cdots \\ \text{H-C} \qquad\qquad \text{Al—O—SiPh}_3 \\ {\searrow}\text{C-O}{\nearrow} \\ \text{Me}{\nearrow} \end{matrix} \tag{149}$$

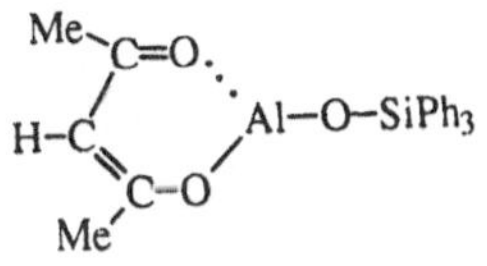

6.3.3
Si-Containing Terminators

Si-containing compounds may also be useful as chain stoppers or terminators not only as initiators. Yet very few examples of such an application were published. As mentioned in Sect. 6.3.1 vinylsilanes (Forms. 88a–c) added in small amounts to a metathesis ROP of cycloolefins act as terminators and yield polyalkeneamers with functional end groups. These end groups are sensitive to hydrolysis and allow a hydrolytical crosslinking of the functionalized polymer chains. Unfortunately this method does not allow a functionalization of all chains. A mixture of three kinds of polymer chains (with regard to their end groups) will result as illustrated for polypenteneamer by Forms. 150–152. One

$$R-CH= (=CH-CH_2-CH_2-CH_2-CH=)_n=CH-SiX_3 \qquad (150)$$

sort of chain is not functionalized and bears two inert endgroups 151. This short-coming is avoidable, when terminators with two Si—X_3 endgroups (eg. 152) are added to the polymerization process, or when these terminators are equilibrated with preformed polyalkeneamers [137].

$$R-CH=(=CH-CH_2-CH_2-CH_2-CH=)=CH-R \qquad (151)$$

$$X_3Si-CH=(=CH-CH_2-CH_2-CH_2-CH=)=CH-SiX_3$$

$$\qquad (152)$$

$$X = Cl\ ,\ OMe \qquad\qquad R = \text{inert chain end}$$

Ketene silyl acetals are well documented as coinitiators in group-transfer-polymerizations (chapter 2). Interestingly a recent report [138] describes their application as terminators in cationic polymerizations of THF, Eqs. (153, 154). The alkaline saponification of the ester end groups, Eq. (155), yields telechelic poly THF with two carboxyl end groups, which are useful for syntheses of a variety of segmented copolymers.

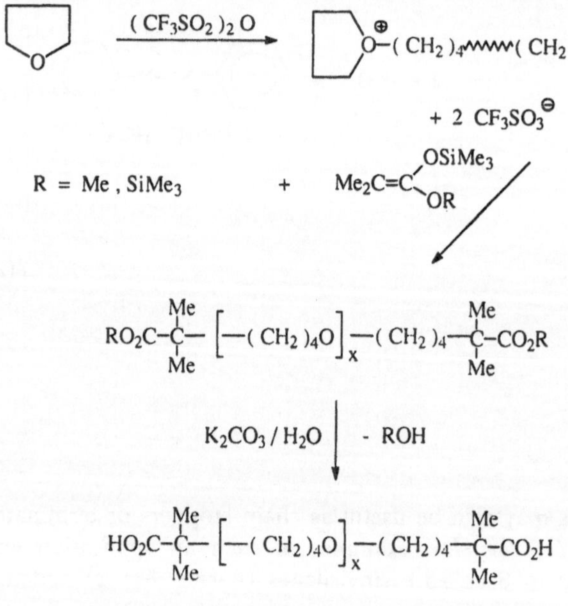

$$PTP = \quad Me_3SiO-\left[-\underset{\underset{\displaystyle OSiMe_3}{|}}{\overset{\overset{\displaystyle O}{\parallel}}{P}}-O-\right]-SiMe_3 \qquad (156)$$

6.4
Various Polymerizations and Polymers

This section summarizes a broad variety of polymerizations and Si-containing polymers which are not mentioned in any other chapter of this handbook. This summary begins with syntheses of Si-free oligomers and polymers and continues with Si-containing polymers which are subdivided into materials with Si- in the backbone and materials with Si-containing substituents.

An interesting dehydrating agent is poly(trimethylsilyl phosphate) (Form. 156). It is soluble in organic solvents in contrast to free poly(phosphoric acid), but an unreacted excess of this agent is also easy to remove by water or alkaline liquids. PTP allows an facile synthesis of thermostable polyamidines either from diamines and carboxylic acids, Eq. (157) or monoacylated diamines Eq. (158).

$$NH_2-\overset{\displaystyle AR}{\bigcirc}-NH_2 \;+\; R-CO_2H \;\xrightarrow{\;PTP\;}\; \left[-\overset{\displaystyle AR}{\bigcirc}-NH-\underset{\underset{\displaystyle R}{|}}{C}=N-\right] \qquad (157)$$

$$R = \quad -\bigcirc \quad , \quad -\bigcirc-CH_3 \quad , \quad -\bigcirc-Cl$$

$$\overset{\displaystyle AR}{\bigcirc} = \quad -\bigcirc- \quad -\bigcirc-X-\bigcirc- \qquad (X = O, CH_2)$$

$$NH_2-\overset{\displaystyle AR}{\bigcirc}-NH-\overset{\overset{\displaystyle O}{\parallel}}{C}-Ph \;\xrightarrow{\;PTP\;}\; \left[-\overset{\displaystyle AR}{\bigcirc}-NH-\underset{\underset{\displaystyle Ph}{|}}{C}=N-\right] \qquad (158)$$

High yields but low viscosity values were reported [140].

The role of silyl groups as OH protecting groups is illustrated by the stepwise synthesis of monodisperse, telechelic oligoesters from phthalic acid and 1,4-butane diol or neopentane diol [141]. Equation (159) illustrates one step of a longer reaction pathway where the monosilylated diol is involved.

The synthesis of Si-containing "starburst-type" dendrimers in a stepwise manner was performed by the reaction sequence outlined in Eqs. (160–162) [142]. Characteristic for this synthetic strategy is (a) the reaction of Si—Cl groups with allyl Grignard reagents and (b) the hydrosilylation of the allyl groups. This strategy is versatile, because alkyl dichlorosilanes may be used instead of trichlorosilane. Up to 5 generations were synthesized corresponding to a

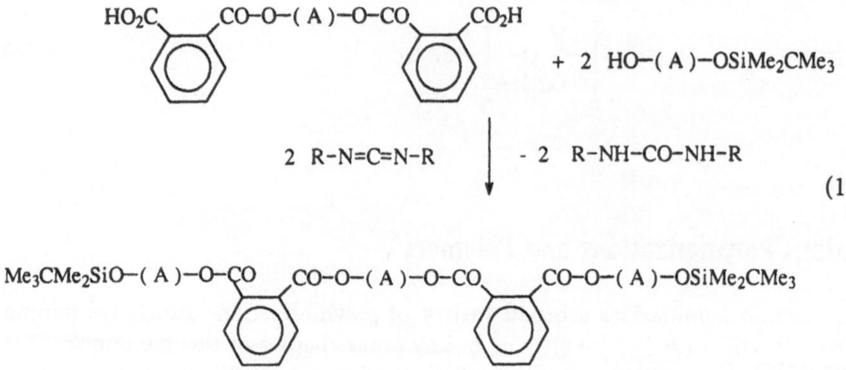

HO$_2$C   CO–O–(A)–O–CO   CO$_2$H

⬡   ⬡   + 2 HO–(A)–OSiMe$_2$CMe$_3$

2 R–N=C=N–R | – 2 R–NH–CO–NH–R

(159)

Me$_3$CMe$_2$SiO–(A)–O–CO   CO–O–(A)–O–CO   CO–O–(A)–OSiMe$_2$CMe$_3$

⬡   ⬡

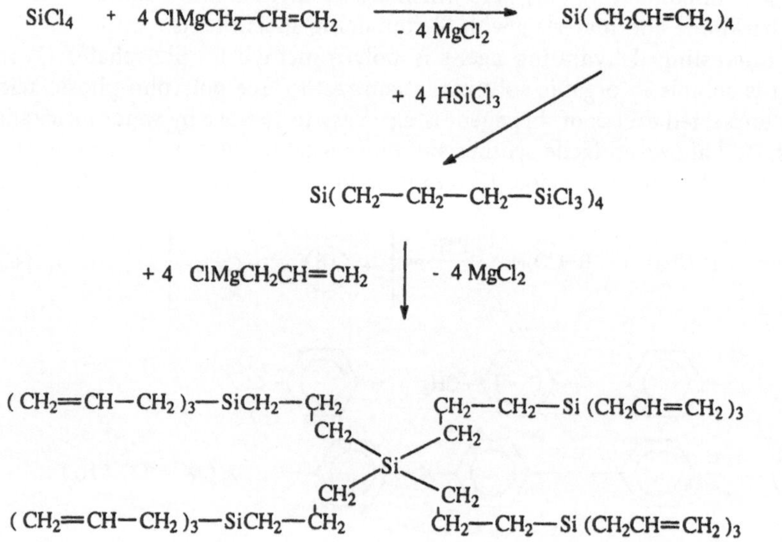

SiCl$_4$ + 4 ClMgCH$_2$–CH=CH$_2$ $\xrightarrow{\text{– 4 MgCl}_2}$ Si(CH$_2$CH=CH$_2$)$_4$ (160)

+ 4 HSiCl$_3$

(161)

Si(CH$_2$—CH$_2$—CH$_2$—SiCl$_3$)$_4$

+ 4 ClMgCH$_2$CH=CH$_2$ | – 4 MgCl$_2$

(CH$_2$=CH—CH$_2$)$_3$—SiCH$_2$—CH$_2$ ⟍ ⟋ CH$_2$—CH$_2$—Si (CH$_2$CH=CH$_2$)$_3$ (162)

CH$_2$ CH$_2$
    ⟍ Si ⟋
CH$_2$ CH$_2$

(CH$_2$=CH—CH$_2$)$_3$—SiCH$_2$—CH$_2$ CH$_2$—CH$_2$—Si (CH$_2$CH=CH$_2$)$_3$

molecular weight of 7400 g/mol. In this connection two previous papers should be mentioned [143, 144] dealing also with synthesis of dendritic or starburst poly(carbosilane)s by Pt-catalyzed addition of Si—H bonds onto allylsilanes.

In contrast to syntheses of Si-containing dendrimers, syntheses of linear poly(alkoxy silane)s, or poly(aryloxy silane)s, Eq. (163) have a long tradition [145–156]. First attempts to prepare poly(alkoxy silane)s by condensation of dimethyldichlorosilane and aliphatic diols were reported in 1947 [145]. However, with the relatively short diols used in these experiments only cyclo-silanes 164a, b were obtained. Yet even when the reaction conditions were varied by addition of pyridin or ammonia and when longer diols were used, these experiments failed to give the desired poly(alkoxy silane)s [146]. Crosslin-ked, gum-like polymers were isolated, when methyltrichlorosilane was reacted with diols or polyvinylalcohol with dimethyldichlorosilane [147, 148]. Somewhat more successful proved polycondensations of diphenols with Me$_2$SiCl$_2$, Et$_2$SiCl$_2$

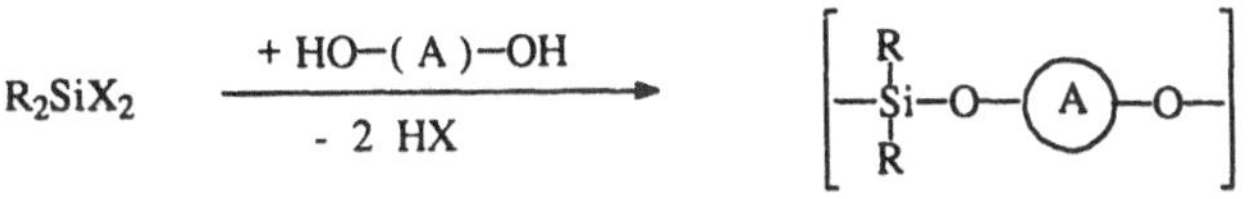

$$R_2SiX_2 \quad \xrightarrow[\text{- 2 HX}]{+ HO-(A)-OH} \quad \left[-\underset{\underset{R}{|}}{\overset{\overset{R}{|}}{Si}}-O-\boxed{A}-O- \right]$$

(163)

R = Me , Et , Ph

X = Cl , OEt , OPh , OAc , NH—Ph , NEt$_2$

$\boxed{A}$ = bivalent aliphatic or aromatic unit

$$Me_2Si\underset{O-CH_2-CH_2-O}{\overset{O-CH_2-CH_2-O}{\diagdown\diagup}}SiMe_2 \qquad\qquad Me_2Si\underset{O}{\overset{O}{\diagup\diagdown}}(\overset{}{C}H_2)_n$$

a b

(164)

or Ph$_2$SiCl$_2$ in the presence of NH$_3$ [146] in solution [146]. However, even in these case the M$_n$'s never exceeded 2000 g/mol.

In addition to dichlorosilanes, diethoxy, diphenoxy and bisacetoxy silanes were studied as monomers [149–151]. Ethoxide catalyzed transacetalization of dimethyldiethoxysilane with methylenebisacetate yielded oligoacetals, Eq. (165)

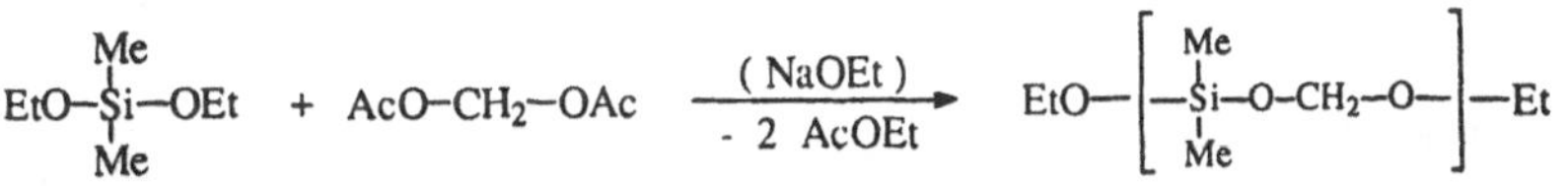

$$EtO-\underset{\underset{Me}{|}}{\overset{\overset{Me}{|}}{Si}}-OEt \ + \ AcO-CH_2-OAc \quad \xrightarrow[\text{- 2 AcOEt}]{(NaOEt)} \quad EtO-\left[\underset{\underset{Me}{|}}{\overset{\overset{Me}{|}}{Si}}-O-CH_2-O-\right]-Et \quad (165)$$

which were isolated by fractionated distillation up to M$_n$'s around 700 g/mol. A high molecular weight poly(alkoxysilane) was prepared by polycondensation of dimethylbisacetoxysilane and 2, 6-di(hydroxymethyl)4-methylanisol Eq. (166),

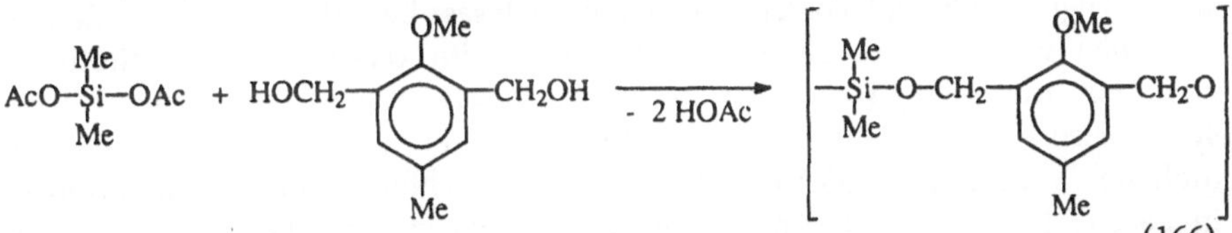

$$AcO-\underset{\underset{Me}{|}}{\overset{\overset{Me}{|}}{Si}}-OAc \ + \ HOCH_2-\underset{\underset{Me}{}}{\overset{\overset{OMe}{}}{\bigcirc}}-CH_2OH \quad \xrightarrow{\text{- 2 HOAc}} \quad \left[-\underset{\underset{Me}{|}}{\overset{\overset{Me}{|}}{Si}}-O-CH_2-\underset{\underset{Me}{}}{\overset{\overset{OMe}{}}{\bigcirc}}-CH_2O\right]$$

(166)

when a two step procedure was applied [150]. The second step was catalyzed by sulfamic acid. Prolonged heating resulted in a crosslinked resin.

A broad variety of high molecular weight poly(aryloxysilane)s were synthesized by polycondensation of diphenols either with diphenoxysilanes or with dianilinosilanes (Forms. 167a–e and 168a–c [151, 152]. These poly(aryloxysilane)s were characterized in much detail. In addition to elemental analyses M$_w$'s (7000–70000) and IR spectra were recorded. Furthermore, the hydrolytic stability and the thermal stability were determined. Moreover, mechanical properties, such as tensile strengths of fibers, heat deflection temperature, elongation at break, impact strength and flexural strength were measured. Polymer

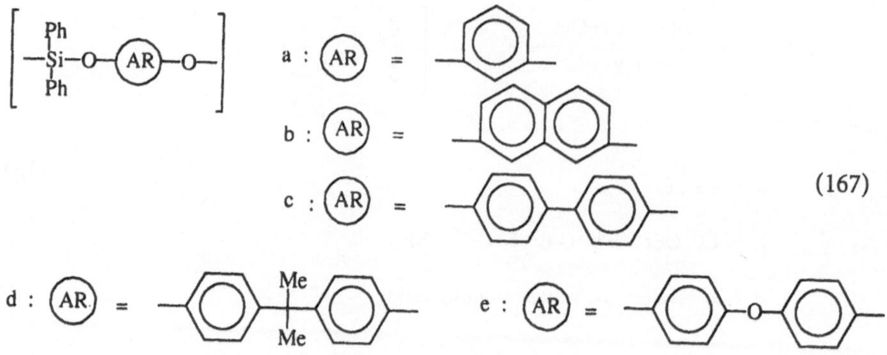

(167)

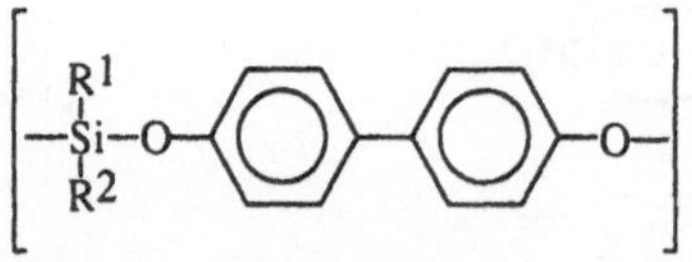

$$a : R^1, R^2 = Ph$$
$$b : R^1, R^2 = Me \qquad (168)$$
$$c : R^1 = Ph, R^2 = Me$$

168a was found to possess the most promising combination of properties. Another research group [153] used the same approach with different diphenols. Again thermostable, amorphous, film-forming poly(aryloxysilane)s were obtained. DSC-measurements (T_g up to 172 °C) and UV-spectra were recorded.

Amorphous and soluble poly(aryloxydisilane)s were prepared by a similar approach Eq. (169) [154, 155]. Twelve different diphenols were used in this

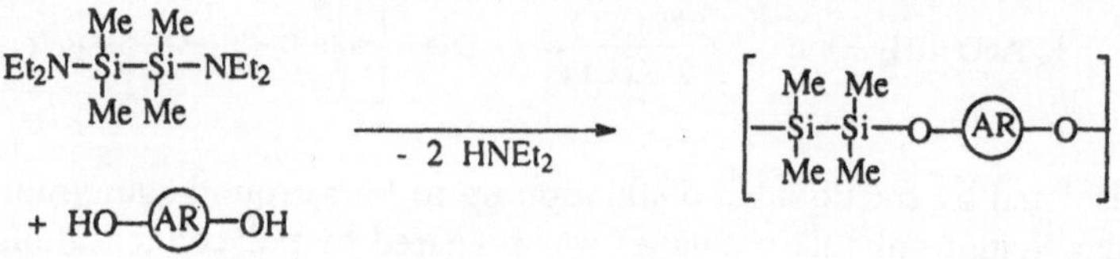

study. High yields and correct elemental analyses, but only low or moderate molecular weights were obtained. Irradiation with UV-light causes photodegradation presumably by radical mechanisms. Poly(aryloxysilane)s can also be synthesized by a quite different approach, namely by transsilylation of silylated diphenols with diphenyldichlorosilane (or other dichlorosilane)s. As mentioned in Sect. (5.10) these polycondensations are sluggish and do not yield high molecular weights despite catalysis with chloride ions. However, they have the advantage that cocondensation with other reactive acid chlorides may be conducted. In this way Si-containing polyesters are accessible, Eq. (170) [156].

Copolymers containing both aryloxysilyl and silazane units were easily prepared by heating neat cyclosilazanes with diphenols, Eq. (171) [157]. Not only the nature of the diphenols but also that of the cyclosilazanes was varied over a broad range (Forms. **172**). Yields above 90% and intrinsic viscosities in the range of 0.4–1.6 were obtained. All materials showed a rubbery character. Their hydrolytic and thermal stability was investigated. Gum-like copolymers of similar structure, but low molecular weight were formed, when hexamethylcyclosilazane was polycondensed with aliphatic diols [158]. Cyclic oligomers were

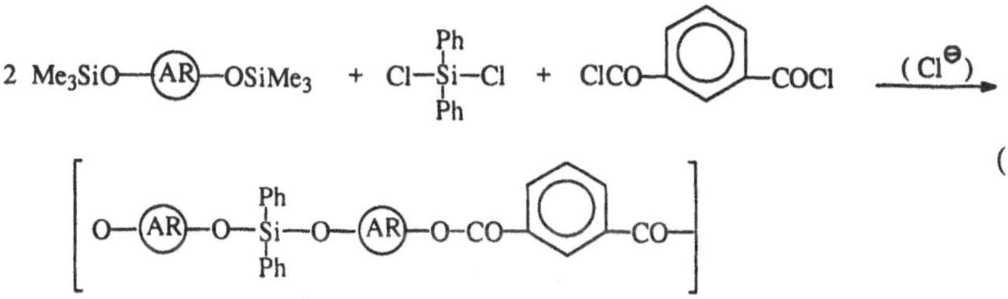

$$2\ Me_3SiO-(AR)-OSiMe_3\ +\ Cl-\underset{\underset{Ph}{|}}{\overset{\overset{Ph}{|}}{Si}}-Cl\ +\ ClCO-\bigcirc-COCl\ \xrightarrow{\ (Cl^{\ominus})\ }$$

$$\left[O-(AR)-O-\underset{\underset{Ph}{|}}{\overset{\overset{Ph}{|}}{Si}}-O-(AR)-O-CO-\bigcirc-CO\right]$$

(170)

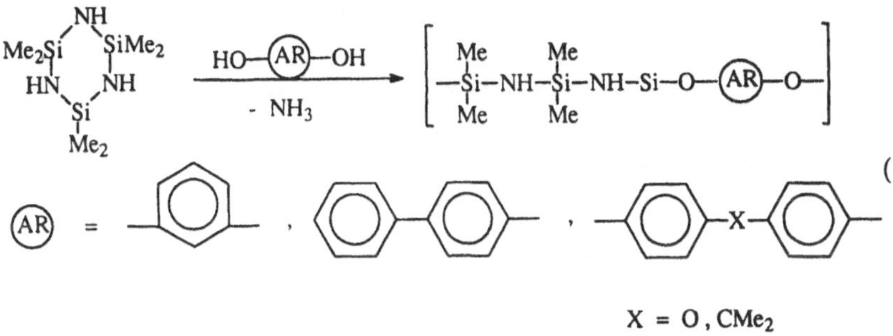

$$\underset{\underset{Me_2}{\overset{|}{Si}}}{\overset{\overset{NH}{/\ \ \backslash}}{Me_2Si\ \ \ \ SiMe_2}}\ \ \underset{HN\ \ \ \ NH}{}\ \ \xrightarrow[-\ NH_3]{HO-(AR)-OH}\ \left[\begin{array}{c} Me\ \ \ \ \ Me \\ -\underset{\underset{Me}{|}}{\overset{\overset{|}{Si}}}-NH-\underset{\underset{Me}{|}}{\overset{\overset{|}{Si}}}-NH-Si-O-(AR)-O- \end{array}\right]$$

(171)

$$(AR)\ =\ -\bigcirc-\ ,\ -\bigcirc-\bigcirc-\ ,\ -\bigcirc-X-\bigcirc-$$

$$X\ =\ O\ ,\ CMe_2$$

the main products of polycondensations involving bis(diethylaminodimethyl-silane) and diaminobenzenes, Eq. (173) [159]. This result is at least surprising for 1,4-diaminobenzene.

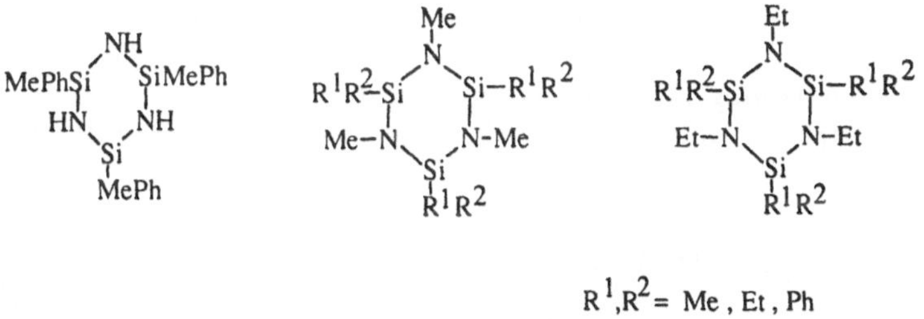

$$\underset{\underset{MePh}{\overset{|}{Si}}}{\overset{\overset{NH}{/\ \ \backslash}}{MePhSi\ \ \ \ SiMePh}}\ \underset{HN\ \ \ \ NH}{}\ \ \ \ \ \underset{\underset{R^1R^2}{\overset{|}{Si}}}{\overset{\overset{\overset{Me}{|}}{N}}{R^1R^2Si\ \ \ \ Si-R^1R^2}}\ \underset{Me-N\ \ \ \ N-Me}{}\ \ \ \ \ \underset{\underset{R^1R^2}{\overset{|}{Si}}}{\overset{\overset{\overset{Et}{|}}{N}}{R^1R^2Si\ \ \ \ Si-R^1R^2}}\ \underset{Et-N\ \ \ \ N-Et}{}$$

(172)

$$R^1,R^2 =\ Me\ ,\ Et\ ,\ Ph$$

$$\underset{NH_2-\bigcirc-NH_2}{\overset{Et_2N-\underset{\underset{Me}{|}}{\overset{\overset{Me}{|}}{Si}}-NEt_2}{}}\ +\ \ \xrightarrow[-\ 2\ HNEt_2]{}\ \left[-\underset{\underset{Me}{|}}{\overset{\overset{Me}{|}}{Si}}-NH-\bigcirc-NH-\right]$$

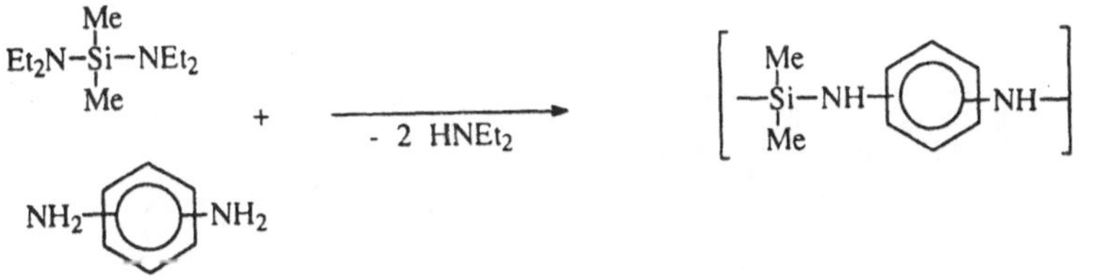

(173)

Aromatic polyamides containing silicon in the backbone were prepared in high yields by the standard solution polycondensation as illustrated by Eq. (174) [160, 161]. The resulting polyamides were characterized by various methods including thermogravimetry in air. In contrast to other polyaramides they are amorphous and soluble in numerous organic solvents. The Si-containing dicarboxylic acids used in this work are, of course, also useful for the synthetic of Si-containing polyesters.

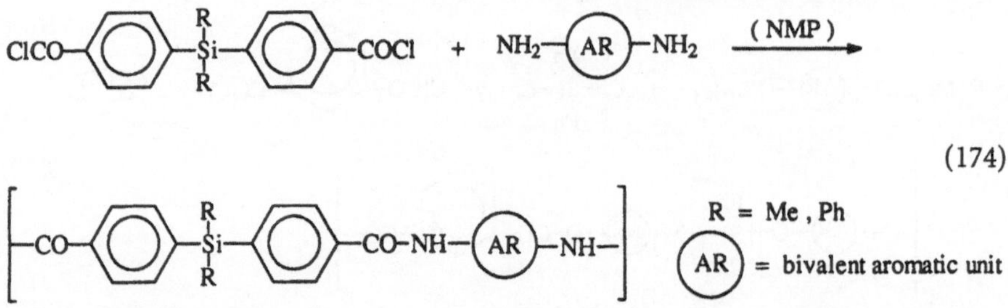

$$\text{ClCO}-\!\!\left\langle \bigcirc \right\rangle\!\!-\!\!\underset{\underset{R}{|}}{\overset{\overset{R}{|}}{Si}}\!\!-\!\!\left\langle \bigcirc \right\rangle\!\!-\!\!\text{COCl} \quad + \quad \text{NH}_2-\!\!\left(\text{AR}\right)\!\!-\!\!\text{NH}_2 \quad \xrightarrow{(\text{NMP})}$$

$$(174)$$

$$\left[-\text{CO}-\!\!\left\langle \bigcirc \right\rangle\!\!-\!\!\underset{\underset{R}{|}}{\overset{\overset{R}{|}}{Si}}\!\!-\!\!\left\langle \bigcirc \right\rangle\!\!-\!\!\text{CO-NH}-\!\!\left(\text{AR}\right)\!\!-\!\!\text{NH}- \right]$$

R = Me , Ph

$\left(\text{AR}\right)$ = bivalent aromatic unit

An approach which has found broad application for the synthesis of hetero-cyclic or fully conjugated polymers is the polycondensation of metallated mono-mers with reactive dihalocompounds (see Chapter 4). A typical example is the polycondensation of lithiated thiophene with dichlorodimethylsilane, Eq. (175)

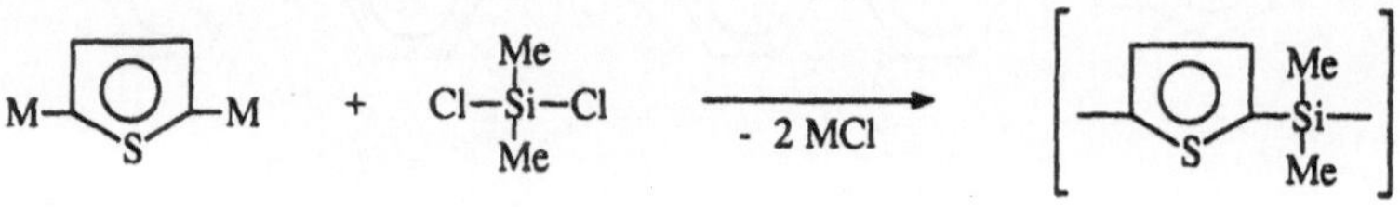

$$\text{M}-\!\!\left\langle\!\!\underset{S}{\bigcirc}\!\!\right\rangle\!\!-\text{M} \quad + \quad \underset{\underset{Me}{|}}{\overset{\overset{Me}{|}}{Cl-Si-Cl}} \quad \xrightarrow{-\,2\,\text{MCl}} \quad \left[-\!\!\left\langle\!\!\underset{S}{\bigcirc}\!\!\right\rangle\!\!-\underset{\underset{Me}{|}}{\overset{\overset{Me}{|}}{Si}}- \right]$$

$$(175)$$

M = Li , MgCl

[162]. Grignard and Inderivatives of thiophenes were used analogously [162, 163]. A somewhat unusual example of this synthetic strategy was described for monomers containing silicon in the main chain and as pending protecting groups (Forms. **176a–d** and **177**) [164]. The dibrominated monomer (**176**)b was

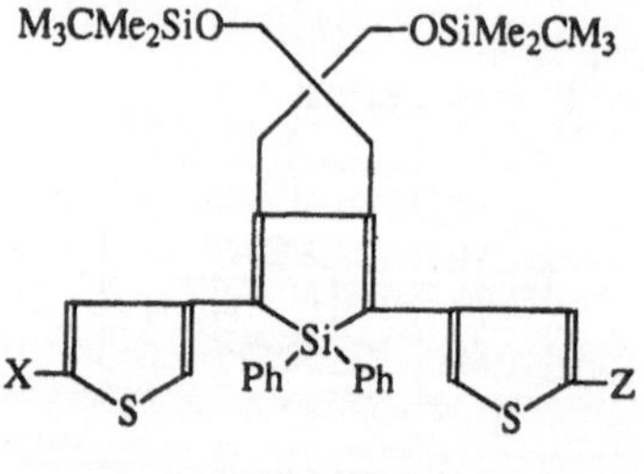

a : X = H , Z = Br
b : X , Z = Br
c : X = H , Z = SnMe₃
d : X , Z = SnMe₃

$$(176)$$

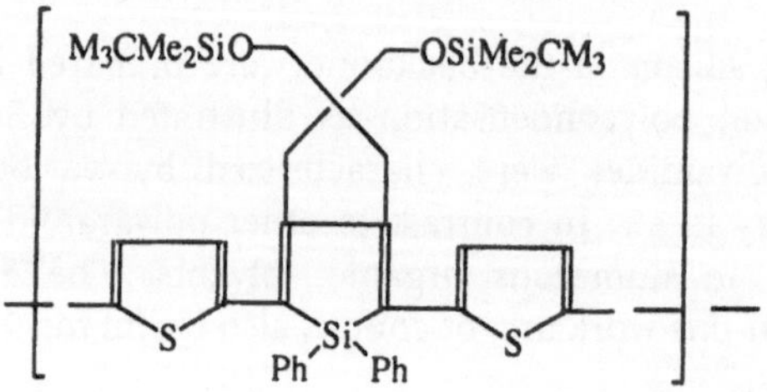

$$(177)$$

reacted with the distannylated monomer **176d** in the presence of Pd complex yielding the polymer **177** with $M_n = 7700$ and $M_w = 12\ 600$. Analogous oligomers were prepared in a stepwise manner, by means of the monomers **176a** and c [164]. All these "polythiophenes" are of interest as potentially electroconducting materials (after doping).

Hydrosilylation of olefinic endgroups is a widely used approach for the synthesis of Si-containing polymers (see also Chapters 3 and 4). The synthesis of an alternating copolymer containing a disiloxane and a ferrocene group may serve here as a special example, because it illustrates the formation of thermostable polymer with a metal complex in the backbone, Eq. (178). High yields but only

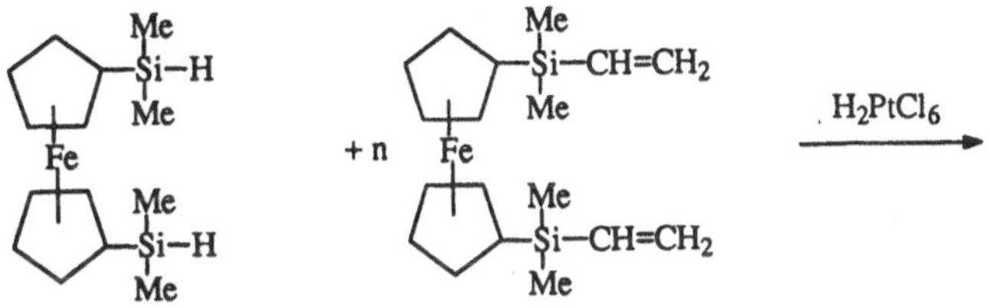

$$\tag{178}$$

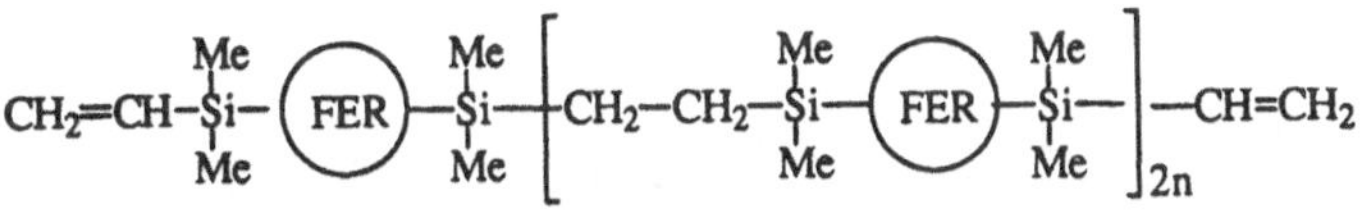

low molecular weights ($M_n < 6000$ g/mol) were reported [165].

Further interesting examples of polymers with metal complexes in the backbone were described in russian patents (Forms. **179a, b**) [166–168]. In this case

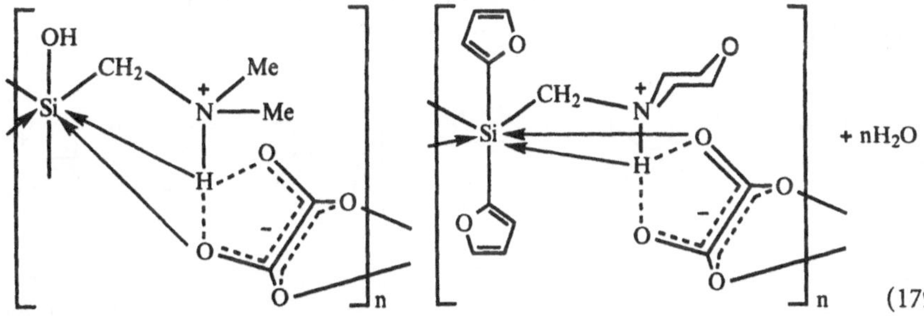

$$\tag{179}$$

a   b

the silicon is the central atom of the complexes, and an ionic bond is also involved in the stabilization of the backbone. An example for the synthesis of polymers with covalent backbone and "pendant" ligands is shown in Eq. (180). The free OH-groups of the silicic units were partially silylated with $ClSiMe_3$ [169]. The resulting polysiloxanes were soluble in $CDCl_3$ and were characterized by ^{1}H NMR and IR-spectroscopy. Polymers on the borderline between organic and inorganic materials were prepared by the polycondensation of NH_2NH_2 with $PhSiH_3$ in the presence of CuCl or CuH catalyst [170]. The repeat unit **181** with trivalent (!) Si-atoms was published. Another unusual material was obtained by

$$
HO-\left(-\underset{\underset{OH}{\overset{Ph}{\mid}}}{Si}-O\right)_n-H \quad \xrightarrow[- \ 2 \ iPrOH]{+ \ iPrO-\underset{\underset{acac}{\mid}}{\overset{acac}{\mid}}{Ti}-OiPr} \quad HO+\left[\left(\underset{\overset{OH}{\mid}}{\overset{\overset{OH}{\mid}}{Si}}-O\right)_n \underset{acac}{\overset{acac}{Ti}}-O\right] \tag{180}
$$

$$
-\underset{\underset{Ph}{}}{Si}\underset{\overset{NH-NH}{\diagdown}}{\overset{NH-NH}{\diagup}} Si-NH-NH-Si\underset{\overset{NH-NH}{}}{\overset{NH-NH}{\diagup}}\diagdown_{\overset{NH-NH}{}}Si- \tag{181}
$$

plasma treatment of ClSiMe$_3$. The films deposed on Si-wafers or glass-plates were modified by heating with TMS azide [171]. No structural formulas were reported.

In addition to polymers with Si-atoms in the backbone polymers with Si-containing substituents need to be mentioned. All polymers prepared from Si-containing vinylmonomers described in Chapter 1 belong to this category. One example, the anionic polymerization of benzyldimethylvinylsilane, is particularly interesting, because an anionic rearrangement occurs which leads to the "migration"of a silyl group into the polymer backbone [172]. In addition to the repeating unit **182a** expected for a clean anionic polymerization, repeat units of structure **182b** were found by spectroscopic methods.

$$
\begin{bmatrix} -\underset{}{CH}-CH_2- \\ \underset{}{SiMe_2} \\ CH_2-Ph \end{bmatrix}_a \qquad \begin{bmatrix} \qquad\qquad Me \\ -CH_2-\underset{}{CH}-\underset{}{Si}-\underset{}{CH}- \\ \qquad CH_2 \ \ Me \ \ Ph \\ \qquad CH_2 \\ \qquad Me_2SiCH_2Ph \end{bmatrix}_b \tag{182}
$$

Aromatic copolyethers resulted from the oxydative coupling of silane substituted phenols with 2,6-dimethylphenol, Eq. (183) [173]. The polymerization

$$
n = 1,2,3 \tag{183}
$$

mechanism was studied kinetically and spectroscopically. A particularly interesting mechanism is the polyaddition of Si-substituted germylenes and stannylenes (the analogs of carbenes) with quinones. The comonomers also play the role of initiators or activators, Eq. (184) [174]. In addition to 1,4-benzoquinone, 2,5-dichloro-, 2,5-bis-*tert*-butyl-, 2,5-diphenyl benzoquinone and 1,4-naphthoquinone were used as comonomers. Germylene also copolymerizes with *N*-phenylquinoneimine, Eq. (185) [175] and with β-unsaturated cyclic ketones, Eq. (186) [176].

Polyimides with short siloxane side chains were prepared by polycondensation of the diamines **187a,b** with various difunctional cyclic

$$M = Ge, Sn \tag{184}$$

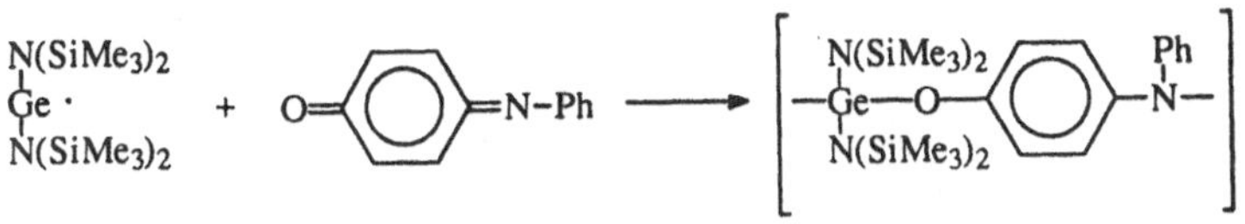

$$(185)$$

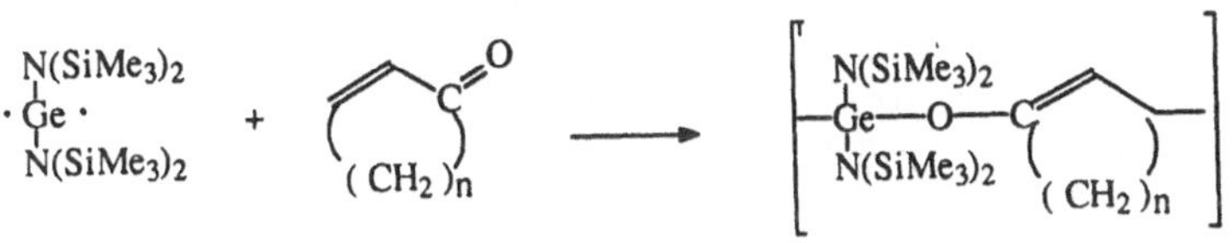

$$(186)$$

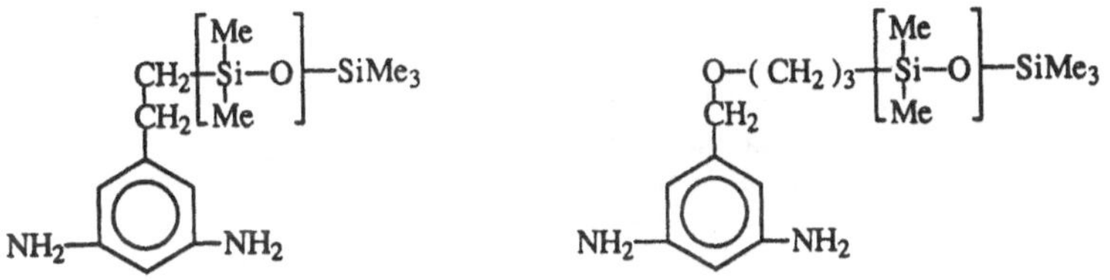

a   b

$$(187)$$

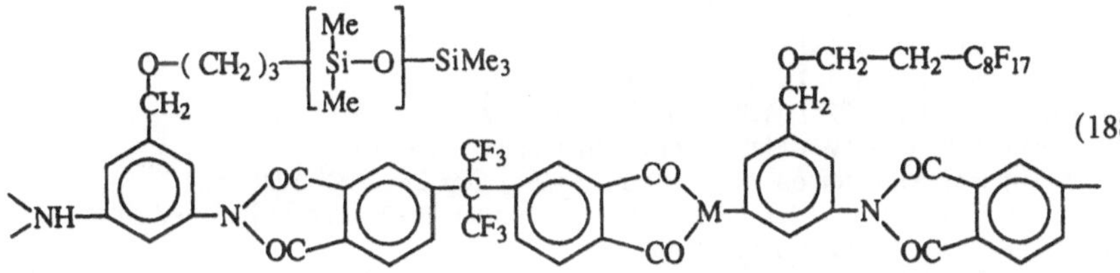

$$(188)$$

anhydrides [177–180]. A typical segment of a polyimide containing both siloxane and fluoro carbon side chains is depicted in Form. **188**. Such polyimides are of interest as membranes with high permeability for gases particularly O_2.

Finally two publications [181, 182] should be mentioned which report on the formation of networks by means of multifunctional Si-compounds. In one case, multifunctional diphenylmethane derivatives such as **189**, which was hydrolyzed together with a Si-terminated polypropylene oxide in the presence of a Sn catalyst [181]. In the second case [182], polystyrene or cationically polymerized

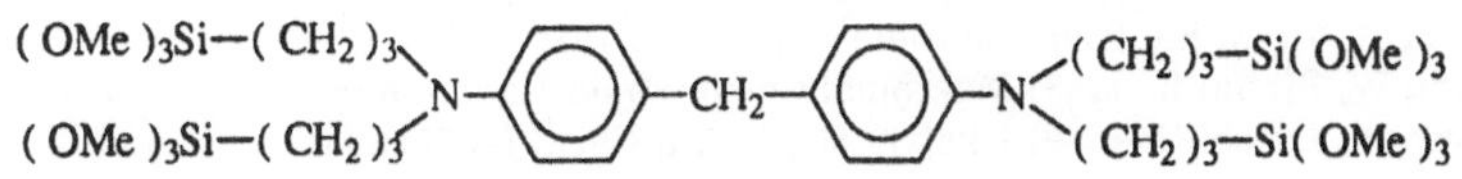

$$(189)$$

vinyl ether were terminated with allyl groups. Hydrosilylation of these allyl end groups with cyclosiloxanes or functional oligosiloxanes followed as second step. The reactive siloxane end groups are then useful for both chain extension and crosslinking. The introduction and hydrolytic crosslinking of trifunctional silyl end groups is also discussed in the preceding section in connection with metathesis ROP.

6.5
References

1. Kricheldorf HR (1972) Synthesis 551
2. Washburne SS, Peterson WR Jr (1972) J Am Oil Chem Soc 49: 694
3. Peterson WR Jr, Radell J, Washburne SS (1973) J Fluorine Chem 2: 437
4. Birkofer L, Ritter A (1965) Angew Chem 77: 414 (1965) Angew Chem Int Ed 4: 417
5. Rühlmann K, Reichert A, Becker M (1965) Chem Ber 98: 1814
6. Wiberg N, Neruda B (1966) Chem Ber 99: 740
7. Kricheldorf HR, (1979) Angew Chem 749 (1979) Angew Chem Int Ed 18: 689
8. Ho TL, Olah G (1976) Angew Chem 88 847 (1976) Angew Chem Int Ed 15: 774
9. Jung ME, Lyster M (1977) J Am Chem Soc 99: 968
10. Kricheldorf HR (1971) Liebigs Ann Chem 748: 101
11. Kricheldorf HR (1975) Angew Chem. 87 517 (1975) Angew Chem Int Ed 14: 502
12. Kricheldorf HR, Fehrle MJ (1980) Makromol Chem 181: 2571
13. Manecke G, Pohl R (1976) Makromol Chem 177: 1999
14. Kricheldorf HR, Leppert E (1976) Synthesis 43
15. Kricheldorf HR, Schultze J (1976) Synthesis 739
16. Schwarz G, Alberts A, Kricheldorf HR (1981) Liebigs Ann Chem 1257
17. Burgert J, Stadler R (1987) Chem Ber 120: 691
18. Mohrmann W, Leukel G (1988) Synth Commun 990
19. Iwakura Y, Uno K, Kang S (1965) J Org Chem 30: 1158
20. Steinbrunn G (1958) Ger Offen 848808 (1949) to BASF AG; C.A. 52: 11116c
21. Kricheldorf HR, Leppert E (1972) Makromol Chem 158: 223
22. Kricheldorf HR (1972) Chem Ber 105: 3958
23. Kricheldorf HR (1973) Chem Ber 106: 3753
24. Kricheldorf HR (1975) Makromol Chem 176: 57
25. Kricheldorf HR (1973) Makromol Chem 173: 13
26. Washburne SS, Peterson WP Jr, Berman DA; J Org Chem 37: 1738
27. Greber G, Kricheldorf HR (1968) Angew Chem 80: 1029 Angew Chem Int Ed
28. Kricheldorf HR (1970) Synthesis 592
29. Kricheldorf HR (1970) Synthesis 649
30. Greber G, Kricheldorf HR (1968) Angew Chem 80: 1028 Angew Chem Int Ed 7: 941
31. Kricheldorf HR (1972) Angew Chem 84: 107
32. Kricheldorf HR (1970) Synthesis 539
33. Kricheldorf HR (1971) Liebigs Ann Chem 748: 101
34. Higashimura T, Kato H, Suzuouki K, Okamura S (1966) Makromol Chem 90: 243 and 254
35. Kricheldorf HR (1973) Makromol Chem 167: 1
36. Kricheldorf HR (1973) Makromol Chem 170: 89
37. Kricheldorf HR, Leppert E (1975) Synthesis 592
38. Kricheldorf HR, Leppert E (1975) Synthesis 591
39. Malamet G, Vernaleken H, Krimm H, Schnell H (1975) Ger Offen 1668076 (1974) to Bayer AG; C A 82: 98 125
40. Mohrmann W, Hißmann (1987) Tetrahedron Lett 3087
41. Mohrmann W, Hohn E (1989) Makromol Chem 190: 1919
42. Mohrmann W, Brahm B (1988) Makromol Chem, Rapid Commun 9: 175
43. Mohrmann W, Leukel G (1989) J Polym Sci, Polym Chem Ed 27: 4341

44. Mohrmann W, Brahm M (1989) Makromol Chem 190: 631
45. Kricheldorf HR, Regel W (1973) Chem Ber 106: 31753
46. Kricheldorf HR (1975) Makromol Chem 176: 2755
47. Kricheldorf HR (1971) Makromol Chem 149: 127
48. Kricheldorf HR (1987) In: α-Amino Acid N-Carboxyanhydrides and Related Heterocycles, Springer, Berlin Heidelberg New York.
49. Kricheldorf HR (1989) In: Allen G. Bevington JC (eds) α-Comprehensive polymer science, vol 3. Pergamon, Oxford, p. 531
50. Kricheldorf HR (1990) In: Penczek S (ed) Models of biopolymers by ring-opening polymerization. CRC, Boca Raton, Fl, chap 1
51. Kricheldorf HR, Mühlhaupt R (1979) Makromol Chem 180: 1419
52. Kricheldorf HR, Mühlhaupt R (1980) J Makromol Sci Chem A 14: 349
53. Kricheldorf HR (1970) Chem Ber 103: 3353
54. Wies R, Pfaender P (1973) Liebigs Ann Chem 1269
55. Hirschmann R, Schwam H, Strachan RG, Schoenewoldt FE, Barkemeyer H, Miller SM, Conn JB, Garky V, Veber DF, Denkewalter RG (1971) J Am Chem Soc 93: 2746
56. Kricheldorf HR (1971) Chem Ber 104: 87
57. Kricheldorf HR (1971) Chem Ber 104: 3168
58. Kricheldorf HR, Greber G (1971) Chem Ber 104: 3131
59. Kricheldorf HR (1070) Angew Chem 82: 550 (1970) Angew Chem Int Ed 9: 526
60. Kricheldorf HR (1971) Chem Ber 104: 3156
61. Kricheldorf HR (1974) Makromol Chem 175: 3343
62. Kricheldorf HR (1970) Synthesis 592
63. Kricheldorf HR (1971) Angew Chem 83: 539 (1971) Angew Chem Int Ed 10: 507
64. Kricheldorf HR, Leppert E, Schilling G (1974) Makromol Chem 175: 1705
65. Kricheldorf HR, Leppert E, Schilling G (1975) Makromol Chem 176: 1629
66. Kricheldorf HR, Schulze J (1977) Makromol Chem 178: 3141
67. Kricheldorf HR, Kaschig J (1978) Eur Polym J 14: 923
68. Kricheldorf HR, Leppert E, Schilling G (1975) Angew Makromol Chem 45: 119
69. Kricheldorf HR (1975) Liebigs Ann Chem 1387
70. Kricheldorf HR, Schilling G (1977) Makromol Chem 178: 3115
71. Kricheldorf HR, Leppert E (1974) Makromol Chem 175: 1731
72. Luderwald I, Kricheldorf HR (1976) Angew Makromol Chem 56: 173
73. Kricheldorf HR, Joshi S, Hull WE (1982) J Polym Sci Chem Ed 20: 2791
74. Kricheldorf HR, Hull WE (1978) J Macromol Sci Chem A 12: 51
75. Kricheldorf HR (1982) Pure & Appl Chem 54: 467
76. Katz TJ, Lee SJ, Shippey MA (1980) J Mol Cat 8: 219
77. Ginsburg EJ, Gorman CB, Marder SR, Grubbs RH (1989) J Am Chem Soc 111: 7621
78. Zimmermann M, Pampus M, Maertens D, Ger Offen 2.460.911 (1974) to Bayer AG; C A 85: 95539
79. Streck R, Weber H (1975) Ger Offen 2.314.543 (1973) to Hüls AG; C A 82: 73848
80. Streck R (1982) J Mol Catal 15: 3
81. Finkelstein ESh, Makovetskii KL, Yampolskii YP, Portnykh EB, Ostrovskaya IY, Kaliuzhnyi NE, Pritula NA, Goldberg AI, Yatsenko MS, Platé NA (1991) Makromol Chem 192: 1
82. Stonich DA, Weber WP (1991) Polym Bull 26: 483
83. Stonich DA, Weber WP (1991) Polym Bull 27: 243
84. Lauters M, Crudden CM, Abd-El-Aziz AS, Wada T (1991) Macromolecules 24: 1425
85. Lammens H, Sartori G, Siffert J, Sprecher N (1971) J Polym Sci Part B 9:341
86. Zhang X, Zhou Q, Weber WP, Horvath RF, Chan TH, Manuel G (1988) Macromolecules 21: 1563
87. Zhou SQ, Park YT, Manuel G, Weber WP (1990) Polym Bull 23: 491
88. Park YT, Manuel G, Weber WP (1990) Macromolecules 23: 1911
89. Zhou SQ, Weber WP (1990) Macromolecules 23: 1915
90. Zhou Q, Manuel G, Weber WP (1990) Macromolecules 23: 1583
91. Stonich DA, Weber WP (1991) Polym Bull 25: 629

92. Liao X, Ko Y-H, Manuel G, Weber WP (1991) Polym Bull 25: 63
93. Park YT, Zhou SQ, Manuel G, Weber WP (1991) Macromolecules 24: 3221
94. Wang L, Ko Y-H, Weber WP (1992) Macromolecules 25: 2828
95. Park YT, Zhou Q, Weber WP (1989) Polym Bull 22: 349
96. Ko Y-H, Weber WP (1991) Polym Bull 26: 487
97. Gilman H, Atwell WH (1964) J Am Chem Soc 86: 2687
98. Vdovin VM, Grinberg PL, Babich ED (1965) Dokt Akad Nauk SSSR 161: 268
99. Nametkin NS, Vdovin VM, Zaryalov VI (1965) Izv Akad Nauk SSSR Ser Khim 1448
100. Theurig M, Weber WP (1992) Polym Bull 28: 17
101. Theurig M, Sargeant JS, Manuel G, Weber WP (1992) Macromolecules 25: 3834
102. Liao CX, Weber WP (1992) Polym Bull 28: 281
103. Liao CX, Weber WP (1992) Macromolecules 25: 1639
104. Finkel'shtein ESh, Ushakov NV, Pritula NA, Andreev EA, Plate NA (1992) Izv Akad Nauk SSSR Ser Khim 223 (1992); CA 116: 256191f
105. Wu HJ, Interrante LV (1991) ACS Polym Prepr 32: 588
106. Wu HJ, Interrante LV (1989) Chem Mater 1: 564
107. Ishikawa M, Hatano T, Hasegawa Y, Horio T, Kunai A, Miyai A, Ishida T, Tsukihara T, Yamanaka T (1992) Organometallics 11: 1604
108. Pitt CG, Gu ZW, Ingram P, Hendren RW (1987) J Polym Sci Part A, Polym Chem 25: 955
109. Uryu T, Yamanaka M, Heumi M, Hatanaka K, Matsuzaki K (1986) Carbohydrate Res 157: 157
110. Allcock HR, Brennan DJ, Graaskamp JM (1986) Organometallics 5: 2434
111. Allcock HR, Brennan DJ, Graaskamp JM (1988) Macromolecules 21: 1
112. Allcock HR, Brennan DJ, Dunn BS (1989) Macromolecules 22: 1534
113. Vandenberg EJ (1985) J Polym Sci, Polym Chem Ed 23: 915
114. Neville RG (1960) J Org Chem 25: 1063
115. Crivello JV, Fan M, Bi D (1992) J Appl Polym Sci 44: 9
116. Gunther P, Haas F, Marwede G, Nützel K, Oberkirch W, Pampus G, Schön N, Witte J (1970) Angew Makromol Chem 14: 87
117. Drapeau A, Leonard J (1985) Macromolecules 18: 144
118. Dreshkin IA, Redkina LI, Kershenbaum IL, Chernenko GM, Makovetsky KL, Tinyakova EI, Dolgoplosk BA (1977) Eur Polym J 13: 447
119. Matsumoto S, Komatsu K, Igarashi K
120. Castner KF, Calderon N (1982) J Mol Catal 15: 47
121. Weiss K, Guthmann W, Maisuls S (1988) Angew Chem 100: 268 (1988); Angew Chem Int Ed 27: 275
122. Weiss K, Lößel G (1989) Angew Chem 101: 75 (1989) Angew Chem Int Ed 28: 62
123. Risse W, Wheeler DR, Canmizzo LF, Grubbs RH (1989) Macromolecules 22: 3205
124. Nambu Y, Endo T (1991) Macromolecules 24: 2127
125. Höcker H, Keul H, Kühling S, Hoverstadt W, Müller AJ (1991) Makromol Chem, Macromol Symp 44: 239
126. Hall HK Jr, Padias AB, Hsumi M, Way TF (1990) Macromolecules 23: 678
127. Dunsing R, Kricheldorf HR (1988) Eur Polym J 24: 145
128. Hrkach JS, Matyjazewski K (1990) Macromolecules 23: 4042
129. Hrkach JS, Matyjazewski K (1990) Macromolecules 25: 2070
130. Xu B, Lillya P, Chien JCW (1992) J Polym Sci Part A, Polym Chem 30: 1899
131. Chalk AJ (1970) Chem Commun 847
132. Crivello JV, Fan M (1992) J Polym Sci Part A, Polym Chem 30: 31
133. Crivello JV, Fan M (1992) J Polym Sci Part A, Polym Chem 30: 1
134. Crivello JV, Fan M (1992) Makromol Chem, Macromol Symp 54/55: 189
135. Crivello JV, Fan M (1991) J Polym Sci Part A, Polym Chem 29: 1853
136. Crivello JV, Fan M (1992) Makromol Chem , Macromol Symp 54/55: 179
137. Hayase S, Onishi Y, Suzuki S, Wada M (1987) J Polym Sci Part A, Polym Chem Vol 25: 753
138. Streck R (1983) Chemtech 758
139. Kobayashi S, Uyama H, Ogaki M, Yoshida T, Saegusa T (1989) Macromolecules 22: 4412

140. Ogata S, Kakimoto M, Imai Y (1985) Makromol Chem, Rapid Commun 6: 835
141. Chen G-F, Jones FN (1991) New Polymeric Mater 2: 329
142. van der Made WA, von Leeuwen PWNM (1992) J Chem Soc Chem Commun 1400
143. Mathias LJ, Carothers TW (1991) J Am Chem Soc 113: 4043
144. Roovers J, Toporowski PM, Zhou L-L (1992) ACS Polym Prepr 33: 182
145. Krieble RH, Burkhard CA (1947) J Am Chem Soc 69: 2689
146. Hopf H, Egli R (1959) Kunststoffe-Plastics 6: 433
147. Hanford WE (1948) US Pat 2.441.066 (1948) to EI DuPont de Nemours & Co, C A 42: 5268
148. Sprung MM (1958) J Org Chem 23: 58
149. Henglein FA, Schmulder P (1954) Makromol Chem 13: 53
150. Martin RW, Frisch KC (1955) US Pat 2.707.191 (1955) CA 49: 13691
151. Dunnavant WR, Markle RA, Stickney PB, Curry JE, Burd JD (1967) J Polym Sci Part A 15: 707
152. Curry JE, Burd JD (1965) J Polym Sci 9: 295
153. Saegusa Y, Kato T, Oshiumi H, Nakamura S (1992) J Polym Sci Part A, Polym Chem 30: 1401
154. Padmanabon M, Kakimoto M, Imai Y, J Polym Sci Part A, Polym Chem 28: 2997
155. Imai Y (1991) J Macromol Sci, Chem A 28: 1115
156. Kricheldorf HR, Stöber O (Manuscript in preparation)
157. Elliot RL, Breed LW (1964) ACS Polym Prepr 5: 587
158. Zhinkin DY, Semenova EA, Morgunova MM, Sobolevski MV (1963) Plaste-Kautschuk, 10 396 (1963); CA 59: 15397
159. Kulpinski J, Lasocki Z (1991) Acta Polymerica 42: 86
160. Mohite SS, Maldar NN, Marvel CS (1988) J Polym Sci Part A, Polym Chem 26: 2777
161. Jadhav AS, Maldar NN, Shinde BM, Vernehav SP (1991) J Polym Sci Part A, Polym Chem 29: 147
162. Chicart P, Corrice R, Moreau J (1991) Fr Dem 2.655.655 (1991) to Rhône-Poulenc; CA 116: 7134u
163. Chicart P, Corrice R, Moreau J (1991) Fr Dem 2.655.655 (1991) to Rhône-Poulenc CA 116: 21728x
164. Tamao K, Yamaguchi S, Shiozaki M, Nakagawa Y, Ito Y (1992) J Am Chem Soc 114 58: 67
165. Greber G, Hallensleben M (1965) Makromol Chem 83: 148
166. Erchak NP, Liepins E(1991) USSR SU 1.680.713 (1991) CA 117: 91078e
167. Erchak NP, Liepins E (1991) USSR SU 1.685.954 (1991) CA 117: 91079f
168. Erchak NP, Liepins E (1991) USSR SU 1.696.440 (1991) CA 117: 91081a
169. Gunji T, Nagao Y, Misono T, Abe Y (1991) J Polym Sci Part A, Polym Chem 29: 941
170. Liu HQ, Harrod JF (1991) ACS Polym Prepr 32: 563
171. Inagaki N, Tano Y (1987) J Polym Sci Part A, Polym Chem 25: 1197
172. Oku J, Haegawa T, Kawakita T, Kondo Y, Takaki M (1991) Macromolecules 24: 1253
173. Hyun SH, Nishide H, Tsuchida E (1986) Polym Bull 16: 395
174. Kobayashi S, Iwata S, Abe M, Yajima K, Kim HJ, Shoda S (1992) Makromol Chem, Macromol Symp 54/55: 225
175. Kobayashi S, Iwata S, Abe M, Shoda S (1990) Polym Prepr Japan 39: 1781
176. Kobayashi S, Shoda S, Iwata S, Yagi K (1991) Polym Prepr Japan 40: 272
177. Akiyama E, Takamura Y, Nagase Y (1992) Makromol Chem 193: 2037
178. Nagase Y, Mori S, Egawa M, Matsui K (1990) Makromol Chem, Rapid Commun 11: 185
179. Nagase Y, Mori S, Egawa M, Matsui K (1990) Makromol Chem 191: 2413
180. Akiyama E, Takamura Y, Nagase Y (1992) Makromol Chem 193: 1509
181. Takai M, Washimi A, Yoshida M, Kimura K (1992) Jap Pat 0436291 (1992) to Toa Gosei Chem Co, CA 117: 27401c
182. Chinami M, Noda K, Kishimoto Y, Fujisawa H, Iwahara T, Yonezawa K (1991) Jap Pat 03.243.603 (1991) to Kanegafuchi Chem Ind Co, CA 116: 84419t

CHAPTER 7

Chemical Modification of Polymers and Surfaces

H.R. Kricheldorf

7.1
Coupling Agents – Introduction and Syntheses

7.1.1
Introduction

Coupling reagents and surfactants are normally bifunctional, sometimes multi-functional compounds with the purpose to modify the surface of silicates in the widest sense. This purpose involves that at least one functional group is a silicon derivative, usually a $-SiCl_3$, $-Si(OMe)_3$ or $-Si(OEt)_3$ group. These groups are designed to develop between one and three stable covalent bonds with OH-groups on the surface of the silicates.

Coupling reagents capable of forming a covalent connection between silicate fillers, in particular glass-fiber, and a surrounding polymer matrix became of great interest when glass-fiber reinforced resins were introduced as high performance composites in the years around 1940. The mechanically weakest point of such composites is the interface between the organic matrix, on the one hand and the silicate on the other. Due to the presence of OH and $Si-O^\ominus$ groups the surface of glass and other silicates is highly polar and hydrophilic, and thus, not directly compatible with most organic polymers. Even, when the composite is prepared under anhydrous conditions, water may creep in along the silicate surface and destablize the cohesion. Therefore, the success of a coupling reagent is based on two synergistic effects: first, it provides a covalent connection between matrix and filler, second, it reduces or eliminates the wetting and destabilizing influence of moisture. A successful application of silicate coupling reagents also requires an optimization of the second functional group (the so-called carbofunctional group) with regard to the matrix polymer. Since glass-fiber reinforcement has proven to be useful for a broad variety of resins, it is not surprising that numerous different coupling reagents were studied. A selection of the most important and interesting compounds is compiled in Tables 7.1–7.3. The most useful and versatile coupling reagents have become commercial (Table 7.4).

An improvement of the mechanical properties of glass-fiber reinforced composites is not the only successful application of silicate coupling reagents. At least for research purposes they were also widely used in the design and preparation of stationary phases for chromatography. In this field the coupling

Table 7.1 Coupling agents based on trichlorosilanes

Chemical Structure	η_D^{20}	b.p. (°C/mm)	References
CH_3CH_2—$SiCl_3$	1.4257	100–101/746	[A] p.66, [C] p. 683
CH_3—$(CH_2)_2$—$SiCl_3$	1.4290	122/740	[A] p.66
CH_4—$(CH_3)_3$—$SiCl_3$	1.4379	146/741	[A] p.66
CH_3—$(CH_2)_{11}$—$SiCl_3$	1.4540	293–294/760	[C] p.625
CH_3—$(CH_2)_{17}$—$SiCl_3$	1.4600	223/ 10	[C] p.1025
$(CH_3)_3C$—$SiCl_3$	m.p. 97–100 °C	132–134/760	[C] p.279
$CH_2{=}CH$—$SiCl_3$		223/ 10	[B] p.108
$CH_2{=}CH$—$CH_2\cdot SiCl_3$	1.4449	117/760	[B] p.112
CH_2—CMe—CH_2—$SiCl_3$	1.4533	139/743	[D] p.4263
CH_3—$CH{=}CH$—CH_2—$SiCl_3$	1.4598	143–144/750	[B] p.111
Cl—$(CH_2)_3$—$SiCl_3$	1.4638	181–182/750	[E]
Br—$(CH_2)_3$—$SiCl_3$		204/760	[E]
C_6H_5—$SiCl_3$		201–202/760	[F] p.113
p—Cl—C_6H_4—$SiCl_3$		88–91/1.5	[G]

[A] W. Noll "Chemieund Technologie der Silicone", VCH Weinheim/ Bergstr. 1968 2nd. Ed.
[B] S. Paulenko in Houben-Weyl, "Methoden der Organischen Chemie", G. Thieme Verlag, Stuttgart-New York 1980, 4. Ed. Vol. XIII/5.
[C] "Katalog Feinchemikalien"; Aldrich Co (Milwaukee, Wisc. USA) 1992–1993.
[D] Beilsteins Handbuch der Organischen Chemie, Springer Verlag Berlin, Heidelberg-New York 1981, 4th Ed. Volume 4.
[E] Ryan JW, Menzie GK, Speier JL (1990) J Am Chem Soc 82: 3601.
[F] Anner H, Grobe J (1980) J Organomet Chem 188(1): 25.
[G] Rosenberg SD, Walburn JJ, Ramsden HE (1957) J Org Chem 22: 1606.

Table 7.2 Coupling agents based on trimethoxysilanes

Chemical Structure	η_D^{20}	b.p. (°C/mm)	References
$CH_3CH_2Si(OMe)_3$	–	123/760	[A] p.73
$CH_3CH_2CH_2Si(OMe)_3$	1.3900	142/760	[B] p.1226
CH_3—$(CH_3)_{17}$—$Si(OMe)_3$	1.4390	m.p. 16–17 °C	[B] p.1090
$CH_2{=}CH$—$(OMe)_3$	1.3930	123/760	[C] p.4256
$CH_2{=}CH$—CH_2—$Si(OMe)_3$	1.4084	136/760	[D]
Cl—$(CH_2)_3$—$Si(OMe)_3$	1.4183	195/750	[E]
NH_2—$(CH_2)_3$—$Si(OMe)_3$	1.4260	91–92/ 150	[B] p.89
NH_2—$(CH_2)_2$—NH—$(CH_2)_3$—$Si(OMe)_3$	1.4416	140–141/ 15	[F]
CH_2—NH—$(CH_2)_3$—$Si(OMe)_3$	1.4462	160/0.5	[F]
CH_2—NH—$(CH_2)_3$—$Si(OMe)_3$			
$CH_2{=}CMeCO_2(CH_2)_3$—$Si(OMe)_3$	1.4310	190/760	[B] p.916
$H_2C\overset{O}{\diagup}CH$—$CH_2$—$O$—$(CH_2)_3Si(OMe)_3$	$\eta_D^{25} = 1.4281$	112/ 1	[G]
C_6H_5—$Si(OMe)_3$	1.4733	130/45	[D] p.73
p—Br—C_6H_4—$Si(OMe)_3$	1.5121	136/13	[H] p.536

[A] W. Noll "Chemieund Technologie der Silicone", VCH Weinheim/ Bergstr. 1968 2nd. Ed.
[B] "Katalog Feinchemikalien"; Aldrich Co. (Milwaukee, Wisc., USA) 1992–1993.
[C] Beilsteins Handbuch der Organischen Chemie, Springer Verlag Berlin, Heidelberg-New York 1981, 4th Ed. Volume 4.
[D] Pola J, Jakoubkova M, Chalovsky V (1978) Collect Czech Chem Commun 43: 3391 (1978).
[E] Ryan JW, Menzie GK, Speier JL (1990) J Am Chem Soc 82: 3601.
[F] Plueddemann EP (1984) U.S.Pat 4.448694 (1984), C.A. 101 76713g.
[G] Lee L-H (1968) J Coll Interf Sci 27 (4): 751.
[H] Beilsteins Handbuch der Organischen Chemie, Springer Verlag Berlin, Heidelberg-New York 1934, 4th Ed. Volume 16.

Table 7.3 Coupling agents based on triethoxysilanes

Chemical Structure	η_D^{20}	b.p. (°C/mm)	References
CH_3—CH_2—$Si(OEt)_3$	1.3853	158–160/760	[A] p.373
CH_3—CH_2—CH_2—$Si(OEt)_3$	1.3956	170–172	[B]
CH_2=CH—$Si(OEt)_3$	1.3966	160–161/760	[C] p.4257
CH_2=CH—CH_2—$Si(OEt)_3$	1.4073	176/760	[C] p.4261
Cl—$(CH_2)_3$—$Si(OEt)_3$	1.4202	98–102/10	[C] p.4243
NH_2—$(CH_2)_3$—$Si(OEt)_3$	1.4198	217/760	[A] p.152
Br—$(CH_2)_3Si(OEt)_3$	–	110/14	[D]
CH_2=$CMeCO_2$—$(CH_2)_3$—$Si(OEt)_3$	–	–	[E]
CH_2—CH=CH_2—O—$(CH_2)_3Si(OEt)_3$	1.4256	122–125/0.3	[A] p.148
C_6H_5—$Si(OEt)_3$	1.4718	235/760	[F]

[A] W. Noll "Chemieund Technologie der Silicone", VCH Weinheim/ Bergstr. 1968 2nd. Ed.

[B] Larsson E (1952) Kg. Fysiograph Sallskap. i. Lund, Hand. 63 (12): 8.

[C] Beilsteins Handbuch der Organischen Chemie, Springer Verlag Berlin, Heidelberg-New York 1981, 4th Ed. Volume 4.

[D] Tundo P (1977) J Chem Soc Chem Commun 18: 641.

[E] Maseiniec B, Gulinski J, Urbaniak W, Nowicka T, Mirecki J (1990) Appl Organomet Chem 4: 27.

[F] Eaborn C (1960) "Organosilicon Compounds", Butterworth Sci Publ London.

reagent may be used in two different functions. First, by analogy with composites it may serve to bind a reactive polymer to the surface of a silicate-type support, such as glass-beads. In this case the functional groups used for the chromatographic process are part of the organic polymer. Second, the coupling reagent itself bears the reactive group which participates in the chromatographic process. Examples for chromatographic applications are: Charge-transfer chromatography, selective extraction of metal ions (via chelating ligands) or immobilization of enzymes.

Finally, the definition and application of surfactants needs to be mentioned. Surfactants are alkyl or arylsilanes designed to modify the surface tension (and thus wettability) of silicate ceramics or oxidized metals without covalent binding of other organic compounds. In most applications, the surfactants lower the surface tension of the substrate and serve as a hydrophobic agent. This function may be of interest to improve the water-resistance of ceramics or the insulating quality of electrical devices and to allow the use of silica in reverse-phase high-performance liquid chromatography. In summary this short introduction should indicate that silicon containing coupling reagents and surfactants are useful for a broad variety of applications. In this connection it should be mentioned that several reviews concerning coupling agents and surface treatment with silicon compounds have been published [1–7]. Particularly informative is the monography of Pluddemann [1].

7.1.2
Synthesis of Coupling Agents

The chemical structure of most coupling agents obeys the general formula: R—SiX_3 (with R = alkyl, vinyl, aryl) for the following reasons.

Table 7.4 Commercial coupling agents and recommended applications (matrix polymers)

Chemical structure	Symbol	Matrix polymers
$CH_2\!=\!CH\!-\!Si(OMe)_3$	A	Polyethylene, Polyethylene, EPDM Styrene-Butadiene-Rubber, EPT, PDAP
$Cl\!-\!(CH_2)_3\!-\!Si(OMe)_3$	B	Polyethylene, Polypropylene, Polystyrene EPDM, Styrene-Butadiene, PVC
$NH_2\!-\!(CH_2)_3\!-\!Si(OMe)_3$	C	Polyethylene, Polypropylene, EPDM, PVC Polyamide, Polycarbonate, Polyurethane
$HS\!-\!(CH_2)_3\!-\!Si(OMe)_3$	D	Epoxide, VF, PF, MF, PSV, FKM, Polyethylene Polypropylene, Polystyrene, ABS, Styrene-Butadiene, PVC, EPDM, PMMA
$\overset{\displaystyle O}{\overset{\diagup\;\diagdown}{CH_2\!-\!CH\!-\!CH_2O\!-\!(CH_2)_3\!-\!Si(OMe)_3}}$	E	Epoxides, Polyethylene, Polypropylene, Polystyrene, ABS, PVC, Polycarbonat, Polyamide, Polyurethane
$\overset{\displaystyle CH_3}{\underset{}{CH_2\!=\!C\!-\!CO\!-\!O\!-\!(CH_2)_3\!-\!Si(OMe)_3}}$	F	Polyethylene, Polypropylene, Polystyrene, Styrene-Acrylonitrile, PMMA, Epoxides, EVA unsaturated Polyesters
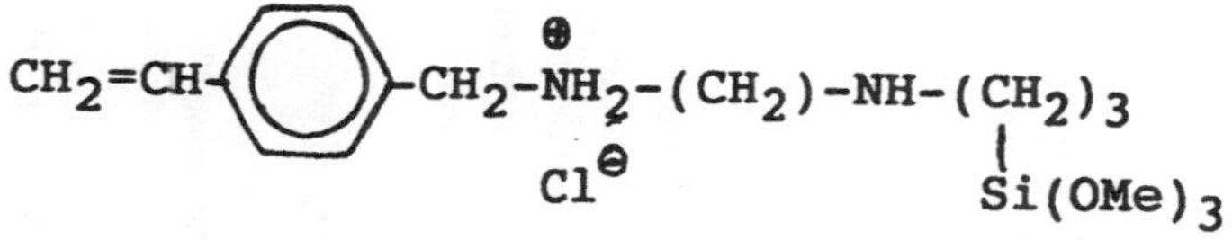 $CH_2\!=\!CH\!-\!\text{(phenyl)}\!-\!CH_2\!-\!\overset{\oplus}{NH_2}\!-\!(CH_2)\!-\!NH\!-\!\underset{\underset{Si(OMe)_3}{\mid}}{(CH_2)_3}$ $Cl^{\ominus}$	G	Polyethylene, Polypropylene, Polystyrene PMMA, PVC, EPDM, Styrene-Acrylnitrile, unsatur. Polyesters, Epoxides

1. When the organic residue bearing the carbofunctional group is attached to the silicon via a heteroatom such as O, S or N this bond is too sensitive to hydrolysis. Thus, a carbon-silicon bond is the only realistic alternative.

2. The X-group is in almost all cases of practical interest a Cl, OMe or OEt group. These groups must be reactive enough under mild conditions to form Si—O—Si bonds with the OH or Si—O$^{\ominus}$ groups on the surface of the silicates. When two or three Si—O—Si bonds are formed, the connection of the coupling agent with the silicate surface is considerably less sensitive to hydrolysis than a connection via a single Si—O—Si bond. Furthermore, the synthesis of coupling agents with R—Si—X3 structure is, in most cases, less expensive than the syntheses of R,R′,SiX$_2$ or R,R′,R″SiX compounds.

3. The position of the carbofunctional group relative to the Si-atom needs to be discussed. A functional group in α-position (i.e. Z—CH$_2$—SiX$_3$) is unfavorable for two reasons. First, a heteroatom in α-position renders the Si—C bond more sensitive to acidic or basic hydrolysis, Eq. (1). For instance, a Cl$_3$C—Si bond is hydrolyzed even by neutral hot water.

$$Z\text{–}CH_2\text{–}Si\equiv \quad + \quad H_2O \quad \xrightarrow{\text{(OH}^{\bullet}\text{)}} \quad Z\text{–}CH_2H \quad + \quad HO\text{–}Si\equiv \qquad (1)$$

Second, the distance between the carbofunctional group and SiX$_3$ is too short for a successful coupling of a matrix polymer to a silicate surface. A heteroatom in the β-position leads to silanes which are unstable against thermal, Eq. (2) or base-catalyzed elimination, Eq. (3). Therefore, most carbofunctional groups are attached to the γ-position or a position further removed from the Si—atom.

$$Z\text{–}CH_2\text{–}CH_2\text{–}Si\equiv \quad \xrightarrow{\Delta T} \quad CH_2{=}CH_2 \quad + \quad Z\text{–}Si\equiv \qquad (2)$$

In principle most synthetic methods that allow the formation of Si—C bonds may be used for the preparation of coupling reagents [8, 9]. Nonetheless, the most widely used procedure for both technical production and research purposes is the addition of siliconhydrides to olefins, Eq. (4). This addition is best catalyzed by hexachloroplanitum acid or other platinum based catalysts [10].

$$Z\text{–}CH_2\text{–}CH_2\text{–}Si\equiv \quad \xrightarrow[-\ NaCl]{+\ NaOH} \quad CH_2{=}CH_2 \quad + \quad HO\text{–}Si\equiv$$

$$Z = \text{any heteroatom} \qquad\qquad\qquad\qquad\qquad\qquad\qquad\qquad (3)$$

$$R\text{–}CH{=}CH_2 \quad + \quad HSiX_3 \quad \xrightarrow{\text{(Cat)}} \quad R\text{–}CH_2\text{–}CH_2\text{–}SiX_3$$

$$R = H,\ CN \qquad\qquad\qquad\qquad\qquad X = Cl,\ OMe \qquad (4)$$

Lowering of the reaction temperature by a catalyst is particularly necessary, when a functional group is present in the olefin. When allyl compounds are used as starting materials the carbofunctional group is finally placed in γ-position of the silicon.

Most commercial coupling agents are prepared on the basis of allyl substrates: for instance γ-chloro or γ-bromopropylsilanes and γ-aminopropyl-silanes Eq. (5). Further coupling agents contain ether groups based on addition reaction of allylethers or esters. It is particularly remarkable that epoxide, oxetane or methacrylate groups are not seriously affected under optimized reaction conditions [11, 12]. In addition to allyl compounds p-substituted styrenes are useful substrates for the silane addition method, Eq. (6). The benzylchloride group is highly useful for further modification by nucle ophilic substitution. 1,4-Divinylbenzene yields styrylethylsilanes, Eq. (7) [13]. Silanes add to conjugated dienes in 1,4-position yielding 3-alkenylsilanes [14].

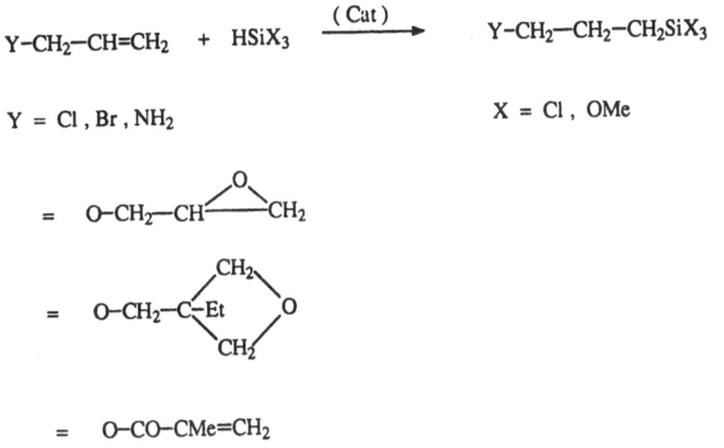

$$Y-CH_2-CH{=}CH_2 \;+\; HSiX_3 \xrightarrow{\;(Cat)\;} Y-CH_2-CH_2-CH_2SiX_3$$

$$Y = Cl, Br, NH_2 \qquad\qquad X = Cl, OMe$$

$$= \quad O-CH_2-CH{-}{-}CH_2 \text{ (epoxide)}$$

$$= \quad O-CH_2-C{-}Et \text{ (oxetane)}$$

$$= \quad O-CO-CMe{=}CH_2$$

$$\tag{5}$$

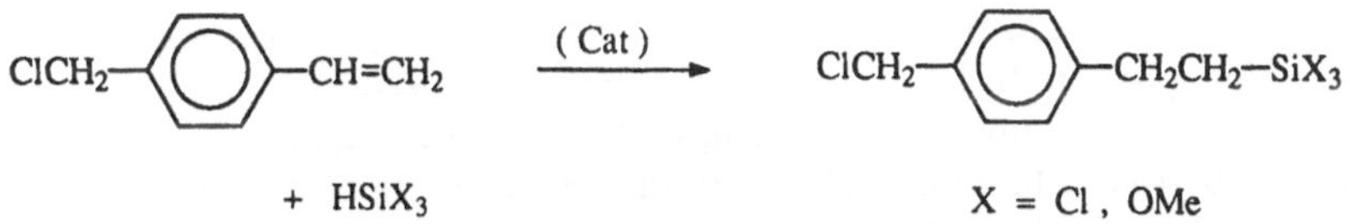

$$ClCH_2{-}\langle\!\langle\bigcirc\!\rangle\!\rangle{-}CH{=}CH_2 \;+\; HSiX_3 \xrightarrow{\;(Cat)\;} ClCH_2{-}\langle\!\langle\bigcirc\!\rangle\!\rangle{-}CH_2CH_2{-}SiX_3$$

$$X = Cl, OMe \tag{6}$$

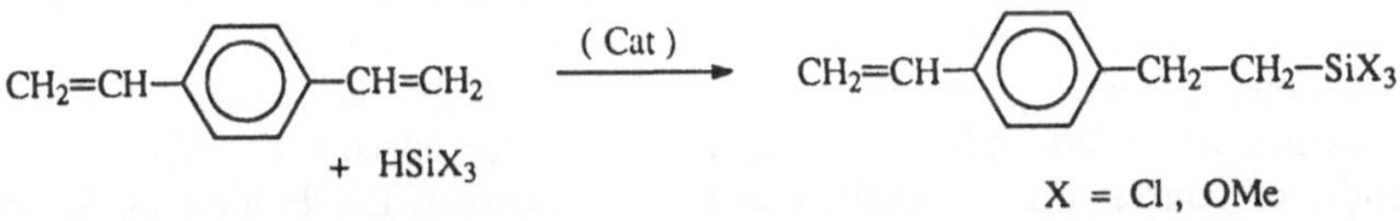

$$CH_2{=}CH{-}\langle\!\langle\bigcirc\!\rangle\!\rangle{-}CH{=}CH_2 \;+\; HSiX_3 \xrightarrow{\;(Cat)\;} CH_2{=}CH{-}\langle\!\langle\bigcirc\!\rangle\!\rangle{-}CH_2{-}CH_2{-}SiX_3$$

$$X = Cl, OMe \tag{7}$$

The addition of silanes to α, ω-diolefins can be conducted in such a way that only one double bond is involved. The remaining free double bond is then available for further modifications, such as oxydation or addition of phosphines [15]. Finally, a noncatalyzed, high temperature addition process based on diisobutylene should be mentioned, Eq. (8). It enables the synthesis of another unsaturated coupling agent [16].

$$Me_3C-CH_2-CMe{=}CH_2 \;+\; HSiCl_3 \xrightarrow{\;500\,°C\;} \underset{\underset{}{}}{CH_2{=}\overset{Me}{C}-CH_2-SiCl_3} \;+\; Me_3CH \tag{8}$$

Unsaturated silanes with a vinyl group directly attached to silicon are conveniently prepared by a base or platinum catalyzed addition of silanes to acetylene or higher alkines, Eq. (9) [17]. Trichlorosilane also reacts with aromatics at high temperatures even in the absence of catalysts yielding aryltrichlorosilanes, Eq. (10). Addition reactions of trimethoxy silane are not attractive for laboratory purposes, because of its high toxicity and low thermal stability. The alkoholysis of R—SiCl$_3$ type starting materials and the further modification of commercial coupling agents (Table 7.4) is thus the most convenient way for a laboratory scale production of coupling agents.

$$R-C{\equiv}C-H \quad + \quad HSiX_3 \quad \xrightarrow{\text{(Cat)}} \quad R-CH{=}CH-SiX_3 \tag{9}$$

$$R = H, Alkyl \qquad\qquad\qquad X = Cl, OMe$$

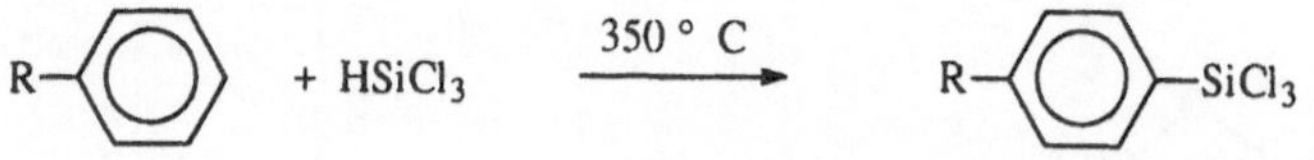

$$R = H, Me \tag{10}$$

The chemical modification of coupling agents will be discussed below in the following order:

I) Modification of silicon-functional groups
II) Modification of carbofunctional groups.

Trimethoxy or higher trialkoxysilanes are easy to prepare by alcoholysis of the corresponding trichlorosilanes, Eq. (11). The only problem is the quantitative removal of large quantities of HCl which may cause side reactions of the carbofunctional group. In laboratory scale experiments addition of a *tert*-amine is certainly the most convenient procedure. An alternative approach consists of the reaction between chlorosilane and orthoformates in the presence of catalytic amounts of alcohols, Eq. (12) [18]. This approach is favorable in the case of methyl or ethylgroups, because methyl and ethylchloride are gaseous byproducts and the boiling points of methyl- and ethylformate are relatively low. For the introduction of higher alkoxygroups in exchange reaction with trimethoxysilanes is an attractive method, Eq. (13). The success of this method depends, of course, on a sufficient difference between the boiling points of

$$R-SiCl_3 \quad + \quad 3\ HOR' \quad \xrightarrow[-\ 3\ HCl]{} \quad R-Si\,(\,OR'\,)_3 \tag{11}$$
$$R' = Me, Et$$

$$R-SiCl_3 \quad + \quad 3\ HC(\,OR'\,)_3 \quad \xrightarrow{\text{(ROH)}} \quad R-Si\,(\,OR'\,)_3 + 3\ RCl \tag{12}$$
$$+ \ 3\ HCO_2R'$$

$$R-Si\,(\,OMe\,) \quad + \quad 3\ R'{-}OH \quad \xrightarrow{\text{(NaOH)}} \quad R-Si\,(\,OR'\,)_3 \quad + \quad 3\ MeOH \tag{13}$$
$$R' = Et\ or\ higher\ Alkyl$$

methanol and the higher alcohol. Sodium hydroxide or tetraalkoxy titanates are required as catalysts to keep the reaction temperature below 100 °C. With such basic or neutral catalysts even epoxycarbofunctional groups are not affected in contrast to acidic catalysis [19]. The amino group of γ-aminopropyl trimethoxy silane is not basic enough to act as catalyst. Trichlorosilanes may also serve as starting materials for the preparation of trisacyloxysilanes. One method consists of the reaction with dry sodium acetate in a dry inert solvent, Eq. (14). The second method, which avoids the formation of salts, is based on acetic anhydride as reaction partner and reaction medium. The liberated acetyl-chloride, Eq. (15), is volatile and easy to remove, whereas the separation of the coupling agent from the excess of acetic anhydride may be difficult.

$$R-SiCl_3 \; + \; 3 \; NaOAc \longrightarrow \qquad R-Si\,(\,OAc\,)_3 \; + \; 3 \; NaCl \qquad (14)$$

$$R-Si\,Cl_3 \; + \; 3 \; Ac_2O \longrightarrow \qquad R-Si\,(\,OAc\,)_3 \; + \; 3 \; AcCl \qquad (15)$$

$$R \; = \; Alkyl\,,\; Vinyl\,,\; Aryl\,,\; in\,Eqs.\;(\,11\,)\;-\;(\,15\,)$$

γ-Chloropropyl-trimethoxy (or triethoxy) silane is a convenient starting material for a broad variation of the carbofunctional group in γ-position. For instance, treatment with sodium iodide in dry acetone yields the more reactive γ-iodopropylsilanes. Analogous nucleophilic substitutions were conducted with potassium cyanide, cyanate or isothiocyanate, Eq. (16) [20]. With sodium methoxide in dry polar solvents the synthesis of methyl ethers is feasible, Eq. (17). This reaction is even more rapid, when a benzylchloride group is involved (Form. 18) [21]. If higher alkoxides are used in combination with trimethoxysilanes a partial exchange of the alkoxygroups at the silicon may take place. Fortunately a relatively clean nucleophilic substitution proceeds with carboxylate ions, such as acetate or methacrylate, Eq. (19) [22]. The methacrylate group is of particular interest, because it enables radical crosslinking with C—C double bonds in a surrounding matrix polymer. The acetate group is

$$Cl-(\,CH_2\,)_3-Si\,(\,OMe\,)_3 \quad \overset{+\;KX}{\underset{-\;KCl}{\longrightarrow}} \quad X-(\,CH_2\,)_3-Si\,(\,OMe\,)_3 \qquad (16)$$

$$X \; = \; J\,,\; CN\,,\; OCN\,,\; SCN$$

$$Cl-(\,CH_2\,)_3-Si\,(\,OMe\,)_3 \quad \overset{+\;NaOMe}{\underset{-\;NaCl}{\longrightarrow}} \quad CH_3O-(\,CH_2\,)_3\;Si\,(\,OMe\,)_3 \qquad (17)$$

$$CH_3O-CH_2-\!\!\left\langle\bigcirc\right\rangle\!\!-CH_2CH_2-Si\,(\,OMe\,)_3 \qquad (18)$$

$$Cl-(\,CH_2\,)_3-Si\,(\,OMe\,)_3 \quad \overset{+\;NaO_2C-CMe=CH_2}{\underset{-\;NaCl}{\longrightarrow}} \quad CH_2=CMe-CO_2^-(\,CH_2\,)_3-Si\,(\,OMe\,)_3 \qquad (19)$$

useful, because acidic methanolysis, allows the introduction of an alcoholic OH-group in α-position of the coupling agent, Eq. (20).

$$AcO-(CH_2)_3-Si(OMe)_3 \xrightarrow[- \ AcOMe]{+ \ MeOH/HCl} HO-(CH_2)_3-Si(OMe)_3 \quad (20)$$

For the synthesis of mercaptanes at least two convenient synthetic methods are known. Treatment of the γ-chloropropylsilane with an excess of H_2S in the presence of 1, 2-diaminoethane yields the γ-mercaptocompound in a direct way, Eq. (21) [23, 24]. A contamination of the alkaline H_2S-solution with oxidation byproducts is a particular risk of this approach. Cleaner mercaptanes can be synthesized by a two-step procedure [25]. The first step consists of the formation of an isothiuronium salt, Eq. (22), which is then cleaved by ammonia yielding the desired mercaptane in addition to urea and ammonium chloride, Eq. (23). The relatively high reactivity of γ-chloropropylsilanes also enables a satisfactory reaction with dimethylmethylphosphonates, Eq. (24) [26].

$$Cl-(CH_2)_3-Si(OMe)_3 \xrightarrow[- \ HCl]{+ \ H_2S} HS-(CH_2)_3-Si(OMe)_3 \quad (21)$$

$$Cl-(CH_2)_3-Si(OMe)_3 \xrightarrow{+ \ NH_2-CS-NH_2} \overset{\oplus}{\underset{NH_2}{\overset{NH_2}{C}-S}}-(CH_2)_3-Si(OMe)_3$$

$$Cl^{\ominus}$$

$$(22)$$

$$+ NH_3 \ \Big| \ \begin{matrix} - \ NH_4Cl \\ - \ NH_2CONH_2 \end{matrix}$$

$$\downarrow$$

$$HS-(CH_2)_3-Si(OMe)_3 \quad (23)$$

$$Cl-(CH_2)_3-Si(OMe)_3 \\ + \ MeP(O)(OMe)_2 \xrightarrow[- \ MeCl]{(Et_3N)} \overset{Me}{\underset{O}{MeO-P-O}}-(CH_2)_3-Si(OMe)_3 \quad (24)$$

Several methods have been described for the introduction of amino groups into coupling agents. Treatment of γ-chloropropyl silanes with an excess of ammonia or amines is one obvious approach, Eq.(25) [27]. Catalytic hydrogenation of the cyanogroup resulting from the silane addition to acrylonitrile (Eq. 4) is an alternative method, Eq. (26) for the introduction of a primary amino group into the γ-position [28]. Aromatic amines were obtained by the treatment of

$$Cl-(CH_2)_3-Si(OMe)_3 \xrightarrow[- \ RNH_2 \cdot HCl]{+ \ RNH_2} R-NH-(CH_2)_3-Si(OMe)_3$$
$$R' = H, Alkyl \qquad\qquad (25)$$

$$NC-CH_2-CH_2-Si(OMe)_3 \xrightarrow{H_2/Pd} NH_2-CH_2-CH_2-CH_2-Si(OMe)_3 \quad (26)$$

4-bromophenyl-trimethoxysilane with NH3 in the presence of copper catalysts, Eq. (27) [29]. Aqueous solutions of nitroarylsilanolates are easily reduced by hydrazine in the presence of Raney nickel, Eq. (28) [29].

$$Br-\!\!\bigcirc\!\!-Si(OMe)_3 + NH_3 \xrightarrow[(110°C)]{Cu/CuCl} NH_2-\!\!\bigcirc\!\!-Si(OMe)_3 \quad (27)$$

$$NO_2-\!\!\bigcirc\!\!-Si(ONa)_3 \xrightarrow[-N_2, -H_2O]{+ N_2H_4/Ni} NH_2-\!\!\bigcirc\!\!-Si(OMe)_3 \quad (28)$$

All these amino-group-containing coupling agents are in turn versatile starting materials for further modifications. For instance, acylation with cyclic anhydrides, such as maleic anhydride yields carboxyl group terminated coupling agents, Eq. (29) [30]. An interesting reaction is the Michael-addition of γ-aminopropyl silanes onto acrylates, Eq. (30) [1, 31]. Normal dialkylamines are nucleophilic enough to react with an excess of acrylates or acrylonitrile (yielding tertiary amino groups) regardless of the solvent. The reactivity of secondary amino groups in γ-position of alkoxysilanes depends largely on the solvent. In less polar, nonprotic solvents (e.g. toluene) the nucleophilicity is comparable with that of other dialkylamines. However, in H-bonding solvents the reactivity is significantly lower. The formation of a cyclic conformation involving the Si(OMe)$_3$ group was considered to be responsible for this change of reactivity. Therefore, the addition of the primary amino group to acrylates stops at the stage of the secondary amino group (without yielding *tert*-amines as byproduct) when conducted in alcohol, Eq. (30).

$$NH_2-(CH_2)_3-Si(OMe)_3$$
$$+ \underset{OC\diagdown_{O}\diagup CO}{\overset{CH=CH}{\diagup\diagdown}} \longrightarrow HO_2C \overset{CH=CH}{\diagup\diagdown} CO-NH-(CH_2)_3-Si(OMe)_3 \quad (29)$$

$$NH_2-(CH_2)_3-Si(OMe)_3$$
$$+ R-O_2C-CH=CH_2 \longrightarrow RO_2C-CH_2-CH_2-NH-(CH_2)_3-Si(OMe)_3 \quad (30)$$

The alkylation of γ-aminopropyltrimethoxysilane with 4-chloromethyl styrene stops again at the stage of the secondary amine, when the liberated HCl is allowed to protonate the secondary amino group which is more basic than the primary one, Eq. (31) [32]. However, a complete alkylation of all primary and secondary amino groups with sodium chloroacetate is desirable and feasible,

$$NH_2-(CH_2)_3Si(OMe)_3$$
$$+ ClCH_2-\!\!\bigcirc\!\!-CH=CH_2 \longrightarrow \overset{+HCl}{CH_2=CH-\!\!\bigcirc\!\!-CH-NH-(CH_2)_3-Si(OMe)_3} \quad (31)$$

Eq. (32) [33]. The resulting diaminotriacetic acid is a useful multidentate ligand for various metal ions (see below). In this connection the formation of quartary ammonium groups by the reaction of tertiary amines with γ-chloropropylsilanes should be mentioned, Eq. (33) [34]. Furthermore, the formation of hydrophilic urea groups by "carbamoylation" of the γ-aminogroup with a urethane is worth noting, Eq. (34) [35].

$$NH_2CH_2CH_2NH-(CH_2)_3\,Si\,(OMe)_3 \quad + \quad 3 \;\; Cl-CH_2CO_2Na$$

$$\downarrow \quad -3 \; NaCl$$

$$\tag{32}$$

$$\begin{array}{c} HO_2C-CH_2 \\ HO_2C-CH_2 \end{array} N-CH_2-CH_2-N-(CH_2)_3-Si\,(OMe)_3$$
$$\underset{\displaystyle CH_2CO_2H}{|}$$

$$Cl-(CH_2)_3-Si\,(OMe)_3 \quad + \quad NR_3 \quad\longrightarrow\quad R_3\overset{\oplus}{N}-(CH_2)_3-Si\,(OMe)_3 \tag{33}$$
$$Cl^{\ominus}$$

$$NH_2-(CH_2)_3-Si\,(OMe)_3$$

$$+ \;\; NH_2-CO_2Et \quad\xrightarrow[-\;EtOH]{}\quad NH_2CONH-(CH_2)_3-Si\,(OMe)_3 \tag{34}$$

$$\underset{O}{\overset{\displaystyle \triangle}{CH_2}}\!\!-\!\!CH-CH_2-O-(CH_2)_3-Si\,(OMe)_3 \;+\; R-OH \quad\xrightarrow{\;(OH)^{\ominus}\;}$$

$$R-O-CH_2-\underset{\underset{\displaystyle OH}{|}}{CH}-CH_2-O-(CH_2)_3-Si\,(OMe)_3 \tag{35}$$

$$R = H, \; C_6H_5, \; CH_3-(O-CH_2CH_2)n$$

In addition to amino groups, epoxide groups are versatile functional groups which allow numerous modifications of the basic coupling agent. Hydrolysis with alkaline water yields a free diol. Nucleophilic attack of phenoxide ions yields arylethers, and the reaction with alkoxides leads to alkyl ethers, Eq. (35). Particularly interesting is the addition of oligoethyleneoxide monoethers, because the resulting coupling agent has now a hydrophilic group with chelating abilities [36]. Finally, it should be mentioned that the γ-mercaptogroups allows numerous alkylation (and arylation) reactions, Eq. (36) [37].

More recently syntheses and applications of oligomeric or polymeric coupling agents with macromeric, telechelic or multifunctional character have found increasing interest [38–51]. These oligomers or polymers may contain one or

$$HS-(CH_2)_3-Si\,(OMe)_3$$

$$\xrightarrow{\qquad} MeO_2C-CH_2-CH_2-S-(CH_2)_3-Si\,(OMe)_3 \tag{36}$$

$$+ \;\; CH_2=CH-CO_2Me$$

two trialkoxysilane endgroups, or they may possess several pending trialkoxy-silane groups. They may be useful as coupling agents, as surfactants as coatings or as monomers (macromers) for the synthesis of starshaped polymers or for polysiloxane networks with pending polymer sides-chains (so-called ceramers).

In a patent claim of Bayer AG [38] telechelic polyurethanes containing two isocyanate endgroups are described. N-Alkylaminomethyltriethoxysilanes were reacted with the isocyanate groups yielding urea end group with triethoxysilane functions. An example of such a telechelic polyurethane is given in Form. 37. The structure of these telechelics was varied over broad range. Aliphatic

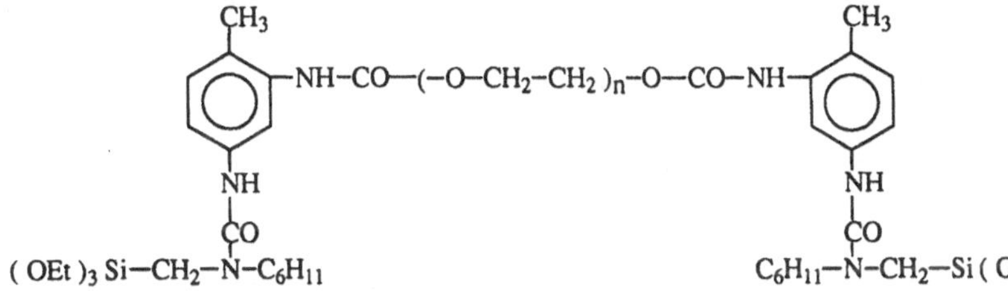

$$(37)$$

$(OEt)_3 Si–CH_2–N–C_6H_{11}$

$C_6H_{11}–N–CH_2–Si(OEt)_3$

oligoesterdiols were used as soft segments in addition to oligoethers, and iso-phoron diisocyanate, biscyclohexylmethane diisocyanate, 1,6-hexa- methylene-diisocyanate and m-xylylene diisocyanate were used for chain extension. Tri-functional triethoxysilanes were derived from the trisisocyanate **38**. In addition to the cyclohexylamine derivative forming part of Form. **37** the amino-methylsilanes **39 a, b** and **40 a, b** were used for the formation of functional end-groups. Another patent claim [39] deals with the synthesis of oligo- and poly-ureas **41, 42** from aliphatic diamines having one or two alkoxysilane groups

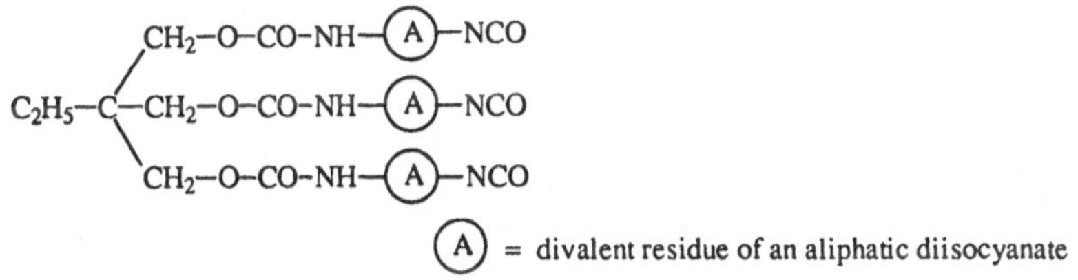

$$(38)$$

(A) = divalent residue of an aliphatic diisocyanate

$C_6H_{11}–NH–CH_2–Si(O–CHMe_2)_3$ $C_2H_5–NH–CH_2–Si(OEt)_3$ (39)

a   b

$C_6H_{11}–NH–CH_2–Si–\overset{OEt}{\underset{O–CMe_3}{|}}–CMe_3$ $CH_3CH_2CH_2–NH–CH_2–Si(OEt)_3$ (40)

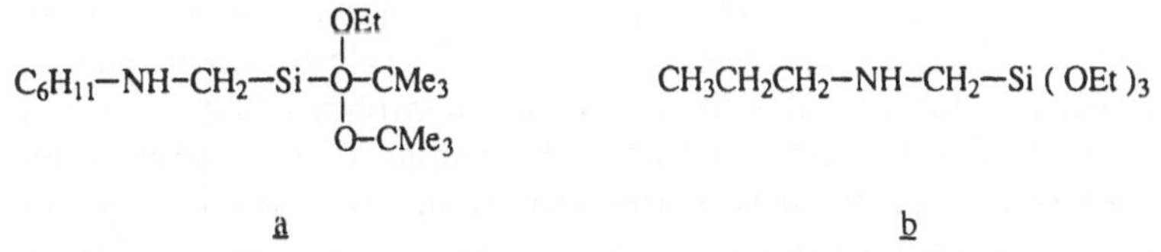

a   b

attached to the nitrogens. Some structural variations of these special diamines are outlined in Formulas **43 a,b** and **44**, Formulas **41** and **42** may also serve as examples for structures of the functionalized polyureas. In addition to the pend-ing alkoxysilane groups, triethoxysilane endgroups were introduced in some

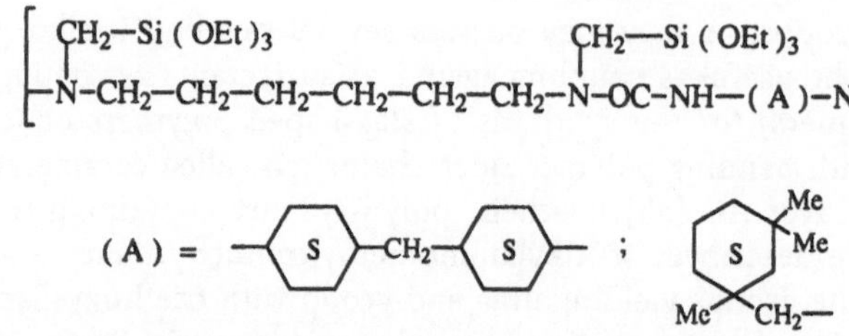

$$
\left[
\begin{array}{c}
\overset{\displaystyle CH_2-Si\,(\,OEt\,)_3}{|} \qquad\qquad \overset{\displaystyle CH_2-Si\,(\,OEt\,)_3}{|} \\
-N-CH_2-CH_2-CH_2-CH_2-CH_2-CH_2-N-OC-NH-(\,A\,)-NH-CO-
\end{array}
\right]
\tag{41}
$$

$$
(\,A\,) = \quad -\!\langle\!\!\bigcirc\!\!\rangle\!-S-\!\langle\!\!\bigcirc\!\!\rangle\!-CH_2-\!\langle\!\!\bigcirc\!\!\rangle\!-S-\!\langle\!\!\bigcirc\!\!\rangle\!- \quad ; \qquad
$$

cases (Form. **42**). More recently a french research group [40] used a similar synthetic approach for the preparation of monomeric or oligomeric urethane-ureas having one triethoxysilane group (e.g. Form. **45**).

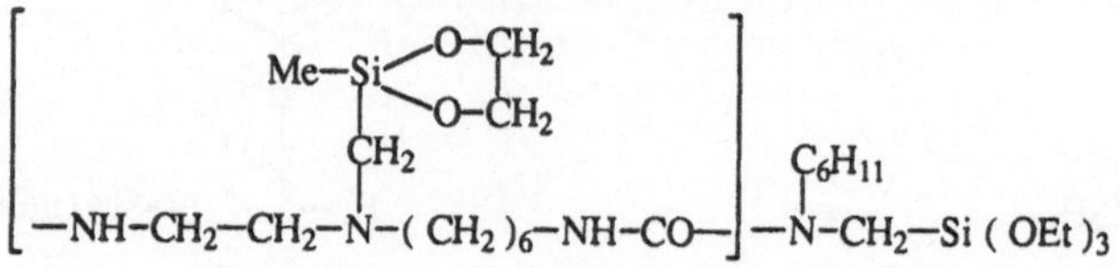

$$
\left[
\begin{array}{c}
Me-Si\!\!\begin{array}{c}O-CH_2\\ \diagdown\\ O-CH_2\end{array} \\
\overset{\displaystyle CH_2}{|} \\
-NH-CH_2-CH_2-N-(\,CH_2\,)_6-NH-CO-
\end{array}
\right]
\overset{\displaystyle C_6H_{11}}{\underset{\displaystyle |}{}}\!-N-CH_2-Si\,(\,OEt\,)_3
\tag{42}
$$

$$
NH_2-CH_2-CH_2-NH-CH_2SiMe\,(\,OEt\,)_2 \qquad\qquad NH_2-CH_2-CH_2-NH-CH_2Si\,(\,OR\,)_3
$$

\qquad\qquad **a** \hspace{7cm} **b**
$$
\tag{43}
$$

$$
R = i-C_3H_7 \quad , \quad C_4H_9
$$

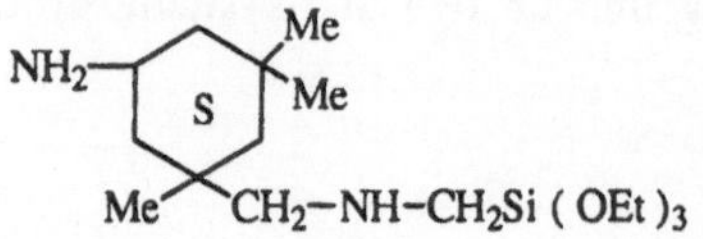

$$
NH_2-\!\langle\!\!\bigcirc\!\!\rangle\!\begin{array}{c}Me\\Me\end{array}
$$
$$
Me-\quad CH_2-NH-CH_2Si\,(\,OEt\,)_3
\tag{44}
$$

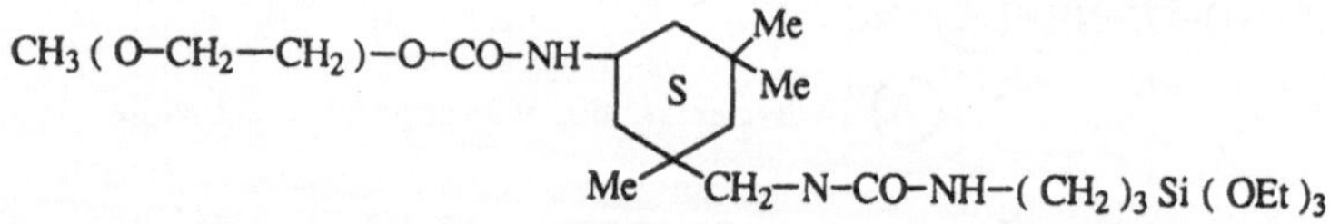

$$
CH_3\,(\,O-CH_2-CH_2\,)-O-CO-NH-\!\langle\!\!\bigcirc\!\!\rangle\!\begin{array}{c}Me\\Me\end{array}
$$
$$
Me-\quad CH_2-N-CO-NH-(\,CH_2\,)_3\,Si\,(\,OEt\,)_3
\tag{45}
$$

A research group of Eastman Kodak [48, 49] studied synthesis and properties of endcapped oligo- and polystyrenes prepared by anionic polymerization of styrene. The anionic chain end was reacted with (α-chloromethylphenyl) trimethoxysilane, Eq. (46), and the silane function hydrolyzed under controlled conditions. Poly(chlorotrifluoroethylene) with pending triethoxysilane groups was prepared from a precurso with free OH-groups by reaction with γ-iso-cyanato-propyl triethoxysilane [50] **47 a**. A polyethylene with pending trimethoxysilane groups **47 b** was obtained by copolymerization of vinyl trimethoxysilane (see Chapter 1) [51]. Both copolymers (**47 a** and **b**) show extremely good adhesion to silica and other oxide-type surfaces.

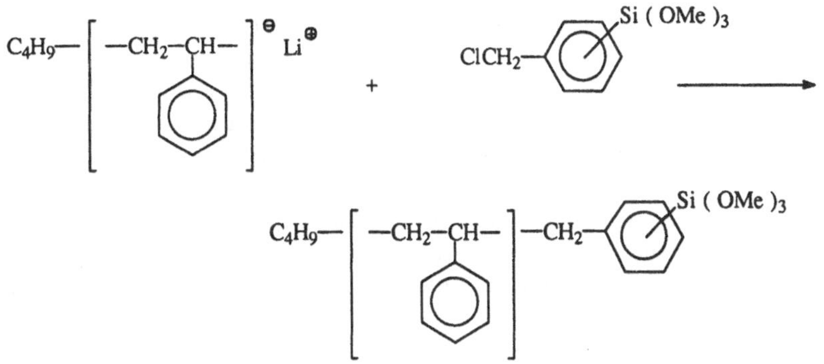

$$C_4H_9 \!-\! \left[-CH_2\!-\!\underset{\substack{| \\ C_6H_5}}{CH}\!- \right]^{\ominus} Li^{\oplus} \quad + \quad ClCH_2 \!-\! \langle\!\!\langle \rangle\!\!\rangle \!-\! Si(OMe)_3 \quad\longrightarrow$$

$$C_4H_9 \!-\! \left[-CH_2\!-\!\underset{\substack{| \\ C_6H_5}}{CH}\!- \right]\!-\!CH_2 \!-\! \langle\!\!\langle \rangle\!\!\rangle \!-\! Si(OMe)_3 \qquad (46)$$

$$\left[\underset{\substack{| \\ O \\ | \\ CO \\ | \\ NH \\ | \\ (CH_2)_3 Si(OEt)_3}}{-CF\!-\!CF_2-} \diagdown -CFCl\!-\!CF_2- \right] \qquad \left[-CH_2CH_2- \diagup \underset{\substack{| \\ Si(OMe)_3}}{-CH_2\!-\!CH_2-} \right]$$

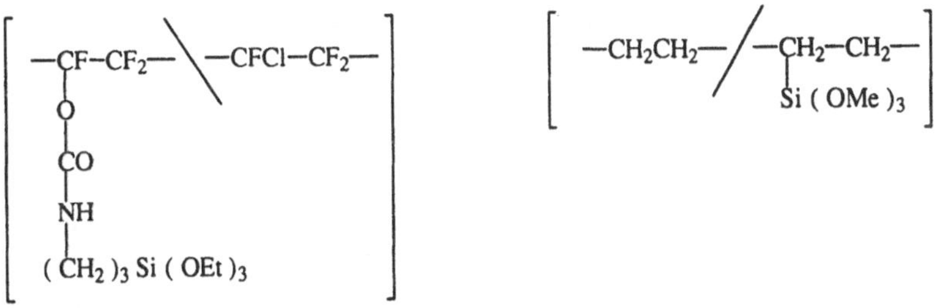

$$\qquad\qquad\qquad \underline{a} \qquad\qquad\qquad\qquad\qquad\qquad\qquad \underline{b} \qquad\qquad\qquad (47)$$

7.2
Application of Coupling Agents in Composites

The application of coupling agents to improve the mechanical properties of silicate-filled composites may be conducted in several ways:

I) Addition of the coupling agent to the matrix polymer before incorporation of the filler.

II) Addition of the coupling agent during the blending process of matrix and filler (so-called in situ addition).

III) Pretreatment of the filler with the neat coupling agent or with its solution in an organic solvent.

IV) Pretreatment of the filler with the aqueous solution of a coupling agent.

The last method is most widely used particularly in connection with glass fibers and will be discussed below in more detail. Pretreatment of silicate powders has the disadvantage that the particulate filler cakes up on drying. In this case methods I or II may be advantageous. However, these methods have the shortcoming that a part of the coupling agent remains dissolved in the organic matrix reducing the viscosity and the glass-transition temperature, but does not contribute to the cohesion between filler and matrix. Thermal posttreatment above T_g (e.g. curing) may improve the result, because then part of the dissolved coupling agent may migrate to the surface of the filler and contribute to the stabilization of the interface. Pretreatment of the filler with the neat coupling agent is only satisfactory, when the coupling agent rapidly migrates to the surface of the filler. Triethoxysilanes and tripropoxysilanes migrate faster than trimethoxysilanes but their reaction with the hydrophilic silicate surface is much slower, and thus, the trimethoxysilanes are usually preferred. Coupling agents with an ionic functional group are reluctant to migrate and in these cases a volatile organic solvent (e.g. alcohol) is helpful. Pretreatment of fillers via the gas-phase or dilute organic solution allows one to generate monolayers on their surface, whereas pretreatment with aqueous solutions produces multilayers of coupling agents.

7.2.1
Aqueous Solutions of Coupling Agents

It is the commercial practice to apply most coupling agents to glass fiber, glass beads and a few other silicate fillers from aqueous solutions of the tri-salkoxysilane. Organofunctional trisalkoxysilanes hydrolyze in contact with water and the resulting silanol (finally a trisilanol) polycondenses by formation of siloxane groups. In general the hydrolysis is considerably more rapid than the polycondensation and under favorable conditions solutions of monomeric trisilanols may be obtained and stored over several hours or days. Several factors influence the rate of hydrolysis and the condensation process: the organic residue, the nature of the alkoxy group, the concentration, the temperature, the presence of organic solvents and the pH. The following discussion will at first concentrate on coupling agents with hydrophobic residues and afterwards on those with hydrophilic groups.

The hydrolysis of higher trisalkoxysilanes is unfavorable compared to that of trimethoxysilanes because of steric hindrance and because of a lower hydrophilicity. Therefore most coupling agents are trimethoxysilanes and most mechanistic and analytical studies concentrate on trimethoxysilanes. Yet, the hydrolysis with neat water is slow even in the case of trimethoxysilanes, because it is initially biphasic when the organic residue is hydrophobic. Detergents and emulsifyers accelerate the hydrolysis, but the hydrolysis in a mixture of an organic solvent (alcohols, acetone etc.) and water is more convenient for commercial and analytical purposes. A quantitative comparison of hydrolysis and stability of the aqueous silanol solution is typically conducted in the following way [52]. 1 g of a coupling agent is dissolved in 2 ml isopropanol and three drops of acetic acid. Water is added dropwise from a burette at such a rate that the reaction mixture maintains a slight haze. When a clear solution is obtained with 50 ml water hydrolysis is assumed to be complete and the resulting trisilanol completely soluble in water. Under these conditions the hydrolysis of methyl or vinyl trimethoxysilane is complete in less than one minute, whereas the g-mercaptoproyl agent requires 7 min, the n-hexyl compound 25 min and 2,5,5-trimethylpentyl Si(OMe)$_3$ 90 min.

Not all silanetriols with hydrophobic residues are entirely soluble in water. The limit of the solubility is given by 6 carbons in an aliphatic and 7 carbons in an aromatic residue. Longer organic residues lead to the formation of micels with a hydrophobic core. The tendency of phase separation increases again when the silanetriol starts to polycondense, because the resulting oligosiloxanes are more or less insoluble in water. The stability of the silanetriol solutions depends strongly on their pH. A pH-range of 2–4 is most favorable for a maximum stability. Furthermore, the stability is enhanced by increasing dilution and by the presence of methanol, whereas fluoride ions favor the poly-condensation. A comparison of various coupling agents (R-Si(OMe)$_3$) in aqueous solution (3% by weight) at pH2 revealed the following order of increasing stability [1]:

n-Butyl < γ-Cl-Propyl < Propyl < Vinyl < Phenyl < Ethyl
< Methyl < CF3CH$_2$CH$_2$—

The polycondensation processes following the hydrolysis of alkyl or aryl trich-lorosilanes and trimethoxysilanes were studied by several research groups [53–64]. Mainly studies of cyclohexyl or phenyl silanetriol which can be isolated in a pure form have contributed to elucidate the nature of the oligomeric condensation products. Scheme (48) presents a simplified illustration of the reaction sequences. The formation of cyclic and cubic siloxanes is favored at the expense of linear and crosslinked products by bulky hydrophobic residues. Intramolecular condensation steps are also favored by entropy, and they may be favored by intramolecular H-bonds. Cubic siloxanes with Si_8O_{12} structure have been isolated and their synthesis and structure has recently attracted much interest [65–68].

Coupling agents with hydrophilic carbofunctional group hydrolyze rapidly even with neat water and their solubility in water in unlimited. γ-Aminopropyl-silane triols can adopt several forms and conformations depending on the pH[1]. [69–71]. In acidic solutions (pH $\geqslant$ 3) a silanetriol with protonated aminogroup **49 a** prevails. A higher pH a zwitter ion **49 b** seems to coexist with a cyclic H-bonded conformation **50 b**. No spectroscopic evidence was found for a cyclic conformation with pentocoordinated silicon analogous to that of **50 a**. Both protonation and intramolecular H-bonds reduce the nucleophilicity of the amino group in γ-position to a significant extent. These effects disappear at a more alkaline pH, but a higher pH also favors rapid polycondensation of the silanol groups.

Coupling agents with stable ionic groups in the side chain are readily hy-drolyzed in water and their silanetriols form rather stable aqueous solutions at a neutral or acidic pH. Typical examples of this group are represented by Forms. **51 a,b** and **52 a,b**.

When silicates are treated with aqueous solutions and dispersions of trisilanols, multilayers of coupling agents rapidly deposit on the silicate surface. The initial steps of this process are certainly governed by the formation of H-bonds. However, it is well recognized that the strength of the cohesion between silicate and layer of coupling agent is mainly a consequence of covalent bonds. These covalent are siloxane groups formed by condensation of SiOH groups, Eq. (53). The trisilanols have the advantage that they can form two

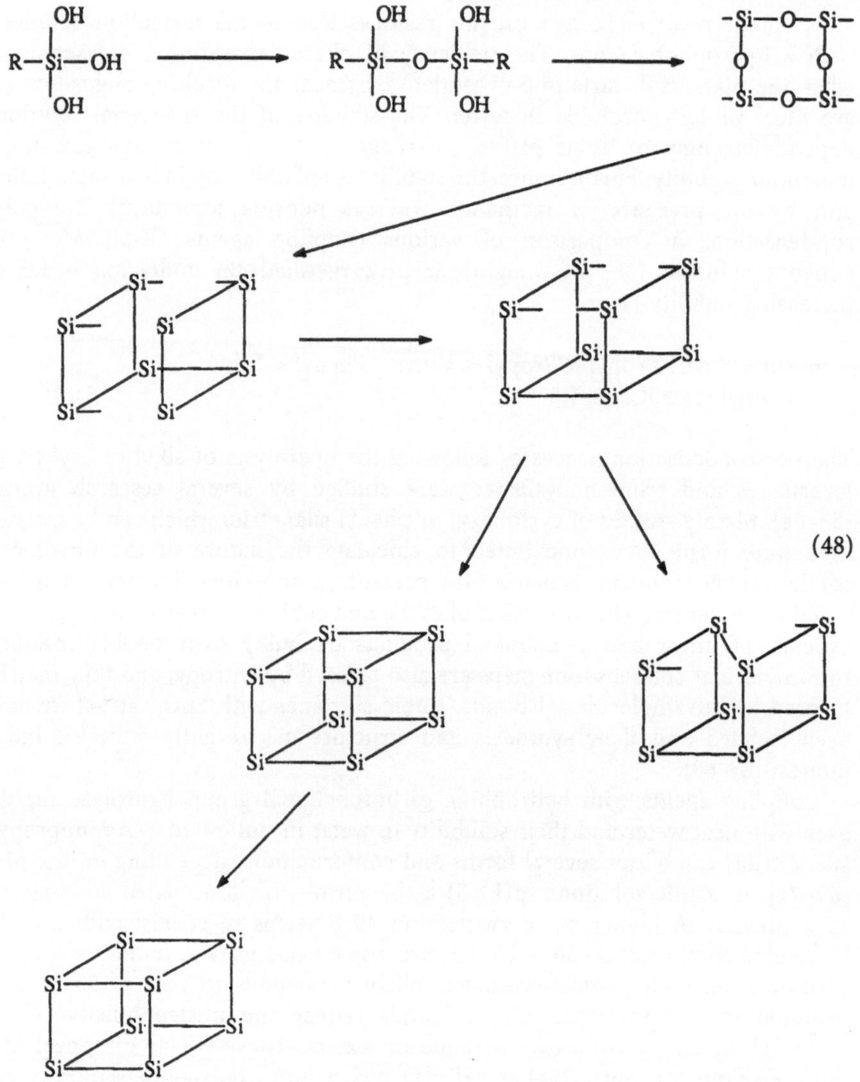

$$(48)$$

$$(\text{HO})_3Si-CH_2CH_2CH_2-\overset{\oplus}{N}H_2-R \qquad\qquad (\text{OH})_2Si-CH_2CH_2CH_2-\overset{\oplus}{N}H_2-R$$

a

$$\qquad\qquad\qquad\qquad\qquad\qquad |\underset{\ominus}{\underline{O}}| \qquad\qquad\qquad (49)$$

b

$$(50)$$

a

b

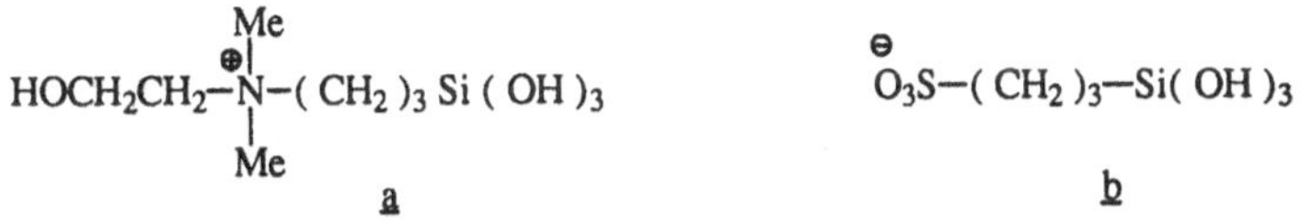

$$
\underset{\underset{\underset{\text{Me}}{|}}{\overset{\overset{\text{Me}}{|}}{\text{HOCH}_2\text{CH}_2\text{—}\overset{\oplus}{\text{N}}\text{—(CH}_2\text{)}_3\text{ Si (OH)}_3}}
\qquad \underset{\underline{b}}{\overset{\ominus}{\text{O}}_3\text{S—(CH}_2\text{)}_3\text{—Si(OH)}_3} \tag{51}
$$

$$
\underline{a}
$$

$$
\begin{array}{l}
\overset{\ominus}{\text{CH}_2\text{—NH—CH}_2\text{CH}_2\text{—CO}_2} \\
\overset{\oplus}{\text{CH}_2\text{—NH}_2\text{—(CH}_2\text{)}_3\text{Si(OH)}_3}
\end{array}
\qquad
\begin{array}{l}
\overset{\ominus}{}\ \overset{\text{Me}}{} \\
| \text{O—P—O—(CH}_2\text{)}_3\text{Si(OH)}_3 \\
\phantom{| \text{O—}}\underset{\text{O}}{\|}
\end{array} \tag{52}
$$

$$
\underline{a} \qquad\qquad\qquad\qquad \underline{b}
$$

$$
\text{R—Si (OMe)}_3 + \text{HO—Si (OR')}_3 \longrightarrow \text{R—Si (OMe)}_2\text{—O—Si (OR')}_3 \tag{53}
$$

covalent links to the silicate surface and one bond to a neighbouring coupling agent or vica versa. Scheme 54 provides a simplified illustration of the covalent bonds an a silicate surface. This formation of three siloxane bonds explains the high stability of the interlayer against hot water or H_2O vapor. Coupling agents based on mono- or disilanol groups form significantly less stable interlayers. When coupling agents are added to silicate surfaces via methods I, II or III, the first reaction step is the hydrolysis of the trialkoxysilane group by the water of the hydrated silicate surface. The following reaction steps are discussed for a pretreatment with aqueous solutions.

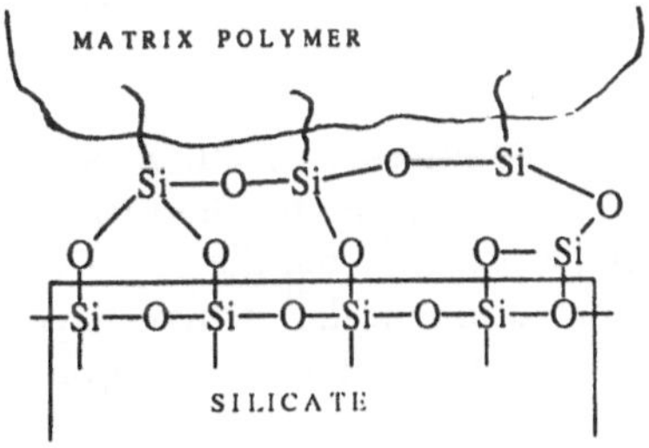

(54)

7.2.2
Structure of the Interlayer and Performance of Composites

The role of the coupling agent in a composite is twofold. First, it has to transmit mechanical forces from the matrix to the filler and from the filler to the matrix. Second, it has to prevent that water separates the matrix from the filler and to interrupt the cohesion. Any oxidized metal, inorganic ceramic, or silicate contains at least a monolayer of water on its surface. In the case of a rough surface and when hygroscopic ions are present, the surface is covered by a multilayer of water even in relatively dry air. This is particularly true for alkali silicate glasses. Drying by heating requires temperatures up to 800 °C for the complete removal of the water layer. Even after such a harsh treatment silanol groups are still detectable by IR-spectroscopy [72].

It has been demonstrated by model reactions that the alkoxy silane groups of coupling agents react not only with water but also with the silanol groups of silicon atoms on the surface of the filler, Eq. (53) and Scheme (54). The bonding of the coupling agent to the filler is catalyzed by aliphatic amines [73]. Therefore, addition of amines to solutions or mixtures containing the coupling agent is recommendable. However, in aqueous solution an alkaline pH accelerates the self-condensation of the trissilanol. Therefore, neutralization of the aqueous solution with a volatile weak acid, such as acetic acid or better CO_2, is required.

The treatment of the filler with a coupling agent reduces considerably the hydrophilicity of the surface. However, it was found that the presence of water is never completely excluded. The water can hydrolize any Si—O bond, and alkaline solutions which are typical for the surface of alkali glass even catalyze the hydrolysis. The success of coupling agents in a wet environment over long periods of time and at elevated temperature is, thus, a phenomenon which deserves a short discussion.

In an early stage of the research on coupling agents monoalkoxysilanes, bisalkoxysilanes and trisalkoxysilanes with identical carbofunctional groups were compared. It was found that composites treated with monoalkoxy or bisalkoxysilanes are by far too sensitive to a moist atmosphere for long periods of service times. Only trisalkoxysilanes yield satisfactory results as illustrated by the results listed in Tables 7.5 and 7.6. Spectroscopic results model reactions and theoretical calculations suggest that most trisalkoxysilanes form only one siloxane bond with the surface of the filler. Only in favorable cases two Si—O—Si-bonds may be formed. However, those alkoxysilane functions that do not react with the surface can form siloxane bonds with neighboring coupling agents. In consequence most silane additives exert three siloxane bonds. A schematic outline of such an interlayer formed by trisalkoxysilanes is illustrated by Scheme (54). Water which may be present from the beginning or which arrives at the interlayer by diffusion through the matrix (e.g. via microcrazes) can certainly hydrolize individual siloxane bonds. However, the

Table 7.5[a]   Fiberglass-reinforced polyester composites

Coupling agent on glass		Flexural strength of composite (MPa)	
Function	Symbol in Table 4	Dry	After 2-h water boil
None		386	234
BJY finish[b]		441	386
Chrome complex (Volan-A$^{R)}$)	–	503	428
Vinyl silane	A	462	414
Methacrylate silane	F	620	586
Cationic methacrylate silane	–	620	566
Cationic vinylbenzyl silane	G	634	566

[a] Data adapted from Ref. [1].
[b] Equimolar CH_2=$CHSiCl_3$ and CH_2=CCl—CH_2OH.

Table 7.6[a]   Mechanical strengths of epoxy laminates (MPa)

Finish on glass	Initial		2-h boil		200-h boil	
	flex	comp	flex	comp	flex	comp
None	448	290	–	–	–	–
Chrome complex (Volan-A$^{R)}$)	503	345	421	282	227	138
$=$SiCH$_2$CH$_2$CH$_2$Cl	586	324	503	310	345	241
$=$SiCH$_2$CH$_2$CH$_2$NH$_2$	544	338	482	303	353	234
$=$Si(CH$_2$)$_3$NHCH$_2$CH$_2$NHCH$_2$C$_6$H$_5$ $\cdot$ HCl	586	365	455	338	324	320
$=$Si(CH$_2$)$_3$NHCH$_2$CH$_2$NHCH$_2$C$_6$H$_4$—CH=CH$_2$ $\cdot$ HCl	586	358	550	310	413	310
$=$Si (CH$_2$)$_3$OCH$_2$—CH$\overset{O}{\overbrace{}}CH_2$	509	339	457	332	322	221

[a] Table from Ref. [1] with courtesy of Pergamon Press.

hydrophobic character of the interlayer prevents the accumulation of large amounts of water. Yet only the hydrolysis of all three siloxane bonds enables the liberation and diffusion of a coupling agent. However, such an event is unlikely for low concentrations of water in the interlayer. Furthermore, the reversibility of the hydrolytic reaction and the gain in entropy favors the "healing process" if only one or two siloxane bonds are hydrolyzed and no diffusion of the silane takes place. This model of an interlayer and its properties is based on numerous spectroscopic results, model reactions, extraction studies and mechanical measurements of serveral research groups (see Refs. 1, 53–68, 74–75 and section 7.2.2). Regardless of details of the interlayer model, numerous experimental results clearly demonstrate that the interlayer, and thus, the cohesion between matrix an filler is best protected against the deteriorating influence of water by

A) a nonpolar hydrophobic matrix polymer
B) a neutral filler which does not allow the diffusion of water (e.g. quartz) and
C) a relatively hydrophobic coupling agent (i.e. free of NH and ionic groups).

The binding of coupling agents to the matrix polymer may be as complex as the formation of the interlayer itself. This complexity is mainly a consequence of the fact that the interlayer of formed by the coupling agents is itself a multilayer (in contrast to the simplified Scheme (54)). The individual molecules of the interlayer are connected by siloxane bonds. Upon annealing or cure networks of siloxane chains may be formed. Furthermore, chlorinated paraffins and radical initiators such as peroxides may be helpful in the case of polyolefins as matrix materials. These additions either cause crosslinking of the polyolefin or formation of covalent bonds via recombination of radicals. In other words, semipenetrating and interpenetrating networks may be formed [76-80]. The formation of covalent bonds via radical reactions is characteristic for coupling agents having S—H methacrylate and styryl carbofunctional groups. Matrix polymers having ester groups can react with amino silanes under formation of amide groups. Matrix polymers which possess OH, NH or SH groups can undergo a nucleophilic addition onto coupling agents with epoxide or methacrylate groups. In summary the cohesion between matrix polymers and interlayer may be based on the following interactions:

1 covalent bonds
2 hydrogen bonds
3 van der Vaals and dipole-dipole forces
4 semipenetrating networks
5 interpenetrating networks (when both interlayer and matrix polymer form crosslinks).

With regard to physical character, mechanical properties and application composites may be subdivided into three groups: filled elastomers, filled thermoplastics and filled thermosetting resins. For all types of composites a significant improvement of the wet-strength retention is a basic advantage of the application of coupling agents. In the case of elastomers the following changes of their properties are expected (property first, change second):

1) elastic modulus: .... increase
2) tensile strength: .... increase
3) abrasion resistance: .... increase
4) Shore A hardness .... little influence
5) Bashore rebound: .... little influence
6) elongation; natural and
   nitrile rubber .... decrease
   styrene-butadiene rubber .... increase
7) cure time: natural and
   nitrile .... increase
   styrene-butadine rubberer .... decrease
8) heat build up .... decrease
9) tension set .... decrease
10) compression set .... decrease

For thermoplastic and thermosetting resins, again an increase of elastic modulus, tensile or flexural strength and abrasion resistance is expected along with a decrease of toughness and elongation at break. A particular advantage for the processing of thermoplastics is a considerable reduction of the melt viscosity. Selected commercial coupling agents and recommended applications are summarized in Table 7.4. The data compiled in Tables 7.5-7.10 illustrate the influence of coupling agents on the mechanical and electrical properties of various thermoplastic and elastomeric matrix materials. Tables 7.5, 7.6 and 7.10

Table 7.7[a]   Performance of polyester resin composites with various fillers and methacrylate silane F (see Table 4)

Filler	% D added	Flexural strength, MPa		% Improvement over nonsilane control	
		Initial	8-h boil	Initial	8-h boil
Quartz	1, 30	138	103	75	130
Wollastonite	0, 80	124	103	38	100
Aluminum silicate	0, 30	103	97	25	100
Clay-chopped glass	2, 25	159	117	29	25–69
Calcined clay	2, 20	103	97	15	27
Hydrous clay	2, 20	83	62	25	22
Hydrated alumina	1, 00	76	55	28	15
Calcium carbonate	2, 00	83	62	20	nil

[a] Data adapted from Ref. [81].

Table 7.8[a]   Properties of poly(methyl methacrylate) containing 60 wt % of untreated or silanized christobalite (grain size 60 μm) and untreated or silanized quartz (grain size 40 μm)

Property	untreated		silanized	
	Christobalite	Quartz	Christobalite	Quartz
Viscosity (mPa.s) (Brookfield 10 min^{-1}	9200	17 700	5 040	4 600
Flexural strength (N/mm^2) initial	73	70	115	123
after 6 h in H$_2$O/100 °C	56	49	99	126
Impact strength kJ/m^2 initial	2.4	2.4	4.8	4.7
after 6 h in H$_2$O/100 °C	2.2	2.0	4.5	4.4

[a] Data adapted from Ref. [5].

Table 7.9[a]   Mechanical properties of clay-filled natural rubber (clay treated with 1 wt% of aqueous trimethoxysilane)

Silane on Clay	None	Chloropropyl-trimethoxy-silane	Mercaptopro-pyltrimethoxy-silane	Aminepropyl-trimethoxy-silane
Peel adhesion, N/cm	nil	1.1	3.5	7.7
300% Modulus, MPa	7.2	6.8	10.2	11.4
Tensile strength, Mpa	23.6	25.4	26.8	27.1
Elongation, %	585	585	540	520
Tension set, %	43	43	41	38
Compression set, %	8	7	6	6
Bashore rebound, %	59	53	64	67
Tear Strength, ppi	N/cm	242	245	207
Peel adhesion, N/cm	nil	0.2	2.1	11.0
300% Modulus, MPa	2.0	2.0	2.8	2.6
Tensile strength, MPa	7.7	9.5	10.4	11.9
Elongation, %	925	1015	885	975
Tension set, %	37	41	35	38
Compression set, %	13	11	10	9
Bashore rebound, %	46	47	48	48
Tear strength, N/cm	247	262	270	275
Flex, JIS 10^{-2}	337	567	259	520
Abrasion resistance	100	130	155	140

[a] Data adapted fro Ref. [1].

Table 7.10ᵃ Electrical properties of mineral-filled polyethyleneᵇ

Silane addedᶜ	Clay				Wollastonite				Quartz			
	Dielectric constant		Dissipation factor		Dielectric constant		Dissipation factor		Dielectric constant		Dissipation factor	
	Initial	Wet	Initial	Wet	Dry	Wet	Dry	Wet	Dry	Wet	Dry	Wet
Filled, no silane	2.7	3.0	0.003	0.082	2.8	4.2	0.009	0.147	2.7	5.2	0.029	0.228
Filled + D	2.7	2.7	0.002	0.003	2.8	2.9	0.007	0.014	–	2.5	0.010	0.012
Filled + C	2.7	2.7	0.002	0.005	2.8	2.9	0.007	0.013	–	–	–	–

ᵃFrom Ref. [1] with courtesy of Pergamon Press.
ᵇResin, filler, and silane mixed in a two-roll mill at 240–250 °C. Composition: Polyethylene resin DYNH (Union Carbide), low density, 50 wt%. Polyethylene resin DMD-7000 (Union Carbide), high density, 50 wt%. filler 50 wt% Electrical properties measured at 1000 Hz per ASTM D-150.
ᶜSilane added to clay at 2 phf; to Wollastonite at 0.8 phf; and to quartz at 0.3 phf.

demonstrate the resistance of silane-treated composites against the destabilizing effect of moisture.

7.3
Coupling Agents for Chromatography, Immobilization of Metal Ions and Enzymes

Alkyl and aryl trialkoxy (or chloro)silanes devoid of carbofunctional groups play a role in chromatography by hydrophobization of the stationary phase. This aspect will be discussed in the next section. The present section deals with applications of true coupling agents which fix a specific substrate via a covalent bond to the stationary phase.

7.3.1
Charge-Transfer and Ligand-Exchange-Chromatography

The characterization of polynuclear or fused-ring aromatic hydrocarbons is important because they may possess carcinogenic properties. These hydrocarbons are difficult to separate and isolate because they lack any functional group. Furthermore, they occur only in low concentrations in sources such as cigarette smoke or in the waste gases of engines. Polynuclear aromatics possess highly polarizable π-electrons and are good donors for π-acceptors such as nitroaromatic compounds.

Silica filled columns treated with γ-aminopropyltrialkoxysilane and 3-(2,4,5,7-tetranitro)fluorenone proved to be a useful stationary phase (Form. 55) in liquid chromatography for the separation of difference polynuclear aromatics. The selectivity proved to be high but the efficiency low [83]. A higher efficiency for the separation of substituted anthracens and chrysens was obtained when the γ-aminopropylsilane was diluted with an alkylsilane (molar ratio 2:8) and when the amino groups were transformed into 3-(2,3,4,7-tetranitrofluorene) imines [84].

In a third study dinitrophenylated amino acids were used as π-acceptor components. They were bound to the stationary silica phase via γ-aminopropyltrisethoxysilane (56). As illustrated by the data of Table 7.10 [1] the tendency to form a CT-complex increases with higher polarizability and size of the fused-

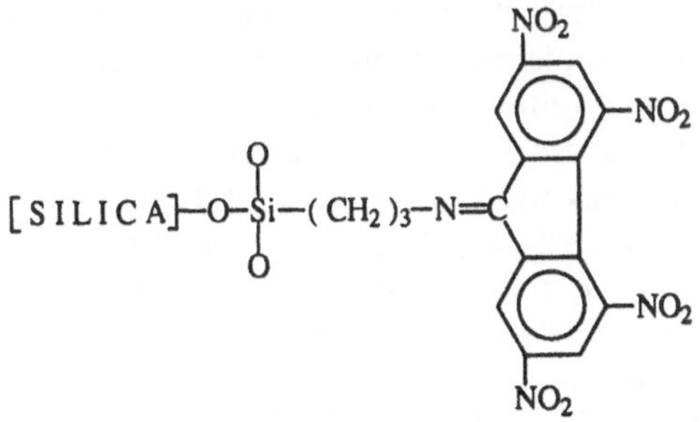

$$[SILICA]-O-\overset{\overset{\displaystyle O}{|}}{\underset{\underset{\displaystyle O}{|}}{Si}}-(CH_2)_3-N=C \qquad (55)$$

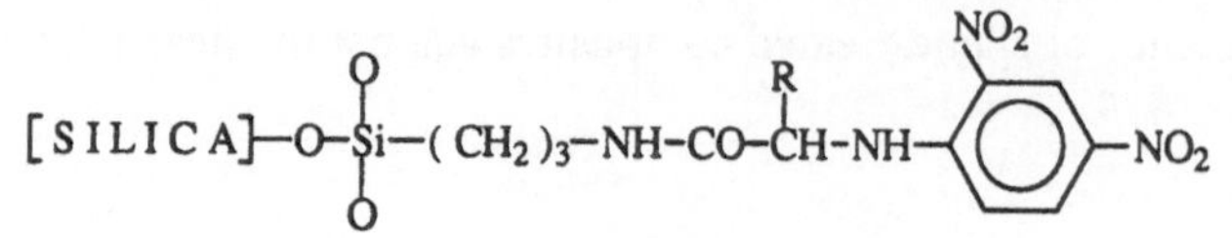

$$[SILICA]-O-\underset{\underset{O}{\overset{O}{||}}}{Si}-(CH_2)_3-NH-CO-\overset{R}{\underset{}{CH}}-NH-\underset{NO_2}{\overset{NO_2}{\bigcirc}}-NO_2$$

(56)

R = H   : Glycine
R = CH$_3$: Alanine
R = CH(CH$_3$)$_2$: Valine
R = CH$_2$—   : Phenylalanine

ring aromatics. Attempts to utilize the chirality of these amino acid derivatives for the separation of 1-AZA-6-helicene racemates into enantiomers were rather successless [1].

Another approach is based on the interaction of aromatic compound such as substituted naphthalenes, anthracenes or fluorenes, with chelated Cu^{2+} ions [85]. The ligands for the copper ions were prepared in two quite different ways. Either the silica treated with γ-aminopropylsilane was reacted with CS$_2$ to form dithiocarbamate (57) or it was acetylated with ethyl benzoylacetate (58). The interaction of the aromatic solutes with the copper ions is partially based on the π-system partially on the substituents (mainly amino groups). Therefore, this chromatographic method may be considered as a kind of ligand-exchange chromatography (LEC).

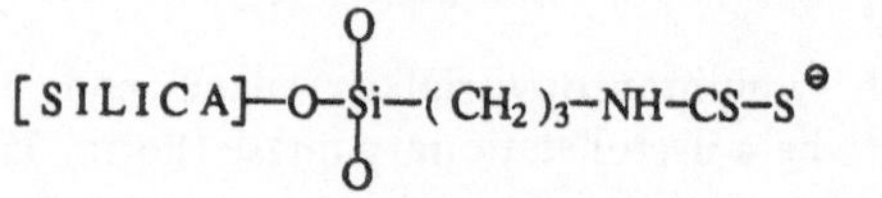

$$[SILICA]-O-\underset{\underset{O}{\overset{O}{||}}}{Si}-(CH_2)_3-NH-CS-S^{\ominus}$$

(57)

$$[SILICA]-O-\underset{\underset{O}{\overset{O}{||}}}{Si}-(CH_2)_3-NH-CO-CH_2-CO-\bigcirc$$

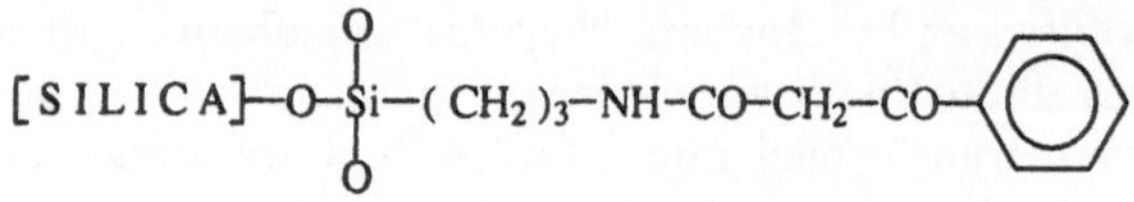

(58)

A classical application of LEC is the separation of different amino acids or amino acid derivatives and the separation of racemic amino acid derivatives into the enantiomers. Several research groups have reported on the synthesis of chiral stationary phases containing chelated metal ions attached to their surface via silane coupling agents. Epoxysilanes [86–90], γ-amino propylsilane [91–93] and γ-chloropropylsilane [94–96] were preferentially used. The formation of a tridentate ligand by addition of proline onto an epoxysilane is illustrated by Eq. (59). In addition to L-proline, L-4-hydroxyproline, L-pipacolic acid and L-azetidine carboxylic acid were successfully used as ligands [90]. Cu^{2+} ions are the most widely used central atoms of such complexes. Such chiral stationary phases not only allow the resolution of racemic amino acids but also the resolution of racemic dipeptides such as Gly-D,L-Ala. Another type of hydrophobic chiral stationary phase based on an amino acid was prepared by hydrosilylation of N-ω-undeceno-yl-L-valine tert.-butylester Eq. (60) [97].

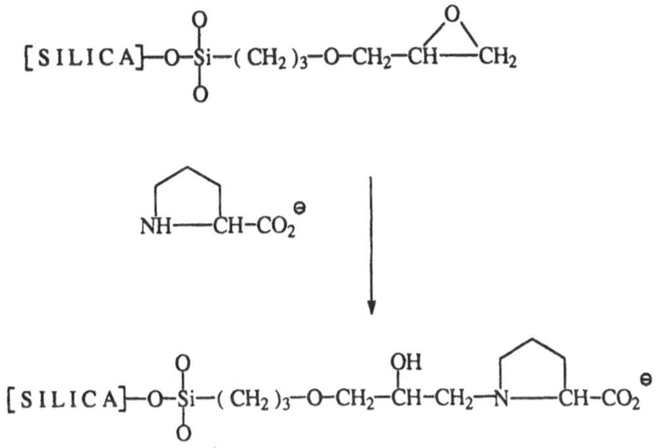

$$[SILICA]-O-\underset{\underset{O}{\|}}{\overset{\overset{O}{\|}}{Si}}-(CH_2)_3-O-CH_2-\overset{\overset{O}{\triangle}}{CH}\underset{}{}CH_2$$

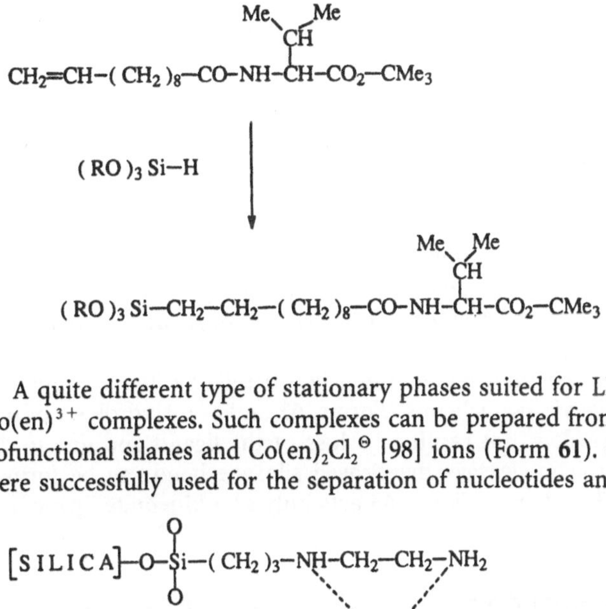

$$\underset{NH\underline{}CH-CO_2^{\ominus}}{\overset{\frown}{}} \qquad \Big\downarrow \tag{59}$$

$$[SILICA]-O-\underset{\underset{O}{\|}}{\overset{\overset{O}{\|}}{Si}}-(CH_2)_3-O-CH_2-\overset{\overset{OH}{|}}{CH}-CH_2-N\underline{}\overset{\frown}{CH}-CO_2^{\ominus}$$

$$CH_2\!=\!CH-(CH_2)_8-CO-NH-\overset{\overset{\displaystyle Me\diagdown\overset{|}{CH}\diagup Me}{}}{CH}-CO_2-CMe_3$$

$$(RO)_3\,Si-H \qquad \Big\downarrow \tag{60}$$

$$(RO)_3\,Si-CH_2-CH_2-(CH_2)_8-CO-NH-\overset{\overset{\displaystyle Me\diagdown\overset{|}{CH}\diagup Me}{}}{CH}-CO_2-CMe_3$$

A quite different type of stationary phases suited for LEC was prepared from Co(en)$^{3+}$ complexes. Such complexes can be prepared from ethylenediamine carbofunctional silanes and Co(en)$_2$Cl$_2{}^{\ominus}$ [98] ions (Form **61**). Such stationary phases were successfully used for the separation of nucleotides and nucleosides [98].

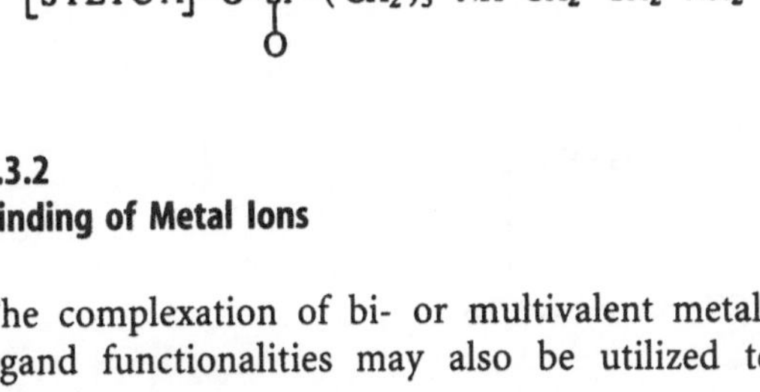

$$[SILICA]-O-\underset{\underset{O}{\|}}{\overset{\overset{O}{\|}}{Si}}-(CH_2)_3-NH-CH_2-CH_2-NH_2$$
$$CoCS_2 \tag{61}$$
$$[SILICA]-O-\underset{\underset{O}{\|}}{\overset{\overset{O}{\|}}{Si}}-(CH_2)_3-NH-CH_2-CH_2-NH_2$$

7.3.2
Binding of Metal Ions

The complexation of bi- or multivalent metal ions by coupling agents with ligand functionalities may also be utilized to collect metals from various aqueous solutions. This application may be useful for several purposes, such as qualitative or semiquantitative analysis of aqueous solutions, recovering of precious metals (e.g. Pt used as catalyst) or purification of waste-water. Glass beads and silica as stationary phases have the advantage that the ion binding and exchange process is not diffusion controlled, and thus, considerably faster. Equilibration with the ligands of the silica surface only requires minutes. Furthermore, the efficiency of all processes does not depend on changes of the pressure in contrast to swollen gels of organic polymers. However, the moderate stability of modified silica and glass surfaces in acidic or alkaline solution may hinder a technical application, if the pH of the aqueous solutions deviates from 7 over long periods of time.

In order to stabilize the layer of ligand-functionalized coupling agents on the silica surface, four methods were proposed [1], but a final evaluation seems to be lacking:

1) The hydrophilic amino groups which are strongly solvated by water are more than three carbon atoms removed from the silicon. An example of this type of couple agents is depicted is the styrene derivative 62.

$$[SILICA]\!-\!O\!-\!\underset{\overset{\displaystyle O}{|}}{\overset{\displaystyle O}{\underset{|}{Si}}}\!-\!CH_2\!-\!CH_2\!-\!\!\left\langle\!\!\bigcirc\!\!\right\rangle\!\!-\!CH_2\!-\!NH\!-\!CH_2\!-\!CH_2\!-\!NH_2 \qquad (62)$$

2) Silanes with free silanol groups which have not completely reacted with the silica surface or neighboring coupling agents are particularly sensitive to any hydrolytic attack. Therefore, additional crosslinking and hydrophobization with difunctional silanes such as $(MeO)_3Si\!-\!CH_2CH_2Si(OMe)_3$ may provide an additional protection.

3) Ligand functionalized silanes containing two trisalkoxysilane groups can bind to the silica surface via two silicons (Forms. 63 and 64). Hydrolysis of one silicon from the surface does not result in a loss of the ligand and due to the reversibility of the hydrolytic cleavage the cleaved siloxane bond can be formed again at a later time. The silane of Form. 63 acts only as a bidentate ligand and

$$[SILICA]\!-\!O\!-\!\underset{\overset{\displaystyle O}{|}}{\overset{\displaystyle O}{\underset{|}{Si}}}\!-\!(CH_2)_3\!-\!NH\!-\!CH_2$$

$$[SILICA]\!-\!O\!-\!\underset{\overset{\displaystyle O}{|}}{\overset{\displaystyle O}{\underset{|}{Si}}}\!-\!(CH_2)_3\!-\!NH\!-\!CH_2 \qquad (63)$$

$$[SILICA]\!-\!O\!-\!\underset{\overset{\displaystyle O}{|}}{\overset{\displaystyle O}{\underset{|}{Si}}}\!-\!(CH_2)_3\!-\!NH\!-\!CH_2\!-\!CH_2\!-\!NH\!-\!CH_2$$

$$[SILICA]\!-\!O\!-\!\underset{\overset{\displaystyle O}{|}}{\overset{\displaystyle O}{\underset{|}{Si}}}\!-\!(CH_2)_3\!-\!NH\!-\!CH_2\!-\!CH_2\!-\!NH\!-\!CH_2 \qquad (64)$$

two closely neighboring silanes of this type are required to form a stable chelate complex. However, such a sterically favorable neighborhood of two silanes of this kind is certainly not a frequent situation. In contrast the silane **64** has an optimum structure for binding of a metal ion.

4) Relatively hydrophobic ligands help to reduce the risk of a hydrolytic attack on the silane-silica surface. An 8-hydroxyquinoline derivative may serve as an example of this strategy. Its synthesis is outlined in Scheme **65**.

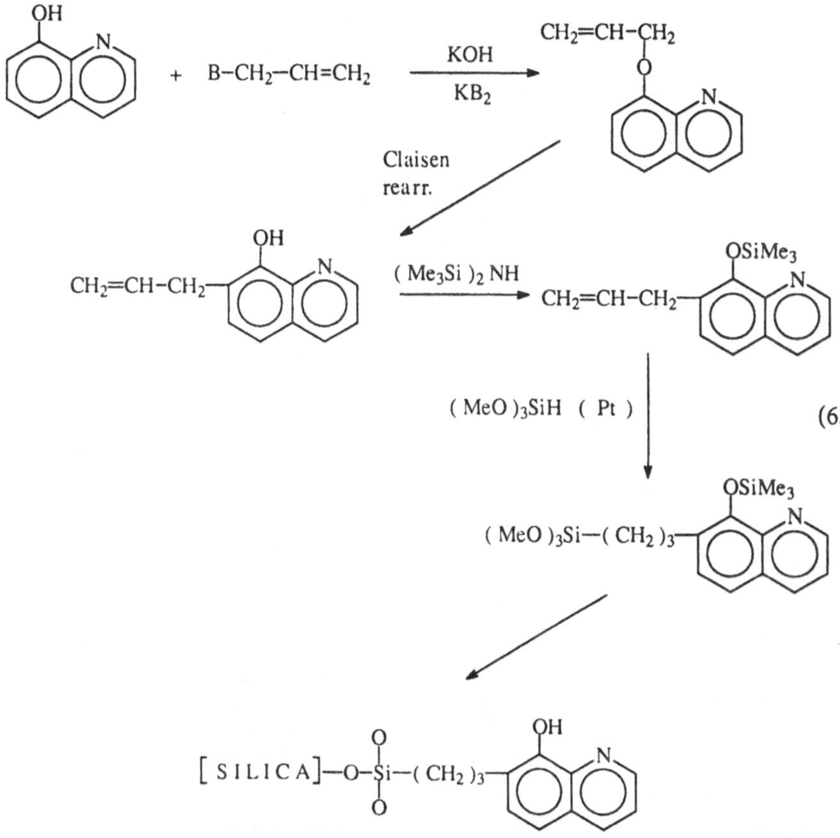

(65)

Metal cations only bind to free amino groups and not to ammonium-ions. Therefore a successful complexation of metal ions with the ligands **62–65** strongly depends on the pH of the solution and on the basicity of the ligand. At a pH above 1 almost all bi- or multivolent metal ions are quantitatively extracted from their aqueous solution. Nonetheless, the minimum pH required for quantitative complexation also depends on the nature of the ion. For instance a $pH \geqslant 7$ is optimum for Zn^{2+}, a $pH \geqslant 5$ for Cu^{2+} and a $pH \geqslant 2.5$ for Hg^{2+}. The cations can be eluted by aqueous solutions of multidentate ligands such as EDTA. In combination with a pH gradient both the binding and the elution process can be used to separate different metals. Whereas the protonated ligands are useless for the complexation of cations, they may be utilized for the extraction of anions from aqueous solutions. Particularly strongly bound are the oxyanions of arsenic, selenium, molybdenium and tungsten. The optimum pH for the extraction of such anions falls into the range of pH 2-5.

Immobilized metal complexes may serve as catalysts. Such immobilized catalysts combine the advantage of a more or less homogeneous catalysis with insolubility in all common solvents. The lack of solubility allows an easy separation of catalyst and substrate and prevents losses of precious catalytic metals such as palladium or platinum. The location of the catalysts on the surface of the stationary phases reduces problems with steric hindrance and slow diffusion of the substrate. Depending on the reactivity of catalyst and substrate free silanol of the coupling agent or on the surface of the support may cause side reactions. Therefore, a pretreatment with silylating agents such as chlorotrimethylsilane or hexamethyldisilazane is recommendable.

Most immobilized catalysts studied so far are based on phosphine ligands. Therefore, three methods for the synthesis of phosphine functionalized silanes are outlined in Eqs. (66), (67) and (68). A list of immobilized catalysts and their catalytic application is compiled in Table 7.11 and reviewed in Ref. [99].

Table 7.11[a] Representative silica-supported metal complex catalysts prepared by use of ligand silane coupling agents[b]

Surface complex catalyst[c] on silica	Method of preparation	Catalytic application
$CpTiCl_2(C_5H_4)-$	B	hydrogenation
$TiCl_2(C_5H_4)_2-$	B	hydrogenation
$RhCl[PPh_2(CH_2)_2]_3-$	A, B	hydrogenation and hydroformation
$RhCl[(CO)PPh_2(CH_2)_2]_2-$	B	hydroformylation
	A	carbonylation
$Rh(accac)(CO)PPh_2(CH_2)_2-$	B	hydroformylation
$PhCl[PPh_2C_6H_4(CH_2)_4]_2-$	B	hydrogenation
$RhCl_3/PPh_2(CH_2)_2-$	A	hydrosilylation
$RhCl_3(COD)[PPh_2(CH_2)_{14}]-$	A, B	hydroformylation
$RhCl_3/NMe_2(CH_2)_3-$	A	hydrosilylation
$PtCl_4^{2-}/SnCl/PPh_2(CH_2)_2-$	A	isomerization
$H_2PtCl_6/NMe_2(CH_2)_3-$	A	hydrosilylation
$Pd(acac)_2/PPh_2(CH_2)_8-$	A	carbonylation
$Co(CO)_2(C_5H_4)-$	A, B	hydroformylation
$Co(acac)_2/PPh_2(CH_2)_2-$	A	hydroformylation
$NiCl_2[PPh_2(CH_2)_2]-$	A	oligomerization
$IrCl[PPh_2C_6H_4(CH_2)_4]-$	A	hydrogenation

[a]Data adapted from Ref. [82].
[b]Abbreviations used in this table as follows: acac: acetyl-acetonate; COD: 1, 5-cyclooctadiene; Cp: cyclopentadienyl.
[c]In cases where undefined metal complexes were formed by Method A, the metal precursor/ligand silica system is specified.

$$(RO)_3 Si-CH=CH_2 + HPPhe_2 \longrightarrow (RO)_3 Si-CH_2-CH_2-PPhe_2 \qquad (66)$$

$$Cl_3SiH \quad + \quad CH_2=CH-CH_2-CH_2-\hspace{-0.3em}\langle\bigcirc\rangle \xrightarrow{\ (\ H_2PtCl_6\ )\ }$$

$$\text{(67)}$$

$$Cl_3Si-(\ CH_2\ )_4-\hspace{-0.3em}\langle\bigcirc\rangle\hspace{-0.3em}-PPhe_2$$

$$[\ SILICA\ ]-O-\underset{\underset{O}{\overset{\displaystyle\|}{}}}{\overset{\displaystyle O}{Si}}-(\ CH_2\ )_3-Cl \quad + \quad \overset{\oplus}{K}\ \overset{\ominus}{PPhe_2} \longrightarrow$$

$$\text{(68)}$$

$$[\ SILICA\ ]-O-\underset{\underset{O}{\overset{\displaystyle\|}{}}}{\overset{\displaystyle O}{Si}}-(\ CH_2\ )_3-PPhe_2$$

When such catalysts are prepared from single metal ions two synthetic strategies with almost identical results can be applied. Strategie I consists of the preparation of a silane/metal complex in solution followed by the binding of this complex to the surface of the support. Strategy II is based on the inverse reaction sequence: the first step is the reaction of coupling agent with the support. Strategy I has the principal advantage that the complex can be purified and characterized in solution. Nonetheless, struture and reactivity of the complexes will be almost identical regardless of the synthetic strategy. This is not necessarily true, when more complex catalysts are involved such as those based on metal clusters. The initial reaction of metal clusters with the oxidized surface of the support may involve redox-reactions, because the metal-metal bonds of the cluster may be strongly reducing agents.

For the immobilization of metal clusters γ-mercaptopropyl silanes **69a** and γ-isocyanidopropylsilanes **69b** proved to be useful coupling agents. The carbofunctional groups react with the clusters in a complex way including redox-reactions and not only donor-acceptor interactions. An example of soluble cluster-silane reaction products is the osmium complex **70a** [100] resulting from the conversion of $[Os_3(CO)_{12}]$ with **69a**. Another example is the ruthemium complex **70b** obtained from $[Ru_3(CO)_{12}]$ and $H_2P-CH_2-CH_2-Si(OEt)_3$ [100]. Numerous other immobilized clusters were studied by Evans and Coworkers [101–106].

$$HS-(\ CH_2\ )_3-Si\ (\ OR\ )_3 \qquad\qquad N{\equiv}C-(\ CH_2\ )_3-Si\ (\ OR\ )_3 \qquad \text{(69)}$$

$$\underline{a} \qquad\qquad\qquad\qquad\qquad\qquad \underline{b}$$

7.3.3
Immobilization of Enzymes and Other Proteins

Chemical processes catalyzed by enzymes are constantly gaining in importance for both research purposes and technical applications. The classical approach,

Os (CO)$_4$

(CO)$_3$ Os = Os (CO)$_3$
H
S
(CH$_2$)$_3$ Si (OR)$_3$

H — Ru — H

(CO)$_3$ Ru —————— Ru (CO)$_3$
P
(CH$_2$)$_3$ Si (OEt)$_3$

a   b   (70)

namely the treatment of a dissolved substrate with a soluble enzyme or entire cell, is unfavorable because the separation from the substrate and its reaction products is difficult and relatively expensive. The immobilization of enzymes, cells and microorganisms on solid insoluble supports is thus, an attractive alternative strategy [107–111]. The immobilization process may have the shortcoming that the efficiency of the enzyme is reduced, but this shortcoming is usually counter-balanced or overcome by a stabilization and longer lifetime of the enzyme.

Compared to crosslinked gels of organic polymers inorganic solid supports have the advantages of being insensitive to variations in pressure and the enzymes are exclusively bound to the surface where they are easily accessible to any substrate. From the view point of efficiency, and high conversion rate particulate supports with a high surface/mass ratio are favorable. Unfortunately, small particles (diameter < 1 mm) have the disadvantage that their dense packing reduces the flow rate. The packing of columns and reactors with particles having a size on the order of 1–5 mm is beneficial for the flow rate, but lowers the surface/mass ratio unless highly porous particles are used. For the immobilization of enzymes the pore size should fall into the range of 20–100 nm. Porous glass beads [112], silica and ceramic particles optimized for this application have become commercially available [113].

The immobilization process may be based on three quite different interactions and synthetic strategies:

(A) Association of the enzyme with the carbofunctional chain of the coupling agent via H-bonds, dipole-dipole-interactions and van der Vaals forces.

(B) Coupling of enzyme and silane via covalent bonds.

(C) Entrapment of enzymes in networks (e.g. polysiloxanes) prepared by the coupling agent and additional reaction partners. This strategy may be particularly useful for the immobilization of cells and microorganisms.

Strategy (A) provides only a relative weak adhesion of the enzyme to the support, and this method is, in general, not well suited for technical application. The formation of covalent bonds between coupling agent and enzyme (method B) is the most widely used strategy for the immobilization of enzymes. The coupling of an enzyme and silane usually involves one of the following four reaction sequences (see Ref. 1 p. 226):

1) Formation of amide bonds:

If the enzyme has free carboxyl groups and no amino groups or at least an excess of carboxyl groups, it can be coupled to amino groups of the silane by means of a water soluble carbodiimide, Eq. (71). If the amino group is directly attached to a silicon (e.g. by aminolysis of chlorinated silica) the resulting amide group is highly sensitive to hydrolysis at the Si—N-bond.

$$EZ–CO_2H \ + \ NH_2–(CH_2)_3 \, Si \, (OMe)_3 \ \xrightarrow{R-N=C=N-R'} \ EZ–CO–NH–(CH_2)_3 \, Si \, (OMe)_3$$

$$EZ = Enzyme \tag{71}$$

If the enzyme has an excess of amino groups, a γ-aminopropylsilane is first reacted with succinic anhydride and then coupled to the enzyme, Eqs. (72, 73).

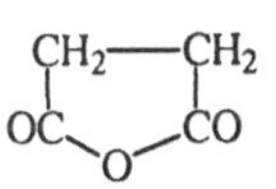

$$\underset{OC\diagdown O \diagup CO}{\overset{CH_2—CH_2}{|\quad\quad|}} \xrightarrow{+ \ NH_2 \, (CH_2)_3 \, Si \, (OMe)_3} HO_2C–(CH_2)_2–CO–NH–(CH_2)_3–Si \, (OMe)_3$$

$$+ \ EZ–NH_2 \tag{72}$$
$$+ \ R–N=C=N–R'$$

$$\downarrow$$

$$EZ–NH–CO–(CH_2)_2–CO–NH–(CH_2)_3–Si \, (OMe)_3$$

$$\tag{73}$$

2) Coupling via epoxide groups

A coupling agent with epoxide (carbofunctional) group is at first reacted with stationary phase and the protein is then coupled with the epoxide group. This method has recently found increasing interest [83–86].

3) Coupling via glutaraldehyde

Glutaraldehyde is a convenient coupling reagent for proteins with free amino groups, Eq. (74). The resulting aldimine groups are sensitive to acidic hydrolysis, but reduction with sodium borhydride, Eq. (75), yields stable alkylamines.

$$EZ–NH_2 \ + \ O=CH–(CH_2)_3–CH=O \ + \ NH_2–(CH_2)_3Si \, (OMe)_3 \ \longrightarrow$$

$$\tag{74}$$

$$EZ–N=CH–(CH_2)_3–CH=N–(CH_2)_3 \, Si \, (OMe)_3$$

$$\downarrow \ Na(BH_4) \tag{75}$$

$$EZ–NH–CH_2–(CH_2)_3–CH_2–NH–(CH_2)_3Si \, (OMe)_3$$

4) Azocoupling

Acylation of γ-aminopropylsilane by 4-nitrobenzoylchloride followed by hydrogenation of the nitro group, Eqs. (76, 77), diazotation and coupling of the diazonium group with the phenol group of tyrosine residues leads to a stable azobond, Eqs. (78, 79).

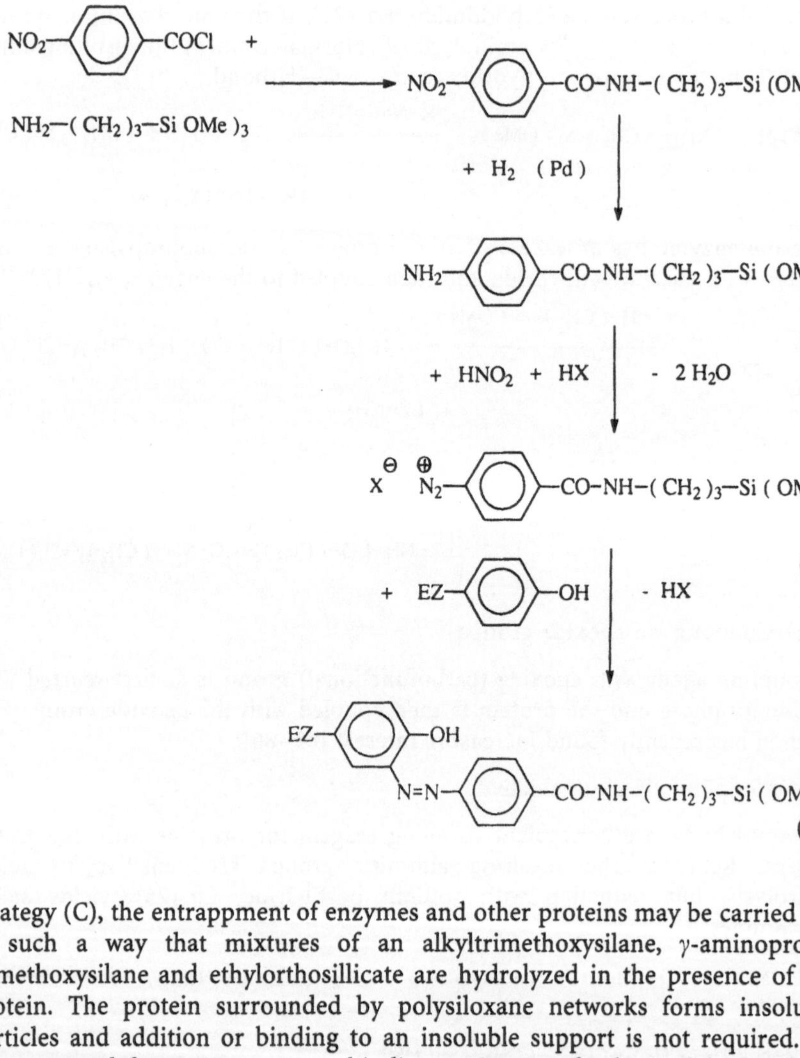

$$NO_2-\langle C_6H_4\rangle-COCl \ + \ NH_2-(CH_2)_3-Si\,OMe)_3 \ \longrightarrow \ NO_2-\langle C_6H_4\rangle-CO-NH-(CH_2)_3-Si(OMe)_3 \tag{76}$$

$$+ \ H_2 \ (Pd)$$

$$NH_2-\langle C_6H_4\rangle-CO-NH-(CH_2)_3-Si(OMe)_3 \tag{77}$$

$$+ \ HNO_2 \ + \ HX \quad -\ 2\,H_2O$$

$$X^\ominus \ \overset{\oplus}{N_2}-\langle C_6H_4\rangle-CO-NH-(CH_2)_3-Si(OMe)_3 \tag{78}$$

$$+ \ EZ-\langle C_6H_4\rangle-OH \quad -\ HX$$

$$EZ-\langle C_6H_3(OH)\rangle-N=N-\langle C_6H_4\rangle-CO-NH-(CH_2)_3-Si(OMe)_3 \tag{79}$$

Strategy (C), the entrappment of enzymes and other proteins may be carried out in such a way that mixtures of an alkyltrimethoxysilane, γ-aminopropyl trimethoxysilane and ethylorthosillicate are hydrolyzed in the presence of the protein. The protein surrounded by polysiloxane networks forms insoluble particles and addition or binding to an insoluble support is not required. As demonstrated for an progesteron binding antiserum, for hemoglobin and for urease [118] no denaturation takes place. The enzymatic activity of the entrapped urease was comparable with that of dissolved urease [118].

Other entrappment methods involve a solid support treated with a functional coupling agent and crosslinking of a multifunctional organic polymer (e.g. poly(ethylene)imine, poly(acrylic acid) or poly(vinylether/maleic anhydride)s) with the carbofunctional groups in the presence of a protein [119].

Immobilized proteins have been used for analytic purposes [120] for affinity chromatography [114–117, 121], and for commercial production [108, 122]. The binding of whole cells has been discussed by Arkles et al. [123].

7.4
Surfactants and Analysis of Silanized Surfaces

7.4.1
Surfactants

Surfactants may be defined as silanes designed to modify the chemical and physical properties of a surface. Coupling reactions by formation of covalent bonds between modified surface and environment are not included in this definition, whereas ionic interactions between the modified surface and the surrounding medium may be desired. The surfaces that need modification by treatment with silanes are not necessarily restricted to silicates. Any kind of ceramics and oxidized metals may be suitable substrates, and even poly-saccharides may be included in this consideration.

Surfactants may be useful for attaching defined cationic (e.g. ammonium) groups or anionic (e.g. carboxylate) groups to the uncharged surface. However the most widely used application is the hydrophobization of more a less hydrophilic, polar surfaces. Alkyl trichlorosilanes, alkyl trialkoxy silane and the corresponding phenylsilanes are typical surfactants used for this application. Alkyl silanes have the advantage that the variation of the alkyl group from methyl to octadecyl allows an easy optimization of the hydrophobization for any special application. A standard criterion for the wettability of a surface is its critical surface tension (γ_{crit}). A liquid with given surface tension will spread over any surface with a higher tension. γ_{crit} is determined in the following way. Several non wetting solvents, covering a wide range of surface tensions, are placed on the surface under investigation and the contact angle θ is measured. These measurements serve for the construction of a plot with cos θ on the ordinate and γ on the abscissa. The extrapolation leads to an interception of the ordinate at γ_{crit} [124–127] (socalled Zisman plot). Unfortunately silanized surfaces rarely yield linear plots and graphical methods were developed to overcome this difficulty. A particular problem for the characterization of silanized surfaces is the sensitivity of γ_{crit} to the thermal history of the surface. In other words, the procedure used for the application of the surfactant and the drying process have a great influence on the final properties of the surface [128, 129]. Nonetheless, a number of γ_{crit} values of silanized and nonsilanized surfaces is listed in Table 7.12.

The hydrophobization of oxidic surfaces have several important applications. Because of the low costs and the low surface tension which prevents wetting by water, methylsilanes are commercial products for improving water-resistance of ceramics, masonry, brick, sandstone, concrete or mortar. The enhancement of the water-resistance of electrical insulators is a similar application, but it is unsuitable for an application of the inexpensive methyl-trichlorosilane, because even traces of HCl remaining after the hydrolysis are a risk owing to corrosion of neighboring metal wires. The methyltrimethoxysilane may be applied as a vapor, in an organic solvent or more conveniently in an alkaline aqueous solution neutralized with CO_2.

Table 7.12 Critical surface tensions (γ_{crit}) of selected surfaces[a]

Surface	γ_{crit}
Silica (fused)	78.0
Glass (soda-lime, dry)	47.0
Nylon-6, 6	46.0
Iron (dry)	46.0
Aluminum (dry)	45.0
Poly(ethylene terephthalate)	43.0
Glycidoxypropyl trimethoxysilane	42.5
γ-Mercaptopropyl (trimethoxysilane)	41.0
γ-Chloropropyl trimethoxysilane	40.5
Phenyltrimethoxysilane	40.0
Polyvinylchloride	39.0
γ-Aminopropyl triethoxysilane	35.0
Polystyrene	34.0
Trifluoropropyl trimethoxysilane	33.5
Polyethylene	33.0
Polypropylene	31.0
Ethyltrimethoxysilane	27.0
Vinyltriethoxysilane	25.0
Methyltrimethoxysilane	22.0
Polytetrafluoroethylene	18.5

[a]Surface tensions of silanes refer to treated silicate-surfaces.

Several metal oxides may be used as semiconductor electrodes. Coating with silanes not only reduces the activity of these electrodes, but also allows one to study the course of electron transfer reaction on their surface in more detail and dyes can be bound to the surface of electrodes to study the redox-reactions in these monolayers. Furthermore, a comparison with the reactivity of the same substrate in solution is feasible. Of particular interest are dyes, such as xanthines or phthalocyanines which can undergo photoinitiated redox-reactions [133, 134].

The so-called water harvesting is a process designed to prevent a dry soil from absorbing rain or dew exclusively on the surface, where it will rapidly evaporate under the influence of sunlight and wind. Hydrophobization of the surface (at least around plants) allows the water to penetrate into deeper layers where it is protected against rapid evaporation in a hot climate. Combinations of silicones with methyl siliconate and a polymer latex (e.g. styrene-butadiene rubber) seems to be the most promising approach to this problem [125, 135–137].

The most wide spread application of hydrophobic surfactants is in so-called reverse phase chromatography [138–154]. This term has been coined, because the normal Gas (GC), Thin Layer (TLC) or Liquid-Chromatography (LC or HPLC) is based on stationary phase wide polar surface and a less polar eluent. After hydrophobization of the stationary phase the eluent is the more polar phase. In the case of TLC and HPLC methanol/H_2O, ethanol/H_2O or actonitrile/H_2O mixtures are typical eluents. This approach goes back to the work of Abel et al. in the mid 1960s [138]. In gas-liquid chromatography the hydrophobization of the silica-filled colums with memthyltrichlorosilane prevents the "tailing" of polar substrates [154].

Most recent studies concern silanized silica used for LC and HPLC. The influence of the following structural parameters were evaluated:

A) Surface area and pore size of the silica.

B) Monofunctional silanes (e.g. dimethyl octadecylchlorosilane) versus trifunctional silanes (e.g. octadecyl trichlorosilane).

C) Length of the alkyl chain (with octadecyl silanes as the longest carbon chain).

Polycyclic aromatic hydrocarbons [139–144], and substituted aromatics [149–151] were the preferred substrates, but proteins were also included in such tests. Trichlorosilanes have the general advantage over monochlorosilanes in that they can form more than one siloxane bond to the surface and an additional siloxane bond with neighboring silanes. The resulting siloxane network (Scheme 54) is more stable against hydrolysis, and thus, more convenient for a routine application in silica packed columns. Longer alkyl chains may show a higher selectivity for hydrophobic substrates than shorter ones, when eluents poor in H_2O are used. A higher water content of the eluent enhances the retention time, for surfaces with long alkyl chains. Yet the different selectivities of short and long alkyl chains are more or less leveled off. Free silanol groups which may be present due to incomplete coating of the surface or due to hydrolysis of trichlorosilanes may interact with the substrate (solute) [155]. This undesired interaction can be eliminated by silylation with a donor of trimethylsilyl groups or better by a difunctional crosslinking agent. A selection of more recent papers (including reviews) dealing with reverse phase chromatography is presented in Refs. [156–170].

Finally, it should be mentioned that silanes can be utilized to attach quaternary ammonium groups to a given surface. These ammonium groups are, of course, useful for the binding (extraction, chromatography) of anionic substrates. However, they also show antimicrobial activities, as discussed by several research groups [125, 171–173].

7.4.2
Analysis of Silanized Surfaces

The structure of silanized surfaces (mainly silica and glass) has been the subject of numerous studies with physical methods regardless whether surfactants or coupling agent were used for the coating process. A detailed description of the analytical methods and their results is beyond the scope of this work, and thus, only a short summary of methods and pertinent references should be given.

In addition to the afore-mentioned measurements of contact angles, IR-spectroscopy has been the most widely used analytical method for the characterization of silanized surfaces. IR-Spectroscopy enables a distinction of siloxane and silanol groups and enables a differentiation between free and H-bonded silanols. A further advantage is the analysis of IR-radiation reflected from the surface, and thus, the characterization of monolayers. The usefulness of IR-spectroscopy has been greatly enhanced by the general availability of the Fourier-transform technique in the years since 1975. FT-IR spectroscopy not only allows a considerable improvement of the signal to noise ratio, it also allows the computerized substraction of background radiation or spectra of model compounds. Results obtained by classical IR-spectroscopy are discussed in Refs. [174–181], and the application of FT-IR-spectroscopy in Refs. [182–191].

NMR spectroscopy has also proved to be a useful analytical tool for the characterization of silanized surfaces, since the cross-polarization/magic angle-spinning method became commercially available. Numerous ^{13}C NMR and ^{29}Si NMR spectroscopic studies of silanized surfaces were meanwhile reported by Manuel and coworkers [192–201]. In contrast to FT-IR and FT-NMR techniques the classical Raman-spectroscopy was of little use. However, the recent development of laser Raman techniques may boost the application of this method as well [202, 203]. Other rarely used analytical methods are small angle neutron scattering (SANS) [204] and inelasting electron tunneling spectroscopy (SIETS) [205, 206]. Several modern analytical methods designed to characterize surfaces are based on a strong irradiation or bombardment of the surface combined with the analysis of electrons or fragments scattered or emitted from the activated surface. To this group of method belongs Auger Electron Spectroscopy (AES), based on the bombardment of a given surface with a electron beam [207, 208]. Ion-Scattering Spectroscopy (ISS) and Secondary Ion-Mass Spectroscopy (SIMS) utilize the analysis of scattered or emitted ions resulting from the bombardment of a surface with highly accelerated ions such as noble gas cations [209].

The most widely used spectroscopic method of this kind is the Electron Scattering Analysis (ESCA) of electrons emitted from the surface upon irradiation with X-ray [210–216]. A rarely used analytical method with different physical basis is optical ellipsometry [217, 218], whereas electron microscopy, particularly in combination with ^{14}C-labeling has been utilized more frequently [219–223].

7.5
Miscellaneous Modifications

7.5.1
Silylation of Polysaccharides

The silylation of polysaccharides has attracted increasing interest in the last two decades because it possesses several useful aspects. Most silylations proceed under mild conditions, so that degradation or oxidation of the polysaccharides is avoided. Crosslinking may occur, if technical chlorosilanes are contaminated with di- or trifunctional silanes. With degrees of substitution (DS) of 1.5–2.5, silylated polysaccharides become soluble in a variety of aprotic organic solvents such as DMF, DMSO, dichloromethane, chloroform or chlorobenzene. With a

DS near 3 (i.e. 100% silylation) trimethylsilylated polysaccharides are soluble in toluene and cyclohexane, but insoluble in DMF, DMSO or pyridine, unless toluene or cyclohexane are added. As compact materials silylated poly-saccharides are hydrophobic and not sensitive to neutral water at moderate temperatures. They can be washed with dry alcohols. However, trimethylsilyl cellulose is easily hydrolyzed by neutral water at 100 °C [224] and by acidic or basic water at lower temperatures. The sensitivity to hydrolysis decreases with increasing bulkiness of the silyl group, i.e. in the following order: [224]

$(CH_3)_3Si > C_6H_5(CH_3)_2Si > (C_6H_5)_2CH_3Si.$

In addition to hydrolysis the trimethylsilyl group allows direct acylation with acid chlorides.

A broad variety of mono- and multifunctional silicon derivatives were used for the silylation of cellulose and starch, Forms 80–85. Yet the most widely used reagents for the silylation of cellulose and other polysaccharides are chlorotrimethylsilane [224–230] and hexamethyldisilazane Form. 80a. Chlorotrimethylsilane is the more reactive and less expensive reagent, yet contamination with dichlorodimethyl or trichloromethylsilane, which are difficult to remove by distillation, may yield crosslinked gels [224]. When silylation of cellulose or starch with pure chlorotrimethylsilane was conducted in pure pyridine, the DS of isolated products never exceeded 2.7 [224, 226, 227]. In contrast quantitative silylation (i.e. DS = 3.0) was found for the silylation of purified cellulose, amylose and pullulan in mixtures of pyridine and ligroin or hexane which are better solvents for completely silylated polysaccharides [228]. Completion of silylation was checked by elemental analyses and IR-spectroscopy, which demonstrated the absence of the OH-bond around 3500cm^{-1}. Contamination of silylated polysaccharides with pyridine hydro-chloride favors desilylation and crosslinking, and thus, washing (or precipitation) of the reaction products with dry methanol or ethanol is recommended [229].

$$Me_3Si-NH-SiMe_3 \qquad CH_3CO-NH-SiMe_3 \qquad CH_3C{\diagup}^{O-SiMe_3}_{\diagdown N-SiMe_3} \qquad (80)$$

$$\qquad a \qquad\qquad\qquad\qquad b \qquad\qquad\qquad\qquad c$$

Even under optimum conditions branched polysaccharides such as amylopectin, glycogen or dextran do not yield persilylated derivatives. In these cases the DS is normally below 2.0 [228], and it is clearly the steric hindrance in the neighborhood of branching points which is responsible for this result. Incomplete silylation of cellulose or starch was also observed [224, 226, 227] when the silyl group was more space filling than the trimethyl-silyl group (e.g. Forms. 81a–c and 82a–c). The DS decreases under identical reaction conditions, when the bulkiness of the substituents attached to silicon increases [224, 228]. Under moderate reaction conditions (T ⩽ 60°C) alkoxy groups attached to silicon (Forms. 81b,c) do not cause side reactions, and soluble starch derivatives were obtained [227]. However, di- or trivalent silicon reagents (Forms. 83a–c

$$
\begin{array}{ccc}
\underset{\overset{|}{\text{Et}}}{\overset{\text{Et}}{\text{Et}-\text{Si}-\text{Cl}}}
&
n\text{-}C_4H_9-O-\underset{}{\overset{}{\text{Si}}}-\text{Cl}\ (\text{with two phenyl groups})
&
\underset{\overset{|}{\text{O}}}{\overset{\overset{\text{CMe}_3}{\text{O}}}{\text{Alkyl}-\text{Si}-\text{Cl}}}\ \underset{\text{CMe}_3}{}
\end{array}
$$

(81)

a   b   c

$$
\begin{array}{ccc}
\text{Me}-(\text{CH}_2)_3-\underset{\overset{|}{\text{Me}}}{\overset{\text{Me}}{\text{Si}}}-\text{X}
&
\text{Ph}-\underset{\overset{|}{\text{Me}}}{\overset{\text{Me}}{\text{Si}}}-\text{X}
&
\text{Ph}_2\overset{}{\text{Si}}\underset{\text{X}}{\overset{\text{Me}}{\diagup}}
\end{array}
$$

(82)

a   b   c

$$
\begin{array}{ccc}
\text{Cl}-\underset{\overset{|}{\text{Me}}}{\overset{\text{Me}}{\text{Si}}}-\text{Cl}
&
\text{Et}_2\text{N}-\underset{\overset{|}{\text{Me}}}{\overset{\text{Me}}{\text{Si}}}-\text{NEt}_2
&
\text{SCN}-\underset{\overset{|}{\text{Me}}}{\overset{\text{Me}}{\text{Si}}}-\text{NCS}
\end{array}
$$

(83)

a   b   c

[227, 231], **84a–c** [227] and **(85)a–d** [227, 231] produce crosslinking under all circumstances.

When the silylation of cellulose was conducted with chlorosilanes **82a–c**, (X=Cl) in pyridine or with silylamides **82a–c**, (X = NH—CO—CH₃) in N-methylpyrrolidone, nearly identical results with regard to both yield and DS were obtained [224]. When regenerated cellulose is treated with an excess of N-trimethylsilylacetamide (Form. **80b**) at 150–170°C, DS up to 2.9 may be obtained [232]. A significantly more powerful silylating reagent is *N,O*-

$$
\begin{array}{ccc}
\text{Ph}-\underset{\overset{|}{\text{Cl}}}{\overset{\text{Cl}}{\text{Si}}}-\text{Ph}
&
\text{Me}_3\text{C}-\text{O}-\underset{\overset{|}{\text{Cl}}}{\overset{\text{Cl}}{\text{Si}}}-\text{O}-\text{CMe}_3
&
\text{Me}_3\text{C}-\text{O}-\underset{\overset{|}{\text{NH}_2}}{\overset{\text{NH}_2}{\text{Si}}}-\text{O}-\text{CMe}_3
\end{array}
$$

(84)

a   b   c

$$
\begin{array}{cccc}
\text{Me}_n\text{Si}(\text{OEt})_m
&
\underset{\text{ClCH}_2}{\overset{\text{Me}}{\diagdown}}\text{Si}-(\text{OEt})_2
&
\text{Ph}\underset{}{\overset{\text{Me}}{\diagup}}\text{Si}-(\text{OEt})_2
&
n\text{C}_8\text{H}_{17}\text{SiCl}_3
\end{array}
$$

d   (85)

a   b   c

bistrimethylsilyl acetamide (Form. **80c**) which allows even the silylation of crude woodpulp cellulose [223]. In this connection it is worth noting that the rate of silylation and the final DS may largely depend on purification and pretreatment (s.c. activation) of the polysaccharides.

A technical process for the production of fibers and films of regenerated cellulose was developed [233–235] based on the silylation of crude cellulose with chlorotrimethylsilane in liquid ammonia. Concentrated solutions of silylated cellulose may be obtained in this way. It is not clear, if the silylating reagent in this system is the chlorosilane itself or a complex of intermediately formed hexamethyldisilazane (HMDS, 80a). Pure HMDS was used by numerous authors as the silylating reagent in particular for cellulose [235–243]. Its reactivity is relatively low, so that activation by addition of trimethylchlorosilane or HCl is recommended [236–238] when a heterogeneous substrate is to be silylated. Yet it is reactive enough under homogeneous reaction conditions, such as solutions of activated cellulose in DMF containing LiCl [239] or in formamide [240].

Silylated cellulose was recently the object of intensive solid state NMR spectroscopic studies [242]. Furthermore, it was demonstrated by Pawlowski et al. [243] that highly trimethylsilyl substituted cellulose shows a thermotropic behavior forming a cholesteric melt upon heating. At high concentration even a lyotropic solution in CH_2Cl_2 was found. With a low degree of substitution (DS = 0.68) of *tert*-buty dimethylsilyl groups neither a thermotropic nor lyotropic mesophase was observed [243]. Another interesting aspect of silylated cellulose is its acylation. Two different methods were studied by Klemm et al. [244–248]. First, free OH-groups of partially silylated cellulose (low DS) were acylated with carboxylic acid chlorides and tertiary amines [244, 245, 247]. Of particular interest are derivatives with pendant cinnamoyl groups, Form. **86**,

$$(86)$$

R = Me , CMe₃   R' = H , NO₂ , (CH₂)₃CH₃

$$(87)$$

which allow photocrosslinking, and thus the synthesis of insoluble membranes. Second, the acylation can be conducted in such a way that the silylgroup itself is exchanged against an acyl residue [246], Eq. (87). Furthermore, the sulfation of silylated cellulose in aprotic solvents was reported [248]. The silylgroups react nearly quantitatively with SO_3 or $ClSO_3H$. Therefore, the DS of the sulfation corresponds to DS of the silylation.

Finally, two publications should be mentioned dealing with silylation of polysaccharide derivatives using TMS-Cl. In the first one [249] the silylation of a partially acetylated cellulose (DS ~ 2.5) is reported along with tests of the resulting polymer in applications for reverse osmosis. The second paper [250] describes the silylation of dextran and dextran derivatives with HMDS/TMS-Cl.

7.5.2
Chemical Modification of Poly-enes and Poly-ynes

During the last twenty years numerous papers dealing with the chemical modification of unsaturated polymers have been published mainly by the group of P. Weber. The poly-enes discussed in this section may be subdivided into two classes:

A) poly-enes with pending silyl or siloxy groups [251–257].
B) poly-yenes with silicon in the main chain [259–264].

When Kunzler and Percec [251] attempted to modify poly(trimethylsilyl-propyne) **88a**, either by bromination with Br_2, by cyclopropanation wit CCl_2, by reduction with $HN{=}NH$ or by hydroboration with BH3 no reaction took place. Both the double bond and the silylgroup proved to be entirely inert. However, the ω-bromobutyl-group of the copolymer **88b** showed an usual sensitivity to

$$
\left[
\begin{array}{c} SiMe_3 \\ -C{=}C- \\ Me \end{array}
\right]
\qquad
\left[
\begin{array}{cc} SiMe_3 & MeSi\,(CH_2)_4X \\ -C{=}C- & -C{=}C- \\ Me & Me \end{array}
\right]
$$

a   b   (88)

$$X = Br\,,\,OH\,,\,HCO_2\,,\,CH_3CO_2$$

$$CH_2{-}CMe{-}CO_2 \qquad CH_2{=}CH{-}\!\!\bigcirc\!\!{-}CO_2$$

nucleophilic substitution, and a variety of derivatives was synthesized without interference of double bonds or silyl groups.

In contrast to poly(trimethylsilylpropyne) both the double bond and the Si—C bond of poly(trialkylsilylbutadiene)s are sensitive to various reagents as demonstrated by Weber and coworkers [252–257]. When a solution of E-poly(2-

$$
\left[
\begin{array}{c} SiEt_3 \\ -CH_2 \quad\diagup\diagdown\quad CH_2- \end{array}
\right]
\xrightarrow[\text{- HOSiEt}_3]{HJ\,+\,H_2O}
\left[
-CH_2 \quad\diagup\diagdown\quad CH_2-
\right]
\qquad (89)
$$

triethylsilyl-2,3-butadiene) in CH_2Cl_2 is treated at 20 °C with 47% HJ complete desilylation occurs, Eq. (89). Treatment of E-poly(2-triethylsilyl-1,3-butadiene) with bromine yields in a highly stereoselective reaction sequence Z-poly(2-bromo-1,3-butdiene). An addition-elimination mechanism (catalyzed by fluoride ions) was formulated [252] Eqs. (90,91). The polymer thus obtained is not obtainable by direct polymerization of 2-bromo-1,3-butadiene. When E-poly(2-triethylsilyl-1,3-butadiene) is reacted analogously with J-Cl and KF·2H₂O a complex copolymer was obtained containing *cis* (E) and *trans* (Z) double bonds with a mixed chlorine and iodine substitution, Eq. (92) [254]. The reverse process the replacement of a halogen by trialkylsilyl-groups was partially

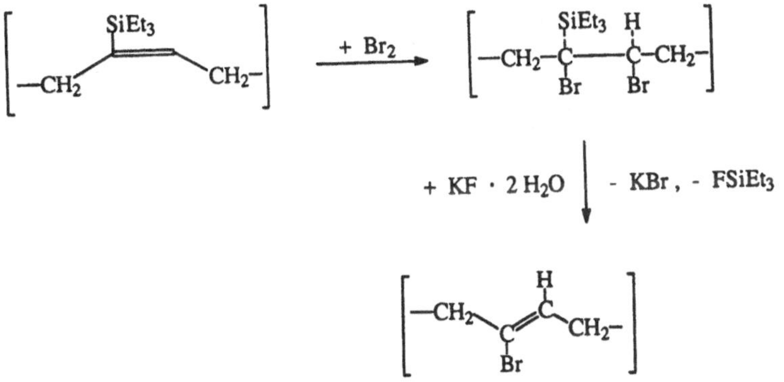

$$\left[-CH_2 \underset{CH_2-}{\overset{SiEt_3}{\diagup}}\right] \xrightarrow{+ Br_2} \left[-CH_2-\underset{Br}{\overset{SiEt_3}{C}}-\underset{Br}{\overset{H}{C}}-CH_2-\right] \tag{90}$$

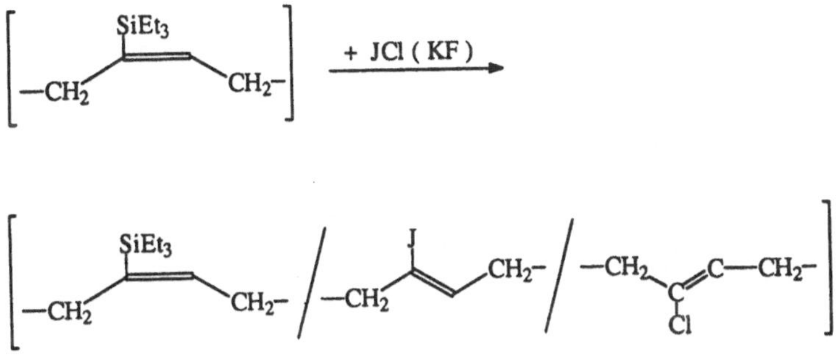

$$+ KF \cdot 2 H_2O \downarrow \quad - KBr, \ - FSiEt_3$$

$$\left[-CH_2 \underset{Br}{\diagup} \overset{H}{\underset{}{C}} CH_2-\right] \tag{91}$$

$$\left[-CH_2 \underset{CH_2-}{\overset{SiEt_3}{\diagup}}\right] \xrightarrow{+ JCl\ (KF)} \tag{92}$$

$$\left[-CH_2 \underset{CH_2-}{\overset{SiEt_3}{\diagup}} \ \middle/ \ -CH_2 \underset{CH_2-}{\overset{J}{\diagup}} \ \middle/ \ -CH_2 \underset{Cl}{\overset{}{\diagup}} C-CH_2-\right]$$

successful, when *E*-poly(2-chloro-1,3-butadiene) was used as substrate. Treatment with chlorotrimethylsilane and sodium in THF yielded a copolymer containing silyl and chloro groups attached to *cis*- and *trans*-double bonds Eq. (93) [255].

The cleavage of the Si—C bond from *E*-poly(2-(trimethylsilylmethyl)-1,3-butadiene) with J_2 was studied in D_2O. ^1H-, ^{13}C-NMR and IR-spectroscopy revealed a highly regioselective reaction with formation of poly(2-methylene-3-deuteriobutane), Eq. (94) [254]. A complex mechanism with an analogous reaction product was observed, when *E*-poly(2-(trimethylsilylmethyl)-1,3-butadiene)

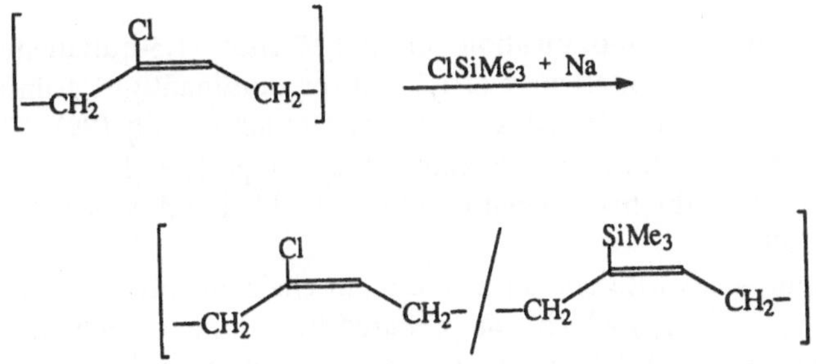

$$\left[-CH_2 \underset{CH_2-}{\overset{Cl}{\diagup}}\right] \xrightarrow{ClSiMe_3\ +\ Na} \tag{93}$$

$$\left[-CH_2 \underset{CH_2-}{\overset{Cl}{\diagup}} \ \middle/ \ -CH_2 \underset{CH_2-}{\overset{SiMe_3}{\diagup}}\right]$$

$$\left[-CH_2 \underset{CH_2-}{\overset{CH_2SiMe}{\diagup}}\right] \xrightarrow{J_2/D_2O} \left[-CH_2 \underset{CH_2-}{\overset{CH_2}{\underset{}{C}}} \overset{D}{\underset{}{CH}}\right] \tag{94}$$

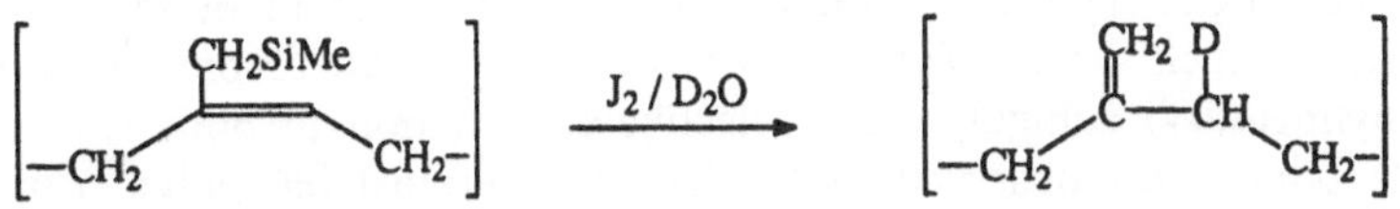

was reacted with phenylsulfenylchloride Eqs. (95, 96). A copolymer was obtained

$$
\left[\begin{array}{c} CH_2SiMe \\ -CH_2 \quad\quad CH_2- \end{array}\right] \xrightarrow{C_6H_5SCl} \left[\begin{array}{c} S-C_6H_5 \\ Me_3SiCH_2 \quad \oplus \quad H \\ -CH_2 \quad\quad CH_2- \end{array}\right] \qquad (95)
$$

$$
Cl^{\ominus}
$$

$$
\left[\begin{array}{c} CH_2 \; SC_6H_5 \\ C-CH \\ -CH_2 \quad\quad CH_2- \end{array}\right] \quad\quad \left[\begin{array}{c} Me_3SiCH_2 \; SC_6H_5 \\ C-CH \\ -CH_2 \; Cl \quad CH_2- \end{array}\right] \qquad (96)
$$

containing a 2-methylene-3-phenylthiobutane unit in a molar fraction around 80% [257]. All these modifications have in common that the reaction of the C—C double bond resulted in a cleavage of the Si—C bond.

A selective reaction of the C—C π-bond has also been reported [258]. With 3-chloroperbenzoic acid epoxydation of E-poly(2-triethylsilyl-1,3-butadiene) or E-poly(2,3-bistriethylsilyl-1,3-butadiene) is feasible without cleavage of the Si—C bonds, Eqs. (97, 98).

$$
\left[\begin{array}{c} SiEt_3 \\ -CH_2 \quad\quad CH_2- \end{array}\right] \xrightarrow{ClC_6H_4-CO_3H} \left[\begin{array}{c} Et_3Si \quad O \quad H \\ -CH_2 \quad\quad CH_2- \end{array}\right] \qquad (97)
$$

$$
\left[\begin{array}{c} SiEt_3 \\ -CH_2 \quad\quad CH_2- \end{array}\right] \xrightarrow{ClC_6H_4-CO_3H} \left[\begin{array}{c} Et_3Si \quad O \quad SiEt_3 \\ -CH_2 \quad\quad CH_2- \end{array}\right] \qquad (98)
$$

$$
\left[\begin{array}{c} OSiMe_3 \\ -CH_2-C=CH-CH_2- \end{array}\right] \xrightarrow{H_2O/AcOH} \left[\begin{array}{c} O \\ -CH_2-C-CH_2-CH_2- \end{array}\right] \qquad (99)
$$

Two papers describe the modification of poly(2-siloxy-1,3-butadiene)s [259, 260]. These polymers are copolymers of 1,2 and (predominantly) 1,4-units. The hydrolysis of the 1,4-units yielded a polyketone structure, Eq. (99). The hydrolysis of neighboring 1,2-units entails ring closure, Eqs. (100, 101) [259]. The second paper describes the hydrogenation of the double bond with H_2 and Pd/coal as catalyst [260].

Modifications of poly-enes with silicon in the main chain mainly concerned poly(1,2-dimethylsila-pent-3-ene) which was prepared by anionic or methatesis ring-opening polymerization of 1,1-dimethylsilacyclopentene (see Chapter 6). Weber and coworkers [261, 264] studied the cyclopropanation of the C—C double bonds in much detail. With difluorocarbene (generated from trifluoro-methyl-phenylmercury) a highly stereoselective *cis* addition to both *cis*- and *trans*-double bonds was observed and relatively thermostable polymer was

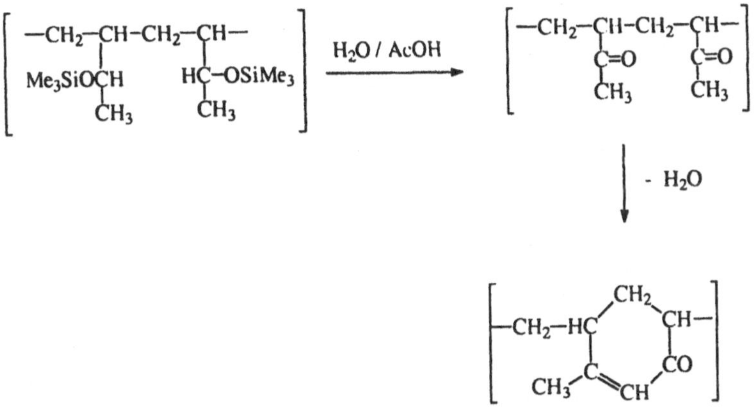

$$\left[\begin{array}{c} -CH_2-CH-CH_2-CH- \\ \underset{\displaystyle CH_3}{Me_3SiOCH} \quad \underset{\displaystyle CH_3}{HC-OSiMe_3} \end{array} \right] \xrightarrow{\;H_2O\,/\,AcOH\;} \left[\begin{array}{c} -CH_2-CH-CH_2-CH- \\ \underset{\displaystyle CH_3}{C=O} \quad \underset{\displaystyle CH_3}{C=O} \end{array} \right] \qquad (100)$$

$$\Big\downarrow \; - H_2O$$

$$\left[\begin{array}{c} CH_2 \\ -CH_2-HC \diagup\quad\diagdown CH- \\ CH_3-C=CH \quad CO \end{array} \right] \qquad (101)$$

obtained, Eq. (102) [261]. In contrast, the polyadducts of dichlorocarbene (from $CHCl_2$ with PTC) to poly(1,1-dimethylsila-pent-3-ene) were considerably less

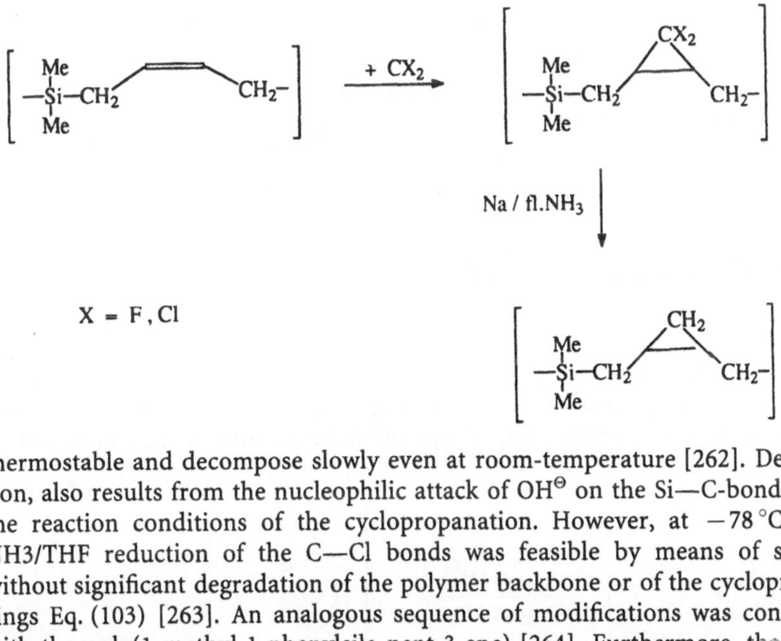

$$\left[\begin{array}{c} Me \\ -Si-CH_2 \diagdown\quad\diagup CH_2- \\ Me \end{array} \right] \xrightarrow{\;+\,CX_2\;} \left[\begin{array}{c} CX_2 \\ Me \triangle \\ -Si-CH_2 \diagdown CH_2- \\ Me \end{array} \right] \qquad (102)$$

$$\Big\downarrow \; Na\,/\,fl.NH_3$$

X = F, Cl

$$\left[\begin{array}{c} CH_2 \\ Me \triangle \\ -Si-CH_2 \diagdown CH_2- \\ Me \end{array} \right] \qquad (103)$$

thermostable and decompose slowly even at room-temperature [262]. Degradation, also results from the nucleophilic attack of $OH^{\ominus}$ on the Si—C-bond under the reaction conditions of the cyclopropanation. However, at $-78\,^{\circ}C$ in fl. NH3/THF reduction of the C—Cl bonds was feasible by means of sodium without significant degradation of the polymer backbone or of the cyclopropane rings Eq. (103) [263]. An analogous sequence of modifications was conducted with the poly(1-methyl-1-phenylsila-pent-3-ene) [264]. Furthermore, the cyclopropanation of this poly-ene was achieved by the Simmons-Smith reaction, Eq. (104). By incomplete conversion of the double bonds with difluorocarbene (generated from trifluoromethylphenyl mercury) a series of copolymers was

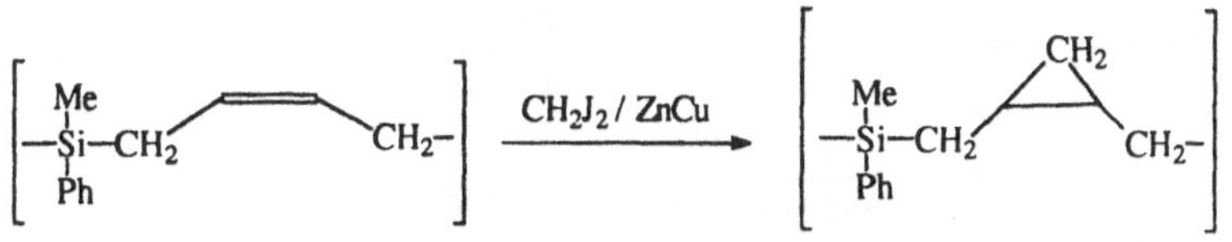

$$\left[\begin{array}{c} Me \\ -Si-CH_2 \diagdown\quad\diagup CH_2- \\ Ph \end{array} \right] \xrightarrow{\;CH_2J_2\,/\,ZnCu\;} \left[\begin{array}{c} CH_2 \\ Me \triangle \\ -Si-CH_2 \diagdown CH_2- \\ Ph \end{array} \right] \qquad (104)$$

synthesized and characterized [265]. The same poly-ene was also successfully hydrogenated by means of diimine, Eq. (105) [266]. Finally, the stereoselective *cis*-addition of dichloroketene onto the double bond of poly(1,1-dimethylsila-pent-3-ene) is worth noting Eq. (106) [257].

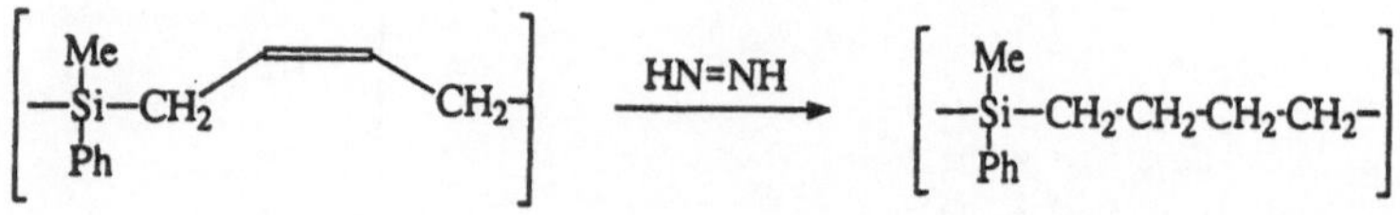

$$
\begin{bmatrix} \overset{Me}{\underset{Ph}{-Si-CH_2}} \diagup CH_2- \end{bmatrix} \xrightarrow{HN=NH} \begin{bmatrix} \overset{Me}{\underset{Ph}{-Si-CH_2 \cdot CH_2 \cdot CH_2 \cdot CH_2-}} \end{bmatrix} \tag{105}
$$

$$
\begin{bmatrix} \overset{Me}{\underset{Me}{-Si-CH_2}} \diagup CH_2- \end{bmatrix} \xrightarrow{Cl_2C=C=O} \begin{bmatrix} \overset{Cl_2C---CO}{\underset{Me}{-Si-CH_2}} CH_2- \end{bmatrix} \tag{106}
$$

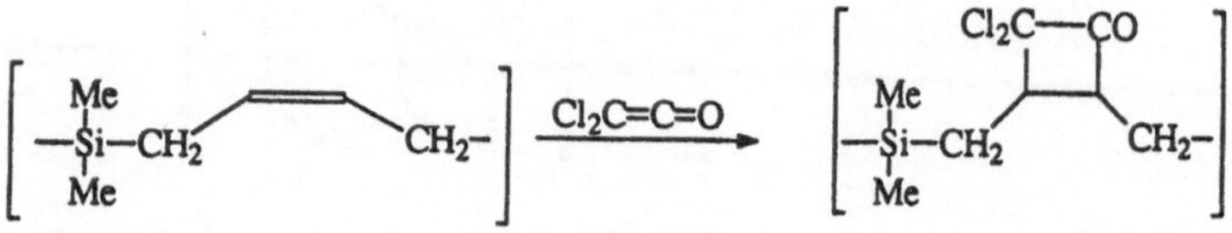

7.5.3
Chemical Modification of Various Polymers

Numerous authors reporting on chemical modification of silicon containing polymers describe the hydrolysis of the silyl group required as protection for the successful polymerization of functional monomers (see Chapter 1). A typical example is the hydrolysis of silyl protected poly(*p*-hydroxyethylstyrene), Eq. (107) [268]. Hirao et al. [268] varied the structure of the silyl group and studied the influence of this variation on the rate of hydrolysis. The following order of decreasing reactivity was found:

$$Me > CMe_3 > OCMe_3.$$

When Higashimura et al. [269] studied the desilylation of silylated poly-(hydroxyethyl vinylalcohol) they also found acidic water sufficient for the trimethylsilyl group, whereas tetraalkylammonium fluorides were more satisfac-

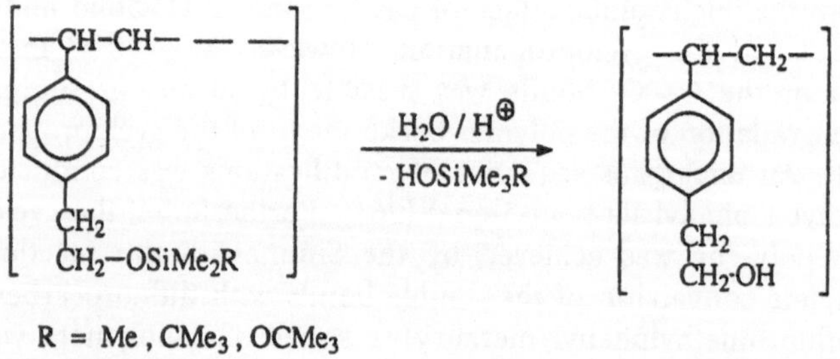

$$
\begin{bmatrix} -CH-CH- --- \\ \bigcirc \\ CH_2 \\ CH_2-OSiMe_2R \end{bmatrix} \xrightarrow[- HOSiMe_3R]{H_2O / H^{\oplus}} \begin{bmatrix} -CH-CH_2- \\ \bigcirc \\ CH_2 \\ CH_2 \cdot OH \end{bmatrix} \tag{107}
$$

R = Me , CMe₃ , OCMe₃

$$
\begin{bmatrix} -CH_2-CH_2- --- \\ O \\ CH_2 \\ CH_2 \\ O-SiMe_2CMe_3 \end{bmatrix} \xrightarrow[- FSiMe_2CMe_3]{R_4N^{\oplus}F^{\ominus}} \begin{bmatrix} -CH-CH_2- \\ O \\ CH_2 \\ CH_2 \\ OH \end{bmatrix} \tag{108}
$$

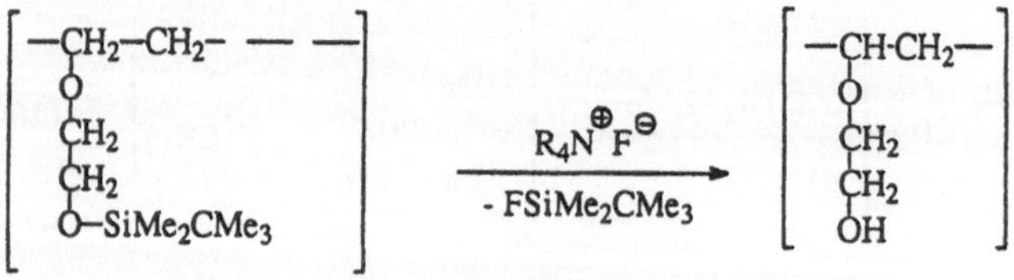

tory in the case of the *tert*-butyldimethylsilyl group, Eq. (108). However, the *tert*-butyldimethylsilyl group is easy to hydrolyze from mercapto groups as demonstrated by Yamaguchi et al. [270] for silylated poly(*p*-mercaptomethyl styrene), Form. 109a. In addition to homopolymers, synthesis and desilylation of A-B blockcopolymers were also investigated [270]. Hydrolysis in combination with potentionmetric titration of silylated poly(fumaric acid), 109b, was

$$
\left[\begin{array}{c} -CH-CH_2- \; -- \; -- \\ \bigcirc \\ CH_2 \\ S-SiMe_2CMe_3 \end{array} \right]
\qquad
\left[\begin{array}{c} CO_2SiMe_3 \\ -C=C- \; -- \; -- \\ CO_2SiMe_3 \end{array} \right]
\tag{109}
$$

a   b

investigated by Kitano et al. [215]. Nagai et al. [259, 272–273] studied the desilylation of copolymers prepared by radical copolymerization of α-trimethylsiloxy styrene. Dry HCl in hot dioxane was used as desilylating agent.

$$
\left[\begin{array}{c} OSiMe_3 \\ -C-CH_2- \\ \bigcirc \end{array} \bigg/ \begin{array}{c} -CH-CH_2- \\ \bigcirc \end{array} \right]_a
\qquad
\left[\begin{array}{c} OSiMe_3 \\ -- -C-CH_2-CH-CH_2- \\ \bigcirc \quad\;\; C \\ \quad\quad\;\; N \end{array} \right]_b
\tag{110}
$$

This agent carries the risk that HCl-catalyzed side reactions occur, but detailed studies of this aspect have not been reported. With styrene as comonomer more or less random copolymers were obtained, Form. 110a, which upon desilylation yielded polystyrenes with pending OH-groups. With maleic anhydride, fumarodinitrile or acrylonitrile as comonomers mainly alternating 1:1 copolymers were isolated, Form. 110b. The hydrolysis has in such cases the consequence that five-membered lactone rings are formed, Eq. (111).

$$
\left[\begin{array}{c} OH \quad C\equiv N \\ -C-CH_2CH-CH_2- \end{array} \right]
\xrightarrow[\; -NH_4^{\oplus} \;]{H_2O/H^{\oplus}}
\left[\begin{array}{c} O----CO \\ -C-CH_2CH-CH_2- \end{array} \right]
\tag{111}
$$

In two further papers Suzuki et al. [274,275] report on the desilation of homo- and copolystyrenes with double silylated amino groups in para-position, Forms. 112a,b. The desilylation was followed by benzylation with benzoylchloride in pyridine. Furthermore, the hydrolytic desilylation of homo- and copolystyrenes with silylated boronic acid in *para*-position should be mentioned Eq. (113) [276]. This hydrolysis proceeds quite easily in aqueous dioxane. Hydrolytic desilylation of trimethoxy silanes followed by crosslinking is an useful strategy for synthesizing well-defined networks [277]. The functional groups

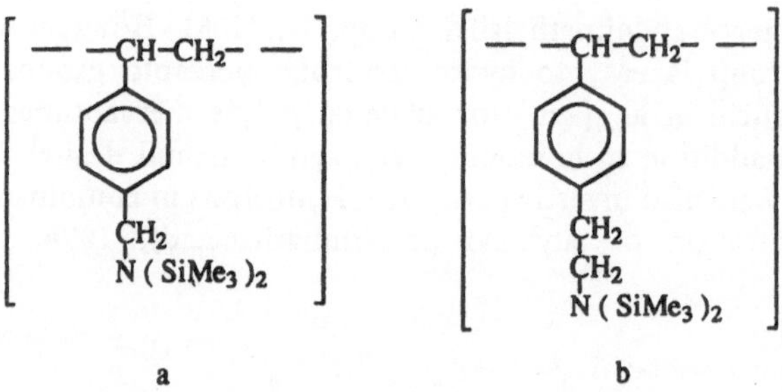

$$
\left[\begin{array}{c} -\ -CH-CH_2-\ - \\ \bigcirc \\ CH_2 \\ N(SiMe_3)_2 \end{array}\right] \qquad \left[\begin{array}{c} -\ -CH-CH_2-\ - \\ \bigcirc \\ CH_2 \\ CH_2 \\ N(SiMe_3)_2 \end{array}\right] \tag{112}
$$

a   b

$$
\left[\begin{array}{c} -\ -CH-CH_2-\ - \\ \bigcirc \\ B(OSiMe_3)_3 \end{array}\right] \xrightarrow{H_2O} \left[\begin{array}{c} -\ -CH-CH_2-\ - \\ \bigcirc \\ B(OH)_2 \end{array}\right] \tag{113}
$$

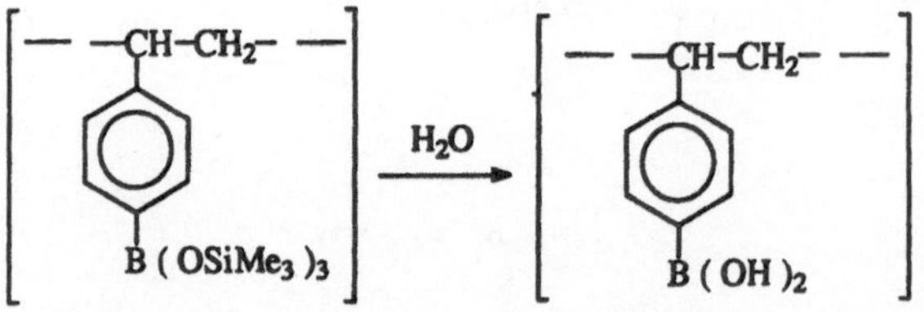

were obtained by radical grafting of vinyltrimethoxysilane onto polypropylene. This approach can be applied to various other polymers containing CH, CH_2 or CH_3 groups. Nonhydrolytic desilylation with tetraalkylammonium fluorides followed by acylation with acetylchloride was reported by another author [278].

Not only desilylation reactions of polymers were reported, but also silylations of hydroxy groups. An interesting example is the reaction of poly(N-(2-hydroxy-ethyl)methacrylamide) with γ-aminopropyl triethoxysilane, Eq. (114). Schott and Britzger [279] prepared in this way a polymer with pendant amino groups which was used as support for the synthesis of oligonucleotides. γ-Aminopropyltrimethoxysilane was also used for the modification of poly(dichlorophosphazene), Eq. (115) and other polyphosphazenes containing OC_6H_5,

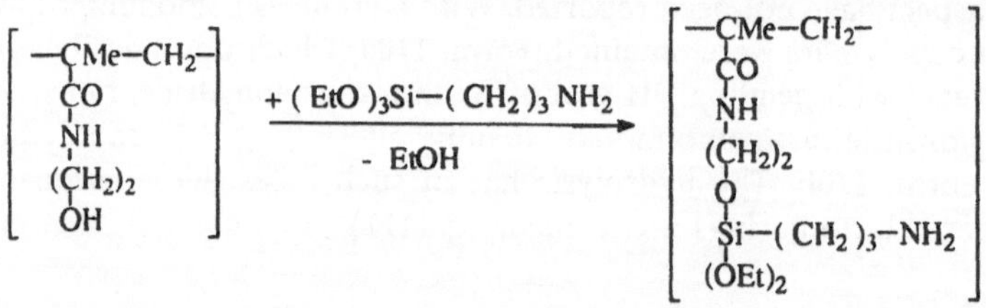

$$
\left[\begin{array}{c} -CMe-CH_2- \\ CO \\ NH \\ (CH_2)_2 \\ OH \end{array}\right] \xrightarrow[\ -\ EtOH\ ]{+\ (EtO)_3Si-(CH_2)_3NH_2} \left[\begin{array}{c} -CMe-CH_2- \\ CO \\ NH \\ (CH_2)_2 \\ O \\ Si-(CH_2)_3-NH_2 \\ (OEt)_2 \end{array}\right] \tag{114}
$$

$$
\left[\begin{array}{c} Cl \\ -N=P- \\ Cl \end{array}\right] \xrightarrow[\ -\ 2\ Et_3N\cdot HCl\ ]{+\ NH_2-(CH_2)_3-Si(OMe)_3(+Et_3N)} \left[\begin{array}{c} (CH_2)_3Si(OMe)_3 \\ NH \\ -N=P- - - - \\ NH \\ (CH_2)_3Si(OMe)_3 \end{array}\right] \tag{115}
$$

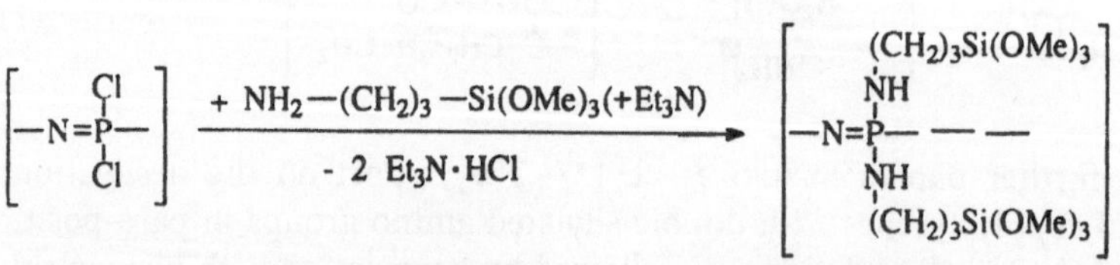

OCH_2CF_3 and $OCH_2CH_2OCH_2CH_2OCH_3$ substituents [280]. Furthermore, γ-Aminopropyl methoxydimethylsilane was used for analogous modifications. All these reactions have in common that the aminogroup of the coupling agents react with the poly(phosphazene)s. Furthermore, the silylation of poly(bis-4-bromophenylphosphazene) should be mentioned, Eq. (116, 117) [281]. Model reactions with cyclophosphazenes were also reported [281].

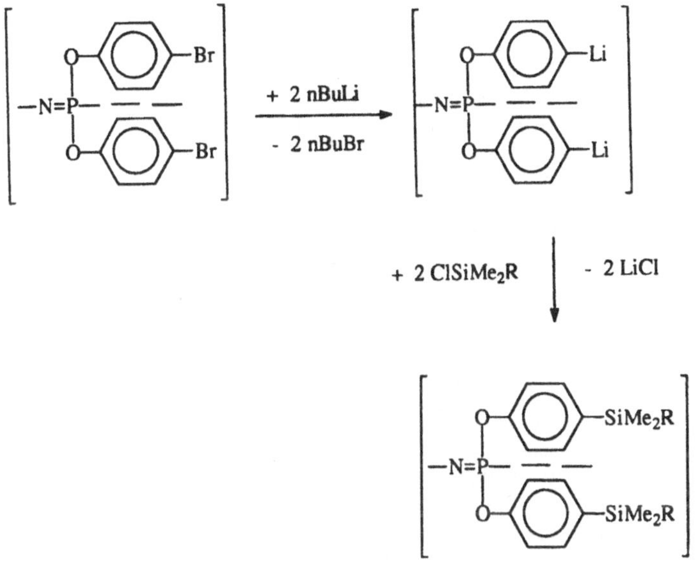

$$(116)$$

$$+ 2\ ClSiMe_2R \quad \Big| \quad -2\ LiCl$$

$$(117)$$

R = Me , Ph , CH=CH$_2$, OSiMe$_2$O tBu

Finally, an example for the modification of end groups with a silicon reagent should be mentioned. Chloro-terminated polyisobutylene was substituted with allyl trimethylsilane in combination with a Lewis acid [282, 283]. Extensive studies of model reactions, Eq. (118), proved that a nearly quantitative "allylation" is feasible under mild conditions. The allyl end groups enable further functionalization, chain extension, synthesis of blockcopolymers or crosslinking.

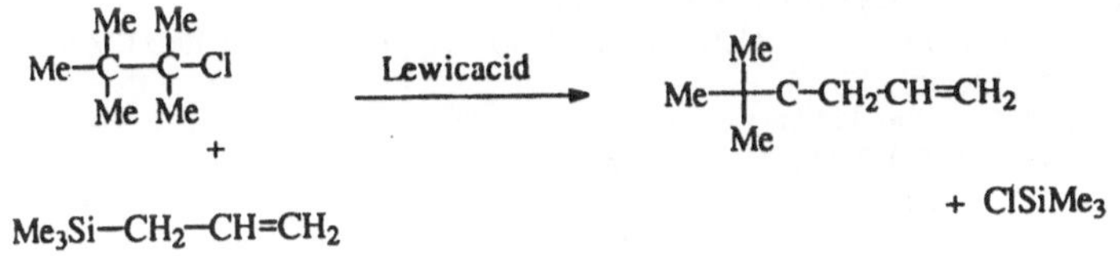

$$(118)$$

7.6
References

1. Plueddemann EP (1982) Silane coupling agents, Plenum, New York
2. Leyden DE (ed) (1985) Silanes, surfaces and interfaces, Gordon and Breach, New York
3. Plueddemann EP (1986) In: Mark H, Bikalis NM, Overberger GC, Menges (eds) 2nd edn Encyclopedia of polymer science and engineering, Wiley, New York, p 284
4. Skudelny E (1978) Kunststoffe 68: 65
5. Skudelny D (1987) Kunststoffe 77: 11
6. Payne R, Theysohn R (1987) Kunststoffe 77: 505
7. Helmdach V, Adler J (1986) Kunststoffe 76: 9
8. Eaborn C (1960) Organosilicon Compounds, Butterworth, London

9. Noll W, (1968) Chemistry and Technology of Silicons, Academic Press, New York
10. Speier JL (1955) USPat 2723987 to Dow Corning Co
11. Plueddemann EP, Clark HA, Nelson LE, Hoffmann KR (1962) Mad Plast 39: 135
12. Plueddemann EP, Clark HA (1966) USPat 3258477 to Dow Corning Co
13. Plueddemann EP (1963) USPat 3079361 to Dow Corning Co
14. Bailey DL, Pines AN (1954) Ind Eng Chem 46: 2363
15. Oswald AA, Murell LL (1978) USPat 4081803 to ET Du Pont
16. Krahnke RH, Michael KW, Plueddemann EP (1971) USPat 3631085 to Dow Corning Co
17. Burkhard CA, Krieble RH (1947) J Am Chem Soc 69 2687
18. Shorr LM (1954) J Am Soc 76: 1390
19. Plueddemann EP, Fanger G (1959) J Am Chem Soc 81: 2635
20. Berger A (1974) USPat 3821218 to General Electric Co
21. Plueddemann EP (1969) USPat 3427340 to Dow Corning Co
22. Merker RL (1957) USPat 2793223 to Dow Corning Co
23. Speier JL (1978) USPat 4082790 to Dow Corning Co
24. Ornietanksi GM, Petty HE, USPat 3849471 to Union Carbide Co
25. LeGrow GE (1971) USPat 3631194 to Dow Corning Co
26. Plueddemann EP (1978) USPat 4093641 to Dow Corning Co
27. Speier JL, Roth CA, Ryan JW (1971) J Org Chem 36: 3120
28. Sommer LH, Pockett J (1951) J Am Chem Soc 73: 5130
29. Bennett EW, Orenski P (1972) USPat 3646087 to Union Carbide Co
30. Plueddemann EP (1976) US Pat 3956353 to Dow Corning Co
31. Plueddemann EP (1972) SPI, 27th Ann Tech Conf Reinf Plast 11-B, 21-B
32. Plueddemann EP (1974) USPat 3819675 to Dow Corning Co
33. Plueddemann EP (1978) USPat 4071546 to Dow Corning Co
34. Plueddemann EP (1973) USPat 3734763 to Dow Corning Co
35. Peppe EJ, Marzden JG (1972) USPat 3671562 to Union Carbide Co
36. Plueddemann EP (1962) USPat 3057901 to Dow Corning Co
37. Hurd CD, Gershbein LL (1947) J Am Chem Soc 69: 2328
38. Wagner K, Oertel G, Gölitz HD, Quiring B (1973) DOS 2155258 to Bayer AG
39. Wagner K, Oertel G, Gölitz HD, Quiring B (1973) DOS 2155259 to Bayer AG
40. Surivet F, Lam TM, Pascault J-P (1991) J Polym Sci Part A, Polym Chem 29: 1977
41. Schmidt H (1984) Mat Res Symp Proc 32: 327
42. Huang H-H, Orler B, Wilkes GL (1987) Macromolecules 20: 1322
43. Glasser RH, Wilkes GL (1989) Polym Bull 22: 527
44. Noell JL, Wilkes GL, Moharty DK, McGrath JE (1990) J Appl Polym Sci 40: 1177
45. Mark EJ (1985) Br Polym J 17: 144
46. Sur GS, Mark EJ (1985) Eur Polym J 21: 1051
47. Clarson SJ, Mark JE (1987) Polym Commun 28: 249
48. Long TE, Kelts LW, Turner SR, Wesson JA, Mourney TH (1990, 1991) Polym Prepr 31: 490 and Macromolecules 24: 1431
49. Moureg TH, Miller SM, Wesson JA, Long TE, Kelts LW (1992) Macromolecules 25: 45
50. Lee K-W, McCarthy TJ (1988) Macromolecules 21: 3353
51. Ulre'n L, Hjertberg T (1989) J Appl Sci 37 1269
52. Plueddemann EP (1969) SPI 24th an Tech Conf Reinf Plast 19-A
53. Lentz CW (1964) Inorg Chem 3: 574
54. Ishida H, Koenig J-L (1978) Appl Spectros 32: 462
55. Ishida H, Koenig J-L (1978) Appl Spectros 32: 469
56. Pearce BW, Mayhan KG, Montle JF (1973) Polymer 14: 420
57. Andrianov KA, Izmaylov BA (1967) J Organomet Chem 8: 435
58. Sprung MM, Guenther FO (1958) J Polym Sci 28: 17
59. Brown JF, jr, Vogt LH (1965) J Am Chem Soc 84: 4313
60. Brown JF, jr (1965) J Am Chem Soc 84: 4317
61. Feher FJ, Budzichowski TA, Rahimian K, Ziller JW (1992) J Am Chem Soc 114: 3859
62. Feher FJ, Weller KJ (1991) Inorg Chem 30: 880
63. Feher FJ, Newman DA (1990) J Am Chem Soc 112: 1931

64. Feher FJ, Newman DA, Walzer FJ (1984) J Am Chem Soc 111: 1741
65. Scott DW (1946) J Am Chem Soc 68: 356
66. Day VW, Klemperer WG, Mainz VV, Millar DM (1985) J Am Chem Soc 107: 8262
67. Agaskar DA, Day VW, Klemperer WG (1987) J Am Chem Soc 109: 5554
68. Calzaferri G (1992) Nachrichten aus Chemie 40: 1106
69. Johnson OK, Stark FO, Vogel GE, Fleischmann RM (1970) J Compos Mater 1: 278
70. Chiang CH, Ishida H, Koenig JC (1980) J Colloid Interface Sci 74: 396
71. Ishida H, Naviroj S, Tripathy SK, Fitzgerald JJ, Koenig JL (1981) SPI 36th Tech Conf Reinf Plast 2-C
72. Hair ML (1967) Infrared Spectroscopy in Surface Chemistry, Marcel Dekker, New York
73. Kaas RL, Kardos JL (1976) SPE, 32nd ANTEC Paper 22
74. Vaughan DJ, McPherson EL (1972) SPI 27th Ann Techn Conf Reinf Plast 21-C
75. Plueddemann EP (1972) SPI 27th Ann Techn Conf Reinf Plast 21-B
76. Lutz MA, Polmanteer KE (1979) J Coat Technol 51: 37
77. Plueddemann EP (1975) In: A Rembaum, E Selequg (eds) Polyelectrolites and their Application D Reidel Dordreckt Holland, p 119
78. Hastlein RC (1971) Ind Eng Chem Prod Res Div 10: 92
79. Meyer FJ, Newman S (1979) SPI 34th Ann Techn Conf Reinf Plast 14: 6
80. Plueddemann EP, Stark GL (1980) SPI 35th Ann Techn Conf Reinf Plast 20-B
81. Sterman S, Marsden RG (1963) SPI 18th Ann Techn Conf Reinf Plast 1-D
82. Pinnavaia TJ, Lee G-S, Abedini M (1980) In: Leyden DE, Collins WT (eds) Silylated Surfaces Gordon and Breach Sci Publ New York, London p 333-344
83. Lochmuller CH, Amoss CW (1975) J Chromatogr 108: 85
84. Lochmuller CH (1980) In: Leyden DE, TW Collins (eds) Silylated Surfaces Gordon & Breach, London
85. Choro FK, Grushka E (1978) Anal Chem 50: 1346
86. Gübitz G, Jellenz W, Löffler G, Santi W (1979) J High Resol Chromatogr Commun 2: 145
87. Gübitz G, Jellenz W, Santi W (1981) J Liquid Chromatogr 4: 701
88. Gübitz G, Jellenz W, Santi W (1981) J Chromatogr 203: 377
89. Gübitz G, Hoffmann F, Jellenz W (1983) Chromatograhia 16: 103
90. Gübitz G (1985) In: Leyden DE (ed) Silanes, surfaces and interfaces Gordon & Breach Sci Publ New York p 391
91. Foucault A, Caude M, Oliveros L (1979) J Chromatogr 185: 345
92. Lindner W (1980) Naturwissenschaften 67: 354
93. Engelhardt H, Kromidas S (1980) Naturwissenschaften 67: 353
94. Sudgen K, Hunter C, Loyd-Jones G (1980) J Chromatogr 192: 228
95. Roumeliotis P, Unger KK, Kurganov AA, Davankov YA (1983) J Chromatogr 255: 51
96. Roumeliotis P, Kurganov AA, Davankov YA (1983) J Chromatogr 266: 439
97. Feibush B, Cohen MJ, Karger BL (1983) J Chromatogr 282: 3
98. Chow FC, Grushka E (1980) In: Leyden DE, Collins WT (eds) Silylated surfaces Gordon & Breach, New York-London, p 301–319
99. Pinnavaia, TJ, Lee JG-S, Abidini M (1980) In: Leyden DE, Collins WT (eds) Silylated Surfaces Gordon & Breach, New York-London, p 333
100. Evans J (1985) In: Leyden DE (eds) Silanes Surfaces and Interfaces Gordon & Breach, New York-London, p 203
101. Evans J, Gracey BP (1983) J Chem Soc, Chem Commun 247
102. Evans J, Gracey BP (1982) J Organomet Chem 228 C 4
103. Evans J, Gracey BP (1982) J Chem Soc, Dalton Trans 1123
104. Evans J, McNulty GS (1984) J Chem Soc, Dalton Trans 587
105. Cook SL, Evans J (1983) J Chem Soc, Chem Commun 713
106. Brown SC, Evans J (1981) J Mol Catal 11: 143
107. Lynn M (1975) In: Weetal HH (ed) Immobilized, Enzymes, Antigens and Peptides Marcel Dekker, New York
108. Messing RA (1975) Immobilized Enzymes for Industrial Reactors, Academic Press New York
109. Ohlson S, Hansson L, Larsson P-O, Mosbach K (1978) FEBS Lett 93: 5

110. Lewis C, Scouten WH (1983) In: Scouten WH (ed) Solid Phase Biochemistry Wiley, New York, p 665
111. Scouten WH (1985) In: Leyden DE (ed) Silanes, Surfaces and Interfaces Gordon & Breach, p 59
112. Haller W (1983) Scouten WH (ed) Solid Phase Biochemistry Wiley, New York, p 535
113. Featon DL (1980) Leyden DE, Collins WT (eds) Silylated Surfaces Gordon & Breach New York , p 201
114. Anselme M, Cholin S, Haquet A, Sibille B, Piquion J (1985) Bull Soc Chim Fr 6: 1115
115. Murakami Y, Mori S (1986,1987) Jap Pat 61104 255; CA 106: 168 211a
116. Ernst-Carrera K, Wilchek M (1987) J Chromatogr 397: 187
117. Hjerten S, Kunguan Y, Liao J (1988) Makromol Chem, Macromol Symp 17: 349
118. Venton DL, Gudipati E (1985) In: Leyden DE (ed) Silanes, Surfaces and Interfaces Gordon & Breach, New York-London, p 73
119. Royer GP, Liberatore FA (1980) In: Leyden DE, Collins WT (eds) Silylated Surfaces Gordon & Breach New York-London, p 189
120. Schrifreen RS, Hanna DA, Bowers LD, Carr PW (1977) Anal Chem 49
121. Cuatrecaras P, Anfinzen CB (1971) Ann Rev Biochem 40 259
122. Skinner KJ (1975) Enzymes Technology in Chem and Eng News 18: 22
123. Arkles BC, Miller AS, Brinigar WS (1980) In: Leyden DE, Collins WT (eds) Silylated Surfaces Gordon & Breach, New York-London, p 363
124. Bascom WD (1968) Adv Chem Ser 87: 38
125. Lee LH (1968) J Colloid Interface Sci 27: 751
126. Kaelble DH, Dynes PJ, Cerlin EH (1974) J Adk 6: 23
127. Lee LH (1975) Adhesion Science and Technology Plenum New York, Vol 9B, p 647
128. Bascom WD (1968) J Colloid Interface Sci 27: 789
129. Owen MJ (1980) Ind Eng Chem Prod Res Devel 19: 97
130. Moses PR, Wier L, Murray RW (1975) Anal Chem 47: 1882
131. Diaz A (1977) J Am Chem Soc 99: 5383
132. Diaz A (1980) In: Leyden DE, Collins WT (eds) Silylated Surfaces Gordon & Breach Sci Publ New York-London, p 137
133. Armstrong NR (1980) In: Leyden DE, Collins WT (eds) Silylated Surfaces Gordon & Breach New York-London, p 159
134. Lenhard JR, Murray RW (1977) J Electroanal Chem 77: 393
135. Plueddemann EP (1974) Proc of the Water Harvesting Symposium, Phoenix March 26, p 76
136. Snyder GH, Ozaki HJ, Hayslip NC (1974) Proc Soil Science Soc of Am 38 678
137. Scholl D (1974) Proc Soil Science Proc of Am 38:
138. Abel FW, Pollard FH, Uden PC, Nickless G (1966) J Chromatogr 22: 23
139. Wise SA, Bonnett WJ, Guenther FR, May WE (1981) J Chromatogr Sci 19: 457
140. Wise SA, May WE (1983) Anal Chem 55: 1479
141. Wise SA, Sandere LC, May WE (1983) J Liqu Chromatogr 6: 2709
142. Sander LC, Wise SA (1984) Anal Chem 56: 504
143. Sander LC, Wisw SA (1984) J Chromatogr 316: 163
144. Wise SA, Sander LC, May WE (1985) In: Leyden DE (ed) Silanes, Surfaces and Interfaces Gordon & Breach, New York -, p 349
145. Ogan K, Katz E (1980) J Chromatogr 188: 115
146. Colusjo A, McDonald JC (1980) Chromatographia 13: 350
147. Amos R (1981) J Chromatogr 204: 469
148. Berendsen GE, Pikaart KA, de Galan L (1980) J Liqu Chromatogr 3: 1437
149. Eengelhardt H, Müller H (1981) J Chromatogr 218: 395
150. Engelhardt H, Dreyer B, Schmidt H (1982) Chromatographia 16: 11
151. Engelhardt H (1985) In: Leyden DE (ed) Silanes, Surfaces and Interfaces Gordon & Breach, New York-London, p 381
152. Engelhardt H, Müller H (1984) Chromatographia 19: 77
153. Cohen SA, Schellenberg K, Benedek K, Karger BL, Grego B, Hearn MTW (1984) Anal Biochem 140: 223

154. Nestrich TJ, Kemparski LL, Stehl RH (1979) Anal Chem 51: 2273
155. Nawrocki J (1986) J Chromatogr 362: 117
156. Bayer E, Paukes A, Peters B, Laupp G, Reiners J, Albert K (1986) J Chromatogr 364: 25
157. Golding RD, Barry AJ, Burke MF (1987) J Chromatogr 384: 105
158. Jones K (1987) J Chromatogr 392: 1
159. Jaeger DA, Clennan MW (1988) J Org Chem 53: 3985
160. Floyd TR, Sagliano N Jr, Hartwick RA (1988) J Chromatogr 45: 43
161. Dawkins JV, Gabbott NP, Loyd LL, Mc Conville JA, Warnere FP (1988) J Chromatogr 452: 145
162. Pharr PY, Uden PC, Siggia S (1988) J Chromatogr Sci 26: 432
163. Gaddines M, MAckemans MT, Everaerts FM, van der Linden PJ, Vader HL, Wiegerinck MAH (1988) J Chromatogr 431: 317
164. Bridge TB, Williams MH, Fell AF (1989) Anal Chim Acta 223: 175
165. De Leon P, Bertram CA, Colin CH (1989) J High Resol Chromatogr 12: 493
166. Matsuda R, Hayashi Y, Ishibashi M, Takeda Y (1989) Annal Chim Acta 222: 301
167. Dorsey JG, Dill KH (1989) Chem Rev 89: 331
168. Hoffmann NE, Liao JC (1990) J Chromatogr Sci 28: 428
169. Phyllis Brown R, Anal Chem 62: 995 A
170. Dorsey JG, Foley JR, Cooper WT, Barford RA, Barth HG (1990) Anal Chem 62: 324 R
171. Asquith AJ, Abott EA, Walters PA (1973) Appl Microbiol 23: 859
172. White WC, Gettings RL (1985) In: Leyden DE (ed) Silanes, Surfaces and Inter-faces Gordon & Breach New York-London, p 107
173. Marthy RS, Caravajal GS, Leyden DE (1985) In: Leyden DE (ed) Silanes, Surfaces and Interfaces Gordon & Breach, New York-London, p 141
174. White TE (1965) SPI, 20th Ann Tech Conf Reinf Plast 3-D
175. Duffy JV (1967) J Appl Chem 17: 35
176. Hertl W (1968) J Phys Chem 72: 1248
177. Alsne VJ, Kronberg VT, Eidus YA (1968) Makkanika Polimerov 4: 182
178. Bascom WD (1972) Macromolecules 5: 792
179. Greenler RG (1966) J Chem Phys 44: 310
180. Boerio FJ, Armogam FJ, Cheng SY (1980) J Colloid Interface Sci 73: 416
181. Boerio FJ, Cheng SY, Armogen L, Williams JW, Gosselim C (1980) SPI 35th Ann Tech Conf Reinf Plast 23-C
182. Ishida H, Koenig JL (1978) Appl Spectros 32: 462
183. Ishida H, Koenig JL (1978) Appl Spectros 32: 467
184. Ishida H, Koenig JL (1978) J Colloid Interface Sci 64: 555
185. Ishida H, Koenig JL (1978) J Colloid Interface Sci 64: 565
186. Ishida H, Koenig JL (1979) J Polym Sci, Polym Phys Ed 17: 615
187. Ishida H, Koenig JL (1980) J Polym Sci, Polym Phys Ed 18: 323
188. Chiang CH, Ishida H, Koenig JL (1980) J Colloid Interface Sci 74: 396
189. Ishida H, Koenig JL (1980) J Polym Sci, Polym Phys Ed 18: 1931
190. Koenig JL (1985) In: Leyden DE (ed) Silanes, surfaces and Interfaces Gordon & Breach, New York-, p 43
191. Müller JD, Ishida H (1985) In: Leyden DE (ed) Silanes, surfaces and interfaces Gordon & Breach, New York-, p 525
192. Maciel GE, Sindorf DW (1980) J Am Chem Soc 102: 7606
193. Maciel GE, Sindorf DE, Bartuska VJ (1981) J Chromatogr 205: 438
194. Sindorf DW, Maciel GE (1981) J Am Chem Soc 103: 4263
195. Sindorf DW, Maciel GE (1982) J Phys Chem 86: 5208
196. Sindorf DW, Maciel GE (1983) J Am Chem Soc 105: 1487 and 3767
197. Sindorf DW, Maciel GE (1983) J Phys Chem 87: 5516
198. Miller ML, Linton RW, Maciel GE, Hawkins BL (1985) J Chromatogr 319: 9
199. Rudzinski WE, Montgomery TL, Trye JS, Hawkins BL, Maciel GE (1985) J Chromatogr 323: 281
200. Gavavajal GS, Leyden DE, Maciel GE (1985) In: Leyden DE (ed) Silanes, surfaces and interfaces Gordon & Breach, New York p 283

201. Maciel GE, Zeigler RC, Taft RK, In: Leyden DE (ed) Silanes, surfaces and Interfaces Gordon & Breach, New York 1985, p 413
202. Koenig JL, Smith PTK (1091) J Colloid Interface Sci 36: 247
203. Ishida H, Koenig JL, Kenney M (1979) SPI, 34th Ann Tech Conf Plast 17-B
204. Sander LC, Glinka GJ, Wise SA (1985) In: Leyden DE (ed) Silanes, surfaces and Interfaces Gordon & Breach, New York p 431
205. Diaz AF, Hetzler U, Kay E (1977) J Amer Chem Soc 99: 6781
206. Werrett CR, Comyn J, Oxley DP, Pritchard RG, Reynolds S (1985) In: Leyden DE (ed) Silanes, surfaces and interfaces Gordon & Breach, New York p 305
207. Cain JF, Sacher E (1978) J Colloid Interface Sci 67: 538
208. Wong R (1972) J Adhes 4: 171
209. DiBenedetto AT, Scola DA (1980) J Colloid Interface Sci 74: 150
210. Nichols GD, Hercules DM, Peek RC, Vaughan DJ (1976) Appl Spectrosc 281: 219
211. Finkler HO, Murray RW (1979) J Phys Chem 83: 353
212. Willmann KW, Greer EW, Murray RW (1979) Nouv J Chim 3: 455
213. Nichols GD, Hercules DM (1979) Appl Spectrosc 28: 219
214. Herecules DM, Cor LE, Onisick S, Nichols GD, Carnes JC (1973) Anal Chem 45: 1973
215. Phillipps LV, Hercules DM (1985) In: Leyden DE (ed) Silanes, surfaces and interfaces Gordon & Breach, New York p 235
216. Sacher E, Klemberg-Sapieha J, Schreiber HP, Wertheimer MR, Intyre NSM In: Leyden DE (ed) Silanes, surfaces and Interfaces Gordon & Breach, New York 985, p 189
217. Tutas DJ, Stromberg R, Passaglia E (1964), SPE Trans 4: 256
218. Bascom WD (1968) Adv Chem Soc 87: 38
219. Sterman S, Bradley HB (1961) SPI Meth Ann Tech Conf Reinf Plast 8-D
220. Vogel GE, Johannson OK, Stark FO, Fleischmann (1967), SPI, 22nd Ann Tech Conf Reinf Plast 13-B
221. Vaughan DJ, Peek RC, SPI, 30th Ann Tech Conf Reinf Plast 22-B
222. Schrader ME, Cernen I, Doria FJ (1967) Mod Plast 45: 195
223. Johannson OK, Stark FO, Vogel GE, Fleischmann RM (1967) J Compos Mater 1: 278
224. Klebe JF, Finkbeiner HL (1969) J Polym Sci A-1, 7: 1947
225. Klemm D, Schnabelrauch M, Stein A, Hinze T, Erler V, Vogt S (1991) Das Papier 43: 773
226. Schuyten HA, Weawer JW, Reid JD, Jürgens JF (1948) J Am Chem Soc 70: 1919
227. Kerr RW, Hobbs KC (1953) Ind Engin Chem 45: 2542
228. Keilich G, Fihlarik K, Husemann E (1968) Makromol Chem 120: 87
229. Klebe JF (1969) US Pat 3418312 (1968) to General Electric CA 70: 59084
230. Klebe JF (1969) USPat 3418313 (1968) to General Electric Co, CA 70 59081a
231. Predvotitelev DA, Rogovin ZA (1967) Vysokomol Soedin 9 611; CA 67 118258
232. Bredereck K, Strunk K, Menrad H (1969) Makromol Chem 126: 139
233. Greber G, Paschinger O (1981) Das Papier 35: 547
234. Greber G, Paschinger O (1981) Fr Pat 2477157 (1982) to Lenzing AG, CA 96: 8425
235. Greber G, Detamble A (1984) Austrian AT 373605 (1984); CA 100 158467a
236. Khalikov KR, Tsagaraeva NA, Yul'chibaeva GS, Turgunov ON (1974) VzbiKhim Zh (1974) 18: 28; CA 81: 123163k
237. Nagu J, Borbely-Kuszman A Becker-Palóssy K, Zimonyi-Hegedüs K (1973) Makromol Chem 165: 335
238. Green JG (1983) US Pat 4390692 (1983) to Dow Chem Co; CA 99: 72434m
239. Schempp W, Krause T, Seifried U, Koura A (1984, 1985) Papier 38: 607; CA 102: 80575d
240. Harmon RE, De KK, Gupta SK (1973) Carbohyd Res 31: 407
241. Cooper GK, Sandberg KP, Hinck JF (1981) J Polym Appl Polym Sci 26: 3827
242. Panolowski WP, Sankar SS, Gilbert RD (1987) J Polym Sci Part A, Polym Chem 25: 3356
243. Panolowski WP, Gilbert RD (1988) J Polym Sci Part B, Polym Phys 26: 1101
244. Klemm D, Schnabelrauch M, Stein A, Philipp B, Wagenknecht, Nehls I (1990) Das Papier 44: 624
245. Klemm D, Schnabelrauch M, Stein A, Niemann M, Ritter H (1990) Makromol Chem 191: 2985
246. Stein A, Klemm D (1988) Makromol Chem, Rapid Commun 9: 569

247. Klemm D, Schuhmann P, Hartmann M (1983) ZChem 24: 62
248. Wagenknecht W, Nehls I, Stein A, Klemm D, Phillipp B (1992) Acta Polymer 43: 266
249. Gatge ND, Skhisti R (1986) Intern J Polymeric Mater 11: 221
250. Hashimoto K, Imanishi S-I, Okada M, Sumitomo H (1991) J Polym Sci Part A, Polym Chem 29: 1271
251. Kunzler J, Percec V (1990) New Polymeric Mater 1: 271
252. Ding Y-X, Weber WP (1988) Macromolecules 21: 530
253. Jiang W, Weber WP (1988) Polym Bull 20: 15
254. Jiang W, Weber WP (1989) Polym Bull 21: 427
255. Jiang W, Weber WP (1989) Polym Bull 21: 335
256. Ding Y-X, Weber WP (1988) Macromolecules 21: 2672
257. Ding Y-X, Weber WP (1988) Polym Bull 20: 7
258. Jiang W, Weber WP (1988) Polym Bull 20: 249
259. Nagai K, Asada K, Kuramoto N (1990) J Polym Sci Part A, Polym Chem 28: 2845
260. Takenaka K, Kato K, Hattori A, Hirao T, Nakaliama S (1990) Macromolecules 23: 3619
261. Zhou Q, Weber WP (1989) Macromolecules 22: 2987
262. Zhou Q, Weber WP (1989) Macromolecules 22: 1300
263. Zhou Q, Weber WP (1989) Polym Bull 21: 173
264. Liao X, Lee HS-J Weber WP (1992) Makromol Chem 191: 2173
265. Lee HS-J, Weber WP (1992) Polymer 33: 4299
266. Liao X, Weber WP (1991) Polym Bull 25: 621
267. Lee HS-J, Weber WP (1992) Polymer 33: 1748
268. Hirao A, Yamamoto A, Takenaka K, Yamaguchi K, Nakahama S (1987) Polymer 28: 303
269. Higashimura T, Ebara K, Aoshima S (1989) J Polym Sci Part A, Polym Chem 27: 2937
270. Yamaguchi K, Kato T, Hirao A, Nakahama S (1987) Makromol Chem, Rapid Commun 8: 203
271. Kitano T, Ishiyaki A, Uimatsu G, Kawaguchi S, Ito K (1987) J Polym Sci Part A, Polym Chem 25: 979
272. Nagai K, Asada K, Chiba K, Kuramoto N (1989) J Polym Sci, Part A, Polym Chem 27: 3779
273. Nagai K, Chiba K Asada K, Masui K, Kuramato N (1990) J Polym Sci Part A, Polym Chem 28: 2195
274. Suzuki K, Yamaguchi K, Hirao A, Nakahama S (1989) Macromolecules 22: 2607
275. Suzuki K, Hirao A, Nakahama S (1989) Makromol Chem 190: 2893
276. Hartmann M, Carlson H, Paals J (1976) Makromol Chem 177: 131
277. Lee Y-D, Wang L-F; J Appl Polym Sci 32; 4639
278. Zeigler M (1984) ACS Polym Prepr 25: 223
279. Schott H, Britzger W (1987) Makromol Chem 188: 2277
280. Allcock HR, Coggio WD (1990) Macromolecules 23: 1626
281. Allcock HR, Coggio WD Archibald RS, Brennan DJ (1989) Macromolecules 22: 3571
282. Wilczek L, Kennedy JP (1987) Polym Bull 17: 37
283. Wilczek L, Kennedy JP (1987) J Polym Sci Part A, Polym Chem 25: 3255

Appendix A: Silicon-Based Thermoset Resins

R. Mülhaupt

A.1
Design of Toughening Agents

Among thermoset materials highly crosslinked epoxy resins are well known to combine excellent dimensional, chemical, and environmental stabilities with low weight, high stiffness, adhesion, low dielectric constants, high creep resistance, and attractive economics [1]. Today epoxy resins have found numerous applications, e.g., as thermosetting binders of coatings, matrix resins of advanced composite materials in aerospace industries, and components of structural and semi-flexible adhesives, sealants, chip encapsulations, electrical insulators and microelectronics. In order to compete with metals and metal alloys in structural applications, bulk as well as filled and fiber-reinforced epoxy resins must resist external mechanical stresses, deformation, and crack propagation, especially when exposed to impact forces. When crosslink densities are increased to improve dimensional and chemical stabilities, most epoxy resins become brittle, i.e., small mechanical stresses can cause premature mechanical failure as reflected by rapid unrestricted crack propagation throughout the epoxy matrix or along the epoxy/filler interface. Therefore, improving toughness/stiffness balance without sacrificing other useful properties represents an important research objective in epoxy resin development.

In order to enhance toughness, energy at the crack tip must be dissipated, preferably by involving a much larger sample volume in crack propagation. This has been achieved by incorporating rubber microphases which are dispersed in the rigid epoxy matrix. Due to the different coefficients of thermal expansion, after thermal cure internal stresses develop at the epoxy/rubber interface. When external mechanical stresses are applied, addition of internal and external stresses leads to localized plastic deformation, shear yielding, crazing ('microvoiding') of the matrix between dispersed particles. Crack front pinning resulting from rubber microparticle bridging contributes to impact energy dissipation, provided that interparticle distance is not too large. As a consequence, stress-concentrating microphases represent efficient energy sinks for absorbing impact energy. Moreover, when rubber microphases are incorporated into the epoxy matrix, mechanical stresses cause cavitation of the rubber and finally formation of additional microvoids due to debonding of epoxy matrix and rubber. Rubber cavitation, voiding, and shear band formation are illustrated in Fig. A1. Important parameters governing stress transfer and impact energy

Fig. A.1. Mechanical failure behavior of crosslinked epoxy resins containing dispersed rubber microphases

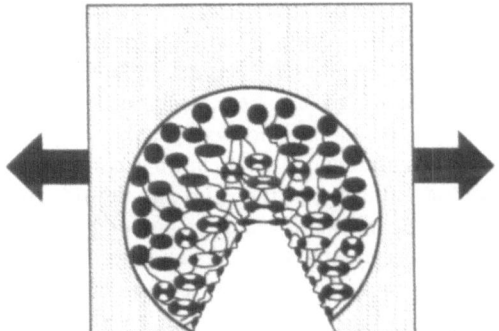

dissipation are microparticle size distribution, interparticle distance (often referred to as "connectivity"), and especially interfacial adhesion [2–16].

Since crosslinked rubbers are difficult to process, especially with respect to formation of micro- and nano-scale dispersions in uncured epoxy resins, telechelic elastomers (reactive liquid rubbers), which comprise oligomers with one or more reactive end groups, are added to the epoxy resin. In principle, either immiscible liquid rubbers are dispersed and crosslinked in the uncured epoxy resin or, in the preferred mode, miscible reactive liquid rubbers and epoxy resins are cured simultaneously. Multiphase simultaneously interpenetrating networks are formed when the rubber components crosslinks and phase separates during cure. Both strategies are discussed below for silicone-toughened epoxy resins. In both strategies it is important to match compatibilities of liquid rubber and uncured epoxy resin containing the epoxy hardener. On one hand, liquid rubber and epoxy resin must be incompatible to afford rubber phase separation during cure. Incorporation of highly flexibile silicone moieties into the rigid epoxy resin would substantially flexibilize the matrix, thus reducing strength, stiffness, and dimensional stabilities. On the other hand, uncured epoxy and liquid rubber must be miscible to afford a single-phase mixture. Moreover, a certain degree of compatibility is a prime requirement for interfacial adhesion after cure. As presented in a previous review in more detail [17], molecular architecture of liquid rubbers can be tailored to adjust rubber/ thermoset resin compatibility balance. Routes to silicone-toughened epoxy resins, involving compatibilized silicone liquid rubbers and compatibilized silicone rubber dispersions, are presented in the following section emphasizing key structure/property relationships. This overview is not mend to be comprehensive but places the major focus on basic concepts and recent advances illustrated with selected examples.

A.2
Difunctional Silicone Liquid Rubbers

Among compatibilized liquid rubbers carboxy-terminated poly(butadiene-co-acrylonitrile) ("CTBN")liquid rubbers, pioneered by Drake, Siebert, and Rowe at B.-F. Goodrich, are widely applied. Compatibility between CTBN and epoxy results both from CTBN carboxy end groups, which are readily converted into an epoxy end group via adduct formation with epoxy resins, and incorporation of polar acrylonitrile units into the non-polar poly(butadiene) backbone [9, 13, 18, 19]. In fact, best results in terms of toughness/stiffness/strength balance of epoxy resins derived from bisphenol A are achieved with acrylontrile contents of approximately 18 wt%. Similar concepts have been applied to the synthesis of compatibilized reactive silicone liquid rubbers useful as epoxy toughening agents. Synthesis of reactive silicone liquid rubbers and copolymers with flexible silicone segments [20–23], and their application as epoxy toughening agents [24–27], applying concepts for enhancing compatibility similar to those developed for CTBN, were reviewed by McGrath. As illustrated in Scheme 1, today a family of difunctional silicone liquid rubbers with a variety of end groups is available, e.g., silicone liquid rubbers with amine (*ATPS* for amine-terminated polysiloxane), piperazine (*PATPS*), carboxylic acid (*CTPS*), phenolic (*PTPS*), dicarboxylic anhydride (*AHTPS*), cycloaliphatic epoxy (*CETPS*), hydroxy (*HTPS*) and epoxy (*ETPS*) end groups.

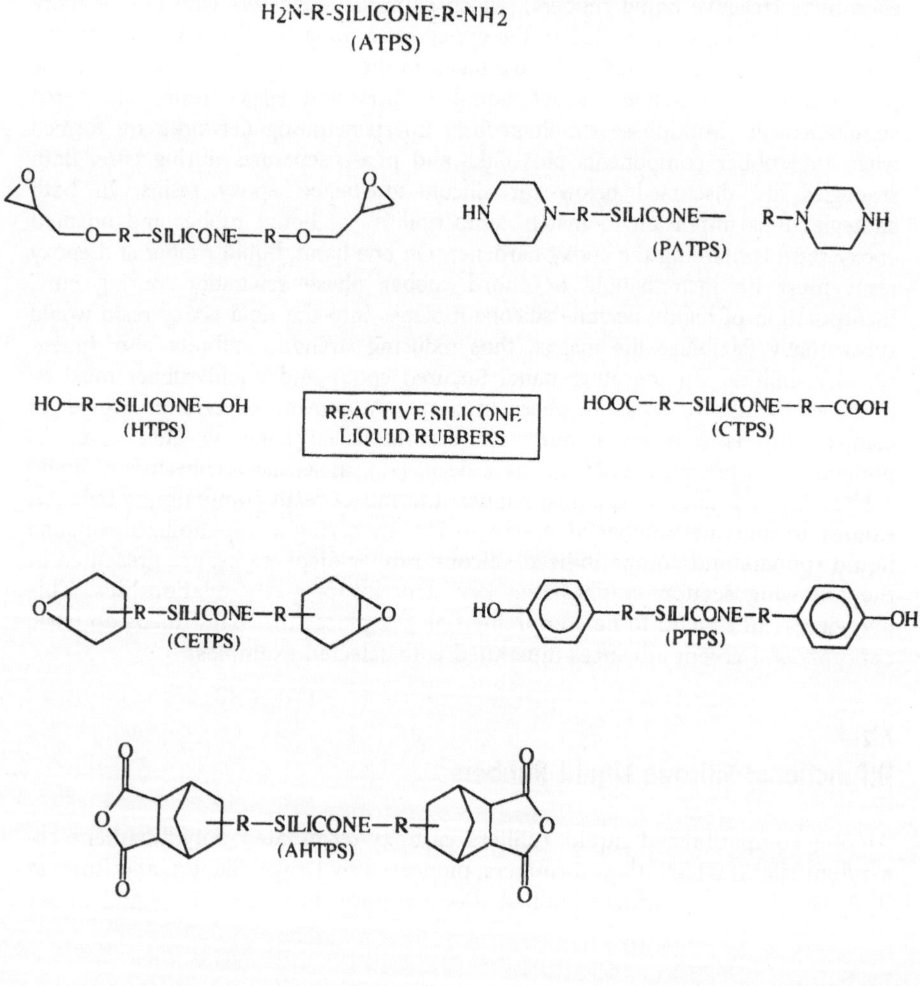

H$_2$N-R-SILICONE-R-NH$_2$
(ATPS)

In comparison to CTBN, silicone toughening agents offer several advantages: (1) much lower glass transition temperatures of dispersed poly(dimethylsiloxane)s (-127 °C relative to -36 °C for nitrile rubber containing 26 wt% acrylonitrile); (2) low water-uptake, (3) low dielectric constant, (4) much better thermooxidative stability due to the absence of unsaturated olefinic groups. Moreover, during cure a small portion of the reactive silicone may migrate to the surface to form in situ a thin flexible protective coating which is also water-repellent and improves friction and wear properties [28, 29].

Two different synthetic routes are commonly applied for manufacturing difunctional silicone liquid rubbers, i.e., hydrosilylation of bis(vinyl)- or bis(allyl)- functional intermediates with bis-(SiR_2H)-terminated oligosiloxanes Eq. (1), and equilibration reaction of difunctional tetramethyldisiloxanes, X—$SiMe_2OSiMe_2$—X, with octamethylcyclotetrasiloxane in the presence of base and acid catalysts, Eq. (2). A third route to reactive silicone liquid rubbers, copolycondensation of dichloro- and functionalized chloro-or alkoxy-alkylsilanes respectively has also been introduced but is less important in view of commercial applications [30, 31].

$$H-SiMe_2OSiMe_2-H \quad + \quad 2\ CH_2=CHCH_2-R-X \quad \longrightarrow \quad X-PDMS-X \quad (1)$$

$$X-SiMe_2OSiMe_2-X \quad + \quad
\begin{array}{c}
O-SiMe_2 \\
Me_2Si \qquad\quad O \\
\qquad\quad SiMe_2 \\
O \\
Me_2Si-O
\end{array}
\quad \longrightarrow \quad X-PDMS-X \quad (2)$$

In a typical example for the hydrosilylation route, Eq. (3), Buchholz [32, 33] reacts bis($SiMe_2H$)-terminated oligo(dimethylsiloxane) of $M_n = 2260$ g/mol, prepared by equilibration of $HSiMe_2OSiMe_2H$ and cyclooctamethyltetrasiloxane in the presence of trifluoroacetic acid, with 2 molar equivalents of O-allylglycdiol in sulfur-free toluene at 80 °C for several hours, using 10 ppm H_2PtCl_6, dissolved in ethanol, as catalyst to form tetrahydroxy-terminated oligo(dimethylsiloxane) $(HO)_2$-PDMS-$(OH)_2$. As apparent from Table A.1 and discussed later in more detail, four hydroxy end groups are not sufficient to compatibilize the oligo(dimethylsiloxane) segment with the epoxy matrix. Hydrosilylation has been applied extensively using SiH-functional silicones with allyl- or vinyl-functional intermediates, e.g., allylacetate to produce *HTPS* via hydrolysis of the acetate [31], allyl ethers of 5-hydroxyisophthalic acid to form *CTPS* [34], 4-allyl-2,6-dimethyl-phenol-glycdiylether to prepare *PTPS* and *ETPS* [35], and 4-vinylcyclohexaneoxide for silicones containing cycloaliphatic epoxy groups (*CETPS*) [36]. Hydrosilylation of Si—H functional silicones and allyl-functional phenolic or epoxy resins has also been carried out in situ during cure. For example, epoxy novolac systems were combined with 2-allylphenol and SiH-terminated oligo(dimethylsiloxane)s to produce silicone-toughened epoxy resins [37] with a glass transition temperature of 253 °C.

$$H-SiMe_2OSiMe_2-H \quad + \quad
\begin{array}{c}
CH_2OH \\
CHCH_2OCH_2CH=CH_2 \\
OH
\end{array}
\quad \longrightarrow \quad (HO)_2-PDMS-(OH)_2$$

$$(3)$$

$$
\begin{array}{l}
\text{HOCH}_2 \\
\quad \text{CHCH}_2\text{O}(\text{CH}_2)_3\!-\!\text{SiMe}_2\!-\!\!\big[\text{OSiMe}_2\big]_x\!-\!\text{OSiMe}_2\!-\!(\text{CH}_2)_3\text{OCH}_2\text{CH} \\
\quad \text{OH} \\
\end{array}
\qquad
\begin{array}{l}
\text{CH}_2\text{OH} \\
\\
\text{OH}
\end{array}
$$

$$(\text{HO})_2\!-\!\text{PDMS}\!-\!(\text{OH})_2$$

The second versatile synthetic route to silicone liquid rubbers with functional end groups, Eq. (2), involves equilibration of octamethylcyclotetrasiloxane with difunctional disiloxanes $XSiMe_2OSiMe_2X$. This equilibration is carried out either in the presence of strong acids, e.g., trifluoroacetic acid and trfluorosulfinic acid, or bases such as NR_4OH or ammonium siloxanolates respectively. As mentioned above, this equilibration reaction provides also the synthetic base for preparation of bis($SiMe_2H$)-terminated oligosiloxanes. Volatile starting materials and cyclic oligomers are readily removed, e.g., using thin-film evaporation. Equilibration reaction mechanisms and kinetics have been studied by McGrath et al. [20, 23, 38]. In a typical *ATPS* synthesis, ring-opening polymerization of octamethylcyclotetrasiloxane is performed at 105–140 °C in the presence of bis(1,3-aminopropyl)tetramethyldisiloxane using a siloxanolate catalysts, which is formed in situ by reacting KOH or NR_4OH with the cyclic siloxane. The disiloxane acts as chain transfer agent controlling both molecular weight and end group composition of the linear difunctional oligosiloxanes [38]. Base catalysis has been applied to prepare *ATPS* [20, 23, 28, 38, 39], piperazine-functional *PATPS* [26], *ETPS* [20, 23, 38, 39], whereas acid catalysis has been useful in syntheses of hydroxy-functional *HTPS* [20, 23, 39–42], carboxy-functional *CTPS* [43], and dicarboxylic-anyhdride-functional *AHTPS* [44].

When *ETPS*, prepared by hydrosilylation of allylglycidylether with bis($SiMe_2H$)-terminated oligo(dimethylsiloxane) of $M_n = 2200$ g/mol, is added to bisphenol A diglycidylether and cured with hexahydrophthalic anhydride, silicone phase separates during cure. Fractography by means of scanning electron microscopic (SEM) imaging of fracture surfaces, depicted in Fig. A.2, reveals the presence of craters with average diameters in the order of magnitude of 10–50 μm. As schematically presented in Fig. A1, these surface structures could result from rubber cavitation and voiding. In fact, at higher magnification (Fig. A2, right), residues of the rubber are detected in the craters. According to transmission electron microscopy (TEM) average domain sizes of the dispersed silicone microphases fall in range of 1–5 μm. In spite of favorable morphologies, according to the properties listed in Table A.1 *ETPS* fails to improve impact energy absorption but impairs stiffness, as expressed by flexural and Young's modulus.

A.3
Compatibilized Silicone Liquid Rubbers

In fact, the majority of oligosiloxanes with two epoxy-reactive end groups and oligosiloxane molecular weights of $M_n > 2000$ g/mol are rather ineffective as epoxy toughening agents, most likely due to poor interfacial adhesion of the

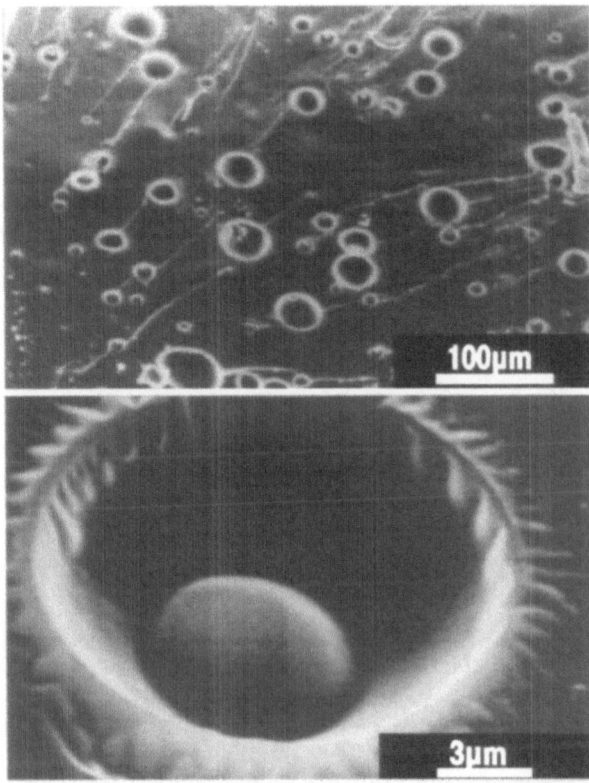

Fig. A.2. SEM fracture surface image of hexahydrophthalic-anhydride-cured bisphenol A-diglycidyl-ether containing 10 wt % *ETPS* with oligo(dimethylsiloxane) of $M_n = 2200$ g/mol

dispersed phases. Although non-bonded dispersed silicone act as stress concentrators, impact energy is not dissipated because crack propagation proceeds very rapidly along the epoxy/silicone interfaces and does not permit stress transfer. In order to enhance compatibility and interfacial adhesion, two approaches toward modification of the silicone rubber proven to be successful: (1) variation of the silicone backbone using more compatible copolysiloxanes instead of highly incompatible poly(dimethylsiloxane)s, and (2) forming segmented silicones containing compatibilizing segments. Saito et al. [45, 46] used tetramine-functional *ATPS* with poly(dimethylsiloxane-co-methylphenylsiloxane) backbone. Increasing the percentage of methylphenylsiloxane units with respect to dimethylsiloxane units enhanced compatibility and simultaneously reactivity of amine end groups and silicone volume fraction dissolved in the epoxy matrix. With increasing phenylmethylsiloxane content, however, glass transition of the silicone rubber increased from -128 °C to -17 °C. In a similar approach, McGrath et al. have introduced piperazine-terminated poly(di-methylsiloxane-co-diphenylmethylsiloxane) and poly(dimethyl-siloxane-co-methyltrifluoro-pro-pylsiloxane) (*PATPS*). Incorporation of trifluoropropyl groups accounts for better compatibility and smaller average diameter of the disperse silicone micro-phases. Highest fracture toughness is obtained when 40% trifluoro-propyl-methylsiloxane are incoporated into the poly(dimethylsiloxane) backbone. Again silicone glass transition temperatures increase with increasing comonomer content.

$$H_2N(CH_2)_2HN(CH_2)_3SiMe_2O-[SiMe_2O]_x[\underset{\underset{\bigcirc}{|}}{Si}O]_y-SiMe_2(CH_2)_3NH(CH_2)_2NH_2$$

(ATPS)

$$HN\underset{}{\bigcirc}N-CH_2CH_2NHCO(CH_2)_3-[SiMe_2O]_x[\underset{\underset{CF_3}{\overset{\overset{Me}{|}}{\underset{CH_2}{|}}}{Si}O]_y-(CH_2)_3CONHCH_2CH_2-N\underset{}{\bigcirc}NH$$

(PATPS)

In order to overcome this drawback of compatibility associated with high silicone glass transition temperature, recent research has been aimed at the development of segmented silicone liquid rubbers comprising poly(dimethyl-siloxane) segments which are coupled with compatibilizing segments. Objective is to achieve colloidal silicone dispersions which are typical for block copoly-mers containing mutually immiscible segments. In conventional processing, rather ill-defined segmented silicones are obtained when silicone end groups react with epoxy resins to form adducts. In a recent approach, Buchholz and Mülhaupt [32, 47, 48] have prepared two families of well-defined linear and branched hydroxy-functional poly(caprolactone)-block-poly(dimethylsiloxane)-block-poly(caprolactone) triblock silicone liquid rubbers as epoxy toughening agents. Similar to procedures first reported by Riffle et al. [49], hydrosilylation chemistry is combined with organotin-catalyzed ring-opening polymerization of caprolactone, using dihydroxy- and tetrahydroxy-functional oligo(dimethyl-siloxane)s as chain transfer agents. According to Eq. (4), linear dihydroxy-terminated oligo(dimethylsiloxane)s are obtained when caprolactone is poly-merized onto undecene-1-ol by heating caprolactone and undecene-1-ol several hours at 220 °C in the presence of $Bu_2Sn(OOCC_{11}H_{23})$ catalyst. In a subsequent step, bis(SiMe$_2$H)-terminated oligo(dimethylsiloxane) with $M_n = 2200$ or 4800 g/mol is coupled with the vinyl-terminated oligo(caprolactone) using Pt-catalyzed hydrosilylation to form linear HO-PCL-PDMS-PCL-OH triblock silicone.

When caprolactone is polymerized onto (HO)$_2$-PDMS-(OH)$_2$ as described above and schematically represented in Eq (5), novel branched segmented sili-cones are obtained. Number average molecular weight of the segmented liquid rubbers varies between 3600 and 10 000 g/mol with PCL content ranging be-tween 38 and 70 wt%. All segmented silicones phase separate when bisphenol A diglycidylether is cured with hexahydrophthalic anhydride (3 h at 150 °C and 1 h

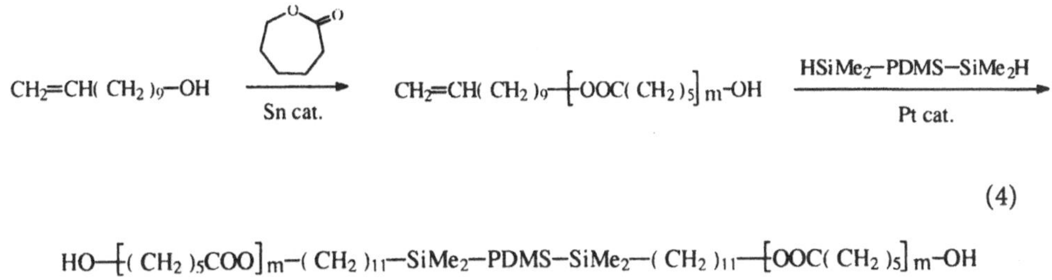

$$CH_2=CH(CH_2)_9{-}OH \quad \xrightarrow{\text{Sn cat.}} \quad CH_2=CH(CH_2)_9{-}[OOC(CH_2)_5]_m{-}OH \quad \xrightarrow{\text{HSiMe}_2{-}\text{PDMS}{-}\text{SiMe}_2\text{H}}_{\text{Pt cat.}}$$

$$(4)$$

$$HO{-}[(CH_2)_5COO]_m{-}(CH_2)_{11}{-}SiMe_2{-}PDMS{-}SiMe_2{-}(CH_2)_{11}{-}[OOC(CH_2)_5]_m{-}OH$$

$$(\ HO{-}PCL{-}PDMS{-}PCL{-}OH\ )$$

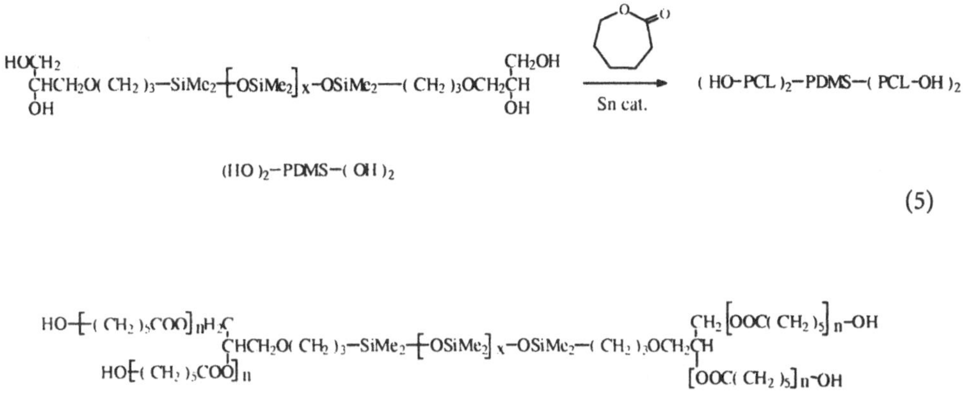

$$\begin{array}{l} HOCH_2 \\ \ \ CHCH_2O(CH_2)_3{-}SiMe_2{-}[OSiMe_2]_x{-}OSiMe_2{-}(CH_2)_3OCH_2CH \\ OH \qquad\qquad\qquad\qquad\qquad\qquad\qquad\qquad\qquad\quad\ \ OH \end{array} \quad CH_2OH \xrightarrow{\text{Sn cat.}} (\ HO{-}PCL\ )_2{-}PDMS{-}(\ PCL{-}OH\ )_2$$

$$(\ HO\ )_2{-}PDMS{-}(\ OH\ )_2$$

$$(5)$$

$$\begin{array}{l} HO{-}[(CH_2)_5COO]_nH_2C \\ \qquad\qquad\qquad CHCH_2O(CH_2)_3{-}SiMe_2{-}[OSiMe_2]_x{-}OSiMe_2{-}(CH_2)_3OCH_2CH \\ HO{-}[(CH_2)_5COO]_n \end{array} \quad \begin{array}{l} CH_2[OOC(CH_2)_5]_n{-}OH \\ \\ \ \ \ [OOC(CH_2)_5]_n{-}OH \end{array}$$

$$(\ HO{-}PCL\ )_2{-}PDMS{-}(\ PCL{-}OH\ )_2$$

at 180 °C) in the presence of 1 wt% N-benzyl-dimethylamine as cure accelerator. While oligo(caprolactone) plasticizes the epoxy matrix, thus reducing modulus and glass transition temperatures, the same oligo(caprolactone) attached to silicone segments improves compatibility of silicone but does not plasticize the matrix, as evidenced by high glass transition temperatures of the silicone-modified matrix. From Table A.1 it is apparent that both linear and branched silicone rubbers give much better impact energy absorption than ETPS with the equivalent oligo(dimethylsiloxane) segment length. With respect to the linear HO-PCL-PDMS-PCL-OH the branched (HO-PCL)$_2$-PDMS-(PCL-OH)$_2$ give markedly higher resistance to stresses, as expressed by higher Young's modulus and tensile and flexural strengths. Also the viscosity of the uncured silicone-containing epoxy resin is much lower in the case of branched silicones. While the matrix glass transition temperature is reduced only by 4 to 6 °C at 10 wt% rubber content, a second low temperature phase transition reveals the presence of silicone dispersed phases with a glass transition temperature near -130 °C. Using element-specific TEM, as shown in Fig A.3, it was possible to image the dispersed silicone phase with average diameters of 10–20 nm. These silicone-nanophase-toughened epoxy resins are optically transparent. It is important to

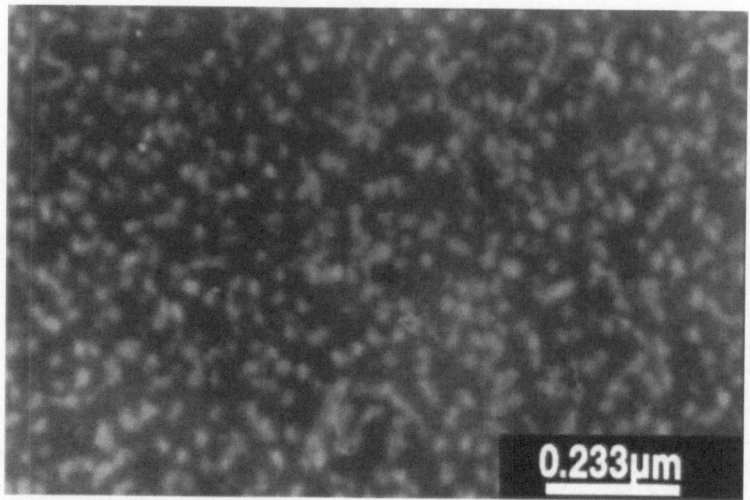

Fig. A.3. Element-specific TEM image of a thin cut prepared from hexahydrophthalic-anhydride-cured bisphenol A-diglycidylether containing 10 wt% (HO-PCL)$_2$-PDMS-(PCL-OH)$_2$ with M$_n$(PDMS) = 2200 g/mol and 47 wt% PCL

note that prime requirement for nanophase separation is the presence of reactive hydroxy groups. In the absence of hydroxy groups, much larger micro-phases are formed and mechanical properties of the resulting silicone-modified epoxy materials are poor. Synergistic properties have been attributed to the formation of nanophase-separated interpenetrating networks.

The SEM fractography of fracture surfaces obtained in fatigue loading test [50], reveal another interesting feature of silicone-nanophase-toughened epoxy resins. While non-modified resin (Fig. A.4 left) exhibit very rough surfaces with parabolic textures relating to secondary crack initiation, the silicone-nanophase-toughened epoxy resin (Fig. A.4, right) give smoother fracture surfaces with much less secondary crack inition [50]. Clearly, the much larger number of silicone nanophases with respect to silicone microphases at the same silicone content, accounts for excellent impact energy dissipation by silicone nano particle bridging.

In an alternative approach to colloidal silicone particle dispersions, silicone liquid rubbers are dispersed and crosslinked in uncured epoxy resins [51, 52]. For example, Pt-catalyzed hydrosilylation reaction of silicone rubbers containing SiH groups with bis(vinyl)-terminated oligo(dimethylsiloxane)s is used. Compatibility can be improved, e.g., by adding allyl-functional epoxy resins which are covalently attached dispersed silicones. Also core/shell-type silicone particles can be formed in situ.

In conclusion, colloidal silicone micro- and nanoparticle dispersions with controlled silicone particle functionality and surface compatibility, either prepared by controlled phase separation of segmented silicones during cure or by formation of prefabricated structured particles, are the key to novel silicone-toughened epoxy resins which exhibit substantially higher impact resistance without sacrificing high strength, stiffness, dimensional, environmental and chemical stability.

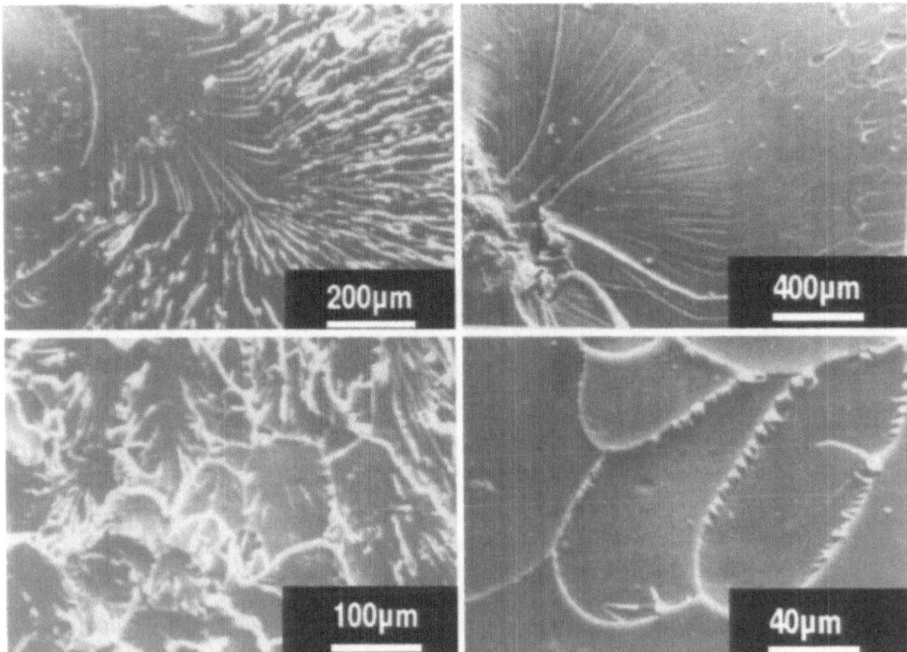

Fig. A.4. SEM image of a fracture surface of a fatigue loading test specimen for non-modified (*left*) and with 5 wt% (HO-PCL)$_2$-PDMS-(PCL-OH)$_2$-modified (*right*) hexahydrophthalic-anhydride-cured bisphenol A-diglycidylether

A.4
References

1. Morawetz G (1988) In: Woebcken W (ed) "Duroplaste",vol.10, 2nd edn., Hanser, Munich, p 338
2. Bucknall CB, Bucknall CB (ed) (1977) In: 'Toughened Plastics', Elsevier London
3. Bucknall CB, Allen G, Bevington JC (eds) (1989) In: 'Comprehensive Polymer Science', vol. 7 Pergamon, Oxford, p. 27–49
4. Kinloch AJ, Kinloch AJ, (ed) (1986) In: 'Structural Adhesives' Elsevier, London, p. 127
5. Riew CK, Gillham JK (ed) (1984) 'Rubber-Modified Thermoset Resins', Adv Chem Ser 208, American Chemical Society, Washington D.C.
6. Riew CK (ed) (1989) 'Rubber-Toughened Plastics', Adv Chem Ser 222, American Chemical Society, Washington, D.C.
7. Allen K (1990) Prog Rubber Plast Technol 6(2): 159
8. Hedrick JC, Hedrick JL, Cecere JA, Liptak SC, McGrath JE (1990) Polym Mater Sci Eng 63: 190
9. Drake R (1990) Polym Mater Sci Eng 63: 802
10. Bauer R (1990) Polym Mater Sci Eng 63: 672
11. Garg AC, Mai YW (1988) Compos Sci Technol 31(3): 179
12. Moloney AC, Kausch HH, Kaiser T, Beer HR (1987) J Mater Sci 22(2): 381
13. Siebert AR (1983) Polym Mater Sci Eng 49: 427
14. Kausch HH (ed) (1987) Polymer fracture Springer, Berlin Heidelberg New York
15. Kinloch AJ, Young RJ (eds) (1983) 'Fracture Behaviour of Polymers', Elsevier Applied Science, London

16. Brostow W, Corneliussen RD (eds) (1986) 'Failure of Plastics', Hanser Publ Co., Munich
17. Mülhaupt R (1990) Chimia 44: 43
18. Rowe EH, Siebert AR, Drake RS (1970) Mod Plat 47: 110
19. Siebert AR, Guiley CD, Eplin AM; in ref. 6), p. 389
20. Yilgor I, McGrath JE (1988) Adv Polym Sci 86: 1
21. Yilgor I, Riffle JS, Steckle WP, Jr., Banthia AK, McGrath JE (1984) Polym Mater Sci Eng 50: 518 (1984)
22. Chujo Y, McGrath JE (1983) Am Chem Soc, Div Polym Chem, Polym Prepr 24(2): 47
23. Yilgor I, Riffle JS, McGrath JE (1985) Am Chem Symp Ser 282 (Reactive Oligomers) 161
24. Yorkgitis EM, Eiss NS, Jr., Tran C, Wilkes GL, Mc Grath JE (1985) Adv Polym Sci 72: 80
25. Yorkgitis EM, Tran C, Eiss NS, Jr., Hu TY, Yilgor I in ref. 5), p. 137
26. Wilkes GL, McGrath JE (1983) Polym Mater Sci Eng 49: 498
27. Riffle JS, Yilgor I, Tran C, Wilkes GL, McGrath JE, Banthia AK, (1983) Am Chem Soc Symp Ser 221 (Epoxy Resin Chemistry 2) 21
28. Riffle JS, Yilgor I, Banthia AK, Tran C, Wilkes GL, McGrath JE (1982) Am Chem Soc Symp Ser 221: 21
29. McGrath JE, In: ref. 24), p. 81
30. Greber G (1971) J Prakt Chem 313(3): 461
31. Boutevin B, Youssef B (1989) Makromol Chem 190(2): 277
32. Buchholz U, Mülhaupt R (1992) Am Chem Soc, Div Polym Chem, Polym Prepr 33(1): 205
33. Buchholz U (1992) Ph.D. thesis, University of Freiburg, Freiburg
34. Kawakami Y, Saibara S, Suzuki F, Abe T (1991) Polym Bull (Berlin) 25(5): 521
35. Zahir SAC (1994) Eur Pat Appl EP 319472 (1988) to Ciba-Geigy AG, Chem Abstr 112: 37309
36. Union Carbide Corp. (1993) Jpn Kokai Tokkyo Koho, JP 59147018 (1984), Chem Abstr 102(8): 63716h
37. Ota M, Yanagisawa K (1993) Jpn Kokai Tokkyo Koho JP 02173034 (1988) to Sumitomo Bakelite Co., Ltd., Chem Abstr 113(22): 193175u
38. Sormani PM, McGrath JE (1985) Am Chem Soc, Div Polym Chem 26(1): 258
39. McGrath JE, Sormani PM, Elsbernd CS, Kilic S (1986) Makromol Chem Mcromol Symp 6: 67
40. Murachashvili DU, Kopylov VM, Khananashvili LM, Shkol'nik MI, Tsomaya NI, Volkova RV, Savitskii AA (1990) Vysokomol Soedin Ser B 32(2): 168
41. Yilgor I, Riffle JS, Steckle WP, Jr., Banthia AK, McGrath JE (1984) Polym Mater Sci Eng 50: 518
42. Steckle WP, Jr., Yilgor E, Riffle JS, Spinu M, Yilgor I, Ward RS (1987) Am Chem Soc Div Polym Chem 28(1): 254
43. Jewel BS, Riffle JS, Allison D, McGrath JE (1989) Am Chem Soc, Div Polym Chem 30(1): 295
44. Rich JD Ger Offen DE 3 542 346 (1980) to General Electric Co.
45. Takahashi T, Nakajima N, Saito N in ref. [6], p. 243
46. Saito N, Nakajima N, Ikushima T, Kanagawa S, Takahashi T (1987) Polym Mater Sci Eng 57: 558
47. Mülhaupt R, Buchholz U In: 'Toughened Plastics II: Science and Engineering', (CK Riew, AJ Kinloch, eds), Adv Chem Ser in preparation San Diego ACS conference 1994 Am Chem Soc Washington DC, submitted
48. Buchholz U, Mülhaupt R, manuscript in preparation
49. Riffle JS, Steckle WP, Jr., White KA, Ward RS (1985) Am Chem Soc, Div Polym Chem, Polym Prepr 26(1): 251
50. Köncöl L, Döll W, Buchholz U, Mülhaupt R, J Appl Polym Sci 54: 815
51. Block H, Pyrlik M (1988) Kunststoffe 78: 12
52. Itho K, Shiobara T, Futatsumori K, Tomiyoshi K (1988) Eur Pat Appl EP 218: 228

Appendix B: Silylation and Silylating Agents

H. R. Kricheldorf

B.1
Reactivity of Silyl Groups

The purpose of this appendix is twofold. Firstly, a short introduction to the basic principles underlying the chemistry of silicon should be given. Since numerous books and reviews cover this aspect [1–40], it was not intended to present a detailed discussion in this handbook. Secondly, a short description of silylation reactions and silylalting agents should be given. Again, numerous reviews exist in this field [12–40]. Nonetheless, silylations are described briefly here again, because they play an extremely important role in analytical and preparative organic chemistry and also a constantly increasing role in preparative polymer chemistry (see Chapters 5–7).

The term silylation is defined as an exchange of a more or less acidic proton against a silyl group. In more than 90% of all practical applications the proton is originally attached to a O-, N- or S-atom, and the silyl group is trimethylsilyl. A silylation entails at least two usually more of the following five effects:

I) It reduces melting points due to the elimination of H-bonds.
II) It improves the volatility due to a reduction of polarity and/or elimination of H-bonds.
III) It improves the solubility in less polar organic solvents.
IV) It protects functional groups (particularly OH).
V) It may activate the nucleophilicity of functional groups (e.g. amide groups).

The electronic configuration of silicon is $1s^2$, $2s^2$, $2p^6$, $3s^2$, $3p^2$. Silicon like carbon prefers a tetracoordinate state with sp^3 hybridization. Because the atom radius of silicon (1.17 Å) is greater than that of carbon (0.77 Å) and because orbitals of the principle quantum number 3 are involved in covalent bonding, the steric demands for substituents and substitution reactions are significantly reduced. For instance, a bimolecular nucleophilic substitution of chloromethyl trimethylsilan, Eq. [1] is more rapid than the analogous substitution of neopentyl

$$Me_3Si-CH_2-Cl + NaJ \longrightarrow NaCl + Me_3Si-CH_2-J \tag{1}$$

chloride [1]). Unlike carbon, silicon possesses energetically favorable vacant 3d-orbitals which are available for binding of ligands and nucleophilic reagents. Hence, the energy of activation in nucleophilic substitution is in general lower than in the case of carbon chemistry. A characteristic illustration of this difference is the rapid hydrolysis of $SiCl_4$ even in neutral water which occurs at 25 °C in a few minutes under homogeneous conditions. In contrast CCl_4 is stable against H_2O for months and years. Furthermore, the nucleophilic substitution of CH_3SiCl is a S_{N2} reaction under almost all circumstances, whereas a S_{N1} course is typical for many substitutions of $(CH_3)_3C-X$ (X = halogen). Nucleophilic substitutions of carbon atoms quite frequently yield the kinetically controlled product, whereas the thermodynamical control is typical for most nucleophilic substitutions of Si-atoms. Therefore, it is of interest to know the stability of Si—X bonds not only for radical reactions (see Table B. 1).

Another, characteristic difference between silicon and carbon concerns their electronegativity. As indicated by the data in Table B.2, silicon is considerably more electropositive than carbon and even more electropositive than hydrogen. Consequently, hydrogen attached to silicon has a "hydride character", but a "proton character" when attached to carbon. The reactions of methyl lithium with triphenylsilane Eq. (2) and triphenylmethane Eq. (3) illustrate this aspect [8]. However, in several reactions the reactivity of the Si—H bond is comparable

$$Ph_3Si-H \ + \ CH_3Li \ \longrightarrow \ Ph_3Si-CH_3 \ + \ Li^{\oplus}H^{\ominus} \tag{2}$$

$$Ph_3C-H \ + \ CH_3Li \ \longrightarrow \ Ph_3C^{\ominus}Li^{\oplus} \ + \ CH_3-H \tag{3}$$

$$R-C{\equiv}CH \ + \ HSiCl_3 \ \xrightarrow{(H_2PtCl_6)} \ \underset{H}{\overset{R}{\diagdown}}C{=}C\underset{SiCl_3}{\overset{H}{\diagup}} \tag{4}$$

with H_2. A typical example is the addition of silanes to alkines under the influence of platinum catalysts [9].

Whereas the reactivity of Si—C bond in $(CH_3)_4Si$ is not much different from the reactivity of C—C bonds in $(CH_3)_4C$, the polarization and reactivity against

Table B.1 Average bond energies of Si—X (kcal/ mol) (Adapted from [11]).	
Si—S	54
Si—I	59
Si—Si	68
Si—C	69
Si—H	70
Si—Br	76
Si—N	75–80
Si—Cl	93
Si—O	112
Si—F	142

Table B.2 Relative electronegativity (non-empirical scale) (Adapted from [7]).	
F	4.0
O	3.52
N	3.16
Cl	2.84
H	2.79
Br	2.52
S	2.52
Se	2.4
C	2.35
P	2.11
Si	1.64

polar reagents becomes evident, when electron withdrawing substituents are attached to the carbon atom. In extreme cases such as $R_3Si-CCl_3$ the Si—C bond is even cleaved by alkaline water Eq. (5). Another example is the easy reaction of trimethylsilylcyamide with ketones (or other electrophiles), Eq. (6). This reaction is quite analogous to the addition of HCN, Eq. (7). Thus, this reaction also illustrates that silyl groups quite often play the role of a "big, soft proton". Other examples of such proton analogous reactions are acylation of silylated SH,

$$R_3Si-CCl_3 \ + \ H_2O \ \xrightarrow{(OH^{\ominus})} \ R_3Si-OH \ + \ HCCl_3 \tag{5}$$

$$\begin{matrix} R \\ {\diagdown} \\ {\diagup} \\ R \end{matrix} C=O \ + \ Mc_3Si-CN \ \longrightarrow \ \begin{matrix} R \\ | \\ NC-C-O-SiMe_3 \\ | \\ R \end{matrix} \tag{6}$$

$$\begin{matrix} R \\ {\diagdown} \\ {\diagup} \\ R \end{matrix} C=O \ + \ H-CN \ \longrightarrow \ \begin{matrix} R \\ | \\ NC-C-OH \\ | \\ R \end{matrix} \tag{7}$$

OH or NH groups as illustrated by Eqs. (8, 9). In contrast to the acylation of the protonated nucleophile, Eq. (8), no HCl acceptor is required for acylation of the silylated nucleophile, Eq. (9). For this and other reasons silylated nucleophiles

$$R-CO-Cl \ + \ H_2N-R' \ \xrightarrow[-\ HCl]{} \ R-CO-NH-R' \tag{8}$$

$$R-CO-Cl \ + \ Mc_3Si-NHR \ \xrightarrow[-\ Me_3SiCl]{} \ R-CO-NH-R' \tag{9}$$

are useful as monomers for a variety of polycondensations (see chapter 5). In this connection it should be mentioned that the transition state of acylations may involve a four membered, Eqs. (10–12), or six membered transition state Eqs. (13–15). In the latter case the acylation does not necessarily result in a substitution.

Finally, it is worth noting that silicon in contrast to carbon stabilizes a negative charge in α-position due to its vacant d-orbitals [5]. Therefore, addition of ions to vinylsilanes usually occur in β-position, Eq. (16). Furthermore, silicon stabilizes positive charges in β-position, and thus, facilitates β-eliminations such as the solvolytic degradation of 2-bromoethyl silanes, Eq. (17) [6]. Also in contrast to carbon, silicon like other elements of higher periods does not form stable multiple bonds. Hence, compounds with Si—C or Si—Si double or triple bonds do not play a role in polymer synthesis. However, an interesting analogy exists between the reactivity of divalent silylenes and carbenes. A few polymer syntheses based on the high reactivity of silylenes have been reported (see chapter 6).

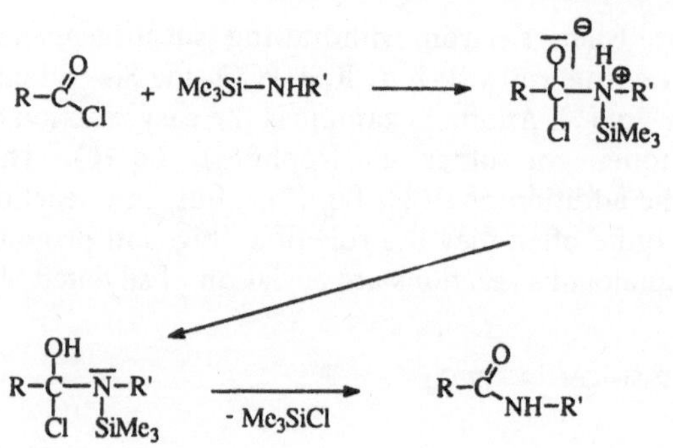

$$R-\overset{\displaystyle O}{\underset{\displaystyle Cl}{C}} \ + \ Me_3Si-NHR' \ \longrightarrow \ R-\overset{\displaystyle \overline{O}|^{\ominus}}{\underset{\displaystyle Cl}{\underset{\displaystyle SiMe_3}{C}}}-\overset{\displaystyle H}{\underset{}{\overset{\oplus}{N}}}-R' \tag{10}$$

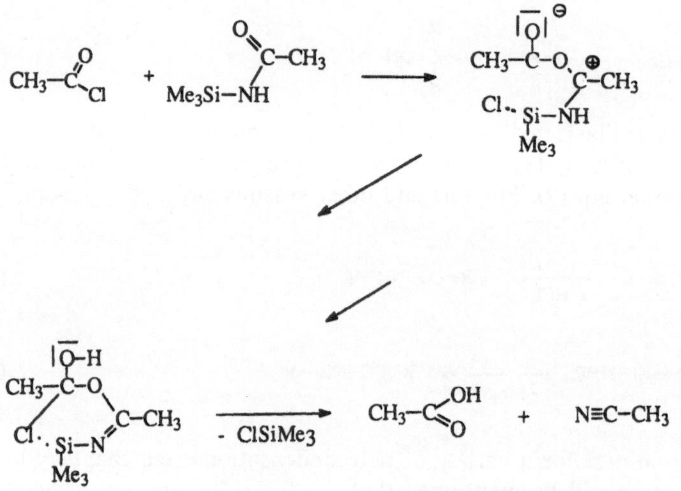

$$\tag{11}$$

$$R-\overset{\displaystyle OH}{\underset{\displaystyle Cl}{\underset{\displaystyle SiMe_3}{C}}}-\overline{N}-R' \ \xrightarrow{\ -Me_3SiCl\ } \ R-\overset{\displaystyle O}{\underset{}{C}}\diagdown_{NH-R'} \tag{12}$$

$$CH_3-\overset{\displaystyle O}{\underset{\displaystyle Cl}{C}} \ + \ \overset{\displaystyle \overset{\displaystyle O}{\overset{\parallel}{C}}-CH_3}{Me_3Si-NH} \ \longrightarrow \ \overset{\displaystyle \overline{|O|}^{\ominus}}{CH_3-\overset{}{C}-O\diagup}\overset{\oplus}{\underset{Cl\cdots\ \underset{\displaystyle Me_3}{Si-NH}}{C}}-CH_3 \tag{13}$$

$$\tag{14}$$

$$\overset{\displaystyle |\overline{O}-H}{CH_3-\overset{}{C}-O\diagdown}\overset{C-CH_3}{\underset{Cl\cdots\ \underset{\displaystyle Me_3}{Si-N}}{}} \ \xrightarrow{\ -ClSiMe_3\ } \ CH_3-\overset{\displaystyle OH}{\underset{\displaystyle O}{C}}\diagdown \ + \ N\equiv C-CH_3 \tag{15}$$

$$Ph_3Si-CH=CH_2 \ + \ Ph-Li \ \longrightarrow \ Ph_3Si-\underset{}{\overset{\ominus}{C}H}-CH_2-Ph \quad \overset{Li^{\oplus}}{} \tag{16}$$

$$\underset{\underset{\displaystyle Br}{H_2C-CH_2}}{\overset{Me_3Si\cdots}{}\ \delta^{\oplus}} \ \xrightarrow{\ (\Delta T, EtOH)\ } \ Me_3Si-OEt \ + \ CH_2=CH_2 \ + \ HBr \tag{17}$$

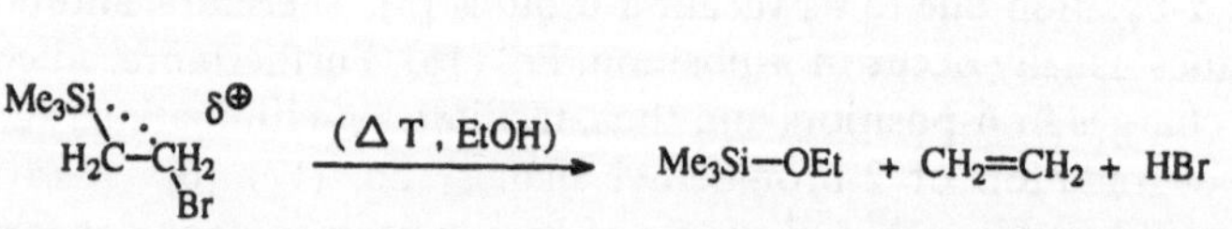

Like most reactions in organic chemistry, reactions of silyl groups are influenced by the electronic and steric effects of substituents. Obviously, reaction rates of any kind of silylation decreases when the steric demands of the reaction partners increase. The influence of substituent effects on the hydrolytic desilylation of methoxy or phenoxysilanes is illustrated by the data compiled in Table B.3.

Table B.3 Relative stabilities ($1/k_{rel}$) of $R^1R^2R^3Si$—OR^4 [a] towards acid or base catalyzed hydrolysis. (adapted from [10])

	Me$_3$Si— 1	PhMe$_2$Si— 1, 2	Et$_3$Si 64	i·PrMe$_2$Si— 86	n·Pr$_3$Si— ~90*
Acid catalyzed	n·Bu$_2$Si 130*	Ph$_3$Si— 400	iPr$_2$MeSi— 5 × 10^3*	t—BuMe$_2$Si— 2 × 10^4*	ThexylMe$_2$Si— 5 × 10^4
	i·Pr$_3$Si— 7 × 10^5	t—BuPH$_2$Si— 5 × 10^6			
	PhMe$_2$Si— 0.4	Ph$_3$Si— 1	Me$_3$Si 1	i·PrMe$_2$Si 600	Et$_3$Si— 1.3 × 10^3 (1.6 × 10^2)*
Base catalyzed	n·PrSi— 5 × 10^2*	n·Bu$_3$Si— 8 × 10^2*	t—BuMe$_2$Si— 2 × 10^4*	i—Pr$_3$Si— 10^5	

[a] R^4 menthyl or phenyl (*)

B.2
Silylating Agents

On the following pages a short description of the most important silylating agents is given (roughly in alphabetical order). The selected references mainly deal with preparative aspects, but it should be emphasized that most publication deal with analytical applications in gas chromatography and mass spectroscopy. A more comprehensive review is available [40].

B.2.1
Allyldimethylchlorosilane (ADMCS) [41 – 45]

ADMCS was almost exclusively used for the silylation of hydroxy groups in combination with analytical applications. Nonetheless, the modification of a substrate with a functional silyl group may also be of preparative interest. For example, the allyl group might be useful for selective crosslinking of silylated polymers. Silylations with ADMCS were usually conducted in the presence of an HCl acceptor, such as imidazole (in DMF) [41] diethylamine (in acetonitrile) [42] or triethylamine (in Et$_2$O) [45].

B.2.2
Allyltrimethylsilane [46 – 50]

The reactivity of this reagent is based on the weakness of the allyl Si—C bond. Nonetheless, most silylations require a strong acid as catalyst. The advantage of allyltrimethylsilane consists of the formation of propene as a nonacidic highly volatile byproduct, Eq. (18). In addition to strong acids [46, 48, 49] Br_2 or J_2 may be used as catalyst [47]. OH-groups are the prefered substrate.

$$Me_3Si-CH_2-CH=CH_2 \quad \xrightarrow{\quad H^{\oplus}\quad} \quad R-O-SiMe_3 \; + \; CH_3-CH=CH_2 \qquad (18)$$
$$+ \; R-OH$$

B.2.3
Bromomethyldimethylchlorosilane (BMDMCS) [51 – 57]

This silylating agent is usually applied in combination with diethylamin or triethylamine. Mild conditions are required to avoid side reactions of the $BrCH_2$ group. The bromomethyldimethylsilyl group may be useful for mass-spectroscopy and for the synthesis of heterocycles [51].

B.2.4
N,O-Bistrimethylsilylacetamide (BSA) [22, 27, 30, 58–80]

BSA [58] in one of the most powerful and most widely used silylating agents and preferentially used for analytical purposes [61]. Depending on the substrate and on the reaction conditions either one or both silyl groups will be transferred. The byproducts acetamide or N-TMS-acetamide (Eqs. 19, 20) are volatile enough to allow direct GC analyses of reaction mixtures. BSA allows an almost quantitative silylation of various types of substrates and particularly useful for the

$$CH_3-C\begin{smallmatrix}OSiMe_3\\ \\NSiMe_3\end{smallmatrix} \; + \; HN\begin{smallmatrix}R^1\\ \\R^2\end{smallmatrix} \longrightarrow CH_3-\overset{O}{\overset{\|}{C}}-NH-SiMe_3 \; + \; Me_3Si-N\begin{smallmatrix}R^1\\ \\R^2\end{smallmatrix} \quad (19)$$

$$CH_3-\overset{O}{\overset{\|}{C}}-NH-SiMe_3 \; + \; HO-R \longrightarrow CH_3-\overset{O}{\overset{\|}{C}}-NH_2 \; + \; Me_3Si-O-R \qquad (20)$$

silylation of multifunctional compounds such as carbohydrates, nucleosides, aminoalcohols and aminoacids or imino acids. It silylates amides, lactams, ureas and sterically hindered phenols or carboxylic acids. In the case of sterically hindered alcohols and phenols small amounts of trifluoroacetic acids may act as catalyst [63]. In other cases chlorotrimethylsilane [62, 65] or traces of HCl were recommended as catalytic additives. Regioselective and stereoselective syntheses of silyl enol ethers from different carbonyl groups were reported to occur in HMPT with catalytic amounts of sodium metal.

B.2.5
N,O-Bistrimethylsilylcarbamate (BSC) [81, 82]

This silylating agent was recommended for the silylation of sterically non-hindered alcohols, phenols and carboxylic acids. It is not powerful enough for the silylation of amines. Its main advantage are the gaseous byproducts NH_3 and CO_2.

B.2.6
Bistrimethylsilyl sulfate (BSS) [83 –87]

This reagent can be used as Lewis acid, as sulfuration and as silylating agents. It is particularly useful for the silylation of metal salts of organic or inorganic acids and phenols. Insoluble metal sulfates are then the only byproducts.

B.2.7
N,O-Bistrimethylilyltrifluoroacetamide (BSTFA) [22, 27, 30, 59, 63, 69, 88 – 96]

This expensive reagent is almost exclusively used for analytical purposes. It is slightly more powerful than BSA and allows silylations of almost any kind of substrates. An additional advantage is the relatively high volatility of the byproducts. Pyridine is the solvent of choice and chlorotrimethylsilane [89–91] or trifluoroacetic acid [63, 92] were recommended as catalytic additives.

B.2.8
N,N-Bistrimethylsilylurea (BSU) [97, 98]

BSU is useful for the silylation of alcohols and carboxylic acids. The neutral urea which is easy to remove from reaction mixtures by filtration is the only byproduct.

B.2.9
Chloromethyldimethylchlorosilane (CMDMCS) [13, 22, 44, 52, 71, 99 – 101]

In combination with diethylamine or triethylamine this chlorosilane is mainly used for analytical applications. However, when reacted with 1,2-difunctional substrates such as 1,2-diols, 1,2-diamines or 1,2-amino alcohols it will form heterocycles [71–100]. Both chloroatoms are substituted when CMDMCS is reacted with the metal salts of carboxylic acids or phenols [101].

B.2.10
1,3-Bis(chloromethyl)-1,1,3,3-tetramethyldisilazane (CMTMDS) [27, 42, 44, 102–104]

This reagent may be used instead of the above mentioned chlorosilane. Most powerful is a combination of both silylating agents. This combination has the additional advantage that NH_4Cl is formed as byproduct and side reactions of NH_3 with the chloromethyl group are avoided.

B.2.11
Dimethylisopropylchlorosilane (DMIPSCl) [105 – 109]

DMIPSCl is usually applied in combination with imidazole (in DMF) or triethylamine as HCl-acceptors. The DMIP-silyl group plays a minor role as protection for alcohols or phenols.

B.2.12
Diphenylmethylchlorosilane (DPMSCl) [110 – 114]

This chlorosilane is preferentially used for the silylation of primary, secondary or tertiary alcohols in combination with imidazole in DMF. Triethylamine in a nonpolar solvent is an alternative. Treatment of ketones with BuLi followed by DPMSCl yields enolethers [110].

B.2.13
1,3-Dimethyl-1,1,3,3-tetraphenyldisilazane (TPDMDS) [115, 116]

TPDMDS was rarely used instead the above mentioned chlorosilane. It is an interesting silylating agent for the hydrophobization of glass or silica surfaces [115, 116].

B.2.14
Ethyltrimethylsilylacetate (ETSA) [117 – 120]

ETSA was used for silylation of alcohols, thiols, acetylenes and ketones. Tetrabutylammonium fluoride served as catalyst. The advantage of this reagent is the formation of the neutral, volatile ethyl acetate as the only byproduct Eq. (21).

$$Me_3Si-CH_2-CO_2Et \quad + \quad HO-R \xrightarrow{(F^{\ominus})} \quad R-O-SiMe_3 \quad + \quad CH_3-CO_2Et \qquad (21)$$

B.2.15
Hexamethyldisilazane (HMDS) [12, 13, 22, 30, 63, 121 – 147]

HMDS is one of the most widely used silylating agents despite its low reactivity. Its advantages are:

1) relatively low price
2) relatively low boiling point
3) it can be used without solvent
4) high volatility of the only byproduct NH_3
5) the silylation power can be increased by addition of (acidic) catalysts.

Neat HMDS is useful for the silylation of sterically nonhindered alcohols, phenols and carboxylic acids [122, 142, 143]. Sterically hindered substrates and amines can be silylated in many cases by the more powerful equimolar mixture of HMDS and trimethylchlorosilane [132]. The crystalline NH_4Cl is then the only byproduct. Small amounts of imidazole, trimethylbromosilane and trifluoro-acetic acid may be more effective catalysts than trimethylchlorosilane [63, 129]. Further catalysts are: HCl, $(NH_4)_2$, SO_4, p-toluene sulfonic acid and saccharin. HMDS is an important silylating agent for the hydrophobization of glass and silica [136 – 138].

B.2.16
Hexamethyldisiloxane (HMDSO) [148 – 156]

HMDSO is the poorest of all silylating agents described in this work. Nevertheless, it is inexpensive and allows the silylation of primary alcohols, phenols and carboxylic acids, when the liberated water is removed by azeotropic distillation, and when an acidic catalyst (H_2SO_4, 4-toluene sulfonic acid) is present [148 – 150]. HMDSO was used for the silylation of inorganic acids and salts including silicates, minerals and boric acid.

B.2.17
Hexamethyldisilthiane [157 – 161]

Hexamethyldisilthiane is a more powerful silylating agent than HMDS or HMDSO, but its terrible smell prevented a broader application [157 – 159]. However, this reagent is particularly useful for the preparation of organic sulfides [159, 160] (poly)thioanhydrides [161], silylated thiocarboxylic acids [161] Eqs. (22, 23) and for the reduction of sulfoxides [40].

$$R-CO-Cl \quad + \quad (Me_3Si)_2S \quad \longrightarrow \quad R-CO-S-SiMe_3 + ClSiMe_3 \qquad (22)$$

$$+ \ R-CO-Cl \ \Bigg| \ \begin{array}{c} - \ ClSiMe_3 \end{array}$$

$$\qquad\qquad\qquad\qquad (23)$$

$$R-CO-S-CO-R$$

B.2.18
N-Methyl-N-trimethylsilylacetamide (MSA) [73, 162–165]

MSA is a strong and useful silylating agent for various substrates. Its silylation power is slightly less than that of BSA. Compared to BSA it has the advantage that only one byproduct, N-methylacetamide, is formed. Worth noting is the easy silylation of amino acids [73] or dipeptides [164], and the transformation of 2-bromoethylamine hydrobromide into N,N-bis TMS-2-bromoethylamine [165].

B.2.19
N-Methyl-N-trimethylsilyl-trifluoroacetamide (MSTFA) [166 – 173]

The reactivity of this reagent is comparable to that of BSA. It is more expensive, but has two advantages; only one byproduct is formed, and this byproduct is more volatile than those of BSA. MSTFA was mainly used for analytical purposes, in particular for steroids. Silylations of multifunctional substrates were frequently conducted with mixtures of MSTFA and $BrSiMe_3$, $JSiMe_3$ or TMS-imidazole.

B.2.20
N-Methyl-N-tert-butyldimethylsilyl trifluoroacetamide (MTBSTFA) [168, 174–181]

In addition to the TMS group the tert-butyldimethylsilyl group (TBDMS) is the most important protecting group for both analytical and preparative applications. Whereas, the cheaper tert-butyl dimethyl chlorosilane is more widely used for preparative applications the expensive TBDSTFA is advantages for analytical purposes. Its silylation power is higher (comparable to that of BSA) and the crude reaction mixture may be injected into a chromatograph.

B.2.21
1-(tert-Butyldimethylsilyl)imidazole (TBDMSIM) [182 – 186]

The silylating power of this reagent is considerably lower than that of MTBSTFA (see above), but this reduced reactivity has the advantage of a higher selectivity, for instance, a discrimination between amino groups in different environment. Its application was limited to analytical purposes.

B.2.22
tert-Butyldimethylchlorosilane (TBDMSCI) [173, 175, 187 – 201]

The TBDMS group has become one of the most useful protecting group in organic synthesis [31, 32, 187 – 189] particularly for hydroxy groups. Yet, its usefulness and importance in analytical chemistry is also permanently increasing [26, 27, 174]. The usefulness of the TBDMS protecting group is based on the following properties:

1) TBDMS derivatives are much more stable to hydrolysis than the corresponding TMS derivatives (see Tables B.3 and B.4).
2) TBDMS alkylethers are relatively sensitive to acidic hydrolysis but rather stable to alkaline water. The opposite is true for the aryl ethers.
3) TBDMS ethers (and some esters or amines) are stable to various reactions, such as most oxidations, several mild reductions, hydrogenolyses, many metal organic agents, all reaction steps of nucleotide synthesis etc.
4) Most TBDMS derivatives are stable to TLC and column chromatography.
5) TBDMS derivatives have a greater tendency to crystallize than TMS derivatives.
6) Ester enolates react with TBDMS-Cl to yield almost exclusively O-silylation.
7) TBDMS derivatives can be selectively cleaved under conditions not affecting the protecting groups.

TBDMS-Cl is the least expensive and most widely used reagent for the introduction of the TBDMS group. The protection of primary and secondary OH-groups is the most common application. However, silylation of *tert*-alcohols (difficult) [196 – 198] phenols, thiols, carboxylic acids, their esters, lactames ketones, nitriles and nitrocompounds has also been reported. Due to the steric demands of the TBDMS group silylations of hydroxy or amino groups in different steric environment may be conducted very selectively [40].

B.2.23
tert-Butyldimethylsilyltriflate (TBDMS triflate) [195, 202 – 208]

This reagent has higher silylation power than TBDMS-Cl and is preferentially used for the silylation of sterically hindered functional groups and for the synthesis of enolethers or ketene acetals.

B.2.24
tert-Butyldiphenylchlorosilane (TBDPS-Cl) [209 – 215]

This reagent was frequently used for the selective silylation of primary alcohol groups. Sterically hindered alcohols or phenols do not react at all or so slowly that a high selectivity for sterically less hindered OH-groups is obtainable. The selectivity of the silylation may be varied by variation of solvent, temperature and base used as HCl acceptor [40].

B.2.25
Thexyldimethylchlorosilane (TDS-Cl) [216 – 218]

This silylating agent is used (in combination with imidazole or triethylamine), when a protecting group with slightly higher steric demands than the TBDMS group is required.

B.2.26
Triethylchlorosilane (TES-Cl) [140, 219 – 224]

The TES group was mainly used for the protection of OH-groups and for syntheses of enol ethers. Silylation procedures and characteristics of TES-derivatives are analogous to those of the TBDMS-derivatives. The sensitivity of TES alkyl or phenyl ethers to hydrolysis is so low that TMS ethers may be quantitatively hydrolyzed in the presence of TES ethers.

B.2.27
Triethylsilyl triflate (TES-triflate) [204, 206, 208, 225 – 229]

The bulkiness of the TES group results in a relatively low reactivity of TES-Cl when compared to TMS-Cl. TES-triflate is a more powerful silylating agent which may be helpful for the silylation of sterically hindered alcohols or phenols.

B.2.28
Triisopropylchlorosilane (TIPS-Cl) [230 – 234]

Because of the bulkiness of the TIPS group metallation of the substrate is recommendable for silylation instead of an amine as HCl-acceptor. Protection of alcohols or pyrrole and syntheses of enolethers were the most common applications.

B.2.29
Trimethylbromosilane (TMBS) [87, 235 – 238, 262]

TMBS (in combination with a base) is a more powerfull silylating agent than trimethylchlorosilane (see below), but it is less reactive than trimethyliodosilane (TMIS) or trimethylsilyltriflate. Of particular interest its potential to cleave cyclic ethers yielding ω-bromoalkoxysilanes, Eq. (31) [235]. Furthermore, it cleaves lactones yielding silylated ω-bromocarboxylic [236] acids, Eq. (32), and it is capable of cleaving cyclic carbonates yielding ω-bromoalkoxysilanes along with CO_2, Eq. (33). However, in contrast to the more reactive trimethyliodosilane it can not cleave noncyclic ethers or esters. Its advantage over TMIS is its chemical and thermal stability. The preparative applications of TMBS are reviewed in Ref. [11] and Ref. [39] (chapter 3).

B.2.30
Trimethylchlorosilane (TMS-Cl) [13, 32, 34, 121, 123, 173, 188, 208, 235–256]

TMS-Cl is the least expensive and most frequently used of all silylating agents. Numerous books and reviews discuss syntheses, reactions and applications of TMS containing compounds [10–40]. In this work chapter 5 and large parts of chapters 1 and 6 deal with synthesis and reactivity of silylated monomers. TMS-Cl alone has a rather low silylation potential. Yet, silylation of carboxylic acids is feasible under reflux and with removal of the liberated HCl in a slow stream of N_2, Eq. (24). In addition to carboxylic acids, sulfuric and sulfonic

$$R-CO_2H + ClSiMe_3 \xrightarrow{\Delta T} R-CO-OSiMe_3 + HCl \tag{24}$$

acids, boric acid and phosphonic acid can be silylated with neat TMS-Cl. Furthermore, amino acids react with TMS-Cl in special reaction mixtures under exclusive formation of their trimethylsilyl ester hydrochlorides, Eq. (25). These highly hygroskopic compounds can be used in situ to introduce various protecting groups to the amino group, Eq. (26). With a double molar amount of TMS-Cl and triethylamine NO-bissilylated amino acids are easily obtainable,

$$NH_2-(CHR)_n-CO_2H \xrightarrow{+ ClSiMe_3} ClH \cdot NH_2-(CHR)_n-CO_2SiMe_3 \tag{25}$$

$$\begin{array}{c} R-COCl \\ + 2\,NEt_3 \end{array} \Bigg| \begin{array}{c} {} \\ - 2\,NEt_3 \cdot HCl \end{array} \tag{26}$$

$$R-CO-NH-(CHR)_n-CO_2SiMe_3$$

$$NH_2-(CHR)_n-CO_2H \xrightarrow[- 2\,NEt_3]{+ 2\,ClSiMe_3\ +\ 2\,NEt_3} Me_3Si-NH-(CHR)_n-CO_2SiMe_3 \tag{27}$$

(Eq. 27). TMS-Cl in combination with pyridine, triethylamine, imidazole or HMDS are the standard reagents for the silylation of sterically less hindered functional groups. In the case of primary or secondary amines a double molar amount of the substrate may be used, Eq. (28). In the case of sterically hindered functional groups metallation (e.g. with BuLi) is necessary prior to the treatment with TMS-Cl, Eqs. (29, 30) [255].

$$R-NH_2 + ClSiMe_3 \longrightarrow R-NH-SiMe_3 + R-NH_2 \cdot HCl \qquad (28)$$

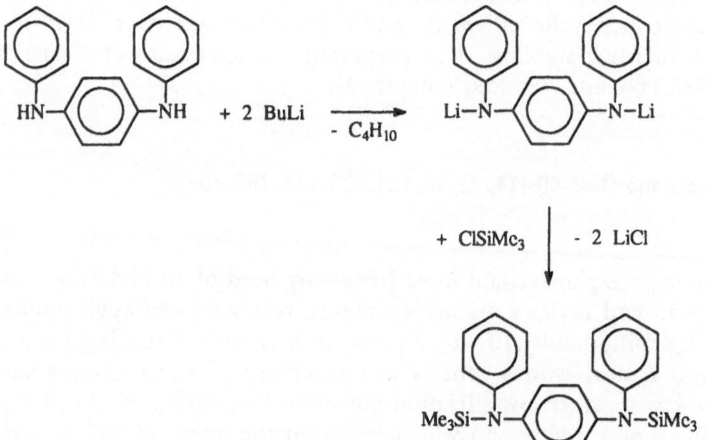

(29)

(30)

B.2.31
Trimethyliodosilane (TMIS) [34, 168, 237, 248, 257 – 262].

TMIS (in combination with HCl acceptors) is one of the strongest silylating agents. Therefore, it was mainly used for the synthesis of enolethers [257 – 262]. Unfortunately, it is expensive, thermally unstable and sensitive to light. Due to this instability TMIS is usually contaminated with J_2 which is unfavorable for analytical applications. However, TMIS is particularly reactive in ring opening reactions involving cyclic ethers, Eq. (31), lactones, Eq. (32), or cyclic carbonates. Eq. (33). Furthermore, it is reactive enough to cleave methylether, Eq. (34), and benzylethers or methylesters Eq. (35), and benzyl esters. These

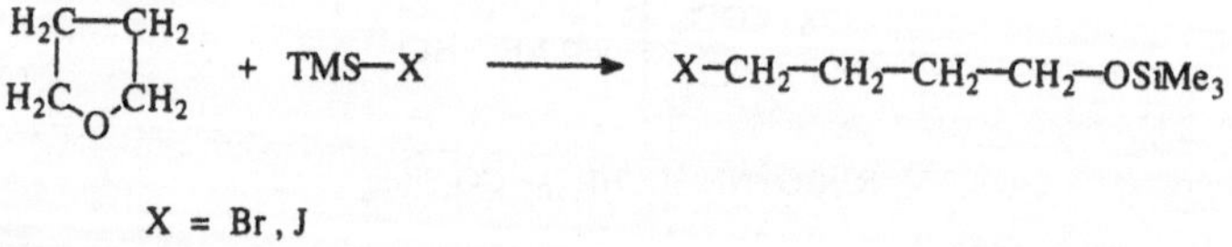

$$X = Br , J \qquad (31)$$

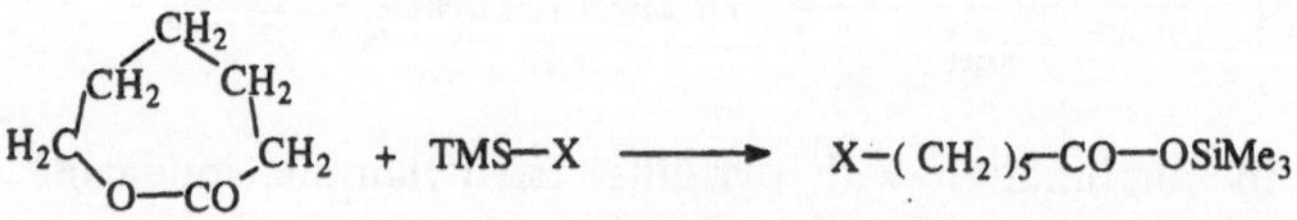

(32)

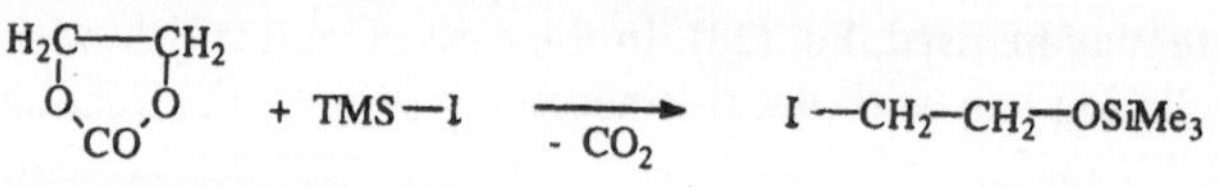

(33)

reactions may be summarized under the title "silylation by dealkylation". These and other useful preparative applications are reviewed in Ref. [11] and in Ref. [34] (chapter 3).

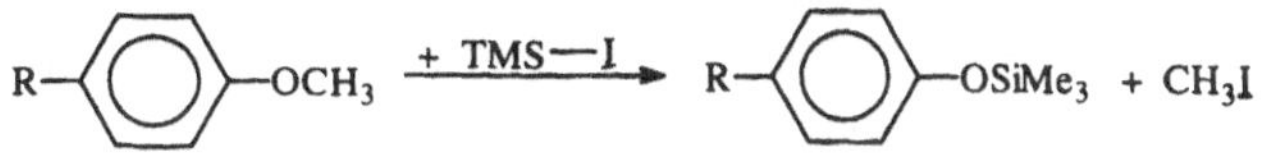

$$R\text{—}\langle\bigcirc\rangle\text{—OCH}_3 \xrightarrow{+ \text{ TMS—I}} R\text{—}\langle\bigcirc\rangle\text{—OSiMe}_3 + \text{CH}_3\text{I} \tag{34}$$

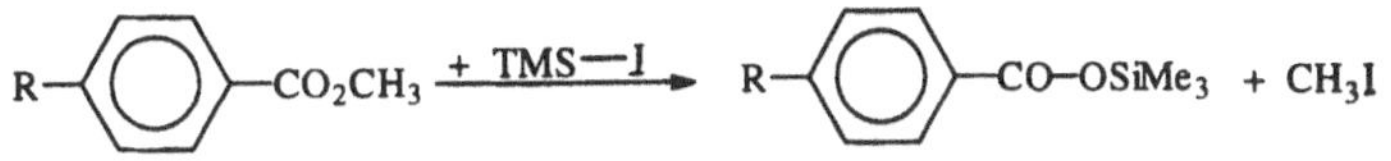

$$R\text{—}\langle\bigcirc\rangle\text{—CO}_2\text{CH}_3 \xrightarrow{+ \text{ TMS—I}} R\text{—}\langle\bigcirc\rangle\text{—CO—OSiMe}_3 + \text{CH}_3\text{I} \tag{35}$$

B.2.32
Trimethylsilylazide [263, 264]

This mild silylating agent may be used for the silylation of phenols and primary or secondary alcohols. Its attractivity suffers from the fact the liberated HN_3 is both poisonous and highly explosive. However, TMS azide is very useful for various preparative applications involving the azide group, particularly for syntheses of isocyanates. These preparative aspects are reviewed in Ref. [11] and in Ref. [34]. Syntheses of difunctional monomers by means of TMS azide are discussed in chapter 6 of this work.

B.2.33
Trimethylcyanide (TMSCN) [265, 266]

TMSCN is a relatively poor and poisonous silylating agent. Its main attractivity of preparative application results from reactions with electrophiles, Eq. (6). These preparative aspects are reviewed in Ref. [11] and in Ref. [34].

B.2.34
Trimethylsilyldiethylamine (TMS-DEA) [267 –275]

TMSDEA is a basic silylating agent with a silylating power in between that of HMDS/TMS-Cl and BSA. Its main advantage is the volatility of the byproduct $HNEt_2$. TMSDEA has been used for silylations of various substrates, particularly for amino acids [267, 268, 273] and dipeptides [271]. Interesting is its selectivity against the steric environment of hydroxy groups. It selectively silylates equatorial but never axial OH-groups [269]. It also silylates α, ω-diamines [275] and in the case of glycine even a N,N,O-tristrimethylsilyl derivative was obtained [276]. Neat TMSDEA can be used as its own solvent, and addition of acids may have a catalytic effect.

B.2.35
Trimethylsilyl dimethylamine (TMSDMA) [276 – 279]

The silylating power and applications of TMSDMA are almost identical with those of TMSDEA. However, its lower boiling point and the higher volatility of $HNMe_2$ may be an advantage for analytical applications in gas-chromatography.

B.2.36
1-(Trimethylsilyl)imidazole (TMSIM) [13, 22, 27, 30, 59, 63, 123, 277, 280 – 283]

TMSIM is a powerful silylating agent for substrates with acidic protons (alcohols, phenols, acids, prim.amides). However, it does not react with basic amines. Therefore it is mainly used for analytical applications. Combinations with other silylating agents such as TMS-Cl, BSA, BSTFA or HMDS are frequently used. When imidazole is combined with TMSCl, TMBS, BSA and other silylating agents TMSIM is prepared and used in situ.

B.2.37
N-Trimethylsilylacetamide [284 – 287]

TMS acetamide is the silylamide with the lowest silylation potential. It is mainly useful for the silylation of primary and secondary OH-groups. Carbohydrates can be silylated in molten TMS acetamide [285]. The byproduct acetamide has the advantage of crystallizing rapidly and almost quantitatively from most nonpolar solvents.

B.2.38
3-Trimethylsilyloxazolidone (TMSO) [288 – 291]

TMSO is potent silylating agent for alcohols, phenols, thiols, carboxylic acids, amino acids and sulfonic acids. However, there is no advantage over MSA.

B.2.39
Trimethylsilyltriflate [87, 208, 228, 248, 292 – 294]

TMS triflate is a highly moisture sensitive and strong silylating agent. It is useful as catalyst for less expensive silylating agents such as HMDS or TMSDEA. Its usefulness as silylating agent is discussed in Ref. [228]. TMS-triflate is a strong electrophile which may catalyze cationic reactions, for example polymerizations of vinyl ethers and heterocycles (see chapters 1 and 5).

B.2.40
4-Trimethylsiloxy-3-penten-2-one (TMS-acac) [295 – 297]

TMS-acac is very potent silylating agent. Particularly noteworthy is the rapid silylation of *tert*-alcohols even in the absence of catalysts and solvents.

B.2.41
Triphenylchlorosilane (TPSCI) [173, 298 – 300]

The triphenylsilyl group is a rarely used protecting group, but it possesses some advantages:

1) due to its bulkiness it allows selective silylation of primary OH-groups
2) the stability of TPS-ethers to acidic hydrolysis is about 400 times higher than that of TMS ethers.
3) The TPS group shows only aromatic protons in ^{1}H NMR spectra.
4) The TPS group can easily be split off by alkaline EtOH, dilute HCl or fluoride ions.

TPS-Cl has been used for the silylation of sterically nonhindered alcohols, thiols, phenols and N-heterocycles with an amine as HCl acceptor. Recommendable is the metallation of the substrate prior to the treatment with TPS-Cl.

For more information on silylating agents Ref. [40] is recommended and Refs. [11] and [34] for preparative applications of silicon reagents.

B.3
References

1. Bott RW, Eaborn C, Swaddle TW (1966) J Organometal Chem 5: 233
2. Gusel'nikov LE, Nametkin NS (1979) Chem Rev 79: 529
3. Walsh R (1981) Accts of Chem Res 14: 246
4. Ebsworth EAV (1968) physical basis of the chemistry of the group IV elements In: A.G. McDiarmid (ed) "Organometallic Compounds of the Group IV Elements" Marcel Dekker p 1–104
5. Peterson JD (1967) Organometal Chem 9: 373
6. Sommer LH, Braughman GA (1961) J Am Chem Soc 83: 3346
7. Simons G, Zandler ME, Talaty ER (1976) J Am Chem Soc 98: 7869
8. Gilman H, Melvin HW Jr (1949) J Am Chem Soc 71: 4050
9. Benkeser RA (1966) Pure and Appl Chemistry 13: 133
10. Sommer LH (1965) Stereochemistry, mechanism and silicon. McGraw Hill, New York p 127
11. Pawlenko S (1980) "Organo Silicum Verbindungen" in Houben-Weyl-Müller "Methoden der organischen chemie" 4th edn. G Thieme Verlag, Stuttgart-New York
12. Birkofer L, Ritter A (1967) In: "Neuere Methoden der präparativen organischen Chemie" Foerst edn, Verlag Chemie, Weinheim, Vol. V, 185
13. Pierce AE (1968) Silylation of organic compounds (a technique for gas-phase analysis), Pierce Chemical Co., Rockford III
14. Klebe JF (1972) In: Tayler EC (ed) Adv in Organic Chemistry, Wiley Intersience, New York, Vol. VII, 97

15. Roth CA (1972) Silylation of organic chemicals. Ind Eng Chem Prod Res Develop 11: 134 (1972)
16. Miller V, Pacakova V (1973) Trimethylsilyl Derivatives in Gas Chromatography, Chem Listy 67: 1121
17. Giesselmann G (1974) "Importance of chemical transformation for analytical purposes"in Method Chim, F Korte Ed, Academic Press, New York, Vol 1, part A: 247
18. Lukevits E et al (1974) The Silyl Method of Synthesis of Nucleo- sides and Nucleotides. Russ Chem Rev 43: 140
19. Kashutina MV et al (1975) Silylation of organic compounds. Usp Khim 44: 1620 (1975), Russ Chem Rev 44: 733
20. Drozd J (1975) Chemical derivatization in gas chromatography. J Chromatogr 113: 303
21. Rasmussen JK (1977) "O-Silylated enolates-versatile intermediates for organic synthesis", Synthesis 91
22. Blau K, King GS (1977) "Handbook of Derivatives for Chromatography", Heyden & Son, London
23. Cooper BE (1978) Silylation as a protective method in organic synthesis, Chem Ind 794
24. Colvin EW (1978) "Silicon in Organic Synthesis, Chem Soc Rev 7: 15
25. Nicholson JD (1978) "Derivative formation in the quantitative gas chromatographic analysis of pharmaceuticals". "Silylation", Analyst 103: 193
26. Poole CF, Zlatkis A, "Trialkylsilyl Ether Derivatives (Other Than TMS) for Gas Chromatography and Mass Spectrometry"J Chromatogr Sci 17: 115
27. Knapp DR (1979) Handbook of Analytical Derivatization Reactions John Wiley & Sons, New York
28. Fleming I (1979) "Organic Silicon Chemistry"In: Jones DN (ed) Comprehensive Organic Chemistry, Pergamon Press, Oxford, Vol. 3
29. Cooper BE (1980) Silylation in organic synthesis Proc Biochem 9
30. Drozd J (1981) Chemical derivatization in gas chromatography In: J Chromatogr Libr., Elsevier, Amsterdam Vol. 19
31. Greene TW (1981) Protective groups in organic synthesis John Wiley & Sons, New York
32. Colvin EW (1981) "Silicon in Organic Synthesis Butterworths, London
33. Plueddemann EP (1982) "Silylating Agents" in Kirk-Othmer Encycl Chem Technol, 3 rd edn, John Wiley & Sons, New York Vol. 20 p 962
34. Weber WP (1983) "Silicon Reagents for Organic Synthesis"Springer-Verlag, Berlin
35. Denney RC (1983) "Silylation Reagents for Chromatography"Spec Chem 6
36. Brownbridge P (1983) "Silyl Enol Ethers in Synthesis"Synthesis 1: 85
37. Schaumann E (1984) "Schutzgruppen der alkoholischen Hydroxy-Funktion"In Houben-Weyl "Methoden der organischen Chemie"Vol. 6, 1b, part 3, p 735
38. Lalonde M, Chan TH (1985) "Use of Organosilicon Reagents as Protective Groups in Organic Synthesis"Synthesis 817
39. Ollson L-I, (1986) "Silicon-based protective groups in organic synthesis", Acta Pharm. Suecia 23: 370
40. van Look G (1988) "Silylating Agents" Fluka Chemie AG
41. Blair IA, Philiipou G (1977) J Chromatogr Sci 15: 478
42. Harvey DJ (1978) J Chromatogr 14A: 291
43. Cella JA (1982) J Org Chem 47: 2125
44. Poole CF, Zlatkis A (1979) J Chromatogr Sci 17: 115
45. Steffenrud S, Borgeat P, Evans MJ, Bertrand MJ (1987) Mass Spectrom. 14: 313
46. Morita T, Okamoto Y, Sakurai H (1980) Tetrahedron Lett. 21: 835
47. Hosomi A, Sakurai H (1981) Chem Lett 85
48. Olah GA, Husain A, Gupta BGB, Salem GF, Narang SC (1981) J Org Chem 46: 5212
49. Olah GA, Husain A, Singh BP (1983) Synthesis 892
50. Yalpani M, Wilke G (1985) Chem Ber 118: 661
51. Simmler W (1963) Chem Ber 96: 349
52. Eaborn C, Holder CA, Walton DRM, Thomas BS (1969) J Chem Soc (C) 2502
53. Chapman JR, Bailey E, (1974) J.Chromatogr. 89: 215
54. Nishiyama N, Kitajima T, Matsumoto M, Itoh K (1984) J Org Chem 49: 2298

55. Stork G, Sofia MJ (1986) J Am Chem Soc 108: 6826
56. Koruda M, George IA (1986) J Am Chem Soc 108: 8098
57. Magnol E, Malaeria M (1986) Tetrahedron Lett. 27: 2255
58. Birkofer L, Ritter A, Gießler W (1963) Angew Chem 75: 93
59. Nicholson JD (1978) Analyst 103: 193
60. Lane TH, Frye CL (1978) J Org Chem 43: 4890
61. Klebe JF (1966) J Am Chem Soc 88: 3390
62. Jolliffe VA (1979) J Chromatogr 179: 333
63. Hoffmann NE, Peteranetz KA (1972) Anal Lett 5: 589
64. Sternson LA, Hincal F, Bannister SJ (1977) J Chromatogr 144: 191
65. Englmaier P (1986) Fresenius Z Anal Chem 324: 338
66. Piekos R, Osmialowski K, Kobylczyk K, Grzybowski J (1976) J Chromatogr 116: 315
67. Garzo' G, Hoebbel D, Ecsery ZJ, Ujszaszi K (1978) J Chromatogr. 167: 321
68. Körtvelyessy G, Szoradi S, Sztrukar I, Ladanyi L (1984) J Chromatogr 303: 370
69. Kawashiro K, Morimoto S, Yoshida H (1984) Bull Soc Chem Jpn 57: 2871
70. Galbraith MN, Horn DHS, Middleton EJ (1968) J Chem Soc Chem Commun 466
71. Lasocki Z (1973) Synth Inorg Metal Org Chem 3: 29
72. de Koning JJ, Kooreman HJ, Tan HS, Verweij J (1975) J Org Chem 40: 1346
73. Kricheldorf HR (1972) Liebigs Ann Chem 763: 17
74. Eggerding G, Wert R (1972) J Am Chem Soc 97: 17
75. Adams JL, Chen T, Metcalf BW (1985) J Org Chem 50: 2730
76. Snatzke G, Vlahov J (1985) Liebigs Ann Chem 439
77. Sharma SC, Torssell K (1979) Acta Chem Seand B 33: 379
78. Colvin EW, Beck AK, Bastani B, Seebach D, Kai Y, Dunitz JD (1980) Helv Chim Acta 63: 697
79. Kozilowski AP (1978) Schmiesing R, Tetrahedron Lett. 4241
80. Dedier J, Gerval P, Frainnet E (1980) J Organomet Chem 185: 183
81. Birkofer L, Sommer P (1975) J Organomet Chem 99 C 1
82. Kozyukov VP, Mironova NV, Mironov VF (1980) Zh Obshch Khim 50: 955 and 2022
83. Sommer LH, Kerr GT, Whitmore FC (1948) J Am Chem Soc 70: 445
84. Voronkov MG, Roman VK, Maletina EA (1982) Synthesis 277
85. Kantlehner W, Hang E, Mergen WW (1980) Synthesis 460
86. Cooper BE, Butler DW (1977) Ger Offen 2.649.536, C.A. 87 135844 m (1977)
87. Hergott HH, Simchen G (1980) Liebigs Ann Chem 1718
88. Stalling DL, Gehrke CW, Zumwalt RW (1968) Biochem Biophys Res Commun 31: 616
89. Gehrke CW, Leimer K (1970) J Chromatogr 53: 201 and 219
90. Valdez D, J Chromatogr Sci 23: 128
91. Valdez D, Iller HD (1986) J Am Oil Chem Soc 63: 119
92. Fell V, Lee CR (1976) J Chromatogr 121: 41
93. Binder H, Ashy AA (1984) J Chromatogr Sci 22: 536
94. Michael G (1985) Z Chem 25: 19
95. Hermann F, Matousek P, Dufka O, Churacek J (1986) J Chromatogr 370: 49
96. Lehtonen K, Ketola M (1986) J Chromatogr 370: 465
97. Verboom W, Visser GW, Reinhoudt DN (1981) Synthesis 807
98. Cooper BE (1978) Chem Ind 794
99. Chapman JR, Bailey E (1973) Anal Chem 45: 1636
100. Wieber M, Schmidt M (1963) J Organomet Chem 1: 22
101. Yoder CH, Tesno SL, Heaney SM, Bohan C (1985) Synth React Inorg Chem 15: 321
102. Stadler J (1978) Anal Biochem 86: 477
103. Morita H, Montgomery WG (1976) J Chromatogr 123: 454
104. Hammar CG (1978) Biomed Mass Spectrom 5: 25
105. Corey EJ, Varma RK (1971) J Am Chem Soc 93: 7319
106. Donaldson RE, Fuchs PL (1977) J Org Chem 42: 2032
107. Conrad PC, Kwiatkowski PL, Fuchs PL (1987) J Org Chem 52: 586
108. Mukaiyama T, Iwasawa N, Stevens RW, Haga T (1984) Tetrahedron. 40: 1381

109. Yamashita K, Watanabe K, Iskibaski M, Miyazaki H, Yokota K, Horie K, Yamamoto S (1987) J Chromatogr 399: 223
110. Denmark SE, Hammer RP, Weber EJ, Habermas KL (1987) J Org Chem 52: 165
111. Tacke R, Wannagat V (1975) Monatsh Chem 106: 1005
112. Friedrich G, Bartsch R, Rühlmann K (1977) Pharmazie 32: 394
113. Muslin DV, Sikyapina N (1984) Izvest Akad Nauk, SSSR, Ser Khim 2433
114. Rutherford KG, Seidewand RJ (1975) Com J Chem 53: 67
115. Grob K, Grob G (1980) J High Res Chromatogr Chromatorg Commun 3: 197
116. Welsch T, Müller R, Eugewald W, Werner G (1982) J Chromatogr 241: 41
117. Nakamura E, Murofushi T, Shimizu M, Kuwajima I (1976) J Am Chem Soc 98: 2346
118. Nakamura E, Hashimoto K, Kuwajima I (1981) Bull Chem Soc, Jpn 54: 805
119. Kuwajima I, Nakamura E, Hashimoto K (1983) Org Synth 61: 122
120. Crimmins MT, Mascorella SW (1986) J Am Chem Soc 108: 3435
121. Langer SH, Connell S, Wender I (1958) J Org Chem 23: 50
122. Mason PS, Smith ED (1966) J Gas Chromatogr 4: 398
123. Birkofer L, Ritter A (1965) Angew Chem 77: 414
124. Torkelson S, Ainsworth C (1976) Synthesis 722
125. Harpp DN, Friedlander BT, Larsen C, Steliou K, Stockton A (1978) J Org Chem 43: 3481
126. Miller RD, McKean DR (1979) Synthesis 730
127. Hässig R et al (1982) Chem Ber 115: 1990
128. Smith AB, Visnick M, Haseltine JN, Sprengeler PA (1986) Tetrahedron 42: 2957
129. Nikovlov ZL, Reilly PJ (1983) J Chromatogr 254: 157
130. Vorbrüggen H, Krolikievicz K (1984) Chem Ber 117: 1523
131. Brieynes CA, Jarriens TK (1982) J Org Chem 47: 3966
132. Sweeley CC, Bentley R, Makita M, Wells WW (1963) J Am Chem Soc 85: 2497
133. Vandenheuvel WJA (1967) J Chromatogr 27: 85
134. Novina R (1984) Chromatographia 15: 704 (1982),17 441 (1983), 18 21
135. Mahmud F, Catterall E, Pakistan (1986) J Sci Ind Res 29: 72
136. Deyhimi F, Coles JA (1982) Helv Chim Acta 65: 1752
137. Sindorf DW, Maciel GE (1983) J Phys Chem 87: 5516
138. Nawrocki J (1985) Chromatographia 20: 308
139. Bassindale AR, Walton DRM (1970) J Organomet Chem 25: 389
140. Stewart RF, Miller LL (1980) J Am Chem Soc 102: 4999
141. Marsmann HC, Horn HG, Naturforsch Z (1972) 27n: 1448
142. Chapman A, Jenkins AD (1977) J Polym Sci 15 3075
143. Larson GL, Ortiz M, Rodriguez de Roca M (1981) Synth Commun 11: 583
144. Vostokov I (1977) Zhur Obshch Khim 48: 2140
145. Wies R, Pfaender P, Liebigs (1973) Ann Chem 1269
146. Appel R, Montenarh M (1975) Chem Ber 108: 1442
147. Su TL, Bennua B, Vorbrüggen H, Lindner HJ (1981) Chem Ber 114: 1269
148. Voronkov MG, Shabarova ZI (1959) Zhur Obshch Khim 29: 1528
149. Matsumoto H, Hoshino Y, Nakabayashi J, Nakano T, Nagai Y, (1980) Chem Lett 1475
150. Pinnih HW, Bal BS, Lajis NH (1978) Tetrahedron Lett 4261
151. Schmidt M, Schmidbauer H (1961) Chem Ber 94: 2446
152. Duflaut N, Calas R, Dunoguès J (1963) Bull Soc Chim Fr 512
153. Jung ME, Lyster MA (1980) Org Synth 59 35
154. Yokoyama M, Yoshida S, Imamoto T (1982) Synthesis 591
155. Aizpurua JM, Polomo C (1984) Bull Soc Chim Fr II, 142
156. Aizpurua JM, Polomo C (1985) Synthesis 206
157. Abel EW (1961) J Chem Soc 4933
158. Fiorenza M, Reginato G, Ricci A, Taddei M (1984) J Org Chem 99: 551
159. Abel EW, Armitage DA, Bush RP (1964) J Chem Soc 2455
160. Harpp DN, Steliou K (1976) Synthesis 721
161. Kricheldorf HR, Leppert E (1972) Makromol Chem 158: 223
162. Birkofer L, Donike M (1967) J Chromatogr 26: 270
163. Donike M (1975) J Chromatogr 103: 91

164. Rogozhin SU, Davidovich YA, Andreev SM, Mironova NV, Yurtanov AI (1974) Izvest Akad Nauk, SSSR Ser Khim 1868
165. Piper F, Rühlmann K (1976) J Organomet Chem 121: 149
166. Donike M (1969) J Chromatogr 42: 103
167. Donike M (1973) J Chromatogr 85: 1
168. Donike M, Zimmermann J (1980) J Chromatogr 202: 483
169. Blum W (1986) High J Res Chromatogr Chromatogr Commun 718 (1985) and 350
170. Page BD, Conacher HBS (1982) In: Frei RW, Lawrence JF (eds) Chemical Derivatization in Analytical Chemistry Plenum Press, New York Vol.2, p 263
171. Donike M (1975) J Chromatogr 115: 591
172. Staab HA, Herz CP (1977) Angew Chem 89: 406
173. Gerlach M, Jutzi P, Stasch J-P, Przuntek H (1983) Z Naturforsch. 38b 237
174. Bazan AC, Knapp DR (1982) J Chromatogr 236: 201
175. Mawhinney TP, Madson MA (1982) J Org Chem 47: 3336
176. McKenzie SL, Tenaschuk D, Fortier G (1987) J Chromatogr 387: 241
177. Ballard KD, Knapp DR, Oatis JE Jr, Walle T (1983) J Chromatogr 277: 333
178. Schwenk WF, Berg PJ, Beaufrere B, Miles JM, Haymond MW (1984) Anal Biochem 141: 101
179. Abbott FS, Kassam J, Acheampong A, Ferguson S, Panesar S, Burton R, Farrell K, Orr J (1986) J Chromatogr 375: 285
180. Corbett ME, Scrimgeour CM, Watt PW (1987) J Chromatogr 419: 263
181. Hwang K-J, Logusch EW, Brannigan LH (1987) J Org Chem 52: 3435
182. Quilliam MA, Westmore JB (1977) Anal Chem 50: 59
183. Harvey DJ (1977) Biomol Mass Spectrom 4: 265
184. Bandi ZL, Amari GAS (1986) J Chromatogr 363: 402
185. Ogilvie KK, Beaucage SL, Entwistle DW, Thompson EA, Quilliam MA, Westmore JB (1976) J Carbohydrates, Nucleosides, Nucleotides 3: 197
186. Reetz MT, Neumeier G (1981) Liebigs Ann Chem 1234
187. Stork G, Hudrlik PF (1968) J Am Chem Soc 90: 4462
188. Corey EJ, Venkatsewarlu A (1970) J Am Chem Soc 94: 6190
189. Lalonde M, Chan TH (1985) Synthesis 817
190. Kendall PM, Johnson JV, Cook CE (1979) J Org Chem 44: 1421
191. Aizpurua JM, Polomo C (1985) Tetrahedron Lett 26: 475
192. Hansen Jr DW, Pilipauskas D (1985) J Org Chem 50: 945
193. Calverley MJ (1983) Synth Commun 13: 601
194. Thomas EJ, Williams AC (1987) J Chem Soc Chem Commun 992
195. Wissner A, Grudzinskas CV (1978) J Org Chem 43: 2102
196. Braish TF, Fuchs PL (1986) Synth Commun 16: 111
197. Olah GH, Gupta BGB, Narang SC, Malhotra R (1979) J Org Chem 44: 4272
198. Dauben WG, Gerdes JM, Look GC (1986) Synthesis 532
199. Kinzy W, Schmidt RR (1987) Liebigs Ann Chem 407
200. Rathke MW, Sullivan DE (1973) Synth Commun 3: 67
201. Mc Donged PG, Rico JG, Oh Y, Condon BD (1986) J Org Chem 51: 3388
202. Riediker M, Graf W (1979) Helv Chim Acta 62: 205
203. Willis JP, Gogins KAZ, Miller LL (1981) J Org Chem 46: 3215
204. Corey EJ, Cho H, Rücker C, Hua DH (1981) Tetrahedron Lett 22: 3455
205. Mander LN, Sethi SP (1984) Tetrahedron Lett 25: 5953
206. Emde H, Simchen G (1983) Liebigs Ann Chem 816
207. Ried W, Reiher U (1987) Chem Ber 120: 657
208. Feger H, Simchen G (1986) Liebigs Ann Chem 428
209. Hanessian S, Lavallee P (1975) Can J Chem 53: 2975
210. Stern A, Swenton JS (1987) J Org Chem 52: 2763
211. Quilliam MA, Yaraskaiitch JM (1985) J Liquid Chromatogr 8: 449
212. Pecquet F, d'Angelo J (1982) Tetrahedron Lett 23: 2777
213. Stewart OA, Williams RM (1985) J Ann Chem Soc 107: 4289
214. Brooks DW, Kellogg RP, Cooper CS (1987) J Org Chem 52: 192

215. Berlage U, Schmidt J, Peters U, Welzel P (1987) Tetrahedron Lett 28: 3091
216. Wetter H, Oertle K (1985) Tetrahedron Lett 26: 5515
217. Walkup RD, Cunningham RT (1987) Tetrahedron Lett 28: 4619
218. Kerscher V, Kreiser W (1987) Tetrahedron Lett 28: 531
219. Hancock RL (1968) J Gas Chromatogr 6: 431
220. Hart TW, Metcalfe DA, Scheinmann F (1979) J Chem Soc Chem Commun 156
221. Andrews DR, Barton DHR, Hesse RH, Pechet MM (1986) J Org Chem 51: 4819
222. Roush WR, Russo-Rodriguez S (1986) J Org Chem 52: 598
223. Smith AB, Rivero RA (1987) J Am Chem Soc 109: 1272
224. Still WC, Schneider MJ (1977) J Am Chem Soc 99: 948
225. Heathoock GH, Young SD, Hagen JP, Pilli R, Badertscher U (1985) J Org Chem 50: 2095
226. Kinoshita M, Arai M, Ohsawa N, Nakata M (1986) Tetrahedron Lett 27: 1815
227. Seebach D, Chow H, Jackson RFW, Sutter MA, Thaisrivongs S, Zimmermann J (1986) Liebigs Ann Chem 1281
228. Emde H, Domsch D, Feger H, Frick U, Götz A, Hergott HH, Hofmann K, Kober W, Krägeloh K, Oesterle T, Steppan W, West W, Simchen G (1982) Synthesis 1
229. Danishefsky S, Harvey DF (1985) J Am Chem Soc 107: 6647
230. Allen AD, Charlton JC, Eaborn C, Modena G (1957) J Chem Soc 3668
231. Cunico RF, Bedell L (1980) J Org Chem 45: 4797
232. Corey EJ, Pan B, Hua DH, Deardorff DR (1982) J Am Chem Soc 104: 6816
233. Kowalski CJ, Lal GS, Haque MS (1986) J Am Chem Soc 108: 7127
234. Ohwa M, Eliel EL (1987) Chem Lett 41
235. Kricheldorf HR (1979) Angew Chem 91: 749 Angew Chem Int Ed 18
236. Kricheldorf HR, Mörber G, Regel W (1981) Synthesis 382
237. Schmidt HA (1980) Chemiker Ztg 104: 253
238. Aizpurua JM, Polomo L (1984) Nouv J Chimie 8: 51
239. Sauer RO (1944) J Am Chem Soc 66: 1707
240. Marsmann AC, Horn HG, Naturforsch Z (1972) 27b: 1448
241. Hils J, Rühlmann K (1967) Chem Ber 200: 1638
242. Kricheldorf HR (1970) Synthesis 592
243. Kricheldorf HR, Greber G (1971) Chem Ber 104: 3131
244. Kricheldorf HR (1971) Liebigs Ann Chem 745: 81
245. Kricheldorf HR (1970) Synthesis 649
246. Kricheldorf HR (1975) Liebigs Ann Chem 13: 1378
247. Schwarz G, Alberts H, Kricheldorf HR (1981) Liebigs Ann Chem 1257
248. Bassindale AR, Stout T (1985) Tetrahedron Lett 26: 3403
249. Nakonieczna L, Chimiak A (1987) Synthesis 418
250. Visser RG, Bos HJT, Brandsma L (1980) Rec Trav Chim Pay Bas 99: 70
251. Birkofer L, Wegner P (1970) Org Synth 50: 107
252. Washburne SS, Peterson WR (1971) J Organomet Chem 33: 153
253. Reetz MT, Chatziiosifidis I (1982) Synthesis 330
254. Hünig S, Wehner G (1979) Synthesis 522
255. Kakimoto M, Oishi Y, Imai Y (1985) Makromol Chem Rapid Commun 6: 557
256. Weisenfeld RB (1986) J Org Chem 51: 2434
257. Hergott HH, Simchen G (1960) Liebigs Ann Chem 1718
258. Miller RD, McKean DR (1979) Synthesis 730
259. Miller RD, McKean DR (1982) Synth Commun 12: 319
260. Kramarova EP, Shipov AG, Artamkina OB, Baukov YI (1984) Zhur, Obshch, Khim 54: 1921
261. Hässig R, Siegel H, Seebach D (1982) Chem Ber 115: 1990
262. Hosomi A, Sekurai H (1981) Chem Lett 85
263. Untze W, Fresenius Z (1972) Anal Chem 259: 212
264. Simon D, Emizane M, Synthesis
265. Fischer K, Hünig S (1987) Chem Ber 120: 325
266. Mai K, Patil G (1986) J Org Chem 51: 3545
267. Rühlmann K (1959) J Pract Chem 9: 315

268. Rühlmann K (1961) Chem Ber 94: 1876
269. Weisz I, Felfoldi K, Kovacs K (1968) Acta Chim Acad Sci Hung 58: 189
270. Mason PS, Smith ED (1966) J Gas Chromatogr 4: 398
271. Rogozhin SV s. Pkt. 264
272. Vostokov IA (1978) Zhur Obshch Khim 48: 2140
273. Cooper BE, Westall S (1978) Ger Offen 2.722.092 (1977)C.A. 88 7048x
274. Taddei M, Tempesti F (1985) Syn Commun 15: 1019
275. Hvidt T, Martin OR, Szarek WA (1986) Tetrahedron Lett 27: 3807
276. Smith ED, Shewbart KL (1969) J Chromatogr Sci 7: 704
277. Chambaz EM, Horning E (1969) Anal Biochem 30: 7
278. Ruppert I (1977) Tetrahedron Lett 1987
279. Hellberg LH, Juarez J (1974) Tetrahedron Lett 3553
280. Ladner W (1983) Chem Ber 116: 3413
281. Torkelson S, Ainsworth C (1976) Synthesis 722
282. Glass RS (1973) J Organomet Chem 61: 83
283. Müller H, Frey V (1981) Ger Offen 2.923.604 (1980), C.A. 94 175249f
284. Klebe JF, Finkbeiner H, White DM (1966) J Am Chem Soc 88: 3390
285. Birkofer L, Ritter A, Bentz F (1964) Chem Ber 97: 2196
286. Richardson JD, Bruice TC, Waraszkiewicz SM, Berchtold GA (1974) J Org Chem 39: 2088
287. Atkins RK, Frazier J, Moore LL, Weigel LO (1986) Tetradron Lett 27: 2451
288. Polomo C (1981) Synthesis 809
289. Ballester A, Polomo AL (1983) Synthesis 571
290. Aizpurua JM, Polomo C (1984) Can J Chem 62: 336
291. Perold GW (1984) J Chromatogr 291: 365
292. Roesky HW, Giere HH (1970) Z Naturforsch 25b: 773
293. Vorbrüggen H, Bennua B (1982) Chem Ber 114: 1279
294. Vorbrüggen H, Krolikiewicz K, Bennua B (1981) Chem Ber 114: 1234
295. Veysoglu T, Mitscher LA (1981) Tetrahedron Lett 22: 1303
296. Taba KM, Dahlhoff WV (1982) Synthesis 652
297. Yalpani M, Wilke G (1985) Chem Ber 118: 661
298. Petersen RC, Ross SD (1963) J Am Chem Soc 85: 3164
299. Allen AD, Lavery SJ (1969) Can J Chem 47: 1263
300. Brandes D (1977) J Organomet Chem 136: 25

Subject Index

F. Ciardelli, E. Tsuchida, D. Wöhrle (Eds.)

Macromolecule-Metal Complexes

1996. XVIII, 318 pp. 94 figs., 19 tabs. Hardcover **DM 198,-**; öS 1445,40; sFr 173,-
ISBN 3-540-59383-7

Macromolecule-Metal Complexes gives the first concise overview on the topic,
both on fundamentals and new application areas. Their synthesis, kinetics and
thermodynamics are detailed; special properties such as gas transport, charge transport, catalysis and lightinduced processes are emphasized. Furthermore, the authors
treat the actual working areas for new application methods. Thus, the book will be a
very helpful tool for Polymer Scientists, Materials Scientists, Organic Chemists, and
Physical Chemists working in these fields.

G. Fink, R. Mülhaupt, H.H. Brintzinger (Eds.)

Ziegler Catalysts

Recent Scientific Innovations and Technological Improvements

1995. XIII, 511 pp. 267 figs., 2 in color, 124 tabs. Hardcover **DM 198,-**; öS 1544,40;
sFr 173,- ISBN 3-540-58225-8

Forty years after Ziegler's discovery of the „Aufbaureaktion" and low-pressure ethene
polymerization, transition metal catalyzed olefin and diolefin polymerization continues to represent one of the most active and exciting areas. Since the 1980s, outstanding scientific innovations and process improvements have revolutionized polyolefin technology and greatly simplified polymerization processes. Well-defined
catalyst systems are now at hand and facilitate the understanding of basic reaction
mechanisms and correlations between catalyst structures, polymer microstructures,
and polymer properties. This book reviews some of the modern approaches in organometallic chemistry, Ziegler-Natta catalysis, polymerization processes, design of novel
materials, and the modelling in catalyst and
process development.

Springer

Preisänderungen vorbehalten.

Springer-Verlag, Postfach 31 13 40, D-10643 Berlin, Fax 0 30 / 82 07 - 3 01 / 4 48   e-mail: orders@springer.de   tmBA96.04.24

L.A. Pilato, M.J. Michno

Advanced Composite Materials

1994. XIII, 208 pp. 50 figs., 49 tabs. Hardcover **DM 138,-**; öS 1076,40; sFr 121,50
ISBN 3-540-57563-4

Advanced composite materials or high performance polymer composites are an unusual class of materials that possess a combination of high strength and modulus and are substantially superior to structural metals and alloys on an equal weight basis. The book provides an overview of the key components that are considered in the design of a composite, of surface chemistry, of analyses/testing, of structure/property relationships with emphasis on compressive strength and damage tolerance. Newly emerging tests, particularly open hole compression tests are expected to provide greater assurance of composite performance. This publication is an „up-to-date" treatment of leading edge areas of composite technology with literature reviewed until recently and includes thermoplastic prepregs/composites and major application areas.

V. Shibaev (Ed.)

Polymers as Electrooptical and Photooptical Active Media

With contributions by **C. Bräuchle, K. Horie, K. Ichimura, S.A. Ivanov, S. Kostromin, F.H. Kreuzer, S. Machida, A. Miller, A. Petri, S.J. Picken, V.P. Shibaev, C.P.J.M.vande Vorst**

1996. XIII, 210 pp. 117 figs., 30 tabs. (Macromolecular Systems - Materials Approach) Hardcover **DM 148,-**; öS 1080,40; sFr 130,50 ISBN 3-540-59486-8

Polymeric materials have special advantages over other materials used for the recording, storage and retrieval of information, telecommunication transmission and visualization of images. The authors describe the synthesis, the physico-chemical behavior and the applications of these highly sensitive macromolecular systems. They discuss the most essential developments in this field.

For scientists and professionals working in the field of electrooptical and photooptical polymeric materials.

Springer

Preisänderungen vorbehalten.

Springer-Verlag, Postfach 31 13 40, D-10643 Berlin, Fax 0 30 / 82 07 - 3 01 / 4 48   e-mail: orders@springer.de   tmBA96.04.24

Springer-Verlag
and the Environment

We at Springer-Verlag firmly believe that an international science publisher has a special obligation to the environment, and our corporate policies consistently reflect this conviction.

We also expect our business partners – paper mills, printers, packaging manufacturers, etc. – to commit themselves to using environmentally friendly materials and production processes.

The paper in this book is made from low- or no-chlorine pulp and is acid free, in conformance with international standards for paper permanency.